Introduction
to
Signals and Systems

McGraw-Hill Series in Electrical and Computer Engineering

Senior Consulting Editor
Stephen W. Director, University of Michigan, Ann Arbor

Circuits and Systems
Communications and Signal Processing
Computer Engineering
Control Theory and Robotics
Electromagnetics
Electronics and VLSI Circuits
Introductory
Power
Antennas, Microwaves, and Radar

Previous Consulting Editors
Ronald N. Bracewell, Colin Cherry, James F. Gibbons, Willis W. Harman, Hubert Heffner, Edward W. Herold, John G. Linvill, Simon Ramo, Ronald A. Rohrer, Anthony E. Siegman, Charles Susskind, Frederick E. Terman, John G. Truxal, Ernst Weber, and John R. Whinnery

Communications and Signal Processing

Senior Consulting Editor
Stephen W. Director, *University of Michigan, Ann Arbor*

Auñón/Chandrasekar: *Introduction to Probability and Random Processes*
Antoniou: *Digital Filters: Analysis and Design*
Bose: *Neural Network Fundamentals with Graphs, Algorithms, and Applications*
Carlson: *Communication Systems: An Introduction to Signals and Noise in Electrical Communication*
Cassandras: *Discrete Event Systems Modeling and Performance Analysis*
Cherin: *An Introduction to Optical Fibers*
Childers: *Probability and Random Processes Using MATLAB*
Collin: *Antennas and Radiowave Propagation*
Collin: *Foundations for Microwave Engineering*
Cooper and McGillem: *Modern Communications and Spread Spectrum*
Davenport: *Probability and Random Processes: An Introduction for Applied Scientists and Engineers*
Drake: *Fundamentals of Applied Probability Theory*
Gardner: *Introduction to Random Processes*
Jong: *Method of Discrete Signal and System Analysis*
Keiser: *Local Area Networks*
Keiser: *Optical Fiber Communications*
Kershenbaum: *Telecommunication Network Design and Algorithms*
Kraus: *Antennas*
Kuc: *Introduction to Digital Signal Processing*

Lee: *Mobile Communications Engineering*
Mitra: *Digital Signal Processing: A Computer-Based Approach*
Papoulis: *Probability, Random Variables, and Stochastic Processes*
Papoulis: *Signal Analysis*
Papoulis: *The Fourier Integral and Its Applications*
Parsons: *Voice and Speech Processing*
Peebles: *Probability, Random Variables, and Random Signal Principles*
Powers: *An Introduction to Fiber Optic Systems*
Proakis: *Digital Communications*
Russell: *Telecommunications Protocols*
Schwartz: *Information Transmission, Modulation, and Noise*
Siebert: *Circuits, Signals, and Systems*
Smith: *Modern Communication Circuits*
Taub and Schilling: *Principles of Communication Systems*
Taylor: *Principles of Signals and Systems*
Taylor: *Hands-On Digital Signal Processing*
Viterbi and Omura: *Principles of Digital Communications and Coding*
Viniotis: *Probability and Random Processes*
Walrand: *Communications Networks*
Waters: *Active Filter Design*

Introduction
to
Signals and Systems

Douglas K. Lindner
Virginia Polytechnic Institute and State University

WCB McGraw-Hill

Boston   Burr Ridge, IL   Dubuque, IA   Madison, WI   New York   San Francisco   St. Louis
Bangkok   Bogotá   Caracas   Lisbon   London   Madrid
Mexico City   Milan   New Delhi   Seoul   Singapore   Sydney   Taipei   Toronto

WCB/McGraw-Hill

A Division of The **McGraw·Hill** *Companies*

INTRODUCTION TO SIGNALS AND SYSTEMS

This book is printed on acid-free paper.

1 2 3 4 5 6 7 8 9 0 DOC/DOC 9 3 2 1 0 9 8

ISBN 0-256-25259-9

Vice president/editor-in-chief: *Kevin T. Kane*
Publisher: *Thomas Casson*
Executive editor: *Elizabeth A. Jones*
Editorial assistant: *Michelle L. Flomenhoft*
Marketing manager: *John T. Wannemacher*
Project manager: *Margaret Rathke*
Production supervisor: *Michael R. McCormick*
Freelance design coordinator: *JoAnne Schopler*
Cover photographer: *Sharon Hoogstaten*
Cover designer: *Matthew Baldwin*
Typeface: *10/12 Times*
Senior supplement coordinator: *Cathy L. Tepper*
Printer: *R. R. Donnelley & Sons Company*

Library of Congress Cataloging-in-Publication Data

Lindner, Douglas K.
     Introduction to signals and systems / Douglas K. Lindner.
       p.   cm.
     Includes index.
     ISBN   0-256-25259-9 (acid-free paper)
     1. Signal processing -- Mathematical models.   2. Transformations
       (Mathematics).   3. Systems analysis.   I. Title
     TK5102.9.155   1999
     621.382'2 -- dc21   98-50542

www.mhhe.com

Preface

Introduction This textbook is designed for a one- or two- semester, sophomore-junior level core course on signals and systems. A typical background of a student would include a course on network analysis, an introduction to physics, and differential equations. The proposed text, however, only assumes a sophomore level maturity.

It is widely recognized that there many common elements in the modeling, analysis, and design in many diverse engineering systems. These elements have been collected under the heading of *system theory*, formalized, and given a systematic treatment. It is the purpose of this book to discuss the most fundamental concepts associated with this theory while stressing their relationship to the engineering problems from which system theory evolved. The mathematical background for the concepts discussed here is also contained in this book.

Structure of the Material The two most basic concepts in this theory are a *signal* and a *system*. The intention of the organization of the text is to present the core signals and systems material in terms of a few well-drawn, tightly interconnected concepts. The development starts from the definition of a signal and a system. The results related to signals are grouped together and the results related to systems are grouped together. One of the major benefits of this organization is that the concepts related to system models (or representations) can now be grouped together. The interrelationships between convolution integrals, transfer functions, and state space equations can be emphasized. These relationships are particularly important in view of modern computational tools which implement these interrelationships. In the same way the concepts related to signals can be grouped together to emphasize their continuity. The frequency domain concepts of the spectral content of a signal and the frequency response of a system then provide a link between the signals material and the systems material. This presentation has the effect of focusing the reader on the engineering aspects of the material instead of emphasizing the transform mathematics.

System theory rests on mathematical transform theory: Fourier, Laplace, and z-transforms. In this text, however, we have chosen to relegate the transform theory to background chapters rather than present it as interwoven with the system theory results.

This book covers both continuous-time and discrete-time signals and systems. It has been noted that continuous-time and discrete-time theory share many mathematical properties. On the other hand, the physical processes that are modeled by continuous-time systems are very different than physical processes that are modeled by discrete-time systems. Because this book emphasizes the connections between physical processes and their models, we have separated the continuous-time theory from the discrete-time theory. Both presentations, however, are structured in a parallel fashion to emphasize the mathematical similarities of the material. In fact, the discrete-time material can be covered before the continuous-time material.

Integration of the Computer Computer aided design has been and continues to be an important component of industrial engineering. In recent years professionally written CAD packages have become available for classroom use. The computer tools reflect the way the practicing engineer organizes the underlying signals and systems concepts to solve everyday engineering problems. The conceptual framework employed by the practicing engineer is embodied in the conceptual framework of the text. In this way classroom discussion is integrated with professional practice and the structure of CAD packages. The signals material as it is presented in existing texts conforms to this standard. The conceptual framework of the text is structured to also integrate the systems material smoothly with the various computer tools currently available for classroom use.

The purpose of the introduction of the computer into a signals and systems course is to give the student the tools and ability to use these tools to analyze complex, realistic examples. We have chosen MATLAB®[1] for integration into this material. MATLAB, however, requires a familiarity with a core set of concepts and terminology in signals and systems. Since this course is the introductory course into the area of signals and systems, the essential issue is to give the students these concepts and terminology to understand the manuals and use MATLAB in a way that also advances their understanding of signal and systems material. Here we are concerned with the compatibility of the computational tool with the concepts that are being taught in the course. The organization of the proposed text is motivated, in part, by the relationship between the students' understanding of the basic concepts, the commands available in the computer package, and the students' ability to use the data structure of the CAD package. MATLAB has been integrated into the text in a structured and systematic way as described in Section 1.7, How to Use MATLAB with This Book.

Continuity with the Network Analysis Course The courses which are prerequisites for a signals and systems course are generally a networks course, a differential equations course, and a course which includes an introduction to dynamics. All of these topics are quite closely related to signals and systems, but the students frequently don't make the connection between these two blocks of material, particularly in the beginning. This loss of continuity between the networks course and the signals and systems course can be traced to the sequence of topics that are found in many introductory treatments of signals and systems. Typically, after signals and systems are defined, the abstract properties of systems are defined followed by the derivation of the convolution integral as a system representation. This discussion is followed by a discussion of Fourier series and transforms. These topics are abstract and mathematically challenging. Furthermore, these topics are not always covered in the preceding networks course. This initial block of mathematically oriented material leaves the students disoriented.

---

[1] MATLAB is a registered trademark of The MathWorks, Inc. 24 Prime Park Way, Natick, MA 01760-1500. Phone: 508-647-7000, http://www.mathworks.com

Table of Contents

Chapter 1

Introduction to Signals and Systems

Chapter Outline

The purpose of this chapter is to explain the basic philosophical concepts that form the basis for the technical concepts presented in the rest of the text. We discuss how these philosophical ideas are evident in the organization of the text. This philosophy also helps to explain how the material in this text is related to the engineering literature as a whole. To readers wholly unfamiliar with the contents of this book, the following discussion may seem a little abstract. Readers are urged to return to this chapter as they progress through the text to develop the big picture as well as the details.

Summary of Sections

Section 1.1:   We discuss the philosophy on which this book is based.

Section 1.2:   We introduce the two most fundamental concepts used in the text: signals and systems.

Section 1.3:   We discuss the principle of mathematical modeling.

Section 1.4:   We discuss more on the concepts of signals and systems.

Section 1.5:   We discuss continuous-time and discrete-time.

Section 1.6:   We discuss the basic organization of the text.

Section 1.7:   We explain how to use MATLAB with this book.

1.1 A LITTLE PHILOSOPHY

As engineers and scientists we are interested in understanding the phenomena in the physical world around us. This knowledge can be used to improve the way we interact with our environment, show us how to improve upon the mechanisms we find in it, and show us how to design and fabricate entirely new devices. There are some underlying principles in the methodologies used for the acquisition of this knowledge for extending our understanding of known concepts. The acquisition of knowledge begins with the observation of a physical process. We use "observation" in a rather general sense meaning not only a direct sensory perception of the physical process, but also indirect perception through a sensor. It is crucial, however, that the process be *observed* in some way.

Once the process has been observed (implying repeatability) the acquisition of a deeper knowledge of the process proceeds in two modes of inquiry. The first mode of inquiry continues with direct observation of the process - experimentation. The physical process is observed in a variety of settings. Its action on other physical processes is documented. Various techniques can be developed to act on the physical process to alter its characteristics. This approach is highly developed, and a vast array of laboratory instrumentation is available for investigation of every type of physical process.

The second mode of inquiry involves developing an abstract description of the physical process. Then this abstract description is used to indirectly investigate the properties of the physical process. The simplest abstract description is a verbal description of the physical process. The statement of Newton's laws gives us a verbal description of these fundamental laws of physics. Verbal descriptions, however, are limited in their ability to accurately describe the physical process. A much more powerful language for the description of a physical process is a mathematical description. Mathematics, in its broadest interpretation, contains a wealth of knowledge that can be brought to bear in the investigation of the properties of the physical process by analyzing its abstract representation. In this mode of inquiry, our understanding of the physical process is developed indirectly by studying the properties of the mathematical description using the tools of mathematics. For example, differential calculus is very useful for understanding and applying Newton's laws.

In recent years the computer has evolved as a new tool for understanding abstract descriptions of physical processes. Our observations of the physical process must be translated into numbers, a mathematical description of the physical process. By processing these numbers possibly in conjunction with a mathematical description of the physical process such as a differential equation, we are able to greatly expand our understanding of the physical process through automated computation. The extension of abstract descriptions of physical processes into the computer environment is having an enormous impact on the way engineering is done today.

Consider, for example, an oil painting. An oil painting is a physical process in that it persists through time. The painting is experimentally created by the artist with paints and a brush. The artist's understanding of the painting is, in part, through the act of painting, an experimental mode of inquiry. It is also possible to develop an abstract description of the painting. At the simplest level such an abstract description may be a verbal description of the color and geometry of the painting. With persistence, a more sophisticated description can be developed using the laws of physics along with a mathematical description of the colors and geometry. This more abstract, sophisticated description is useful in that it allows us to reproduce the painting on our computer screen. We can also use the computer model of the painting to gain insight into its historical origins.

At this early point in our discussion we emphasize that we are primarily interested in developing tools that can be used in the understanding of physical processes. Furthermore, there are two separate, but complementary approaches for understanding the physical process: experimentation and abstraction. Neither one of these approaches is satisfactory by itself, but depends on the other to guide it. This basic fact provides the foundation and orientation for the material in this book.

1.2 BASIC CONCEPTS

1.2.1 Introduction

In this book we will discuss the abstract description of a physical process. It turns out that diverse physical processes have mathematical descriptions that are similar in their mathematical properties. Furthermore, the same analytical tools can be used for the analysis of the mathematical descriptions of many of these processes. Therefore, the organization of the material in the book tends to emphasize the mathematical aspects of the subject. While this organization underscores the power and usefulness of this material, it should be remembered that the ultimate goal is to use these concepts to further our understanding of the physical processes. Some attempt has been made in the text to keep the readers tuned in to this objective. The fact that we don't discuss the experimental aspects of the analysis of physical processes doesn't imply that this knowledge is less useful than the concepts contained in this text. Its just that space is limited.

This book is concerned with the mathematical descriptions of a physical process and the analytical tools used to analyze these descriptions. To a lesser extent we will address the problem of design: synthesizing a mathematical description with the ultimate goal of constructing a physical device that matches the abstract description. The readers are undoubtedly familiar with this approach, it being the way of science. One of the primary goals of a first networks course is to introduce the mathematical tools used to describe the operation of a electric network composed of a resistor, capacitor, and inductor along with a voltage or current source. The voltages and currents in a network are represented by functions. The relationship between these voltages and currents is shown to be given by differential equations. Later it is shown how these mathematical objects can be analyzed using Laplace and Fourier transforms. These transforms simplify the calculations through the use of complex impedance, source transformations, etc. More importantly, these transforms expose properties of the voltages, currents, and circuits that allow deep insight into their behavior that is not readily apparent from the functions or differential equations or from experimentation. All of these results are simply abstract, mathematical descriptions of the physical processes which allow deep insight into the behavior of the circuit.

It is the purpose of this book to formalize and extend many of the results related to mathematical descriptions of physical processes obtained in the context of network analysis. In the course of this development it becomes apparent that the techniques involved in network analysis can be applied to a wide variety of physical phenomena. There are obvious parallels to mechanical systems, in particular vibration analysis and dynamics. Less obvious parallels appear in chemical processes, hydraulics, robotics, optics, and electric machines, to name a few. Even more obscure connections turn up in systems with decidedly different observed behavior such as economics, biological populations, image processing, and computers. It is exactly the fact that these techniques can be applied to such a wide spectrum of physical processes that gives them their power and beauty.

1.2.2 Two Fundamental Concepts

The challenge, of course, is to develop descriptive concepts that apply to these many diverse physical processes. There are two fundamental concepts that play a central role throughout the book: a signal and a system. A major purpose of this book is to develop these two concepts and their interaction in sufficient depth that they can be applied to the analysis of advance problems in engineering. As such these two concepts motivate the underlying organization of this book.

A signal, the starting point for our investigation, is the formal definition of an abstract mathematical description of an observed physical process. A signal[1] is a function that is used to describe an observed physical variable of a physical process; it is an abstract mathematical description of the observation. Hence, a signal provides the most basic connecting link between the physical process, in which we are ultimately interested, and the mathematical analysis techniques developed in this text. Everything grows from a signal.

Armed with the concept of a signal, we can begin to describe the physical world we see around us. Familiar signals include the functions used to describe the voltages and currents in circuits. Position, velocity, and acceleration of a mass (such as an automobile) are also readily described by signals. Images on the computer screen become signals. Newspapers boast signals describing the fluctuations of the Dow Jones average.

Once we begin to recognize the many signals around us, we may note that certain signals always appear when another signal is present. When we switch on a flashlight (motion of the switch) a light appears (electromagnetic waves). When a voltage is applied to a circuit, voltages and current appear at the circuit terminals. When the Fed changes the prime interest rate, the Dow Jones average changes. In fact there is a physical interrelationship between many physical processes. This interrelationship is the second fundamental component in the abstract mathematical

---

[1] A signal is defined formally in Chapter 5.

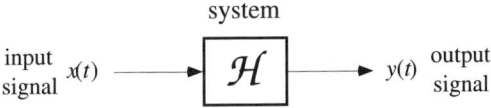

Figure 1.2.1 A System

language developed here to describe observed physical phenomena. The terms of the mathematical language, one signal, called the input signal, causes the appearance of a second signal, called the output signal. The mathematical relationship between the input signal and the output signal is called a system.[2] The abstract concept of a system is frequently described with a cartoon as shown in Figure 1.2.1. Because of Figure 1.2.1, this definition of a system is often called a black box definition of a system. A system is described by its external connections to the world; we don't care what is in the black box.

A system can be generally identified with a physical process, device, or (large, complex) interconnection of devices. As implied in the description above, a system is identified by defining the input signal and the output signal. A network becomes a system by identifying the voltage across the power source as the input signal and the voltage across the load resistor as the output signal. An automobile becomes a system by identifying the pressure on the accelerator as the input signal and the velocity of the car as the output signal. A stereo becomes a system by identifying the laser light reflecting off the tracks of the compact disk as the input signal, and the sound coming out of the speakers as the output signal. This definition of a system may appear somewhat novel, but we hope to show over the course of the book that it is extremely useful.

1.3 MATHEMATICAL MODELING

Signals and systems are mathematical objects that describe the physical world we observe around us. We call these mathematical objects representations or models of the signal or system. If signals and systems are to be a useful descriptive language, they must satisfy two criteria. First, the signal or system must accurately reflect the observed physical process. That is, the signal or system must be compared to the observations of the physical process, and they must agree to an acceptable level of accuracy. It is exactly here that the experimental characterization of the physical process is critical. The less precisely the signal or system matches the observations, the less precise our understanding of the physical process will be. Conversely, the functional form of the signal or system often motivates the laboratory experiments and establishes the acceptable ranges of parameter values. Hence, there is a close relationship between the abstract mathematical description of the physical process and the experimental understanding of that process.

Second, the mathematical model must be of a form which provides useful information. If the physical process is complex, then it is often easy to develop a complicated model which seems to accurately describe the experimental

---

[2] A system is defined formally in Chapter 6.

observations. Such a complex model, however, may not be tractable by any known analysis tools. Hence, we won't be able to develop any useful information about the physical process from the abstract description. If the model is oversimplified, then it won't retain any of the interesting behavior of the physical process. Again, the model serves no useful purpose. Therefore, it is important that the model fall into the category where the complexity of the physical process is represented with a model that is amenable to our analysis tools.

The development of mathematical models has been significantly impacted by the emergence of sophisticated computer simulation packages. These computer tools dramatically extend the usefulness of mathematical models. Some models that previously were too complex for analysis with pencil and paper become almost trivial with current computer technology. Furthermore, the range and depth of the analysis that can be performed within an acceptable period of time is dramatically increased. The advances in hardware and software technology allow much more of the system analysis and design to be transferred from experimental hardware into the computer laboratory.

New computational tools, which can be thought of as an extension of an abstract mathematical description of the physical process, have made computer simulation a full partner to laboratory experimentation. It must be remembered that these two modes of inquiry into the nature of the physical process are still complementary and mutually supportive. The advancements in computer technology, however, are shifting the relative importance of these two approaches in engineering analysis and design. In general, laboratory experiments can be very complex, expensive, difficult to construct, and time-consuming. In the most extreme cases, it may impossible to duplicate the physical process in the lab. In these situations, the computer simulations offer an attractive alternative physical experiments. Of course, for the computer simulations to be successful, accurate mathematical models must exist of the physical process. These models must also be numerically tractable. If the models can't be simulated even on the largest computers, then even complex, expensive experiments are necessary.

Many electronic circuits are easy to build and test in the lab, but they pose difficult problems for numerical simulations. Switching power supplies are one such class of circuits in which the transistors are operated in an on-off mode. This operation introduces discontinuities into the numerical simulations which are problematic. For these devices, it is often easier and faster to test their behavior in the lab rather than spend many hours on computer simulation.

NASA is currently building the space station. This structure is quite large, but it must be constructed of lightweight materials. It is a complicated truss structure with many joints and large point masses. It is not known how this structure will respond dynamically in space. It is impossible to experimentally investigate the dynamic response because a zero gravity environment of the required size can't be fabricated on earth. Therefore, computer simulation remains the only feasible method for characterization of the dynamic response of such large space structures.

Understanding how sound propagates in complex enclosures is very important to many engineering designs. For example, there is a lot of interest currently in developing systems that reduce the sound levels in aircraft cabins. It is possible to develop models that describe the propagation of acoustic waves in these environments, but these models require an inordinate amount of computer time to simulate. Therefore, computer simulations are of limited value in developing noise reduction systems for realistic enclosures at this time.

There has been a lot of interest in recent years in developing devices to reduce vibrations in flexible structures. These devices typically include a sensor to measure the vibrations in the structure, an actuator to apply forces to the structure, and a digital computer to generate an input signal into the actuator in response to the electronic output of the sensor. Unfortunately, if these devices are not configured correctly, they can cause the vibrations to increase, rather than decrease, leading to damage to the structure. In fact, the whole system can destruct rather rapidly. Nor is it always clear how to set the parameters of the vibration suppression device to obtain the desired performance. For these devices simulation models can be quite useful, not only for determining the device parameters for optimal performance from design methodologies, but also for evaluating those parameter values where the whole system would go into failure.

It may have occurred to the reader that a model of the stock market could be financially beneficial. To date no (published) model of this institution exists. The main problem is that while the proposed models match the observed data, apparently they don't predict the future behavior of the market. That is, the mathematical modeling of this system is inadequate. Experimentation remains the best method for understanding this system.

Engineers developing a model must understand what analysis tools are available as well as the physical process they are trying to model. It is the purpose of this book to describe some of the basic models that are used to describe a physical process, introduce some of the analysis tools that are available to analyze these models, and touch on some of the design procedures for synthesizing a device from a mathematical description.

The computer simulation tools are developed out of the abstract mathematical description language used to model physical processes. It is not surprising, then, that these computer tools have influenced the growth of the mathematical language and the analysis tools for signals and systems and vice versa. This interplay between the mathematical language and the computer simulation tools is part of an ongoing process in the growth of engineering analysis and design tools. These tools have reached a stage of development where it is appropriate to include them in a text at this level. Accordingly, MATLAB has been integrated into the material. Indeed, some thought has gone into presenting the material in the text to emphasize the relationship between the theoretical concepts and their implementation in MATLAB.

1.4 SIGNALS AND SYSTEMS

1.4.1 Introduction

As we discussed above, there are two fundamental concepts in our abstract mathematical descriptive language: a signal and a system. Because these two concepts play such a central role in this book we will discuss them further here. We will explain the various components of the analysis of signals and systems. We will discuss the ways in which simple models of signals and systems can be combined to represent complicated systems. We will also discuss how signals and systems interact with each other. Throughout the discussion we will give examples that provide motivation for studying these concepts.

1.4.2 Signals

Signals are functions that describe the time variation of a physical variable of a physical process.[3] This concept is rather easy to assimilate because we observe these physical variables around us on a daily basis. A good example of an observed physical variable is the oscilloscope trace of a voltage waveform. The form of the signal is easy enough to determine if we can observe the oscilloscope, but suppose we wish to explain the shape of the signal to someone who can't see the oscilloscope? Or suppose we wish to infer properties of the circuit from the shape of the waveform? In these cases a more sophisticated description of the oscilloscope trace is needed. At this point signal theory enters the picture.

There are three components to signal theory: modeling, analysis, and design. Signal modeling is concerned with developing an abstract description of the time history of an observed physical variable. Quite often signal modeling is an ad hoc process which relies heavily on the past experience of the engineer doing the modeling. In such situations, the physics of the underlying process often plays a prominent role in the form of the signal with the parameters of the signal being physically meaningful. A second way to construct a signal model is to sample the signal and then enter the sample values into a computer. The computational power of the computer allows very sophisticated and powerful models of signals to be developed. We will discuss a few simple computer-based techniques for developing signal models from observed experimental data.

The second component of signal theory is signal analysis. Signal analysis is concerned with extracting information about the underlying physical process from the signal. If a signal is developed as a description of an observed physical variable, then we can associate characteristics of the signal with the physical variable. In fact, this concept is central to our understanding of physical processes. For example, suppose we observe a sinusoidal oscilloscope trace. Then we would describe this observation with a sinusoidal function. In particular, we would choose an amplitude and frequency for the sinusoidal function so that it matches the observed trace. (The construction of the sinusoid is signal modeling.) Then we would say that we observed a signal of a certain frequency. This frequency, a parameter of the signal, could well turn out to characterize an important property of the circuit. In fact, we use this type of analysis routinely in the lab. The key point here is that the frequency is a property of the signal, the function, not the underlying physical variable. Only by attaching a mathematical description to the observation are we able to characterize it. One of the goals of signal theory is to develop this idea into a sophisticated and powerful modeling and analysis tool.

The third component of signal theory is signal design. Signal design is the reverse of signal modeling and analysis. Here we start with a signal, which is an abstract mathematical description, and we proceed to synthesize a physical process that is described by the signal. For example, suppose we want to generate a sinusoidal voltage of a specific frequency and amplitude. In merely stating the problem, we have begun with the signal. We would then proceed to construct a

---

[3] In advanced theory the definition of a signal can be more abstract.

circuit, a physical process, that would generate a voltage which when viewed on the oscilloscope screen would be described by a sinusoid of the specified amplitude and frequency.

There are several purposes for signal design. The most important reason is that by specifying the functional form of the signal we can associate an information content with the signal. For example, by specifying the functional form of the signal to be a series of positive and negative pulses, we can associate binary "1" and "0" with the positive and negative pulses, respectively. In this way we can encode into the signal a binary sequence. Clearly, this binary sequence can carry an information content. By synthesizing a voltage waveform, say, which is described by this signal we have embedded abstract information into the physical process, the waveform. This concept is fundamental to both communication and computers.

The second purpose of signal design is to determine the shape of the signal so that it will propagate through a system in a specified manner, usually with minimal distortion. For example, consider transmitting an electromagnetic wave from one antenna to another. We can consider the voltage supplied to the antenna that is used to generate the electromagnetic wave leaving the transmitting antenna as the input signal. The output signal is the voltage generated by the electromagnetic wave at the receiving antenna. These two signals clearly describe a system. The job of the antenna designer is to generate the voltage supplied to the transmitter such that the electromagnetic wave will propagate to the receiver with minimal distortion. This task begins by choosing a signal whose shape describes an electromagnetic wave that will propagate with minimal distortion. Typically this signal would also carry some information content. Here the signal design is also driven by the characteristics of the system.

1.4.3 Systems

Introduction The second fundamental concept associated with the material in this book is a system. A system is defined in terms of the relationship between two signals. The input signal into a system generates an output signal. In that signals are part of the mathematical language, a system is also part of that language. A system is typically used to describe a physical process, a device, or (large, complex) interconnection of devices. We interpret *device* in the broadest possible sense. For example, an aircraft can be a device that is modeled as a system. More abstractly, a computer algorithm can also be a system, so we also label it a device. The definition of a system implies that the input signal is qualitatively different from the output signal. The input signal represents a physical process that is generated independently from the system. The output signal, however, is generated by the physical process represented by the system when the input signal is present. The reverse process is not necessarily true; the output signal will not necessarily cause the appearance of the input signal.

Systems represent physical processes that can perform a variety of functions. One class of systems represent physical processes that transform energy from one form to another. These systems are often defined by identifying the input signal to the system as the signal that controls the energy transformation. Transportation vehicles are examples of this type of physical process. Some kind of chemical energy (fuel) is transformed into mechanical energy in the form of motion. The input signal into the system controls the conversion of chemical energy into mechanical energy. Another example of this type of physical process are electronic amplifiers. A low power voltage controls the power flow from the power bus to the load. The amplifier is represented as a system when the reference bias voltage is identified as the input signal and the voltage to the load is identified as the output signal.

A second class of systems represent physical processes that perform a function. An example of this type of physical process is a robot arm. The robot arm becomes a system when the command signal to the robot arm is identified as an input signal and the position of the gripper is identified as the output signal. The function of the gripper is to grip an object.

A third class of systems represent physical processes that in some way process the input signal. (These systems are called filters.) These systems enhance or remove characteristics of the input signal. A digital image is often enhanced using a computer algorithm. If the original digital image is identified as the input signal and the processed digital image is identified as the output signal, then the computer algorithm becomes the system. In this case the system restores a degraded signal. Another type of system extracts parameters or information from a signal. We explained above how information can be embedded into an electronic waveform. If this waveform is transmitted over some distance it can become corrupted. In order to extract the information, the received waveform is used as the input signal to a circuit whose output waveform is a clean version of the received waveform. Here the circuit is a system that extracts the information (output signal) from the corrupted waveform (input signal). Another very common function of a system is to pass certain signals through to the output signal while blocking the transmission of other signals. The equalizer on a stereo works on this principle. The input signal from the storage medium contains signals of all frequencies. The equalizer as a system attenuates some of these signals more than others. How the system acts on the signal depends on the frequency of the input signal.

System Modeling, Analysis, and Design   A system as a mathematical object is a model or representation. Again we emphasize that if this model is to be useful in the engineering design process it must satisfy two criteria. It must accurately describe the physical process or device it represents. This model must also lend itself to analysis being neither too simplistic nor too complicated. The study of these system models is called system analysis. System analysis consists of three main areas: mathematical modeling, analysis, and design.

Mathematical modeling is concerned with the development of different forms of equations that can be used to represent a system. The mathematical modeling of systems has long been studied, and there are many specialized methods for obtaining system models. These methods can be divided into two approaches. The first approach uses physical laws to develop the model. This approach predominates in the introductory courses on networks and dynamics. The second approach, known as system identification, attempts to back calculate the model of the system from knowledge of the input and output signals. In this approach, a known input signal is applied to the physical process and the corresponding output signal is measured experimentally. Then a model of the system is computed so that the simulated response of the model matches the experimental data. This approach to modeling is rooted in the basic concepts of signals and systems developed in this book.

System analysis consists of analyzing the system model to uncover properties and characteristics of the system. There are a large number of analysis tools for investigating the properties of a system, many of which are discussed in this text. In addition, computer analysis tools extend our ability to use a system model to understand the underlying physical process.   System analysis can support the development of a new device in several ways.   Through system analysis we can examine the effect on the performance of the system as one or more parameters of the system change. If each of these parameters corresponds to a component in the system, then this analysis can aid in the determination of the performance specifications within the tolerances of the components or in the selection of the components. In another interpretation, the changes in the parameter values may represent variations due to changes in the environment such as temperature and pressure, or changes due to aging. System analysis can determine whether a system will continue to meet its performance specifications throughout its operating envelope.   Very complicated systems often exhibit unusual and/or unexpected behavior. System analysis often lends insight into the cause or source of this behavior. Conversely, system analysis allows the system behavior to be thoroughly probed in an efficient manner to expose system deficiencies. System analysis can replace lengthy and potentially costly laboratory experimentation.   It should be remembered that the viability of system analysis depends on the accuracy of the model. There is a close relationship between the model development and the system analysis.

System design is the converse process of the mathematical modeling. System design begins by postulating a mathematical model with desirable properties and then constructing a device which matches the mathematical model. In this way we ensure that the device has the properties we desire.   System design in general requires ingenuity on the part of the designer. Hence, there are many approaches to system design which depend on the underlying device. We may distinguish two classes of design, however. In the first class we seek to modify an existing system by the addition of hardware, or by attaching another system to the given system. Often the reason for modifying the existing system is to improve its response to given input signals. This type of system design falls into the domain of control theory. In the second class of system design we seek to fabricate an entirely new system. The most important example of this type of design is filter design. Historically, this type of design is concerned with building a circuit with a given transfer function. That is, we want the circuit to respond in a certain way to sinusoidal input signals with specified frequencies. More recently, the electronic circuit has been replaced by a microprocessor. Now the filter design focuses on finding an appropriate algorithm to achieve the desired processing of the input signal.

1.4.4 Interconnections of Signals and Systems

Signals and systems form the basic building blocks in the mathematical language that we use to describe physical processes. Obviously many physical processes are very complex and consist of many interrelated components.   To describe these complicated physical processes it is often useful to think of them as being composed of many interrelated signals and systems. By establishing interconnections between simple signals and systems we can build up complex systems that represent very complicated physical processes.

There are several ways to combine signals together. Two of the most useful ways are by adding two signals together and by multiplying two signals together. These operations are shown in Figure 1.4.1.

A system is defined as having an input signal and an output signal. It is entirely possible that the output signal of one system may be the input signal to another system. Or the output signals of two systems may be combined and used as the input signal to another system. In this way the larger systems are formed from the interconnection of smaller subsystems. Several interconnection patterns are shown in Figure 1.4.2. The importance of breaking a complex system into interconnected subsystems for analysis can't be overstated. The behavior of complex systems can be puzzling and difficult to understand. By breaking the system into smaller pieces, and analyzing each piece separately much insight is gained into the collective behavior of the entire system.

A very familiar example of an interconnected system is a stereo system. Consider the input signal to be the light reflected from the CD and the output signal to be the sound from the speakers. If the CD player is separate from the amplifier which in turn is separate from the speakers, we have an interconnected system. The first system would be the CD player. The input signal to this system is the light reflected from the CD. The output signal is the voltage at the output terminals of the CD player. This output signal forms the input signal into the amplifier, the second system. The output signal from the amplifier is the voltage at its output terminals. This output signal forms the input signal to the speakers, the third system. The output signal from the speakers is the sound, the output signal of the whole system. Here we have a composite system of three subsystems hooked together in a serial fashion.

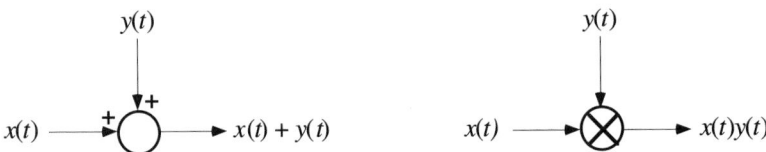

Figure 1.4.1 Combining Two Signals

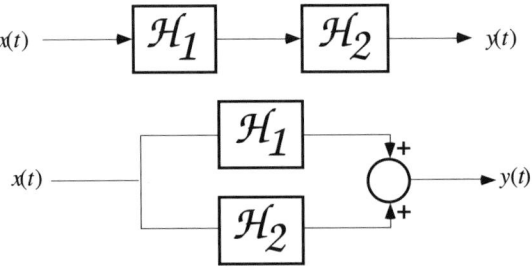

Figure 1.4.2 Interconnections of Systems

1.4.5 Interaction Between Signals and Systems

Signals and systems are closely interrelated concepts. Each can be studied separately and we shall do so. We are also interested in the relationship between signal concepts and systems concepts, however. This relationship can best be understood by thinking of the input signal propagating through the system and emerging as the output signal. From this point of view, the system changes the input signal as it propagates through the system. The question to which we address ourselves is characterizing the change in the signal properties using properties of the system. This question is of fundamental importance in many physical processes of interest.

We have noted that a signal can carry information by properly selecting the waveform. These signals are the foundation of communication systems where the signal carries information over long distances. In effect the communication signal is the input signal into a system (the communication channel). This signal is sent by the transmitter. The output signal is the signal received by the receiver. The input signal (the transmitted signal) is changed (distorted) by the system (the communication channel) so that the output signal (the received signal) only vaguely resembles the input signal. In order to properly design a communication signal it is important to understand how the system changes the communication signal so that the amount of distortion introduced by the channel can be minimized. Indeed, pulse shaping, used to reduce the distortion, is one of the standard results in communication theory.

In aircraft the flaps on the wings are controlled by the pilot. The pilot pushes on the "stick" a specified amount to change the flap deflection by a specified angle. One way this system is implemented is by actuating the flap with electric motors. The movement of the stick generates an electric signal that causes the motor shaft to rotate through a specified angle. A gear linkage then moves the flap. If we denote the input signal as the position of the stick and the output signal as the position of the flap, we have a system. In this system we want the output signal to follow exactly the input signal; i.e., we want them to have the same functional form, although they may be scaled differently because of the units. A direct linkage between the stick and the flap is not desirable because it will require a considerable amount of energy to move the flap because of the aerodynamic forces. The function of the system is to change a low-energy signal into a high-energy signal. We must understand the relationship between the input signal and the system so that we can design the system properly.

There are many applications where the information in a signal is carried in the frequency of the signal. That is, the signal is a sinusoid and the different frequencies are correlated to different information. To extract the information in the signal it is necessary to determine the frequency of the signal. One way to solve this problem is to input the signal into a system which is specially designed to pass only input signals with a given frequency. All other signals are severely attenuated. It turns out that the key concept is contained in the way a sinusoidal signal propagates through a system. The Fourier and Laplace transform also play a central role in the relationship. Hence, this interaction between the signal and the system is said to be "in the frequency domain." These results are described in detail in the text. These results are very deep and of fundamental importance.

1.5 CONTINUOUS-TIME AND DISCRETE-TIME SIGNALS AND SYSTEMS

We have defined a signal as a function which describes a physical variable as it evolves in <u>time</u>. In this text we will use two different conceptions of time. The first conception of time is the most common usage of the word. Here time is modeled using the real numbers. Signals become functions of a real variable. We say these signals are <u>continuous-time</u> signals. Systems with continuous-time input and output signals are called <u>continuous-time systems</u>.

Almost all introductory courses in networks, physics and mechanics are devoted to continuous-time signals and systems. These signals represent physical processes readily observed in the world around us. This concept of time is most directly part of our reality. Voltage, current, position, velocity, pressure, etc. are all physical processes represented by continuous-time signals.

The second conception of time is associated frequently with computers and economics. Here time is modeled using the integers. The signals, which are functions of the integers, effectively become sequences. We say these signals are <u>discrete-time</u> signals. Systems with discrete-time input and output signals are called <u>discrete-time systems</u>.

Discrete-time signals are not unfamiliar to us. A glance at the market page of the newspaper will show a graph of a recent trend in the stock market. Typically, one price is given for each day. By associating an integer with each day we readily obtain a discrete signal. Certainly economic theory makes extensive use of discrete-time signals, but an even more important example for electrical engineers occurs in the computer. Because computers inherently process data sequentially in discrete steps, discrete-time signals naturally model this process.

It turns out that the models of discrete signals and systems share many of the same mathematical properties. Therefore, many of the same modeling, analysis, and design tools can be applied to both types of systems with only minor modifications. The text is written to exploit this similarity.

1.6 ORGANIZATION OF THE MATERIAL

1.6.1 Organization of the Chapters

The concepts that are presented in this book are highly interrelated. Their relationship could best be presented by a flow chart that is constructed as a matrix. The layout of a book, however, requires that the concepts be presented in a linear sequential structure. Because the matrix relationship between the concepts can be reorganized into a column in several different ways, there are several organizations of this material that can be adopted for presentation in book form. In this section we will briefly explain the overall organization of this book. It must be emphasized, however, that a deep understanding of this material recognizes the cross-chapter relationship of the concepts.

Given the scope and depth of this material the many cross relationships between the concepts can be very confusing to the beginning student. Therefore, this text has been organized so that it can be read straight through. For the advanced student, however, the interrelationships between the concepts are quite important. Therefore, the text has been organized to exploit these interrelationships. The following discussion is an aid for the advanced students to guide them through the more subtle organizational aspects of this text.

There are several fundamental divisions of the material as it is presented in this book which we will explain next. The primary division of the presentation is the between continuous-time and discrete-time signals and systems. Continuous-time signals and systems are discussed in Part I of the text and discrete-time signals and systems are discussed in Part II. This basic division of the text is shown in Figure 1.6.1. It must be emphasized, however, that these two sets of results are closely related. In fact, they are essentially parallel to each other. It has been widely recognized that mathematically, these results are the same (on most points). (Of course, continuous-time and discrete-time systems are used to model very different physical processes.) The close relationship between these two blocks of material can be observed in the Table of Contents by comparing the continuous-time chapters to the discrete-time chapters. Part II of this text has been written so that it can be read independently of Part I (aside from the introductory chapters in Part I).

Within Part I and Part II of the text, the material has been further subdivided according to the concepts introduced in this chapter. The discussion begins with a formal introduction to the definitions of a signal and system. The concepts directly related to signals are separated from the concepts related to systems. Finally, the relationship between signals and systems is developed. This division is accomplished through the chapter designations. This organization of the material is repeated in both the continuous-time and discrete-time material. The breakdown of the chapters is summarized in Figure 1.6.1.

In the introductory chapters we discuss basic functions that can be used for signals, and various equations that can be used for modeling systems. The chapters on signals focus on developing frequency domain descriptions of signals. The main tools in this discussion are the Fourier series, Fourier transforms, and discrete-time Fourier transform. The chapters on systems focus on the development of system models and their interrelationship. The transfer function, convolution integral, and state space representation are discussed. The properties of these system models are also discussed, including linearity and time invariance. The chapters on the relationship between signals and systems build on the concepts introduced in the previous chapters.

The signals and systems concepts in this book require a certain amount of mathematical background. There are several chapters which cover this background. A concerted effort has been made to separate this background material from the concepts which have an explicit engineering orientation. The chapters devoted to background material can be covered as necessary. Here is a summary of this background material.

Chapters 2 - 4	This material is commonly found in the prerequisite math and networks courses of a signals and systems course.
Chapter 7	The Fourier series and Fourier transform.
Chapter 9	The Laplace transform.
Chapter 18	The discrete-time transforms that parallel the Laplace and Fourier transform.

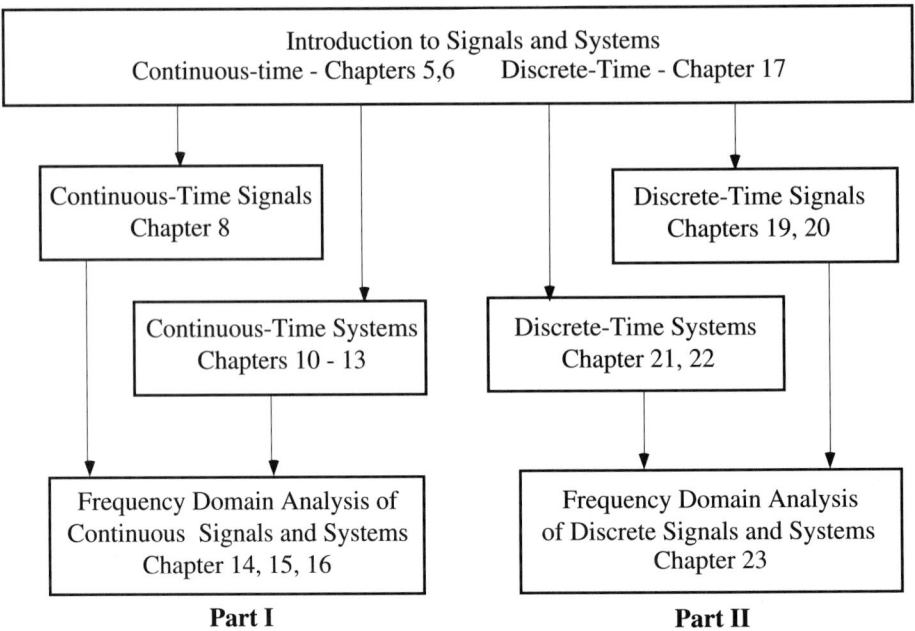

Figure 1.6.1 Outline of the Chapter Structure

1.6.2 Coverage of the Text

The many concepts in this text are related in many ways. The relationship between the chapters can best be described by a matrix structure shown in Figure 1.6.2. Figure 1.6.2 can be used as a road map for reading selected portions of this book. As can be seen from Figure 1.6.2 the later chapters don't necessarily require all of the results from the previous chapters. The interrelationships between the chapters in Figure 1.6.2 is also summarized in the chapter introduction in the subsections entitled *Coverage of the Text*. These subsections give a detailed description of the prerequisite material that is required for each section of the chapter.   This information, which can be quite confusing to the beginning student, is provided primarily for the advanced student as a guide for tracing the specific relationships between concepts.

1.7 HOW TO USE MATLAB WITH THIS BOOK

The primary purpose of this text is to provide a comprehensive framework for the basic concepts of the theory of signals and systems. This presentation of this theory is self-contained, and it is illustrated by numerous examples in the text. It must be recognized, however, that some of the theoretical results derive their practical importance from their application to numerical computation.   Furthermore, an understanding of the examples and the underlying theory can be enhanced by

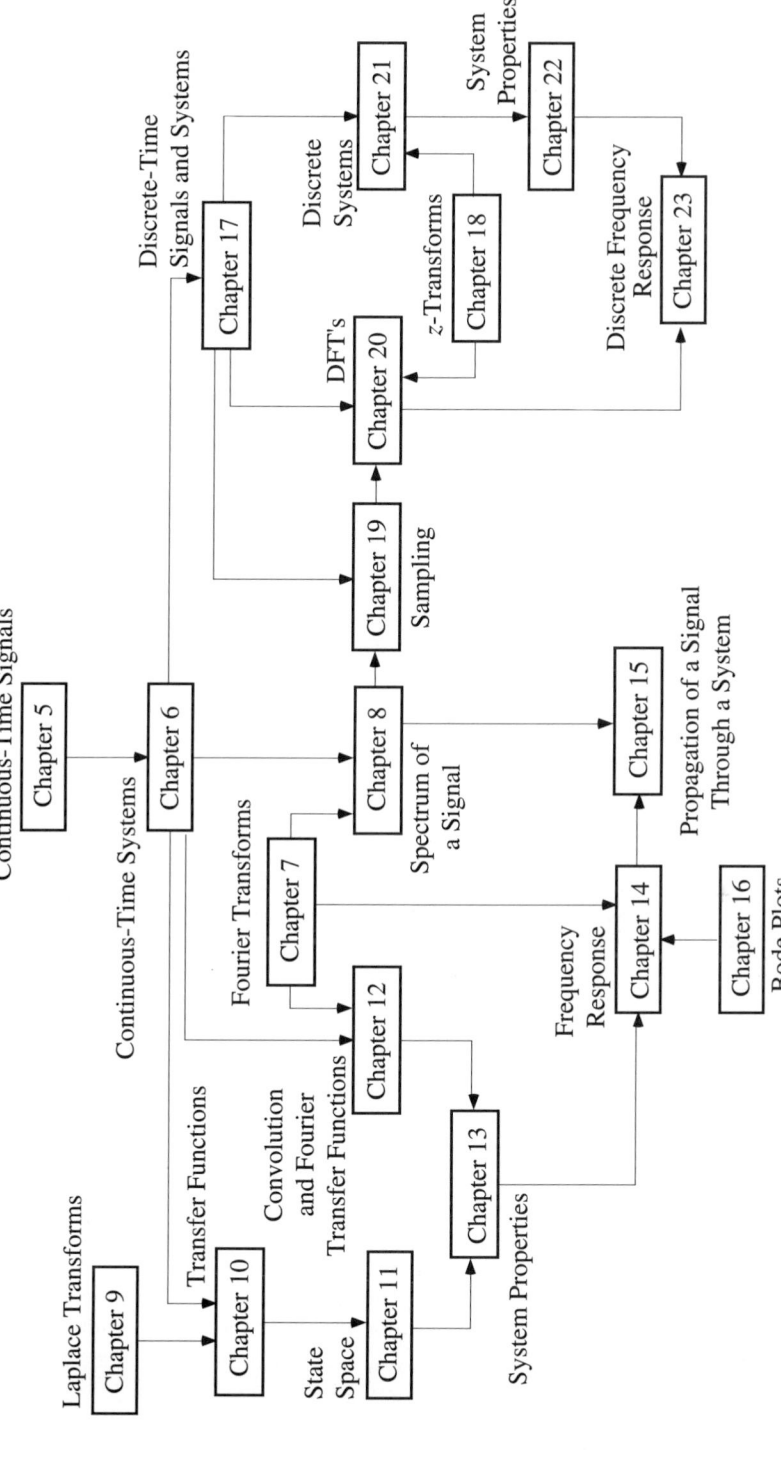

Figure 1.6.2 Relationship Between Chapters in Text with Topical Coverage

investigation through numerical simulation. Therefore, a computational software tool, MATLAB, has been integrated into the text as an extension of the existing material. The discussion of MATLAB in relation to the material generally follows the discussion of the main theoretical concepts and examples in each section or chapter, so the availability of MATLAB is not strictly required for self-study.

As a computational tool MATLAB is well-suited to extend the readers' understanding of the theory beyond the traditional confines of the written page. As each new theoretical concept is introduced, the relationship between the concept and the MATLAB command structure is explained. Then a MATLAB M-file is given illustrating a calculation using the concept. Generally, this M-file recreates the supporting example in the text. The example in the text gives the readers a point of reference for the interpretation of the output from the M-file. From this clearly defined starting point, readers can build their understanding by modifying the M-file to probe deeper into the meaning of the example. The Exploratory Exercises serve as a guide to readers in this investigation.

MATLAB has evolved into a standard computational tool that is widely used in industry as well as academia. Therefore, there is merit in gaining a working knowledge of the software itself. This knowledge will allow the readers to use MATLAB in conjunction with the theory of signals and systems to solve real problems. This basic philosophy guides the integration of MATLAB into the material. The way this philosophy is implemented in the text structure is that most sections contain a subsection at the end titled "MATLAB Experiments." This subsection contains a brief description of the MATLAB commands that pertain to the concepts introduced into the section. Then a MATLAB M-file is given which illustrates the relationship between the concept presented in the section and the MATLAB command. It is suggested that the readers type in each M-file. (They are usually short.) In this way the readers will learn the syntax and structure of MATLAB through example. This experience will build the expertise of readers to write their own M-files to solve their own problems. By entering the code of the particular M-file the readers will understand how this M-file solves the particular problem. This understanding will allow the readers to modify the M-file. The readers can then use the M-file to explore the implications of the concept through the example with the aid of the Exploratory Exercises.

The previous remarks apply to the integration of MATLAB into the main body of the text. The problems at the end of the chapters present a more in-depth challenge in the application of MATLAB. Because of the comprehensive integration of MATLAB into this material, it is generally assumed that MATLAB is available for problem solving. Hence, computer problems are not identified as such.

The description of the MATLAB commands in this text are not intended to substitute for the MATLAB manual or help files. It is assumed that readers are familiar with the basics of MATLAB as provided in the introductory chapters of the MATLAB manual. It is also assumed that the readers will make extensive use of the MATLAB help files, particularly if there is some question about the command syntax in the M-files given in the text. Rather, the descriptions of the MATLAB commands are intended to provide an interface between the material in the text and MATLAB, explaining how MATLAB can be used to further the understanding of the material and how it is applied to problem solving.

Chapter 2

Real Functions

Chapter Outline

This chapter has two primary goals. First, we review basic mathematical background required for the rest of the text. Second, we establish the mathematical notation, particularly with respect to functions, which we will use throughout the text.

In Sections 2.1 and 2.2 we review the concept of a function, and we introduce the notation of a number of basic functions including singularity functions. Because these functions are used extensively, it is suggested that readers familiarize themselves with the notation.

Section 2.3 introduces the notation of a discrete function. This material provides the starting point for the discussion of discrete signals and systems.

Summary of Sections

Section 2.1:   We introduce some basic concepts of real functions.

Section 2.2:   We define several common functions including singularity functions and some special functions.

Section 2.3:   We introduce discrete functions.

Coverage of the Text

This chapter is self-contained.

2.1 CONTINUOUS-TIME FUNCTIONS

In this text we assume that the reader is familiar with elementary set theory.

Notation: Let $\mathcal{R}$ be the set of real numbers.

Definition 2.1.1: An <u>interval</u> is a set with one of the forms

(i) $I_1 = \{t \in \mathcal{R} \mid t_- < t < t_+\} = (t_-, t_+)$

(ii) $I_2 = \{t \in \mathcal{R} \mid t_- < t \leq t_+\} = (t_-, t_+]$

(iii) $I_3 = \{t \in \mathcal{R} \mid t_- \leq t < t_+\} = [t_-, t_+)$

(iv) $I_4 = \{t \in \mathcal{R} \mid t_- \leq t \leq t_+\} = [t_-, t_+]$

We say that the interval (i) is an <u>open</u> interval. We say that the interval (iv) is a <u>closed</u> interval if $-\infty < t_- < t < t_+ < \infty$.

▲▲

Intervals (ii) and (iii) are neither open nor closed.

Notation: If $t_- = -\infty$ or $t_+ = \infty$, then the strict inequality applies. When we write

$$I_1 = \{t \in \mathcal{R} \mid t_- < t < \infty\} \tag{2.1.1}$$

we mean that t can take on arbitrarily large positive values, but not infinity.

Definition 2.1.2: Let I_1 and I_2 be two intervals. Let f be a rule which assigns a member of I_2, $t_2 \in I_2$, to each member of I_1, $t_1 \in I_1$; write $f(t_1) = t_2$. Then we say f is a (real) <u>function</u>; write $f(t)$. The interval on which a function $f(t)$ is defined, I_1, is called the <u>domain</u> of the function. The interval from which the function takes values, I_2, is called the <u>range</u>.

▲▲

A function can take on only one value. Obviously, functions can be defined for any set, not just intervals. A function is not defined outside its domain. The domain of definition plays an important role in the definitions of the Laplace and Fourier transforms below. Functions are distinguished, in part by their domains of definition. Let us consider the function $\sin t$. Suppose we define the interval

$$I_1 = \{t \in \mathcal{R} \mid 0 \leq t < \infty\} \tag{2.1.2}$$

Now define the function

$$f_1 : I_1 \to \mathcal{R}, \quad f_1(t) = \sin t. \tag{2.1.3}$$

This function is shown in Figure 2.1.1. In particular, the negative real axis is not in Figure 2.1.1, because it is not in the domain of definition. Consider next the interval

$$I_2 = \{t \in \mathcal{R} \mid -\infty < t < \infty\} = \mathcal{R}. \tag{2.1.4}$$

Define a second function

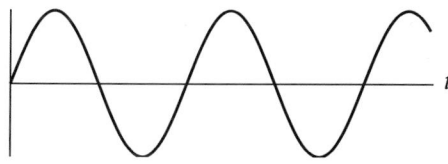

Figure 2.1.1 Function $f_1(t) = \sin t, \quad t \geq 0$

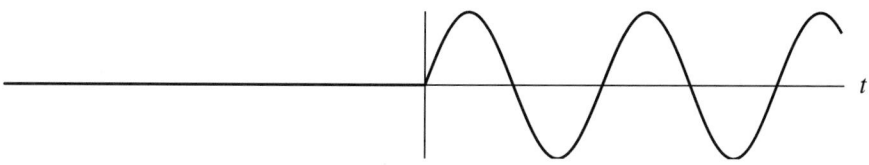

Figure 2.1.2 Function $f_2(t) = \begin{cases} \sin t, & t \geq 0 \\ 0, & t < 0 \end{cases}$

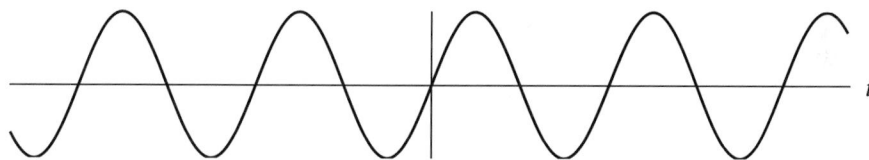

Figure 2.1.3 Function $f_3(t) = \sin t, \quad -\infty < t < \infty$

$$f_2: \mathcal{R} \to \mathcal{R}, \quad f_2(t) = \begin{cases} \sin t, & t \geq 0 \\ 0, & t < 0 \end{cases} \tag{2.1.5}$$

This function, shown in Figure 2.1.2, is defined on the whole real axis; it is zero for negative values of t. The function $f_1(t)$ can't be compared to function $f_2(t)$ because they are defined on different domains. Finally, consider a third function

$$f_3: \mathcal{R} \to \mathcal{R}, \quad f_3(t) = \sin t. \tag{2.1.6}$$

This function is shown in Figure 2.1.3. The functions $f_2(t)$ and $f_3(t)$ are defined on the entire real line, but they are clearly not the same function since they are not equal for $t < 0$.

The properties of even and odd functions are useful in the discussion of Fourier series and Fourier transforms.

Definition 2.1.3:   (a) The function x(t) is <u>even</u> if $x(t) = x(-t)$.
                          (b) The function x(t) is <u>odd</u> if $x(t) = -x(-t)$.

▲▲

Example 2.1.4: The functions $\cos t$ and $|x(t)|$ are even functions. The functions $\sin t$ and $x(t) = t$ are odd functions.

▲▲

Remark 2.1.5: Let $x_o(t)$ be an odd function and $x_e(t)$ be an even function. Then these two functions have the following properties:

(a) $x_e(t)x_e(t) = $ even function,

(b) $x_o(t)x_e(t) = $ odd function,

(c) $x_o(t)x_o(t) = $ even function,

(d) $\displaystyle\int_{-a}^{a} x_0(t)\,dt = 0$,

(e) $\displaystyle\int_{-a}^{a} x_e(t)\,dt = 2\int_{0}^{a} x_e(t)\,dt$.

▲▲

According to properties (d) and (e) even and odd functions can significantly reduce the amount of computation required for evaluating an integral. These two properties can be easily established by interpreting the integral as the area under the function.

2.2 COMMON FUNCTIONS

In the following section we introduce a number of functions which are used frequently throughout this text and we discuss some of their properties. This section is used primarily to establish notation.

Definition 2.2.1: The <u>unit impulse function</u> (or <u>Delta function</u>) $\delta(t)$ satisfies

$$\int_{-\infty}^{\infty} f(t)\,\delta(t)\,dt = f(0)$$

if $f(t)$ is continuous at $t = 0$.

▲▲

The unit impulse "function" is not, technically speaking, a function. Therefore, the graph of this function does not exist. It is useful, however, to represent $A\delta(t - t_0)$ by a vertical arrow as shown in Figure 2.2.1. The arrow is located at t_0. The height of the arrow is shown as A, but this graph can't be justified mathematically. It is to be used as a memory aid.

Remark 2.2.2: From the definition we can deduce that

$$\delta(t) = 0, \quad t \neq 0. \tag{2.2.1}$$

It is clear from the definition that the impulse function is not zero at the origin. On the other hand it can be shown that a function, as defined mathematically, can't satisfy (2.2.1) and also satisfy Definition 2.2.1. We conclude that the impulse

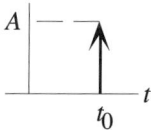

Figure 2.2.1 Impulse Function $A\delta(t - t_0)$

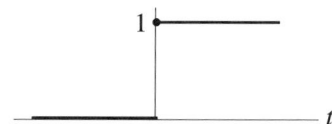

Figure 2.2.2 Unit Step Function $u_s(t)$

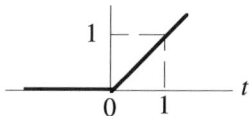

Figure 2.2.3 Unit Ramp Function $r_p(t)$

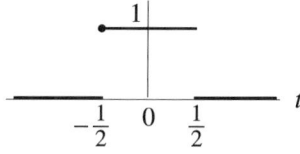

Figure 2.2.4 Unit Pulse Function $\Pi(t)$

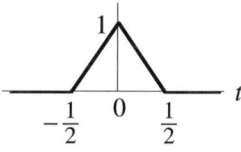

Figure 2.2.5 Unit Triangle Function $\Lambda(t)$

"function" is not really a function. Hence, some care should be exercised in its use. Using (2.2.1) we have

$$f(t)\delta(t - t_0) = f(t_0)\delta(t - t_0) \tag{2.2.2}$$

if $f(t)$ is continuous at t_0. Using (2.2.2) in Definition 2.2.1 we have

$$\int_{\infty}^{\infty} f(t)\,\delta(t-t_0)\,dt = \int_{-\infty}^{\infty} f(t_0)\,\delta(t-t_0)\,dt = f(t_0). \tag{2.2.3}$$

This property of the impulse function is sometimes called the <u>sifting</u> property.

If we take

$$f(t) = \begin{cases} 0, & t \le t_1 \\ 1, & t_1 < t < t_2 \\ 0, & t_2 \le t \end{cases} \tag{2.2.4}$$

and use (2.2.1) we conclude

$$\int_{t_1}^{t_2} f(t)\,\delta(t-t_0)\,dt = \int_{t_1}^{t_2} \delta(t-t_0)\,dt = \begin{cases} 1, & \text{if } t_1 < t_0 < t_2 \\ 0, & \text{if } t_0 < t_1,\ \ t_0 > t_2 \end{cases} \tag{2.2.5}$$

In (2.2.5) if $t_1 = t_0$ or $t_2 = t_0$, the integral is undefined. The nonexistence of the integral is really a consequence of (2.2.1) where we noted that the impulse function is not defined at the point where the argument of the impulse function is zero. It is here that problems with the practical application of the impulse function arise. If one of the limits of the integral falls where the argument of the impulse function is zero, then the limit of the integral must be redefined.   ▲▲

Definition 2.2.3: The <u>unit step function</u> $u_s(t)$ is defined as

$$u_s(t) = \begin{cases} 1, & t \ge 0 \\ 0, & t < 0 \end{cases}$$

▲▲

This function is shown in Figure 2.2.2.

Notation: When it is important to identify the value of a function at a discontinuity, the function value is denoted by a dot as shown in Figure 2.2.2 at $t = 0$.

Notation: The notation for the unit step function is not standardized. Several other symbols are commonly used including $u(t)$, $1(t)$, and $\mu(t)$. When reading other materials the notation should always be verified. The notation for the unit impulse function is standard.

Definition 2.2.4: The <u>unit ramp function</u> $r_p(t)$ is defined as

$$r_p(t) = \begin{cases} t, & t \ge 0 \\ 0, & t < 0 \end{cases}$$

▲▲

This function is shown in Figure 2.2.3.

Definition 2.2.5: The <u>unit pulse function</u> $\Pi(t)$ is defined as

$$\Pi(t) = \begin{cases} 1, & -\frac{1}{2} \le t < \frac{1}{2} \\ 0, & \text{elsewhere} \end{cases}$$

▲▲

This function is shown in Figure 2.2.4. The following function is occasionally useful in the discussion of Fourier transforms.

Definition 2.2.6: The <u>unit triangle function</u> $\Lambda(t)$ is defined as

$$\Lambda(t) = \begin{cases} t + \dfrac{1}{2}, & -\frac{1}{2} \le t \le 0 \\ \dfrac{1}{2} - t, & 0 \le t \le \frac{1}{2} \\ 0, & \text{elsewhere} \end{cases}$$

▲▲

This function is shown in Figure 2.2.5. The unit pulse function is not usually classified as a singularity function as its usefulness beyond this text is limited. We shall find frequent use for it, however. The unit triangle function is also not widely used.

   The following functions are widely used.

Definition 2.2.7: The function $f(t)$ is called an <u>exponential</u> function if

$$f(t) = Ae^{at}.$$

▲▲

Notation: For convenience the exponential function is sometimes written as

$$e^{at} = \exp(at). \tag{2.2.6}$$

Definition 2.2.8: If $t > 0$, the <u>natural logarithm</u> of t is any number y such that

$$t = e^{y}.$$

We write

$$y = \ln(t).$$

▲▲

Definition 2.2.9: If $t > 0$, the <u>logarithm base 10</u> of t is any number y such that

$t = 10^y$.

We write

$y = \log(t)$.

▲▲

Notation: To distinguish a base 10 logarithm from a logarithm in another base we write $\log_{10}(y)$.

Definition 2.2.10: The function $f(t)$ is called a <u>sinusoidal</u> function if

$f(t) = A \sin(\omega t + \phi)$.

The <u>amplitude</u> of this function is A. The <u>frequency</u> of this function is

$\omega_0 = 2\pi f_0$

where ω_0 is in radians/second and f_0 is in hertz. The <u>period</u> T_0 of this function is

$$T_0 = \frac{1}{f_0} = \frac{2\pi}{\omega_0}$$

where T_0 is in seconds. The <u>phase</u> of this function is ϕ. All of these terms are identified in Figure 2.2.6.

▲▲

In Figure 2.2.6 at $t = 0$ the value of $f(t)$ is

$f(0) = A \sin(\phi).$ \hfill (2.2.7)

From (2.2.7) we see that the phase of a sinusoid controls the time shift of a sinusoid.

Terminology: Note that

$\cos \omega t = \sin\left(\omega t - \frac{\pi}{2}\right).$ \hfill (2.2.8)

The term "sinusoidal function" is used to refer to both sine and cosine functions.

The following two functions occur commonly in Fourier transforms, signal processing, and communication systems.

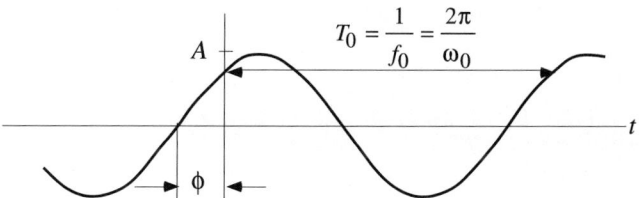

Figure 2.2.6 The Sinusoidal Function $f(t) = A\sin(\omega_0 t + \phi)$

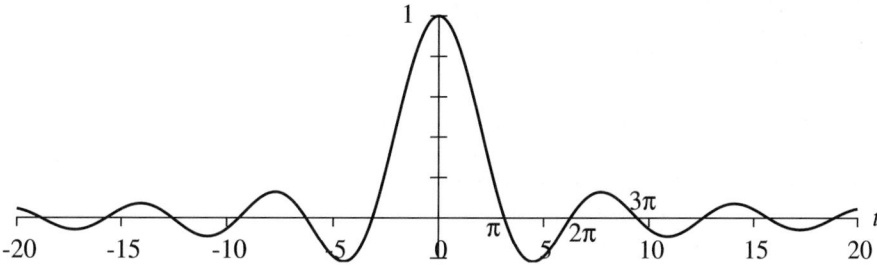

Figure 2.2.7 The Function Sa(t)

Definition 2.2.11: (a) The <u>Sa function</u> Sa(t) is defined as

$$Sa(t) = \begin{cases} \dfrac{\sin t}{t}, & t \neq 0 \\ 1, & t = 0 \end{cases}$$

(b) The <u>sinc function</u> sinc(t) is defined as

$$sinc(t) = \begin{cases} \dfrac{\sin \pi t}{\pi t}, & t \neq 0 \\ 1, & t = 0 \end{cases}$$

▲▲

The graph of the Sa function is shown in Figure 2.2.7. The appearance of the variable t in the denominator of this function means that this function is not defined at $t = 0$. It is logical to define the value at $t = 0$ to be 1, however. To see why, we

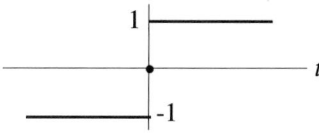

Figure 2.2.8 Sign Function sgn(t)

apply L'Hopital's rule. Taking the derivative of the numerator and denominator we get

$$\lim_{t \to 0} \frac{\sin \pi t}{\pi t} = \lim_{t \to 0} \frac{\frac{d \sin \pi t}{dt}}{\frac{d \pi t}{dt}} = \lim_{t \to 0} \frac{\pi \cos \pi t}{\pi} = 1. \qquad (2.2.9)$$

This calculation shows that an appropriate way to define the sinc function at $t = 0$ is to set it equal to 1, as shown in Figure 2.2.7.

When $t \neq 0$, the zeros of these two functions are at those values of the argument where the sine function is zero. For the sinc function the zeros are at

$$t = k, \quad \text{for } k \text{ integer}, \quad k \neq 0. \qquad (2.2.10)$$

For the Sa function, the zeros appear at $t = k\pi$.

Definition 2.2.12: The <u>sign</u> function is defined as

$$\text{sgn}(t) = \begin{cases} 1, & t > 0 \\ 0, & t = 0 \\ -1, & t < 0 \end{cases}$$

▲▲

A graph of the sign function is shown in Figure 2.2.8.

2.3 DISCRETE-TIME FUNCTIONS

We have been concerned with functions that depend on a real variable t. In this text we will also consider functions that are defined only on the integers.

Notation: Let the set of integers be denoted by I. We shall reserve κ to denote a fixed integer.

Definition 2.3.1: Let $f: I \to \mathcal{R}$ be a function that maps the integers into the real number; write $f(n)$, $n \in I$. Then we say $f(n)$ is a <u>discrete function</u>.

▲▲

Terminology: A sequence of numbers is represented by

$$a_n, \quad n = 1, 2, 3, \dots . \qquad (2.3.1)$$

A sequence is also a discrete function if we set

$$f(n) = a_n . \qquad (2.3.2)$$

So there is no real difference between a sequence and a discrete function.

Definition 2.3.2: The sequence $f(n)$ is a <u>right-handed</u> sequence if

$$f(n) = 0, \quad \text{for all} \quad n < n_0 < \infty.$$

The sequence $f(n)$ is a <u>left-handed</u> sequence if

$$f(n) = 0, \quad \text{for all} \quad -\infty < n_1 < n.$$

The sequence $f(n)$ is a <u>finite length</u> sequence if

$$f(n) = 0, \quad \text{for all} \quad n < n_0 \leq n_1 < n.$$

▲▲

Terminology: The nonzero values of a right-hand sequence appear on the right. Right-handed sequences are sometime called *causal* sequences if $n_0 = 0$.

The definitions of the continuous-time singularity functions as defined in Section 2.2 motivate the definitions of discrete singularity functions.

Definition 2.3.3: The (<u>discrete</u>) <u>unit impulse function</u> $\delta(n)$ is defined as

$$\delta(n) = \begin{cases} 1, & n = 0 \\ 0, & n \neq 0 \end{cases}$$

▲▲

Terminology: The same terminology is used for the continuous-time impulse function and the discrete-time equivalent of that function. The impulse function in continuous-time is distinguished from the impulse function in discrete-time by the discrete-time argument n. Usually it is quite clear from the context which function is being discussed.

Remark 2.3.4: The discrete unit impulse function $\delta(n)$ is a perfectly well-defined function, whereas the continuous impulse response function $\delta(t)$ is not really a function at all. See Remark 2.2.2. Therefore, to rigorously define the continuous impulse response function $\delta(t)$ we had to resort to an integral as in Definition 2.2.1, and we have to be careful in its usage. The discrete unit impulse function $\delta(n)$ we can use without concern.

▲▲

Definition 2.3.5: The <u>unit step function</u> $u_s(n)$ is defined as

$$u_s(n) = \begin{cases} 1, & n \geq 0 \\ 0, & n < 0 \end{cases}$$

▲▲

Definition 2.3.6:   The (discrete) exponential function is defined as

$$f(n) = a^n.$$

▲▲

The exponential function is in some ways analogous to the exponential function $f(t) = e^{at}$. But notice that

$$f(n) \to 0 \quad \text{as} \quad n \to \infty \quad \text{if and only if } |a| < 1. \tag{2.3.3}$$

Definition 2.3.7: The Sa function $\text{Sa}(n)$ is defined as

$$\text{Sa}(n) = \begin{cases} \dfrac{\sin(n)}{n}, & n \neq 0 \\ 1, & n = 0 \end{cases}$$

The sinc function $\text{sinc}(n)$ is defined as

$$\text{sinc}(n) = \begin{cases} \dfrac{\sin(\pi n)}{\pi n}, & n \neq 0 \\ 1, & n = 0 \end{cases}$$

▲▲

Definition 2.3.8: The pulse function $\Pi_\kappa(n)$ is defined as

$$\Pi_\kappa(n) = \begin{cases} 1, & -\kappa \leq n \leq \kappa, \quad \kappa \in I \\ 0, & \text{otherwise} \end{cases}$$

▲▲

Notation: The pulse function in continuous-time is distinguished from the pulse function in discrete-time by the discrete-time argument n. Furthermore, the discrete-time pulse function must carry a subscript to identify its width.

Definition 2.3.9: The sinusoidal function is defined as

$$f(n) = A \sin(\Omega_0 n + \theta_0).$$

▲▲

Terminology: The number Ω_0 in the sinusoidal function is called the frequency of the function. Because the sinusoid is a discrete function, we call Ω_0 a discrete frequency to distinguish it from the frequency of a continuous-time sinusoid. The reason for this distinction will become clear when we discuss discrete signals and systems.

Notation: We use a capital Ω to stand for discrete frequency.

Remark 2.3.10: Discrete sinusoidal functions have several special properties not found in continuous-time sinusoids. Let

$$f(n) = \cos(\Omega_0 n). \tag{2.3.4}$$

Using a trigonometric formula it follows that

$$\cos\big((\Omega_0 + 2\pi k)n\big) = \cos(\Omega_0 n)\cos(2\pi k n) - \sin(2\pi k n)\sin(\Omega_0 n) = \cos\Omega_0 n. \tag{2.3.5}$$

The relationship between the two frequencies in (2.3.5) is shown in Figure 2.3.1. Equation (2.3.5) shows that it is not possible to distinguish between discrete sinusoids the difference of whose frequencies is a 2π multiple. Therefore we need only consider discrete sinusoids whose frequency is between 0 and 2π, or between $-\pi$ and π.

If in (2.3.4) $0 < \Omega_0 < \pi$, then

$$\cos\big((2\pi - \Omega_0)n\big) = \cos(2\pi n)\cos(\Omega_0 n) + \sin(2\pi n)\sin(\Omega_0 n) = \cos(\Omega_0 n). \tag{2.3.6}$$

The relationship between the frequencies of the sinusoids in (2.3.6) is shown in Figure 2.3.2. Equation (2.3.6) says that the sinusoids $\cos(\Omega_0 n)$ and $\cos\big((2\pi - \Omega_0)n\big)$ are identical. Hence, we can't distinguish between these two frequencies in discrete time. Therefore we need only consider sinusoids whose frequency is between 0 and π.

▲▲

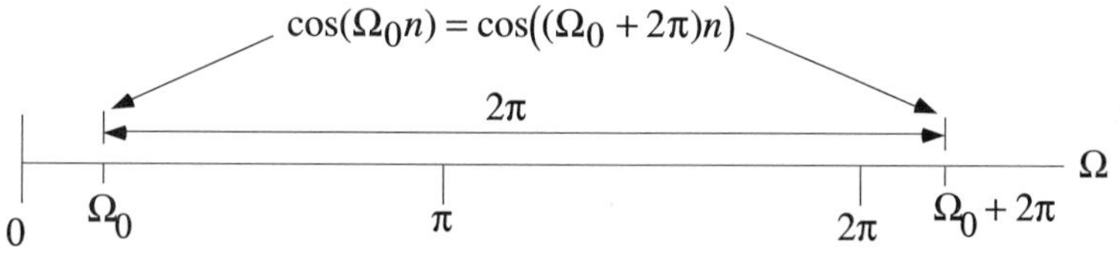

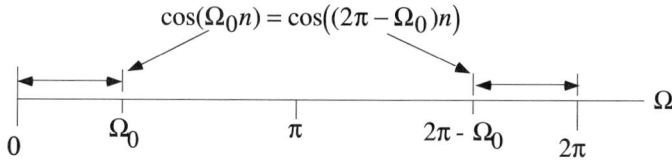

Figure 2.3.1 Relationship Between the Frequencies in (2.3.5)

Figure 2.3.2 Relationship Between the Frequencies in (2.3.6)

2.4 HOMEWORK FOR CHAPTER 2

Homework Problems for Section 2.1

2.1.1   Sketch the following functions.

(i) $f(t) = e^{-2t}, \quad -1 < t < 1$

(ii) $f(t) = \begin{cases} 0, & t < 0 \\ e^{-2t}, & t \geq 0 \end{cases}$

(iii) $f(t) = e^{-2t}, \quad t \geq 0$

2.1.2   Determine if the following functions are even or odd or neither.

(i) $f(t) = \tan^{-1}(t)$ (iv) $f(t) = |\sin(3t)|$

(ii) $f(t) = \cos(\pi t)$ (v) $f(t) = e^{at}$

(iii) $f(t) = \sin(3t)$ (vi) $f(t) = |t|$

2.1.3   Evaluate the following integrals.

(i) $\displaystyle\int_{-m\pi}^{m\pi} \tan^{-1}(t)\,dt$ (iii) $\displaystyle\int_{-\pi}^{\pi} \sin(3t)\,dt$

(ii) $\displaystyle\int_{-2}^{2} \cos(\pi t)\,dt$ (iv) $\displaystyle\int_{-5}^{5} |t|\,dt$

2.1.4   Consider a signal $x(t)$.

(a)   Show that the following signal is even.

$$x_e(t) = \frac{x(t) + x(-t)}{2}$$

(b)   Show that the following signal is odd

$$x_e(t) = \frac{x(t) + x(-t)}{2}$$

(c)   Show that

$$x_e(t) = \frac{x(t) + x(-t)}{2}$$

2.1.5   Prove each of the properties in Remark 2.1.6.

Homework Problems for Section 2.2

2.2.1   Show that

$$\text{Sa}\left(\frac{\omega\varepsilon}{2}\right) = \text{sinc}(f\varepsilon), \quad \omega = 2\pi f.$$

2.2.2   Given a function $f(t)$ show that

$$\sum_{k=1}^{N} f(t-kT) = \int_{-\infty}^{\infty} f(t-\lambda) \left[\sum_{k=1}^{N} \delta(\lambda - kT) \right] d\lambda$$

if $f(t)$ is continuous at the points $t = kT$.

2.2.3   Plot the following signals.
    (i) $x(t) = [\sin(t)]u_s(t)$ (iv) $x(t) = \Pi(t)\sin(20t)$
    (ii) $x(t) = [\sin(t)]\text{sgn}(t)$ (v) $x(t) = \Lambda(t)\sin(20t)$
    (iii) $x(t) = \sin(2t + \phi)$, $\phi = 0, \frac{\pi}{4}, -\frac{\pi}{4}$ (vi) $x(t) = \ln\left(10^t\right)$

2.2.4   Show that

$$\delta(\varepsilon t) = \frac{1}{\varepsilon}\delta(t), \quad \varepsilon > 0$$

Hint: Introduce a change of variables in the defining integral.

2.2.5   Evaluate the following integrals, if possible. If not, give a reason why.
    (i) $\displaystyle\int_{-\infty}^{\infty} e^{3\lambda}\,\delta(\lambda)\,d\lambda$ (iv) $\displaystyle\int_{-5}^{5} 4\lambda^2\,\delta(\lambda+1)\,d\lambda$

    (ii) $\displaystyle\int_{-\infty}^{\infty} 4\cos(\pi\lambda)\,\delta(\lambda)\,d\lambda$ (v) $\displaystyle\int_{0}^{5} \lambda^3\,\delta(\lambda+1)\,d\lambda$

    (iii) $\displaystyle\int_{0}^{\infty} 4\cos(\pi\lambda)\,\delta(\lambda-\pi)\,d\lambda$ (vi) $\displaystyle\int_{0}^{\infty} 4\cos(\pi\lambda)\,\delta(\lambda)\,d\lambda$

Homework Problems for Section 2.3

2.3.1   Sketch the following functions.
    (i) $f(n) = \delta(n) + 2\delta(n-1) + \delta(n-2)$ (iv) $f(n) = \text{Sa}(\pi n)$
    (ii) $f(n) = (0.8)^{n-1}u_s(n)$ (v) $f(n) = -2\Pi_3(n)$
    (iii) $f(n) = (0.8)^{n-1}u_s(-n-1)$ (vi) $f(n) = \cos(0.4\pi n)$

2.3.2   In Problem 2.3.1 classify each sequence as a right-hand sequence, a left-hand sequence, or a finite length sequence.

2.3.3   Show the following equalities.
    (i) $\cos(2.4\pi n) = \cos(0.4\pi n)$
    (ii) $\cos(1.6\pi n) = \cos(0.4\pi n)$
    (iii) $\cos(8.4\pi n) = \cos(0.4\pi n)$

2.3.4   Consider the following discrete functions.

    (i) $f_1(n) = \cos(\Omega_0 n), \quad 0 < \Omega_0 < \pi$

    (ii) $f_2(n) = \cos\!\big((\Omega_0 + 2\pi k)n\big), \quad 0 < \Omega_0 < \pi$

    (iii) $f_3(n) = \cos\!\big((2\pi - \Omega_0)n\big), \quad 0 < \Omega_0 < \pi$

Write a MATLAB program to plot each of these functions on the same graph using different symbols for various values of Ω_0 and k.

2.3.5   Show that

    (i) $\sin\!\big((\Omega_0 + 2\pi k)n\big) = \sin(\Omega_0 n)$

    (ii) $\sin\!\big((2\pi - \Omega_0)n\big) = \sin(\Omega_0)$

Chapter 3
Review of Complex Variables

Chapter Outline

This chapter contains a review of complex numbers and complex rational functions. This material is required throughout the text. For the most part, this material is found in all introductory treatments of complex variables. The notation is standard.

The algebraic and geometric representation of complex numbers and complex arithmetic are routinely used throughout the text. The exponential representation is fundamental to the algebra of complex numbers. Complex exponentials also play a central role in Fourier series. This material is discussed in Section 3.1.

Rational functions of a complex variable play a central role for Laplace and z-transforms. Insofar as these two subjects play a central role in continuous and discrete systems, this material is absolutely essential to obtain a firm grasp of the concepts developed in the remainder of the text. Because of the commonality of the concepts of Laplace and z-transforms, basic terminology of rational functions are presented in Section 3.2. In particular, the indexing convention of the coefficients of the rational function is followed throughout the discussion of continuous-time systems. This section also contains a discussion of second-order polynomials that is particular to certain engineering subjects.

Summary of Sections

Section 3.1:   We present a review of complex numbers.

Section 3.2:   We discuss rational functions.

Coverage of the Text

This chapter is self-contained.

3.1 COMPLEX NUMBERS

3.1.1 Introduction

We begin with the definition of a complex number along with the simple arithmetic properties of these numbers. We present two representations of a complex number which are used interchangeably. These representations of a complex number have a graphical representation as well, which is developed below.

3.1.2 Definitions

Definition 3.1.1: Let j ($= \sqrt{-1}$) be the root of the equation

$$s^2 + 1 = 0.$$

A <u>complex number</u> s is defined as

$$s = \sigma + j\omega, \quad \sigma \in \mathcal{R}, \quad \omega \in \mathcal{R}.$$

Notation: The set of complex numbers is denoted by C.

▲▲

   The components parts of a complex number are as follows.

Definition 3.1.2:   The <u>rectangular representation</u> of the complex number s is
$s = \sigma + j\omega$.
The <u>real part</u> of s is σ, write $\text{Re}(s) = \sigma$.
The <u>imaginary part</u> of s is ω, write $\text{Im}(s) = \omega$.

▲▲

The algebraic manipulations of complex numbers are defined as follows.

Definition 3.1.3: We define <u>addition</u> (and subtraction) of two complex numbers as

$$s_1 + s_2 = (\sigma_1 + j\omega_1) + (\sigma_2 + j\omega_2) = (\sigma_1 + \sigma_2) + j(\omega_1 + \omega_2).$$

We define the <u>multiplication</u> of two complex numbers as

$$(s_1)(s_2) = (\sigma_1 + j\omega_1)(\sigma_2 + j\omega_2) = (\sigma_1\sigma_2 - \omega_1\omega_2) + j(\omega_1\sigma_2 + \sigma_1\omega_2).$$

We define the <u>division</u> of two complex numbers as

$$\frac{s_1}{s_2} = \frac{\sigma_1 + j\omega_1}{\sigma_2 + j\omega_2} = \frac{\sigma_1\sigma_2 + \omega_1\omega_2}{\sigma_2^2 + \omega_2^2} + j\frac{\omega_1\sigma_2 - \sigma_1\omega_2}{\sigma_2^2 + \omega_2^2}.$$

▲▲

Example 3.1.4: Consider the complex number $s_0 = 1 + j4$. The real part and imaginary part of s_0, respectively, are

$$\text{Re}(s_0) = 1 \quad \text{and} \quad \text{Im}(s_0) = 4. \tag{3.1.1}$$

Consider a second complex number $s_1 = 2 - j3$. Then

$$s_0 + s_1 = (1 + j4) + (2 - j3) = (1 + 2) + j(4 - 3) = 3 + j1. \tag{3.1.2}$$

$$s_0 s_1 = (1 + j4)(2 - j3) = [(1)(2) - (4)(-3)] + j[(4)(2) + (1)(-3)] = 14 + j5,$$

$$\frac{s_0}{s_1} = \frac{1 + j4}{2 - j3} = \frac{[(1)(2) + (4)(-3)]}{(2)^2 + (3)^2} + j\frac{[(4)(2) - (1)(-3)]}{(2)^2 + (3)^2} = -\frac{10}{13} + j\frac{11}{13}.$$

▲▲

3.1.3 Graphical Representation

We will routinely use the graphical representation of a complex number, which we introduce next. To each complex number s we can associate an order pair of real numbers

$$s = \sigma + j\omega \leftrightarrow (\sigma, \omega). \tag{3.1.3}$$

This ordered pair of numbers is, in turn, associated with a point in the Cartesian coordinate system shown in Figure 3.1.1a. The ordered pair is associated with the vector shown in Figure 3.1.1b. So we see how a complex number can be represented with a vector. The rules of complex number addition and subtraction correspond to the rules of vector addition and subtraction. These algebraic operations are often conceptually more clear graphically. When we use a graphical representation of a complex number, we will use the vector representation rather than the ordered pair of numbers.

There are two ways to represent a vector: rectangular form or polar form. The rectangular form of the complex vector is used in the definition of a complex number and in Figure 3.1.1b. The polar representation is derived from the rectangular representation by simple geometry. Denoting the magnitude of the vector in Figure 3.1.1b by ρ and the angle by θ, we arrive at the relationship

$$\sigma = \rho\cos\theta \quad \text{and} \quad \omega = \rho\sin\theta. \tag{3.1.4}$$

and

$$\rho = \sqrt{\sigma^2 + \omega^2} \quad \text{and} \quad \theta = \tan^{-1}\left(\frac{\omega}{\sigma}\right). \tag{3.1.5}$$

There is a compact notation for the polar representation of a complex number based on the notion of the complex exponential function.

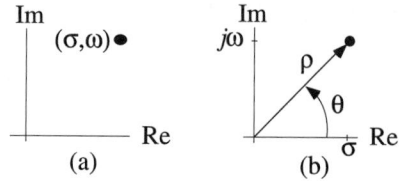

Figure 3.1.1 Graphical Representation of the Complex Number s

Definition 3.1.5: The <u>exponential function</u> of a complex number $r = \tau + j\theta$ is defined as

$$e^r = e^{(\tau + j\theta)} = e^\tau (\cos\theta + j\sin\theta).$$

where e^τ is the real exponential function.

▲▲

The special form of the exponential function when $r = j\theta$ carries a name.

Definition 3.1.6: The equation

$$e^{j\theta} = \cos\theta + j\sin\theta$$

is called <u>Euler's identity</u>.

▲▲

Using Euler's identity we have a second representation of a complex number.

Definition 3.1.7: The <u>polar representation</u> of a complex number s is $s = \rho e^{j\theta}$.
The <u>magnitude</u> of s is ρ; write $|s| = \rho$.
The <u>angle</u>, or <u>phase</u> of s is θ; write $\angle s = \theta$.

▲▲

The magnitude of the complex number also carries another name.

Definition 3.1.8: The <u>absolute value</u> or <u>modulus</u> $|s|$ of a complex number s is defined as

$$|s| = \sqrt{\sigma^2 + \omega^2} = \rho.$$

▲▲

Example 3.1.9: Consider the complex number

$$s_2 = 5e^{j(0.52)}. \tag{3.1.6}$$

The magnitude and phase, respectively, of this complex number are

$$|s_2| = 5 \quad \text{and} \quad \angle s_2 = 0.52. \tag{3.1.7}$$

The rectangular form of this number is determined using Euler's identity. We have

$$s_2 = 5e^{j(0.52)} = 5[\cos(0.52) + j\sin(0.52)] \tag{3.1.8}$$
$$= 5[0.86 + j0.5] = 4.3 + j2.5.$$

Consider the complex number $s_0 = 1 + j4$. The magnitude and phase, respectively, of this complex number are

$$|s_0| = \rho = \sqrt{1^2 + 4^2} = \sqrt{17} \quad \text{and} \quad \angle s_0 = \theta = \tan^{-1}\left(\frac{4}{1}\right) = 1.33. \tag{3.1.9}$$

Hence, the polar representation is

$$s_0 = 1 + j4 = \left(\sqrt{17}\right)e^{j1.33}. \tag{3.1.10}$$

▲▲

Example 3.1.10: Consider the complex number $s = -1$. The polar representation of this number is

$$s = -1 = e^{j\pi} = e^{-j\pi}. \tag{3.1.11}$$

If we try to calculate this number using the formula (3.1.5) we obtain

$$\tan^{-1}\left(\frac{0}{-1}\right) \tag{3.1.12}$$

which is multi-valued. Here we must select π rad to obtain the correct answer. In using the angle formula (3.1.5) care must be taken to interpret the inverse tangent as this function is not single valued.

▲▲

Example 3.1.11: The polar representation simplifies the multiplication and division of two complex numbers. Given two complex numbers we have

$$s_0 s_1 = \left(A_0 e^{j\theta_0}\right)\left(A_1 e^{j\theta_1}\right) = A_0 A_1 e^{j(\theta_0 + \theta_1)} \tag{3.1.13}$$

and

$$\frac{s_0}{s_1} = \frac{A_0 e^{j\theta_0}}{A_1 e^{j\theta_1}} = \frac{A_0}{A_1} e^{j(\theta_0 - \theta_1)}.$$

▲▲

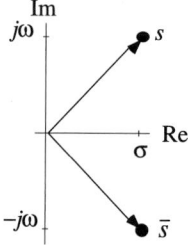

Figure 3.1.2 Complex Conjugate of a Complex Number

The complex conjugate of a complex number plays a significant role in the theory of complex numbers.

Definition 3.1.12: The <u>complex conjugate</u> of a complex number s, denoted $\bar{s}$, is obtained by replacing j by $(-j)$ in s. That is,

$$\bar{s} = \sigma - j\omega = \rho e^{-j\theta}.$$

▲▲

Notation: There are several other notations for the complex conjugate that are in general use including s^* and s'.

The graphical representation of the conjugate of a complex number is just a reflection of this number about the real axis as shown in Figure 3.1.2. The complex conjugate also satisfies the following equalities:

$$|s| = \sqrt{s\bar{s}}, \quad \overline{s_1 + s_2} = \bar{s}_1 + \bar{s}_2, \quad \overline{s_1 s_2} = (\bar{s}_1)(\bar{s}_2). \tag{3.1.14}$$

Using the complex conjugate, the division of two complex numbers becomes

$$\frac{s_0}{s_1}\frac{\bar{s}_1}{\bar{s}_1} = \frac{\sigma_1 + j\omega_1}{\sigma_2 + j\omega_2} \cdot \frac{\sigma_2 - j\omega_2}{\sigma_2 - j\omega_2} = \frac{(\sigma_1\sigma_2 + \omega_1\omega_2) + j(\omega_1\sigma_2 - \sigma_1\omega_2)}{\sigma_2^2 + \omega_2^2}. \tag{3.1.15}$$

Euler's identity can be used to express sine and cosine in terms of complex exponentials. Taking the complex conjugate of Euler's identity we obtain

$$e^{-j\theta} = \cos\theta - j\sin\theta. \tag{3.1.16}$$

Now taking the sum and difference of Euler's identity and (3.1.16) we have

$$\cos\theta = \frac{e^{j\theta} + e^{-j\theta}}{2} \tag{3.1.17}$$

and

$$\sin\theta = \frac{e^{j\theta} - e^{-j\theta}}{2j}. \tag{3.1.18}$$

The formulas in (3.1.17) and (3.1.18) are used often throughout this text.

3.1.4 Sets in the Complex Plane

Given a complex number, it can be associated with a point in the plane. Hence, the set of complex numbers corresponds to the plane. The representation of the set of complex numbers as a plane is extremely useful, and we will use it routinely. In fact we give this set a name.

Definition 3.1.13: The complex plane in Figure 3.1.1a is known as the _s-plane_.   ▲▲

Notation: The name s-plane is derived from the notation for the Laplace transform variable, s.

There are several sets of complex numbers that play an important role in some of the chapters that follow.

Definition 3.1.14: The two sets of complex numbers

$$\text{open RHP} = \left\{ s \in C \mid \sigma > 0 \right\},$$

and

$$\text{closed RHP} = \left\{ s \in C \mid \sigma \geq 0 \right\},$$

are called the open and closed right-half plane (RHP), respectively.   ▲▲

These sets are identified in Figure 3.1.3. Roughly, the right-half plane is the set of numbers to the right of the imaginary axis. If the RHP is open, it doesn't include the imaginary axis. If the RHP is closed, it does include the imaginary axis. We have similar definitions for the left-half plane.

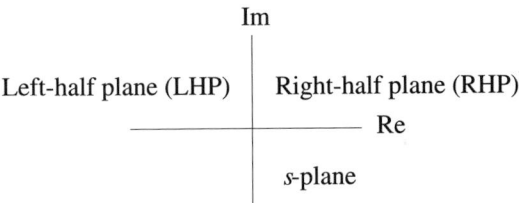

Figure 3.1.3 Left- and Right-Half Planes

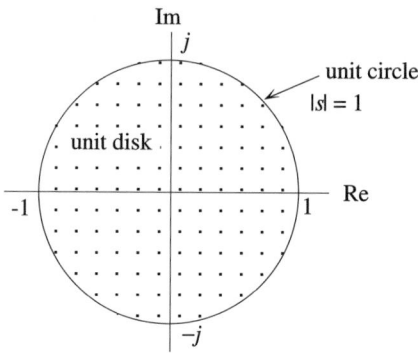

Figure 3.1.4 Unit Circle and Unit Disk in the Complex Plane

In the study of discrete systems we will be interested in the sets of complex numbers shown in Figure 3.1.4.

Definition 3.1.15: The set of complex numbers

$$\text{unit circle} = \left\{ s \in C \,\middle|\, |s| = 1 \right\}$$

is called the <u>unit circle</u>. The sets of complex numbers which satisfy

$$\text{open unit disk} = \left\{ s \in C \,\middle|\, |s| < 1 \right\},$$

and

$$\text{closed unit disk} = \left\{ s \in C \,\middle|\, |s| \leq 1 \right\}.$$

are called the <u>open</u> and <u>closed unit disk</u> respectively.

▲▲

The open unit disk contains all points inside the unit circle but not the unit circle itself. The closed unit disk contains the unit circle as well as all points inside the unit circle. Using Euler's identity the unit circle can also be described as

$$s = e^{j\theta}, \quad \text{for} \quad 0 \leq \theta < 2\pi. \tag{3.1.19}$$

3.2 COMPLEX FUNCTIONS

3.2.1 Definitions

In this section we introduce complex functions. For the most part we will restrict ourselves to rational functions, except for the exponential function which is used pervasively.   This section introduces the definitions, notation (which is very important), and basic concepts of rational functions.

Definition 3.2.1: A <u>complex function</u> $X(s)$ is a function that maps the set of complex numbers into the set of complex numbers, $X: C \to C$.
▲▲

A special form of a complex function is a function that maps the real numbers into the complex numbers.   Let $X(s)$ be any complex function.   The following two characteristics of a complex function play an important role throughout the text.

Definition 3.2.2: The complex function $X(s)$ has a <u>pole</u> at $s = p_0$ if

$$\lim_{s \to p_0} X(s) = \infty.$$
▲▲

Definition 3.2.3: The complex function $X(s)$ has a <u>zero</u> at $s = z_0$ if

$$\lim_{s \to z_0} X(s) = 0.$$
▲▲

The next example illustrates poles and zeros.

Example 3.2.4: Consider the function

$$X(s) = \frac{s - b}{s - a} \qquad\qquad (3.2.1)$$

where $a \neq b$. This function has a pole at $p_0 = a$, and a zero at $z_0 = b$.
▲▲

3.2.2 Rational Functions

In this text we will be primarily concerned with one type of complex function.

Definition 3.2.5: A <u>rational function</u> is a complex function which is the ratio of two polynomials with real coefficients; i.e.,

$$X(s) = \frac{b_m s^m + b_{m-1} s^{m-1} + \cdots + b_0}{s^n + a_{n-1} s^{n-1} + \cdots + a_0} = \frac{b(s)}{a(s)}.$$

We define the <u>order</u> of this function to be n. The polynomial $b(s)$ is called the <u>numerator</u> polynomial. The polynomial $a(s)$ is called the <u>denominator</u> polynomial.
▲▲

A polynomial is a rational function with a denominator of 1.

Notation: We will follow the notation convention in Definition 3.2.5 for writing rational functions when we are discussing Laplace transforms. The indexing on the coefficients always matches the power of s. The leading coefficient of the denominator is always 1.

We will discuss z-transforms which are rational functions in the complex variable z. The indexing for these functions will be reversed, in deference to custom. This change in notation is discussed in Chapter 17.

Definition 3.2.6: We say that the rational function $X(s)$ is:
   (1) <u>strictly proper</u> if $m < n$,
   (2) <u>proper</u> if $m \leq n$, or,
   (3) <u>not proper</u> if $m > n$.

▲▲

Terminology: Not proper rational functions are sometimes called <u>noncausal functions</u>.

It is well known that if we allow the roots of a polynomial to be complex, then any polynomial can be factored into linear factors. This result is stated for rational functions.

Theorem 3.2.7: Let $X(s)$ be a rational function with real coefficients. Then we can factor $b(s)$ and $a(s)$ into linear factors as

$$X(s) = \frac{K(s - z_1)(s - z_2)\cdots(s - z_m)}{(s - p_1)(s - p_2)\cdots(s - p_n)}.$$

If the coefficients of $X(s)$ are real, then the complex roots of the $b(s)$ and $a(s)$ will appear in complex conjugate pairs.

▲▲

The form of the rational function in Theorem 3.2.7 clearly displays the poles and zeros of $X(s)$. Usually, we will be concerned with proper or strictly proper rational functions. If the rational function is strictly proper, $m \leq n$, then the number of zeros m is less than the number of poles n. In this case it is sometimes useful to consider such a function to have $n - m$ zeros at infinity.

Suppose we have a rational function with a common factor in the numerator and the denominator, say,

$$X(s) = \frac{(s+1)(s+2)}{(s+3)(s+2)}. \tag{3.2.2}$$

Because we can cancel the common factors, this function is equivalent to a reduced function

$$X_1(s) = \frac{(s+1)}{(s+3)}. \tag{3.2.3}$$

The greatest common factor of numerator and denominator polynomials of $X(s)$ is $(s+2)$. The greatest common factor of numerator and denominator polynomials of $X_1(s)$, $b_1(s)$, and $a_1(s)$ is 1.

Terminology: When the greatest common factor of two polynomials, $b(s)$ and $a(s)$, is 1 we say $b(s)$ and $a(s)$ are <u>coprime</u>.

Notation: Many of the results below are stated in terms of the poles and zeros of the function. These results apply only to poles and zeros that can't be canceled out of the function. Hence, we must always work with reduced rational functions. We will *always* assume that the numerator and denominator polynomials are coprime.

We now have two ways to represent a rational function: (1) as a ratio of polynomials as in Definition 3.2.5, or, (2) in terms of the poles, zeros, and the gain K of the factored polynomials as in Theorem 3.2.7. The representation of the rational function as poles and zeros is used frequently in this text, partially because it lends itself to graphical representation.

Definition 3.2.8: We can represent $X(s)$ graphically by marking in the s-plane each zero z_i by an o and each pole p_i by an x. The resulting plot is called a <u>pole-zero</u> plot.

▲▲

Example 3.2.9: The pole-zero plot of the rational function

$$X(s) = \frac{K(s-2)}{(s+1)(s+2+j2)(s+2-j2)} \tag{3.2.4}$$

is shown in Figure 3.2.1. The pole-zero plot defines the rational function (3.2.4) up to a constant K.

▲▲

3.2.3 Second-Order Polynomials

It turns out that we will often encounter a second-order polynomial with two complex roots in the LHP. Next we will introduce some notation concerning these

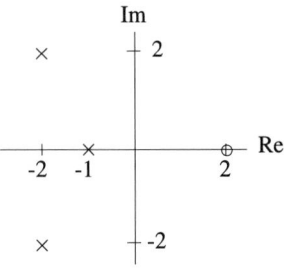

Figure 3.2.1 Pole-Zero Plot of Example 3.2.9

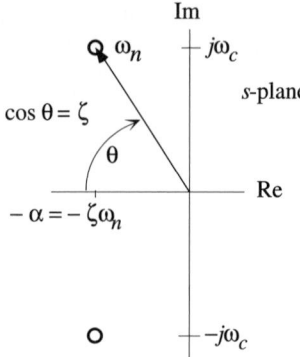

Figure 3.2.2 Roots of the Polynomial of a Second-Order Polynomial

polynomials. They generally have the form

$$b(s) = s^2 + b_1 s + b_0, \quad b_1 > 0, \quad b_0 > 0. \tag{3.2.5}$$

Suppose $b(s)$ has two complex roots

$$b(s) = (s + \alpha + j\omega_c)(s + \alpha - j\omega_c) = (s + \alpha)^2 + \omega_c^2, \quad \alpha > 0, \quad \omega_c > 0. \tag{3.2.6}$$

From the form of the polynomial in (3.2.6) we can read off the roots of the polynomial directly in rectangular form. The roots of this polynomial are shown in Figure 3.2.2.

Notation: The notation in this subsection, as in (3.2.6), is standard in the engineering literature. It will be used throughout the text.

If we complete the square in (3.2.6), we obtain

$$b(s) = (s + \alpha)^2 + \omega_c^2 = s^2 + 2\alpha s + (\alpha^2 + \omega_c^2) = s^2 + 2\zeta\omega_n s + \omega_n^2. \tag{3.2.7}$$

It is clear from (3.2.7) that

$$\omega_n = \sqrt{\alpha^2 + \omega_c^2}. \tag{3.2.8}$$

If we represent the complex root of this equation with a vector, then ω_n is the length of this vector, as shown in Figure 3.2.2. The angle that vector makes with the *negative* real axis is defined to be θ. Then the component of this vector in the direction of the negative real axis is

$$\omega_n \cos\theta. \tag{3.2.9}$$

Now we define

$$\cos\theta = \zeta. \tag{3.2.10}$$

It follows that

$$\zeta\omega_n = \alpha. \tag{3.2.11}$$

We have now established the relationship between the coefficients in (3.2.7). Having defined ζ via (3.2.10), it follows that

$$\sin\theta = \sqrt{1-\zeta^2}. \tag{3.2.12}$$

Then from Figure 3.2.2 we have

$$\omega_c = \omega_n\sqrt{1-\zeta^2}. \tag{3.2.13}$$

The beauty of this parameterization is that the parameters (ζ,ω_n) can be simply computed directly from (3.2.7) and the corresponding root locations determined accordingly. This calculation doesn't require the quadratic formula.

Notation: Sometimes these root locations are abbreviated (ζ,ω_n).

The second-order polynomial in (3.2.7) is used *so* frequently we give it a name.

Definition 3.2.10: A second-order polynomial defined in terms of the parameters (ζ,ω_n) and written as

$$s^2 + 2\zeta\omega_n s + \omega_n^2$$

is said to be in <u>standard second-order form</u>.

▲▲

Example 3.2.11: In this example we illustrate the relationship between the coefficients of a polynomial and its roots using the results above. Consider the polynomial

$$s^2 + 2s + 2. \tag{3.2.14}$$

From (3.2.7) (ζ,ω_n) can be determined from the coefficients of (3.2.14) as follows. From the constant coefficient we have

$$\omega_n^2 = 2, \quad \text{or,} \quad \omega_n = \sqrt{2}. \tag{3.2.15}$$

Then it follows that

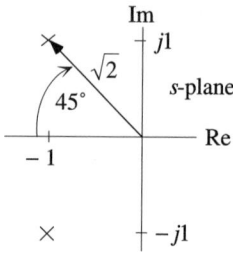

Figure 3.2.3 Roots of the Polynomial in Example 3.2.11

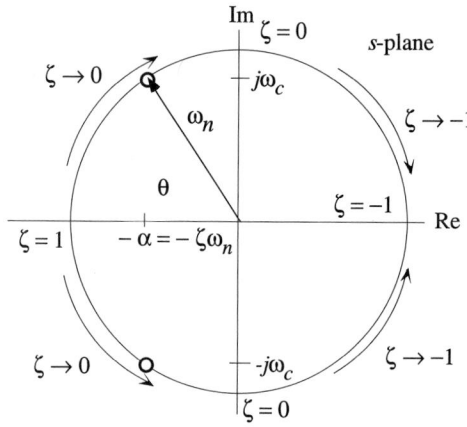

Figure 3.2.4 Roots of a Second-Order Polynomial with Constant ω_n

$$2\zeta\omega_n = 2, \quad \text{or,} \quad \zeta = \frac{2}{2\sqrt{2}} = 0.707. \tag{3.2.16}$$

From (3.2.16) we find that

$$\cos\theta = 0.707, \quad \text{or,} \quad \theta = 45°. \tag{3.2.17}$$

These roots are shown in Figure 3.2.3. We can determine the real and imaginary parts of the roots using (3.2.11) and (3.2.13). From (3.2.16) we have

$$\zeta\omega_n = 1 = \alpha, \tag{3.2.18}$$

and from (3.2.13)

$$\omega_c = \omega_n \sqrt{1-\zeta^2} = (\sqrt{2})\left(\sqrt{1-\left(\frac{1}{\sqrt{2}}\right)^2}\right) = 1. \tag{3.2.19}$$

These numbers are also shown in Figure 3.2.3.

▲▲

Suppose we start with the polynomial in (3.2.7). First, fix ω_n. Now let ζ vary for $1 \geq \zeta \geq -1$. The roots of the polynomial that are traced out are shown in Figure 3.2.4. When ζ is negative, the roots of the polynomial lie in the RHP. This result is in accord with (3.2.10). Suppose that the coefficients of the polynomial (3.2.7) are such that $\zeta \leq -1$ or $\zeta \geq 1$. The roots of such a polynomial are both real. In this case, (3.2.10) is meaningless.

3.3 HOMEWORK FOR CHAPTER 3

Homework Problems for Section 3.1

3.1.1 Using Euler's formula, convert the following complex numbers into polar form.

(i) $s = -.1 + j10$ (iv) $s = -1$ (vii) $s = 2 + j\pi$
(ii) $s = 3 + j6$ (v) $s = 1 + j2$ (viii) $s = -1 + j3$
(iii) $s = 1 + j3$ (vi) $s = -1 + j2$ (ix) $s = -2 - j4$

3.1.2 Using Euler's formula, convert the following complex numbers into rectangular form.

(i) $s = 0.5e^{j2.4}$ (iv) $s = 4e^{-j1.5}$

(vii) $s = 2e^{j\frac{k\pi}{6}}$

(ii) $s = 2e^{-j.2}$ (v) $s = 2e^{j\pi}$

(viii) $s = \frac{1}{3k}e^{j\frac{k\pi}{3}}$

(iii) $s = 2e^{j\frac{2\pi}{5}}$ (vi) $s = 2e^{j\frac{\pi}{3}}$ (ix) $s = 5e^{j4}$

3.1.3 Find the complex conjugate of each of the complex numbers in Problems 3.1.1 and 3.1.2.

3.1.4 Plot all of the complex numbers in Problems 3.1.1 and 3.1.2 using MATLAB. Use equal scaling on each of the axis to obtain a rectangular grid.

3.1.5 Carry out the indicated multiplication.

(i) $s = (-.1 + j10)(3 + j6)$ (v) $s = \left(0.25e^{j0.2}\right)\left(4e^{-j0.2}\right)$

(ii) $s = (1 + j3)(1 + j2)$

(vi) $s = \left(5e^{j\frac{\pi}{2}}\right)\left(2e^{j\frac{\pi}{2}}\right)$

(iii) $s = (-2 - j4)(-2 + j4)$ (vii) $s = (3 + j4)\left(4e^{-j0.3}\right)$

(iv) $s = \left(3e^{j0.2}\right)\left(4e^{-j0.3}\right)$ (viii) $s = (3 + j4)\left(4e^{-j5\pi}\right)$

3.1.6   Carry out the indicated division.

(i) $s = \dfrac{2 + j3}{1 - j4}$

(iv) $s = \dfrac{2e^{j2}}{4e^{j3}}$

(vii) $s = \dfrac{e^{j3}}{2k + j}$

(ii) $s = \dfrac{2 + j2\pi k}{j2\pi k}$

(v) $s = \dfrac{2e^{j2}}{2e^{-j2}}$

(viii) $s = \dfrac{1 + 2j}{0.3e^{j.2}}$

(iii) $s = \dfrac{1}{2 - j3k}$

(vi) $s = \dfrac{6e^{-j3\pi}}{12e^{j5\pi}}$

(ix) $s = \dfrac{e^{j3\pi}}{2 + j2\pi k}$

3.1.7   Show that

$$s\bar{s} = |s|^2.$$

Homework Problems for Section 3.2

3.2.1   Consider the following polynomials.

(i) $s^2 + 2s + 100$

(v) $s^2 + 2s + 81$

(ii) $(s - 3)^2 + 25$

(vi) $(s + 6)^2 + 36$

(iii) $s^2 + 2(.3)(6)s + (6)^2$

(vii) $s^2 + 2(.9)(8)s + (8)^2$

(iv) $s^2 + 0.8s + 16$

(viii) $s^2 + 9$

(a) By hand find the roots of each of these polynomials in rectangular form and using (ζ, ω_n). Express each polynomial in standard second-order form.

(b) Enter each polynomial into MATLAB. Use MATLAB to confirm the answers to part (a). Show the vector you entered. Show the command you used for each calculation. (The **damp** command will calculate (ζ, ω_n) for complex roots.)

(c) Plot the roots of these equations in the s-plane. Label (ζ, ω_n).

3.2.2   Find the poles and zeros of the following rational functions. Express each complex pair in terms of (ζ, ω_n) and in terms of (α, ω_c). Plot the pole-zero diagram and label all quantities.

(i) $H(s) = \dfrac{16(s^2 + 1)}{(s + 2)(s^2 + 5.2s + 16)}$

(iii) $H(s) = \dfrac{s + 2}{s^2 + 3s + 9}$

(ii) $H(s) = \dfrac{4}{s^2 + 2s + 25}$

(iv) $H(s) = \dfrac{s^2 + s + 1}{s^2 + 1.02s + 1.01}$

3.2.3   Consider the zero plots of the second-order polynomials shown in Figure P3.2.3.

(i)   Find (ζ, ω_n).

(ii)   Find the corresponding polynomial.

(iii) Enter this polynomial into MATLAB and verify the root locations.

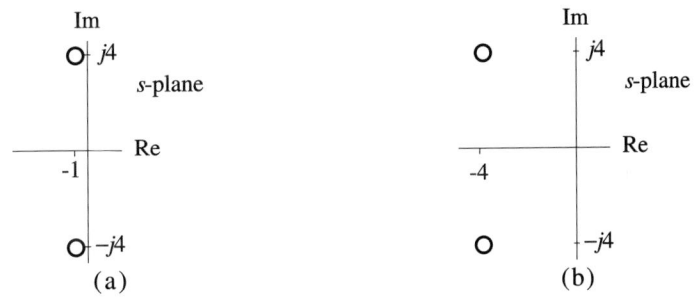

Figure P3.2.3

3.2.4   Suppose that the roots of a second-order polynomial are characterized by (ζ, ω_n). Find the coefficients of the corresponding polynomial. Also find (α, ω_c) for these roots. Plot the roots in the s-plane.

(i)   (ζ, ω_n) = (0.3, 2)   (iv)   (ζ, ω_n) = (0.01, 1)
(ii)   (ζ, ω_n) = (0.9, 2)   (v)   (ζ, ω_n) = (0.95, 6)
(iii) (ζ, ω_n) = (0.707, 1)   (vi)   (ζ,ω_n) = (0.8,4)

3.2.5   Suppose that the roots of a second-order polynomial are characterized by (α, ω_c). Find the coefficients of the corresponding polynomial. Also find (ζ,ω_n) for these roots. Plot the roots in the s-plane.

(i)   (α, ω_c) = (−2.5, 2)   (iv)   (α, ω_c) = (20, 2)
(ii)   (α, ω_c) = (2, 2)   (v)   (α, ω_c) = (0.5, 2)
(iii) (α, ω_c) = (0.02, 2)   (vi)   (α, ω_c) = (4, 2)

3.2.6   Consider the polynomial

$$a(s) = s^2 + a_1 s + a_0.$$

(a) Show that if $a_0 < 0$, then the polynomial has two real roots.
(b) Show that if $a_0 > 0$, but $a_1 < 0$, then the polynomial has roots in the RHP.

3.2.7   (a) Consider the polynomial expressed in factored form

$$a(s) = (s - a + jb)(s + c)$$

Show that this polynomial has complex coefficients if $b \neq 0$.
(b) Consider the polynomial expressed in factored form

$$a(s) = (s - a + jb)(s + c)$$

Show that this polynomial has real coefficients.
(c) Consider the polynomial expressed in factored form

$$a(s) = (s - a + jb)(s - a - jb)(s + a + jb)(s + a - jb)$$

Show that this polynomial has real coefficients. Also show that it has no odd powers of s.

3.2.8   Consider the two rational functions

$$\frac{b_i(s)}{a_i(s)}, i = 1,2.$$

Suppose that the order of the numerator and denominator of each rational function is m_i and n_i, $i = 1,2$ respectively. Find the order of the numerator and denominator of the following rational functions assuming no pole-zero cancellations.

(i) $\dfrac{b_1(s)}{a_1(s)} \cdot \dfrac{b_2(s)}{a_2(s)}$

(ii) $\dfrac{b_1(s)}{a_1(s)} + \dfrac{b_2(s)}{a_2(s)}$

(iii) $\dfrac{b_1(s)}{a_1(s)} + d$

3.2.9   Consider the polynomial

$$a(s) = s^2 + 2\zeta\omega_n s + \omega_n^2$$

Fix ζ. Plot the poles of this polynomial for $0 < \omega_n < \infty$.

Chapter 4

Review of Matrix Theory

Chapter Outline

In this chapter we review some of the elementary definitions and results from matrix theory. The intention is to provide the background for concepts involving state space representations. The required concepts are mainly the elementary operations of matrices, vectors, and eigenvalues.

Computer Usage: Many CAD packages for system analysis are based on matrices. This format has shown itself to be both flexible and powerful. This chapter was written to provide some of the basic background needed to use these CAD packages, in particular MATLAB. Hence, the discussion of partitioned matrices, which normally wouldn't be included with this material.

Summary of Sections

Section 4.1:   We define the basic quantities associated with matrices and we introduce elementary matrix operations.

Section 4.2:   We introduce vectors. Eigenvalues and eigenvectors are also defined.

Coverage of the Text

This chapter only requires the notion of a complex number from Chapter 3.

4.1 BASIC DEFINITIONS AND ELEMENTARY OPERATIONS

4.1.1 Definitions

We begin by defining an array of numbers.

Definition 4.1.1: An $n \times m$ matrix A is the two-dimensional, $n \times m$ array of numbers

$$A = \begin{bmatrix} a_{11} & a_{12} & \cdots & a_{1m} \\ a_{21} & a_{22} & \cdots & a_{2m} \\ \vdots & \vdots & \ddots & \vdots \\ a_{n1} & a_{n2} & \cdots & a_{nm} \end{bmatrix}.$$

Each entry a_{ij} is called an element of A. We say the dimensions of A are $n \times m$.

▲▲

Notation: The elements a_{ij} of the matrix are located by their row index i and column index j. Sometime for convenience, we denote the matrix by $A = [a_{ij}]$. With this notation it is assumed that dimensions of the matrix, n rows and m columns, are known.

Note: The dimensions of the matrix play an important role in matrix theory, and in the implementation of matrix calculations on a computer.

Definition 4.1.2: A square matrix is a matrix where the number of rows equals the number of columns. The dimensions of a square matrix are $n \times n$. We say the matrix has order n.

▲▲

In this text we will be most often concerned with matrices filled with real or complex numbers. However, we need not confine the elements of the matrix to real or complex numbers. The elements of A can be chosen from any set of objects. The elements can also be functions. We will use both matrices whose elements are polynomials and those matrices whose elements are real functions.

Terminology: It is commonplace to refer to the matrix by type of elements in the matrix. Matrices with real numbers as elements are called real matrices. Matrices with complex elements are called complex matrices. Matrices with polynomials as entries are called polynomial matrices. And so on.

A given matrix A may be subdivided, or partitioned, into smaller matrices. For example, let

$$A = \begin{bmatrix} a_{11} & a_{12} & a_{13} \\ a_{21} & a_{22} & a_{23} \\ a_{31} & a_{32} & a_{33} \\ a_{41} & a_{42} & a_{43} \end{bmatrix} = \left[\begin{array}{cc|c} a_{11} & a_{12} & a_{13} \\ a_{21} & a_{22} & a_{23} \\ a_{31} & a_{32} & a_{33} \\ a_{41} & a_{42} & a_{43} \end{array} \right] = \begin{bmatrix} A_{11} & A_{12} \\ A_{21} & A_{22} \end{bmatrix} \tag{4.1.1}$$

where

$$A_{11} = \begin{bmatrix} a_{11} & a_{12} \end{bmatrix}, \quad A_{12} = \begin{bmatrix} a_{13} \end{bmatrix}, \quad A_{21} = \begin{bmatrix} a_{21} & a_{22} \\ a_{31} & a_{32} \\ a_{41} & a_{42} \end{bmatrix}, \quad A_{22} = \begin{bmatrix} a_{23} \\ a_{23} \\ a_{23} \end{bmatrix}. \tag{4.1.2}$$

This process of partitioning a matrix can be reversed, and a larger matrix can be created from several smaller matrices. The new matrix that is created from the smaller matrices must be rectangular, of course.

Example 4.1.3: Consider the matrices

$$A_1 = \begin{bmatrix} 1 & 2 \\ 3 & 4 \end{bmatrix}, \quad A_2 = \begin{bmatrix} 5 & 6 & 7 \\ 8 & 9 & 10 \end{bmatrix}. \tag{4.1.3}$$

These matrices can be joined together to form a new matrix

$$B = \begin{bmatrix} A_1 & A_2 \end{bmatrix} = \left[\begin{array}{cc|ccc} 1 & 2 & 5 & 6 & 7 \\ 3 & 4 & 8 & 9 & 10 \end{array} \right]. \tag{4.1.4}$$

▲▲

Definition 4.1.4: A <u>diagonal</u> matrix is a matrix whose off-diagonal elements are zero. That is, the elements satisfy $a_{ij} = 0, i \neq j$.

▲▲

Sometimes we are interested in extracting certain elements from the matrix.

Definition 4.1.5: The <u>diagonal</u> of a $n \times m$ matrix A, denoted diag(A), is the vector

$$\text{diag}(A) = \begin{bmatrix} a_{11} & a_{22} & \cdots & a_{kk} \end{bmatrix}^T, \quad k = \min(n, m).$$

The elements $a_{i,i+1}$ are called the <u>(first) superdiagonal</u> of A. The elements $a_{i-1,i}$ are called the <u>(first) subdiagonal</u> of A.

▲▲

Example 4.1.6: Consider the matrix

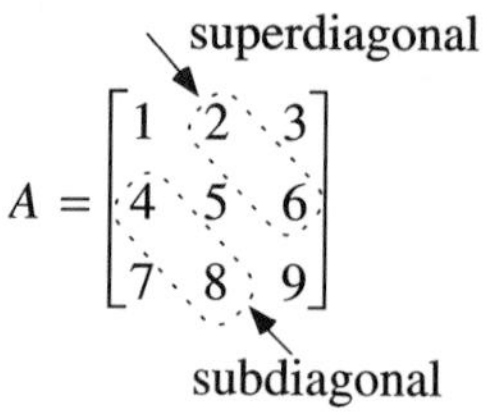

superdiagonal

$$A = \begin{bmatrix} 1 & 2 & 3 \\ 4 & 5 & 6 \\ 7 & 8 & 9 \end{bmatrix}$$

subdiagonal

Figure 4.1.1 Subdiagonal and Superdiagonal of A in Example 4.1.6

$$A = \begin{bmatrix} 1 & 2 & 3 \\ 4 & 5 & 6 \\ 7 & 8 & 9 \end{bmatrix}. \tag{4.1.5}$$

The diagonal of A is given by

$$\text{diag}(A) = \begin{bmatrix} 1 & 5 & 9 \end{bmatrix}. \tag{4.1.6}$$

The subdiagonal and superdiagonal are shown in Figure 4.1.1.

▲▲

Definition 4.1.7: A <u>zero</u> matrix is a matrix whose elements all zeros. A <u>unit</u> matrix is a matrix whose elements are all ones.

▲▲

Notation: We denote a zero matrix by a bold 0; i.e., **0**. If we wish to explicitly identify the dimensions of a zero matrix, we do so with subscripts. For example,

$$\begin{bmatrix} 0 & 0 & 0 \\ 0 & 0 & 0 \end{bmatrix} = \mathbf{0}_{23}. \tag{4.1.7}$$

Definition 4.1.8: An <u>identity</u> matrix I is a square, diagonal matrix whose diagonal elements are one; $a_{ii} = 1$; $a_{ij} = 0$, $i \neq j$.

▲▲

Notation: We identify the order of the identity matrix by a subscript, write I_n.

4.1.2 Elementary Operations with Matrices

Next we consider several elementary operations on matrices.

Definition 4.1.9: Given the $n \times m$ matrix A, the <u>transpose</u> of A is $A^T = [a_{ji}]$. Write A^T.

▲▲

Notation: Another notation for the transpose of a matrix is A' (which is used by MATLAB).

Note: If A is an n x m matrix, the transpose of A, A^T, is an m x n matrix.

Example 4.1.10: The transpose of the matrix A is

$$A^T = \begin{bmatrix} a_{11} & a_{12} & a_{13} \\ a_{21} & a_{22} & a_{23} \end{bmatrix}^T = \begin{bmatrix} a_{11} & a_{21} \\ a_{12} & a_{22} \\ a_{13} & a_{23} \end{bmatrix}. \tag{4.1.8}$$

The matrix B and its transpose are

$$B = \begin{bmatrix} 1 & 2 \\ 3 & 4 \\ 5 & 6 \end{bmatrix}, \quad B^T = \begin{bmatrix} 1 & 3 & 5 \\ 2 & 4 & 6 \end{bmatrix}. \tag{4.1.9}$$

▲▲

Definition 4.1.11:   Let A and B be n x m matrices.   Then the sum of these two matrices is defined as

$$A + B = \left[a_{ij} + b_{ij} \right], \quad i = 1, \ldots, n, \quad j = 1, \ldots, m.$$

▲▲

Note: To add two matrices, they must have the same dimensions.

Example 4.1.12: (a) The sum of the matrices A and B is

$$\begin{aligned} A + B &= \begin{bmatrix} a_{11} & a_{12} & a_{13} \\ a_{21} & a_{22} & a_{23} \end{bmatrix} + \begin{bmatrix} b_{11} & b_{12} & b_{13} \\ b_{21} & b_{22} & b_{23} \end{bmatrix} \\ &= \begin{bmatrix} a_{11} + b_{11} & a_{12} + b_{12} & a_{13} + b_{13} \\ a_{21} + b_{21} & a_{22} + b_{22} & a_{23} + b_{23} \end{bmatrix}. \end{aligned} \tag{4.1.10}$$

(b) Consider the matrices

$$A = \begin{bmatrix} 1 & 2 \\ 3 & 4 \\ 5 & 6 \end{bmatrix}, \quad B = \begin{bmatrix} -1 & 1 \\ -2 & 4 \\ 2 & -3 \end{bmatrix}, \quad \text{and} \quad C = \begin{bmatrix} 10 & 11 \\ 12 & 13 \end{bmatrix}. \tag{4.1.11}$$

The sum of the matrices $A + B$ is

$$A + B = \begin{bmatrix} 1 & 2 \\ 3 & 4 \\ 5 & 6 \end{bmatrix} + \begin{bmatrix} -1 & 1 \\ -2 & 4 \\ 2 & -3 \end{bmatrix} = \begin{bmatrix} 1-1 & 2+1 \\ 3-2 & 4+4 \\ 5+2 & 6-3 \end{bmatrix} = \begin{bmatrix} 0 & 3 \\ 1 & 8 \\ 7 & 3 \end{bmatrix}. \tag{4.1.12}$$

The matrix C can't be added to either A or B because the dimensions aren't compatible.

▲▲

Definition 4.1.13: Let A be an n x m matrix and B be an m x p matrix. Then the product of these matrices is defined as

$$C = AB = [c_{ij}] = \left[\sum_{k=1}^{m} a_{ik}b_{kj}\right], \quad i = 1,\ldots,n, \quad j = 1,\ldots,p.$$

The product of a matrix A with a constant α, called <u>scalar multiplication</u>, is defined as

$$\alpha A = \alpha \begin{bmatrix} a_{11} & a_{12} \\ a_{21} & a_{22} \end{bmatrix} = \begin{bmatrix} \alpha a_{11} & \alpha a_{12} \\ \alpha a_{21} & \alpha a_{22} \end{bmatrix}.$$

▲▲

The definition of the multiplication of two matrices says that the ijth element of C is found by multiplying the ith row of A by the jth row of B.

Remark 4.1.14: We make the following observations about multiplication of two matrices, AB.

1. The dimensions of the matrices must be compatible as given in Definition 4.1.13. The number of columns of A must be the same as the number of rows of B. The dimension of the product of these two matrices, $C = AB$, is the number of rows of A by the number of columns of B; C is n x p.
2. If AB is defined, the product BA may *not* exist. If BA is defined, it may not be true that $AB = BA$.   Since in general $AB \neq BA$, we say that matrix multiplication is <u>not</u> commutative.
3. The product of A with an identity matrix is A; i.e., $AI_m = I_nA = A$.
4. The transpose of the product of two matrices is given by $(AB)^T = B^TA^T$.

▲▲

Example 4.1.15: (a) The product of the matrices A and B is

$$\begin{aligned}
AB &= \begin{bmatrix} a_{11} & a_{12} \\ a_{21} & a_{22} \end{bmatrix}\begin{bmatrix} b_{11} & b_{12} & b_{13} \\ b_{21} & b_{22} & b_{23} \end{bmatrix} \\
&= \begin{bmatrix} a_{11}b_{11} + a_{12}b_{21} & a_{11}b_{12} + a_{12}b_{22} & a_{11}b_{13} + a_{12}b_{23} \\ a_{21}b_{11} + a_{22}b_{21} & a_{21}b_{12} + a_{22}b_{22} & a_{21}b_{13} + a_{22}b_{23} \end{bmatrix}.
\end{aligned}$$

(4.1.13)

(b) Using the matrices in (4.1.11), the product AC is computed as

$$AC = \begin{bmatrix} 1 & 2 \\ 3 & 4 \\ 5 & 6 \end{bmatrix} \begin{bmatrix} 10 & 11 \\ 12 & 13 \end{bmatrix} = \begin{bmatrix} 1(10)+2(12) & 1(11)+2(13) \\ 3(10)+4(12) & 3(11)+4(13) \\ 5(10)+6(12) & 5(11)+6(13) \end{bmatrix} = \begin{bmatrix} 34 & 37 \\ 78 & 85 \\ 122 & 133 \end{bmatrix}. \qquad (4.1.14)$$

(c) The matrices A and B in (4.1.11) can't be multiplied together because their dimensions aren't compatible.

(d) Using s as a complex number, we have

$$sI_2 = s \begin{bmatrix} 1 & 0 \\ 0 & 1 \end{bmatrix} = \begin{bmatrix} s & 0 \\ 0 & s \end{bmatrix} \qquad (4.1.15)$$

by the property of scalar multiplication. If A is a 2 x 2 matrix we can subtract A from sI_2. We get

$$sI_2 - A = \begin{bmatrix} s & 0 \\ 0 & s \end{bmatrix} - \begin{bmatrix} a_{11} & a_{12} \\ a_{21} & a_{22} \end{bmatrix} = \begin{bmatrix} s - a_{11} & -a_{12} \\ -a_{21} & s - a_{22} \end{bmatrix}. \qquad (4.1.16)$$

Note that $(sI - A)$ is a polynomial matrix.

▲▲

Elementary matrix computations can be carried out on partitioned matrices as long as the laws of matrix addition and multiplication are observed.

Example 4.1.16: Consider the matrices

$$B_1 = \begin{bmatrix} A_1 & A_2 \end{bmatrix} = \begin{bmatrix} 1 & 2 & 5 & 6 & 7 \\ 3 & 4 & 8 & 9 & 10 \end{bmatrix}, \qquad (4.1.17)$$

$$B_2 = \begin{bmatrix} C_1 & C_2 \end{bmatrix} = \begin{bmatrix} -10 & -9 & -6 & -5 & -4 \\ -8 & -7 & -3 & -2 & -1 \end{bmatrix}.$$

Then the sum of these matrices is

$$B_1 + B_2 = \begin{bmatrix} A_1 & A_2 \end{bmatrix} + \begin{bmatrix} C_1 & C_2 \end{bmatrix} = \begin{bmatrix} A_1 + C_1 & A_2 + C_2 \end{bmatrix} \qquad (4.1.18)$$

$$= \begin{bmatrix} -9 & -7 & -1 & 1 & 3 \\ -5 & -3 & 5 & 7 & 9 \end{bmatrix},$$

$$A_1 + C_1 = \begin{bmatrix} -9 & -7 \\ -5 & -3 \end{bmatrix}, \quad A_1 + C_1 = \begin{bmatrix} -1 & 1 & 3 \\ 5 & 7 & 9 \end{bmatrix}.$$

The matrices B_1 and B_2 can't be multiplied because their dimensions don't agree. We can, however, multiply the transpose of one matrix by the other. That is,

$$(4.1.19)$$

$$B_1 B_2^T = \begin{bmatrix} A_1 & A_2 \end{bmatrix} \begin{bmatrix} C_1^T \\ C_2^T \end{bmatrix} = \begin{bmatrix} 1 & 2 & | & 5 & 6 & 7 \\ 3 & 4 & | & 8 & 9 & 10 \end{bmatrix} \begin{bmatrix} -10 & -8 \\ -9 & -7 \\ -6 & -3 \\ -5 & -2 \\ -4 & -1 \end{bmatrix} = \begin{bmatrix} -116 & -56 \\ -199 & -104 \end{bmatrix}$$

$$= A_1 C_1^T + A_2 C_2^T = \begin{bmatrix} -28 & -22 \\ -66 & -52 \end{bmatrix} + \begin{bmatrix} -88 & -34 \\ -133 & -52 \end{bmatrix}.$$

In (4.1.19) the dimensions between (A_1, C_1) are compatible, as are the dimensions between (A_2, C_2).

▲▲

4.2 VECTORS

4.2.1 Definition

It is sometimes useful to designate n x 1 arrays by a special name.

Definition 4.2.1: A <u>vector</u> is an n x 1 array, write

$$\vec{q} = \begin{bmatrix} q_1 \\ \vdots \\ q_n \end{bmatrix}.$$

The <u>dimension</u> of this vector is n.

▲▲

Terminology: A vector as we have defined it is sometimes called a *column* vector. A *row* vector can also be defined, but we will not need this concept in this text.

Note: A vector is rigorously defined by introducing the appropriate abstract concepts from linear algebra. This treatment of vector spaces is beyond the scope of this text.

Vectors obey the rules of matrix algebra. Vectors can be multiplied by a constant. Vectors of the same dimension can be summed. Vectors can't be multiplied, because their dimensions are not compatible. Vectors can be left multiplied by matrices. The result is a new vector. Let $\vec{q}$ be an m-dimensional vector and A an n x m matrix. The product of this matrix with the vector yields

$$\vec{y} = A\vec{q}. \qquad (4.2.1)$$

The vector $\vec{y}$ has dimension n by the rules of matrix multiplication. The dimensions of the equation can be visualized as

$$n\left\{\left[\begin{array}{c}\vec{y}\end{array}\right]=\left[\begin{array}{c}A\end{array}\right]\left[\begin{array}{c}\vec{q}\end{array}\right]\right\}m. \tag{4.2.2}$$

$$\underbrace{\qquad\qquad}_{m}$$

Next we consider an n dimensional vector $\vec{q}$ and an $n \times n$ matrix A. If we multiply this vector by an $n \times n$ matrix we get a new vector of the same dimension

$$\vec{q}_2 = A\vec{q}_1. \tag{4.2.3}$$

Suppose that we wish to partition the two vectors as

$$\begin{array}{c}n_1\{\\n_2\{\end{array}\left[\begin{array}{c}\vec{q}_{21}\\\hline\vec{q}_{22}\end{array}\right]=A\left[\begin{array}{c}\vec{q}_{11}\\\hline\vec{q}_{12}\end{array}\right]\begin{array}{c}\}n_1\\\}n_2\end{array} \tag{4.2.4}$$

This partition of the vectors induces a partition of the matrix A,

$$\begin{array}{c}n_1\{\\n_2\{\end{array}\left[\begin{array}{c}\vec{q}_{21}\\\hline\vec{q}_{22}\end{array}\right]=\left[\begin{array}{c|c}A_1 & A_2\\\hline A_3 & A_4\end{array}\right]\left[\begin{array}{c}\vec{q}_{11}\\\hline\vec{q}_{12}\end{array}\right]\begin{array}{c}\}n_1\\\}n_2\end{array} \tag{4.2.5}$$

where

$$\left[\begin{array}{cc}A_1 & A_2\\A_3 & A_4\end{array}\right]\begin{array}{c}\}n_1\\\}n_2\end{array} \tag{4.2.6}$$

$$\underbrace{\quad}_{n_1}\ \underbrace{\quad}_{n_2}$$

Notation: When we require the vectors in (4.2.5) to be partitioned compatibly, we impose a partition of A where the diagonal blocks are *square*.

4.2.2 Eigenvectors and Eigenvalues

Eigenvectors and eigenvalues play an important role in the study of systems represented by first-order linear differential equations. While a detailed study of these equations is beyond the scope of this text, we will use a few simple results. A brief summary of eigenvectors and eigenvalues is given next.

Again, we confine ourselves to square matrices. In general, we must allow for complex matrices, although we usually will be concerned with real matrices.

Definition 4.2.2: Let A be a square matrix. Let $\vec{v}$ be a complex vector of dimension n, and let λ be a complex number. If $\vec{v}$ and λ satisfy

$$A\vec{v} = \vec{v}\lambda$$

we say that $\vec{v}$ is an <u>eigenvector</u> and λ is an <u>eigenvalue</u> of A.

▲▲

The eigenvalues can be computed as follows. From the definition of the eigenvalues we have

$$0 = \lambda\vec{v} - A\vec{v} = \lambda I_n \vec{v} - A\vec{v} = (\lambda I_n - A)\vec{v}. \tag{4.2.7}$$

This equation can be interpreted as a system of linear equations. This equation will have a solution only if the determinant of $(\lambda I - A)$ is zero; i.e.,

$$\det(\lambda I - A) = 0. \tag{4.2.8}$$

If we expand this determinant we will obtain a polynomial in λ.

Definition 4.2.3: Let A be an n x n matrix. The polynomial

$$\det(\lambda I - A) = \lambda^n + a_{n-1}\lambda^{n-1} + a_{n-2}\lambda^{n-2} + \cdots + a_1\lambda + a_0$$

is called the <u>characteristic (polynomial) equation</u> of A.

▲▲

We make the following observations:

1. The zeros of the characteristic equation are eigenvalues of A because of (4.2.8).
2. It can be verified from the algorithm for the calculation of the determinant that the characteristic equation is always of order n, the same as the dimensions of A. In particular, the coefficient of the highest power of characteristic equation is always one as shown in Definition 4.2.3.
3. It follows that there are n roots of the characteristic equation, and n eigenvalues of A.
4. The eigenvalues of a real matrix can be either real or complex. If there is a complex eigenvalue, than the complex conjugate of that eigenvalue also appears as a eigenvalue because the coefficients of the characteristic equation are real. The eigenvalues can be repeated.
5. If the matrix has complex eigenvalues, it will also have complex eigenvectors, even if the matrix is real.
6. An eigenvector can be determined only up to multiplication by a scalar. That is, if $\vec{v}$ is an eigenvector of A, then $\alpha\vec{v}$ is also an eigenvector of A.

Example 4.2.4: Consider the matrix

$$A = \begin{bmatrix} 0 & -4 \\ 1 & -5 \end{bmatrix}. \tag{4.2.9}$$

To compute the characteristic equation we first form the matrix

$$\left(\lambda I - A\right) = \begin{bmatrix} \lambda & 4 \\ -1 & \lambda + 5 \end{bmatrix}. \tag{4.2.10}$$

Taking the determinant of (4.2.10) we obtain

$$\det\left(\lambda I - A\right) = \lambda(\lambda + 5) + 4 = \lambda^2 + 5\lambda + 4. \tag{4.2.11}$$

The eigenvalues of A are the roots of the characteristic equation in (4.2.11). We have

$$\lambda_1 = -1, \quad \lambda_2 = -4. \tag{4.2.12}$$

Once the eigenvalues have been determined, we can solve for the eigenvectors. The eigenvalue/eigenvector pairs satisfy the equation

$$\left(\lambda_i I - A\right)\vec{v}_i = 0 \tag{4.2.13}$$

From this equation we can calculate the eigenvectors. We have

$$\left(\lambda_1 I - A\right)v_1 = \begin{bmatrix} \lambda_1 & 4 \\ -1 & \lambda_1 + 5 \end{bmatrix}\begin{bmatrix} v_{11} \\ v_{21} \end{bmatrix} = \begin{bmatrix} -1 & 4 \\ -1 & 4 \end{bmatrix}\begin{bmatrix} v_{11} \\ v_{21} \end{bmatrix} = 0 \quad \Rightarrow v_1 = \begin{bmatrix} 4 \\ 1 \end{bmatrix} \tag{4.2.14}$$

$$\left(\lambda_2 I - A\right)v_2 = \begin{bmatrix} \lambda_2 & 4 \\ -1 & \lambda_2 + 5 \end{bmatrix}\begin{bmatrix} v_{12} \\ v_{22} \end{bmatrix} = \begin{bmatrix} -4 & 4 \\ -1 & 1 \end{bmatrix}\begin{bmatrix} v_{12} \\ v_{22} \end{bmatrix} = 0 \quad \Rightarrow v_2 = \begin{bmatrix} 1 \\ 1 \end{bmatrix}$$

▲▲

Remark 4.2.5: When a matrix has n distinct eigenvalues, it is possible to associate a unique eigenvector with each eigenvalue. Unfortunately, when a matrix has repeated eigenvalues, it may not be possible to identify a unique eigenvector for each eigenvalue. There are generalizations of the eigenvector concept that lead to n "generalized" eigenvectors, but they are beyond the scope of this text. We will confine ourselves to matrices that have n eigenvectors.

▲▲

The question arises "Are the eigenvalues unique to the matrix A?" This question is answered in part by the following calculation. Let T be an n x n, nonsingular[1] matrix. Consider the matrix

$$\tilde{A} = TAT^{-1}. \tag{4.2.15}$$

The characteristic equation of A and $\tilde{A}$ are the same. To see this, first form the equation

$$\lambda I - \tilde{A} = \lambda I - TAT^{-1} = \lambda TT^{-1} - TAT^{-1} = T(\lambda I - A)T^{-1}. \tag{4.2.16}$$

---

[1] The inverse of T exists.

Taking the determinant of (4.2.16) we find that

$$\det(\lambda I - \tilde{A}) = \det\left[T(\lambda I - A)T^{-1}\right] = \det(T)\det\left(T^{-1}\right)\det(\lambda I - A) \qquad (4.2.17)$$
$$= \det(\lambda I - A).$$

Hence, the eigenvalues of A and $\tilde{A}$ are the same.

The eigenvectors of the matrices A and $\tilde{A}$ are not the same, but they are related. To find this relationship start with the basic eigenvalue/eigenvector relationship in Definition 4.2.2. Using T, modify this relationship as

$$ATT^{-1}\vec{v} = \vec{v}\lambda. \qquad (4.2.18)$$

Now multiply (4.2.18) on the left by T^{-1}. We obtain

$$T^{-1}ATT^{-1}\vec{v} = T^{-1}\vec{v}\lambda. \qquad (4.2.19)$$

Compare (4.2.19) to (4.2.15). Now define the new eigenvector

$$\vec{\tilde{v}} = T^{-1}\vec{v}. \qquad (4.2.20)$$

Using (4.2.20), (4.2.19) can be rewritten as

$$\tilde{A}\vec{\tilde{v}} = \vec{\tilde{v}}\lambda. \qquad (4.2.21)$$

Hence, the eigenvectors of the matrix $\tilde{A}$ in (4.2.15) are related to the eigenvectors of the matrix A by the relationship in (4.2.20).

4.3 HOMEWORK FOR CHAPTER 4

Homework Problems for Section 4.1

4.1.1   Consider the following matrices.

$$A_1 = \begin{bmatrix} 3 & 1 \\ 4 & 5 \end{bmatrix}, \quad A_2 = \begin{bmatrix} -2 \\ -4 \end{bmatrix}, \quad A_3 = \begin{bmatrix} 6 & 7 \end{bmatrix}, \quad A_4 = \begin{bmatrix} 10 \end{bmatrix}.$$

(a) Perform each of the following calculations.
(b) Using MATLAB perform each of these calculations.

(i) A_3^T (iv) $A_3 A_2$

(ii) $A = \begin{bmatrix} A_1 & A_2 \\ A_3 & A_4 \end{bmatrix}$ (v) $A_2 A_3$

(iii) $A_2 + A_3^T$ (vi) $A_4 A_1 A_2$

4.1.2 Consider the following matrices.

(a) $A_1 = \begin{bmatrix} 3 & 1 \\ 4 & 5 \end{bmatrix}$, $A_2 = \begin{bmatrix} -2 & 0 & 8 \\ -4 & 2 & 0 \end{bmatrix}$, $A_3 = \begin{bmatrix} 6 & 7 \end{bmatrix}$.

(b) $A_1 = \begin{bmatrix} 3 & 1 \\ 4 & 5 \\ 0 & 1 \end{bmatrix}$, $A_2 = \begin{bmatrix} -2 & 0 & 8 \\ -4 & 2 & 0 \\ 0 & -3 & 1 \end{bmatrix}$, $A_3 = \begin{bmatrix} 6 & 7 & 0 \end{bmatrix}$

Perform the following calculations in MATLAB if the operations are defined. If the operations are not defined, explain why.

(i) A_3^T (iv) $A_1 A_2$

(ii) $A = \begin{bmatrix} A_1 & A_2 \end{bmatrix}$ (v) $A_1 A_3^T$

(iii) $A_2 + \begin{bmatrix} A_1 & A_3^T \end{bmatrix}$ (vi) $A_3 A_2$

4.1.3 Consider the following matrices.

(i) $A = \begin{bmatrix} s-1 & 4 & 3 \\ 7 & s & 5 \\ -1 & 6 & s+8 \end{bmatrix}$ (ii) $A = \begin{bmatrix} 0 & 0 & -a_0 \\ 1 & 0 & -a_1 \\ 0 & 1 & -a_2 \end{bmatrix}$

(a) Find the diagonal of each matrix.
(b) Find the subdiagonal of each matrix.
(c) Find the superdiagonal of each matrix.

Homework Problems for Section 4.2

4.2.1 Find the eigenvalues and eigenvectors of the following matrices.

(i) $A = \begin{bmatrix} 0 & 1 \\ 0 & -2 \end{bmatrix}$ (iii) $A = \begin{bmatrix} 0 & 1 \\ -2 & -3 \end{bmatrix}$

(ii) $A = \begin{bmatrix} -2 & 0 \\ 0 & -6 \end{bmatrix}$ (iv) $A = \begin{bmatrix} 0 & 1 \\ -4 & -1 \end{bmatrix}$

4.2.2 Find the eigenvalues and eigenvectors of the following matrices.

(i) $A = \begin{bmatrix} \sigma & \omega \\ -\omega & \sigma \end{bmatrix}$

(ii) $A = \begin{bmatrix} \lambda & 0 & 0 \\ 0 & \sigma & \omega \\ 0 & -\omega & \sigma \end{bmatrix}$

4.2.3   Consider the matrix

$$A = \begin{bmatrix} a & 0 & b \\ c & d & 0 \\ 0 & 0 & f \end{bmatrix}$$

Assume that $a \neq d$ and $a \neq f$. Find the eigenvalues and eigenvectors of A.

4.2.4   Determine the characteristic polynomial for each of the following matrices.

(i) $A = \begin{bmatrix} 0 & 1 & 0 \\ 0 & 0 & 1 \\ -a_0 & -a_1 & -a_2 \end{bmatrix}$ (ii) $A = \begin{bmatrix} 0 & 0 & -a_0 \\ 1 & 0 & -a_1 \\ 0 & 1 & -a_2 \end{bmatrix}$

Chapter 5

Introduction to Signals

Chapter Outline

In Chapter 1 we informally introduced the concept of a signal. It is the purpose of this chapter to formally introduce the definition of a signal. We will then give a number of examples of signals. We will also discuss several ways of combining two or more signals to form more complicated signals. This discussion marks the beginning of the discussion of the main topics in this text.

As we discussed in Chapter 1, a signal is a function that is used to describe physical observations. In order to use the signal to enhance our understanding of the physical phenomenon we must develop a representation of the signal. That is to say, we must write an explicit expression for the function that can be analyzed. This chapter is devoted to introducing several elementary methods for developing a representation of the signal. These methods are primarily ad hoc, but they are routinely used in system theory. First we discuss several mathematical techniques for systematically reshaping known functions to match observations. These techniques are pervasive in signal modeling. Then we discuss modeling signals by breaking the signal into time intervals and developing a representation on each time interval. This technique leads to a model of a digital waveform that is widely used in digital communications and in computers. Finally, we introduce the idea of modeling signals with sinusoids. This concept, which is fundamental to much of the analysis in this text, takes on its most subtle form in the Fourier series. This important topic is taken up in Chapter 7.

Summary of Sections

Section 5.1:   The definition of a signal is introduced.

Section 5.2:   We discuss time scaling and time shifting as well as signals defined as limits

Section 5.3:   Signal modeling based on intervals is introduced.

Section 5.4:   We discuss digital waveforms.

Section 5.5:   We discuss signals that are sums of sinusoids.

Section 5.6:   Chapter summary section.

Coverage of the Text

This chapter requires primarily the material on functions in Chapter 2. The MATLAB files require elementary knowledge of matrices given in Chapter 4.

5.1 DEFINITION OF A SIGNAL

5.1.1 Definitions

We are interested in describing and understanding the physical processes we observe in the world around us. Examples of physical variables are current and voltage at a given node in an electric network; position, velocity, and acceleration of a mass; the concentration of a certain chemical in a batch process; the height of the water in a reservoir; the intensity of light at the endpoint of an optical fiber; etc. In order to discuss and analyze these observations we need to provide a vocabulary that allows us to explore the deeper properties of these physical phenomena in a systematic way. It is the purpose of this text to introduce a language for describing and analyzing physical phenomena. The following definition is the first step down this road.

Definition 5.1.1: A <u>signal</u> is a function which represents the time variation of a physical variable.

▲▲

The essence of Definition 5.1.1 is that we attach a *mathematical* description to the observations of the physical process. Through this mathematical description we hope to increase our understanding of the physical process. We must emphasize that signals are *functions* which describe the evolution in time of physical variables, not the physical variables themselves. The signal model is an abstraction of reality.

Definition 5.1.1 of a signal is abstract in that we have not specified the function that is used to describe the signal. To proceed with the investigation we must develop a specific function that is amenable to analysis.

Definition 5.1.2: The function which is chosen to describe a signal is called the <u>representation</u> of the signal. The process of building a representation of the signal is called signal <u>modeling</u>.

▲▲

Terminology: The signal representation is also called the signal <u>model</u>.

There are many kinds of signal models which yield different kinds of information about the underlying physical process. Functions of time are the most obvious kind of signal model. For the simplest signal model we simply select a real function to describe the observations. This type of signal modeling is very familiar. The functions used to describe voltages and currents in networks are examples of this type of signal. When the observations don't appear to be exactly repeatable, we can include probability concepts. We call this class of signal models random processes. (We will not discuss this class of signals in this text.)

We have made several references to the observation of physical processes that evolve as a function of time. In fact we will investigate signals that depend on two different concepts of time. The first concept of time corresponds to the standard concept that is used everyday. Here time is modeled by a real variable.

Definition 5.1.3: A signal that depends on a real variable t and that models a physical variable that evolves in real time is called a <u>continuous (-time)</u> signal.

▲▲

Notation: Continuous-time signals are denoted by a function with the real variable t as an argument as in: $x(t)$.

In the second part of this text we will have occasion to consider another concept of time. This concept of time, which is discussed in more detail in Chapter 17, is based on the fact that many observations are not made continuously. In fact, the observations are frequently made at discrete instants of time. Then we model the signal only at the discrete observation instants. We can enumerate these observation instants with integers. Hence, time can be modeled with the integers rather than the real numbers. Then the signals are defined using real functions that depend on the integers rather than the real numbers.

Definition 5.1.4: A signal that depends on a discrete variable n and that models a physical variable that evolves in discrete time is called a <u>discrete (-time)</u> signal.
▲▲

Notation: Discrete-time signals are denoted by a function with an integer n as an argument as in: $x(n)$.

Discrete-time signals are widely used to model the operation of computer algorithms and economic systems.

When we are trying to model actual experimental data, we generally represent the signal with a *real* function of time. Sometimes it is convenient to abstract this definition and allow the function to be *complex*. Of course, a complex function doesn't relate directly to a physical process so we lose the physical interpretation of a signal as the model of a physical process. Complex discrete signals can be used effectively to model the operation of computer algorithms, however. Most of the results in this text extend directly to complex signals with only minor modifications. Some of the results will be stated using complex signals because that is the way they are often used in advanced analysis.

Another kind of signal model employs a transform of the function of time. The idea here is to start with a function of time which describes the observations of the physical process. Then this function is transformed into another domain using a mathematical transform. The most common transforms used are Fourier and Laplace transforms for continuous signals and z-transforms for discrete signals. The signal analysis is then carried out by analyzing the transform of the signal. The Fourier and Laplace transforms are widely used in network analysis, for example, as representations for signals. We will investigate this type of signal representation in detail.

In Definition 5.1.1 we define a signal as a function of *time*. This definition can be extended by replacing *time* with *space*. These signals occur in image processing. A two-dimensional image is represented by a signal by discretizing the image and assigning a number to each discrete point. The signal is defined by considering these points to be a function of spatial location. Thus the signal is a function of space. Most of the concepts we will develop for signals that depend on time extend directly to signals that depend on a spatial variable.

When analyzing a specific signal the representation plays a crucial role in extracting information about the physical process. Given a specific function, we generally have three ways we can represent that function: (1) with an analytical expression, (2) with a graph, and (3) with a computer file. In advanced signal analysis, analytical expressions have many limitations. We will find graphical representations to be extremely useful, however, and we will use them frequently. Visual analysis of a graphical representation of a signal is widely used in the laboratory where the signals are routinely captured using digital collection electronics or analog chart recorders. Finally, if a signal can be represented in the computer, the analysis possibilities are greatly expanded.

5.1.2 Examples

In this subsection we will present two examples which illustrate the concept of a signal.

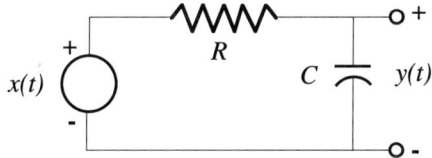

Figure 5.1.1 *RC* Network

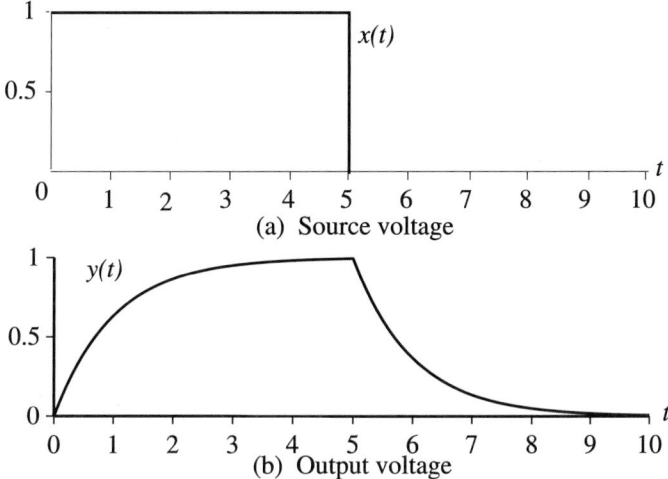

(a) Source voltage

(b) Output voltage

Figure 5.1.2 Input and Output Voltage Applied to the Network in Figure 5.1.1

Example 5.1.5: Consider the *RC* network in Figure 5.1.1. The applied voltage is a signal. It is denoted by $x(t)$. The voltage across the resistor is a signal as is the voltage across the capacitor. The current flowing through the resistor is also a signal.

Suppose that at time $t = 0$ we switch on the applied voltage and at time $t = 1$ we switch it off. While the voltage is on, this source applies a constant one volt to the network. This applied voltage is measured, and it is shown in Figure 5.1.2a. Figure 5.1.2a is a graphical representation of the voltage signal. An analytical representation of the source voltage signal is

$$x(t) = u_s(t) - u_s(t - t_0). \tag{5.1.1}$$

Suppose we also measure the voltage across the capacitor. This measured voltage is shown in Figure 5.1.2.b as a graphical representation of the signal. One analytical signal model of this voltage is

$$y(t) = \begin{cases} 1 - e^{-\frac{t}{RC}}, & 0 \le t < t_0 \\ e^{-\frac{(t-t_0)}{RC}} - e^{-\frac{t}{RC}}, & t > t_0 \end{cases} \tag{5.1.2}$$

This type of analysis is commonplace in network textbooks. Here we are simply giving a name to the representations of the physical processes.

The new technologies of the future are not so well understood as resistors, capacitors and inductors, however. We need analytical tools to investigate, understand and use these new devices and concepts. Therefore, we are going to generalize the analysis tools of networks in an attempt to put them to new uses. As this example suggests, we are just rephrasing network analysis in a more general framework.

▲▲

Example 5.1.6: Consider a mass on a frictionless plane shown in Figure 5.1.3. Suppose that this mass is being acted on by a force f. As the result of this force, the mass begins to move in a direction determined by direction of action of the force. The displacement and velocity are physical variables associated with this mass. We can model these physical variables by signals. We choose a coordinate reference frame along the line of motion of the mass. Then the displacement of the mass from the origin is a signal. The velocity of the mass is also a signal.

A somewhat more interesting signal is the force imparted to the mass by the hammer. Clearly, the force acts on the mass for a finite (nonzero) time interval with a finite magnitude. It is often convenient to represent this force with an impulse function, however. This type of signal model is discussed further in Section 5.2.

▲▲

5.1.3 Interconnections of Signals

Often complicated signals can be constructed from the interconnection of simpler signals. These composite signals may be the result of the action of a physical process. Indeed, this type of signal model is often part of a large complex model of interconnected systems. Or these composite signals may be just a convenient way for the signal modeler to construct a signal to represent a complicated physical variable.

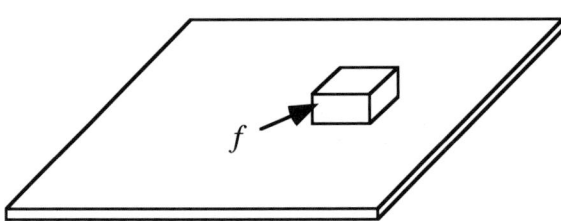

Figure 5.1.3 Mass on a Frictionless Plane Being Acted on by a Force

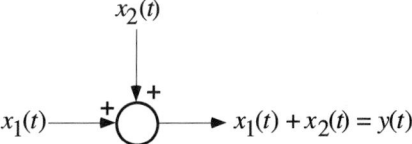

Figure 5.1.4 Addition of Two Signals

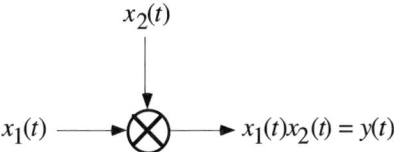

Figure 5.1.5 Multiplication of Two Signals

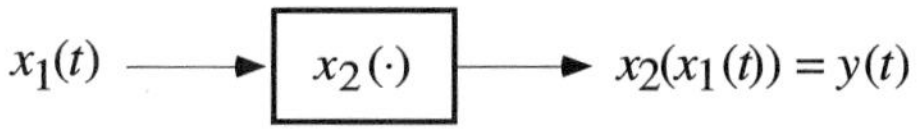

$$x_1(t) \longrightarrow \boxed{x_2(\cdot)} \longrightarrow x_2(x_1(t)) = y(t)$$

Figure 5.1.6 Composition of Two Signals

We will discuss three ways to combine two signals: addition, multiplication, and composition. In the following discussion we assume that we are given two signals: $x_1(t)$ and $x_2(t)$ which are combined to form the new signal $y(t)$. The first way to generate a new signal is to add these two signals together. This process is shown in Figure 5.1.4. This figure will frequently appear in this text. The second way these two signals can be combined is through multiplication. This process is shown in Figure 5.1.5. The third way two signals can be combined is through the formal mathematical process of composition. In equations this process is given by

$$y(t) = x_2(x_1(t)). \tag{5.1.3}$$

This process is shown in Figure 5.1.6. All three of these techniques are used to generate signals throughout the text, and, in particular, in the remaining sections of this chapter. The following example illustrates these three types of signal interconnections.

Example 5.1.7: Consider the most elementary model of a communication system shown in Figure 5.1.7. The basic idea in a communication system is to encode information into a signal called a <u>message</u> signal $m(t)$. This signal is then transmitted from the one place to another through the communication <u>channel</u>. Quite often the message signal can't be transmitted directly, but it must be combined with a <u>carrier</u> signal $x_c(t)$ to create the <u>transmitted</u> signal $s(t)$ that can be transmitted over long distances through the channel. The function of the <u>transmitter</u> (in part) is to combine the message signal with the carrier signal. As the transmitted signal

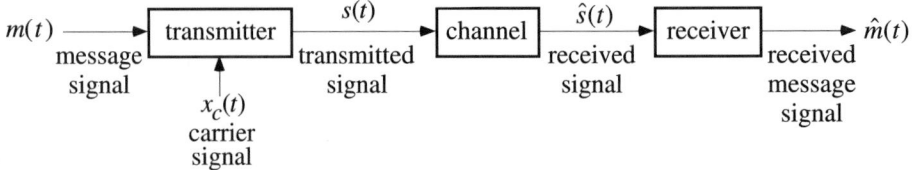

Figure 5.1.7 Components of a Communication System

propagates through the communication channel, it is often corrupted by noise, distorted by nonlinearities of the channel, and suffers interference with the other transmitted signals in the channel. This corrupted signal, the <u>received</u> signal $\hat{s}(t)$ arrives at the <u>receiver</u>. The function of the <u>receiver</u> (in part) is to extract the corrupted message signal from the carrier signal. Finally, the information is extracted from the corrupted message signal $\hat{m}(t)$.

Most radio stations transmit radio signals as electromagnetic waves. The electromagnetic waves are the carrier signals. The frequency of these waves is selected for optimal propagation through the atmosphere and so the various stations don't interfere with each other. A simple model of the carrier signal, the electromagnetic wave, is

$$x_c(t) = A_c \cos(\omega_c t). \tag{5.1.4}$$

Suppose that the message signal $m(t)$ is a representation of a voice signal. More precisely, it would represent the voltage from a microphone. This message signal can be impressed into either the amplitude A_c or frequency ω_c of the carrier signal. This process of varying one of the parameters of the carrier wave according to a modulating signal $m(t)$ is called <u>modulation</u>. Because the carrier wave is a sinusoid, we call this type of modulation <u>continuous-wave</u> modulation.   The transmitter accomplishes the modulation. At the receiver, the message signal is extracted from the received signal. This process is called <u>demodulation</u>.

If the message signal is impressed into the amplitude of the carrier signal, we say the signal is <u>amplitude modulated</u> (AM). In practice, this modulation takes the form

$$s(t) = \left[1 + k_a m(t)\right] A_c \cos(\omega_c t). \tag{5.1.5}$$

The constant k_a is called the <u>amplitude sensitivity</u>. The construction of the transmitted signal from the message signal and the carrier signal is shown in Figure 5.1.8. Clearly, the transmitted signal is constructed through both addition and multiplication of two signals. An amplitude modulated signal is shown in Figure 5.1.9 where the message signal is a cosine function.

If the message signal is impressed into the frequency of the carrier signal, we say the signal is <u>frequency modulated</u> (FM). Writing the frequency of the carrier wave as $\omega_i = 2\pi f_i$, the message signal is impressed into the frequency as

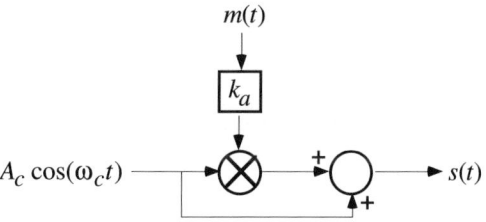

Figure 5.1.8 Amplitude Modulation

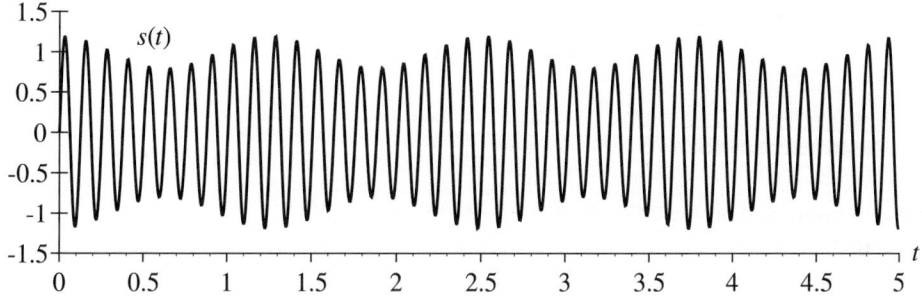

Figure 5.1.9 Amplitude-Modulated Signal

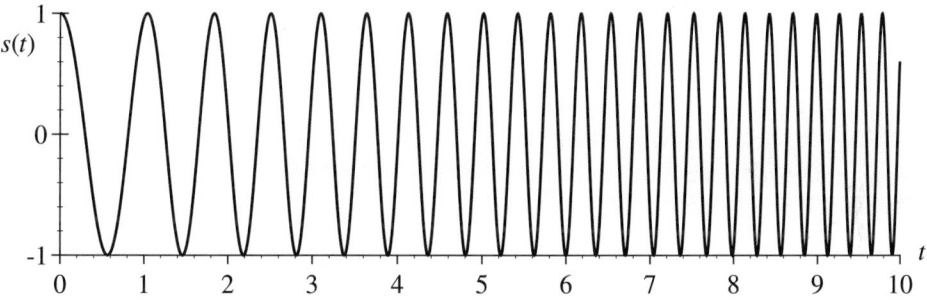

Figure 5.1.10 Frequency-Modulated Signal

$$f_i(t) = f_c + k_f m(t). \tag{5.1.6}$$

The constant k_f is called the <u>frequency sensitivity</u>. In order to interpret a "time varying frequency" we note that the frequency of the carrier wave in (5.1.4) is the derivative of the argument of the cosine. Therefore, if we integrate (5.1.6) and multiply by 2π we obtain

$$2\pi \int_0^t f_i(\lambda)\, d\lambda = 2\pi f_c t + k_f \int_0^t m(\lambda)\, d\lambda. \tag{5.1.7}$$

The frequency modulated signal is then given by

$$s(t) = A_c \cos\left(2\pi f_c t + k_f \int_0^t m(\lambda) \, d\lambda\right).$$

(5.1.8)

Here the transmitted signal is constructed by the composition of two signals. A frequency modulated signal is shown in Figure 5.1.10 where the message signal is given by $m(t) = 2r_p(t)$.

▲▲

5.2 TIME SCALING, TIME SHIFTING, AND LIMITS OF SIGNALS

5.2.1 Introduction

The number of functions that have been introduced so far could be used to model only a small fraction of the number of signals in which we are interested. In this chapter we discuss several ways of modifying an existing function so that it can be used as a signal model. Two basic methods for changing a standard function into a representation of an observed variable are time scaling and time shifting. These transformations are discussed in this section.

5.2.2 Time Scaling

New functions can be created by replacing the argument of the function with the argument multiplied by a constant. Given the signal $x(t)$, let

$$t = \varepsilon\tau.$$

(5.2.1)

Substituting (5.2.1) into $x(t)$ we obtain $x(\varepsilon\tau) = \tilde{x}(\tau)$. This action can be viewed as "stretching" or "shrinking" the signal as well as flipping it around the ordinate axis.

Definition 5.2.1: Given the signal $x(t)$, the signal $x(\varepsilon\tau) = \tilde{x}(\tau)$ is said to be <u>time scaled</u>.

▲▲

First suppose that $\varepsilon = -1$. Then (5.2.1) becomes

$$t = -\tau.$$

(5.2.2)

The time scale of the signal is <u>reversed</u>. Time reversal causes the function to be flipped about the ordinate axis. The following example illustrates time reversal.

Example 5.2.2: Consider the unit step function $u_s(t)$. Using the time reversal in (5.2.2) we obtain

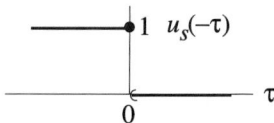

Figure 5.2.1 Time Reversal for a Unit Step Function

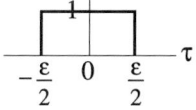

Figure 5.2.2 Scaled Unit Pulse Signal

$$x(\tau) = u_s(-\tau) = \begin{cases} 1, & \tau \le 0 \\ 0, & \tau > 0 \end{cases} \tag{5.2.3}$$

This signal is shown in Figure 5.2.1.

▲▲

Anytime ε is negative in (5.2.1), the new signal undergoes time reversal. In the following we will concentrate on the effect of a positive ε in (5.2.1). The following example illustrates the basic idea.

Example 5.2.3: Consider the unit pulse function $\Pi(t)$ in Definition 2.1.8. If we time-scale this function using

$$t = \frac{\tau}{\varepsilon} \tag{5.2.4}$$

we obtain

$$x(\tau) = \Pi\left(\frac{\tau}{\varepsilon}\right) = \begin{cases} 1, & |\tau| \le \frac{\varepsilon}{2} \\ 0, & |\tau| > \frac{\varepsilon}{2} \end{cases} \tag{5.2.5}$$

The signal in (5.2.5) is shown in Figure 5.2.2. We see from Figure 5.2.2 the pulse signal goes from 0 to 1 when the argument of the pulse function is $-1/2$. To find this transition set

$$\frac{\tau}{\varepsilon} = -\frac{1}{2}, \quad \tau = -\frac{\varepsilon}{2}. \tag{5.2.6}$$

We can find the second transition from 1 to 0 in the same manner.

▲▲

We will frequently have occasion to use the scaled version of the unit pulse. For the signal (5.2.5) in Figure 5.2.2 we see that the pulse width is ε. As ε gets larger, the pulse width has expanded. It's as if the time scale has expanded. Conversely, as ε gets smaller, the pulse width gets smaller. For this case, the time scale shrinks. This idea of expanding and shrinking time scales is further illustrated in the following example.

Example 5.2.4: Consider the RC network in Figure 5.2.3. Assume that the voltage $x(t)$ is a unit step function and the initial voltage across the capacitor is zero. The signal which represents the voltage across the capacitor $y(t)$ is given by

$$y(t) = \left(1 - e^{-\frac{t}{RC}}\right) u_s(t). \tag{5.2.7}$$

If $R = 1 \text{ k}\Omega$ and $C = 1 \text{ }\mu\text{F}$, then the time constant is $RC = 10^{-3}$ sec. This voltage is shown in Figure 5.2.4. (The purpose of the discrete points is explained below.) The top time scale in Figure 5.2.4 is in seconds. One way to rescale the graph in Figure 5.2.4 is to rescale (5.2.7) using

$$t = RC\tau = 10^{-3}\tau. \tag{5.2.8}$$

The original time scale is in units of seconds while the new time scale is in units of milliseconds. The output voltage (5.2.7) in the new time scale becomes

$$\tilde{y}(\tau) = \left(1 - e^{-\frac{1}{RC}(RC\tau)}\right) u_s(\tau) = \left(1 - e^{-\tau}\right) u_s(\tau). \tag{5.2.9}$$

This time scale is also shown in Figure 5.2.4. The scaling in (5.2.9) normalizes the time scale by the system parameter RC. Hence, Figure 5.2.4 shows the qualitative behavior of all networks of the form in Figure 5.2.3. We say the output signal has been normalized by the RC time constant.

Which time scale we chose depends on the motivation for the signal model. Suppose we scale the time axis by

$$\tau = 5t_1. \tag{5.2.10}$$

In this new time scale (5.2.9) becomes

$$y_1(t_1) = \left(1 - e^{-5t_1}\right) u_s(t_1). \tag{5.2.11}$$

In this time scale every unit of t_1 corresponds to 5 msec. This voltage in this time scale is shown in Figure 5.2.5a. We might consider a second time scale

$$\tau = 0.5t_2. \tag{5.2.12}$$

In this time scale (5.2.9) becomes

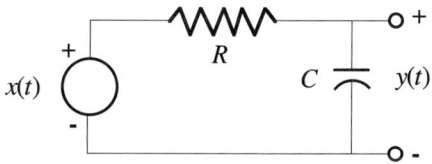

Figure 5.2.3 *RC* Network

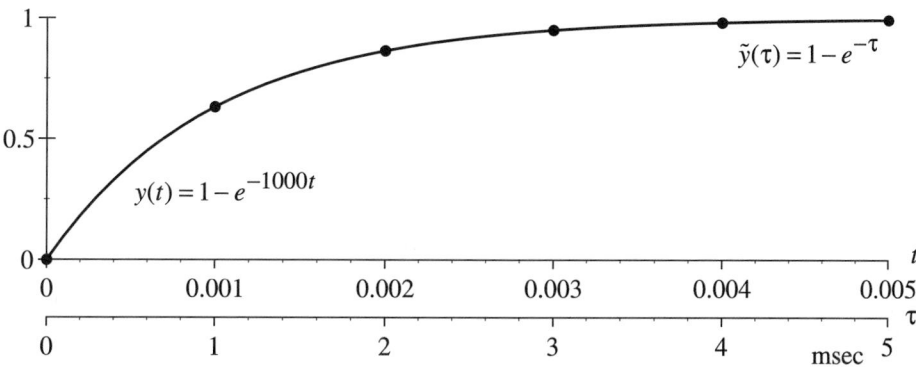

Figure 5.2.4 Output Signal for the Network in Figure 5.2.3

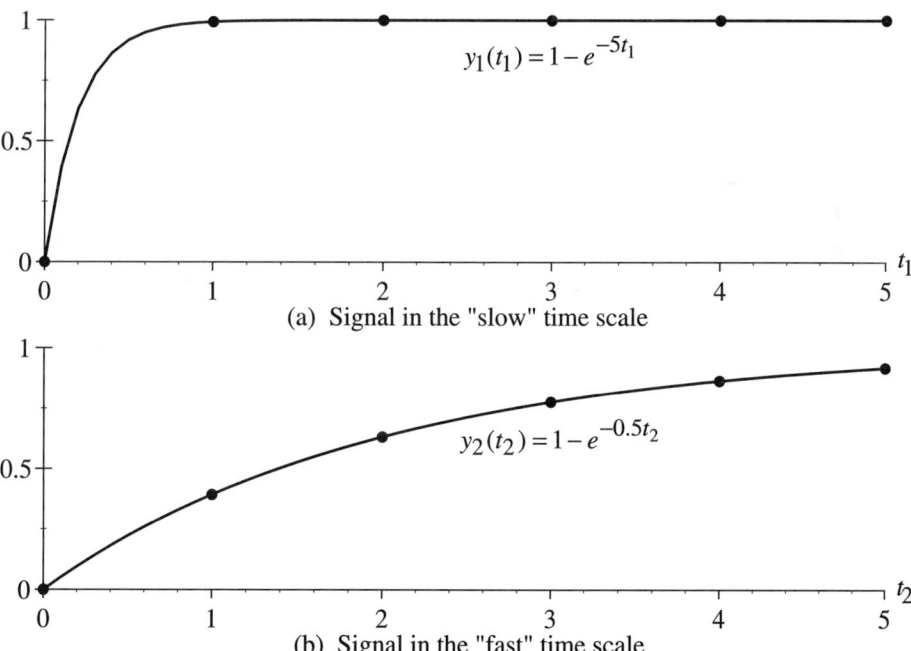

(a) Signal in the "slow" time scale

(b) Signal in the "fast" time scale

Figure 5.2.5 Two Time Scales of the Output Signal of the Network in Figure 5.2.3

$$y_2(t_2) = \left(1 - e^{-0.5t_2}\right)u_s(t_2).$$

(5.2.13)

This voltage in the new time scale is shown in Figure 5.2.5b. In this time scale every unit of t_2 corresponds to 0.5 msec.

Signal Modeling for Analysis The signal representation we chose depends on our purpose for analyzing the signal. If the capacitance in the RC network in Figure 5.2.3 represents stray capacitance, then the RC time constant will be small compared to the long time behavior of the voltage $y_1(t_1)$. If we compare the graphs in Figure 5.2.5a and Figure 5.2.5b it appears as if the transient in Figure 5.2.4 has shrunk. As a result, the long time behavior of the signal $y_1(t_1)$ is emphasized. This long time behavior is sometimes referred to as the "slow" time scale.

On the other hand, suppose we were interested in developing a model of the capacitor for a resonant circuit. For such applications, the input voltage is constantly switching from 0 V to 1 V every 2 msec, say, in Figure 5.2.3. In this context, the discharge time of the capacitor is an important factor in the operation of the circuit. For this analysis we would select the model in (5.2.13) and Figure 5.2.5b. If we compare the graphs in Figure 5.2.5a and Figure 5.2.5b it appears as if the transient in Figure 5.2.5b has been stretched. As a result, the short time behavior of the signal $y_2(t_2)$ is emphasized. This short time behavior is sometimes referred to as the "fast" time scale.

Signal Modeling for Computer Analysis Time scales have important ramifications for computer-aided design. Suppose that we were asked to plot the signal in Figure 5.2.5a, but we are allowed to only plot one point every unit of time uniformly spaced. Furthermore, we are restricted to a vector of finite length because of memory requirements. If we elect to use the original time scale, then these points are shown in Figure 5.2.4. These points capture much of the behavior of this signal over five seconds.

If we elect to plot this signal in the slow time scale, the resulting points are shown in Figure 5.2.5a. In Figure 5.2.5a we clearly see that the transient behavior is lost. On the other hand, we have plotted this signal for $(5)(5) = 25$ sec; i.e., this signal has been plotted over a much longer time using the same number of points.

If we elect to use the fast time scale, the points would be plotted are shown in Figure 5.2.5b. Plotting in this time scale captures the fast transient behavior, but it would take five times more points to plot the signal in Figure 5.2.4 for 5 msec.

So we see that the choice of time scale depends on the behavior of interest and the capability of the computer. The time step and the time scale are closely related. Each must be chosen carefully if the appropriate computer results are to be obtained. ▲▲

5.2.3 Time Shift

A second way to create a new signal is to move the origin. For the signal $x(t)$ we make the substitution

$$t = \tau - b.$$

(5.2.14)

This change in (5.2.14) causes the origin of the original signal to be shifted to $t = b$.

Definition 5.2.5: Given the signal $x(t)$, we say that the signal $x(\tau - b) = x_s(\tau)$ has been <u>time-shifted</u>. If $b > 0$ we say that the time shift is a <u>right</u> shift. If $b < 0$ we say that the time shift is a <u>left</u> shift.

▲▲

Example 5.2.6: Consider the time-shifted pulse signal $\Pi(\tau - b)$. This signal is shown Figure 5.2.6. The pulse is now centered on $\tau = b$. This shift is a right shift because the center of the shifted pulse is to the right of the ordinate axis.

▲▲

Of course, time shifting and time scaling can be combined to create new functions as illustrated in the next example.

Example 5.2.7: The graph of

$$\Pi\left(\frac{\tau - b}{\varepsilon}\right) = \Pi\left(\frac{\tau}{\varepsilon} - \frac{b}{\varepsilon}\right) \tag{5.2.15}$$

is shown in Figure 5.2.7. The pulse in Figure 5.2.7 is the same shape as the pulse in the basic definition of the Π function. It has been modified in two ways, however. First, the width of the pulse has been changed to ε due to a change in time scale as discussed in Example 5.2.3. Then this scaled pulse has been time shifted to the right so that it is centered at $\tau = b$. We will frequency encounter this signal in this text.

▲▲

5.2.4 Limits of Signals

In some applications it is useful to represent a signal as the limit of a set of parameterized signals.

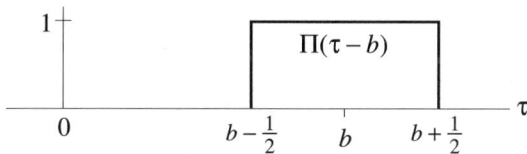

Figure 5.2.6 Right Time-Shifted Unit Pulse Signal $\Pi(t - b)$

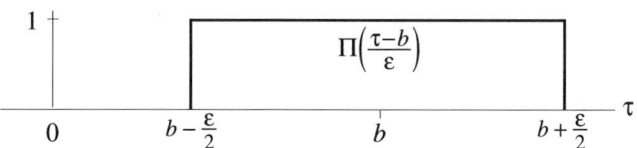

Figure 5.2.7 Graph of the Pulse Function in Example 5.2.7

Example 5.2.8: Consider the scaled pulse function

$$\delta_\varepsilon(t) = \frac{1}{\varepsilon} \Pi\left(\frac{t}{\varepsilon}\right), \quad \varepsilon > 0. \tag{5.2.16}$$

This signal is shown for two different values of ε in Figure 5.2.8. The limit of this sequence of signals as $\varepsilon \to 0$ can be considered to approximate an impulse signal. To justify this observation, we note that

$$\int_{-\infty}^{\infty} \delta_\varepsilon(t)\, dt = 1 \quad \varepsilon > 0. \tag{5.2.17}$$

The signals in (5.2.16) satisfy

$$\delta_\varepsilon(t) = 0 \quad \text{for} \quad t < -\frac{\varepsilon}{2}, \quad t > \frac{\varepsilon}{2}. \tag{5.2.18}$$

So the approximating signals have the properties of the impulse function in Remark 2.2.2 in the limit. Because the area is constrained to be one, the height of the signals tends to infinity as $\varepsilon \to \infty$, $\varepsilon \neq 0$. It can be shown that

$$\lim_{\varepsilon \to 0} \int_{-\infty}^{\infty} \delta_\varepsilon(t)x(t)\, dt = \int_{-\infty}^{\infty} \delta(t)x(t)\, dt = x(0) \tag{5.2.19}$$

provided that $x(t)$ is continuous at $t = 0$. It follows that

$$\lim_{\varepsilon \to 0} \delta_\varepsilon(t) = \delta(t). \tag{5.2.20}$$

The limit of the approximating sequence of signals in (5.2.20) is a signal that has the properties of an impulse function. It can be shown that any sequence of signals with the properties in (5.2.17) and (5.2.18) results in the same limit. So the impulse function is uniquely defined as a limit of these approximation signals. This analysis suggests that for ε small we think of the impulse function as a pulse of short duration with unit area.   ▲▲

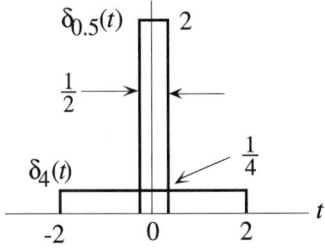

Figure 5.2.8 Signals of Example 5.2.8

Example 5.2.9: The approximation for the impulse function in Example 5.2.8 is commonly used in dynamics where large forces act over a short time interval. For example, when a bat contacts a ball, the contact time is short as far as an outfielder is concerned. The force applied to the ball can be modeled as an impulse function.

We should keep in mind that the signal model depends on the intended use of that model. The baseball manufacturers, for example, are concerned with constructing the ball of materials that give it a desirable response on contact with the bat. Baseballs have a very hard core that results in long flights with proper contact with the bat. For the local softball league, such dynamic properties are undesirable because the fences are short and the players are not as agile as professional athlete. A change in the core material, for example, can lead to shorter flight distance appropriate for the local league. So the baseball manufacturers are only interested modeling the behavior of the bat and ball in the time interval of the contact of the ball with the bat. For the manufacturers of baseballs, the impulse function is not a good model for the force applied to the ball by the bat.

▲▲

Example 5.2.10: Consider the flow of charge into a capacitor. When a capacitor is connected to a current source, charge flows into the capacitor until the potential across the terminals of the capacitor is equal to the source. From physical considerations this flow can't change discontinuously, so the charging of the capacitor must take place in a finite amount of time. For many applications, however, an instantaneous charging of the capacitor is suitable for network analysis. From this viewpoint the impulse function as a model of the current used in charging of the capacitor is appropriate.

▲▲

5.2.5 MATLAB Experiments

Plotting signals is a basic skill in MATLAB that will be used routinely through the text. An introduction to this subject can be found in the MATLAB manuals. Basically, a signal is represented by a vector, called the signal vector, that represents values of the signal at specified points in time. These time points are stored in a separate vector, called the time vector. MATLAB plots the signal by plotting the points in the signal vector vs. the points in the time vector on an x–y grid and then connecting the points.

To perform operations on the signal in MATLAB it is necessary to perform operations on the signal vector. Often these operations take the form of arithmetic operations on elements of the signal vector. For example, to square a signal it is necessary to square each element of the signal vector. Since the ordinary arithmetic operations in MATLAB are *matrix* operations, the usual arithmetic symbols must be modified. To perform arithmetic operations on each element of a matrix, the operator is preceded by a period. For example, to square a matrix A the MATLAB command is "A^2". Of course, A must be square. To square each element in the matrix B the MATLAB command is "B.^2". The matrix B doesn't have to be square. The following example illustrates the comments above.

Example 5.2.11: Consider the signal

$$x(t) = t^2 \sin(10t), \quad 0 \le t \le 3. \tag{5.2.21}$$

We would like to plot this signal. To generate the signal in (5.2.21) in MATLAB two intermediate signals are defined

$$x_1(t) = t^2, \quad \text{and} \quad x_2(t) = \sin(10t). \tag{5.2.22}$$

Then the signal in (5.2.21) is the product of the signals in (5.2.22). The following M-file plots the signal in (5.2.21).

```
clear
% generate signal
t = linspace(0,3,200);      % time vector
x1 = t.^2;                   % signal, x1
x2 = sin(10*t);             % signal, x2
x = x1.*x2;                  % product of signals
% plot signal
plot(t,x)
xlabel('time')
title('Product of Functions')
```

▲▲

5.3 SIGNALS DEFINED ON INTERVALS

5.3.1 Pulse Definition

Many times a complicated signal can be represented in terms of simple signals defined on intervals. Then the signal modeling proceeds by constructing a signal model on each interval using the techniques discussed in this chapter. Finally, these intervals are assembled to complete the signal model. This method is very flexible and often used.

A representation of a signal on an interval has two components: (1) a pulse that selects the interval on which the signal is modeled, and (2) a signal representation for that interval. The pulse is introduced in the first example.

Example 5.3.1: Consider the pulse shown in Figure 5.3.1. This pulse can be represented using the unit pulse or using step functions as shown in (5.3.1)

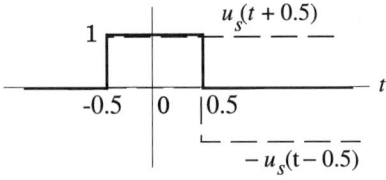

Figure 5.3.1 Two Representations of the Unit Pulse

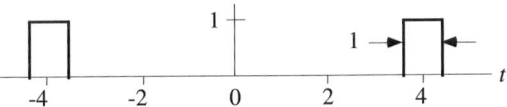

Figure 5.3.2 Signals in Example 5.3.2

$$\Pi(t) = u_s\left(t + \frac{1}{2}\right) - u_s\left(t - \frac{1}{2}\right) \tag{5.3.1}$$

We will have occasion to use both representations.

▲▲

The next example shows how the pulse in Example 5.3.1 can be used to model a signal.

Example 5.3.2: Suppose we wish to construct a model of the signal in Figure 5.3.2. For the pulse centered at $t = -4$ we have

$$x_-(t) = \Pi(t + 4). \tag{5.3.2}$$

The signal in (5.3.2) is defined on the entire real line, not just on the interval where it is nonzero. Similarly, we define the signal

$$x_+(t) = \Pi(t - 4). \tag{5.3.3}$$

Now we can represent the signal in Figure 5.3.2 as

$$x(t) = x_-(t) + x_+(t). \tag{5.3.4}$$

This representation is convenient because the nonzero parts of the two signals (5.3.2) and (5.3.3) don't overlap.

▲▲

5.3.2 General Signals

This technique can be extended to signals divided into many intervals.   The following procedure is used.

Procedure 5.3.3: Let $x(t)$ be the given signal.
Step 1. Divide the signal into intervals, such that the signal on each interval has a simple representation.
Step 2. Write the pulse function for each interval. This function is one over the interval of interest and zero elsewhere.
Step 3.   Develop a signal representation on each interval.   Multiply this representation by the appropriate pulse function.
Step 4. Sum the representations for each interval to form the signal model.

▲▲

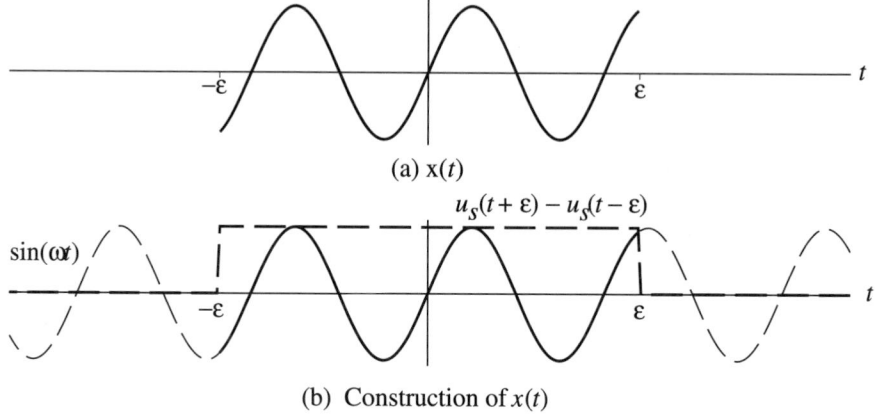

(a) x(t)

(b) Construction of $x(t)$

Figure 5.3.3 Signal of Example 5.3.4

The following examples illustrate the use of Procedure 5.3.3 to construct signal representations.

Example 5.3.4: On several occasions we will have cause to consider a truncated sinusoid as shown in Figure 5.3.3a. We can develop a representation of this signal using Procedure 5.3.3.   First, note that there is only one interval of interest, $I_1 = \{t \mid -\varepsilon \leq t < \varepsilon\}$.   Second, the pulse function for this interval is shown in Figure 5.3.3b. The signal representation over the interval I_1 is just $\sin(\omega t)$.   Now the signal representation is obtained by multiplying the pulse by $\sin(\omega t)$. We obtain

$$x(t) = (\sin \omega t)\big[u_s(t+\varepsilon) - u_s(t-\varepsilon)\big]. \tag{5.3.5}$$

▲▲

Example 5.3.5: Consider the signal shown in Figure 5.3.4a. To find a representation of this signal we follow the steps in Procedure 5.3.3.
Step 1. First we break the signal up into the intervals

$$I_1 = \{t \mid t < 0\}, \tag{5.3.6}$$
$$I_2 = \{t \mid 0 \leq t < 1\},$$
$$I_3 = \{t \mid 1 \leq t < 2\},$$
$$I_4 = \{t \mid t \geq 2\}.$$

Step 2. Next we construct a pulse function for each interval. For interval I_1 the appropriate pulse would be

$$1 - u_s(t). \tag{5.3.7}$$

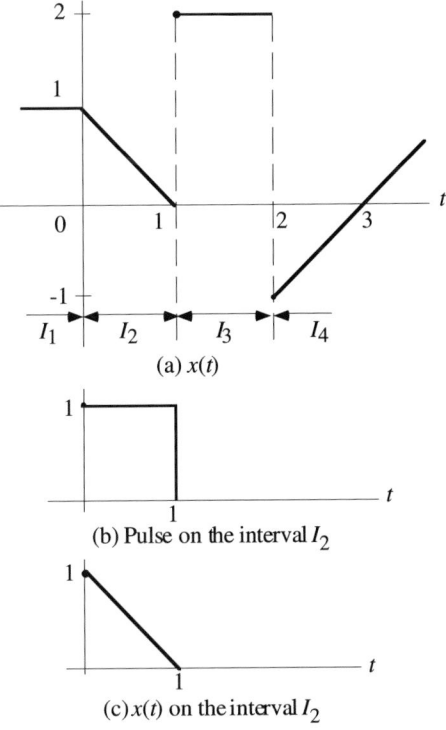

(a) $x(t)$

(b) Pulse on the interval I_2

(c) $x(t)$ on the interval I_2

Figure 5.3.4 Signal of Example 5.3.5

For interval I_2 the appropriate pulse would be

$$u_s(t-0) - u_s(t-1). \qquad (5.3.8)$$

This pulse is shown in Figure 5.3.4b. And so on.

Step 3. Next we represent the signal on each interval. For interval I_1, the function is a constant value of 1. Now using (5.3.7) we represent $x(t)$ on the interval I_1 as

$$(1)\big[1 - u_s(t)\big]. \qquad (5.3.9)$$

Next we consider the signal on the interval I_2. The signal on this interval can be expressed as

$$(1-t). \qquad (5.3.10)$$

Now using (5.3.8) and (5.3.10) we represent $x(t)$ on the interval I_2 as

$$(1-t)\big[u_s(t-0) - u_s(t-1)\big]. \qquad (5.3.11)$$

The signal in (5.3.11) is shown in Figure 5.3.4c. The signal in (5.3.11) corresponds with $x(t)$ on the interval I_2 but is zero elsewhere. The representation of the signal in Figure 5.3.4a on the remaining two intervals is

Interval I_3 $(2)\left[u_s(t-1)-u_s(t-2)\right],$ (5.3.12a)

Interval I_4 $(t-3)\left[u_s(t-2)\right].$ (5.3.12b)

Combining (5.3.9), (5.3.11), and (5.3.12), we find the representation of $x(t)$ to be

$$x(t) = (1)\left[1-u_s(t)\right]+(1-t)\left[u_s(t)-u_s(t-1)\right] \tag{5.3.13}$$
$$+(2)\left[u_s(t-1)-u_s(t-2)\right]+(t-3)\left[u_s(t-2)\right].$$

Note that the final expression of $x(t)$ in (5.3.13) is a sum of step and ramp functions. Such expressions are useful when analyzing the response of a system.

▲▲

Of course, we need not restrict ourselves to a finite number of intervals.

Example 5.3.6: Consider the signal in Figure 5.3.5. To find a representation of this signal, first note that this signal is a string of shifted pulse signals. For the pulse signal centered at $2m$ the representation is

$$\Pi(t-2m). \tag{5.3.14}$$

Now the signal in Figure 5.3.5 can be represented by summing up the signals in (5.3.14). We have

$$x(t) = \sum_{m=-\infty}^{\infty} \Pi(t-2m). \tag{5.3.15}$$

▲▲

We will use pulse trains to illustrate many of the concepts in signal analysis.

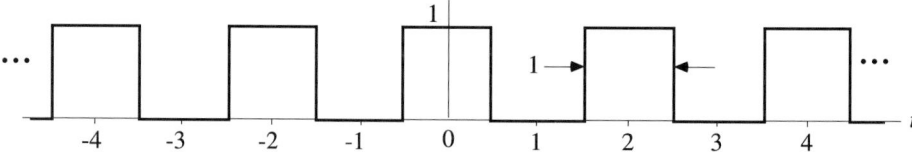

Figure 5.3.5 Pulse Train of Example 5.3.6

5.3.3 MATLAB Experiments

The computer analysis of signals and systems using MATLAB routinely calls for the representation of complicated signals. Readers are referred to the MATLAB manual for a complete discussion of the representation of functions in MATLAB. There are several methods for the representation of a function. Typically, however, the representation contains two vectors: a time vector and a signal vector. The time vector contains the points at which the signal is evaluated. The signal vector contains the values of the signal at the times contained in the time vector.

There follow three M-files for the representation of the signal in Example 5.3.5. Each of these M-files is basically a computer implementation of Procedure 5.3.3. In each case a time vector is constructed and partitioned into the appropriate intervals. Then the signal is constructed on each time interval.

The first M-file constructs the signal on separate time intervals and then joins both the time and signal vector. Note that the time vector on each interval must be constructed such that there is no overlap of points when the vectors are joined.

```
clear
dt = 0.05;                  % time increment for plotting
t1 = [-1:dt:-dt];           % first interval
x1 = ones(size(t1));        % signal on first interval
t2 = [0:dt:1];              % second interval
x2 = 1 - t2;                % signal on second interval
t3 = [1+dt:dt:2];           % third interval
x3 = 2*ones(size(t3));      % signal on third interval
t4 = [2+dt:dt:4];           % fourth interval
x4 = (t4 - 3);              % signal on fourth interval
t = [t1,t2,t3,t4];          % total time interval
x = [x1,x2,x3,x4];          % total signal
plot(t,x)
```

The second M-file constructs the time vector at once, and then constructs the signal vector on the various intervals of the time vector by knowledge of the indices of the entries of the time vector.

```
clear
t = [-1:.05:4];                       % total time interval
x(1:21) = ones(size(t(1:21)));        % signal on first interval
x(22:41) = 1 - t(22:41);              % signal on second interval
x(42:61) = 2*ones(size(t(42:61)));    % signal on third interval
x(62:101) = t(62:101) - 3;            % signal on fourth interval
plot(t,x)
```

The third M-file constructs the signal in basically the same way as the second M-file, but it uses the **find** command to locate the indices of the elements of the time vector that demarcate the time intervals. The advantage of this method over the second M-file is that the time increment can be adjusted without rewriting the commands that create the signal vector.

```
clear
dt = 0.05;                                  % time increment
t = [-1:dt:4];                              % total time interval
i1 = min(find(t >= 0));                     % find index of t = 0
x(1:i1) = ones(size(t(1:i1)));             % signal on first interval
i2 = min(find(t >= 1));                     % find index of t = 1
x(i1+1:i2) = 1 - t(i1+1:i2);               % signal on second interval
i3 = min(find(t >= 2));                     % find index of t = 2
x(i2+1:i3) = 2*ones(size(t(i2+1:i3)));     % signal on third interval
i4 = max(size(t));
x(i3+1:i4) = t(i3+1:i4) - 3;               % signal on fourth interval
plot(t,x)
```

5.4 DIGITAL WAVEFORMS

5.4.1 Introduction

One of the fundamental concepts in information processing is the ability to impress information into an electronic waveform and to extract that information after it has been processed. There are two basic ways to impress information into a waveform. The first way is to modulate the parameter of a carrier wave with a continuous information signal. This process is discussed in Section 5.1. The second way to impress information into a waveform is to divide the waveform into intervals in time and then associate a specific unit of information with each interval. These units of information are usually designated by a 0 and a 1. Hence, the information is binary in nature and each unit of information is called a bit. The resulting waveform will carry the digital information.

In digital communication systems the information is encoded into the waveform at one location, the waveform is transmitted to another location, and then the information is extracted from the waveform. The waveform could be transmitted halfway around the world via a satellite system, or the waveform could be transmitted from a PC to a printer standing beside it.

Encoding information into an electronic waveform in a computer for digital processing is fundamental to the operation of the computer, of course. In recent years it has become economically and technically feasible to use computer processing chips, or DSP chips, in many applications. To get the information into and out of the processor requires that the bit be impressed into an electronic waveform.

We will discuss the anatomy of electronic waveforms that carry digital information in this section. The idea is to divide the waveform into intervals and specify a specific waveform for each interval. These particular waveforms carry the digital information. These waveforms can be seen as an extension of the signal modeling ideas in the last section.

5.4.2 Symbols

In order to impress binary information into a waveform, the signal must be divided into uniform time intervals with a bit being assigned to each time interval. If we are to extract this information, we must also know the time origin of the signal so we can accurately locate the beginning and the end of each interval. Each of these time intervals is defined by a clock which also establishes the time origin.

Definition 5.4.1: A <u>clock</u> is an electronic device that generates a narrow electronic pulse, called a <u>timing pulses</u>, every T_c sec. The spacing between the pulses, $T_c > 0$, is called the <u>clock period</u>. The clock period divides the time axis into intervals

$$I_c = \left\{ t \mid nT_c \le t < (n+1)T_c \right\}.$$

Each interval I_c is called a <u>clock cycle</u>. The <u>clock frequency</u>, $f_c > 0$, is

$$f_c = \frac{1}{T_c}.$$

If an electronic signal is divided into clock intervals with timing pulses between the intervals, we say the signal is <u>clocked</u>.

▲▲

The waveform generated by the clock, the timing pulses, is shown in Figure 5.4.1. In the waveform analysis of high speed digital electronics it is important to consider the effects of timing pulses on the waveforms in the circuit. For our purposes, it is enough to assume that these pulses demarcate the clock intervals.

Terminology: The timing pulses that are generated by a clock can be used to synchronize all aspects of the digital electronics. When the signals in the electronic devices are coordinated in this way, it is called <u>synchronous</u> operation. Otherwise, the operation is called <u>asynchronous</u> operation.

The timing pulses are used to accurately define a larger time interval to which we will assign a bit.

Definition 5.4.2: A <u>symbol period</u> T_b is a fixed, integer number of clock cycles. A <u>symbol</u> is a given signal which is zero outside the symbol period. A <u>binary pair</u> of symbols are two symbols $b_0(t)$ and $b_1(t)$ that are distinct from each other.

▲▲

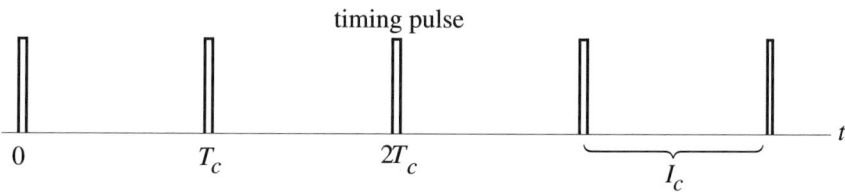

Figure 5.4.1 Clock Cycles

Notation: To a binary pair of symbols we make the assignment of a logical 0 and 1,

$$b_0(t) \leftrightarrow 0, \quad b_1(t) \leftrightarrow 1. \tag{5.4.1}$$

Example 5.4.3: One obvious choice of a binary symbol pair is

$$b_0(t) = 0, \quad b_1(t) = A\Pi\left(\frac{t}{T_b}\right). \tag{5.4.2}$$

▲▲

5.4.3 Message Signals

Using a clock, we can divide a waveform into symbol periods. To each of these symbol periods we can assign a symbol. Next we will show how information can be encoded into this signal. Define a binary sequence by

$$\{a_k = 0 \text{ or } 1\} \quad \text{for} \quad k = \ldots, -2, -1, 0, 1, 2, \ldots \tag{5.4.3}$$

where each element of the sequence can take on a logical value of 1 or 0. Each element of the sequence is a bit. This binary sequence of bits is our information. We assign one bit to each symbol period. On each symbol period we assign a symbol according to the bit assigned to that symbol period. In this way we can encode binary information into a waveform.

Definition 5.4.4: A <u>digital message</u> signal is a signal that has been divided into symbol intervals where the origin of the signal is fixed and known. One of the binary symbols $b_0(t)$ or $b_1(t)$ is assigned to each symbol interval according to the binary sequence we are encoding.

▲▲

This terminology is taken from communication theory where these signals are frequently found.

Example 5.4.5: A digital message signal using the symbol pair in (5.4.2) is shown in Figure 5.4.2.

▲▲

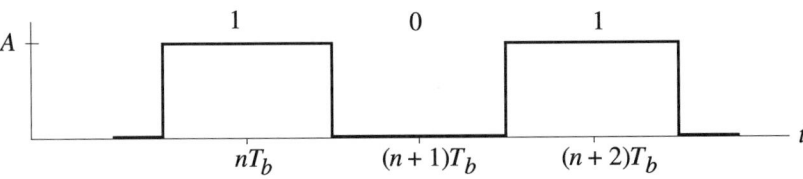

Figure 5.4.2 A Digital Message Signal

Next we will develop a model for digital message signals. Let the symbol centers be spaced by a symbol period T_b. Let the two symbols be $b_0(t)$ and $b_1(t)$. Define the continuous-time signal

$$m(t) = \sum_{k=-\infty}^{\infty} b_{a_k}(t - kT_b). \tag{5.4.4}$$

The signal $m(t)$ is a digital message signal composed of a sequence of symbols that corresponds to a binary sequence defined by (5.4.3).

Example 5.4.6: Consider the five-bit sequence $\{10110\}$. Then the binary sequence in (5.4.3) is

$$a_1 = 1, \quad a_2 = 0, \quad a_3 = 1, \quad a_4 = 1, \quad a_5 = 0. \tag{5.4.5}$$

Let the symbols be

$$b_1(t) = \Pi\left(\frac{t}{T_b}\right), \quad b_0(t) = -b_1(t). \tag{5.4.6}$$

The message signal $m(t)$ for this example is

$$m(t) = b_1(t - T_b) + b_0(t - 2T_b) + b_1(t - 3T_b) + b_1(t - 4T_b) + b_0(t - 5T_b). \tag{5.4.7}$$

This message signal is shown in Figure 5.4.3.   ▲▲

In principle, it is obvious how to extract the information from the digital waveform. On each symbol interval we simply determine which symbol is present. Then we determine which bit corresponds to the symbol. Consider, for example, the digital waveform in Figure 5.4.3. For this waveform we could determine which symbol is present by sampling the waveform at the times nT_b. If the sample is positive, a 1 is represented by that symbol. If the sample is negative, a 0 is represented by that symbol. In this way the binary sequence contained in the waveform can be reconstructed.

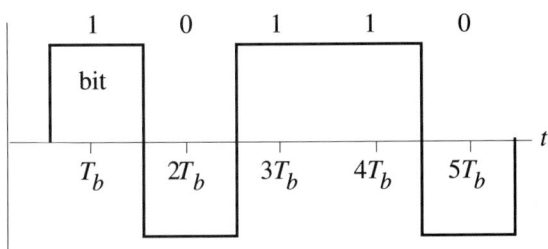

Figure 5.4.3 Message Signal in Example 5.4.5

5.4.4 Time Division Multiplexing

It is often desirable to impress more than one digital message signal into a single waveform. This is easily done with digital message signals. Suppose we are given two digital message signals $m_1(t)$ and $m_2(t)$. The symbols from the first message signal $m_1(t)$ are simply interleaved with the symbols from the second message signal $m_2(t)$. That is, the first symbol period is assigned to the first message signal. The second symbol period is assigned to the second message signal. The third symbol period is assigned to the first message signal. And so on. This process is illustrated in Figure 5.4.4. Clearly, this idea can be extended to more than two signals.

Definition 5.4.7: The concept of combining two signals into a single waveform in such a way that the two signals can be separated at a future time is called <u>multiplexing</u>. The separation of the two signals from the multiplexed signal is called <u>demultiplexing</u>. The multiplexing of the two signals as shown in Figure 5.4.4 is called <u>time division</u> multiplexing.

▲▲

The motivation for the name "time division" multiplexing is clear from Figure 5.4.4. The two signals are separated by dividing the signals into time intervals. The demultiplexing proceeds by determining which symbol interval belongs to which signal.

The time division multiplexing of several signals comes with a cost: the amount of time required to transmit a binary sequence of fixed length is increased. If a single binary sequence is impressed into a digital message signal, then the amount of time required to transmit M bits is just MT_b. That is, it requires M symbol periods. If K binary sequences are impressed into a multiplexed message signal, however, only one bit of each binary sequence is contained in every K symbols. Therefore, to transmit M bits of one binary sequence requires $K(MT_b)$ symbol intervals.

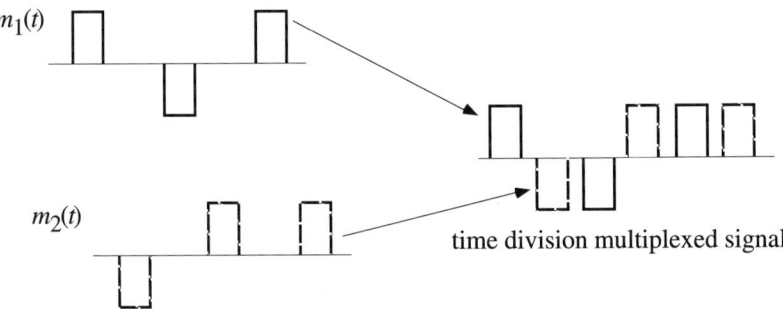

Figure 5.4.4 Time Division Multiplexing

5.5 SIGNALS AS SUMS OF SINUSOIDS

In the previous sections we have discussed signals that are defined on intervals. An essentially unique signal representation is assigned to each interval on the time axis. In this section we introduce a different approach to signal representation. The basic building block of this type of signal representation is the sinusoid. A single sinusoid is defined by a frequency, an amplitude, and a phase. Complicated signals are represented by sums of sinusoids with different frequencies, amplitudes, and phases. This approach is fundamentally different from the approach of the previous sections because each sinusoid is defined over the entire time axis rather than over just a finite length interval. While it may appear that this approach is unnecessarily complicated, it turns out that this approach to signal representations plays a fundamental role in signal modeling and system analysis. In many ways sinusoidal signals are the most important signals we use in system theory.

Example 5.5.1: This example illustrates how a sum of sinusoids can be used to represent a complicated signal. Figure 5.5.1 shows a complicated graph of a voltage that could have been measured in the lab. We wish to develop a signal model for this graph. Close inspection shows that this signal is composed of two sinusoids: one large amplitude low frequency sinusoid and one low amplitude, high frequency sinusoid. From the graph we can write

$$x(t) = A\sin\omega_1 t + B\sin(\omega_2 t + \theta). \tag{5.5.1}$$

From inspection of the graph we can see that the period T_1 of the low frequency component is

$$T_1 \approx 6.2 = \frac{2\pi}{\omega_1} \text{ sec} \quad \text{or} \quad \omega_1 \approx 1 \text{ rad / sec.} \tag{5.5.2}$$

We can also see that the amplitude of the low frequency sinusoid is $A \approx 1$.

Further inspection of the graph in Figure 5.5.1 shows that the high frequency sinusoid has approximately 10 cycles per cycle of the low frequency component. This component of the signal can be represented as

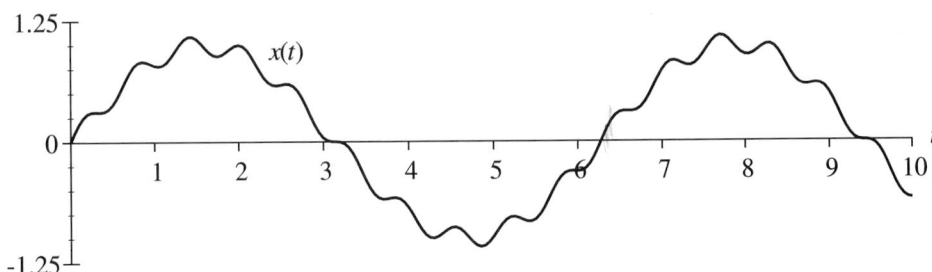

Figure 5.5.1 Signal of Example 5.5.1

$\omega_2 \approx 10\omega_1 = 10$ rad / sec. $\hspace{4cm}$ (5.5.3)

It is also easily seen that the amplitude of the high frequency component is $B \approx 0.1$.

Finally, the relative phase between the two signals must be determined. Because the high frequency is an integer multiple of the low frequency and $x(t) = 0$ at $t = 0$, we conclude that $\theta = 0$. So the signal model of the graph in Figure 5.5.1 is

$$x(t) = (1\sin t) + (0.1\sin 10t). \hspace{3cm} (5.5.4)$$

Typically, the phase of a sinusoidal signal is difficult to determine.

The signal representation in (5.5.4) consists of two components, both sinusoids. Each one of these components is nonzero over the entire time axis. This signal representation is fundamentally different from the representations we developed in previous sections that were based on representing the signals on finite length intervals.

▲▲

The last example presented a signal as a sum of two sinusoids. In fact, by varying the parameters of the sinusoids the signal shape can be varied significantly. The following example examines the effects of phase on the signal shape.

Example 5.5.2: Consider the signal

$$x(t) = \sin(t) + 0.2\sin(2t + \theta). \hspace{3cm} (5.5.5)$$

This signal is shown in Figure 5.5.2a for $\theta = 0$ and in Figure 5.5.2b for $\theta = 90°$.

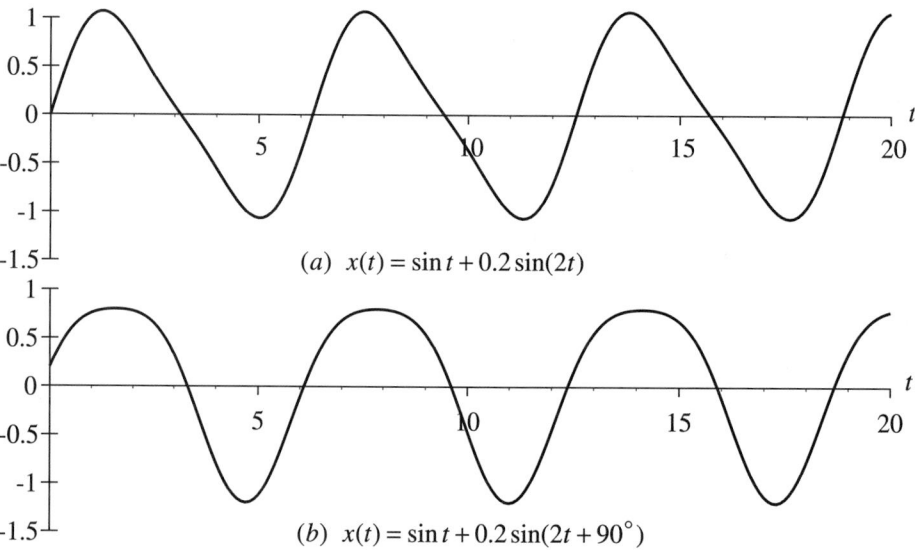

(a) $x(t) = \sin t + 0.2\sin(2t)$

(b) $x(t) = \sin t + 0.2\sin(2t + 90°)$

Figure 5.5.2 Effects of Phase

Comparison of the signals in Figure 5.5.2a and 5.5.2b shows that both the amplitude and the shape of the signal can be varied significantly by only varying the phase of the second sinusoid in (5.5.5). There is a significant variation between the signals even though the amplitude of the second signal is small compared to the amplitude of the first signal. This example implies that a wide variety of waveforms can be modeled with sinusoids.

▲▲

In the previous two examples we have considered a finite sum of sinusoids. By increasing the number of sinusoids in the sum more complicated signals can be modeled. In the limit we can consider an infinite sum of sinusoids as long as the parameters of the sinusoids satisfy certain conditions. These infinite sums of sinusoids are called <u>Fourier series</u>, which we will study extensively. The following example shows how a Fourier series can be used to represent a signal that seems to have nothing in common with a sinusoid.

Example 5.5.3: Consider the periodic signal in Figure 5.5.3. It will be shown that the Fourier series for this signal is given by

$$x(t) = \frac{1}{2} + \sum_{\substack{m=1 \\ m \text{ odd}}}^{\infty} \frac{2}{m\pi} \cos(m\pi t + \theta_m) \qquad (5.5.6)$$

where

$$\theta_m = \begin{cases} \pi, & m = 3, 7, 11, \ldots \\ 0, & m = 1, 5, 9, \ldots \end{cases}$$

This same signal is shown in Figure 5.3.5 where it is represented as a sum of shifted pulse functions. Using (5.5.6) and the previous signal representation we can write

$$x(t) = \frac{1}{2} + \sum_{\substack{m=1 \\ m \text{ odd}}}^{\infty} \frac{2}{m\pi} \cos(m\pi t + \theta_m) = \sum_{k=-\infty}^{\infty} \Pi(t - 2k). \qquad (5.5.7)$$

We have two very different signal representations in (5.5.7) for the signal in Figure 5.5.3. The first signal representation is an infinite sum of sinusoids. The second representation is based on representing the signal by a specific function over each

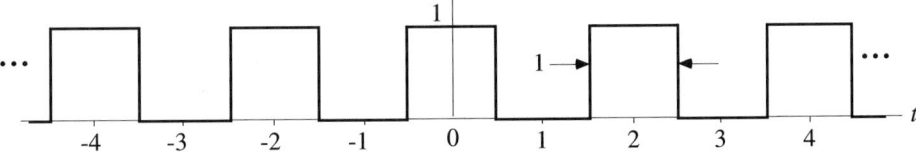

Figure 5.5.3 Periodic Signal of Example 5.5.3

interval where it is nonzero. Yet the equal sign means that they are essentially the same signal. In particular, this example shows that very complicated signals can be represented using sinusoids. This observation motivates our study of the relationship between sinusoidal signals and systems in Chapters 14 and 15.

▲▲

5.6 CHAPTER SUMMARY

5.6.1 Definitions

In this chapter we have introduced the most fundamental concept of this text: a signal.

Definition 5.1.1: A <u>signal</u> is a real function which represents the time variation of a physical variable.

▲▲

For a signal to be useful to us, we need a way to write down the signal in a way in which we can analyze the signal.

Definition 5.1.2: The function which is chosen to describe a signal is called the <u>representation</u> of the signal. The process of building a representation of the signal is called signal <u>modeling</u>.

▲▲

We will consider two classes of signals in this text.

Definition 5.1.3: A signal that depends on a real variable t and that models a physical variable that evolves in real time is called a <u>continuous (-time)</u> signal.

▲▲

Definition 5.1.4: A signal that depends on a discrete variable n and that models a physical variable that evolves in discrete time is called a <u>discrete (-time)</u> signal.

▲▲

We also discussed the interconnection of two signals to form a new signal by addition, multiplication, and composition.

5.6.2 Signal Models

In the remaining sections of the chapter introduced several ways of modeling signal which we shall use through the text. These methods include:

(1) Time Scaling   Section 5.2
(2) Time Shifting   Section 5.2
(3) Signals Defined by Limits   Section 5.2
(4) Signals Defined on Intervals   Section 5.3
(5) Digital Message Signals   Section 5.4
(6) Fourier Series   Section 5.5

5.7 HOMEWORK FOR CHAPTER 5

Homework Problems for Section 5.1

5.1.1   (a) Give an example of a continuous-time signal not discussed in the text.
(b) Give an example of a discrete-time signal not discussed in the text.

5.1.2   Plot the following signals using MATLAB.
(i) $x(t) = [\sin(2\pi t)]\cos(50\pi t)$ (iii) $x(t) = [1 + 0.2\sin(2\pi t)]\cos(50\pi t)$

(iv) $x(t) = [1 + 1.2\sin(2\pi t)]\cos(50\pi t)$
(ii) $x(t) = \Lambda\!\left(\dfrac{t}{4}\right)\cos(50\pi t)$

5.1.3   Plot the following signals using MATLAB.
(i) $x(t) = \cos(10\pi t + \beta\sin(\pi t)),\quad \beta = 3,8$

(ii) $x(t) = \cos\!\left[10\pi t + \Lambda\!\left(\dfrac{t}{4}\right)\right]$

(iii) $x(t) = \cos\!\left[10\pi t - \Lambda\!\left(\dfrac{t}{4}\right)\right]$

(iv) $x(t) = \cos\!\left[10\pi t + \dfrac{\pi}{2}(u_s(t+1) - 2u_s(t-1))\right],\quad -1 \le t \le 1$

Homework Problems for Section 5.2

5.2.1   (a) Sketch the following signals.
(b) Plot the following signals using MATLAB.
(i) $x(t) = r_p(-t)$ (iii) $x(t) = \text{sgn}(-t)$
(ii) $x(t) = [\sin(-\pi t)]u_s(-t)$
(iv) $x(t) = \Pi\!\left(\dfrac{0.01 - t}{0.005}\right)$

5.2.2   (a) Sketch the following signals.
(b) Plot the following signals using MATLAB.
(i) $x(t) = \text{sinc}(t - c),\quad c = 0, 1, \pi$ (iii) $x(t) = \text{Sa}(\varepsilon t),\quad \varepsilon = 1, 2, \pi$

(ii) $x(t) = \text{sinc}(t) + \text{sinc}(t - \pi)$
(iv) $x(t) = \Pi\!\left(\dfrac{t}{\varepsilon}\right)\text{Sa}(t),\quad \varepsilon = \dfrac{\pi}{2}, \pi, 4\pi$

5.2.3   (a) Sketch the following signals.
(b) Plot the following signals using MATLAB.
(i) $x(t) = \Pi\!\left(\dfrac{t}{10}\right)$ (iii) $x(t) = \Pi\!\left(\dfrac{t - 0.01}{0.005}\right)$
(ii) $x(t) = \Pi(t + 8)$
(iv) $x(t) = \Pi\!\left(\dfrac{t}{0.005} - 2\right)$

5.2.4   (a) Sketch the following signals.
(b) Plot the following signals using MATLAB.

$$(i) \quad x(t) = e^{-2(t-3)} u_s(t-3) \qquad (ii) \quad x(t) = e^{-2(3-t)} u_s(3-t)$$

5.2.5   Consider the signal

$$x_\varepsilon(t) = \frac{2}{\varepsilon} \Lambda\left(\frac{t}{\varepsilon}\right).$$

(a) Sketch this signal for $\varepsilon = 1, 0.5$.
(b) Determine the signal that is the limit of these signals as $\varepsilon \to 0$.

Homework Problems for Section 5.3

5.3.1   (a) Sketch each of the following signals.
(b) Plot these signals using MATLAB.

$$(i) \quad x_1(t) = A\Pi\left(\frac{t}{\varepsilon}\right)$$

$$(ii) \quad x_2(t) = B\left(1 + \cos\frac{2\pi t}{\varepsilon}\right)\Pi\left(\frac{t}{\varepsilon}\right)$$

$$(iii) \quad x_3(t) = C\left(\cos\frac{2\pi t}{\varepsilon}\right)\Pi\left(\frac{t}{\varepsilon}\right)$$

5.3.2   Consider the signal in Figure P5.3.2.
(a) Find a representation for this signal.
(b) Plot this signal using MATLAB.

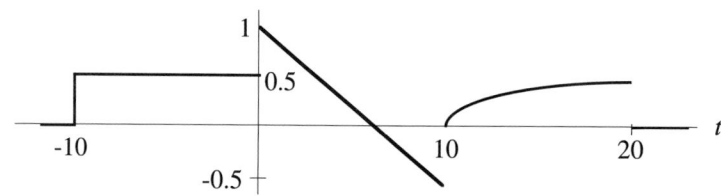

Figure P5.3.2

5.3.3   (a) Sketch the following signals.
(b) Plot the following signals using MATLAB.

$$(i) \quad x(t) = \Pi\left(\frac{t+3}{2}\right) + t^2\left[u_s(t-1) - u_s(t-3)\right]$$

$$(ii) \quad x(t) = r(t)\left[u_s(t+2) - u_s(t-1)\right] + \left[u_s(t-1) - u_s(t-3)\right] - e^{(t-3)}u_s(t-3)$$

(iii) $x(t) = 2\Pi\left(\dfrac{t}{0.5}\right) + \displaystyle\sum_{\substack{n=-2 \\ n\neq0}}^{2} \dfrac{1}{n}\Pi\left(\dfrac{t-2n}{0.5}\right)$

(iv) $x(t) = \displaystyle\sum_{n=1}^{\infty} \Lambda\left(\dfrac{t-n}{\frac{1}{n^2}}\right)$

(v) $x(t) = \Pi\left(\dfrac{t}{2}\right)\sin(2\pi t) + \delta(t+5) + \delta(t-5)$

(vi) $x(t) = (t)\Pi\left(\dfrac{t-3}{2}\right) + 2\delta(t+2)$

(vii) $x(t) = \displaystyle\sum_{k=0}^{3} (k+1)\left[u_s(k) - u_s(k+1)\right]$

(viii) $x(t) = \displaystyle\sum_{k=-2}^{2} (-1)^{-n}\Lambda(t-2k)$

5.3.4   Sketch the following signals.

(i) $x(t) = \displaystyle\sum_{m=-\infty}^{\infty} \Pi\left(\dfrac{t-0.1m}{0.02}\right)$ (ii) $x(t) = \displaystyle\sum_{m=-\infty}^{\infty} \Pi\left(\dfrac{t-0.1m}{0.02}\right)\cos(2\pi t)$

5.3.5   Consider the signal in Figure P5.3.5.
(a) Find a representation for this signal.
(b) Plot this signal using MATLAB.

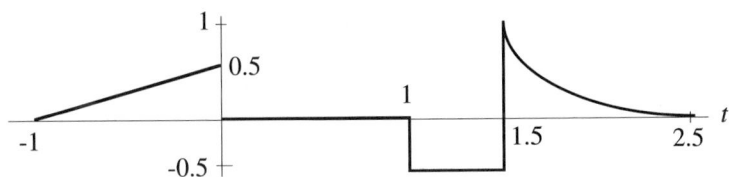

Figure P5.3.5

5.3.6   Consider the signal $x(t)$ shown in Figure P5.3.6. Find a representation of this signal.

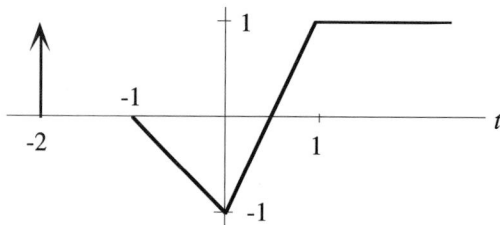

Figure P5.3.6

5.3.7   Consider the signal

$$x(t) = \frac{\text{sinc}\left(\dfrac{t}{T_b}\right)\cos\left(\dfrac{\alpha\pi t}{T_b}\right)}{1 - 4\alpha^2\left(\dfrac{t}{T_b}\right)^2}$$

(a) Using L'Hospital's rule show that

$$\lim_{t \to \frac{T_b}{2\alpha}} \frac{\cos\left(\dfrac{\alpha\pi t}{T_b}\right)}{1 - 4\alpha^2\left(\dfrac{t}{T_b}\right)^2} = \frac{\pi}{4}$$

(b) Plot this signal using MATLAB for $\alpha = 0, 0.5, 1$.

Homework Problems for Section 5.4

5.4.1   Define the sequence $\tilde{a} = \{a_1 \quad a_2 \quad a_3\} = \{0.2 \quad 0.8 \quad 0.5\}$.   Using this sequence, define the signal

$$x(t) = \sum_{k=1}^{3} a_k \Pi(t - k)$$

Sketch this signal.
Note: The sequence $\tilde{a}$ which defines this signal can be recovered from the amplitude of the pulses in the signal.   This signal is said to be pulse amplitude modulated.

5.4.2   Define the sequence $\tilde{a} = \{a_1 \quad a_2 \quad a_3\} = \{0.2 \quad 0.8 \quad 0.5\}$.   Using this sequence, define the signal

$$x(t) = \sum_{k=1}^{3} \Pi\left(\frac{t-k}{a_k}\right)$$

Sketch this signal.
Note: The sequence $\tilde{a}$ which defines this signal can be recovered from the signal by the width of the pulses (as long as $|a_k| < 1$). This signal is said to be pulse width modulated.

5.4.3 Define the sequences $\tilde{a} = \{a_1 \ a_2 \ a_3\} = \{0.2 \ 0.8 \ 0.5\}$ and $\tilde{b} = \{b_1 \ b_2 \ b_3\} = \{-0.1 \ 0.6 \ -0.4\}$. Using these sequences, sketch the signal

$$x(t) = \sum_{k=1}^{3} \Pi\left(\frac{t-(2k-1)}{a_k}\right) + \sum_{k=1}^{3} \Lambda\left(\frac{t-2k}{b_k}\right)$$

5.4.4 Define the sequence $a = [1 \ 0 \ 1 \ 1]$. Now define the signal

$$s(t) = \sum_{m=0}^{3} a(m)b(t-mT_b)$$

(a) Sketch $s(t)$ for each of the following signals $b(t)$ when $T_b = 10^{-2}$.
(b) Repeat part (a) for the sequence $\tilde{a} = [-1 \ 1 \ -1 \ 1]$.
(c) Plot the signals in parts (a) and (b) in MATLAB.

(i) $\quad b(t) = \Pi\left(\dfrac{t}{T_b}\right)$

(ii) $\quad b(t) = \Pi\left(\dfrac{t-\frac{T_b}{4}}{\frac{T_b}{2}}\right) - \Pi\left(\dfrac{t+\frac{T_b}{4}}{\frac{T_b}{2}}\right)$

(iii) $\quad b(t) = \Pi\left(\dfrac{t}{T_b}\right)\cos\left(\dfrac{\pi t}{T_b}\right)$

5.4.5 Consider the binary sequences $\tilde{a} = \{0 \ 1 \ 1 \ 0\}$ and $\tilde{b} = \{1 \ 0 \ 1 \ 1\}$. Suppose these two sequences are to be time division multiplexed into a digital waveform using the symbols in Problem 5.4.4. Sketch the multiplexed waveform for each symbol.

Homework Problems for Section 5.5

5.5.1   Find a representation for the signal shown in Figure P5.5.1.

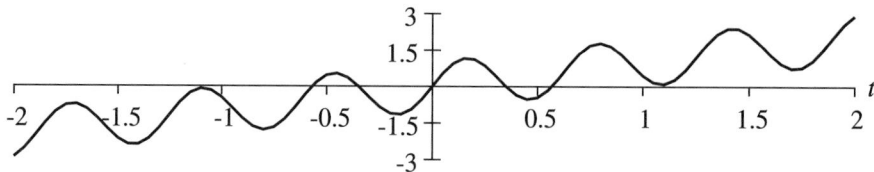

Figure P5.5.1

5.5.2   Find a representation for the signal shown in Figure P5.5.2.

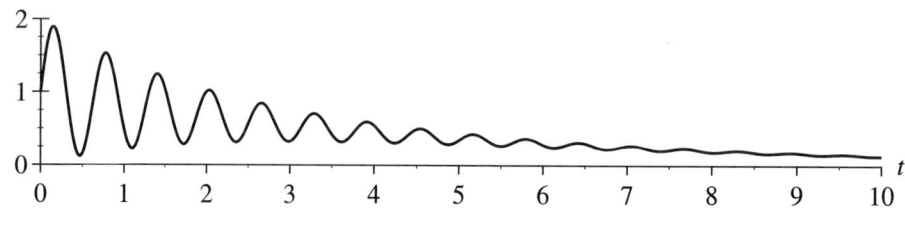

Figure P5.5.2

5.5.3   For a sinusoidal signal show that a phase shift is the same as time delay.

5.5.4   Consider the signal

$$x(t) = A\sin(t) + B\sin(\omega t + \theta)$$

(a) Write a MATLAB M-file to plot this signal.   Initially chose the parameters of the sinusoids to match the parameters of Example 5.5.1.
(b) Investigate the waveforms obtained by varying the phase as $0 \le \theta \le 2\pi$.
(c) Reverse the amplitudes of the two sinusoids and plot the signal.
(d) Change the frequency of the second sinusoid as $2 \le \omega \le 20$ and describe the effect on the signal.

5.5.5   Consider the signal

$$x(t) = \sin(t) + A\sin(\omega t + \theta)$$

(a) Write a MATLAB M-file to plot this signal.   Chose the parameters of the second sine to match the parameters of Example 5.5.2.
(b) Investigate the waveforms for $0 \le \theta \le 2\pi$.
(c) For this variation in phase investigate the effects of changing the amplitude A.

Chapter 6

Introduction to Systems

Chapter Outline

In Chapter 5 we introduced the concept of a signal. This concept is one of the most fundamental concepts in system theory. In this chapter we use the concept of a signal to define the second fundamental concept of system theory: a system. As described in Chapter 1, the input signal to a system produces an output signal.

This definition is very general, and it allows us great flexibility in applying it to a variety of physical processes. At the same time there are a number of very powerful system theoretic tools available that we can use to analyze very complex systems. It is the purpose of this text to begin the development of the concepts and tools related to systems.

A system is given concrete mathematical expression by an explicit mathematical description of the system called a _model_. The premise of system theory is that a mathematical model can be used to gain insight into a physical processes. Implied in this statement is that: (1) we can find a model that adequately describes the physical process, (2) this model is suitable for analysis, and (3) this model is suitable for design (if applicable). It is important to recognize that a model can accurately describe the physical observations, but be too complicated for analysis or design. Or a model that lends itself to analysis and design can be too simplistic to accurately describe the observations. In order to choose an appropriate model, it is important to understand what system analysis techniques exist, and to what kinds of system models these analysis techniques apply. These analysis techniques include both mathematical tools and computer-aided design packages. It is the purpose of this text to develop this background. It is the purpose of this chapter to introduce several different kinds of models, called representations, that we will study in the coming chapters.

In order to firmly fix the concept of a system, we will also introduce several systems that we will use to illustrate the concepts in the coming chapters. These systems serve to connect the concepts in this text with other disciplines. These examples also serve to focus the discussion of the many diverse ideas in this text, illustrating how these many concepts each yield insight into the behavior of a particular system.

The first section of this chapter defines a system, and introduces several different classes of systems. This material is fundamental to the rest of the text. The rest of the chapter discusses system models with which many readers may be familiar. The primary motivation of the last sections of this chapter is to introduce a number of examples for material that is often covered in previous courses, but which is explained in the terminology of this text. These discussions rely on mathematical concepts that many readers may have seen, but which are formally introduced in this text in later chapters. None of the material in Sections 6.2 - 6.5 is formally required for the coming chapters. It is a sneak preview of sorts.

Summary of Sections

Section 6.1:   We define a system and its representation.

Section 6.2:   We introduce three examples of system representations.

Section 6.3:   We discuss electrical networks as systems.

Section 6.4:   We develop mass-spring-damper systems.

Section 6.5:   We discuss a particular type of motor, a linear motor.

Section 6.6:   Chapter summary section.

Coverage of the Text

This chapter requires the definition of a signal from Section 5.1. Several essential concepts are defined in Sections 6.1-6.2 which are required throughout the rest of the text.   Sections 6.3 - 6.5 serve as an introduction to the systems frequently encountered in later chapters, but these sections can be skipped at this time.

The concept of a system introduced in Section 6.1 is developed in two parts of this text. Continuous-time systems are developed in Chapters 10 - 13. Discrete-time systems are developed in Chapters 21 - 22 after an introduction to discrete-time theory in Chapter 17. The reader may proceed directly in either of these directions.

6.1 DEFINITION OF A SYSTEM

6.1.1 Definitions

In this chapter we formally introduce the second fundamental concept of this text: a system.   The idea of a system was first outlined in Chapter 1.   Through our observations of physical processes we note that when one process is present, we also observe another process. In effect the first process causes the second process. This interrelationship between processes is a system. It is the purpose of this chapter to formalize this concept. We will also give several examples of a system.

The concept of a system begins with a formal description of the physical processes that give rise to a system. The physical process is usually characterized by the observation of a physical variable. The description of the physical variable is, of course, a signal. Since a signal is fundamental to the definition of a system we repeat the definition here.

Definition 5.1.1: A <u>signal</u> is a function which represents the time variation of a physical variable.

▲▲

Now the relationship between two signals, a system, can be formally defined.

Definition 6.1.1: A <u>system</u> generates a response, or output signal, for a given input signal.

▲▲

A system is a relationship between signals. Since signals are mathematical objects, functions, a system must also be a mathematical object.[1] The idea of a system, as expressed by Definition 6.1.1,[2] is represented in (abstract) mathematical symbols as

$$y(t) = \mathcal{H}[x(t)]. \tag{6.1.1}$$

The concept of a system expressed by (6.1.1) is shown in Figure 6.1.1 by a cartoon.

Notation: We will denote the input signal by $x(t)$ and the output signal by $y(t)$.

---

[1] These objects are called operators.

[2] A more mathematically precise definition of a system might be as follows. "A system is a rule that assigns an output signal to an input signal." This definition, however, doesn't have quite the engineering flavor of the given definition.

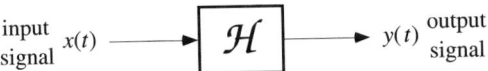

Figure 6.1.1 Cartoon of a System

At this very early stage of our discussion it is important to focus on the main point of the definition of a system without getting lost in the abstraction. Figure 6.1.1 reminds us that a system is defined by its input and output signals. The figure expresses one of the most important precepts of system theory: we view systems from an input-output perspective. This concept underlies almost all of the material to follow in this text.

The other important point in Figure 6.1.1 is highlighted by the arrows. The arrows signify that the *input* signal leads to the *output* signal. In particular, the reverse direction is not automatically defined. For many systems it doesn't make physical sense that the output signal will produce an input signal. This point is shown by the examples below. For other systems we can define another system in which the output signal will produce an input signal. This second system, however, must be independently defined. To summarize, Figure 6.1.1 very nicely reminds us that the *input signal* to the *system* will produce an *output signal*. The mathematical abstraction in (6.1.1) also brings this message.

Terminology: The word <u>system</u> is very overworked. It has been used to mean several things in several different contexts. We, however, have defined system quite precisely. We will always use it as specified by Definition 6.1.1.

The idea of a system is very intuitive. We define a system by first identifying the input signal and then identifying the output signal. The system is the implied relationship between these signals. The following examples illustrate the point.

Example 6.1.2: Consider the temperature in a room as a signal. When the thermostat setting is changed, the temperature in the room changes. We select the setting of the thermostat as the input signal. We select the temperature in the room as the output signal. These two signals define a system, the heating system responsible for the room temperature.   ▲▲

Example 6.1.3: As a second example, consider a car traveling along a highway. For simplicity, assume the car is traveling in a straight line at a certain speed. If we press on the accelerator, the car's speed will increase. Suppose we select the input signal to be the pressure on the accelerator pedal, and the speed of the car as the output signal. These two signals define a system, in this case the car.   ▲▲

Systems are defined in terms of signals. In Section 5.1 we introduced two major classes of signals: continuous-time signals and discrete-time signals. Based on these two classes of signals we can define two classes of systems by distinguishing to which class the input and output signals belong. To emphasize the definition of these two classes of systems we begin by recalling the definitions of the signals.

Definition 5.1.3: A signal that depends on a real variable t and that models a physical variable that evolves in real time is called a <u>continuous (-time)</u> signal.

▲▲

Definition 5.1.4: A signal that depends on a discrete variable n and that models a physical variable that evolve in discrete-time is called a <u>discrete (-time)</u> signal.

▲▲

First, suppose that continuous-time signals are used to define a system.

Definition 6.1.4: If the input signal and output signal of a system are continuous-time signals, then the system is called a <u>continuous(-time)</u> system.

▲▲

Terminology: The terminology "continuous system" is used very loosely. The definition appears to refer to a mathematical concept, but, in fact, only refers to the underlying signals that define the system.

A second class of system arises when discrete-time signals are used to define the system.

Definition 6.1.5: If the input signal and output signal of a system are discrete-time signals, then the system is called a <u>discrete-time</u> system.

▲▲

This text is divided into two parts: one on continuous-time signals and systems (Chapters 7–16) and one on discrete-time signals and systems (Chapters 17–23). These two types of systems have many properties in common. Through the text we will emphasize the similarities and differences between these two classes of systems.

A system can have more than one input and/or output signal. We can classify a system by the number of input and output signals that define the system.

Definition 6.1.6: If the system has one input signal and one output signal, we call that system a <u>single-input-single-output</u> (SISO) system. If the system has more than one input and/or output signal, then the system is called a <u>multiple-input-multiple-output</u> (MIMO) system.

▲▲

The following system is an example of a MIMO system.

Example 6.1.7: Consider an automobile traveling down a highway discussed in Example 6.1.3. The system defined in that example doesn't completely describe the system. We can make that model more complete by including more input and output signals. The input signals can be identified as the pressure on the accelerator and brake. The output signals can be identified as the position and velocity of the auto. Now we can think of the auto as a system. Since this system has more than one input and output signal, it is a MIMO system.

▲▲

Note: In this text we will consider only SISO systems unless specifically noted otherwise.

6.1.2 Representations of Systems

The implication in Figure 6.1.1 is that the system is a mathematical relationship between the two signals. In order to analyze this system we must have an explicit expression for this system.

Definition 6.1.8: An explicit mathematical expression for the system in (6.1.1) is called a system <u>representation</u>.

▲▲

Terminology: A system representation is also called a system <u>model</u>.

The premise of system theory is that insight into a physical process can be obtained by studying a mathematical representation for that system. The insight gained is explicitly or implicitly contained in the representation we chose for the system. Typically, one system can have several representations. Each representation is useful in its own way and gives a different insight into the physical process. System representations are the subject of Chapters 10 - 13. We then proceed to show how these representations can be used to analyze and synthesize systems in Chapters 14 and 15. A similar development is undertaken for discrete systems in Chapters 7 - 16.

We have defined a system as a mathematical relationship between two signals. Of course, given a particular system, we must find a representation of that system.

Definition 6.1.9: The process of deriving a system representation is called <u>modeling</u>.

▲▲

There are two ways for constructing a system model: (1) directly from the laws of physics, and (2) from empirical observations. To construct a system model from the laws of physics, the physical processes must be well-defined and understood. Furthermore, the system representation that is derived from the modeling process must be simple enough to analyze and simulate. Network analysis is a good example of how physical laws are used to derive models to analyze the behavior of an existing network and synthesize new circuits. This analysis is now embedded into CAD packages.

There are many other systems which are either too complex for direct analysis or their dynamics are not well understood. To study these systems all that is required is a clearly defined input and output signal. A carefully chosen input signal is used to stimulate the system and the corresponding output signal is measured. A system representation is than back-calculated from these two signals.

Terminology: The development of a system model from measured input and output signals is called <u>system identification</u>.

There exist many books on identification. In this text we will usually assume that we have a system model, leaving system identification to more advanced treatments.

The development of a model for a given system is often closely tied to the physics of a system. The techniques for developing the model are often motivated by the physics of the process. Quite frequently a combination of the two modeling techniques described above is used to develop the system representation. In this text we will not focus on modeling methodologies, but rather concentrate on system analysis techniques used to analyze a given system model.

6.2 SYSTEM REPRESENTATIONS

6.2.1 Introduction to System Representations

In the last section we introduced the definition of a system. A system is a mathematical relationship which assigns an output signal to an input signal. This definition is concisely summarized in the equation

$$y(t) = \mathcal{H}\big[x(t)\big]. \tag{6.2.1}$$

The system which corresponds to a particular physical process has an explicit mathematical expression called a representation. The idea is that we can gain insight into the system by studying the representation of the system. The representation of a system is not unique. The representation must be selected to serve the intended purpose.

Generally, the representation of the system serves two main purposes. First, the representation must describe the observed physical behavior of the system. Given an input signal we can compute the output signal using the system representation. We can then compare both the input signal and the computed output signal to the corresponding observations of the physical process. They should match, of course. The accuracy with which the output signal models the observations will depend on the complexity of the representation. Typically as the representation becomes more complex the output signal will more accurately match the observations. Second, the representation must be simple enough to allow for analysis. Equivalently, there must exist analysis tools which can be applied to the representation to deduce its properties, thus lending insight into the physical process. These tools might include analytical tools from calculus, and they might include computer analysis software. The basic insight we gain into the system, however, will depend on the representation we select for the system. Many times a complex system representation simply is too complex for analysis. So while a complex representation of a system will accurately describe the physical observations, it will be too complex to yield insight into the physical process because there are no analytical tools to analyze the representation. Thus, a representation must be selected to balance these two demands.

One of the purposes of this text is to introduce several system representations. We will explain the relationships between these representations, the similarities and differences. We will also introduce a number of fundamental analysis tools for each representation and explain the insight each representation provides into the behavior of the system. These analysis tools include both analytical (mathematical) tools and computer tools. This background will provide a basis for selecting the appropriate system representation for a given task.

There is an extensive literature on system representations and analysis methodologies for these representations. In this text we will focus on the most fundamental of these representations. We will be primarily concerned with four specific types of models: (1) a differential equation, (2) a Laplace transfer function, (3) a convolution integral, and (4) a Fourier transfer function.[3] In this section we will briefly introduce each of these models and give the most basic of relationships between them. A more in-depth discussion is given in the following chapters.

It is anticipated that many readers are already familiar with these system representations, if by a different name. This chapter is written with these readers in mind. We will introduce formally the above system representations and standardize our terminology. This discussion requires some mathematical concepts that are introduced in later chapters. If the reader is not familiar with these concepts, the rest of this chapter can be skipped (or skimmed) at this time. For those readers who are familiar with this background, this chapter will provide orientation and a point of departure for the discussion of system representations. The following discussion gives concrete expression to the abstract discussion above.

6.2.2 Differential Equations

The first system representation we shall consider is the differential equation

$$y^{(n)}(t) + a_{n-1}y^{(n-1)}(t) + \cdots + a_1\dot{y}(t) + a_0 y(t) \tag{6.2.2}$$

$$= b_m x^{(m)}(t) + b_{m-1}x^{(m-1)}(t) + \cdots + b_0 x(t),$$

$$y^{(n-1)}(0) = y_{n-1}, y^{(n-2)}(0) = y_{n-2}, \ldots, \dot{y}(0) = y_1, y(0) = y_0.$$

Notation: In (6.2.2) we assume that the coefficients a_k and b_k are real numbers; i.e., they are constants that don't depend on time. We also assume that $m \le n$.

The input to this system is the signal $x(t)$ and the output of this system is $y(t)$. The *system* is the relationship between $x(t)$ and $y(t)$ implied by the differential equation. The order of the highest derivative in (6.2.2) is n. We say that the differential equation in (6.2.2) is an <u>nth order</u> differential equation. In Chapter 10 and 11 we develop an alternative form of the differential equation (6.2.2) (state space equations) which is widely used.

Differential equations are widely used to model physical processes. They occur frequently in networks, dynamics, physics, etc.

6.2.3 Laplace Transfer Functions

Consider (6.2.2) when all of the initial conditions are zero. If we take the Laplace transform of (6.2.2) and solve for the ratio of the output signal over the input signal we get

---

[3] These models are not discussed in this order. In the following chapters they are presented in a modular structure as discussed in Chapter 1.

$$\frac{Y(s)}{X(s)} = \frac{b_m s^m + b_{m-1} s^{m-1} + \cdots + b_0}{s^n + a_{n-1} s^{n-1} + \cdots + a_1 s + a_0} = H(s). \qquad (6.2.3)$$

In (6.2.3) the input signal $x(t)$ is represented by its Laplace transform $X(s)$. Similarly, the output signal $y(t)$ of this system is also represented by its Laplace transform $Y(s)$.[4] The ratio $H(s)$ of the transform of the input signal $X(s)$ and the transform of the output signal $Y(s)$ in (6.2.3) is called the <u>transfer function.</u> We will study transfer functions formally in Chapter 10.

The *system* in (6.2.3) includes the input and output signals, $X(s)$ and $Y(s)$, as well as the transfer function $H(s)$. The transfer function in (6.2.2) is defined as the ratio of the transform of the output signal over the transform of the input signal. This definition is a reiteration of the definition of a system in terms of the input and output signals. The definition of the Laplace transform contains three rational functions.[5] Two of these rational functions $X(s)$ and $Y(s)$ represent signals and the third rational function $H(s)$ represents the system. Thus, these rational functions have very different physical interpretations.

The Laplace transform is used in mathematics and network analysis as a convenient tool to solve differential equations. In this text we will use it in a greatly expanded role.

6.2.4 Convolution Integral

The third system representation is derived from the transfer function (6.2.3) by expressing the output signal as a product of the transfer function and the input signal

$$Y(s) = H(s)X(s). \qquad (6.2.4)$$

Taking the inverse Laplace transform of (6.2.4) yields

$$y(t) = \int_{0^-}^{\infty} h(t - \lambda)x(\lambda)\, d\lambda \qquad (6.2.5)$$

where

$$h(t) = \mathcal{L}^{-1}\{H(s)\}. \qquad (6.2.6)$$

The convolution integral in (6.2.5) is the third system representation we will discuss, but we will use it in a slightly more general form. The representation

$$y(t) = \int_{-\infty}^{\infty} h(t - \lambda)x(\lambda)\, d\lambda \qquad (6.2.7)$$

---

[4] When we compute the transfer function from the differential equation, we <u>always</u> set the initial conditions of the differential equation to zero.

[5] See Definition 3.2.6.

of the system (6.2.1) is called the <u>convolution representation</u> of the system.   The convolution representation is discussed extensively in Chapter 12.

The convolution representation of a system (6.2.1) depends on two functions: the input signal $x(t)$ and the function $h(t)$. From a purely mathematical viewpoint, these functions are indistinguishable. From an engineering viewpoint, a sharp distinction is drawn between these two functions: $x(t)$ is a signal and $h(t)$ is associated with the physical process that generates the output signal $y(t)$.

The convolution integral would appear to be unnecessarily complicated mathematically. This representation is simpler to use for some common systems than the more familiar representations given above. It is also useful in some theoretical analysis.

6.2.5 Fourier Transfer Function

From the convolution integral we can derive the fourth system representation by taking the Fourier transform of (6.2.7). Using the convolution property of the Fourier transform we obtain

$$Y(\omega) = H(\omega)X(\omega) \tag{6.2.8}$$

where

$$H(\omega) = \mathcal{F}\{h(t)\} \tag{6.2.9}$$

and $h(t)$ is from the convolution integral in (6.2.7). Taking the ratio of the output signal over the input signal in (6.2.8) we obtain the <u>Fourier transfer function</u>

$$\frac{Y(\omega)}{X(\omega)} = H(\omega). \tag{6.2.10}$$

The Fourier transfer function is, in concept, almost identical to the Laplace transfer function. Here the Fourier transform is employed instead of the Laplace transform.

The Fourier transform is very closely related to the Laplace transform. There are certain physical situations in which the Fourier transform is more appropriate than the Laplace transform (and vice versa). Therefore, both the Fourier and Laplace transfer functions are useful.

All of the above system representations are closely related. There are also some important differences between them. Understanding these similarities and differences between these representations is important for selecting the appropriate system representation for the particular engineering task at hand. Over the course of the text we will elaborate at length on the relationships between these four system representations.

6.2.6 Response to Standard Inputs

A system is defined by specifying the input and output signals. It is not surprising, therefore, that a system can be characterized by the output signal when the input signal is a specific test signal. In this subsection we will discuss two important test input signals in the context of the system representations introduced above.[6]

The first important test input signal is an impulse function applied at time $t = t_0$

$$x(t) = \delta(t - t_0).$$ (6.2.11)

Suppose the system is modeled by a convolution integral (6.2.5). Then the output to the system (6.2.5) is

$$y_i(t) = \int_{-\infty}^{\infty} h(t - \lambda)\delta(\lambda - t_0)\,d\lambda$$ (6.2.12)

$$= h(t - t_0) = \begin{cases} \text{the system output signal at time } t \\ \text{to a unit impulse input signal at time } t_0 \end{cases}$$

We give this output signal a name.

Definition 6.2.1: The <u>impulse response</u> of a system is the output signal of the system when the input signal is an impulse function applied at time $t = 0$.
▲▲

Terminology: The word *response* refers to the output signal of the system for a given input signal. Here we see that the output signal of the system is the same function that is used to define the convolution representation. Often the *impulse response* is referred to as the *impulse response function* of the system.

Another important test input signal is the unit step function.

Definition 6.2.2: The <u>step response</u> of a system is the output signal of the system when the input signal is a unit step function.
▲▲

The step response is easily calculated using the transfer function. If the input signal is a unit step function, then its Laplace transform is

$$X(s) = \frac{1}{s}.$$ (6.2.13)

When the input signal to the system is the unit step function in (6.2.13), then the output signal can be found by taking the inverse Laplace transform of

$$Y_s(s) = H(s)\frac{1}{s}.$$ (6.2.14)

---

[6] In particular, the initial conditions of the system are zero.

That is to say, the step response of a system is found by multiplying the transfer function by $1/s$ and taking the inverse Laplace transform to obtain $y_s(t)$.

The impulse response function is related to the step response of the system. To take a derivative in the Laplace transform domain, we multiply the transform by s. Performing this operation on (6.2.14), we obtain

$$sY_s(s) = s\left[H(s)\frac{1}{s} \right] = H(s).$$

(6.2.15)

Taking the inverse Laplace transform of (6.2.15) we have

$$\frac{dy_s}{dt} = h(t).$$

(6.2.16)

Hence, the derivative of the step response is the impulse response function. This fact is sometimes used to experimentally measure the impulse response. Of course, we can reverse this process and obtain the step response from the impulse response function by integrating the impulse response function.

In the remaining three sections of this chapter we will introduce three systems that will be discussed throughout the text. For each of these three systems we will develop the three system representations introduced in this section. We will also calculate and plot the impulse and step response of each of these three systems.

6.3 ELECTRICAL NETWORKS

6.3.1 System Representations

Much of the system theory discussed in this text originated in the theory of electrical networks. From this foundation the theory has grown until it includes many other physical processes other than just networks. Nonetheless, the ideas discussed here are extremely useful for the analysis and design of networks, and we will use networks to illustrate the ideas.

The general approach followed in the discussion below is to develop the concept, illustrate the concept with a simple example amenable to hand calculation, and then show how the concept can be used for the computer analysis of more complex systems. It is in this spirit that we chose the RC network in Figure 6.3.1 for our example network. The idea is to use this network to give concrete expression to the theoretical results we will develop below. In most cases we will just be

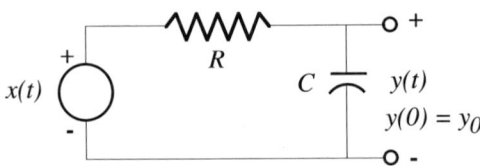

Figure 6.3.1 RC Network

formulating the network analysis problem in the language of system theory. Once the connection is made between the system theory results and network analysis, the analysis of complicated networks can be undertaken. It is not the purpose of this text to develop new network analysis techniques. Hence, the simple network in Figure 6.3.1 will be sufficient for our purposes of establishing the link between system theory and network analysis.

The purpose of this section is to derive the four models identified in the last section for the network in Figure 6.3.1. For those readers familiar with network analysis, this discussion will establish the link between the terminology of system theory presented thus far in the text and the basic results of network analysis. This network will be used repeatedly throughout the text.

In order to develop the system representations we must first identify the system. We take the input signal to be the input voltage. We chose the output signal to be the voltage across the capacitor. The system is the relationship between the input signal and the output signal as determined by the network.

The first representation we will develop for the network in Figure 6.3.1 is the differential equation. This representation is derived from the relationships between the voltage and current through the circuit elements. The currents and voltages are labeled in Figure 6.3.1. The current through the capacitor is given by

$$i(t) = C\frac{dy(t)}{dt}. \tag{6.3.1}$$

This current is equal to the current through the resistor

$$C\dot{y}(t) = i(t) = \frac{x(t) - y(t)}{R}. \tag{6.3.2}$$

For reasons that will become obvious later in the text, we will always write our differential equation models in the form of (6.2.2). Rewriting (6.3.2) into this format we have

$$\dot{y}(t) + \frac{y(t)}{RC} = \frac{x(t)}{RC}, \quad y(0) = y_0. \tag{6.3.3}$$

The initial condition in (6.3.3) is just the voltage across the capacitor at $t = 0$. Given the initial voltage across the capacitor, say, $y(0) = 0$, and the input voltage $x(t)$ we can solve for the voltage across the capacitor $y(t)$. Hence, the relationship between the input and output signals implied by the system is expressed by the differential equation (6.3.3).

The second important system representation is the transfer function. It turns out that the transfer function of the system identified in Figure 6.3.1 can be derived by using complex impedances. This approach assumes that all currents and voltages are Laplace transformable. First, we assign complex impedances as shown in Figure 6.3.2. The voltage across the capacitor can be determined using a voltage divider. We have

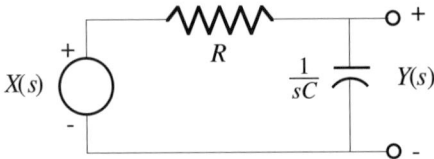

Figure 6.3.2 *RC* Network with Complex Impedances

$$Y(s) = \frac{\frac{1}{sC}}{R + \frac{1}{sC}} X(s). \tag{6.3.4}$$

The transfer function representation of (6.3.4) is expressed as the ratio of the output signal over the input signal. We also write the transfer function such that the leading coefficient of the denominator polynomial is 1. Then (6.3.4) becomes

$$\frac{Y(s)}{X(s)} = \frac{\frac{1}{RC}}{s + \frac{1}{RC}} = H(s). \tag{6.3.5}$$

The transfer function in (6.3.4) can also be derived from the differential equation in (6.3.3). This approach requires that the initial voltage across the capacitor is zero.

The third system representation is the convolution integral. The simplest way to obtain this representation is to take the inverse Laplace transform of the transfer function (6.3.5). We find that

$$\mathcal{L}\{H(s)\} = h(t) = \frac{1}{RC} e^{-\frac{t}{RC}} u_s(t). \tag{6.3.6}$$

Using the function in (6.3.6), the convolution representation is given by

$$y(t) = \int_{-\infty}^{\infty} \left(\frac{1}{RC} e^{-\frac{(t-\lambda)}{RC}} u_s(t-\lambda) \right) x(\lambda)\, d\lambda. \tag{6.3.7}$$

The fourth system representation is the Fourier transfer function. The Fourier transfer function of the system in Figure 6.3.1 can be found through steady state analysis of the network. We assign the complex impedances as shown in Figure 6.3.3. The voltage across the capacitor can be determined using a voltage divider. We have

$$Y(\omega) = \frac{\frac{1}{j\omega C}}{R + \frac{1}{j\omega C}} X(\omega). \tag{6.3.8}$$

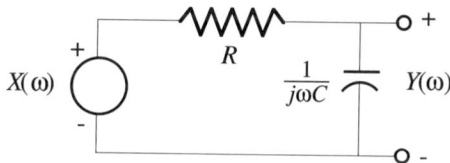

Figure 6.3.3 *RC* Network with Complex Impedances

As with the Laplace transfer function, the Fourier transfer function is written in standard form. From (6.3.8) we obtain

$$\frac{Y(\omega)}{X(\omega)} = H(\omega) = \frac{\frac{1}{RC}}{j\omega + \frac{1}{RC}}. \tag{6.3.9}$$

In summary the four system representations for the *RC* network are: (1) the differential equation in (6.3.3), (2) the Laplace transfer function in (6.3.5), (3) the convolution integral in (6.3.7), and (4) the Fourier transfer function in (6.3.9).

6.3.2 The Step and Impulse Responses

It is often useful to know the response of the system to a step or impulse input signal. For the network in Figure 6.3.1 the impulse response and the step response can be calculated from the system representations derived in this section.

The step response of this network is most easily found from the Laplace transfer function. If the input signal is a step function, then the output signal of the system is given by

$$Y_s(s) = \frac{\frac{1}{RC}}{s + \frac{1}{RC}} \frac{1}{s} = H(s)\frac{1}{s}. \tag{6.3.10}$$

Taking the inverse Laplace transform of (6.3.10) we obtain

$$y_s(t) = \left(1 - e^{-\frac{t}{RC}}\right) u_s(t). \tag{6.3.11}$$

If the input signal is an impulse function, then the Laplace transfer function of this signal is just $X(s) = 1$. So the Laplace transform of the impulse response is just

$$Y_i(s) = H(s). \tag{6.3.12}$$

The inverse Laplace transfer function of (6.3.12) is given in (6.3.6).

The impulse and step responses of the system in Figure 6.3.1 are shown in Figure 6.3.4 where we have assumed that $RC = 1$. The form of both the step and impulse responses is determined from the system representation (the transfer function). The exact shape of the curves is determined by the parameters of the system *RC*.

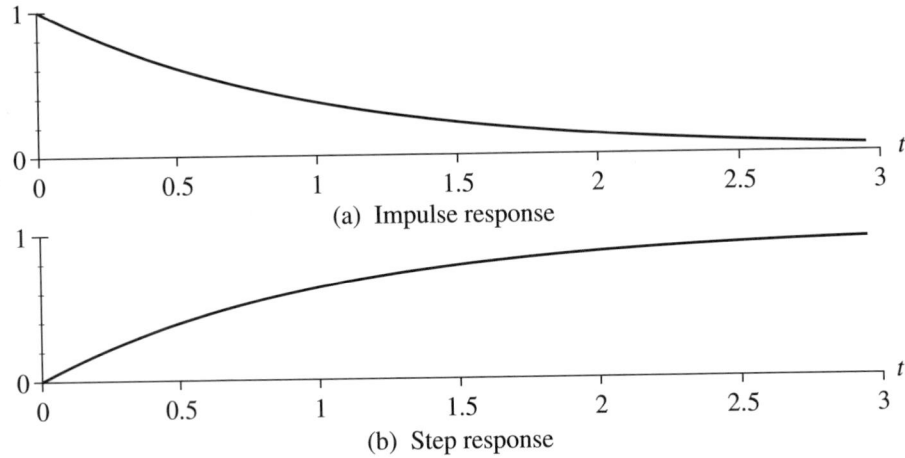

Figure 6.3.4 Impulse and Step Responses of an RC Network

In network analysis we can discern two approaches to analyzing a circuit. The first approach is to assign complex impedances to each element and then compute the relationship between the voltage and/or currents of interest. The particular configuration of circuit elements is the focus of attention. The second approach also begins with the construction of the equations that link a particular voltage to another voltage. Having obtained this relationship, we then focus on the relationship between the voltages; i.e., we ignore the particular configuration which gave rise to this equation. This later approach is sometimes called a <u>black box</u> approach to the network analysis. More formally, we call it system analysis from an input-output viewpoint.

6.4 MASS-SPRING-DAMPER SYSTEM

6.4.1 System Representations

A second class of systems is offered by a mass attached to a fixed support with a spring and a damper. An idealization of such a configuration is shown in Figure 6.4.1. The mass is m_{st}, the spring constant is k_{st}, and the damping coefficient is c_{st}. The force applied to the mass is $f_{st}(t)$ and the displacement of the mass due to this force is $y_{st}(t)$.[7] We assume that the mass moves on a frictionless surface. For simplicity, we will refer to this system as a mass-spring-damper system.

First, we will identify a system in the configuration in Figure 6.4.1. The input signal for this system is taken as the force $f_{st}(t)$ applied to the mass. Next, the output signal is defined as the displacement of the mass $y_{st}(t)$. The relationship between these two signals defines a system. We shall use this system frequently in

---

[7]The subscripts will be explained below.

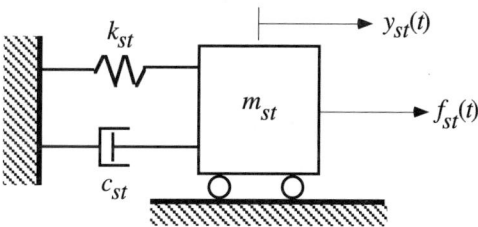

Figure 6.4.1 Mass-Spring-Damper System

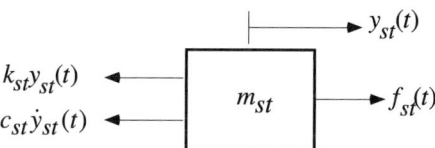

Figure 6.4.2 Free-Body Diagram of a Mass-Spring-Damper System

this text. The purpose of this section is to derive the four system representations introduced in Section 6.2 for the system in Figure 6.4.1.

The first system representation is the differential equation. The derivation of the differential equation for the mass-spring-damper system in Figure 6.4.1 is a standard topic in dynamics texts. We can derive the differential equation using the free body diagram shown in Figure 6.4.2. By Newton's second law, the sum of the forces must be equal to the mass times the acceleration, or

$$m_{st}\ddot{y}_{st}(t) = f_{st}(t) - k_{st}y_{st}(t) - c_{st}\dot{y}_{st}(t). \tag{6.4.1}$$

For our purposes, the equations in (6.4.1) are reformatted so that the leading coefficient is 1, and the input and output signals and their derivatives are collected on each side of the equals sign. Using this formatting (6.4.1) becomes

$$\ddot{y}_{st}(t) + \frac{c_{st}}{m_{st}}\dot{y}_{st}(t) + \frac{k_{st}}{m_{st}}y_{st}(t) = \frac{1}{m_{st}}f_{st}(t), \tag{6.4.2}$$

$$\dot{y}_{st}(0) = \dot{y}_0, \quad y_{st}(0) = y_0.$$

The initial conditions are the initial velocity and displacement of the mass.

The Laplace transfer function is obtained by setting the initial conditions in the differential equation (6.4.2) to zero and taking the Laplace transform. Solving for the ratio of the input signal over the output signal we obtain

$$\frac{Y_{st}(s)}{F_{st}(s)} = \frac{\dfrac{1}{m_{st}}}{s^2 + \dfrac{c_{st}}{m_{st}}s + \dfrac{k_{st}}{m_{st}}} = H_{st}(s). \tag{6.4.3}$$

The third system representation is the convolution integral. The convolution integral requires the impulse response function of the system. This function is obtained by taking the inverse Laplace transform of the transfer function of the mass-spring-damper in (6.4.3). Using a table of Laplace transforms we obtain

$$h(t) = \frac{1}{k_{st}} \left\{ \frac{\omega_n e^{-\zeta \omega_n t}}{\sqrt{1-\zeta^2}} \sin\left[\left(\omega_n \sqrt{1-\zeta^2} \right) t \right] \right\} u_s(t) \tag{6.4.4}$$

where

$$\omega_n = \sqrt{\frac{k_{st}}{m_{st}}} \quad \text{and} \quad \zeta = \frac{c_{st}}{2m_{st}} \sqrt{\frac{m_{st}}{k_{st}}} = \frac{c_{st}}{2\sqrt{k_{st}m_{st}}}. \tag{6.4.5}$$

(In stating the impulse response function in (6.4.4) we have made use of the results on polynomials in standard second-order form discussed in Section 3.2.) Using the impulse response function in (6.4.5), the convolution representation of the mass-spring-damper system in Figure 6.4.1 is

$$y_{st}(t) = \int_{-\infty}^{\infty} \left[c e^{-\zeta \omega_n (t-\lambda)} \sin\left((t-\lambda)\omega_n \sqrt{1-\zeta^2} \right) \right] u_s(t-\lambda) x(\lambda) \, d\lambda, \tag{6.4.6}$$

$$c = \frac{1}{k_{st}} \frac{\omega_n}{\sqrt{1-\zeta^2}}.$$

Finally, the fourth system representation is the Fourier transfer function. This system representation is most easily obtained from the Laplace transfer function using the formula

$$H_{st}(\omega) = H_{st}(s)\big|_{s=j\omega}. \tag{6.4.7}$$

Applying (6.4.7) to (6.4.3) we obtain

$$\frac{Y_{st}(\omega)}{F_{st}(\omega)} = \frac{\dfrac{1}{m_{st}}}{\dfrac{k_{st}}{m_{st}} - \omega^2 + \dfrac{c_{st}}{m_{st}} j\omega} = H_{st}(\omega). \tag{6.4.8}$$

In summary the four system representations for the mass-spring-damper system are in Figure 6.4.1: (1) the differential equation in (6.4.2), (2) the Laplace transfer function in (6.4.3), (3) the convolution integral in (6.4.6), and (4) the Fourier transfer function in (6.4.8).

6.4.2 The Step and Impulse Responses

It is interesting to observe the step and impulse responses of this system. The step response of this system is found by multiplying the transfer function in (6.4.3) by $(1/s)$ and computing the inverse Laplace transform (with the help of a table). This calculation yields

$$y_S(t) = \frac{1}{k_{st}} \left\{ 1 - \frac{e^{-\zeta \omega_n t}}{\sqrt{1-\zeta^2}} \sin\left[\left(\omega_n \sqrt{1-\zeta^2} \right) t + \cos^{-1}(\zeta) \right] \right\} u_s(t). \tag{6.4.9}$$

The impulse response of this system is given (6.4.4). The step and impulse responses of the mass-spring-damper system are shown in Figure 6.4.3 with $\zeta = 0.3$ and $\omega_n = 8.165$.

It is interesting to note that the step and impulse responses in Figure 6.4.3 are significantly different from the step and impulse responses of the RC network in Figure 6.3.4. In particular the output signals in Figure 6.4.3 exhibit a damped oscillation. We can investigate the impulse response further by rewriting the impulse response in (6.4.4) as

$$h(t) = \frac{1}{k_{st}} \left[\frac{\omega_n e^{-\alpha t}}{\sqrt{1-\zeta^2}} \sin(\omega_c t) \right] u_s(t) \tag{6.4.10}$$

where

$$\alpha = \zeta \omega_n \quad \text{and} \quad \omega_c = \omega_n \sqrt{1-\zeta^2}. \tag{6.4.11}$$

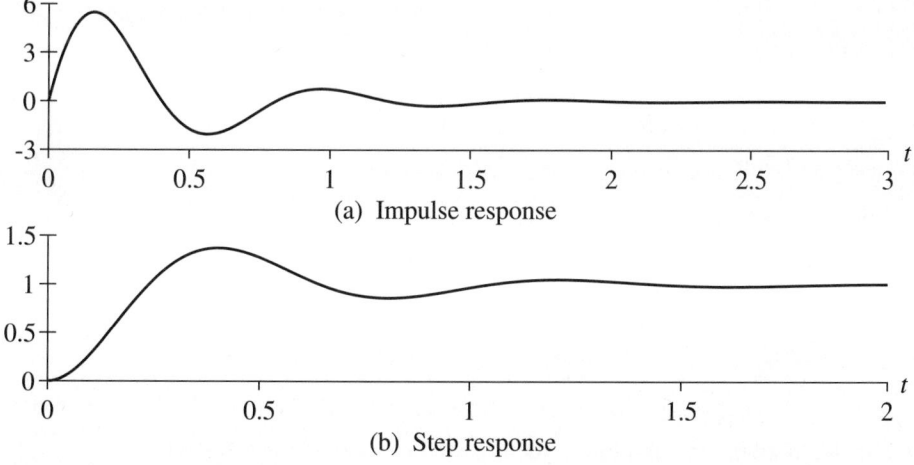

(a) Impulse response

(b) Step response

Figure 6.4.3 Impulse and Step Responses of the Mass-Spring-Damper System

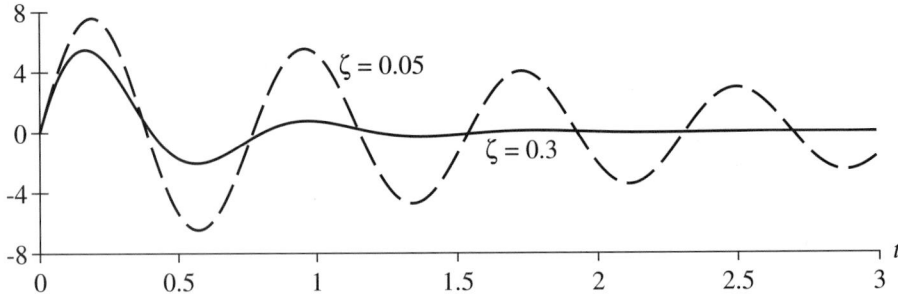

Figure 6.4.4 Impulse Response Function for Two Different Damping Ratios

The impulse response functions for two different values of ζ are shown in Figure 6.4.4. Thus, we see that the amount of damping in the oscillations is determined by the ζ. The number ζ is directly related to damping coefficient c_{st} as shown in (6.4.5). Inspection of (6.4.10) also shows that the decay of the amplitude of the sinusoid is controlled by the coefficient α, which depends on ζ.

Assume that $\zeta \approx 0$. Then from (6.4.4) the frequency of the oscillation is

$$\omega_n \approx \sqrt{\frac{k_{st}}{m_{st}}}. \tag{6.4.12}$$

This frequency is the frequency at which this system will oscillate if an impulsive force is applied to the mass. If ζ is not almost zero, then (6.4.4) shows that the frequency of oscillation is given by ω_c in (6.4.11). Based on these observations we introduce names for these parameters.

Definition 6.4.1: The frequency ω_n is called the <u>natural frequency</u>. The frequency ω_c is called the <u>critical frequency</u>. The parameter ζ is called the <u>damping ratio</u>. The parameter α is called the <u>damping factor</u>.
▲▲

These parameters, particularly ζ and ω_n, are widely used in many different engineering fields. We will use them frequently.

6.5 PROOF-MASS ACTUATORS

6.5.1 Linear Motors as Actuators

A third example we will use frequently in this text is provided by a linear motor. A linear DC motor operates on the same principles as an ordinary DC motor except the electromagnetic coupling imparts a rectilinear motion to a mass instead of a rotational motion. A simplified picture of one possible configuration of a linear motor is shown in Figure 6.5.1. This linear DC motor consists of two basic components: (1) a motor base and (2) a proof-mass. The proof-mass is a bar which

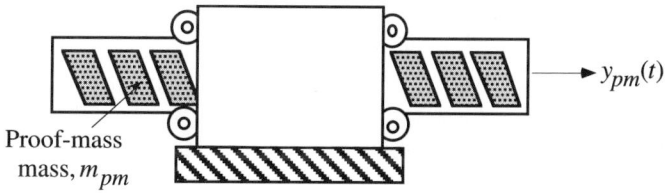

Figure 6.5.1 A Proof-Mass Actuator

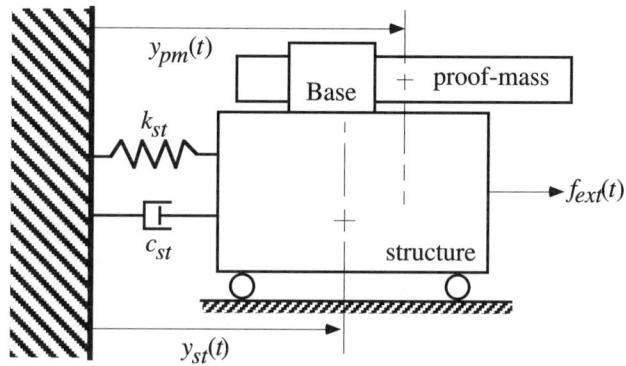

Figure 6.5.2 Linear Motor on a Mass-Spring-Damper System

contains permanent magnets and is free to move in a linear direction. The motor base contains the motor coils and the power electronics. The input signal to the motor is the input voltage $v_{pm}(t)$ to the power electronics.

Suppose that the linear motor in Figure 6.5.1 is attached to the mass-spring-damper system, described in the last section, as shown in Figure 6.5.2. If an external force f_{ext} is applied to the mass, the mass will oscillate. The linear motor can be used to suppress these vibrations. The principle of operation is as follows. When an input voltage is given to the device, the motor coils impart a force on the proof-mass. An equal and opposite force is generated on the base of the motor. This force is transmitted to the mass through the rigid coupling of the actuator to the mass. If the mass is vibrating then, in principle, the vibrations could be decreased by choosing the input voltage in such a way that the force imparted to the mass always opposes the motion of the mass. During the course of this text we will explore this idea in some detail.

The primary purpose for the linear motor is to impart a force that is proportional to a commanded electrical input signal. The input signal is called "commanded" because we are free to generate this signal ourselves. Such a device is called an actuator.

Terminology: When a linear DC motor is employed for vibration suppression, it is called a proof-mass actuator. Henceforth, we shall refer to this device by this name.

6.5.2 System Representations

The purpose of this section is to develop three system representations for the proof-mass actuator. First, we define the system. The input signal to this system is the voltage input to the motor, $v_{pm}(t)$, (not shown in Figure 6.5.1). The output signal of this system is the position of the proof-mass, $y_{pm}(t)$. The system is the relationship between these two signals.

The first representation is the differential equation representation of the system. To derive this representation first consider the electronics of the motor. We will assume that the force on the proof-mass, $f_{pm}(t)$, is proportional to the input control voltage, $v_{pm}(t)$ i.e.,

$$f_{pm}(t) = K_{ef} v_{pm}(t). \tag{6.5.1}$$

This model ignores the dynamics of the power electronics and the nonlinearities of the flux linkages. The dynamics of this system is due to the dynamics of the proof-mass. A free-body diagram of the proof-mass is shown in Figure 6.5.3. The proof-mass has a mass m_{pm} and length $2d$. We say that the actuator has a <u>stroke</u> of d. The position of the center of the proof-mass with respect to an inertial reference frame is denoted by $y_{pm}(t)$ as shown in Figure 6.5.3. Neglecting friction between the motor base and the proof-mass, the model of the proof-mass motion is just

$$m_{pm} \ddot{y}_{pm}(t) = f_{pm}(t) = K_{ef} v_{pm}(t). \tag{6.5.2}$$

Rewriting (6.5.2) into our standard form we obtain

$$\ddot{y}_{pm}(t) = \frac{K_{ef}}{m_{pm}} v_{pm}(t), \tag{6.5.3}$$

$$\dot{y}_{pm}(0) = y_{v0}, \quad y_{pm}(0) = y_{p0}.$$

We get the second system representation by taking the Laplace transform of (6.5.2) with zero initial conditions. The result is the Laplace transform transfer function

$$\frac{Y_{pm}(s)}{V_{pm}(s)} = \frac{K_{ef}}{m_{pm}s^2} = H_{pm}(s). \tag{6.5.4}$$

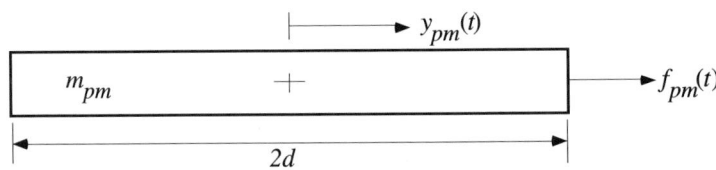

Figure 6.5.3 Free-Body Diagram of the Proof-Mass of the Actuator

The third system representation is the convolution integral. To form this system representation we need the impulse response function. This function is obtained from the inverse Laplace transform of the transfer function as

$$h_{pm}(t) = \mathcal{L}^{-1}\left\{\frac{K_{ef}}{m_{pm}s^2}\right\} = \left(\frac{K_{ef}t}{m_{pm}}\right)u_s(t). \tag{6.5.5}$$

Using (6.5.5), the convolution representation of the proof-mass actuator is

$$y_{pm}(t) = \int_0^\infty \left(\frac{K_{ef}}{m_{pm}}(t-\lambda)u_s(t-\lambda)\right)v_{pm}(\lambda)\,d\lambda. \tag{6.5.6}$$

The fourth system representation that we introduced in Section 6.2 is the Fourier transfer function. It turns out that this system doesn't have a Fourier transfer function, as explained in Section 13.5.

In summary, the three system representations are: (1) the differential equation in (6.5.2), (2) the Laplace transfer function in (6.5.4), and (3) the convolution integral in (6.5.6).

6.5.3 The Step and Impulse Responses

We can also characterize this system by its step and impulse responses. The step response of this system is found by multiplying the transfer function in (6.5.4) by $(1/s)$. We have

$$Y_s(s) = \frac{K_{ef}}{m_{pm}s^2}\frac{1}{s} = H_{pm}(s)\frac{1}{s}. \tag{6.5.7}$$

Taking the inverse Laplace transform of (6.5.7) we obtain the step response

$$y_s(t) = \frac{K_{ef}}{2m_{pm}}t^2 u_s(t). \tag{6.5.8}$$

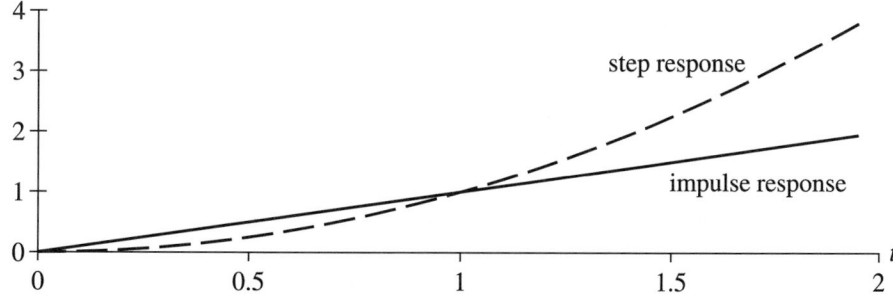

Figure 6.5.4 Impulse and Step Responses of a Proof-Mass Actuator

The impulse response is given in (6.5.5). The impulse and step responses of a proof-mass actuator are shown in Figure 6.5.4. It is worth noting that these step and impulse responses are qualitatively different from those of the systems in the previous two sections. In particular, the responses in Figure 6.5.4 do not tend to zero as time goes to infinity.

6.6 CHAPTER SUMMARY

6.6.1 Definitions

In the last chapter we introduced the first fundamental concept of this text: a signal. In this chapter we introduced the second fundamental concept of this text.

Definition 6.1.1: A <u>system</u> generates a response, or output signal, for a given input signal.

▲▲

Over the course of the text we will introduce several classifications of systems. The next classifications are based on the type of input and output signals that are used to define the system.

Definition 6.1.4: If the input signal and output signal of a system are continuous-time signals, then the system is called a <u>continuous(-time)</u> system.

▲▲

Definition 6.1.5: If the input signal and output signal of a system are discrete-time signals, then the system is called a <u>discrete-time</u> system.

▲▲

Systems are also classified by the number of input and output signals of the system.

Definition 6.1.6: If the system has one input signal and one output signal, we call that system a <u>single-input-single-output</u> (SISO) system. If the system has more than one input and/or output signal, then the system is called a <u>multiple-input-multiple-output</u> (MIMO) system.

▲▲

Since signals are functions, systems are mathematical relationships between functions.

Definition 6.1.8: An explicit mathematical expression for a system is called a system <u>representation</u>.

▲▲

A system representation is also called a system model.

Definition 6.1.9: The process of deriving a system representation is called <u>modeling</u>.

▲▲

In this text we will focus on four main system representations:

(1) differential equations,

$$y^{(n)}(t) + \cdots + a_1 \dot{y}(t) + a_0 y(t) = b_m x^{(m)}(t) + \cdots + b_0 x(t), \qquad (6.2.2)$$

$$y^{(n-1)}(0) = y_{n-1}, \ldots, \dot{y}(0) = y_1, y(0) = y_0.$$

(2) Laplace transfer functions

$$\frac{Y(s)}{X(s)} = \frac{b_m s^m + b_{m-1} s^{m-1} + \cdots + b_0}{s^n + a_{n-1} s^{n-1} + \cdots + a_1 s + a_0} = H(s). \qquad (6.2.3)$$

(3) convolution integrals

$$y(t) = \int_{-\infty}^{\infty} h(t - \lambda) x(\lambda) \, d\lambda \qquad (6.2.7)$$

and (4) Fourier transfer functions

$$\frac{Y(\omega)}{X(\omega)} = H(\omega). \qquad (6.2.10)$$

Another way to characterize a system is to apply a test signal and record the output signal. The following two test signals are widely used.

Definition 6.2.1: The <u>impulse response function</u> $h(t)$ of a system is the output signal of the system when the input signal is an impulse function applied at time $t = 0$.
▲▲

Definition 6.2.2: The <u>step response</u> of a system is the output signal of the system when the input signal is a unit step function.
▲▲

6.6.2 Comparison of Impulse and Step Responses for the Three Systems

In this chapter we have introduced three prototypical continuous-time systems: an *RC* network, a mass-spring-damper system, and a proof-mass actuator. These systems will be used extensively in the following chapters to illustrate various system concepts.

It is interesting to compare the impulse response functions for the three systems introduced in the last three sections. These functions are shown in Figure 6.3.4 for the *RC* network, Figure 6.4.3 for the mass-spring-damper system, and Figure 6.5.4 for the proof-mass actuator.

The impulse response function of the *RC* network exhibits a uniform decay to zero, while the mass-spring-damper system oscillates as it decays to zero. The proof-mass actuator, however, never decays to zero. In fact, the impulse response of the proof-mass actuator increases for increasing time. This observation suggests that the proof-mass actuator has different properties than the other two systems.

The step response function of the *RC* network and the mass-spring-damper system both approach one as time goes to infinity. The *RC* network step response never exceeds one, while the mass-spring-damper system oscillates and attains values greater than one. These two systems respond in different ways to the same input signal as suggested by the different impulse response functions. The step response of the proof-mass actuator is completely different; it drifts off to infinity rather than approaching a finite value. Again, we see that the properties of this system are different than the properties of the other two systems. In the chapters to come we will quantify the differences between these systems.

6.7 HOMEWORK FOR CHAPTER 6

Homework Problems for Section 6.1

6.1.1   (a) Give an example of a continuous-time system that is not given in the text.
   (b) Give an example of a discrete-time system that is not given in the text.
   (c) Give an example of a system that is neither a continuous-time nor discrete-time system that is not given in the text.
   (d) Give an example of a multivariable system that is not given in the text.

Homework Problems for Section 6.2

6.2.1   Suppose that the step response of a system is given as follows.
   (a) Find the impulse response of each system.
   (b) Find the transfer function of each system.
   (i) $y_s(t) = \left(1 - e^{-at}\right) u_s(t)$ (ii) $y_s(t) = e^{-at} u_s(t)$

6.2.2   For the following systems:
   (a) Find the step response of each of these systems.
   (b) Qualitatively compare the step responses of these two systems.
   (i) $\dot{y}(t) + ay(t) = ax(t), \quad a > 0$ (ii) $y(t) = x^2(t)$

6.2.3   Find the impulse response of the following systems.
   (i) $\dot{y}(t) - ay(t) = x(t), \quad a > 0$ (v) $\ddot{y}(t) + \omega_0 y(t) = bx(t)$
   (ii) $\dfrac{Y(s)}{X(s)} = \dfrac{1}{(s^2 + 1)(s^2 + 1.01)}$ (vi) $\dfrac{Y(s)}{X(s)} = e^{-sT}$
   (iii) $\dfrac{Y(\omega)}{X(\omega)} = \dfrac{1}{1 + j\omega}$ (vii) $\dfrac{Y(\omega)}{X(\omega)} = \delta(\omega - 2\pi) + \delta(\omega + 2\pi)$
   (iv) $y(t) = \displaystyle\int_{-\infty}^{\infty} \Pi(t - \lambda) x(\lambda)\, d\lambda$ (viii) $y(t) = \displaystyle\int_{-\infty}^{\infty} \sin(t - \lambda) u_s(t - \lambda) x(\lambda)\, d\lambda$

6.2.4   Suppose the impulse response of a dynamic system is as follows.
(a) Find the step response.
(b) Find the transfer function.
  (i) $y_i(t) = \sin(\omega t)u_s(t)$ (ii) $y_i(t) = e^{-t}u_s(t)$

Homework Problems for Section 6.3

6.3.1   Consider the network in Figures P6.3.1a and b.
(a) Find the differential equation that corresponds to this network.
(b) Find the transfer function of this network.
(c) Find the impulse response of this network.
(d) Find the convolution representation of this network.
(e) Find the step response of this network.

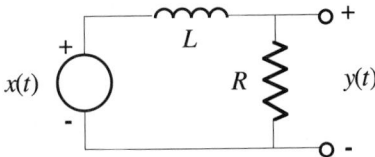

Figure P6.3.1a

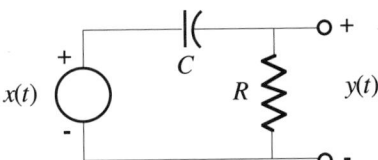

Figure P6.3.1b

6.3.2   Consider the network shown in Figure P6.3.2.
(a) Find the transfer function in terms of L, R, and C.
(b) Find (ζ,ω_n) for the poles.
(c) Plot the impulse response.
(d) Plot the step response.

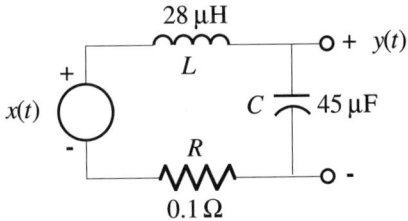

Figure P6.3.2

Homework Problems for Section 6.4

6.4.1   Consider the mass-spring system in Figures P6.4.1a and b.
(a) Find the differential equation that corresponds to this system.
(b) Find the transfer function of this system.
(c) Find the impulse response of this system.
(d) Find the step response of this system.

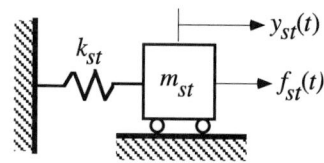

Figure P6.4.1a

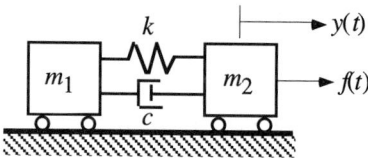

Figure P6.4.1b

6.4.2   Repeat Problem 6.4.1 if the output signal is defined as the velocity of the mass. Compare your answers to those for Problem 6.4.1.

Homework Problems for Section 6.5

6.5.1   Many proof-mass actuators are constructed with centering springs. This concept is shown in Figure P6.5.1. Use the same parameters and variables as defined for the proof-mass actuator in Section 6.5.
(a) Derive the differential equation for this actuator.
(b) Find the transfer function of this actuator.
(c) Find the impulse response function of this actuator.
(d) Find the step response of this actuator.

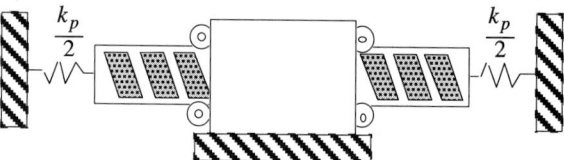

Figure P6.5.1

6.5.2   Consider the proof-mass actuator discussed in Section 6.5. Suppose that the output signal of this system is defined to be the velocity of the proof-mass (rather than the position of the proof-mass). Repeat Problem 6.5.1.

Chapter 7

Fourier Series and Fourier Transforms

Chapter Outline

In this chapter we introduce two very important mathematical tools for analyzing signals and systems: the Fourier series and the Fourier transform. Both of these concepts come from classical mathematics, and this chapter is devoted to developing the required mathematical background to use these concepts in this text.

In keeping with the philosophy of this text, this chapter concentrates on the mathematics of these two concepts. Subsequent chapters apply these ideas to the analysis of signals and systems.

In the first three sections of this chapter we discuss Fourier series. A Fourier series is the representation of a periodic signal by an infinite sum of sinusoids. In this chapter we define the Fourier series and discuss how to compute the Fourier series for a given periodic signal. We also give three alternative representations of the Fourier series, each of which is frequently used, and develop their interrelationship. The focus of this chapter is on the analytical interpretations of the Fourier series. For most applications of the Fourier series in signals and systems, the Fourier series of a signal is represented graphically. This graphical representation and its interpretation is developed in Chapter 8. The Fourier series is also very useful in explaining how a signal propagates through a system. This discussion is presented in Chapter 15.

The Fourier transform can be viewed as an extension of the Fourier series to aperiodic signals. In the last two sections of this chapter we define the Fourier transform and develop several of its properties. The Fourier transform (and its extension to discrete-time) is one of the most useful tools for doing signal and system analysis. It will play a fundamental role throughout the rest of the text including the representation of signals in Chapter 8, the representation of systems in Chapter 12, and the analysis of a signal propagating through a system in Chapters 14 and 15. The discrete-time Fourier transform plays an analogous role in the discussion of discrete-time signals and systems.

Summary of Sections

Section 7.1:   We give an example of a Fourier series to introduce this concept.

Section 7.2:   We formally define three forms of the Fourier series and develop their interrelationship.

Section 7.3:   We give the computational formulas for the Fourier series.

Section 7.4:   We define the Fourier transform.

Section 7.5:   We develop several properties of the Fourier transform.

Section 7.6:   Chapter summary section.

Coverage of the Text

This chapter requires only Chapter 5.

7.1 INTRODUCTION TO FOURIER SERIES

7.1.1 Introduction

This section introduces our study of signal representations. We begin with signals that repeat in time: periodic signals. Periodic signals are commonly encountered in engineering practice and they are important theoretically as well. It turns out that periodic signal can be written as an infinite sum of sinusoids. This infinite sum is called a Fourier series. The purpose of this section is to give a heuristic introduction to the concept of the Fourier series for periodic signal representation. Using a square wave we will show how to compute its Fourier series. We will also give an interpretation of the Fourier series. In the following sections we will give a systematic development of this signal representation, and show how to construct it for any periodic signal.

7.1.2 Periodic Signals

In this chapter we will consider signals of the type shown in Figure 7.1.1. This signal is called a pulse train or a square wave. These signals are characterized by the repetition of a waveform over the entire real line. We have a name for such signals.

Definition 7.1.1: The signal $x(t)$ is <u>periodic</u> if there exists a constant $T_0 > 0$ such that

$$x(t + T_0) = x(t) \quad \text{for all} \ \ t.$$

The smallest T_0 for which this is true is called the (<u>fundamental</u>) <u>period</u>. All other signals are <u>aperiodic</u>.   ▲▲

Notation: We denote a periodic signal with period T_0 by $x_{T_0}(t)$.

Example 7.1.2: The pulse train in Figure 7.1.1 is a periodic signal with a period of $T_0 = 2$. The signals $\sin(t)$ and $\cos(t)$ are periodic signals with period $T_0 = 2\pi$.   ▲▲

There are many physical processes which repeat themselves. Sinusoidal voltages in power systems are one example. Light in an optical fiber (waveguide) is

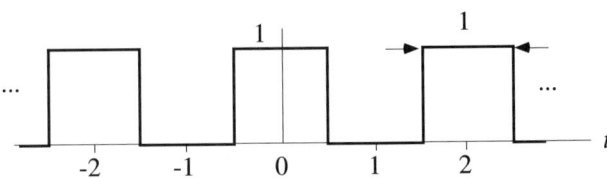

Figure 7.1.1 Pulse Train

another. We can take advantage of their periodic nature to gain insight into the representations for these signals and the structure of the processes themselves. The analysis of periodic signals plays a central role in signal analysis.

An easy way to describe and generate periodic signals is by repetition of a signal defined on an interval.

Definition 7.1.3: Let the signal $x_1(t)$ be an aperiodic signal defined on the interval $t_0 \leq t < t_0 + T_0$. Then define a signal $x_{T_0}(t)$ by repeating the signal $x_1(t)$ on all the intervals $mt_0 \leq t < mt_0 + T_0$, where m is an integer which is not zero. The signal $x_{T_0}(t)$ is said to be created by <u>periodic extension</u>.

▲▲

Example 7.1.4: Define the signal

$$x_1(t) = \begin{cases} 1, & -\dfrac{1}{2} \leq t < \dfrac{1}{2} \\ 0, & \dfrac{1}{2} \leq t < \dfrac{3}{2} \end{cases} \tag{7.1.1}$$

This signal is shown in Figure 7.1.2. The periodic extension of the signal in Figure 7.1.2 is the pulse train shown in Figure 7.1.1.

▲▲

Example 7.1.5: Consider again the signal in Example 7.1.4. Define a new signal by multiplying $x_1(t)$ in (7.1.1) by $\cos(\omega_c t)$. We get

$$x_c(t) = \begin{cases} \cos(\omega_c t), & -\dfrac{1}{2} \leq t < \dfrac{1}{2} \\ 0, & \dfrac{1}{2} \leq t < \dfrac{3}{2} \end{cases} \tag{7.1.2}$$

This signal is shown in Figure 7.1.3a. We can define a new signal $x_p(t)$ by periodic extension of (7.1.2). This signal is shown in Figure 7.1.3b.

At first glance, the signal in Figure 7.1.3 appears to be a pulse train multiplied by a cosine. This signal is not a pulse train multiplied by a cosine, however, because the cosine in (7.1.2) only multiplies the signal (7.1.1) over one period. That is,

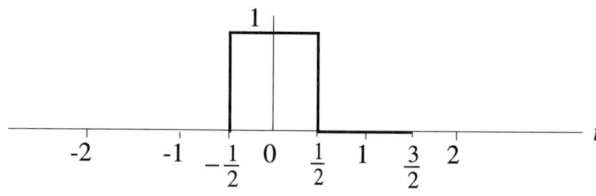

Figure 7.1.2 Signal Defined By Periodic Extension

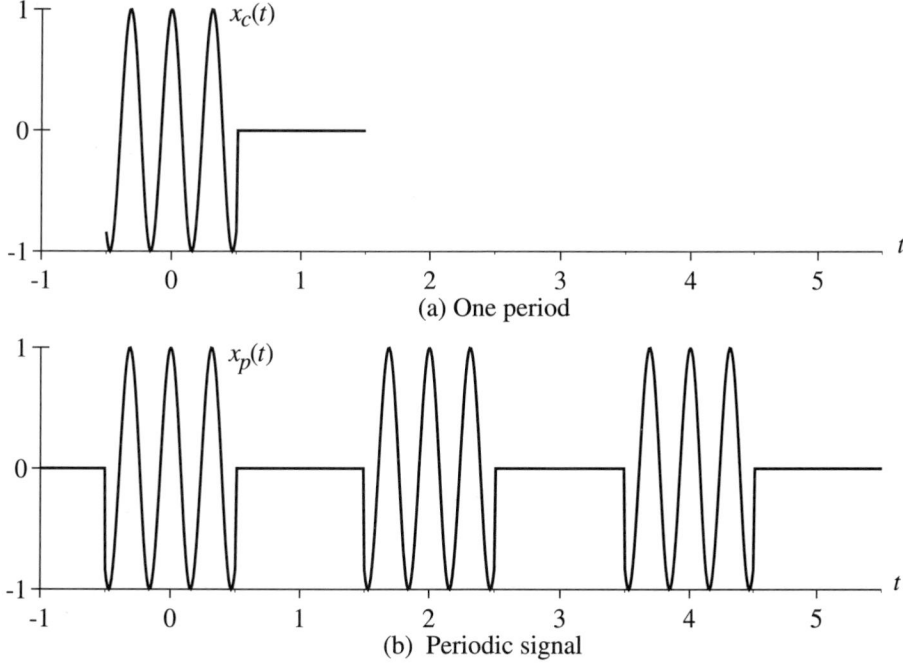

(a) One period

(b) Periodic signal

Figure 7.1.3 Signal and Its Periodic Extension for Example 7.1.5

$$x_p(t) \neq \left[\sum_{k=-\infty}^{\infty} \Pi(t-2k) \right] \cos(\omega_c t). \tag{7.1.3}$$

▲▲

7.1.3 Fourier Series of a Pulse Train

Derivation of the Fourier Series It is the purpose of this section to introduce the concept of the Fourier series by presenting an example of a Fourier series of the pulse train shown in Figure 7.1.1. In the next section it will be formally shown that the signal in (7.1.1) can be represented as an infinite sum of sinusoids

$$x(t) = a_0 + \sum_{m=1}^{\infty} a_m \cos(m\omega_0 t) + \sum_{m=1}^{\infty} b_m \sin(m\omega_0 t), \quad \omega_0 = \frac{2\pi}{T_0}. \tag{7.1.4}$$

The coefficients of (7.1.4) are given by

$$a_0 = \frac{1}{T_0} \int_{T_0} x(t)\, dt, \tag{7.1.5}$$

$$a_m = \frac{2}{T_0} \int_{T_0} x(t) \cos(m\omega_0 t)\, dt, \qquad m = 1, 2, 3, \ldots$$

$$b_m = \frac{2}{T_0} \int_{T_0} x(t) \sin(m\omega_0 t)\, dt, \qquad m = 1, 2, 3, \ldots$$

Notation: The notation in (7.1.5) means that the integration in (7.1.6) can be performed over any interval of length T_0. That is,

$$\int_{T_0} x(t)\, dt = \int_{t_0}^{t_0 + T_0} x(t)\, dt \quad \text{for any } t_0. \tag{7.1.6}$$

To compute the coefficients in (7.1.4) first we note that the period of the signal in Figure 7.1.1 is $T_0 = 2$. This period determines the fundamental frequency ω_0 in (7.1.4) by

$$\omega_0 = \frac{2\pi}{T_0} = \frac{2\pi}{2} = \pi. \tag{7.1.7}$$

The constant term in (7.1.4) is just the average of the signal over one period as shown in (7.1.5). This constant is given by

$$a_0 = \frac{1}{T_0} \int_{-\frac{1}{2}}^{\frac{3}{2}} x(t)\, dt = \frac{1}{2} \int_{-\frac{1}{2}}^{\frac{1}{2}} 1\, dt + \frac{1}{2} \int_{\frac{1}{2}}^{\frac{3}{2}} 0\, dt = \frac{1}{2}(1) = \frac{1}{2}. \tag{7.1.8}$$

Next the coefficients of the cosine terms are computed by multiplying (7.1.4) by $\cos(m\omega_0 t)$ and integrating over the period as stipulated by (7.1.5). We have

$$a_m = \frac{2}{T_0} \int_{T_0} x(t) \cos(m\omega_0 t)\, dt = \frac{2}{T_0} \left[\int_{-\frac{1}{2}}^{\frac{1}{2}} \cos(m\omega_0 t)\, dt - \int_{\frac{1}{2}}^{\frac{3}{2}} 0\, dt \right] \tag{7.1.9}$$

$$= \frac{2}{T_0 m \omega_0} \left(\sin(m\omega_0 t) \Big|_{t=-\frac{1}{2}}^{\frac{1}{2}} \right).$$

Evaluating the limits in (7.1.9) we obtain

$$a_m = \frac{2}{T_0 m \omega_0} \left[\sin\left(\frac{m\pi}{2}\right) - \sin\left(-\frac{m\pi}{2}\right) \right] = \frac{2}{T_0 m \omega_0} \left[2\sin\left(\frac{m\pi}{2}\right) \right]. \tag{7.1.10}$$

Using (7.1.7) in the first factor in (7.1.10) we have that

$$\omega_0 T_0 = \left(\frac{2\pi}{T_0}\right) T_0 = 2\pi. \tag{7.1.11}$$

Substituting (7.1.11) into (7.1.10) we get

$$a_m = \frac{2}{m\pi}\sin\left(\frac{m\pi}{2}\right) = \begin{cases} \dfrac{2}{m\pi}, & m = 1,5,9,\dots \\[2ex] -\dfrac{2}{m\pi}, & m = 3,7,11,\dots \end{cases} \tag{7.1.12}$$

$$= \left(\frac{2}{m\pi}\right)(-1)^{\frac{m-1}{2}}, \quad m \text{ odd.}$$

The coefficients b_m in (7.1.4) are computed from (7.1.5) as follows

$$b_m = \frac{2}{T_0}\int_{T_0} x(t)\sin(m\omega_0 t)\,dt = \frac{2}{T_0}\left[\int_{-\frac{1}{2}}^{\frac{1}{2}}\sin(m\omega_0 t)\,dt - \int_{\frac{1}{2}}^{\frac{3}{2}} 0\,dt\right] \tag{7.1.13}$$

$$= \frac{2}{T_0 m\omega_0}\left[\cos\left(\frac{m\omega_0}{2}\right) - \cos\left(-\frac{m\omega_0}{2}\right)\right] = 0.$$

Combining (7.1.8), (7.1.12), and (7.1.13) we arrive at the Fourier series for the signal in Figure 7.1.1,

$$x(t) = \frac{1}{2} + \sum_{\substack{m=1 \\ m \text{ odd}}}^{\infty}\left((-1)^{\frac{m-1}{2}}\right)\frac{2}{m\pi}\cos(m\pi t). \tag{7.1.14}$$

Remark 7.1.6: Indexing of the Coefficients In (7.1.14) the coefficients of the Fourier series are expressed in terms of the index of summation m. This relationship can be rewritten without changing the essential content of the Fourier series. Suppose that we redefine the index m as

$$m = 2k - 1, \quad \text{for} \quad -\infty < k < \infty. \tag{7.1.15}$$

Using (7.1.15) in (7.1.14) we obtain

$$x(t) = \frac{1}{2} + \sum_{k=1}^{\infty}\left((-1)^{k-1}\right)\frac{2}{(2k-1)\pi}\cos(k\pi t). \tag{7.1.16}$$

The Fourier series in (7.1.16) is equivalent to the Fourier series in (7.1.14) even thought the terms in the summation are indexed differently.

▲▲

Alternative Representations of the Fourier Series We have seen that a signal can have several representations. The same is true of the Fourier series. To introduce these alternative representations we will rewrite the Fourier series of the pulse train in (7.1.14). This first alternative representation relies on the trigonometric identity

$$\cos(\omega t - \pi) = -\cos(\omega t). \tag{7.1.17}$$

Substituting (7.1.17) into (7.1.14) the Fourier series becomes

$$x(t) = \frac{1}{2} + \sum_{\substack{m=1 \\ m \text{ odd}}}^{\infty} \frac{2}{m\pi} \cos(m\pi t + \theta_m) \tag{7.1.18}$$

where

$$\theta_m = \begin{cases} -\pi, & m = 3, 7, 11, \ldots \\ 0, & m = 1, 5, 9, \ldots \end{cases}$$

Comparing (7.1.18) to (7.1.14) we see that the coefficients of (7.1.18) are all positive whereas the coefficients of (7.1.14) alternate in sign. Furthermore, the terms in the series in (7.1.18) carry a phase angle while the terms in (7.1.14) do not have a phase angle associated with them. If the sign of the amplitude coefficient in (7.1.14) is negative, then this sign is absorbed into the phase angle in (7.1.18). This assumption unambiguously defines the phase angles θ_m.

The Fourier series in (7.1.18) is an example of a another representation of the Fourier series in (7.1.4). The general representation of the series in (7.1.18) is

$$x(t) = A_0 + \sum_{m=1}^{\infty} A_m \cos(m\omega_0 t + \theta_m), \quad \omega_0 = \frac{2\pi}{T_0}. \tag{7.1.19}$$

In the next section we will show that the Fourier series in (7.1.19) is equivalent to the Fourier series in (7.1.4).

7.1.4 Interpretation of a Fourier Series

The Fourier series is of central importance to signal analysis. The importance of the Fourier series is expressed in the representation in (7.1.19). On the left side of the equality is a periodic signal. On the right side of the equality is an infinite sum of sinusoids. This equality is the essential point. Almost any periodic signal[1] can be expressed as a sum of sinusoids. So by studying the sum of sinusoids we can learn about signals that don't appear to be sinusoidal.

---

[1] The exact conditions are given in Theorem 7.2.6.

But all signals are not the same. Hence, the Fourier series on the right side of (7.1.19) must be different for different signals. The Fourier series is distinguished by three sets of parameters: ω_0, A_m, and θ_m.

Terminology: The frequency ω_0 is called the <u>fundamental</u> frequency. The coefficients A_m are called the <u>amplitude</u> coefficients and the coefficients θ_m are called the <u>phase</u> coefficients.

The Fourier series itself is composed of cosines whose frequencies are integer multiples of each other. Cosine terms with other frequencies are excluded. The fundamental frequency is determined by the period of the periodic signal as shown in (7.1.19). Note that all other frequencies are an integer multiple of the fundamental frequency. These other frequencies carry no more information about the signal $x(t)$.

The Fourier series itself is composed of cosines whose frequencies are integer multiples of each other. The existence of the Fourier series depends on the special relationship between these functions as expressed by (7.1.5). This special relationship depends on the integer relationship between the frequencies of sinusoidal functions. There are many other functions that also have this property. In particular we will study Fourier series based on complex exponentials.

The other two sets of parameters that define a Fourier series for a given signal are the amplitude coefficients A_m and the phase coefficients θ_m. These parameters vary from signal to signal. In Chapter 8 we will develop methods for interpreting these coefficients to identify properties of the signal. To begin, we see that for the signal in Figure 7.1.1 the nonzero coefficients of the Fourier series in (7.1.14) tend to zero as the summation index tends to infinity, $m \to \infty$. This fact points to the observation that some terms in the Fourier series are more "important" than others. This importance is determined by the amplitude coefficients A_m. We will quantify this observation in the next chapter. For the time being we will demonstrate this observation graphically.

The Fourier series in (7.1.18) says that we can plot the signal $x(t)$ by plotting the sinusoidal signals on the right side of (7.1.18). Since the sum in (7.1.18) is an infinite series, we can't plot all of the terms. We truncate this sum to

$$x_{0:m_1}(t) = \frac{1}{2} + \sum_{\substack{m=1 \\ m \text{ odd}}}^{m_1} \frac{2}{m\pi} \cos(m\pi t + \theta_m).$$

(7.1.20)

In Figure 7.1.4 we have shown the signal (7.1.1) and the partial sums in (7.1.20) for $m_1 = 3$, and $m_1 = 11$. From Figure 7.1.2 we see that just a few terms of the Fourier series do approximate the original pulse train in Figure 7.1.1. Moreover, as we include more terms in the partial sum in (7.1.20), the approximation to the original signal $x(t)$ improves. In the next chapter we will quantify all of these observations. In the following chapters we will use this knowledge to study how a signal propagates through a system.

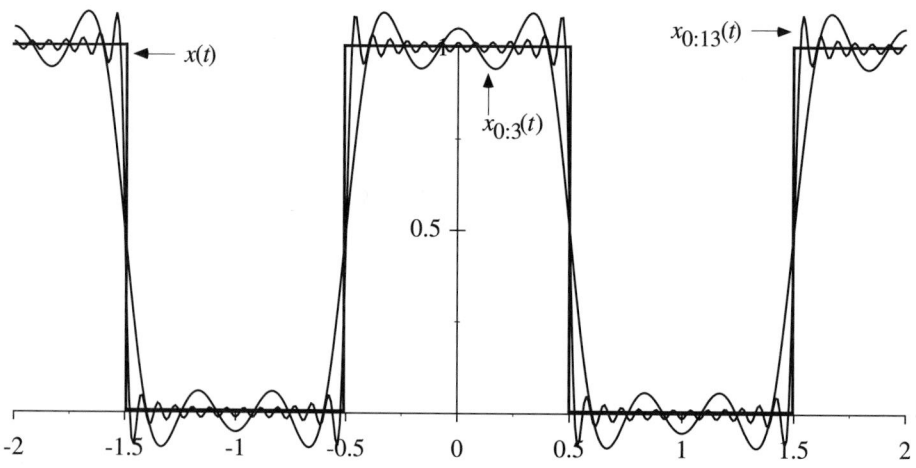

Figure 7.1.4 Partial Sums of the Fourier Series in (7.1.20)

7.2 THREE REPRESENTATIONS OF A FOURIER SERIES

7.2.1 Introduction

In the last section we discussed the Fourier series of a particular periodic signal, a pulse train. In fact, we gave two different forms of the Fourier series. In this section we will formally introduce and define the Fourier series for any periodic signal $x(t)$ with period T_0. Then we will show that there are three equivalent representations of the Fourier series as shown in Table 7.2.1. All of these expressions are useful, and they will be used throughout the text. In this section we will derive the relationship between these representations. In the next section we will introduce the computational formulas that allow us to derive a Fourier series for a given periodic signal.

Table 7.2.1 Fourier Series Representations

$x(t)$, periodic with period T_0	Fourier series, $\omega_0 = \dfrac{2\pi}{T_0}$
Cosine Representation	$\displaystyle\sum_{m=0}^{\infty} A_m \cos(m\omega_0 t + \theta_m)$
Trigonometric Representation	$\displaystyle a_0 + \sum_{m=1}^{\infty} a_m \cos(m\omega_0 t) + \sum_{m=1}^{\infty} b_m \sin(m\omega_0 t)$
Exponential Representation	$\displaystyle\sum_{m=-\infty}^{\infty} X_m e^{jm\omega_0 t}$

7.2.2 Cosine Representation

Of the three representations of the Fourier series in Table 7.2.1, we begin with the cosine Fourier series.

Definition 7.2.1: Let $x(t)$, $-\infty < t < \infty$, be a periodic signal with period T_0. If there exists a convergent series of the form

$$x(t) = \sum_{m=0}^{\infty} A_m \cos(m\omega_0 t + \theta_m), \quad \omega_0 = \frac{2\pi}{T_0}, \quad A_m > 0,$$

then this series is called a <u>(cosine) Fourier series</u>. The numbers A_m are called the <u>(one-sided) amplitude coefficients</u> of the Fourier series. The numbers θ_m are called the <u>(one-sided) phase coefficients</u> of the Fourier series.

▲▲

This form is very useful for signal analysis as will be shown below. The main features of this form of the Fourier series are:

1. All terms are cosine functions
2. The amplitude coefficients are all positive, $A_m \geq 0$, and each term contains a phase angle θ_m.
3. The summation index m runs from 0 to ∞.
4. There is no easy formula to calculate this form of the Fourier series directly. It is easily determined from one of the other representations, however.

7.2.3 Trigonometric Representation

The Definition The second representation of the Fourier in Table 7.2.1 is the trigonometric Fourier series. This representation follows from the cosine representation by a simple trigonometric identity. Consider the general term in the Fourier series in Definition 7.2.1. Using a trigonometric identity on this term we obtain

$$A_m \cos(m\omega_0 t + \theta_m) \tag{7.2.1}$$
$$= A_m (\cos\theta_m)\cos(m\omega_0 t) + (-A_m \sin\theta_m)\sin(m\omega_0 t)$$
$$= a_m \cos(m\omega_0 t) + b_m \sin(m\omega_0 t).$$

Substituting (7.2.1) in the cosine Fourier series in Definition 7.2.1 and splitting apart the summation results in the second standard form of the Fourier series.

Definition 7.2.2: Suppose $x(t)$, $-\infty < t < \infty$, is a periodic function with fundamental period T_0. If there exists a convergent series of the form

$$x(t) = a_0 + \sum_{m=1}^{\infty} a_m \cos(m\omega_0 t) + \sum_{m=1}^{\infty} b_m \sin(m\omega_0 t), \quad \omega_0 = \frac{2\pi}{T_0}$$

then this series is called a <u>trigonometric Fourier series</u>.

▲▲

The main features of this form of the Fourier series are:

1. The series contains both sine and cosine terms.
2. The amplitude coefficients can be both positive and negative, and there are no phase angles associated with these terms.
3. The summation index m runs from 0 to ∞.
4. There exist simple formulas for computing the amplitude coefficients which are given in the next section.

Relationship to Cosine Representation Next we will derive explicit formulas between the coefficients of the cosine and trigonometric Fourier series. Equating coefficients in (7.2.1) we obtain

$$a_0 = A_0 \cos\theta_0,$$ (7.2.2)
$$a_m = A_m \cos\theta_m,$$
$$b_m = -A_m \sin\theta_m.$$

Given the cosine representation, these formulas give the coefficients for the trigonometric Fourier series.

To translate the trigonometric Fourier series back into a cosine Fourier series, we square both sides of the last two rows of (7.2.2) and add them together to get

$$a_m^2 + b_m^2 = A_m^2\left(\cos^2\theta_m + \sin^2\theta_m\right) = A_m^2.$$ (7.2.3)

Thus (7.2.3) yields

$$A_m = +\sqrt{a_m^2 + b_m^2}.$$ (7.2.4)

Taking the ratio of the last two lines of (7.2.2) we have

$$\frac{a_m}{b_m} = -\frac{A_m \sin\theta_m}{A_m \cos\theta_m} = -\tan\theta_m.$$ (7.2.5)

From (7.2.5) we obtain

$$\theta_m = -\tan^{-1}\!\left(\frac{b_m}{a_m}\right). \tag{7.2.6}$$

Equations (7.2.2) and (7.2.4) and (7.2.6) allow for easy conversion between the trigonometric and cosine form of the Fourier series. These results are summarized in Table 7.6.1 in the chapter summary section.

Example 7.2.3: This example illustrates how to translate a cosine series into a trigonometric series. The cosine Fourier series in (7.1.18) is

$$x(t) = \frac{1}{2} + \sum_{\substack{m=1 \\ m \text{ odd}}}^{\infty} \frac{2}{m\pi}\cos(m\pi t + \theta_m) \tag{7.2.7}$$

where

$$\theta_m = \begin{cases} -\pi, & m = 3,7,11,\ldots \\ 0, & m = 1,5,9,\ldots \end{cases}$$

To translate this series into the trigonometric form, we use (7.2.2). The constant term is

$$a_0 = A_0 \cos\theta_0 = \frac{1}{2}. \tag{7.2.8}$$

The coefficients for the cosine terms are

$$a_m = A_m \cos\theta_m = \begin{cases} \dfrac{2}{m\pi}\cos(-\pi), & m = 3,7,11,\ldots \\ \dfrac{2}{m\pi}\cos 0, & \text{otherwise} \end{cases} \tag{7.2.9}$$

$$= (-1)^{\frac{m-1}{2}}\frac{2}{m\pi}, \quad m \text{ odd}$$

The coefficients for the sine terms are

$$b_m = -A_m \sin\theta_m = \begin{cases} \dfrac{2}{m\pi}\sin(-\pi), & m = 3,7,11,\ldots \\ \dfrac{2}{m\pi}\sin 0, & \text{otherwise} \end{cases} \tag{7.2.10}$$

$$= 0$$

Substituting (7.2.8) - (7.2.10) into the Fourier series (7.2.7) we recover the Fourier series in (7.1.16)

$$x(t) = \frac{1}{2} + \sum_{\substack{m=1 \\ m \text{ odd}}}^{\infty} \left((-1)^{\frac{m-1}{2}} \right) \frac{2}{m\pi} \cos(m\pi t). \tag{7.2.11}$$

▲▲

7.2.4 Exponential Representation

The Definition   There is a third form of the Fourier series in Table 7.2.1, the exponential representation. This form of the Fourier series is obtained by expanding the general term of the cosine Fourier series in Definition 7.2.1 using Euler's identity. We have

$$A_m \cos(m\omega_0 t + \theta_m) = A_m \left(\frac{e^{j(m\omega_0 t + \theta_m)} + e^{-j(m\omega_0 t + \theta_m)}}{2} \right). \tag{7.2.12}$$

Substituting (7.2.12) into the definition of the cosine Fourier series and regrouping terms yields

$$x(t) = A_0 + \sum_{m=1}^{\infty} \left(\frac{A_m}{2} e^{j\theta_m} e^{jm\omega_0 t} + \frac{A_m}{2} e^{-j\theta_m} e^{-jm\omega_0 t} \right) \tag{7.2.13}$$

$$= A_0 + \sum_{m=1}^{\infty} \left(\frac{A_m}{2} e^{j\theta_m} \right) e^{jm\omega_0 t} + \sum_{m=1}^{\infty} \left(\frac{A_m}{2} e^{j(-\theta_m)} \right) e^{j(-m)\omega_0 t}.$$

In the second summation of (7.2.13) we replace the negative sign on the summation index m by a negative index

$$x(t) = A_0 + \sum_{m=1}^{\infty} \left(\frac{A_m}{2} e^{j\theta_m} \right) e^{jm\omega_0 t} + \sum_{k=-1}^{-\infty} \left(\frac{A_k}{2} e^{j\theta_k} \right) e^{jk\omega_0 t}. \tag{7.2.14}$$

Comparing the second summation in (7.2.14) to the second summation in (7.2.13) we see that

$$A_m = A_k, \quad (-\theta_m)_{m>0} = (\theta_k)_{k<0}. \tag{7.2.15}$$

Now define

$$X_0 = A_0, \quad X_m = \frac{A_m}{2} e^{j\theta_m}, \quad m > 0. \tag{7.2.16}$$

Note that (7.2.15) implies that

$$X_m = \overline{X}_{-m}. \tag{7.2.17}$$

Using (7.2.16), the two summations in (7.2.14) can be combined into a single sum

$$x(t) = A_0 + \sum_{m=1}^{\infty} \left(\frac{A_m}{2} e^{j\theta_m} \right) e^{jm\omega_0 t} + \sum_{m=-1}^{-\infty} \left(\frac{A_m}{2} e^{j\theta_m} \right) e^{jm\omega_0 t} \qquad (7.2.18)$$

$$= \sum_{m=-\infty}^{\infty} X_m e^{jm\omega_0 t}.$$

Now we can rewrite the Fourier series in Definition 7.2.1.

Definition 7.2.4: Suppose $x(t)$, $-\infty < t < \infty$, is a periodic function with fundamental period T_0. If there exists a convergent series of the form

$$x(t) = \sum_{m=-\infty}^{\infty} X_m e^{jm\omega_0 t}, \quad \omega_0 = \frac{2\pi}{T_0}$$

this series is called the <u>exponential</u> Fourier series. The numbers $|X_m|$ are called the <u>(two-sided) amplitude coefficients</u> of the Fourier series.   The numbers $\angle X_m$ are called the <u>(two-sided) phase coefficients</u> of the Fourier series.

▲▲

Relationship to Cosine and Trigonometric Representation   There are simple relationships between the coefficients of the exponential Fourier series and the other two representations of the Fourier series in Table 7.2.1.   The relationship to the cosine Fourier series is derived above.   To derive the trigonometric Fourier series from the exponential Fourier series, we use Euler's formula to expand the relationship in (7.2.16) to get

$$X_m = \frac{A_m}{2}(\cos\theta_m + j\sin\theta_m) = \frac{A_m}{2}\cos\theta_m - j\left(-\frac{A_m}{2}\sin\theta_m\right) \qquad (7.2.19)$$

$$= \frac{1}{2}(a_m - jb_m).$$

From (7.2.19) we easily obtain the desired relationships. The relationships between the coefficients of all of the representations of the Fourier series are given in Table 7.6.1.

Example 7.2.5: In this example we show how to compute the exponential Fourier series from the cosine Fourier series in Example 7.2.3. To find the exponential Fourier series of this signal substitute the coefficients in (7.2.7) into (7.2.16). We have

$$X_0 = A_0 = \frac{1}{2}, \quad m = 0. \tag{7.2.20}$$

For $m > 0$ (7.2.16) yields

$$X_m = \frac{A_m}{2} e^{j\theta_m} = \begin{cases} \dfrac{1}{2} \dfrac{2}{m\pi} e^{j(-\pi)}, & m = 3, 7, 11, \ldots \\ \dfrac{1}{2} \dfrac{2}{m\pi} e^{j0}, & \text{otherwise} \end{cases} \tag{7.2.21}$$

$$= \frac{1}{m\pi} (-1)^{\frac{m-1}{2}}, \quad m \text{ odd}$$

Finally, inserting (7.2.20) and (7.2.21) into Definition 7.2.4 we have

$$x(t) = \frac{1}{2} + \sum_{\substack{m=-\infty \\ m\neq0 \\ m \text{ odd}}}^{\infty} (-1)^{\frac{m-1}{2}} \frac{1}{m\pi} e^{jm\pi t} \tag{7.2.22}$$

which is the exponential Fourier series of this square wave.

▲▲

7.2.5 Existence Theorem

We have shown that a periodic signal $x(t)$ can be expressed as an infinite sum of sinusoids. Before the Fourier series can be used, three important questions need to be answered.

1. How do we compute the coefficients of the Fourier series?
   If for a given signal we can't find the Fourier series, it would be of little use. Therefore, we will present the computational formulas in the next section for finding the coefficients of the series.

2. Does the infinite series converge?
   In general, a computational algorithm will usually give an answer. But we must guarantee that the answer makes sense if we are to use it for analysis or design. The computational formulas for the coefficients of the Fourier series will produce numbers. We must then make sure that the resulting Fourier series converges.

3. Is the representation unique?
   It could be that a signal has more than one Fourier series. In that case we would draw different conclusions depending on which Fourier series we used. Again, the Fourier series would be of little practical value if it is not unique.

The answer to all three questions is in the affirmative. The exact conditions for the existence of a Fourier series are given in the following theorem.

Theorem 7.2.6: Let $x(t)$, $-\infty < t < \infty$, be a periodic signal with period T_0. Assume that:

(a) $x(t)$ is absolutely integrable over one period, $\int_{T_0} |x(t)|\, dt < \infty$

(b) the signal $x(t)$ has only a finite number of minima and maxima over any period, and

(c) the signal $x(t)$ has only a finite number of discontinuities over any period.

Then the Fourier series representation

$$x(t) = \sum_{m=0}^{\infty} A_m \cos(m\omega_0 t + \theta_m), \quad \omega_0 = \frac{2\pi}{T_0}$$

exists and is unique.   ▲▲

We have the following observations about Theorem 7.2.6.

1. Uniqueness in Theorem 7.2.6 means that if $x_1(t) = x_2(t)$ for all but a finite number of points, then they have the same Fourier series.
2. At jump discontinuities of $x(t)$ the Fourier series converges to the average of the right- and left-hand limits. This behavior can be observed in Figure 7.1.4.
3. Roughly, Theorem 7.2.6 says that the Fourier series exists for any reasonable signal encountered in engineering practice.

7.3 COMPUTATIONAL FORMULAS FOR THE FOURIER SERIES COEFFICIENTS

7.3.1 Introduction

In this section we will present the computational formulas for the coefficients of the trigonometric and exponential Fourier series. For some signals the trigonometric formulas are preferred while for other signals the exponential series formulas are easier to use. So both sets of formulas are presented below.

It will become clear in Chapter 8 that the analysis of the characteristics of a signal may require a different series representation than the representation used for computation of the coefficients of the series. In this case the formulas of the last section are used to translate between the various series representations. In this text we will emphasize the interpretation of the Fourier series rather than the computation of its coefficients.

7.3.2 Trigonometric Representation

We begin with the trigonometric representation of the periodic signal $x(t)$ with period T_0. From Definition 7.2.2 we recall that the expression for the trigonometric Fourier series is

$$x(t) = a_0 + \sum_{k=1}^{\infty} a_k \cos(k\omega_0 t) + \sum_{k=1}^{\infty} b_k \sin(k\omega_0 t), \quad \omega_0 = \frac{2\pi}{T_0}. \tag{7.3.1}$$

Since ω_0 is calculated from the period of $x(t)$, all that is not known in (7.3.1) are the coefficients a_k and b_k.

The computation of the coefficients of Fourier series relies on the following integrals which can be found in any calculus text.

$$I_1 = \int_{T_0} \sin(m\omega_0 t) \sin(k\omega_0 t)\, dt = \begin{cases} 0, & m \neq k \\ \dfrac{T_0}{2}, & m = k \neq 0 \end{cases} \tag{7.3.2a}$$

$$I_2 = \int_{T_0} \cos(m\omega_0 t) \cos(k\omega_0 t)\, dt = \begin{cases} 0, & m \neq k \\ \dfrac{T_0}{2}, & m = k \neq 0 \end{cases} \tag{7.3.2b}$$

$$I_3 = \int_{T_0} \sin(m\omega_0 t) \cos(k\omega_0 t)\, dt = 0, \quad \text{for all } m, k \tag{7.3.2c}$$

Notation: The notation $\int_{T_0} dt$ means the integral can be evaluated over any interval of length T_0.

These formulas rely on the fact that the period of the sinusoids in the integrand is exactly an integer multiple of the interval of integration.

Formula for a_m To find the Fourier coefficients a_k, multiply each side of (7.3.1) with $\cos(m\omega_0 t)$ and integrate over one period. The resulting expression is

$$\int_{T_0} \left[x(t) = a_0 + \sum_{k=1}^{\infty} a_k \cos(k\omega_0 t) + \sum_{k=1}^{\infty} b_k \sin(k\omega_0 t) \right] \cos(m\omega_0 t)\, dt. \tag{7.3.3}$$

Taking the integral inside the summation we have

$$\int_{T_0} x(t)\cos(m\omega_0 t)\,dt = \int_{T_0} a_0 \cos(m\omega_0 t)\,dt \qquad (7.3.4)$$

$$+ \sum_{k=1}^{\infty} a_k \int_{T_0} \cos(k\omega_0 t)\cos(m\omega_0 t)\,dt + \sum_{k=1}^{\infty} b_k \int_{T_0} \sin(k\omega_0 t)\cos(m\omega_0 t)\,dt.$$

Next we consider the integrals in (7.3.4) for each m.

$m = 0$ For $m = 0$, (7.3.4) becomes

$$\int_{T_0} x(t)\,dt = \int_{T_0} a_0 \,dt + \sum_{k=1}^{\infty} a_k \int_{T_0} \cos(k\omega_0 t)\,dt + \sum_{k=1}^{\infty} b_k \int_{T_0} \sin(k\omega_0 t)\,dt. \qquad (7.3.5)$$

Each of the integrals in the summation signs in (7.3.5) is over an integral multiple of the period of the sine or cosine. Hence, they are all zero. We are left with

$$\int_{T_0} x(t)\,dt = \int_{T_0} a_0 \,dt = a_0 T_0. \qquad (7.3.6)$$

Rewriting (7.3.6) we obtain an expression for the constant coefficient

$$a_0 = \frac{1}{T_0} \int_{T_0} x(t)\,dt. \qquad (7.3.7)$$

The coefficient a_0 is the average value of the signal over one period. This coefficient can often be determined by inspection.

$m > 0$ For $m > 0$ the first integral on the right-hand side of (7.3.4) is zero because we are integrating over a integer multiple of periods. Next note that all of the integrals in the last summation are zero because of (7.3.2c). Finally, all the integrals in the middle summation drop out except $m = k$ because of (7.3.2b). We are left with

$$\int_{T_0} x(t)\cos(m\omega_0 t)\,dt = a_m \frac{T_0}{2}. \qquad (7.3.8)$$

Re-arranging (7.3.8) we arrive at the formula

$$a_m = \frac{2}{T_0} \int_{T_0} x(t)\cos(m\omega_0 t)\,dt. \qquad (7.3.9)$$

Formula for the b_m The derivation of the formulas for the Fourier coefficients b_m follows along the lines for the a_m. Multiplying both sides of (7.3.1) by $\sin(m\omega_0 t)$ we obtain

$$\int_{T_0} x(t)\sin(m\omega_0 t)\,dt = \int_{T_0} a_0 \sin(m\omega_0 t)\,dt \tag{7.3.10}$$

$$+ \sum_{k=1}^{\infty} a_k \int_{T_0} \cos(k\omega_0 t)\sin(m\omega_0 t)\,dt + \sum_{k=1}^{\infty} b_k \int_{T_0} \sin(k\omega_0 t)\sin(m\omega_0 t)\,dt.$$

This time the only nonzero integral is when $m = k$ in the last summation of (7.3.10). (Use (7.3.2b).) Thus, we have

$$b_m = \frac{2}{T_0}\int_{T_0} x(t)\sin(m\omega_0 t)\,dt. \tag{7.3.11}$$

From (7.3.7), (7.3.9), and (7.3.11) all of the coefficients for the trigonometric Fourier series can be computed. These results are summarized in Remark 7.3.1.

Remark 7.3.1: Computation of the Trigonometric Fourier Series   Given the trigonometric Fourier series

$$x(t) = a_0 + \sum_{m=1}^{\infty} a_m \cos(m\omega_0 t) + \sum_{m=1}^{\infty} b_m \sin(m\omega_0 t) \tag{7.3.1}$$

the coefficients can be computed with the following formulas.

$$a_0 = \frac{1}{T_0}\int_{T_0} x(t)\,dt \tag{7.3.7}$$

$$a_m = \frac{2}{T_0}\int_{T_0} x(t)\cos(m\omega_0 t)\,dt \tag{7.3.9}$$

$$b_m = \frac{2}{T_0}\int_{T_0} x(t)\sin(m\omega_0 t)\,dt \tag{7.3.11}$$

▲▲

Example 7.3.2: This example illustrates the use of the formulas in Remark 7.3.1. Consider the signal

$$x_1(t) = \begin{cases} A, & 0 \le t < \varepsilon \\ 0, & \varepsilon \le t < T_0 \end{cases} \tag{7.3.12}$$

Define a periodic signal by periodic extension of (7.3.12). This signal is shown in Figure 7.3.1. To find the trigonometric Fourier series for this signal, first use (7.3.7) to find the constant term. We have

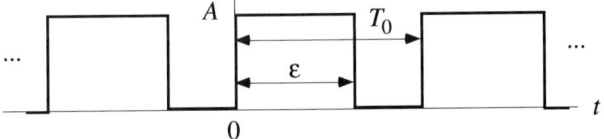

Figure 7.3.1 Signal of Example 7.3.2

$$a_0 = \frac{1}{T_0} \int_0^{T_0} x(t)\, dt = \frac{1}{T_0} \int_0^{\varepsilon} A\, dt + \frac{1}{T_0} \int_\varepsilon^{T_0} 0\, dt = \frac{A\varepsilon}{T_0}. \tag{7.3.13}$$

Clearly, the constant term is the average of the signal over one period. Next the coefficients of the cosine terms are computed from (7.3.9). For the signal in (7.3.12) we have

$$a_m = \frac{2}{T_0} \int_{T_0} x(t)\cos(m\omega_0 t)\, dt = \frac{2}{T_0}\left[\int_0^{\varepsilon} A\cos(m\omega_0 t)\, dt \right] \tag{7.3.14}$$

$$= \frac{2A}{T_0 m\omega_0}\left[\sin(m\omega_0 t)\big|_{t=0}^{\varepsilon} \right] = \frac{2A}{T_0 m\omega_0}\sin(m\omega_0\varepsilon).$$

The coefficients of the sine terms in the Fourier series (7.3.1) are

$$b_m = \frac{2}{T_0} \int_{T_0} x(t)\sin(m\omega_0 t)\, dt = \frac{2}{T_0}\left[\int_0^{\varepsilon} A\sin(m\omega_0 t)\, dt \right] \tag{7.3.15}$$

$$= \frac{2A}{T_0 m\omega_0}\left[-\cos(m\omega_0 t)\big|_{t=0}^{\varepsilon} \right] = \frac{2A}{T_0 m\omega_0}\left[1 - \cos(m\omega_0\varepsilon) \right].$$

These formulas can be simplified by using

$$m\omega_0 T_0 = m\left(\frac{2\pi}{T_0} \right)T_0 = m2\pi. \tag{7.3.16}$$

The Fourier series of this signal is

$$x(t) = \frac{A\varepsilon}{T_0} + \sum_{m=1}^{\infty}\left[\frac{A}{m\pi}\sin(m\omega_0\varepsilon) \right]\cos(m\omega_o t) \tag{7.3.17}$$

$$+ \sum_{m=1}^{\infty}\left[\frac{A}{m\pi}(1 - \cos(m\omega_0\varepsilon)) \right]\sin(m\omega_o t).$$

▲▲

7.3.3 Exponential Representation

In this subsection we will develop formulas for the calculation of the coefficients of the exponential Fourier series

$$x(t) = \sum_{m=-\infty}^{\infty} X_m e^{jm\omega_0 t}, \quad \omega_0 = \frac{2\pi}{T_0}. \tag{7.3.18}$$

The derivation of this formula parallels the development of the formulas for the trigonometric representation in the last subsection. The derivation relies on the following integral which is easily verified by direct calculation:

$$\int_{T_0} e^{jk\omega_0 t} e^{-jm\omega_0 t} \, dt = \begin{cases} 0, & k \neq m \\ T_0, & k = m \end{cases} \tag{7.3.19}$$

To derive the formulas for the X_m first, multiply (7.3.18) by $e^{-jm\omega_0 t}$ and integrate over one period. We have

$$\int_{T_0} x(t) e^{-jm\omega_0 t} \, dt = \int_{T_0} \left[\sum_{k=-\infty}^{\infty} X_k e^{jk\omega_0 t} \right] e^{-jm\omega_0 t} \, dt \tag{7.3.20}$$

$$= \sum_{k=-\infty}^{\infty} X_k \int_{T_0} e^{jk\omega_0 t} e^{-jm\omega_0 t} \, dt = X_m T_0$$

where the last line follows from (7.3.19). From (7.3.20) we obtain the following formula for the coefficients of (7.3.18).

Remark 7.3.3: **Computation of the Exponential Fourier Series**   Given the exponential Fourier series

$$x(t) = \sum_{m=-\infty}^{\infty} X_m e^{jm\omega_0 t}, \quad \omega_0 = \frac{2\pi}{T_0}. \tag{7.3.18}$$

the coefficients of this series can be computed from

$$X_m = \frac{1}{T_0} \int_{T_0} x(t) e^{-jm\omega_0 t} \, dt. \tag{7.3.21}$$

▲▲

Example 7.3.4: This example illustrates the use of the formulas for the coefficients of the exponential Fourier series representation in Remark 7.3.3. Consider the signal

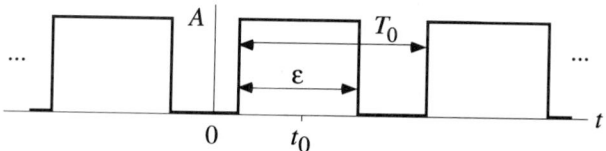

Figure 7.3.2 Generic Square Wave

$$x(t) = \sum_{k=-\infty}^{\infty} A\Pi\left(\frac{t-t_0-kT_0}{\varepsilon}\right) \tag{7.3.22}$$

which is shown in Figure 7.3.2. To find the exponential Fourier series we use (7.3.21). For $m = 0$ we have

$$X_0 = \frac{1}{T_0}\int_{T_0} x(t)\,dt = \frac{1}{T_0}\int_{t_0-\frac{\varepsilon}{2}}^{t_0+\frac{\varepsilon}{2}} A\,dt = \frac{A\varepsilon}{T_0}. \tag{7.3.23}$$

For other values of m we have

$$X_m = \frac{1}{T_0}\int_0^{T_0} x(t)e^{-jm\omega_0 t}\,dt = \frac{1}{T_0}\int_{t_0-\frac{\varepsilon}{2}}^{t_0+\frac{\varepsilon}{2}} Ae^{-jm\omega_0 t}\,dt \tag{7.3.24}$$

$$= \frac{A(-1)}{jmT_0\omega_0}\left[e^{-jm\omega_0 t}\right]_{t=t_0-\frac{\varepsilon}{2}}^{t_0+\frac{\varepsilon}{2}} = \frac{2A(-1)e^{-jm\omega_0 t_0}}{mT_0\omega_0}\left(\frac{e^{-jm\omega_0\frac{\varepsilon}{2}}-e^{jm\omega_0\frac{\varepsilon}{2}}}{2j}\right).$$

Recognizing the last factor of (7.3.24) as Euler's formula for a sine function, we rewrite (7.3.24) as

$$X_m = \frac{2Ae^{-jm\omega_0 t_0}}{m2\pi}\sin\left(\frac{m\omega_0\varepsilon}{2}\right) = \frac{A}{m\pi}e^{-jm\omega_0 t_0}\sin\left(\frac{m\omega_0\varepsilon}{2}\right). \tag{7.3.25}$$

The coefficients in (7.3.25) can also be expressed in terms of an Sa function. We have

$$X_m = \frac{A\varepsilon}{T_0}e^{-jm\omega_0 t_0}\frac{\sin\left(\frac{m\omega_0\varepsilon}{2}\right)}{m\varepsilon\left(\frac{2\pi}{T_0}\right)} = \frac{A\varepsilon}{T_0}e^{-jm\omega_0 t_0}\,\mathrm{Sa}\left(\frac{m\omega_0\varepsilon}{2}\right). \tag{7.3.26}$$

The exponential Fourier series for the signal in Figure 7.3.2 is

$$x(t) = \sum_{m=-\infty}^{\infty} \left[\frac{A\varepsilon}{T_0} e^{-jm\omega_0 t_0} \, \text{Sa}\left(\frac{m\omega_0\varepsilon}{2} \right) \right] e^{jm\omega_0 t}. \tag{7.3.27}$$

A MATLAB M-file to simulate this Fourier series is given in Section 8.8.

▲▲

Example 7.3.5: By inspection of Figure 7.3.2 we see that the signal shown there is the signal in Figure 7.1.1 if we take

$$A = 1, T_0 = 2, t_0 = 0, \text{ and } \varepsilon = 1. \tag{7.3.28}$$

Using the parameters in (7.3.28), the exponential Fourier series becomes

$$x(t) = \sum_{m=-\infty}^{\infty} \left[\frac{A}{m\pi} e^{-jm\omega_0 t_0} \sin\left(\frac{m\omega_0\varepsilon}{2} \right) \right] e^{jm\omega_0 t} \tag{7.3.29}$$

$$= \sum_{m=-\infty}^{\infty} \left[\frac{(1)}{m\pi} e^{-jm\omega_0 (0)} \sin\left(\frac{m\omega_0}{2} \frac{T_0}{2} \right) \right] e^{jm\omega_0 t}$$

$$= \sum_{m=-\infty}^{\infty} \left[\frac{1}{m\pi} \sin\left(\frac{m\pi}{2} \right) \right] e^{jm\pi t} = \frac{1}{2} + \sum_{\substack{m=-\infty \\ m\neq 0 \\ m \text{ odd}}}^{\infty} (-1)^{\frac{m-1}{2}} \frac{1}{m\pi} e^{jm\pi t}.$$

This result agrees with Example 7.2.5.

▲▲

7.3.4 Use of Symmetry

Sometimes it is possible to use the symmetry of a signal to reduce the computations of the Fourier coefficients. Let $x(t)$ be a periodic signal created by periodic extension of the signal

$$x_1(t), \quad -\frac{T_0}{2} \leq t < \frac{T_0}{2}. \tag{7.3.30}$$

If $x_1(t)$ is an odd function then the coefficients of the trigonometric series satisfy

$$a_m = \frac{2}{T_0} \int_{-\frac{T_0}{2}}^{\frac{T_0}{2}} x_1(t) \cos(m\omega_0 t)\, dt = 0, \quad m \geq 0. \tag{7.3.31}$$

If $x_1(t)$ is an even function then the coefficients of the trigonometric series satisfy

$$b_m = \frac{2}{T_0} \int_{-\frac{T_0}{2}}^{\frac{T_0}{2}} x_1(t) \sin(m\omega_0 t)\, dt = 0, \quad m \geq 1. \tag{7.3.32}$$

For example, recall that the signal in Figure 7.1.1 is an even signal. Hence, the b_m coefficients of the Fourier series are zero. The following example also illustrates the use of symmetry in the calculation of the Fourier coefficients.

Example 7.3.6: Consider the signal

$$x_c(t) = \begin{cases} A\cos(\omega_c t), & -\frac{\varepsilon}{2} < t \leq \frac{\varepsilon}{2} \\ 0, & \frac{\varepsilon}{2} < t \leq T_0 - \frac{\varepsilon}{2} \end{cases} \tag{7.3.33}$$

Let the signal $x_{pc}(t)$ be defined from (7.3.33) by periodic extension. This signal is shown in Figure 7.3.3. We will derive the Fourier series using the trigonometric formulas in Remark 7.3.1. Observe from Figure 7.3.3 that this signal is even. We immediately conclude that the coefficients $b_m = 0$. It remains to calculate the coefficients a_m.

The constant terms given by

$$a_0 = \frac{1}{T_0} \int_{-\frac{T_0}{2}}^{\frac{T_0}{2}} x_c(t)\, dt = \frac{1}{T_0} \int_{-\frac{\varepsilon}{2}}^{\frac{\varepsilon}{2}} A\cos(\omega_c t)\, dt. \tag{7.3.34}$$

Next observe that the integrand of the second integral is an even function over the interval of integration as shown in Figure 7.3.3. Hence, by Remark 2.1.5 this integral is given by

$$a_0 = \frac{1}{T_0} \int_{-\frac{\varepsilon}{2}}^{\frac{\varepsilon}{2}} A\cos(\omega_c t)\, dt = \frac{2}{T_0} \int_0^{\frac{\varepsilon}{2}} A\cos(\omega_c t)\, dt \tag{7.3.35}$$

$$= \frac{2A}{T_0 \omega_c} \left[\sin(\omega_c t) \right]_{t=0}^{\frac{\varepsilon}{2}} = \frac{2A}{T_0 \omega_c} \sin\left(\frac{\omega_c \varepsilon}{2} \right).$$

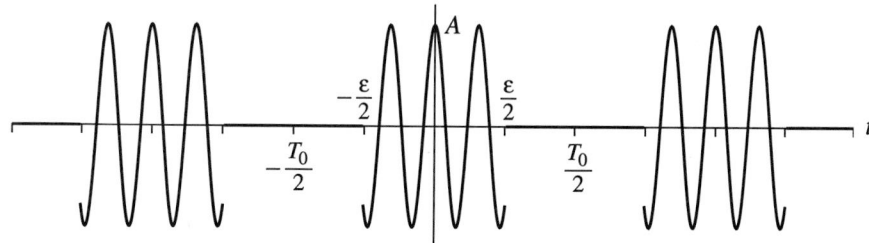

Figure 7.3.3 Signal of Example 7.3.6

When $\omega_c \ne m\omega_0$, the coefficients a_m are given by

$$a_m = \frac{2}{T_0} \int_{T_0} x_c(t) \cos(m\omega_0 t)\, dt = \frac{2}{T_0} \int_{-\frac{\varepsilon}{2}}^{\frac{\varepsilon}{2}} A \cos(\omega_c t) \cos(m\omega_0 t)\, dt. \tag{7.3.36}$$

Again we make use of the fact that the integrand of (7.3.36) is even. Employing a table of integrals we obtain

$$a_m = \frac{2}{T_0} \int_{-\frac{\varepsilon}{2}}^{\frac{\varepsilon}{2}} A \cos(\omega_c t) \cos(m\omega_0 t)\, dt \tag{7.3.37}$$

$$= \frac{4A}{T_0} \left[\frac{\sin(\omega_c - m\omega_0)t}{2(\omega_c - m\omega_0)} + \frac{\sin(\omega_c + m\omega_0)t}{2(\omega_c + m\omega_0)} \right]_{t=0}^{\frac{\varepsilon}{2}}$$

$$= \frac{A\varepsilon}{T_0} \left[\mathrm{Sa}\!\left(\frac{(\omega_c - m\omega_0)\varepsilon}{2} \right) + \mathrm{Sa}\!\left(\frac{(\omega_c + m\omega_0)\varepsilon}{2} \right) \right]$$

Therefore, the trigonometric Fourier series representation of the signal (7.3.33) is

$$x_{pc}(t) = \frac{2A}{\omega_c T_0} \sin\!\left(\frac{\omega_c \varepsilon}{2} \right) \tag{7.3.38}$$

$$+ \frac{A\varepsilon}{T_0} \sum_{m=1}^{\infty} \left[\mathrm{Sa}\!\left(\frac{(\omega_c - m\omega_0)\varepsilon}{2} \right) + \mathrm{Sa}\!\left(\frac{(\omega_c + m\omega_0)\varepsilon}{2} \right) \right] \cos(m\omega_0 t).$$

The exponential representation of this signal is easily derived from the trigonometric representation (7.3.38) using (7.2.19). The complex coefficients of the exponential Fourier series are given by

$$X_m = \frac{a_m - jb_m}{2} = \frac{a_m}{2} = \frac{A\varepsilon}{2T_0} \left[\mathrm{Sa}\!\left(\frac{(\omega_c - m\omega_0)\varepsilon}{2} \right) + \mathrm{Sa}\!\left(\frac{(\omega_c + m\omega_0)\varepsilon}{2} \right) \right], \tag{7.3.39}$$

$$X_0 = a_0 = \frac{2A}{T_0 \omega_c} \sin\!\left(\frac{\omega_c \varepsilon}{2} \right),$$

where we have used (7.3.35), (7.3.37), and the fact that $b_m = 0$ for this series. Now the exponential form of the Fourier series of the signal in (7.3.33) is

$$x_{pc}(t) = \frac{2A}{\omega_c T_0} \sin\!\left(\frac{\omega_c \varepsilon}{2} \right) \tag{7.3.40}$$

$$+ \frac{A\varepsilon}{2T_0} \sum_{\substack{m=-\infty \\ m \ne 0}}^{\infty} \left[\mathrm{Sa}\!\left(\frac{(\omega_c - m\omega_0)\varepsilon}{2} \right) + \mathrm{Sa}\!\left(\frac{(\omega_c + m\omega_0)\varepsilon}{2} \right) \right] e^{jm\omega_0 t}.$$

A MATLAB M-file to simulate this Fourier series is given in Section 8.8.

▲▲

7.3.5 Summary

In the last three sections we have shown that *any* periodic signal can be represented as a sum of sinusoids. This idea is quite powerful, as we will show in later chapters. It is also clear from the examples above that the analytical expressions for coefficients of the Fourier series can be quite complicated. It would be useful if we had a simpler method for displaying these coefficients of a Fourier series. In Chapter 8 we will develop a graphical method for representing the Fourier series. We will show how this graphical representation of the Fourier series lends insight into the structure of the signal. Then in Chapter 15 we will show how the Fourier series is used to analyze the propagation of a signal through a system.

7.4 DEFINITION OF THE FOURIER TRANSFORM

7.4.1 Introduction

The idea that a periodic signal can be represented as a sum of cosines with Fourier series is very powerful. We would like to extend this notion to aperiodic signals. To begin this section we present a heuristic description of such an extension of Fourier series to aperiodic signals. This discussion leads from Fourier series to the Fourier transform, which is defined in this section. While these arguments could be made mathematically rigorous, the intention is to give a description which lends insight into the engineering interpretations of the Fourier series and Fourier transform.

The extension of the Fourier series to aperiodic signals can be motivated by generating an aperiodic signal from a periodic signal by extending the period to infinity. Consider the signal

$$x_1(t) = \begin{cases} 0, & -\dfrac{T_0}{2} \le t < -\dfrac{\varepsilon}{2} \\[2mm] 1, & -\dfrac{\varepsilon}{2} \le t < \dfrac{\varepsilon}{2} \\[2mm] 0, & \dfrac{\varepsilon}{2} \le t < \dfrac{T_0}{2} \end{cases} \tag{7.4.1}$$

which was introduced in Section 7.1. This signal is shown in Figure 7.4.1a and its periodic extension $x_{T_0}(t)$ is shown in Figure 7.4.1b. If we let $T_0 \to \infty$ then in the limit the periodic signal will be reduced to the aperiodic signal shown in Figure 7.4.1c, i.e.,

$$x(t) = \lim_{T_0 \to \infty} x_{T_0}(t). \tag{7.4.2}$$

The periodic signal $x_{T_0}(t)$ has an exponential Fourier series that is given by

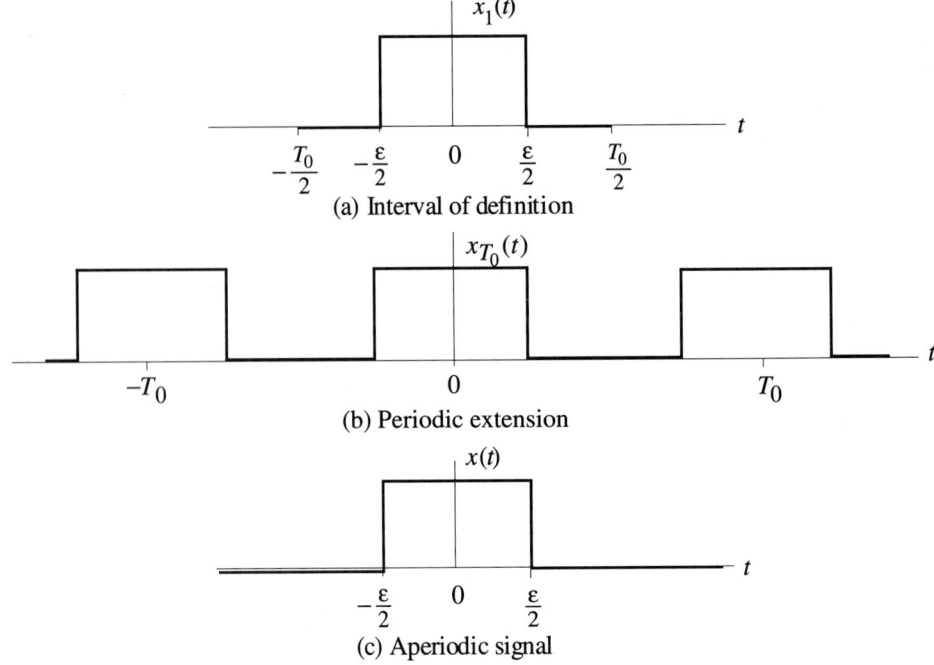

Figure 7.4.1 Aperiodic Signal as Limit of a Periodic Signal

$$x_{T_0}(t) = \sum_{m=-\infty}^{\infty} X_m e^{jm\omega_0 t}, \quad \omega_0 = \frac{2\pi}{T_0}$$

(7.4.3)

where

$$X_m = \frac{1}{T_0} \int_{-\frac{T_0}{2}}^{\frac{T_0}{2}} x_{T_0}(t) e^{-jm\omega_0 t}\, dt.$$

(7.4.4)

As far as the integration is concerned in (7.4.4), the integrand on this integral can be rewritten as

$$X_m = \frac{1}{T_0} \int_{-\frac{T_0}{2}}^{\frac{T_0}{2}} x_{T_0}(t) e^{-jm\omega_0 t}\, dt = \frac{1}{T_0} \int_{-\infty}^{\infty} x(t) e^{-jm\omega_0 t}\, dt.$$

(7.4.5)

Note the limits on the integrals. The second equality holds because on the interval $-\frac{T_0}{2} \le t \le \frac{T_0}{2}$ the signal $x_{T_0}(t)$ and $x(t)$ agree and outside this interval $x(t)$ is zero. Of course, outside of this interval the signals are different.

What we would like to do is generalize the formula for the Fourier series coefficient in (7.4.5). This generalization is accomplished by making the integral a function of frequency. In view of (7.4.5) we define

$$X(\omega) = \int_{-\infty}^{\infty} x(t) e^{-j\omega t} \, dt. \tag{7.4.6}$$

Using (7.4.6) we can rewrite (7.4.5) as

$$X_m = \frac{1}{T_0} X(m\omega_0). \tag{7.4.7}$$

Equation (7.4.7) shows that the function $X(\omega)$ is the envelope of the coefficients X_m of the Fourier series scaled by T_0. Substituting (7.4.7) into (7.4.3) we get

$$x_{T_0}(t) = \sum_{m=-\infty}^{\infty} \frac{X(m\omega_0)}{T_0} e^{jm\omega_0 t}. \tag{7.4.8}$$

Define

$$\Delta\omega = \frac{2\pi}{T_0}. \tag{7.4.9}$$

Using (7.4.9), (7.4.8) can be rewritten as

$$x_{T_0}(t) = \frac{2\pi}{2\pi} \sum_{m=-\infty}^{\infty} \frac{X(m\omega_0)}{T_0} e^{jm\omega_0 t} = \frac{1}{2\pi} \sum_{m=-\infty}^{\infty} X(m\omega_0) e^{jm\omega_0 t} \Delta\omega. \tag{7.4.10}$$

Taking the limit in (7.4.10) and using (7.4.2) we get

$$x(t) = \lim_{T_0 \to \infty} x_{T_0}(t) = \lim_{T_0 \to \infty} \frac{1}{2\pi} \sum_{m=-\infty}^{\infty} X(m\omega_0) e^{jm\omega_0 t} \Delta\omega. \tag{7.4.11}$$

Noting that (7.4.9) is the spacing between the frequencies of the cosines in the Fourier series, in the limit the spacing becomes dense, $m\omega_0 \to \omega$, $\Delta\omega \to d\omega$, and the summation goes over to an integral. From (7.4.11) we have

$$x(t) = \frac{1}{2\pi} \int_{-\infty}^{\infty} X(\omega) e^{j\omega t} \, d\omega. \tag{7.4.12}$$

7.4.2 The Definition

In (7.4.6) we defined $X(\omega)$ and (7.4.12) established $X(\omega)$'s relationship to $x(t)$. Equations (7.4.6) and (7.4.12) represent the generalization of Fourier series to aperiodic signals. We summarize these observations with the following definition.

Definition 7.4.1: Let $x(t)$ be a signal such that:

   (a) $x(t)$, $-\infty < t < \infty$, and

   (b) $\displaystyle\int_{-\infty}^{\infty} |x(t)|\, dt \le M < \infty$,   for $0 < M < \infty$

Then the <u>Fourier transform</u> of $x(t)$ is defined as

$$X(\omega) = \int_{-\infty}^{\infty} x(t)e^{-j\omega t}\, dt = \mathcal{F}\{x(t)\}.$$

The <u>inverse Fourier transform</u> is defined as

$$x(t) = \frac{1}{2\pi}\int_{-\infty}^{\infty} X(\omega)e^{j\omega t}\, d\omega = \mathcal{F}^{-1}\{X(\omega)\}.$$

▲▲

There is a one-to-one relationship between a signal and its Fourier transform.

Definition 7.4.2: The pair

$$\mathcal{F}\{x(t)\} = X(\omega) \leftrightarrow x(t) = \mathcal{F}^{-1}\{X(\omega)\}$$

is called a <u>Fourier transform pair</u>.

▲▲

Notation: Sometimes the integrals in the Fourier transform are expressed using frequency in terms of hertz instead of radian frequency. In that case the defining integral becomes

$$X(f) = \int_{-\infty}^{\infty} x(t)e^{-j2\pi ft}\, dt.$$

(7.4.13)

Using the frequency variable in hertz rescales the frequency axis in the second integral according to

$$d\omega = 2\pi\, df$$

(7.4.14)

so that

$$x(t) = \int_{-\infty}^{\infty} X(f) e^{j2\pi ft} \, df.$$
(7.4.15)

Rescaling the frequency variable in the second integral causes the cancellation of the factor 2π.

▲▲

The following example illustrates the calculation of the Fourier transform.

Example 7.4.3: Consider the pulse

$$x(t) = \Pi\!\left(\frac{t}{\varepsilon}\right).$$
(7.4.16)

The graph of this function is shown in Figure 7.4.1. From the definition of the Fourier transform we have

$$X(\omega) = \int_{-\infty}^{\infty} x(t) e^{-j\omega t} \, dt = \int_{-\frac{\varepsilon}{2}}^{\frac{\varepsilon}{2}} 1 e^{-j\omega t} \, dt = \left[\frac{-e^{-j\omega t}}{j\omega}\right]_{t=-\frac{\varepsilon}{2}}^{\frac{\varepsilon}{2}}$$
(7.4.17)

$$= \frac{2\varepsilon}{\omega\varepsilon}\left(\frac{-e^{\frac{-j\omega\varepsilon}{2}} + e^{\frac{j\omega\varepsilon}{2}}}{2j}\right) = \varepsilon\left[\frac{\sin\!\left(\frac{\omega\varepsilon}{2}\right)}{\frac{\omega\varepsilon}{2}}\right] = \varepsilon\,\mathrm{Sa}\!\left(\frac{\omega\varepsilon}{2}\right).$$

Thus we have the Fourier transform pair

$$\mathcal{F}\!\left\{\Pi\!\left(\frac{t}{\varepsilon}\right)\right\} = \varepsilon\,\mathrm{Sa}\!\left(\frac{\omega\varepsilon}{2}\right) = X(\omega).$$
(7.4.18)

▲▲

Example 7.4.4: Using the result of Example 7.4.3 we can reconstruct the Fourier series of the periodic signal in Figure 7.4.1 following the analysis leading to (7.4.7). Using (7.4.18) we have

$$X_m = \frac{1}{T_0} X(m\omega_0) = \frac{1}{T_0}\left[\varepsilon\,\mathrm{Sa}\!\left(\frac{m\omega_0\varepsilon}{2}\right)\right] = \frac{\varepsilon}{T_0}\,\mathrm{Sa}\!\left(\frac{m\omega_0\varepsilon}{2}\right).$$
(7.4.19)

▲▲

7.4.3 Existence of the Fourier Transform

The definition of the Fourier transform relies on the existence of infinite integrals in Definition 7.4.1. Before we use the Fourier transform we should verify that the infinite integrals in the definition exist for a class of signals. From Definition 7.4.1, the Fourier transform of $x(t)$ is computed from

$$X(\omega) = \int_{-\infty}^{\infty} x(t)e^{-j\omega t} \, dt. \qquad (7.4.20)$$

To show that this integral exists we apply the integral inequality in Theorem 2.4.17. We have

$$\left| \int_{-\infty}^{\infty} x(t)e^{-j\omega t} \, dt \right| \leq \int_{-\infty}^{\infty} \left| x(t)e^{-j\omega t} \right| dt \leq \int_{-\infty}^{\infty} |x(t)| \left| e^{-j\omega t} \right| dt \leq \int_{-\infty}^{\infty} |x(t)| \, dt. \qquad (7.4.21)$$

The calculations in (7.4.21) show that $x(t)$ will have a Fourier transform if $x(t)$ is absolutely integrable, i.e., the last integral in (7.4.21) is finite. The exact conditions for existence for $x(t)$ to have a Fourier transform are given next. We require the following definition.

Definition 7.4.5: The signal $x(t)$ is <u>piecewise smooth</u> if it can be divided into a finite number of intervals $a_i < t < b_i$ such that $x(t)$ has continuous derivatives on these intervals. ▲▲

The next theorem gives a mathematical statement of the existence of the Fourier transform.

Theorem 7.4.6: Suppose that $x(t)$ is piecewise smooth and it satisfies

$$\int_{-\infty}^{\infty} |x(t)| \, dt \leq M < \infty$$

for some finite, nonnegative constant M. Then

$$X(\omega) = \int_{-\infty}^{\infty} x(t)e^{-j\omega t} \, dt$$

exists and is a continuous function[2] of ω. Furthermore,

$$x(t) = \frac{1}{2\pi} \int_{-\infty}^{\infty} X(\omega)e^{j\omega t} \, d\omega$$

exists and the equality holds at every point where $x(t)$ is continuous. If $x(t)$ is discontinuous at $t = t_0$, then the integral converges to the average of the left- and right-hand limits. ▲▲

Theorem 7.4.6 explicitly identifies those signals that have a Fourier transform in the sense of Definition 7.4.1. These signals must be absolutely integrable; they must

---

[2] A continuous function in the mathematical sense, i.e., no step discontinuities.

satisfy the integral constraint.   While many signals that commonly occur in engineering practice do satisfy this constraint, there is a significant class of signals that don't satisfy Theorem 7.4.6.   Sinusoidal signals, for example, don't satisfy Theorem 7.4.6.   In the next section we will extend the definition of the Fourier transform to some signals that don't satisfy Theorem 7.4.6. The extended definition makes the Fourier transform widely useful.   In some applications, however, it is important to distinguish between the signals that satisfy Theorem 7.4.6 and those signals whose Fourier transform exists only by extension.

7.5   PROPERTIES OF THE FOURIER TRANSFORM AND THE GENERALIZED FOURIER TRANSFORM

7.5.1   Introduction

There are several properties of the Fourier transform which make it useful for signal and system analysis.   In this section we will enumerate these properties and illustrate them with examples.   These properties will be used extensively in the following chapters.

In the last section we noted that not all signals of interest have a Fourier transform according to Definition 7.4.1.   In this section we will extend the notion of the Fourier transform to these signals by introducing the *generalized* Fourier transform.   With the generalized Fourier transform we can analyze most signals of interest.

7.5.2   Properties of the Fourier Transform

In the discussion that follows we assume that the Fourier transform exists for the signals $x(t)$ and $h(t)$.

Property 7.5.1:   (Linearity) The Fourier transform satisfies

$$\mathcal{F}\{a_1 x_1(t) + a_2 x_2(t)\} = a_1 X_1(\omega) + a_2 X_2(\omega)$$

for the signals $x_i(t)$ and all constants a_i for $i = 1,2$.

▲▲

Terminology: Because of Property 7.5.1, we say that the Fourier transform is linear.

The proof of this property relies on the linearity of the integral in the definition of the Fourier transform.

Property 7.5.2:   (Duality) If $x(t) \leftrightarrow X(\omega)$ is a Fourier transform pair, then

$$X(t) \leftrightarrow 2\pi x(-\omega)$$

is also a Fourier transform pair.

▲▲

Duality follows from the symmetry between the integrals defining the Fourier transform and its inverse. The following example illustrates this property.

Example 7.5.3: Consider again the Fourier transform pair

$$\Pi\!\left(\frac{t}{\varepsilon}\right) \leftrightarrow \varepsilon\,\mathrm{Sa}\!\left(\frac{\omega\varepsilon}{2}\right) \qquad (7.5.1)$$

that was derived in Example 7.4.3. Using duality, Property 7.5.2, we obtain a second Fourier transform pair

$$\varepsilon\,\mathrm{Sa}\!\left(\frac{t\varepsilon}{2}\right) \leftrightarrow 2\pi\Pi\!\left(\frac{-\omega}{\varepsilon}\right) = 2\pi\Pi\!\left(\frac{\omega}{\varepsilon}\right). \qquad (7.5.2)$$

The second equality in (7.5.2) follows because the pulse function is an even function.
▲▲

The following property of the Fourier transform plays a fundamental role in the analysis of signals. We will discuss it extensively in Chapter 8.

Property 7.5.4: Parseval's Theorem Let $x(t)$ and $y(t)$ be signals with a Fourier transform. Then we have

$$\int_{-\infty}^{\infty} x(t)\bar{y}(t)\,dt = \frac{1}{2\pi}\int_{-\infty}^{\infty} X(\omega)\bar{Y}(\omega)\,d\omega.$$
▲▲

Often when we apply Parseval's Theorem we take $y(t) = x(t)$. Then we obtain

$$\int_{-\infty}^{\infty} |x(t)|^2\,dt = \frac{1}{2\pi}\int_{-\infty}^{\infty} |X(\omega)|^2\,d\omega. \qquad (7.5.3)$$

Property 7.5.5: (Time and Frequency Scaling)   For any positive number a, we have

$$\mathcal{F}\{x(at)\} = \frac{1}{a}X\!\left(\frac{\omega}{a}\right).$$
▲▲

The next example illustrates this property of the Fourier transform.

Example 7.5.6: Consider the Fourier transform pair (7.5.1) and the signal

$$\hat{x}(t) = \Pi\!\left(\frac{t}{\frac{\varepsilon}{2}}\right) = \Pi\!\left(\frac{2t}{\varepsilon}\right). \qquad (7.5.4)$$

Comparing (7.5.4) to (7.5.1) we see that the time axis has been scaled by a factor of $a = 2$. Applying Property 7.5.5 we find that the Fourier transform of (7.5.4) is

$$\mathcal{F}\{\hat{x}(t)\} = \mathcal{F}\left\{\Pi\left(\frac{2t}{\varepsilon}\right)\right\} = \left(\frac{1}{2}\right)X\left(\frac{\omega}{2}\right) = \left(\frac{1}{2}\right)\varepsilon \operatorname{Sa}\left(\frac{\omega}{2}\frac{\varepsilon}{2}\right) = \frac{\varepsilon}{2}\operatorname{Sa}\left(\frac{\varepsilon\omega}{4}\right). \tag{7.5.5}$$

▲▲

The following property forms the basis for system analysis using the Fourier transform. As such it is the cornerstone of much of the analysis of this text.

Property 7.5.7: (Convolution in the Time Domain)   If $x(t)$ and $h(t)$ are Fourier transformable, then

$$\mathcal{F}\left\{\int_{-\infty}^{\infty} h(t-\lambda)x(\lambda)\,d\lambda\right\} = H(\omega)X(\omega).$$

▲▲

Example 7.5.8:   The convolution integral can be used to derive the Fourier transform of a signal expressed as a convolution of two other signals. Consider the unit pulse signal in (7.5.1). If we convolve $x(t)$ with itself, we obtain

$$\int_{-\infty}^{\infty} \Pi\left(\frac{t-\lambda}{\frac{\varepsilon}{2}}\right)\Pi\left(\frac{\lambda}{\frac{\varepsilon}{2}}\right)d\lambda = \frac{\varepsilon}{2}\Lambda\left(\frac{t}{\varepsilon}\right). \tag{7.5.6}$$

Using Property 7.5.7, the Fourier transform of (7.5.6) is

$$\mathcal{F}\left\{\int_{-\infty}^{\infty} \Pi\left(\frac{t-\lambda}{\frac{\varepsilon}{2}}\right)\Pi\left(\frac{\lambda}{\frac{\varepsilon}{2}}\right)d\lambda\right\} = \mathcal{F}\left\{\frac{\varepsilon}{2}\Lambda\left(\frac{t}{\varepsilon}\right)\right\} = \left[\frac{\varepsilon}{2}\operatorname{Sa}\left(\frac{\omega\varepsilon}{2}\right)\right]^2 = \frac{\varepsilon^2}{4}\operatorname{Sa}^2\left(\frac{\omega\varepsilon}{2}\right) \tag{7.5.7}$$

where we have used the Fourier transform of $x(t)$ in Example 7.5.3. Canceling the common factors between (7.5.6) and (7.5.7) (the Fourier transform is linear) we obtain the Fourier transform pair

$$\Lambda\left(\frac{t}{\varepsilon}\right) \leftrightarrow \frac{\varepsilon}{2}\operatorname{Sa}^2\left(\frac{\omega\varepsilon}{4}\right). \tag{7.5.8}$$

▲▲

Applying the duality property to Property 7.5.7 we obtain the following property.

Property 7.5.9: (Convolution in the Frequency Domain)   For two signals $x(t)$ and $h(t)$ we have

$$\mathcal{F}\{h(t)x(t)\} = \frac{1}{2\pi}\int_{-\infty}^{\infty} H(\omega-\lambda)X(\lambda)\,d\lambda.$$

▲▲

The following two properties of the Fourier transform relate the Fourier transform of a signal to its derivative and integral.

Property 7.5.10: (Differentiation) If $x(t)$ has an nth derivative, then we have

$$\mathcal{F}\left\{x^{(n)}(t)\right\} = (j\omega)^n X(\omega).$$

▲▲

Property 7.5.11: (Integration) For the signal $x(t)$ we have

$$\mathcal{F}\left\{\int_{-\infty}^{t} x(\lambda)\,d\lambda\right\} = \frac{1}{j\omega} X(\omega) + \pi X(0)\delta(\omega).$$

▲▲

The next two properties of the Fourier transform describe shifts in the time and frequency variables.

Property 7.5.12: (Time Shift) If the time scale of $x(t)$ is shifted, then we have the Fourier transform pair

$$\mathcal{F}\left\{x(t-t_0)\right\} = X(\omega)e^{-j\omega t_0}.$$

▲▲

Duality applied to the time shift property yields the next property.

Property 7.5.13: (Frequency Shift) If $x(t)$ is multiplied by a complex exponential, then

$$\mathcal{F}\left\{x(t)e^{j\omega_0 t}\right\} = X(\omega - \omega_0).$$

▲▲

Property 7.5.13 is useful in establishing other properties of the Fourier transform. The following property plays a central role in the analysis of signals used in communication systems.

Property 7.5.14: (Modulation) If a signal $x(t)$ is multiplied by a cosine, we have

$$\mathcal{F}\left\{x(t)\cos(\omega_0 t)\right\} = \frac{1}{2}\left[X(\omega + \omega_0) + X(\omega - \omega_0)\right].$$

If a signal $x(t)$ is multiplied by a sine, we have

$$\mathcal{F}\left\{x(t)\sin(\omega_0 t)\right\} = \frac{j}{2}\left[X(\omega + \omega_0) - X(\omega - \omega_0)\right].$$

▲▲

The following two examples illustrate the use of the modulation property.

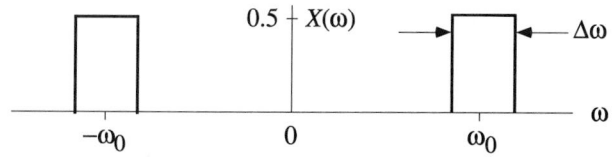

Figure 7.5.1 Fourier Transform of the Signal in Example 7.5.15

Example 7.5.15: Consider the Fourier transform shown in Figure 7.5.1. We would like to find the corresponding signal $x(t)$. From Figure 7.5.1 we can express this Fourier transform as

$$X(\omega) = \frac{1}{2}\left[\Pi\left(\frac{\omega - \omega_0}{\Delta\omega}\right) + \Pi\left(\frac{\omega + \omega_0}{\Delta\omega}\right)\right]. \tag{7.5.9}$$

Applying the modulation property to (7.5.9) we see that

$$x(t) = \mathcal{F}^{-1}\left\{\Pi\left(\frac{\omega}{\Delta\omega}\right)\right\}\cos(\omega_0 t). \tag{7.5.10}$$

Next using the result of Example 7.5.3, we have

$$\frac{\Delta\omega}{2\pi}\,\mathrm{Sa}\left(\frac{\Delta\omega t}{2}\right) \leftrightarrow \Pi\left(\frac{\omega}{\Delta\omega}\right). \tag{7.5.11}$$

Combining the above we have

$$x(t) = \mathcal{F}^{-1}\{X(\omega)\} = \frac{\Delta\omega}{2\pi}\,\mathrm{Sa}\left(\frac{\Delta\omega t}{2}\right)\cos(\omega_0 t). \tag{7.5.12}$$

This signal is shown in Figure 7.5.2.

▲▲

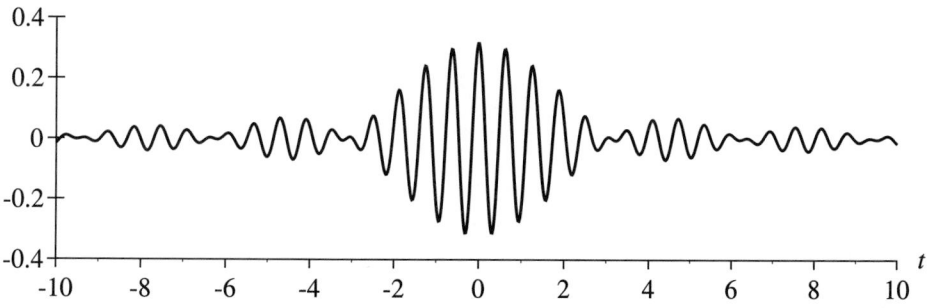

Figure 7.5.2 Signal of Example 7.5.15

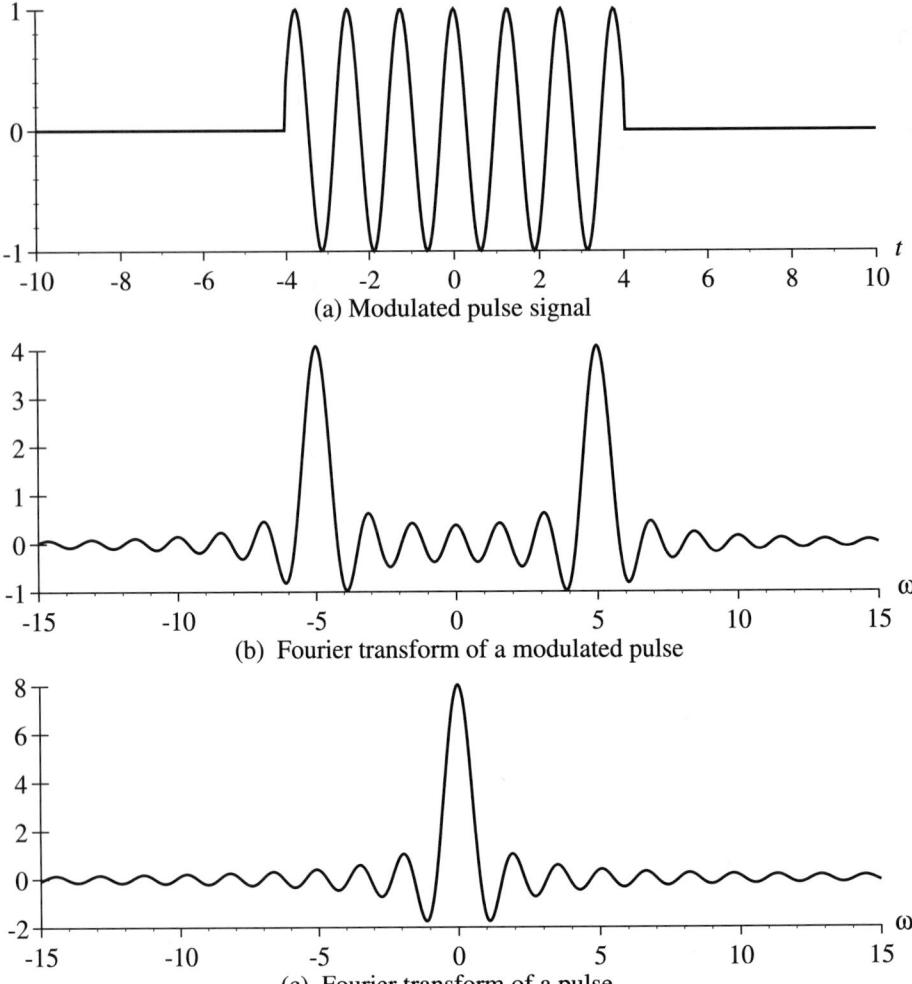

Figure 7.5.3 A Modulated Pulse Signal

Example 7.5.16: Consider the signal

$$x(t) = \Pi\left(\frac{t}{\varepsilon}\right)\cos(\omega_0 t). \tag{7.5.13}$$

The graph of this signal is shown in Figure 7.5.3a. To find the Fourier transform of this signal we use the modulation property along with the Fourier transform pair

$$\Pi\left(\frac{t}{\varepsilon}\right) \leftrightarrow \varepsilon \operatorname{Sa}\left(\frac{\omega\varepsilon}{2}\right). \tag{7.5.14}$$

The Fourier transform of the signal (7.5.13) is

$$\mathcal{F}\left\{\Pi\left(\frac{t}{\varepsilon}\right)\cos(\omega_0 t)\right\} = \frac{1}{2}\left[\varepsilon\,\text{Sa}\left(\frac{(\omega - \omega_0)\varepsilon}{2}\right) + \varepsilon\,\text{Sa}\left(\frac{(\omega + \omega_0)\varepsilon}{2}\right)\right].$$

(7.5.15)

The graph of the Fourier transform in (7.5.15) is shown in Figure 7.5.4b.

The graph of the Fourier transform of the pulse in (7.5.14) is shown in Figure 7.5.3c.   In comparing the Fourier transforms in Figure 7.5.3, we see that multiplication by $\cos(\omega_0 t)$ causes the Fourier transform of the pulse to "duplicate and split," the center of each copy is shifted $\omega = \pm \omega_0$, the frequency of the cosine.

▲▲

7.5.3 Generalized Fourier Transform

Next we turn our attention to signals that aren't Fourier transformable according to Definition 7.4.1. Consider the signal

$$x(t) = \cos(\omega_0 t), \quad -\infty < t < \infty.$$

(7.5.16)

We would like to take the Fourier transform of this signal, but it isn't absolutely integrable; it doesn't satisfy condition (b) of the definition of the Fourier transform, Definition 7.4.1. Nonetheless, it would be useful to have a "Fourier transform" for these signals. An appropriate extension of the Fourier transform is suggested by the modulated pulse function

$$x(t) = \Pi\left(\frac{t}{\varepsilon}\right)\cos(\omega_0 t)$$

(7.5.17)

that was discussed in Example 7.5.16. As $\varepsilon \to \infty$, $x(t)$ approaches a pure cosine. Yet for each ε, the signal (7.5.17) does have a Fourier transform. At the same time, the Fourier transform of $x(t)$ approaches a pair of impulse functions, $\delta(\omega - \omega_0)$ and $\delta(\omega + \omega_0)$, as suggested by comparing Figure 7.5.3b and Figure 7.5.4. This example suggests that we could find the Fourier transform of (7.5.16) from the Fourier transform of the impulse function.

If the impulse function is substituted into the definition of the Fourier transform we get

$$X(\omega) = \int_{-\infty}^{\infty} \delta(t)e^{-j\omega t}\,dt = 1$$

(7.5.18)

which is perfectly well defined. If the constant function of frequency in (7.5.18) is substituted into the definition of the inverse Fourier transform we get

$$\int_{-\infty}^{\infty}(1)e^{j\omega t}\,d\omega$$

(7.5.19)

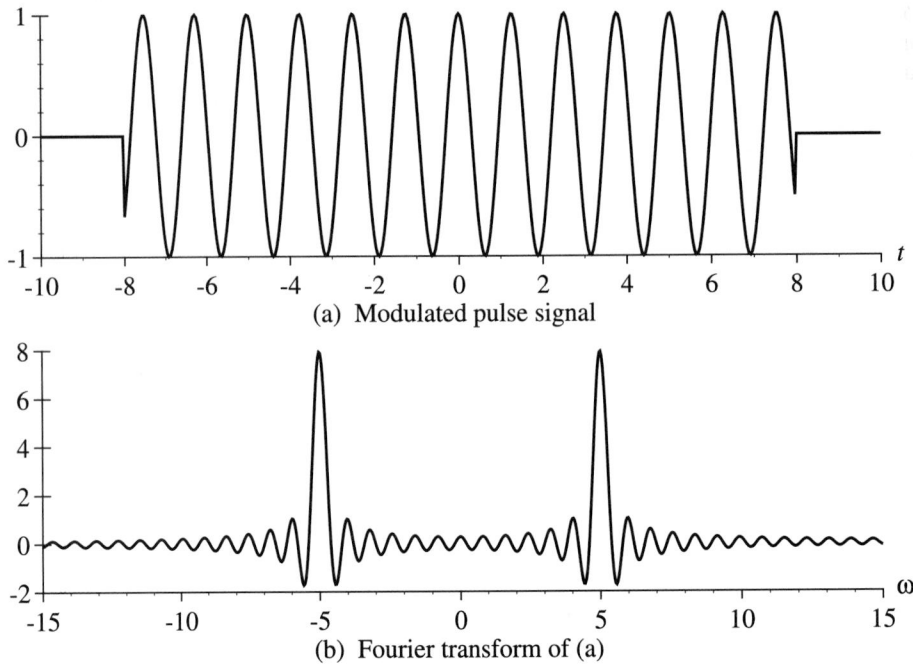

(a) Modulated pulse signal

(b) Fourier transform of (a)

Figure 7.5.4 Fourier Transform for (7.5.17) as ε Grows Large

which is not defined. To make sense of this integral we take the following approach. We construct a sequence of functions whose limit is the impulse function. If we multiply the pulse function by 1/ε then, in the limit as ε approaches 0, the pulse approaches an impulse function (see Example 5.2.8); i.e.,

$$\lim_{\varepsilon \to 0} \frac{1}{\varepsilon} \Pi\left(\frac{t}{\varepsilon}\right) = \delta(t).$$

(7.5.20)

The Fourier transform of the pulse function is

$$\mathcal{F}\left\{\frac{1}{\varepsilon}\Pi\left(\frac{t}{\varepsilon}\right)\right\} = \frac{1}{\varepsilon}\left[\varepsilon \, \mathrm{Sa}\left(\frac{\omega\varepsilon}{2}\right)\right] = \mathrm{Sa}\left(\frac{\omega\varepsilon}{2}\right).$$

(7.5.21)

As the pulse becomes narrower in time, the first zero crossing of the Fourier transform occurs at larger and larger frequencies. The limit of the corresponding Fourier transform is

$$\lim_{\varepsilon \to 0} \mathcal{F}\left\{\frac{1}{\varepsilon}\Pi\left(\frac{t}{\varepsilon}\right)\right\} = \lim_{\varepsilon \to 0} \frac{1}{\varepsilon}\left[\varepsilon \, \mathrm{Sa}\left(\frac{\omega\varepsilon}{2}\right)\right] = 1.$$

(7.5.22)

So while Fourier transform of (7.5.20) exists for all values of $\varepsilon > 0$, it does not exist for the limiting function. These limiting functions, however, point to the Fourier transform pair we would like to have.

Based on these observations we propose the following definition.

Definition 7.5.17: The <u>generalized</u> Fourier transform of the impulse function is the constant function,

$$x(t) = \delta(t) \leftrightarrow 1 = X(\omega).$$

▲▲

Having postulated this Fourier transform pair, we can use this pair just as we would use any other Fourier transform pair. Based on Definition 7.5.17, we generate several other pairs of generalized Fourier transforms using the properties of the Fourier transform introduced above. This process is illustrated by the following example.

Example 7.5.18: Using duality, we also define the Fourier transform pair

$$x(t) = 1 \leftrightarrow 2\pi\,\delta(\omega) = X(\omega). \tag{7.5.23}$$

Applying the modulation property, Property 7.5.14, to (7.5.23) we obtain the pair

$$\mathcal{F}\{1\cos(\omega_0 t)\} = 2\pi\left\{\frac{1}{2}[\delta(\omega - \omega_0) + \delta(\omega + \omega_0)]\right\} \tag{7.5.24}$$

$$= \pi\{[\delta(\omega - \omega_0) + \delta(\omega + \omega_0)]\}.$$

▲▲

Example 7.5.19: The Fourier transform of the impulse function can also be used to find the Fourier transform of a Fourier series

$$x(t) = \sum_{m=-\infty}^{\infty} X_m e^{jm\omega_0 t}. \tag{7.5.25}$$

Using Property 7.5.13 and Definition 7.5.17 the Fourier transform of each term of the Fourier series (7.5.25) is

$$\mathcal{F}\{1 e^{jm\omega_o t}\} = 2\pi\,\delta(\omega - m\omega_0). \tag{7.5.26}$$

Now the Fourier transform of the Fourier series (7.5.25) is

$$\mathcal{F}\{x(t)\} = X(\omega) = \sum_{m=-\infty}^{\infty} X_m 2\pi\,\delta(\omega - m\omega_0). \tag{7.5.27}$$

▲▲

The generalized Fourier transform employs the Fourier transform of the impulse function. The presence of the impulse function serves to distinguish a generalized Fourier transform from the Fourier transform defined in the last section. Very often it is not important to discriminate between signals that have Fourier transforms and signals that have generalized Fourier transforms. On some occasions, however, this distinction is important.

7.6 CHAPTER SUMMARY

7.6.1 Fourier Series

In this chapter we have introduced the Fourier series as a representation of periodic signals.

Definition 7.1.1: The signal $x(t)$ is periodic if there exists a constant $T_0 > 0$ such that $x(t + T_0) = x(t)$ for all t. The smallest T_0 for which this is true is called the (fundamental) period. All other signals are aperiodic.   ▲▲

The three different representations of the Fourier series are summarized in Table 7.2.1. The relationships between these representations are summarized in Table 7.6.1. The fundamental frequency of the these Fourier series is computed from the period of the signal.

There are two sets of computational formulas for the coefficients of the trigonometric and exponential Fourier series.

Table 7.6.1 Relationship Between the Three Fourier Series Representations

$x(t) = \displaystyle\sum_{m=0}^{\infty} A_m \cos(m\omega_0 t + \theta_m)$ $A_m > 0$	$a_0 = A_0 \cos\theta_0$ $a_m = A_m \cos\theta_m$ $b_m = -A_m \sin\theta_m$	$X_m = \dfrac{A_m}{2} e^{j\theta_m}, m > 0$ $X_0 = A_0, \quad m = 0$
$A_m = +\sqrt{a_m^2 + b_m^2}$ $\theta_m = -\tan^{-1}\left(\dfrac{b_m}{a_m}\right)$	$x(t) = a_0 + \displaystyle\sum_{m=1}^{\infty} a_m \cos(m\omega_0 t)$ $+ \displaystyle\sum_{m=1}^{\infty} b_m \sin(m\omega_0 t)$	$2\,\mathrm{Re}\,X_m = a_m$ $-2\,\mathrm{Im}\,X_m = b_m$ $m > 0$
$A_m = 2|X_m|, \quad m > 0$ $\theta_m = \angle X_m, \quad m > 0$ $A_0 = X_0, \quad m = 0$	$2\,\mathrm{Re}\,X_m = a_m$ $-2\,\mathrm{Im}\,X_m = b_m$ $m > 0$	$x(t) = \displaystyle\sum_{m=-\infty}^{\infty} X_m e^{jm\omega_0 t}$

Remark 7.3.1:   Computation of the Trigonometric Fourier Series   The coefficients of the trigonometric Fourier series can be computed with the following formulas.

$$a_0 = \frac{1}{T_0} \int_{T_0} x(t)\,dt. \tag{7.3.7}$$

$$a_m = \frac{2}{T_0} \int_{T_0} x(t)\cos(m\omega_0 t)\,dt. \tag{7.3.9}$$

$$b_m = \frac{2}{T_0} \int_{T_0} x(t)\sin(m\omega_0 t)\,dt. \tag{7.3.11}$$

▲▲

Remark 7.3.3: Computation of the Exponential Fourier Series The coefficients of the exponential Fourier series can be computed from

$$X_m = \frac{1}{T_0} \int_{T_0} x(t)e^{-jm\omega_0 t}\,dt. \tag{7.3.21}$$

▲▲

7.6.2 Fourier Transform

For periodic signals we have Fourier series. For aperiodic signals we use the Fourier transform.

Definition 7.4.1: Let $x(t)$ be a signal such that:

  (a) $x(t)$, $-\infty < t < \infty$, and

  (b) $x(t)$ satisfies

$$\int_{-\infty}^{\infty} |x(t)|\,dt \leq M < \infty$$

for some finite, nonnegative constant M. Then the <u>Fourier transform</u> of $x(t)$ is defined as

$$X(\omega) = \int_{-\infty}^{\infty} x(t)e^{-j\omega t}\,dt = \mathcal{F}\{x(t)\}.$$

The <u>inverse Fourier transform</u> is defined as

$$x(t) = \frac{1}{2\pi} \int_{-\infty}^{\infty} X(\omega)e^{j\omega t}\,d\omega = \mathcal{F}^{-1}\{X(\omega)\}.$$

▲▲

The properties of the Fourier transform are summarized in Table 7.6.2. Several Fourier transform pairs are given in Table 7.6.3.

Table 7.6.2 Fourier Transform Properties

Property	Description	Mathematical Description
7.5.1	Linearity	$\mathcal{F}\{a_1 x(t) + a_2 y(t)\} = a_1 X(\omega) + a_2 Y(\omega)$
7.5.2	Duality	$\mathcal{F}\{X(t)\} = 2\pi x(-\omega)$
7.5.4	Parseval's Theorem	$\displaystyle\int_{-\infty}^{\infty} x(t)y(t)\,dt = \frac{1}{2\pi}\int_{-\infty}^{\infty} X(\omega)\overline{Y}(\omega)\,d\omega$
7.5.5	Time and Frequency Scaling	$\mathcal{F}\{x(at)\} = \dfrac{1}{a} X\left(\dfrac{\omega}{a}\right) \quad a > 0$
7.5.7	Convolution in the Time Domain	$\mathcal{F}\left\{\displaystyle\int_{-\infty}^{\infty} h(t-\lambda)x(\lambda)\,d\lambda\right\} = H(\omega)X(\omega)$
7.5.9	Convolution in the Frequency Domain	$\mathcal{F}\{h(t)x(t)\} = \dfrac{1}{2\pi}\displaystyle\int_{-\infty}^{\infty} H(\omega-\lambda)X(\lambda)\,d\lambda$
7.5.10	Differentiation	$\mathcal{F}\left\{x^{(n)}(t)\right\} = (j\omega)^n X(\omega)$
7.5.11	Integration	$\mathcal{F}\left\{\displaystyle\int_{-\infty}^{t} x(\lambda)\,d\lambda\right\} = \dfrac{1}{j\omega} X(\omega) + \pi X(0)\delta(\omega)$
7.5.12	Time Shift	$\mathcal{F}\{x(t-t_0)\} = X(\omega)e^{-j\omega t_0}$
7.5.13	Frequency Shift	$\mathcal{F}\left\{x(t)e^{j\omega_0 t}\right\} = X(\omega-\omega_0)$
7.5.14	Modulation	$\mathcal{F}\{x(t)\cos(\omega_0 t)\} = \dfrac{1}{2}[X(\omega+\omega_0) + X(\omega-\omega_0)]$ $\mathcal{F}\{x(t)\sin(\omega_0 t)\} = \dfrac{j}{2}[X(\omega+\omega_0) - X(\omega-\omega_0)]$

Table 7.6.3 Fourier Transform Pairs

Signal	Fourier Transform	
$\Pi\left(\dfrac{t}{\varepsilon}\right)$	$\varepsilon\,\mathrm{Sa}\left(\dfrac{\omega\varepsilon}{2}\right)$	Example 7.4.3
$\varepsilon\,\mathrm{Sa}\left(\dfrac{t\varepsilon}{2}\right)$	$2\pi\Pi\left(\dfrac{\omega}{\varepsilon}\right)$	Example 7.5.3
$\Lambda\left(\dfrac{t}{\varepsilon}\right)$	$\dfrac{\varepsilon}{2}\mathrm{Sa}^2\left(\dfrac{\omega\varepsilon}{4}\right)$	Example 7.5.8
$\dfrac{\varepsilon}{2}\mathrm{Sa}^2\left(\dfrac{t\varepsilon}{4}\right)$	$2\pi\Lambda\left(\dfrac{\omega}{\varepsilon}\right)$	
$\delta(t)$	1	Definition 7.5.17
1	$2\pi\delta(\omega)$	Example 7.5.18
$e^{j\omega_0 t}$	$2\pi\delta(\omega-\omega_0)$	Example 7.5.19
$\cos(\omega_0 t)$	$\pi\big[\delta(\omega+\omega_0)+\delta(\omega-\omega_0)\big]$	Example 7.5.18
$\sin(\omega_0 t)$	$j\pi\big[\delta(\omega+\omega_0)-\delta(\omega-\omega_0)\big]$	
$u_s(t)$	$\dfrac{1}{j\omega}+\pi\delta(\omega)$	
$\mathrm{sgn}(t)$	$\dfrac{2}{j\omega}$	
$e^{-at}u_s(t),\quad a>0$	$\dfrac{1}{a+j\omega}$	
$\dfrac{t^n}{n!}e^{-at}u_s(t),\quad a>0$	$\dfrac{1}{(a+j\omega)^{n+1}}$	
$e^{-a|t|},\quad a>0$	$\dfrac{2a}{a^2+\omega^2}$	
$e^{-\frac{t^2}{2\sigma^2}}$	$\left(\sigma\sqrt{2\pi}\right)e^{-\frac{\sigma^2\omega^2}{2}}$	

7.7 HOMEWORK FOR CHAPTER 7

Homework Problems for Section 7.1

7.1.1   The signal $x_c(t)$ shown in Figure P7.1.1 is defined on the interval $-1 \leq t < 3$. Using $x_c(t)$ define a periodic signal $x(t)$ by periodic extension.
(a) Sketch $x(t)$.
(b) What is the period of $x(t)$?

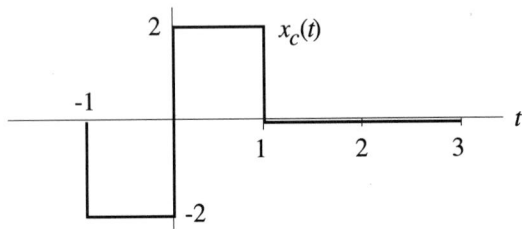

Figure P7.1.1

7.1.2   Suppose the signal $x(t)$ is defined as the periodic extension of

$$x_c(t) = t, \quad 0 \leq t < 2$$

(a) What is the fundamental frequency of this signal?
(b) Sketch the signal $x(t)$.

7.1.3   For what values of ω_c is the following signal periodic?

$$x(t) = \left[\sum_{k=-\infty}^{\infty} \Pi(t - 2k) \right] \cos(\omega_c t)$$

7.1.4   Consider the signal

$$x(t) = \sum_{m=-\infty}^{\infty} \Pi\left(\frac{t - 0.1m}{0.02} \right)$$

Find the Fourier series of $x(t)$.

7.1.5   Let $x(t)$ be a periodic signal with period T_0. What is the period of the signal $x(\alpha t)$?

7.1.6   Which of the following signals are periodic?
  (i) $x(t) = \sin(3t) + \sin(8t)$
  (iii) $x(t) = e^{3|t|}$
  (ii) $x(t) = \cos(2t) + \cos(3\pi t)$
  (iv) $x(t) = (\cos(4\pi t))u_s(t)$

Homework Problems for Section 7.2

7.2.1   Which of the following signals are in the format of a Fourier series? If the signal is a Fourier series, what is its fundamental period?
  (i) $x(t) = \sin(\pi t)$
  (v) $x(t) = \cos(t) + \cos(3t) + \cos(\pi t)$
  (ii) $x(t) = \cos(t) + \cos(3t)$
  (vi) $x(t) = 2 + \cos(t + 0.2) + 3\sin(2t - 0.1)$
  (iii) $x(t) = 1 + \cos(t) + \sin(2\pi t)$
  (vii) $x(t) = 0.2e^{j2t} + 0.2e^{-j2t}$
  (iv) $x(t) = \cos(t) + \sin(3t)$
  (viii) $x(t) = 3e^{-j\pi t} + 3e^{j\pi t} + 2e^{-jt} + 2e^{jt}$

7.2.2   Find the cosine and trigonometric representations of the following Fourier series.

  (i) $$x(t) = \sum_{\substack{m=-\infty \\ m \neq 0}}^{\infty} \left[\frac{2}{jm\pi} \left(1 - \cos\left(\frac{m\pi}{2}\right) \right) \right] e^{j\frac{m\pi}{2}t}$$

  (ii) $$x(t) = \sum_{m=-\infty}^{\infty} \frac{0.5}{(1 + j\pi m)} e^{j\frac{\pi m t}{3}}$$

  (iii) $$x(t) = \sum_{\substack{m=-\infty \\ m \neq 0}}^{\infty} \frac{1}{j2\pi m} \left(1 + e^{-j\frac{2\pi m}{3}} - 2e^{-j\frac{4\pi m}{3}} \right) e^{j\frac{2\pi m t}{3}}$$

  (iv) $$x(t) = 1 + \sum_{\substack{m=-\infty \\ m \neq 0}}^{\infty} \left(\frac{1}{m\pi} e^{j\frac{\pi}{2}} \right) e^{jm\pi t}$$

  (v) $$x(t) = 1 + \sum_{\substack{m=-\infty \\ m \neq 0}}^{\infty} \frac{1}{m} e^{j\left(\frac{m\pi}{4}\right)} e^{j2mt}$$

  (vi) $$x(t) = \sum_{m=-\infty}^{\infty} \frac{1}{2m^2} e^{-j\frac{\pi m}{2}} e^{jmt}$$

  (vii) $$x(t) = 1 + \sum_{\substack{m=-\infty \\ m \neq 0}}^{\infty} \left[\frac{3}{m\pi} \sin\left(\frac{2m\pi}{3}\right) e^{-j\frac{2m\pi}{3}} \right] e^{j\frac{2m\pi t}{3}}$$

7.2.3   Find the trigonometric and exponential representations of the following cosine Fourier series.

(i) $x(t) = \sum\limits_{m=1}^{\infty} \left(\dfrac{2}{2m^2 + 1} \right) \cos(2mt - 0.5m)$

(ii) $x(t) = \sum\limits_{m=1}^{\infty} \dfrac{3}{4m^2} \cos\left(2mt - \dfrac{\pi m}{4} \right)$

7.2.4   Find the cosine and exponential representations of the following trigonometric Fourier series.

(i) $x(t) = \sum\limits_{m=1}^{\infty} \left(\dfrac{2}{m\pi} \right) \left[1 + 2\cos\left(\dfrac{2m\pi}{3} \right) - 3\cos(m\pi) \right] \sin\left(\dfrac{m\pi t}{3} \right)$

(ii) $x(t) = \dfrac{8}{\pi^2} \sum\limits_{n=1}^{\infty} \dfrac{1}{(2m-1)^2} \cos\left(\dfrac{2\pi(2m-1)t}{T_0} \right)$

(iii) $x(t) = \dfrac{2}{\pi} \sum\limits_{m=1}^{\infty} \dfrac{(-1)^{m+1}}{m} \sin\left(\dfrac{2\pi m t}{T_0} \right)$

(iv) $x(t) = \dfrac{2}{\pi} + \sum\limits_{\substack{m=2 \\ m \text{ even}}}^{\infty} \dfrac{4}{\pi(1 - m^2)} \cos(\omega_0 t)$

7.2.5   Using MATLAB plot each of the signals in Problems 7.2.2, 7.2.3, and 7.2.4.

7.2.6   Find a signal that violates each of the assumptions in Theorem 7.2.6.

7.2.7   Let $x(t)$ be a periodic signal with a period of T_0. Suppose that this periodic signal has an exponential Fourier series with coefficients X_m. Verify the following properties of this Fourier series.

Signal	Fourier Coefficients
$x(t - t_0)$	$X_m e^{-jm\omega_0 t_0}$
$x(-t)$	X_{-m}
$x^*(t)$	X_{-m}^*
$e^{jk\omega_0 t} x(t)$	X_{m-k}
$x(\alpha t), \quad \alpha > 0$	X_m
$\dfrac{dx(t)}{dt}$	$jm\omega_0 X_m$
$\displaystyle\int_{-\infty}^{t} x(\lambda)\, d\lambda, \quad \text{if } X_0 = 0$	$\dfrac{1}{jm\omega_0} X_m$

Homework Problems for Section 7.3

7.3.1   Consider the signal in Figure P7.3.1.

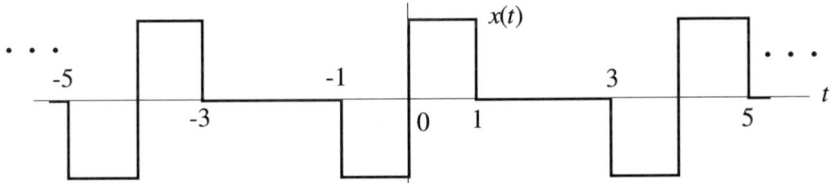

Figure P7.3.1

Show that the Fourier series of this signal is

$$x(t) = \sum_{\substack{m=-\infty \\ m\neq 0}}^{\infty} \left[\frac{2}{jm\pi}\left(1 - \cos\left(\frac{m\pi}{2}\right)\right)\right] e^{j\frac{m\pi}{2}t}$$

7.3.2   Consider the signal

$$x(t) = \begin{cases} 0, & -2 < t \leq -1 \\ e^{-|t|}, & -1 < t \leq 1 \\ 0, & 1 < t \leq 2 \end{cases}$$

Define the signal $y(t)$ as the periodic extension of $x(t)$.
(a) Sketch $x(t)$.
(b) Sketch $y(t)$.
(c) Find the Fourier series of $y(t)$.

7.3.3   Consider the signal $x_c(t)$ defined in the Figure P7.3.3 on the interval $-T_0/2 < t < T_0/2$.   This function is odd and symmetric about $T_0/4$ on the interval $0 < t < T_0/2$.   Let $x(t)$ be defined by periodic extension of $x_c(t)$. Show that the exponential form of the Fourier series is

$$x(t) = \sum_{\substack{m=-\infty \\ m\neq 0}}^{\infty} X_m e^{jm\omega_0 t}$$

where

$$X_m = \frac{j(A-B)}{m\pi}\cos(m\omega_0\tau_2) + \frac{j(B-A)}{m\pi}\cos\left(m\omega_0\left(\tau_2 - \frac{T_0}{2}\right)\right)$$

$$+ \frac{jA}{m\pi}\left[\cos\left(m\omega_0\left(\tau_1 - \frac{T_0}{2}\right)\right) - \cos(m\omega_0\tau_1)\right]$$

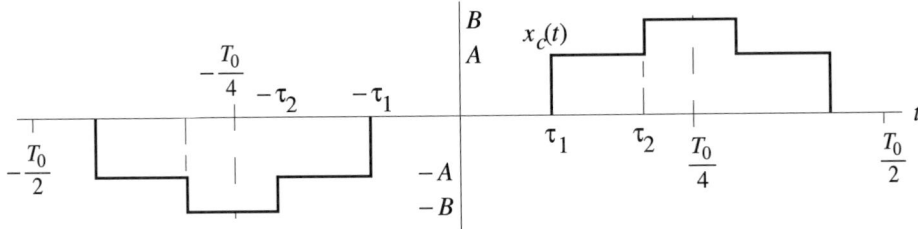

Figure P7.3.3

7.3.4 Consider the signal shown in Figure P7.3.4. Find the exponential Fourier series of this signal.

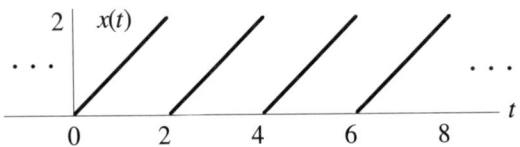

Figure P7.3.4

7.3.5 Consider the signal shown in Figure P7.3.5. Using this signal, define a periodic signal by periodic extension.
(a) What is the fundamental frequency?
(b) Find the exponential Fourier series.

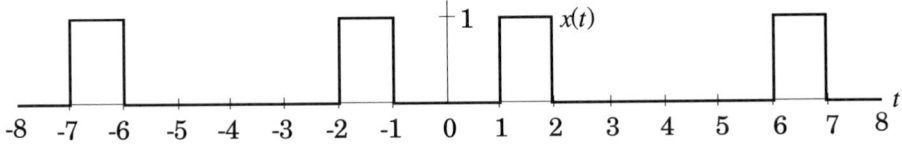

Figure P7.3.5

7.3.6 Consider the periodic signal shown in Figure P7.3.6. Show that the exponential Fourier series of this signal is given by

$$x(t) = \sum_{\substack{m=-\infty \\ m\neq 0}}^{\infty} \frac{1}{j2\pi m}\left(1 + e^{-j\frac{2\pi m}{3}} - 2e^{-j\frac{4\pi m}{3}}\right)e^{j\frac{2\pi mt}{3}}$$

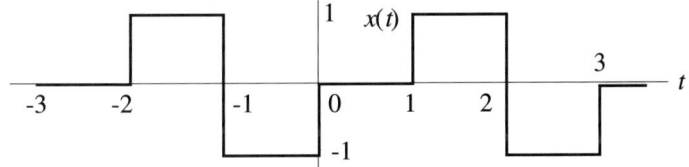

Figure P7.3.6

7.3.7   Consider the periodic signal shown in Figure P7.3.7. Find the Fourier series of this signal.

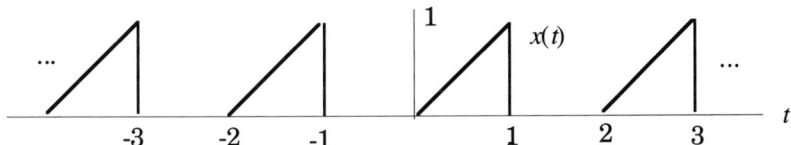

Figure P7.3.7

7.3.8   Using MATLAB verify your answers to Problems 7.3.1 - 7.

7.3.9   Verify the integral relationship in (7.3.19).

Homework Problems for Section 7.4

7.4.1   The Fourier transform $X(\omega)$ of a signal is shown in Figure P7.4.1. Find $x(t)$ using the definition of the inverse Fourier transform.

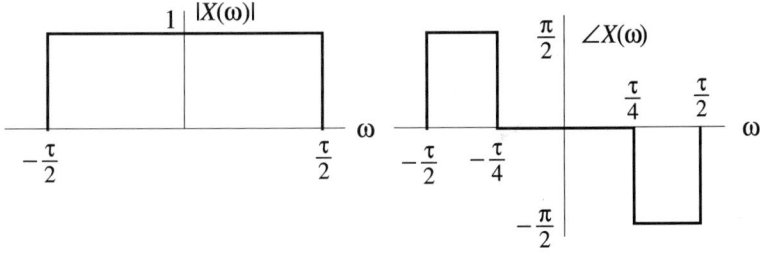

Figure P7.4.1

7.4.2 Which of the following signals is Fourier transformable? Why?

(i) $x(t) = e^{-at} u_s(t), \quad a > 0$

(ii) $x(t) = e^{at}$

(iii) $x(t) = 1$

(iv) $x(t) = [\cos(10t)] \sin(2t)$

(v) $x(t) = \displaystyle\sum_{m=-\infty}^{\infty} \Pi(t - 2m)$

(vi) $x(t) = \ln(t)$

7.4.3 Suppose that $x(t)$ is a real, even function.

(a) Show that

$$X(\omega) = 2 \int_0^\infty x(t) \cos(\omega t) \, d\omega.$$

(b) Show that $X(\omega)$ is a real even function of ω.

7.4.4 Suppose that $x(t)$ is a real, odd function.

(a) Show that

$$X(\omega) = -j2 \int_0^\infty x(t) \sin(\omega t) \, d\omega.$$

(b) Show that $X(\omega)$ is a imaginary, odd function of ω.

7.4.5 Let $X(\omega)$ be the Fourier transform of the signal $x(t)$. If $x(t)$ is written in terms of its real and imaginary parts $x(t) = x_e(t) + x_o(t)$, show that

$$x_e(t) \leftrightarrow \mathrm{Re}[X(\omega)] \quad \text{and} \quad x_o(t) \leftrightarrow j\,\mathrm{Im}[X(\omega)].$$

7.4.6 Find the inverse Fourier transform of each of the Fourier transforms shown in Figures P7.4.6a to d.

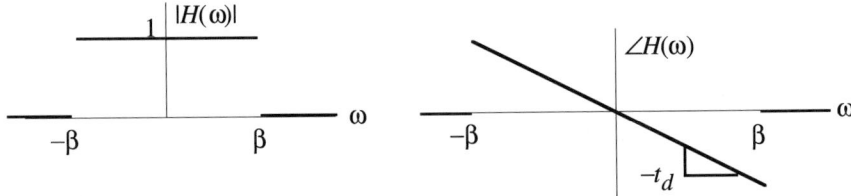

Figure P7.4.6a

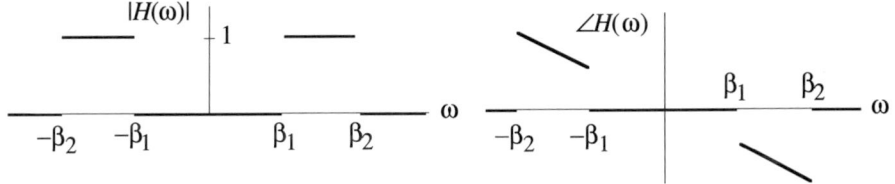

Figure P7.4.6b

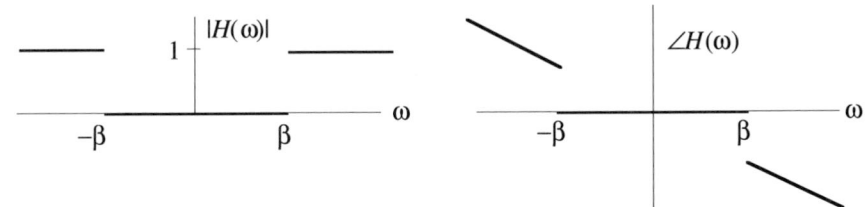

Figure P7.4.6c

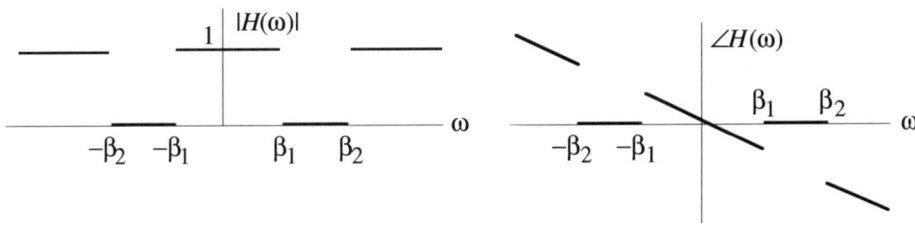

Figure P7.4.6d

7.4.7   Plot the Fourier transform in (7.4.18) with $\varepsilon = 1$ using MATLAB. Superimposed on this graph, plot the Fourier series coefficients in Example 7.4.4 using $T_0 = 2$.

Homework Problems for Section 7.5

7.5.1   Find the Fourier transform of each of the following signals using the definition and properties of the Fourier transform.

(i) $x(t) = \operatorname{sgn}(t)$

(ii) $x(t)\sin(\omega_0 t)$

(iii) $x(t) = \varepsilon\operatorname{Sa}^2\left(\dfrac{\varepsilon t}{4}\right)$

(iv) $x(t) = e^{-at}u_s(t),\quad a > 0$

(v) $x(t) = te^{-at}u_s(t),\quad a > 0$

(vi) $x(t) = e^{-a|t|},\quad a > 0$

(vii) $x(t) = \left[\operatorname{Sa}(3t)\right]\cos(7t)$

(viii) $x(t) = \displaystyle\sum_{m=-\infty}^{\infty} \Pi\left(\dfrac{t - 0.1m}{0.02}\right)$

(ix) $x(t) = \left(1 + k_a \Lambda(t)\right)\cos(10t)$

(x) $x(t) = \delta(t + t_0) + \delta(t - t_0)$

7.5.2   The duality property of Fourier transforms can be stated in terms of the frequency variable f in hertz. If $x(t) \leftrightarrow X(f)$ is a Fourier transform pair, then $X(t) \leftrightarrow x(-f)$ is also a Fourier transform pair. Prove this result using the definition of the Fourier transform.

7.5.3   Find the inverse Fourier transform of the following functions.

(i) $X(\omega) = \text{Sa}(4\omega - 5) + \text{Sa}(4\omega + 5)$

(iv) $X(\omega) = \Pi\left(\dfrac{\omega}{\beta}\right)e^{-j\omega t_0}$

(ii) $X(\omega) = \Lambda(100\omega)$

(v) $X(\omega) = -j\,\text{sgn}(\omega)$

(iii) $X(\omega) = \delta(\omega) + 2\Pi(4\omega)$

(vi) $X(\omega) = \delta(\omega + 5) + \delta(\omega - 5)$
$$+\Lambda(\omega + 5) + \Lambda(\omega - 5)$$

7.5.4   Show that

$$\mathcal{F}\left\{\Pi\left(\frac{t-\varepsilon}{\varepsilon}\right)\sin(\tfrac{\pi}{\varepsilon}t)\right\} = \mathcal{F}\left\{\Pi\left(\frac{t}{\varepsilon}\right)\cos(\tfrac{\pi}{\varepsilon}t)\right\}e^{-j\frac{\pi}{\varepsilon}}.$$

7.5.5   Suppose that we use the definition of the Fourier transform based on the frequency variable f in (7.4.13). Show that

$$\mathcal{F}\left\{\cos(2\pi f_0 t)\right\} = \frac{1}{2}\left[\delta(f + f_0) + \delta(f - f_0)\right].$$

Chapter 8

Spectral Content of a Signal

Chapter Outline

Signals, as defined in Chapter 5, are functions of time which represent a physical variable. To understand the physical variable, we analyze the function of time, that is, the signal. Signal analysis is the systematic investigation of the properties of a mathematical model of a physical variable and the underlying physical process. The purpose of this investigation is to uncover properties of the physical process from the signal that are not immediately apparent from the physical process itself. In this chapter we begin the investigation of the properties of these representations of signals.

The tools we have to analyze a signal depend on how we represent the signal. For example, a signal could be represented by an oscilloscope trace, or an analytical expression. Taking the derivative of the analytical expression may be easy, but taking the derivative of the oscilloscope trace may be nearly impossible. The central idea in signal analysis is to introduce a signal representation that gives insight into the characteristics of the signal. Generally, that means the signal representation should lend itself to numerical computation and graphical representation in such a way that the essential characteristics of the signal are exposed. This is a rather tall order, but several very successful signal representations have been introduced in the last chapter. These two representations are the Fourier series and the Fourier transform. We will study these two representations in this chapter.

The Fourier series represents the signal as a sum of cosines with given frequencies. The Fourier transform of a signal is a function of a frequency variable ω. It is the purpose of this chapter to develop the idea that these signal representations depend on frequency. This concept is known as the spectral content of a signal. This concept plays two important roles in signal analysis. First, the frequencies that play a dominant role in the composition of the signal gives us clues to the properties of the signal. These properties are usually exposed by sampling the signal and using computer processing. An introduction to this subject is given in Chapters 18 - 20. Second, this characterization of the signal can be coupled with system analysis to explain how a signal propagates through a system. This analysis is undertaken in Chapter 15. The results in this chapter provide the background for this analysis.

Two important concepts in this chapter are the energy and power in a signal. The energy (power) in a signal is abstracted from physics. It gives us a measure of the "strength" of a signal as compared to other signals. Through the use of Parseval's theorem we are able to couple the energy in a signal to its Fourier transform. This crucial result enables us to characterize the distribution of the energy across frequency. This result is the essence of the concept of the spectral content of a signal. We develop a parallel set of results for power signals. Extensive examples are given to illustrate these concepts.

The results in this chapter are routinely used to analyze complicated, real world signals. These signals don't admit simple analytical representations. However, we can use graphical representations of these signals very effectively. These graphical representations are based on the Fourier series and Fourier transform of the signal. In particular, they are graphical representations of the energy or power distribution in the signal as a function of frequency. These graphs are often the result of numerical processing of samples of the signal and so they can be realistically computed for real world signals. These computational algorithms are described in Chapter 20. In this chapter we define the basic concepts and illustrate them with simple examples.

Often a result of passing a signal through a real world system is that the signal becomes distorted. This signal distortion is due to nonlinearities in the system. In this chapter we describe how this distortion can be characterized using the Fourier series. This topic is of great practical importance; however, it is independent of the rest of the topics in the text.

Summary of Sections

Section 8.1:   We introduce amplitude and phase spectra as graphical representations of signals.

Section 8.2:   We define the energy and power of a signal.

Section 8.3:   We define the energy spectral density and signal bandwidth.

Section 8.4:   We define the power spectral density.

Section 8.5:   We discuss power calculations for periodic signals.

Section 8.6:   We develop the concept of the spectral content of a signal for periodic signals.

Section 8.7:   We discuss the harmonic distortion introduced by a static nonlinearity.

Section 8.8   We discuss several MATLAB experiments that illustrate the results in this chapter.

Section 8.9:   Chapter summary section.

Coverage of the Text

The results in this chapter require the results of Chapter 7. The results in Sections 8.1 - 8.5 are widely used in the rest of the text. Sections 8.6 and 8.7 are basically extended examples to illustrate the previous concepts.

8.1 AMPLITUDE AND PHASE SPECTRA

8.1.1 Introduction

Signal modeling is one of the three main topics in the theory of signals along with signal analysis, and signal design. In Chapter 5 we introduced signal modeling in its simplest form. The basic idea behind the signal modeling in Chapter 5 is to simply try to match the observed physical variable with a function of time. To that end we introduced a number of techniques for manipulating functions so that they can be used for suitable signal models. There is no question that these signal models play a

Table 8.1.1 Signal Representations

Signal	Signal Representation	
$x(t)$, periodic	$\displaystyle\sum_{m=0}^{\infty} A_m \cos(m\omega_0 t + \theta_m) = \sum_{m=-\infty}^{\infty} X_m e^{jm\omega_0 t}$	Fourier Series
$x(t)$, aperiodic	$\mathcal{F}\{x(t)\} = X(\omega)$	Fourier Transform

fundamental role in signal analysis and design.   This approach is somewhat simplistic, however.

In Chapter 7 we introduced two signal representations that are shown in Table 8.1.1. These two signal representations give us an alternative to simply modeling a physical variable with a function of time. In this chapter we will show that these new representations lend great insight into the properties of the signal which can't be discerned from the function of time directly. Hence, this chapter is really about a second class of signal representations, and how they can be used for signal modeling and analysis.

The two representations in Table 8.1.1 are extremely useful for signal analysis, but most signals are not amenable to analytical representations. Therefore, it is the purpose of this section to develop two widely used graphical representations of the Fourier series and Fourier transform. The graphical representation of the Fourier series and Fourier transform plays a central role in the signal analysis. These two graphical representations will be used extensively in the later chapters to show how a signal propagates through a system.

Computer Usage: Today most graphs are generated by the computer.   Strictly speaking, a computer-generated graph is a representation of a discrete signal. From this technical viewpoint, graphical representations of signals are a stepping stone to digital signal processing discussed in Chapters 17 - 23. Herein lies the importance of these graphical representations. Using digital electronics, samples of signals can be collected for computer processing and graphical display.

8.1.2 Periodic Signals

First we consider Fourier series. Let $x(t)$ be a periodic signal with a Fourier series as shown in Table 8.1.1. Consider the cosine Fourier series. This representation is always a summation of cosines. To define the Fourier series we need to specify: (1) the frequency $m\omega_0$, (2) the amplitude A_m, and (3) the phase θ_m for $m = 0, 1, 2, \ldots$. Furthermore, the amplitude and phase coefficients are indexed to the frequency of each term in the Fourier series. This information essentially defines the analytical representation of the Fourier series in Table 8.1.1. By expressing this information as a function we can concisely represent the Fourier series.   The next definition formally introduces this idea.

Definition 8.1.1: Let $x(t)$ be a periodic signal with a cosine Fourier series

$$x(t) = \sum_{m=0}^{\infty} A_m \cos(m\omega_0 t + \theta_m).$$

The function

$$A_m \text{ vs. } m\omega_0, \quad m = 0, 1, 2, \dots$$

is called the (one-sided) amplitude spectrum of $x(t)$. The function

$$\theta_m \text{ vs. } m\omega_0, \quad m = 0, 1, 2, \dots$$

is called the (one-sided) phase spectrum of $x(t)$.

▲▲

The amplitude and phase spectra is frequently presented graphically by plotting the amplitude vs. frequency and the phase vs. frequency, which are defined next. The usefulness of these two graphs will be apparent by the end of the chapter.

We can also define the amplitude and phase spectra for a Fourier series in exponential form. Again we observe that the series is completely defined by: (1) the frequencies $m\omega_0$, and (2) the coefficients X_m which are usually displayed in graphical form. There are two changes in the spectra for the exponential Fourier series. First, the frequencies start from negative infinity rather than zero. Second, the coefficients of the Fourier series are complex numbers. So these coefficients are defined in terms of their magnitude and phase. These spectra are defined next.

Definition 8.1.2: Let $x(t)$ be a periodic signal with an exponential Fourier series

$$x(t) = \sum_{m=-\infty}^{\infty} X_m e^{jm\omega_0 t}.$$

The function

$$|X_m| \text{ vs. } m\omega_0, \quad m = \dots, -2, -1, 0, 1, 2, \dots$$

is called the (two-sided) amplitude spectrum of $x(t)$. The function

$$\angle X_m \text{ vs. } m\omega_0 \quad m = \dots, -2, -1, 0, 1, 2, \dots$$

is called the (two-sided) phase spectrum of $x(t)$.

▲▲

Terminology: The *one-sided* amplitude and phase spectra are defined on the positive frequency axis while the *two-sided* amplitude and phase spectra start at negative infinity. This terminology is frequently used in signal and system analysis.

The trigonometric Fourier series representation is generally not expressed in graphical form.

Terminology: The amplitude and phase spectra completely define the Fourier series, but time t doesn't explicitly appear as a parameter. In fact, the independent variable in these graphs is frequency ω. Hence, these graphs are said to be in the <u>frequency domain</u>.

Part of the significance of the amplitude and phase spectra is that these representations are of the same form for *all* signals independent of the particular waveform. Hence, these graphs can be used to compare the properties of different periodic signals whose waveforms are seemingly quite different. The following examples illustrate the concept of amplitude and phase spectra.

Example 8.1.3: Consider the (trivial) cosine Fourier series

$$x(t) = A\cos(\omega_0 t + \theta). \tag{8.1.1}$$

The one-sided amplitude and phase spectra for this Fourier series is shown in Figure 8.1.1.

We can also plot the two-sided amplitude spectrum of the exponential representation of the Fourier series in (8.1.1). We rewrite (8.1.1) using Euler's formulas as

$$x(t) = A\cos(\omega_0 t + \theta) = \frac{A}{2}e^{j(-\omega_0 t - \theta)} + \frac{A}{2}e^{j(\omega_0 t + \theta)} \tag{8.1.2}$$

$$= \frac{A}{2}e^{-j\theta}e^{-j\omega_0 t} + \frac{A}{2}e^{j\theta}e^{j\omega_0 t} = X_{-1}e^{-j\omega_0 t} + X_1 e^{j\omega_0 t}.$$

The two-sided amplitude and phase spectra for this exponential Fourier series is shown in Figure 8.1.2. From Figure 8.1.1 we see that a cosine term of frequency ω_0 in the cosine Fourier series corresponds to two terms in the exponential Fourier series in Figure 8.1.2, one term at positive frequency ω_0 and one term at negative frequency $-\omega_0$. This observation is just a graphical illustration of the original derivation of the exponential Fourier series from the cosine Fourier series in (8.1.2).

The amplitude spectrum of an exponential Fourier series has nonzero values for negative frequencies. Of course, "negative" frequencies don't really exist. The amplitude spectrum is an even function so the graph on negative frequencies contains no new information. Therefore, we are just plotting a mathematical artifice by including the negative frequency axis. Nonetheless, some properties of signals are conveniently explained using the entire frequency axis. The two-sided amplitude spectrum is widely used.

▲▲

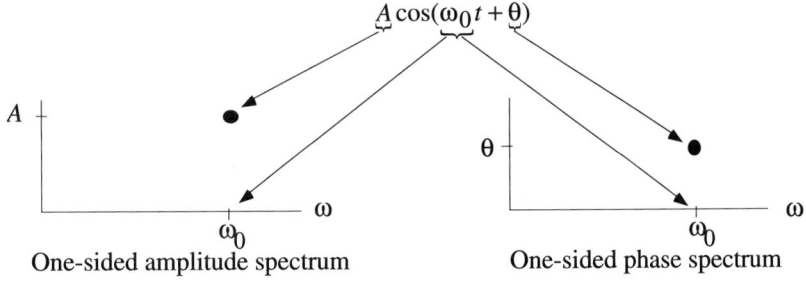

One-sided amplitude spectrum   One-sided phase spectrum

Figure 8.1.1 One-Sided Amplitude and Phase Spectra for a Single Cosine

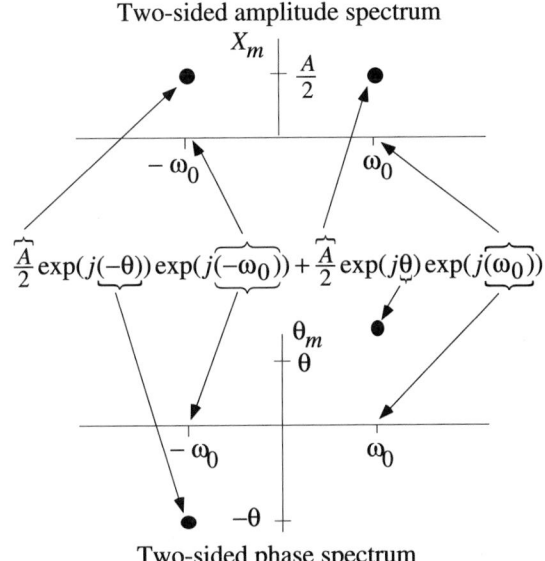

Figure 8.1.2 Two-Sided Amplitude and Phase Spectra for a Single Cosine

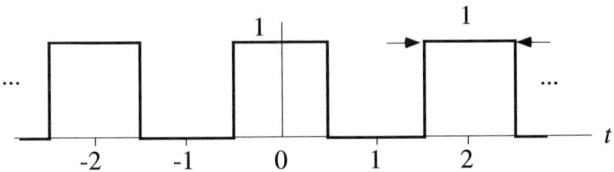

Figure 8.1.3 Pulse Train

Example 8.1.4: Consider again the pulse train $x(t)$ in Figure 8.1.3. In Section 7.1.1 its cosine Fourier series representation is derived as (see Remark 7.1.6)

$$x(t) = \frac{1}{2} + \sum_{m=1}^{\infty} \frac{2}{(2m-1)\pi} \cos\left((2m-1)\pi t + \theta_m\right)$$ (7.1.18)

where

$$\theta_m = \begin{cases} 0, & m = 1,3,5,\dots \\ -\pi, & m = 2,4,6,\dots \end{cases}$$

The one-sided amplitude coefficients are

$$A_m = \begin{cases} \dfrac{1}{2}, & m = 0 \\ \dfrac{2}{(2m-1)\pi}, & m \neq 0 \end{cases}$$ (8.1.3)

The corresponding amplitude and phase spectra are shown in Figure 8.1.4. Recall that the amplitude coefficients of the cosine representation are required to be positive. The phase angles in the phase spectrum represent negative signs on the amplitude coefficients.

▲▲

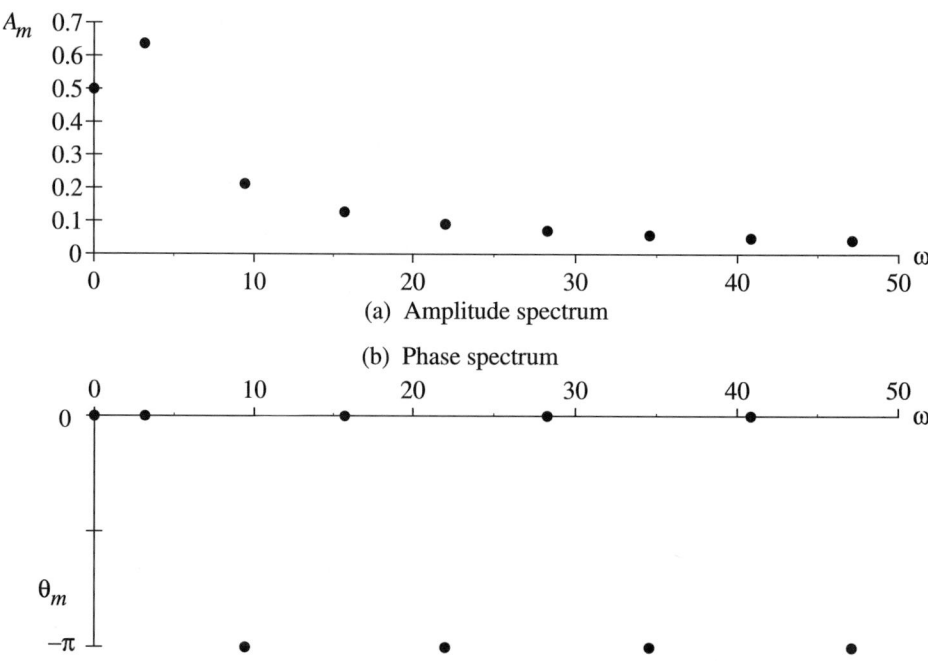

(a) Amplitude spectrum

(b) Phase spectrum

Figure 8.1.4 Amplitude and Phase Spectra of the Signal in Figure 8.1.3

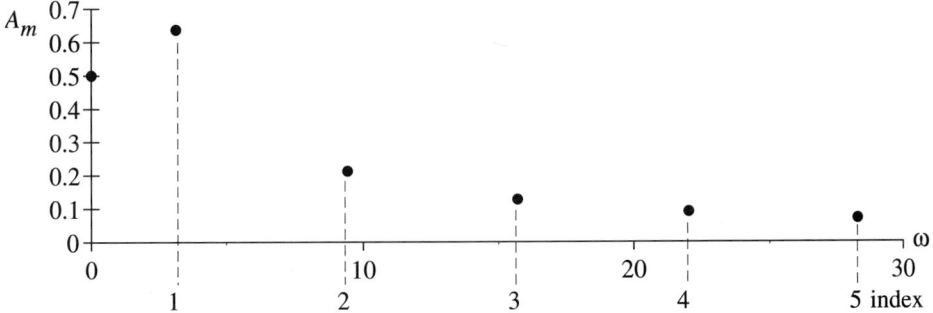

Figure 8.1.5 Amplitude Spectrum of Example 8.1.4

Notation: When graphing the line spectra of a signal, there are two scales commonly chosen for the abscissa. These scales are shown in Figure 8.1.5 for the amplitude spectrum in Figure 8.1.4. One scale is in absolute rad/sec as shown in Figure 8.1.5. From this graph we can extract both the amplitude coefficients and the corresponding frequencies of the cosine terms in the Fourier series. The second scaling takes advantage of the fact that the frequencies of the cosine terms are all integer multiples of the fundamental frequency.   In this scheme the absolute frequency scaling is repressed and only the integer multiple is displayed as in Figure 8.1.5. To extract the absolute frequencies from this graph requires prior knowledge of the fundamental frequency. In this text, only the absolute scaling will be used.

Example 8.1.5:   Consider again the signal in Example 8.1.4.   The exponential Fourier series for this signal, as derived in Example 7.3.5, is

$$x(t) = \frac{1}{2} + \sum_{\substack{m=-\infty \\ m \neq 0}}^{\infty} \left[\frac{1}{(2m-1)\pi} e^{j(m-1)\pi} \right] e^{j(2m-1)\pi t}. \tag{8.1.4}$$

The two-sided amplitude coefficients are

$$|X_0| = \frac{1}{2}, \quad |X_m| = \left| \frac{1}{(2m-1)\pi} e^{j(m-1)\pi} \right| = \left| \frac{1}{(2m-1)\pi} \right|. \tag{8.1.5}$$

The two-sided phase coefficients are

$$\angle X_m = \begin{cases} 0, & m = 1, 3, 5, \ldots \\ -\pi, & m = 2, 4, 6, \cdots \end{cases} \tag{8.1.6}$$

The two-sided line spectra for this signal are shown in Figure 8.1.6.

▲▲

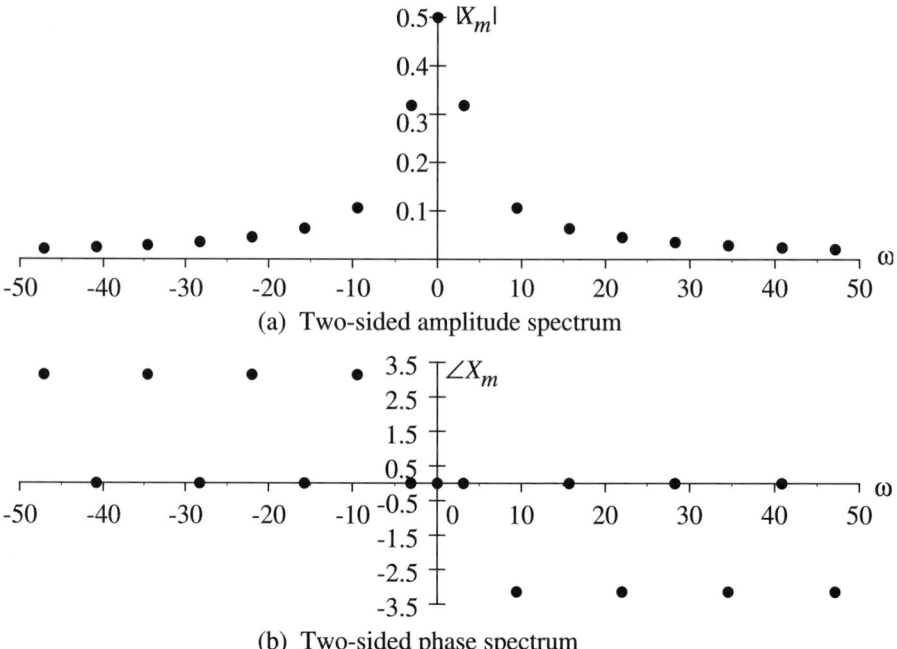

(a) Two-sided amplitude spectrum

(b) Two-sided phase spectrum

Figure 8.1.6 Two-Sided Amplitude and Phase Spectra for the Signal
in Example 8.1.5

Notation and Terminology: There is another format of the amplitude and phase spectra that is widely used. This format consists of using lines instead of discrete points. We call this type of graph a <u>line spectrum</u> for obvious reasons. It turns out that line spectra are very inconvenient for our discussions in later chapters. Therefore, we will not use them.

The two-sided line spectra in Figure 8.1.6 makes it obvious that the amplitude spectrum of a signal is an even function of ω, and the phase spectrum is an odd function of ω when $x(t)$ is a real signal. This fact can be established analytically from the integral that is used to calculate the coefficients of the complex Fourier series X_m in (7.3.22). It is merely necessary to show that X_m and X_{-m} are complex conjugates.

The amplitude and phase spectra of a cosine Fourier series completely defines the signal with the pairs of numbers

$$m\omega_0 \leftrightarrow (A_m, \theta_m), \quad m = 0, 1, 2, \ldots \tag{8.1.7}$$

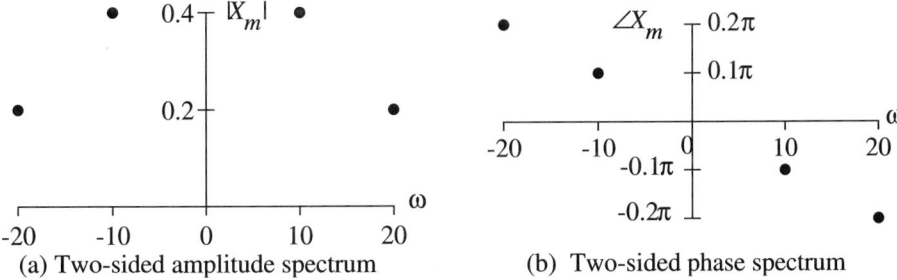

Figure 8.1.7 Amplitude and Phase Spectra of a Fourier Series

Thus, we can reconstruct the signal from the graph. A similar comment applies to the exponential Fourier series. The following example illustrates the construction of a Fourier series from its amplitude and phase spectra.

Example 8.1.6: The two-sided amplitude and phase spectra of a Fourier series is shown in Figure 8.1.7. We can construct the Fourier series from these line spectra. The first nonzero term occurs at the frequency $\omega = 10$ rad/sec, and the second nonzero term occurs at $\omega = 20$ rad/sec. We conclude that the fundamental frequency is $\omega_0 = 10$ rad/sec. At the frequency $\omega = 10$ rad/sec the $m = 1$ term in the Fourier series is

$$X_1 = 0.4e^{-j0.1\pi}e^{j10t}. \tag{8.1.8}$$

The (0.4) factor in (8.1.8) came from the amplitude plot and the exponent came from the phase plot, both of these numbers being read from the frequency $\omega_0 = 10$ rad/sec. Continuing in this manner we obtain the terms

$$X_{-2} = 0.2e^{j0.2\pi}e^{-j20t}, \quad X_{-1} = 0.4e^{j0.1\pi}e^{-j10t}, \quad X_2 = 0.2e^{-j0.2\pi}e^{j20t}. \tag{8.1.9}$$

Combining all of these terms together using Euler's identity we obtain

$$x(t) = 0.2e^{j0.2\pi}e^{-j20t} + 0.4e^{j0.1\pi}e^{-j10t} + 0.4e^{-j0.1\pi}e^{j10t} + 0.2e^{-j0.2\pi}e^{j20t} \tag{8.1.10}$$
$$= 0.8\cos(10t - 0.1\pi) + 0.4\cos(20t - 0.2\pi).$$

▲▲

We will continue the investigation of the relationship between the amplitude and phase spectra and the structure of a periodic signal after we have introduced the notion of energy and power in Sections 8.2 and 8.3.

8.1.3 Aperiodic Signals

The period and the coefficients of the exponential Fourier series completely define a periodic signal. This information is contained in the graphical representation of the signal via the amplitude and phase spectra. This idea can be extended to aperiodic signals through the Fourier transform. The Fourier transform completely defines an aperiodic signal $x(t)$. We can also functionally and graphically represent the Fourier transform in a manner similar to the amplitude and phase spectra for Fourier series. In fact, we employ the same terminology. In general the Fourier transform $X(\omega)$ of a signal $x(t)$ is complex-valued. Therefore, two plots are presented, one for the magnitude and one for the phase of $X(\omega)$.

Definition 8.1.7: Let the Fourier transform of the signal $x(t)$ be $X(\omega)$. The function

$$|X(\omega)| \text{ vs. } \omega, \quad -\infty < \omega < \infty$$

is called the <u>amplitude spectrum</u> of $x(t)$. The function

$$\angle X(\omega) \text{ vs. } \omega, \quad -\infty < \omega < \infty$$

is called the <u>phase spectrum</u> of $x(t)$.

▲▲

As we noted for the amplitude and phase spectra for Fourier series, the amplitude and phase spectra are of the same form for *all* aperiodic signals. Hence, they can be used as a basis for comparison of signals with seemingly quite different waveforms.

The following example illustrates the amplitude and phase spectra.

Example 8.1.8: Consider the pulse shown in Figure 8.1.8. The Fourier transform of this signal is

$$\mathcal{F}\left\{\Pi\left(\frac{t}{\varepsilon}\right)\right\} = \varepsilon \operatorname{Sa}\left(\frac{\omega\varepsilon}{2}\right). \tag{8.1.11}$$

The amplitude spectrum of the Fourier transform of the pulse function in Figure 8.1.8a is

$$|X(\omega)| = \left|\varepsilon \operatorname{Sa}\left(\frac{\omega\varepsilon}{2}\right)\right|. \tag{8.1.12}$$

The amplitude spectrum is shown in Figure 8.1.8b. The phase spectrum, shown in Figure 8.1.8c, of this Fourier transform is

$$\angle X(\omega) = \begin{cases} 0, & 0 \le \varepsilon \operatorname{Sa}\left(\dfrac{\omega\varepsilon}{2}\right) \\[3mm] \pm\pi, & 0 > \varepsilon \operatorname{Sa}\left(\dfrac{\omega\varepsilon}{2}\right) \end{cases} \tag{8.1.13}$$

▲▲

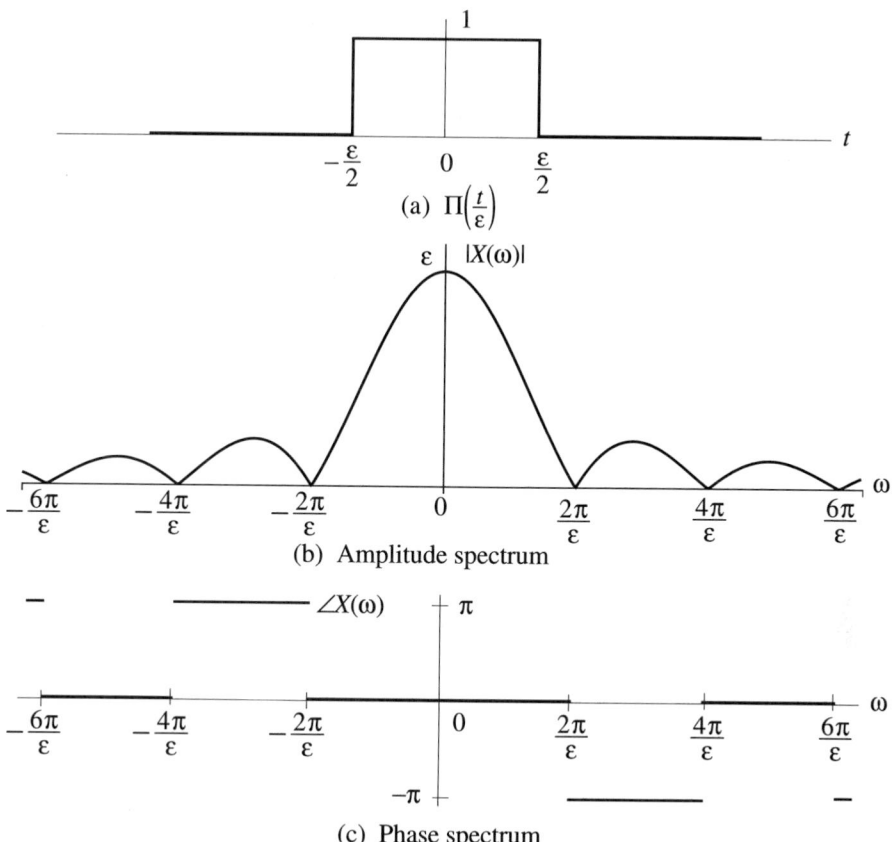

Figure 8.1.8 Pulse and Its Amplitude and Phase Spectra

If $x(t)$ is a real signal, then amplitude spectrum is an even function of ω, and the phase spectrum is an odd function of ω as shown in Figure 8.1.8. This fact can be established from the definition of the Fourier transform by showing that $X(-\omega)$ is the complex conjugate of $X(\omega)$.

Notation: It is common practice to plot the amplitude and phase spectra on the entire frequency axis. The graph on negative frequencies carries no new information because of the symmetries of the amplitude and phase spectra. Of course, negative frequencies have no physical meaning, but this format of the graph is faithful to the definition of the Fourier transform.

8.1.4 Fourier Series and the Generalized Fourier Transform

Consider the (trivial) Fourier series

$$x(t) = A\cos(\omega_0 t) = \frac{A}{2}\left(e^{j\omega_0 t} + e^{-j\omega_0 t}\right). \tag{8.1.14}$$

(a) Amplitude spectrum of the Fourier series   (b) Amplitude spectrum of the Fourier transform

Figure 8.1.9 Amplitude Spectrum of $A\cos(\omega_0 t)$

This signal also has a generalized Fourier transform

$$\mathcal{F}\{x(t)\} = A\pi[\delta(\omega - \omega_0) + \delta(\omega + \omega_0)].$$

(8.1.15)

We now have an option as to whether to represent the amplitude spectrum of this signal as a Fourier series or a Fourier transform. The amplitude spectrum of the Fourier series in (8.1.14) is shown in Figure 8.1.9a. The amplitude spectrum of the Fourier transform in (8.1.15) is shown in Figure 8.1.9b. Both of these graphs carry the same information. Most importantly, the frequency of the cosine is identified on the frequency axis. The amplitude of the signal is scaled by $(1/2)$ in the amplitude spectrum of the exponential Fourier series. The amplitude is scaled by π in the amplitude spectrum of the Fourier transform. Also note that the amplitude spectrum of the Fourier transform is an impulse function. In practice, the most convenient form is used.

8.2 ENERGY AND POWER SIGNALS

8.2.1 Introduction

Energy and power play an important role in the analysis of networks and other physical processes. We have introduced the notion of a signal, Definition 5.1.1, in Chapter 5. In this section we extend the notion of energy and power to the signals we defined in Chapter 5. This notion is necessarily an abstraction of energy and power commonly defined based on physical concepts. Nonetheless it is consistent with those definitions.

8.2.2 Definitions

The following two definitions are abstractions from physics.

Definition 8.2.1: The <u>energy</u> E of the signal $x(t)$ is defined as

$$E = \lim_{\gamma \to \infty} \int_{-\gamma}^{\gamma} |x(t)|^2 dt$$

if the limit exists and it is nonzero.

▲▲

Definition 8.2.2: The <u>power</u> P of the signal $x(t)$ is defined as

$$P = \lim_{\gamma \to \infty} \frac{1}{2\gamma} \int_{-\gamma}^{\gamma} |x(t)|^2 \, dt$$

if the limit exists and it is nonzero.

▲▲

Notation: The energy in the signal $x(t)$ will be denoted by E_x. The power in the signal $y(t)$ will be denoted by P_y.

The next two examples illustrate the difference between energy and power signals. The first example illustrates the calculation of energy and the second example illustrates the calculation of power.

Example 8.2.3: In this example we calculate the energy of the signal

$$x(t) = \Pi\left(\frac{t}{\varepsilon}\right). \tag{8.2.1}$$

Using Definition 8.2.1 the energy in this signal is

$$E = \lim_{\gamma \to \infty} \int_{-\gamma}^{\gamma} |x(t)|^2 \, dt = \lim_{\gamma \to \infty} \int_{-\gamma}^{\gamma} \left|\Pi\left(\frac{t}{\varepsilon}\right)\right|^2 \, dt. \tag{8.2.2}$$

Evaluating (8.2.2) we find that

$$E = \lim_{\gamma \to \infty} \int_{-\gamma}^{\gamma} \left|\Pi\left(\frac{t}{\varepsilon}\right)\right|^2 \, dt = \int_{-\frac{\varepsilon}{2}}^{\frac{\varepsilon}{2}} 1^2 \, dt = \varepsilon. \tag{8.2.3}$$

The power in this signal is

$$P = \lim_{\gamma \to \infty} \frac{1}{2\gamma} \int_{-\gamma}^{\gamma} |x(t)|^2 \, dt = \lim_{\gamma \to \infty} \frac{1}{2\gamma} \int_{-\gamma}^{\gamma} \left|\Pi\left(\frac{t}{\varepsilon}\right)\right|^2 \, dt = \lim_{\gamma \to \infty} \frac{1}{2\gamma}(\varepsilon) = 0. \tag{8.2.4}$$

This example shows that if a signal has finite duration and it is bounded, then the power of the signal is zero.

▲▲

Example 8.2.4 In this example we calculate the power of the signal

$$x(t) = A\cos(\omega_0 t + \theta), \quad -\infty < t < \infty. \tag{8.2.5}$$

This signal plays a central role in much of the analysis in this text. As such this power calculation is of fundamental importance. We note that the energy in this signal is infinite. Using Definition 8.2.2 the power of (8.2.5) is given by

$$P = \lim_{\gamma \to \infty} \frac{1}{2\gamma} \int_{-\gamma}^{\gamma} [A\cos(\omega_0 t + \theta)]^2 \, dt. \tag{8.2.6}$$

Equation (8.2.6) can be integrated by using a trigonometric identity. We have

$$P = \lim_{\gamma \to \infty} \frac{1}{2\gamma} \int_{-\gamma}^{\gamma} A^2 \frac{1}{2}[1 - \cos 2(\omega_0 t + \theta)] \, dt \tag{8.2.7}$$

$$= \lim_{\gamma \to \infty} \frac{1}{2\gamma} \left[\int_{-\gamma}^{\gamma} A^2 \frac{1}{2} \, dt \right] - \lim_{\gamma \to \infty} \frac{1}{2\gamma} \left[\int_{-\gamma}^{\gamma} \frac{1}{2} \cos 2(\omega_0 t + \theta) \, dt \right].$$

The first integral in (8.2.7) leads to

$$\lim_{\gamma \to \infty} \frac{1}{2\gamma} \left[\int_{-\gamma}^{\gamma} A^2 \frac{1}{2} \, dt \right] = \lim_{\gamma \to \infty} \frac{1}{2\gamma} \left[\frac{A^2}{2}(2\gamma) \right] = \frac{A^2}{2}. \tag{8.2.8}$$

Evaluating the second integral we find that

$$\lim_{\gamma \to \infty} \frac{1}{2\gamma} \int_{-\gamma}^{\gamma} \frac{1}{2} \cos 2(\omega_0 t + \theta) \, dt = \lim_{\gamma \to \infty} \frac{1}{2\gamma} \left[\frac{1}{2\omega_0} \sin[2(\omega_0 t + \theta)] \right]_{t=-\gamma}^{\gamma} \tag{8.2.9}$$

$$= \lim_{\gamma \to \infty} \frac{1}{2\gamma} \left[\frac{1}{2\omega_0} [\sin(2\omega_0 \gamma + 2\theta) - \sin(-2\omega_0 \gamma + 2\theta)] \right].$$

Since the sine function is bounded we have

$$\lim_{\gamma \to \infty} \frac{1}{2\gamma} \left| \frac{1}{2\omega_0} [\sin(2\omega_0 \gamma + 2\theta) - \sin(-2\omega_0 \gamma + 2\theta)] \right| \leq \lim_{\gamma \to \infty} \frac{1}{2\gamma} \left(\frac{1}{\omega_0} \right) = 0. \tag{8.2.10}$$

So we find that the power in the signal (8.2.5) is

$$P = \frac{A^2}{2}. \tag{8.2.11}$$

Note: The power of a sinusoidal signal in (8.2.5) is independent of the phase θ of the signal (8.2.5).

▲▲

From these two examples we make the following observations:

1.   If the energy of the signal $x(t)$ is finite, then the power in this signal is zero.

This observation follows from Definition 8.2.1 and Definition 8.2.2.   Rewriting the integral in the definition of the power in the signal we have

$$P = \lim_{\gamma \to \infty} \frac{1}{2\gamma} \int_{-\gamma}^{\gamma} |x(t)|^2 \, dt = \lim_{\gamma \to \infty} \frac{1}{2\gamma} (E).$$

(8.2.12)

Hence, if E is finite, then P must be zero.

2.   If the power in the signal $x(t)$ is nonzero, then the energy in the signal is infinite $E = \infty$.

This observation also follows from (8.2.12).   Example 8.2.3 shows that a bounded signal must be nonzero over an infinite interval, if it is to be a power signal.

3.   There exist signals for which neither energy nor power is finite.

An example of a signal that is neither a power nor an energy signal is

$$x(t) = e^{-at}, \quad -\infty < t < \infty.$$

(8.2.13)

We will not frequently encounter these types of signals.

Definitions 8.2.1 and 8.2.2 above divide signals into mutually exclusive classes. Based on these definitions we can classify signals as follows.

Definition 8.2.5: The signal $x(t)$ is an <u>energy signal</u> if $0 < E < \infty$ for that signal.

▲▲

Definition 8.2.6: The signal $x(t)$ is a <u>power signal</u> if $0 < P < \infty$ for that signal.

▲▲

Examples 8.2.3 and 8.2.4 lend a certain amount of insight into classifying a signal as an energy or power signal based on the waveform.   Energy signals are pulse like.   If the signal is bounded and of finite duration then it is probably an energy signal.   Energy signals can have infinite duration but they tend to zero rapidly in both positive and negative time.
Power signals usually have infinite duration; they don't tend to zero as time goes to infinity.   Mathematically, a bounded signal that is a power signal must exist for an infinite interval of time.   Obviously, this property is physically impossible to verify. For many applications, however, the definition of power is very useful for analysis and design purposes.   In this case we define a signal to be a power signal, if it exists for a sufficiently long time interval, "sufficiently long" being defined by the application at hand.   Power signals are encountered in power systems and communication systems, for example.

8.3 ENERGY SPECTRAL DENSITY

8.3.1 Parseval's Theorem

In the last section we introduced a definition of the energy in a signal $x(t)$. The energy of a signal is related to its Fourier transform by Parseval's theorem. Parseval's theorem was introduced as a property of the Fourier transform (Property 7.5.4). Observing the definition of energy introduced in the last section, we see an immediate connection to Fourier transforms. Here we will restate the theorem and prove it. Parseval's theorem is a cornerstone in signal analysis.

Theorem 8.3.1: Parseval's Theorem Let $x(t)$ be an energy signal which has a Fourier transform $X(\omega)$. Then the energy in this signal is given by

$$E = \int_{-\infty}^{\infty} |x(t)|^2 \, dt = \frac{1}{2\pi} \int_{-\infty}^{\infty} |X(\omega)|^2 \, d\omega.$$

▲▲

Proof: The energy in the signal $x(t)$ is given by

$$E = \int_{-\infty}^{\infty} |x(t)|^2 \, dt = \int_{-\infty}^{\infty} x(t)\bar{x}(t) \, dt. \tag{8.3.1}$$

The second equality follows from complex variables. Substituting the definition of the Fourier transform of $x(t)$ into (8.3.1) we obtain

$$E = \int_{-\infty}^{\infty} \bar{x}(t)x(t) \, dt = \frac{1}{2\pi} \int_{-\infty}^{\infty} \bar{x}(t) \int_{-\infty}^{\infty} X(\omega)e^{j\omega t} \, d\omega \, dt. \tag{8.3.2}$$

Reversing the order of integration yields

$$E = \frac{1}{2\pi} \int_{-\infty}^{\infty} X(\omega) \int_{-\infty}^{\infty} \bar{x}(t)e^{j\omega t} \, dt \, d\omega = \frac{1}{2\pi} \int_{-\infty}^{\infty} X(\omega) \left[\overline{\int_{-\infty}^{\infty} x(t)e^{-j\omega t} \, dt} \right] d\omega. \tag{8.3.3}$$

The inner integral is just the complex conjugate of the Fourier transform of the $x(t)$. So (8.3.3) yields

$$E = \frac{1}{2\pi} \int_{-\infty}^{\infty} X(\omega)\bar{X}(\omega) \, d\omega = \frac{1}{2\pi} \int_{-\infty}^{\infty} |X(\omega)|^2 \, d\omega. \tag{8.3.4}$$

▲▲

Terminology: The definition of the energy of a signal relies on the *time domain* representation of the signal $x(t)$; just integrate the square of the signal. Parseval's theorem gives a second way of computing the energy based on the Fourier transform of the signal; integrate the square of the magnitude of the Fourier transform. Since the Fourier transform is a function of the frequency variable ω, we say this integral is in the *frequency domain*. Hence, Parseval's theorem relates a time domain of the energy in a signal to the frequency domain description.

The version of Parseval's theorem stated in Theorem 8.3.1 requires that the signal be an energy signal with a Fourier transform in the sense of Definition 7.4.1; generalized Fourier transforms are excluded by this version of Parseval's theorem. In particular, sinusoids have a generalized Fourier transform, but they are power signals, not energy signals. We will discuss Parseval's theorem for power signals in the next section.

Parseval's theorem describes how the energy in the signal is distributed along the frequency axis by the function $|X(\omega)|^2$. This theorem gives rise to the following definition.

Definition 8.3.2: The function

$$D(\omega) = |X(\omega)|^2$$

is called the <u>energy spectral density</u>.

▲▲

Notation: If we wish to indicate to which signal the energy spectral density is attached, we attach a subscript to the energy spectral density. That is, the energy spectral density of $x(t)$ is $D_x(\omega) = |X(\omega)|^2$.

We make the following two observations about the energy spectral density.

1. The energy spectral density is determined from the square of the amplitude spectrum of the signal. Hence, the energy spectral density is always a nonnegative, real function.
2. The energy spectral density is determined purely from the amplitude spectrum of $x(t)$. All phase information is lost. So the energy spectral density of a energy signal is unlike the amplitude and phase spectra, which completely determine the signal. That is, many energy signals can have the same energy spectral density.

The next example illustrates these concepts.

Example 8.3.3: Consider the signal

$$x(t) = e^{-\alpha t} u_s(t), \quad \alpha > 0. \tag{8.3.5}$$

The Fourier transform of $x(t)$ is given by

$$X(\omega) = \frac{1}{\alpha + j\omega}. \tag{8.3.6}$$

The amplitude spectrum is computed as follows. Converting the Fourier transform to polar notation we find

$$X(\omega) = \frac{1}{\alpha + j\omega} = \frac{1}{\sqrt{\alpha^2 + \omega^2}\, e^{j\tan^{-1}\frac{\omega}{\alpha}}} = \frac{1}{\sqrt{\alpha^2 + \omega^2}}\, e^{-j\tan^{-1}\frac{\omega}{\alpha}}. \qquad (8.3.7)$$

From (8.3.7) the amplitude spectrum is

$$|X(\omega)| = \frac{1}{\sqrt{\alpha^2 + \omega^2}}. \qquad (8.3.8)$$

The energy spectral density of this signal is

$$|X(\omega)|^2 = \frac{1}{\alpha^2 + \omega^2}. \qquad (8.3.9)$$

A comparison of the amplitude spectrum and energy spectral density is shown in Figure 8.3.1. Using Parseval's theorem, Theorem 8.3.1, we can compute the total energy in this signal. We have

$$E = \frac{1}{2\pi} \int_{-\infty}^{\infty} |X(\omega)|^2\, d\omega = \frac{1}{2\pi} \int_{-\infty}^{\infty} \frac{1}{\alpha^2 + \omega^2}\, d\omega = \frac{1}{2\pi}\left[\frac{\tan^{-1}\left(\frac{\omega}{\alpha}\right)}{\alpha}\right]_{\omega=-\infty}^{\infty} = \frac{1}{2\alpha}. \qquad (8.3.10)$$

Furthermore, we can use Parseval's theorem to compute the energy in a certain frequency band. Let us select the frequency band $0 < \omega < m\alpha$. The energy in this frequency band is

$$E_{m\alpha} = \frac{1}{2\pi} \int_{-m\alpha}^{m\alpha} |X(\omega)|^2\, d\omega = \frac{1}{2\pi\alpha}\left[\tan^{-1}\left(\frac{\omega}{\alpha}\right)\right]_{\omega=-m\alpha}^{m\alpha} = \frac{\tan^{-1}(m)}{\pi\alpha}. \qquad (8.3.11)$$

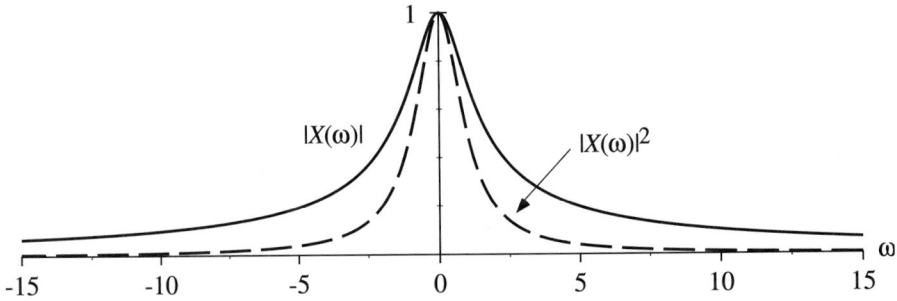

Figure 8.3.1 A Comparison of the Amplitude Spectrum and
Energy Spectral Density of (8.3.6)

The percentage of the energy of this signal in this frequency band is

$$\%E = \frac{\dfrac{\tan^{-1}(m)}{\pi\alpha}}{\dfrac{1}{2\alpha}} \times 100 = \left(\frac{2}{\pi}\right)\tan^{-1}(m) \times 100. \qquad (8.3.12)$$

For $m = 1$, the percentage of energy in the frequency band $0 < \omega < \alpha$ is $\%E = 50\%$. For $m = 2$, the percentage of energy is $\%E = 70\%$.

In this example we computed the energy in the signal contained in the frequency interval $0 < \omega < m\alpha$ using Parseval's theorem. When we carried out the integration, however, we integrated over the frequency interval $-m\alpha < \omega < m\alpha$ as shown in (8.3.11). This convention is generally followed when working with frequency domain representations of signals. We speak of *positive* intervals of frequencies, but when we do calculations, we include the mirror image on the negative axis. In particular, this convention is followed in the discussion of signal bandwidth in the next subsection.

The energy spectral density is an even function. Often this fact can be exploited to reduce the complexity of the integration when computing the energy. In particular, the energy of a signal in a particular frequency band can be found by integrating on the positive frequency interval and multiplying by 2.

▲▲

8.3.2 Signal Bandwidth

Parseval's theorem relates the energy in a signal to the energy spectral density of the signal. Since these functions are a function of frequency, we can relate the distribution of the energy or power of a signal to certain frequency intervals in the frequency domain. This concept of the energy distribution of a signal in terms of frequency plays a central role in filtering, signal processing, and communication with ramifications in many other fields. The fundamental concept is captured by the following definition.

Definition 8.3.4: Let $x(t)$ be a signal with an energy spectral density $D(\omega)$. Then we say this signal is <u>bandlimited</u> if $D(\omega) = 0$ for $|\omega| > BL$. The number BL is called the <u>bandlimit</u>.

▲▲

If a signal is bandlimited, then all of its energy is confined to a particular frequency band. Next we identify this frequency band. Let

$$\omega_{min} = \min_{\omega \geq 0}\{\omega \mid D(\omega) \neq 0\} \qquad (8.3.13)$$

$$\omega_{max} = \max_{\omega > 0}\{\omega \mid D(\omega) \neq 0\}$$

All of the energy is contained in the frequency band $\omega_{min} < \omega < \omega_{max}$.

Definition 8.3.5: The <u>bandwidth</u> B of the signal $x(t)$ is defined as

$$B = \omega_{max} - \omega_{min}.$$

▲▲

Note: The bandlimit and the bandwidth are defined for positive frequencies only. See the comments in Example 8.3.3.

Since the energy spectral density is defined in terms of amplitude spectrum of the signal, the amplitude spectrum can also be used to determine the bandlimit and bandwidth of a signal.

Example 8.3.6: (a) The Fourier transform of the energy signal

$$X(\omega) = \Pi\left(\frac{\omega}{\varepsilon}\right) \tag{8.3.14}$$

is shown in Figure 8.3.2a. The bandlimit of this signal is $\dfrac{\varepsilon}{2} = BL$. The bandwidth of this signal is $B = \dfrac{\varepsilon}{2}$. The signal itself is

$$x(t) = \frac{\varepsilon}{2\pi}\,\mathrm{Sa}\left(\frac{\varepsilon t}{2}\right) \tag{8.3.15}$$

and it is shown in Figure 8.3.2b.

(b) The Fourier transform

$$X(\omega) = \Pi\left(\frac{\omega - \omega_0}{\varepsilon}\right) + \Pi\left(\frac{\omega + \omega_0}{\varepsilon}\right) \tag{8.3.16}$$

of an energy signal $x(t)$ is shown in Figure 8.3.3a. This signal is bandlimited to the frequency band $|\omega| < \omega_0 + \varepsilon = BL$. The bandwidth of this signal is $B = \varepsilon$. The signal itself is

$$x(t) = \frac{\varepsilon}{2\pi}\,\mathrm{Sa}\left(\frac{\varepsilon t}{2}\right)\cos(\omega_0 t), \tag{8.3.17}$$

(see Example 7.5.15) and it is shown in Figure 8.3.3b.

Note that while the energy spectral densities of these two signals are bandlimited (zero everywhere except on a certain frequency band), the signals themselves are nonzero over the entire time axis. It is generally true that signals that are limited in frequency are not limited in time while signals that are limited in time are not limited in frequency.

▲▲

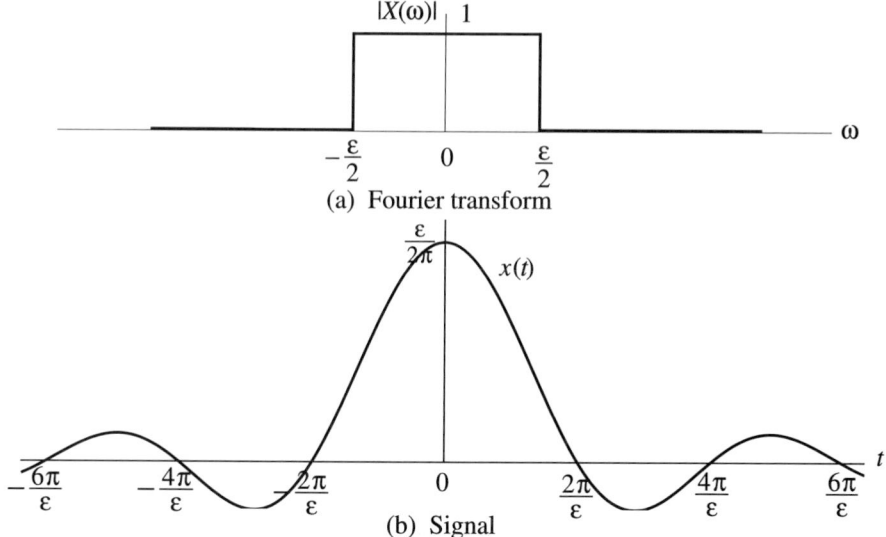

Figure 8.3.2 Fourier Transform (8.3.14) and the Signal

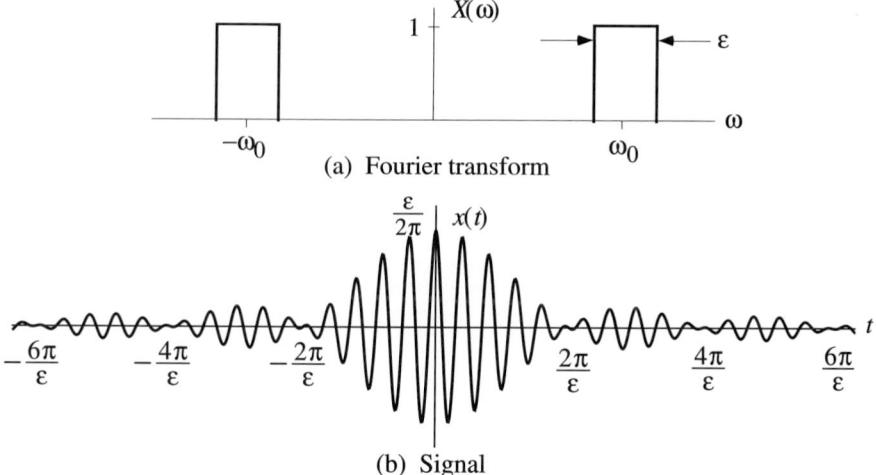

Figure 8.3.3 Fourier Transform (8.3.16) and the Signal (8.3.17)

In this example the energy of the signal shown in Figure 8.3.2a is concentrated near the origin while the energy in the signal shown in Figure 8.3.3a is concentrated away from the origin. These two classes of signals play an important role in communication theory, and we give a name to them.

Definition 8.3.7: If the energy of the signal $x(t)$ is contained in the frequency band $0 < \omega < BL$, the signal is said to be a <u>baseband</u> signal. If the energy is contained in the frequency band $0 < \omega_{min} < \omega < BL$, the signal is said to be a <u>passband</u> signal. If a signal is a passband signal and the frequencies defining the bandwidth satisfy

$$(10)(\omega_{max} - \omega_{min}) < \frac{\omega_{max} + \omega_{min}}{2} \tag{8.3.18}$$

then the signal is said to be <u>narrowband</u>. Otherwise the signal is <u>wideband</u>.

▲▲

Definition 8.3.7 says that if the bandwidth of the signal is ten times the center frequency, then the signal is narrowband.

Example 8.3.8: The signal in Figure 8.3.2 is a baseband signal, and the signal in Figure 8.3.3 is a passband signal.

▲▲

Example 8.3.9: Consider the signal in Example 8.3.6. The energy spectral density is shown in Figure 8.3.3. Suppose $f_0 = 10$ kHz, or $\omega_0 = (2\pi)10$ krad. If $\varepsilon = 100$ Hz, then this signal is a narrowband signal because

$$(10)(f_{max} - f_{min}) < \frac{f_{max} + f_{min}}{2}, \tag{8.3.19}$$

or

$$(10)(100) < \frac{20 \text{ kHz}}{2}.$$

▲▲

In practice, no signal is strictly bandlimited as defined Definition 8.3.4. Some signals, however, have a "high" percentage of their energy concentrated in a certain frequency band. Several definitions are available for the bandwidth of such signals. The following definition is sometimes used. Consider an energy signal $x(t)$ whose energy spectral density is $D(\omega)$. Suppose the maximum value of the energy spectrum is

$$M = \max_{\omega} D(\omega). \tag{8.3.20}$$

Given a number $0 < \alpha < 1$, let N_α satisfy

$$\alpha = \frac{N_\alpha}{M}. \tag{8.3.21}$$

If the signal is a passband signal, define the frequencies ω_{min} and ω_{max} by

$$\omega_{min} = \min_{\omega \geq 0} \left\{ N_\alpha = D(\omega) \right\} \tag{8.3.22}$$

$$\omega_{max} = \max_{\omega \geq 0} \left\{ N_\alpha = D(\omega) \right\}$$

If the signal is a baseband signal, then define these frequencies as

$$\omega_{min} = 0 \tag{8.3.23}$$

$$\omega_{max} = \max_{\omega \geq 0} \left\{ N_\alpha = D(\omega) \right\}$$

Definition 8.3.10: The $\underline{\alpha \text{ energy bandwidth }} B_\alpha$ of the signal $x(t)$ is defined as

$$B_\alpha = \omega_{max} - \omega_{min}.$$

The frequencies ω_{min} and ω_{max} are called the $\underline{\alpha \text{ energy frequencies}}$.

▲▲

Terminology: One common selection for α is $\alpha = 0.5$. In this case the frequencies in Definition 8.3.10 are called the $\underline{\text{half-energy}}$ frequencies, or $\underline{3 \text{ dB}}$ frequencies.

Example 8.3.11: Consider the signal of Example 8.3.3,

$$x(t) = e^{-t} u_s(t). \tag{8.3.24}$$

The energy spectral density of this signal is

$$|X(\omega)|^2 = \frac{1}{1 + \omega^2} \tag{8.3.25}$$

and it is shown in Figure 8.3.4. For this signal $M = 1$ in (8.3.20). If we chose $\alpha = 0.5$, then $N_\alpha = \alpha = 0.5$ in (8.3.21). This signal is, by inspection, a baseband signal. Therefore, $\omega_{min} = 0$ and $\omega_{max} = 1$. Therefore, the half-energy bandwidth is $B_{0.5} = 1$ rad/sec.

▲▲

Here is another way to define the signal bandwidth.

Definition 8.3.12: The $\underline{E\% \text{ bandwidth }} B_{E\%}$ of the signal is defined as

$$B_{E\%} = \omega_{max} - \omega_{min}$$

where $\omega_{min} < \omega < \omega_{max}$ is the frequency interval that contains $E\%$ of the energy of the signal.

▲▲

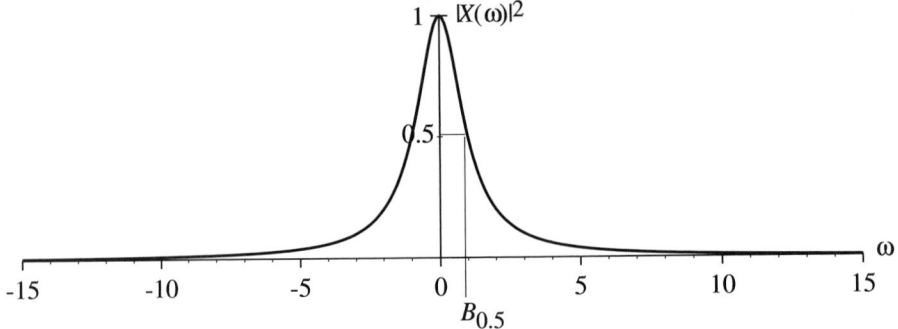

Figure 8.3.4 Half Energy Bandwidth of the Signal (8.3.24)

Example 8.3.13: Consider again the signal in Example 8.3.3. It was also shown in (8.3.12) that the percentage of the energy of this signal in the frequency band $0 < \omega < (1)m$ is

$$E\% = \left(\frac{2}{\pi}\right) \tan^{-1}(m) \times 100. \tag{8.3.26}$$

If we define the bandwidth of the signal to be that frequency band which contains 95% of the energy of the signal, then the bandwidth of this signal is $B_{95\%} = 12.7$ rad/sec. This bandwidth is illustrated in Figure 8.3.5. This method of computing signal bandwidth relies on the area under the energy spectral density. The previous method of determining signal bandwidth uses the energy spectral density curve directly.

▲▲

Note: Throughout this discussion of signal bandwidth, the word "energy" can be replaced by the word "power" if the energy concepts introduced in this section are replaced by the power concepts introduced in the next section.

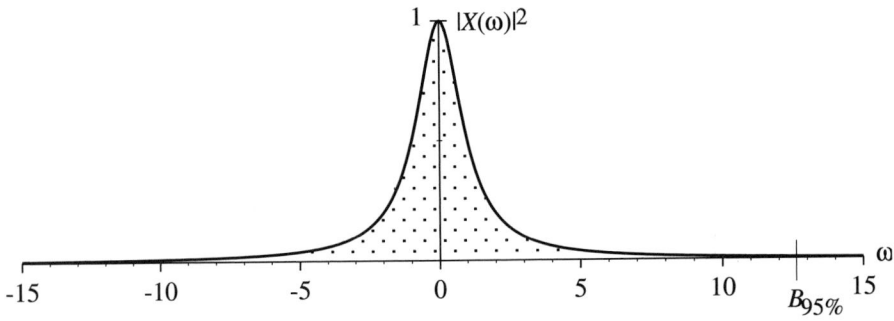

Figure 8.3.5 95% Energy Bandwidth of the Signal

8.4 POWER SPECTRAL DENSITY

8.4.1 Definition

The concept of energy spectral density is very powerful, but as defined it doesn't apply to power signals. Power signals, however, are frequently encountered and we would like to have a notion similar to the energy spectral density. In this section we will develop the analog of energy spectral density for power signals.

Let $x(t)$ be a power signal. The development of the power spectral density relies on the following concept.

Definition 8.4.1: Given a power signal $x(t)$, the <u>truncated version</u> of $x(t)$, $x_\gamma(t)$, is defined as

$$x_\gamma(t) = \begin{cases} x(t), & |t| \leq \gamma \\ 0, & |t| > \gamma \end{cases}$$

▲▲

Note that

$$\lim_{\gamma \to \infty} x_\gamma(t) = x(t). \tag{8.4.1}$$

The truncated signal is an energy signal because it is defined on a finite length interval. The energy signal $x_\gamma(t)$ has a Fourier transform[1]

$$X_\gamma(\omega) = \mathcal{F}\{x_\gamma(t)\} = \frac{1}{2\pi} \int_{-\infty}^{\infty} x_\gamma(t) e^{-j\omega t}\, dt, \tag{8.4.2}$$

because it is only defined on a bounded interval. The energy in the truncated signal is

$$E = \int_{-\infty}^{\infty} |x_\gamma(t)|^2\, dt = \int_{-\gamma}^{\gamma} |x_\gamma(t)|^2\, dt = \frac{1}{2\pi} \int_{-\infty}^{\infty} |X_\gamma(\omega)|^2\, d\omega \tag{8.4.3}$$

using Parseval's theorem, Theorem 8.3.1. Using (8.4.3) the power in the original signal (energy per unit time) is given by

$$P = \lim_{\gamma \to \infty} \frac{1}{2\gamma} \int_{-\gamma}^{\gamma} |x_\gamma(t)|^2\, dt = \lim_{\gamma \to \infty} \frac{1}{2\gamma} \frac{1}{2\pi} \int_{-\infty}^{\infty} |X_\gamma(\omega)|^2\, d\omega. \tag{8.4.4}$$

Under certain conditions the limiting process in the second term in (8.4.4) can be interchanged with the integration. Then we have

---

[1] That is, a generalized Fourier transform is not required.

$$P = \lim_{\gamma \to \infty} \frac{1}{2\gamma} \frac{1}{2\pi} \int_{-\infty}^{\infty} \left| X_\gamma(\omega) \right|^2 d\omega = \frac{1}{2\pi} \int_{-\infty}^{\infty} \lim_{\gamma \to \infty} \frac{1}{2\gamma} \left| X_\gamma(\omega) \right|^2 d\omega. \qquad (8.4.5)$$

Comparing (8.4.5) to the definition of the energy spectral density, we see that the quantity under the integral on the right is what we would like to call the power spectral density. This development leads to the following definition.

Definition 8.4.2: If $x(t)$ is a power signal, let $x_\gamma(t)$ be the truncated version of $x(t)$. Then the function $S(\omega)$

$$S(\omega) = \lim_{\gamma \to \infty} \frac{1}{2\gamma} \left| X_\gamma(\omega) \right|^2$$

is called the <u>power spectral density</u> of $x(t)$.

▲▲

Notation: If we wish to indicate to which signal the power spectral density is attached, we attach a subscript to the power spectral density. That is, $S_x(\omega) \leftrightarrow x(t)$.

The analysis above justifies the following theorem.

Theorem 8.4.3: Let $S(\omega)$ be the power spectral density of the power signal $x(t)$. Then the power in the signal can be determined as

$$P = \lim_{\gamma \to \infty} \frac{1}{2\gamma} \int_{-\gamma}^{\gamma} \left| x_\gamma(t) \right|^2 dt = \frac{1}{2\pi} \int_{-\infty}^{\infty} S(\omega) \, d\omega.$$

▲▲

Theorem 8.4.3 plays the role of Parseval's theorem for power signals.

The power spectral density is determined from the square of the amplitude spectrum of the truncated signal. Hence, the power spectral density is always a nonnegative real function. Since the power spectral density is determined from the amplitude spectrum it doesn't contain any phase information. So the power spectral density doesn't uniquely identify the signal. That is, many power signals can have the same power spectrum. Nonetheless, it is of interest to determine the power spectral density from the direct measurement of physical processes. This area is called spectral analysis.

8.4.2 Examples

It turns out that power spectral densities are most easily calculated from autocorrelation functions, a subject that is beyond the scope of this text. Therefore, we will simply give the power spectral density of several signals which we will use frequently.

Consider the signal that is a constant,

$$x(t) = A. \qquad (8.4.6)$$

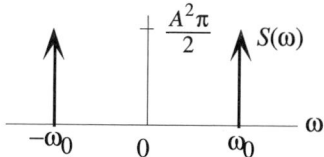

Figure 8.4.1 Power Spectral Density of a Cosine

The power spectral density of this signal is

$$S(\omega) = 2\pi A^2 \delta(\omega). \tag{8.4.7}$$

Consider the signal that is the sum of two cosines,

$$x(t) = A_1 \cos(\omega_1 t + \theta_1) + A_2 \cos(\omega_2 t + \theta_2) = x_1(t) + x_2(t). \tag{8.4.8}$$

The power spectral density of this signal is

$$S(\omega) = \frac{A_1^2 \pi}{2}\left[\delta(\omega - \omega_1) + \delta(\omega + \omega_1)\right] + \frac{A_2^2 \pi}{2}\left[\delta(\omega - \omega_2) + \delta(\omega + \omega_2)\right]. \tag{8.4.9}$$

This result shows that the power spectral density of a cosine is basically a pair of impulse functions located at the frequency of the cosine as shown in Figure 8.4.1. The constant multiplier of the impulse function is proportional to the square of the amplitude of the cosine. (The power spectral density in Figure 8.4.1 looks like the amplitude spectrum shown in Figure 8.1.9b, but its constant multiplier is different.) *The power spectral density is independent of the phase angle of the cosine.*
      The power spectral density in (8.4.9) also shows that the power spectral density of the sum of two cosines is the sum of the power spectral densities of each of the individual terms. That is, the power spectral density of the cross terms is zero. This fact is really a consequence of Parseval's theorem, and it has far-reaching implications for signal analysis.

8.5 POWER CALCULATIONS FOR PERIODIC SIGNALS

8.5.1 Introduction

In the last section we defined the power spectral density for power signals. One of the most important classes of power signals are periodic signals. In this section we will discuss power calculations for periodic signals. Recall from Definition 7.1.1 that a periodic signal $x(t)$ is a signal that satisfies

$$x(t) = x(t + T_0), \quad T_0 > 0. \tag{8.5.1}$$

The power calculations will be carried out in terms of Fourier series representation of this periodic signal

$$x(t) = \sum_{m=-\infty}^{\infty} X_m e^{jm\omega_0 t} = \sum_{m=0}^{\infty} A_m \cos(m\omega_0 t + \theta_m).$$

(8.5.2)

We will extend Parseval's theorem, Theorem 8.3.1, to these signals and develop the power spectrum of the Fourier series in two forms. In the next section we will show how these concepts lend insight into the structure of the signal (8.5.2).

8.5.2 Power of Periodic Signals

To calculate the power in a signal from the definition requires that a limit be evaluated. If the power signal is a periodic signal, then that calculation can be simplified as the following theorem shows.

Theorem 8.5.1: Suppose that $x(t)$ is a periodic signal with period T_0. Then the power in $x(t)$ is given by

$$P = \lim_{\gamma \to \infty} \frac{1}{2\gamma} \int_{-\gamma}^{\gamma} |x(t)|^2 \, dt = \frac{1}{T_0} \int_{t_0}^{t_0+T_0} |x(t)|^2 \, dt.$$

▲▲

Theorem 8.5.1 says that the power in a periodic signal can be calculated by averaging the energy over one period of the signal. This computation is illustrated in the next example.

Example 8.5.2: A representation of the periodic signal in Figure 8.5.1 is

$$x(t) = \sum_{m=-\infty}^{\infty} \Pi(t - 2m).$$

(8.5.3)

The period of this signal is $T_0 = 2$. The power in this signal is given by

$$P = \frac{1}{T_0} \int_{t_0}^{t_0+T_0} |x(t)|^2 \, dt = \frac{1}{2} \int_{-\frac{1}{2}}^{\frac{3}{2}} |x(t)|^2 \, dt = \frac{1}{2} \int_{-\frac{1}{2}}^{\frac{1}{2}} 1^2 \, dt = \frac{1}{2}.$$

(8.5.4)

▲▲

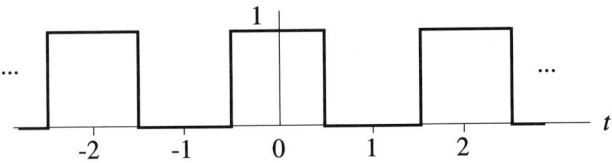

Figure 8.5.1 Periodic Signal

8.5.3 Parseval's Theorem

Next we consider periodic signals that have a Fourier series representation (8.5.2). It turns out that the power in the periodic signal can be calculated by summing the power in each term. The power calculation in the next theorem is a variation on Parseval's theorem.

Theorem 8.5.3: Parseval's Theorem Suppose that $x(t)$ is a periodic signal with period T_0. Assume also that this series has a Fourier series which is given by

$$x(t) = \sum_{m=-\infty}^{\infty} X_m e^{jm\omega_0 t}.$$

Then the power P in this signal can be expressed as

$$P = \lim_{\gamma \to \infty} \frac{1}{2\gamma} \int_{-\gamma}^{\gamma} |x(t)|^2 \, dt = \sum_{m=-\infty}^{\infty} |X_m|^2.$$

▲▲

Proof:[2] From Theorem 8.5.1 we have

$$P = \frac{1}{T_0} \int_{T_0} |x(t)|^2 \, dt = \frac{1}{T_0} \int_{T_0} x(t) \bar{x}(t) \, dt \tag{8.5.5}$$

where we have used a result from complex variables. Substituting the Fourier series of $x(t)$ into (8.5.5) and rearranging terms we have

$$P = \frac{1}{T_0} \int_{T_0} x(t) \left[\sum_{m=-\infty}^{\infty} \bar{X}_m e^{-jm\omega_0 t} \right] dt = \sum_{m=-\infty}^{\infty} \bar{X}_m \left[\frac{1}{T_0} \int_{T_0} x(t) e^{-jm\omega_0 t} \, dt \right]. \tag{8.5.6}$$

The quantity in the brackets is just the expression for the coefficient of the Fourier series of $x(t)$. Hence, (8.5.6) yields

$$P = \sum_{m=-\infty}^{\infty} \bar{X}_m X_m = \sum_{m=-\infty}^{\infty} |X_m|^2. \tag{8.5.7}$$

▲▲

Technically, the interchange of the summation and integration must be justified. This operation can be justified, but the proof is beyond the scope of this text. Also we must be sure that the infinite sum exists. This condition is guaranteed by the existence of the Fourier series.

---

[2] Compare with the proof of Theorem 8.3.1.

Parseval's theorem says that the power in a periodic signal can be computed by computing the power in each term of the Fourier series and summing the results. In particular, there is no power associated with the cross terms when the signal is squared.

This theorem can be stated using the cosine Fourier series. The basic idea can be seen by considering a Fourier series with one term in both exponential and cosine form

$$x(t) = \frac{A_1}{2} e^{-j\theta_0} e^{-j\omega_0 t} + \frac{A_1}{2} e^{j\theta_0} e^{j\omega_0 t} = A_1 \cos(\omega_0 t + \theta_0). \tag{8.5.8}$$

Using Theorem 8.5.3 the power in this signal is

$$P = |X_{-1}|^2 + |X_1|^2 = \left(\frac{A_1}{2}\right)^2 + \left(\frac{A_1}{2}\right)^2 = \frac{A_1^2}{2}. \tag{8.5.9}$$

Hence, the calculation of the power in a single sinusoid using Theorem 8.5.3 agrees with the direct calculation in Example 8.2.4. This calculation also shows that Parseval's theorem can be stated using the cosine form of the Fourier series.

Corollary 8.5.4: Parseval's Theorem Suppose that $x(t)$ is a periodic signal with period T_0. Assume also that this series has a cosine Fourier series representation which is given by

$$x(t) = \sum_{m=0}^{\infty} A_m \cos(m\omega_0 t + \theta_m).$$

Then the power in this signal is given by

$$P = \frac{1}{T_0} \int_{T_0} |x(t)|^2 \, dt = A_0^2 + \sum_{m=1}^{\infty} \frac{A_m^2}{2}.$$

▲▲

The physical interpretation of Corollary 8.5.4 is somewhat clearer than Theorem 8.5.3 because each sinusoid in the Fourier series is a real function. Furthermore, the total power in the signal turns out to be the sum of the powers of each individual term. This interpretation is easily visualized. Theorem 8.5.3, however, is more widely used at the advanced level for reasons that will become clear below.

These two theorems suggest a graphical representation of the power in a periodic signal. From Theorem 8.5.3 the power in each term of the exponential Fourier series is

$$|X_m|^2 \quad \leftrightarrow \quad X_m e^{jm\omega_0 t}, \tag{8.5.10}$$

and from Corollary 8.5.4 the power in each term of the cosine Fourier series is

$$\frac{A_m^2}{2} \leftrightarrow A_m \cos(m\omega_0 t + \theta_m). \tag{8.5.11}$$

Observe that the power in each term is indexed by the frequency of the corresponding function of time $m\omega_0$. Hence, we are led to the following definition.

Definition 8.5.5: Let $x(t)$ be a periodic signal. If $x(t)$ is represented with an exponential Fourier series

$$x(t) = \sum_{m=-\infty}^{\infty} X_m e^{jm\omega_0 t},$$

then the function

$$|X_m|^2 \quad \text{vs.} \quad m\omega_0, \quad m = \dots, -2, -1, 0, 1, 2, \dots$$

is called the (two-sided) power spectrum of $x(t)$.

If $x(t)$ is represented with an cosine Fourier series

$$x(t) = \sum_{m=0}^{\infty} A_m \cos(m\omega_0 t + \theta_m),$$

then the function

$$\frac{A_m^2}{2} \quad \text{vs.} \quad m\omega_0, \quad m = 0, 1, 2, \dots$$

is called the (one-sided) power spectrum of $x(t)$. ▲▲

The power spectra provide fundamental insight into the structure of the signal because they identify the distribution of power in a signal as a function of the frequency of its sinusoidal components. This idea is developed in detail in the next section. It is applied to the analysis of a signal propagating through a system in Chapter 15.

Example 8.5.6: Consider again the signal in Example 8.5.2. The exponential Fourier series for this signal, derived in Example 7.3.5, is

$$x(t) = \frac{1}{2} + \sum_{\substack{m=-\infty \\ m \neq 0 \\ m \text{ odd}}}^{\infty} \left[\frac{1}{m\pi} e^{j\left(\frac{m-1}{2}\right)\pi} \right] e^{jm\pi t}. \tag{8.5.12}$$

The power in each term of the Fourier series of this signal is

$$|X_0|^2 = \frac{1}{4}, \quad |X_m|^2 = \left| \frac{1}{m\pi} \right|^2 = \frac{1}{\pi^2 m^2}, \quad m \neq 0. \tag{8.5.13}$$

The two-sided power spectrum is shown in Figure 8.5.2. The cosine Fourier series of this signal is given by

$$x(t) = \frac{1}{2} + \sum_{\substack{m=1 \\ m \text{ odd}}}^{\infty} \left(\frac{2}{m\pi} \right) \cos\left[m\pi t - \left(\frac{(m-1)\pi}{2} \right) \right]. \tag{8.5.14}$$

The power in each term of the Fourier series of this signal is

$$A_0^2 = \frac{1}{4}, \quad \frac{A_m^2}{2} = \frac{1}{2}\left| \frac{2}{m\pi} \right|^2 = \frac{2}{\pi^2 m^2}. \tag{8.5.15}$$

The one-sided power spectrum is shown in Figure 8.5.3.   ▲▲

8.5.4 Power Spectral Density of a Fourier Series

There is a second approach to representing the power in a periodic signal that is based on the Fourier transform of the Fourier series (8.5.2). It can be shown that the power spectral density for the cosine Fourier series in (8.5.2) is

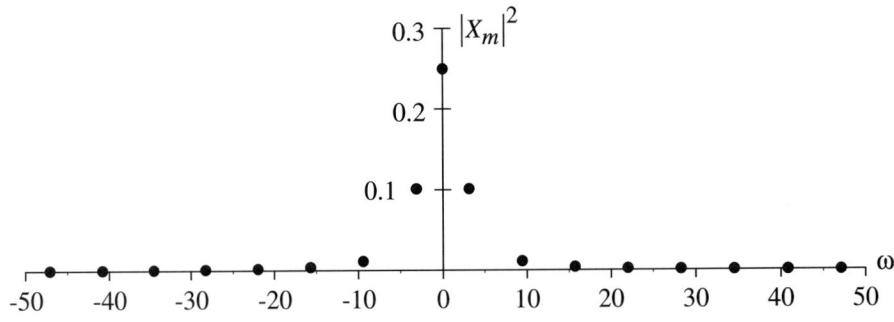

Figure 8.5.2 Two-Sided Power Spectrum of the Fourier Series in (8.5.12)

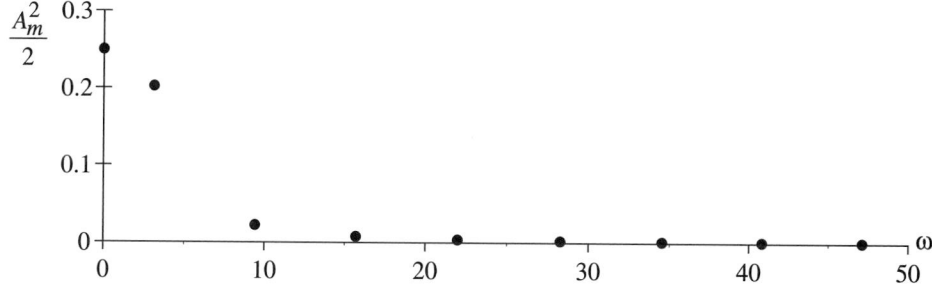

Figure 8.5.3 One-Sided Power Spectrum of the Fourier Series in (8.5.14)

$$S(\omega) = 2\pi A_0^2 \delta(\omega) + \sum_{m=1}^{\infty} \frac{\pi A_m^2}{2}\left(\delta(\omega - m\omega_i) + \delta(\omega + m\omega_i)\right). \tag{8.5.16}$$

If we split the summation in (8.5.16) we obtain

$$S(\omega) = 2\pi A_0^2 \delta(\omega) + \sum_{m=1}^{\infty} \frac{\pi A_m^2}{2}\left(\delta(\omega - m\omega_i) + \delta(\omega + m\omega_i)\right) \tag{8.5.17}$$

$$= 2\pi A_0^2 \delta(\omega) + \sum_{m=-\infty}^{-1} \frac{\pi A_m^2}{2}\delta(\omega + m\omega_i) + \sum_{m=1}^{\infty} \frac{\pi A_m^2}{2}\delta(\omega - m\omega_i).$$

This sum can be rewritten by noting that

$$\frac{A_m^2}{2} = 2\left(\frac{A_m^2}{4}\right) = 2|X_m|^2 \tag{8.5.18}$$

where we have used the relationship between the cosine and exponential Fourier series, Table 7.6.1. Using (8.5.18) in (8.5.17) we obtain

$$S(\omega) = \sum_{m=-\infty}^{\infty} 2\pi|X_m|^2 \delta(\omega - m\omega_i). \tag{8.5.19}$$

The summation in (8.5.19) is the power spectral density of the periodic signal $x(t)$ in terms of the exponential representation. The power spectrum is most conveniently expressed in terms of a graph.

Definition 8.5.7: Let $x(t)$ be a periodic signal with an exponential Fourier series

$$x(t) = \sum_{m=-\infty}^{\infty} X_m e^{jm\omega_0 t}.$$

The graph

$$2\pi |X_m|^2 \delta(\omega - m\omega_0) \quad \text{vs.} \quad \omega$$

is called the <u>power spectral density</u> of $x(t)$.

▲▲

The following example illustrates concept of the power spectral density.

Example 8.5.8: Consider again the signal in Example 8.5.6. The exponential Fourier series for this signal is

$$x(t) = \frac{1}{2} + \sum_{\substack{m=-\infty \\ m \neq 0 \\ m \text{ odd}}}^{\infty} \left[\frac{1}{m\pi} e^{j\left(\frac{m-1}{2}\right)\pi} \right] e^{jm\pi t}. \tag{8.5.20}$$

The power spectrum of this signal is

$$2\pi |X_m|^2 \delta(\omega - m\pi) = 2\pi \left|\frac{1}{m\pi}\right|^2 \delta(\omega - m\pi) = \frac{2}{\pi m^2} \delta(\omega - m\pi) \tag{8.5.21}$$

$$|X_0|^2 \delta(\omega) = (2\pi)\frac{1}{4}\delta(\omega) = \frac{\pi}{2}\delta(\omega).$$

This power spectrum is shown in Figure 8.5.4.

If we compare the power spectral density of a Fourier series in Definition 8.5.7 to the power spectrum of the same Fourier series in Definition 8.5.5, then we see that while these two graphs are technically different, they carry the same information. The power spectral density is defined in terms of impulse functions, and the magnitude is scaled by 2π. But the essential magnitude and frequency information is in both graphs. In fact, the same terminology is used interchangeably for the two graphs. The power spectral density should also be compared to the amplitude spectrum of a cosine as discussed in Section 8.1.4. Note that the amplitude spectrum and power spectral density are two different functions.

▲▲

We shall use primarily the power spectrum in Definition 8.5.5 because of the convenience of the form of the graphs, particularly for our discussion in Chapter 15.

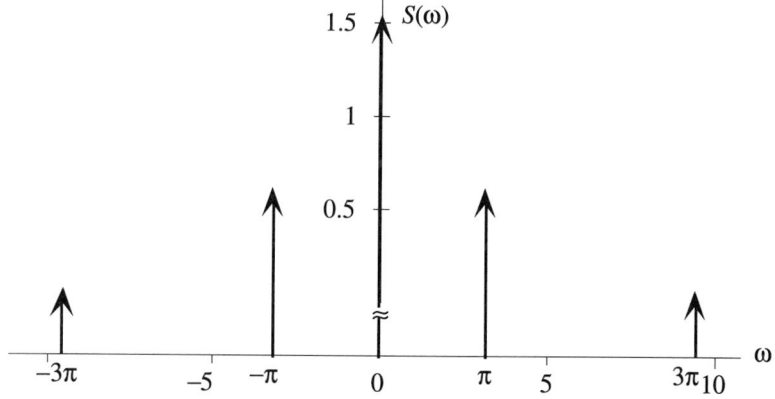

Figure 8.5.4 Power Spectral Density of the Fourier Series in (8.5.20)

The power spectral density in Definition 8.5.7 is often used in advanced work because of the continuity with the theory of stochastic processes.

8.6 SPECTRAL CONTENT OF A SIGNAL: AN EXAMPLE

8.6.1 Introduction

In this section we will consider a periodic signal $x(t)$ with period T_0 which has a Fourier series representation

$$x(t) = \sum_{m=0}^{\infty} A_m \cos(m\omega_0 t + \theta_m), \qquad \omega_0 = \frac{2\pi}{T_0}. \tag{8.6.1}$$

As we have seen the amplitude and phase spectra can be used to characterize the Fourier series and so the original signal. In this section we will discuss how to interpret the amplitude spectrum in light of the Fourier series. The amplitude spectrum gives us deep insight into the structure of this signal. The insight gained is of fundamental importance in analyzing signals and their propagation through systems. Furthermore, the concepts elaborated in this section extend to aperiodic signals and random processes (although this discussion is left to advanced treatments). Yet the fundamental ideas are most easily seen in the Fourier series (8.6.1).

To begin, we make a fundamental observation about (8.6.1). On the left side of the equal sign we have a periodic signal. On the right side of the equal sign we have a sum of cosines. *This equation says that a periodic signal can be written as a sum of sinusoids.* This observation is the essence of analysis, the key to all. First, the Fourier series provides a uniform basis for comparison of waveforms with different shapes. Second, the Fourier series allows us to couple signal analysis with system analysis to determine how a signal with an arbitrary waveform will propagate through a system.

From this starting point we observe that the cosines in the summation are not arbitrary. First and foremost, the frequencies of the sinusoids are fixed by the period of the signal $x(t)$. That is, the frequencies of the cosines are given by $m\omega_0, m = 0, 1, 2, \ldots$ where ω_0 is given in (8.6.1). Therefore, the periodic signal is composed of cosines with frequencies that are uniquely determined by the signal. These frequencies identify the signal (in part).

Once the frequencies of the cosines have been determined, the signal is uniquely determined by the amplitude and phase coefficients A_m, and $\theta_m, m = 0, 1, 2, \ldots$. All of the cosines that make up the sum in (8.6.1) are not equally important. The importance of a particular cosine to the signal is determined primarily by the amplitude coefficient A_m. Very crudely, the bigger the coefficient A_m, the more important the term $A_m \cos(m\omega_0 t + \theta_m)$ is to the signal. The structure of the signal, then, is determined by those sinusoidal terms, identified by their frequency, whose amplitude coefficients are the most significant. This information is readily apparent in the amplitude spectrum (or the power spectrum). It is the purpose of this section to quantify the contribution of each cosine in the sum (8.6.1) to the periodic signal $x(t)$. We reiterate that each cosine term is identified by its frequency $m\omega_0$. That is, the information in the amplitude spectrum is frequency dependent.

Terminology: The information in the amplitude spectrum or power spectrum is known as the spectral content of the signal or the frequency content of a signal.

8.6.2 Approximation of Periodic Signals Using Fourier Series

How do we determine which cosines in (8.6.1) are important to the signal $x(t)$? First, we will choose the terms which we think are important. For simplicity, we denote these terms by

$$\hat{x}_{m_1:m_2}(t) = \sum_{m=m_1}^{m_2} A_m \cos(m\omega_0 t + \theta_m). \tag{8.6.2}$$

Notation: The subscript on (8.6.2) $m_1{:}m_2$ denotes which terms are in the sum (8.6.2) extracted from the Fourier series in (8.6.1).

We shall refer to (8.6.2) as the approximating signal. To make the following discussion interesting, we will fix the number of terms in the approximating signal at $m_0 = m_2 - m_1$. Then we subtract the terms in (8.6.2) from the Fourier series (8.6.1). We obtain the error signal

$$e(t) = x(t) - \hat{x}_{m_1:m_2}(t) = \sum_{m=0}^{m_1-1} A_m \cos(m\omega_0 t + \theta_m) + \sum_{m=m_2+1}^{\infty} A_m \cos(m\omega_0 t + \theta_m). \tag{8.6.3}$$

If the terms in (8.6.2) really are important to the signal $x(t)$ in (8.6.1), then the error signal should be "small." Now we must establish a meaning to the word "small." One way to measure the error signal is to use the power in that signal. If the power in the error signal is small as compared to the power in the original signal, then the terms in (8.6.2) must be important to the signal $x(t)$. The goal of this section is to identify the terms in (8.6.2) that are most important for a given signal $x(t)$ by making the error signal small.

Parseval's theorem, Theorem 8.5.3, tells us how to choose the approximating signal to minimize the power in the error signal. The power in $x(t)$ in terms of the Fourier series components is

$$P_x = A_0^2 + \sum_{m=1}^{\infty} \frac{A_m^2}{2}. \tag{8.6.4}$$

(Here we are using two crucial facts from the previous sections. First, the power is a sinusoid is independent of its amplitude and phase. Second, the total power in a Fourier series can be computed by summing the powers of the individual terms.) The power in the approximating signal (8.6.2) is

$$P_{\hat{x}} = \sum_{m=m_1}^{m_2} \frac{A_m^2}{2}. \tag{8.6.5}$$

Using (8.6.4) and (8.6.5), the power in the error signal is

$$P_e = P_x - P_{\hat{x}} = \left(A_0^2 + \sum_{m=1}^{\infty} \frac{A_m^2}{2} \right) - \left(\sum_{m=m_1}^{m_2} \frac{A_m^2}{2} \right) = A_0^2 + \sum_{m=1}^{m_1-1} \frac{A_m^2}{2} + \sum_{m=m_2+1}^{\infty} \frac{A_m^2}{2}. \tag{8.6.6}$$

In words, the power in the error signal is the power in the signal $x(t)$ minus the power in the terms contained in the approximation of $x(t)$, $\hat{x}_{m_1:m_2}(t)$. We can make this statement because of Parseval's theorem. Also note that the power in the approximating signal is always less than the power in the signal $x(t)$, $P_{\hat{x}} < P_x$. By minimizing the power in the error signal, we achieve a better approximation of $x(t)$.

It is clear from (8.6.6) that in order to minimize the power in the error signal in (8.6.6), we should construct the approximating signal from the terms of the Fourier series of $x(t)$ that contain the most power. These terms then form the approximating signal. The cosine terms in the Fourier series (8.6.1) that contain the most power are easily identified; they have the largest amplitude coefficients A_m. Given a periodic signal with a Fourier series (8.6.1), if we want to approximate this signal with a sum of m_0 cosine terms with the most power, we simply select the m_0 terms with the largest amplitude coefficients. This selection is easily made from the power spectrum. The following examples illustrate these observations.

8.6.3 Examples

Example 8.6.1: Consider the signal pulse train in Figure 8.6.1. The Fourier series for this signal is derived in Section 7.1. It is

$$x(t) = \frac{1}{2} + \sum_{m=1}^{\infty} \frac{2}{(2m-1)\pi} \cos((2m-1)\pi t + \theta_m) \tag{8.6.7}$$

where

$$\theta_m = \begin{cases} -\pi, & m \text{ even} \\ 0, & m \text{ odd} \end{cases}$$

The power in each term of the Fourier series, as derived in Example 8.5.6, is

$$A_0^2 = \frac{1}{4}, \quad \frac{A_m^2}{2} = \frac{2}{(2m-1)^2 \pi^2}. \tag{8.6.8}$$

The one-sided power spectrum is shown in Figure 8.6.2.

Now for a fixed integer m_0, we want to minimize the power in the error signal. Inspection of the power spectrum in Figure 8.6.2 shows that we should retain the lowest frequency terms because their powers are the largest. Therefore, the approximating signals (8.6.2) will be of the form

$$x_{0:m_2}(t) = \sum_{m=m_1=0}^{m_2} A_m \cos((2m-1)\pi t + \theta_m). \tag{8.6.9}$$

The graph of (8.6.9) for $m_2 = 3$ and $m_2 = 7$ is shown in Figure 8.6.3. The quality of the approximation signal now depends on the number of terms retained in (8.6.9). As expected, the approximation gets better as we increase the number of terms in the approximating signal.

To reinforce the idea, suppose instead we chose the 7th - 10th terms for the approximating sum. In this case the approximating signal would be

$$x_{7:10}(t) = \sum_{m=7}^{10} \frac{2}{(2m-1)\pi} \cos((2m-1)\pi t + \theta_m). \tag{8.6.10}$$

The partial sum in (8.6.10) is shown in Figure 8.6.4.

Note: Compare the scales of Figure 8.6.3 and Figure 8.6.4.

This approximation bears little resemblance to the original signal shown in Figure 8.6.1. We see that the terms in (8.6.10) contribute less to the signal (8.6.7) than the terms in (8.6.9). (Note that the signal in Figure 8.6.4 is largest near the discontinuities of the original signal where the approximating signals are the worst.)

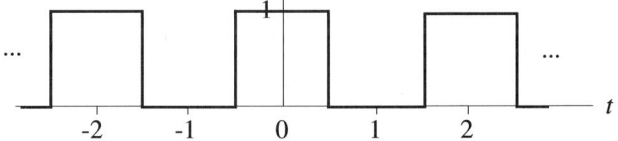

Figure 8.6.1 Pulse Train

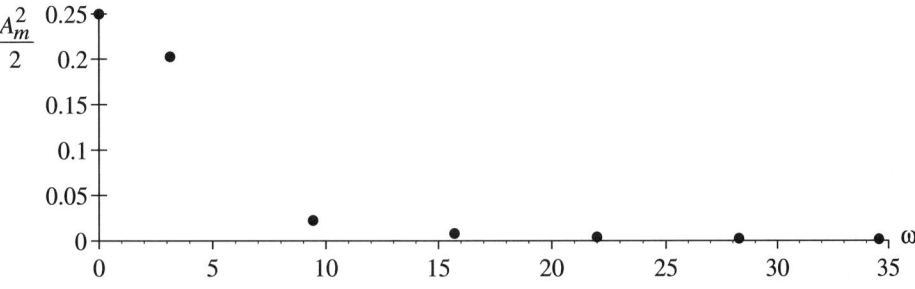

Figure 8.6.2 One-Sided Power Spectrum of (8.6.7)

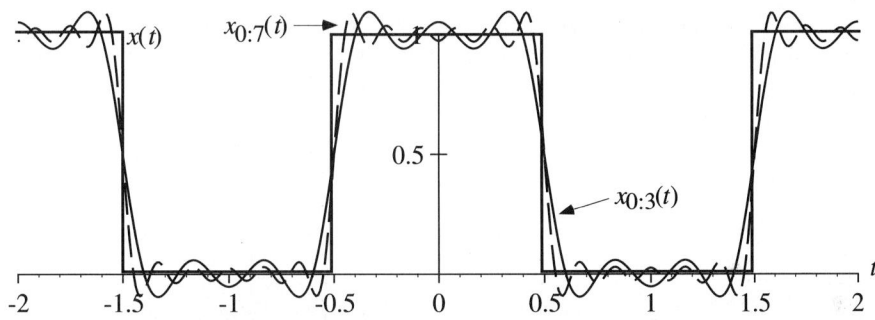

Figure 8.6.3 Approximations to the Signal (8.6.9)

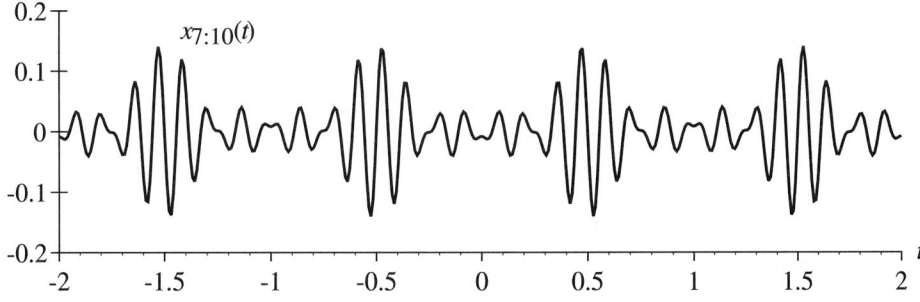

Figure 8.6.4 Approximation of the Signal (8.6.7) Using (8.6.10)

Table 8.6.1 Power in the Signal (8.6.7) and Its Approximating Partial Sums

Terms in the Approximation	Power	Percent of Total Power
all	0.5	100
0 - 3	0.4833	96.7
0 - 7	0.4928	98.6
7 - 10	0.0045	0.9

The graphs in Figures 8.6.3 and 8.6.4 can be compared by computing the power in each approximation with the power in $x(t)$. To compute the power in $x(t)$ we use a direct calculation from Theorem 8.5.1. We have

$$P_x = \frac{1}{T_0} \int_{T_0} |x(t)|^2 dt = \frac{1}{2} \int_{-\frac{1}{2}}^{\frac{1}{2}} (1)^2 dt = 0.5. \tag{8.6.11}$$

The power in the approximating signals is computed using (8.6.5) where we take $m_1 = 0$. A comparison of power in the various signals in Figures 8.6.3 and 8.6.4 shown in Table 8.6.1. The first two approximating signals in Figure 8.6.3 have a large percentage of the total signal power in them, while the last approximating signal in Figure 8.6.4 has only a small percentage of the total power. We see that the power in the approximating partial sums (8.6.9) rapidly approaches the total power in the signal (8.6.11) as m_2 increases. That is, the terms on the tail of the Fourier series don't carry much of the power in the signal. That is why the terms in (8.6.10) contribute so little to the signal in Figure 8.6.1.

▲▲

In this first example, we have used the cosine representation of the Fourier series to emphasize the relationship of a periodic signal and its equivalent representation as a sum of cosines. The cosine representation is a good representation for this conceptualization in the time domain. In practice, however, the two-sided amplitude and phase spectra are usually employed when the Fourier series is represented in the frequency domain by its amplitude or power spectrum. When the exponential representation is used, the approximating signal is chosen by selecting the term with the negative index whenever the term with the positive index is chosen. That is, to form the approximating signal from the terms m_1 to m_2 we would write

$$x_{m_1:m_2}(t) = \sum_{m=-m_2}^{-m_1} X_m e^{jm\omega_0 t} + \sum_{m=m_1}^{m_2} X_m e^{jm\omega_0 t} = \sum_{m=m_1}^{m_2} A_m \cos(m\omega_0 t + \theta_m). \tag{8.6.12}$$

We will use the exponential representation and the two-sided power spectrum to analyze the signal in the next example. This example will reinforce the concepts introduced in the last example.

Example 8.6.2: Consider the signal

$$(8.6.13)$$

$$x_c(t) = \begin{cases} A\cos(\omega_c t), & -\dfrac{\varepsilon}{2} < t \le \dfrac{\varepsilon}{2} \\[2mm] 0, & \dfrac{\varepsilon}{2} < t \le T_0 - \dfrac{\varepsilon}{2} \end{cases}$$

Let the signal $x_{pc}(t)$ be defined from (8.6.13) by periodic extension. This signal is shown in Figure 8.6.5. Here we intend to find the power distribution among the terms of the Fourier series in the signal $x_c(t)$ and compare it to the signal in Example 8.6.1.

The exponential form of the Fourier series, as derived in Example 7.3.6, is

$$(8.6.14)$$

$$x_{pc}(t) = \frac{2A}{\omega_c T_0}\sin\left(\frac{\omega_c \varepsilon}{2}\right)$$

$$+\frac{A\varepsilon}{2T_0}\sum_{\substack{m=-\infty \\ m\ne 0}}^{\infty}\left[\mathrm{Sa}\left(\frac{(\omega_c - m\omega_0)\varepsilon}{2}\right)+\mathrm{Sa}\left(\frac{(\omega_c + m\omega_0)\varepsilon}{2}\right)\right]e^{jm\omega_0 t}.$$

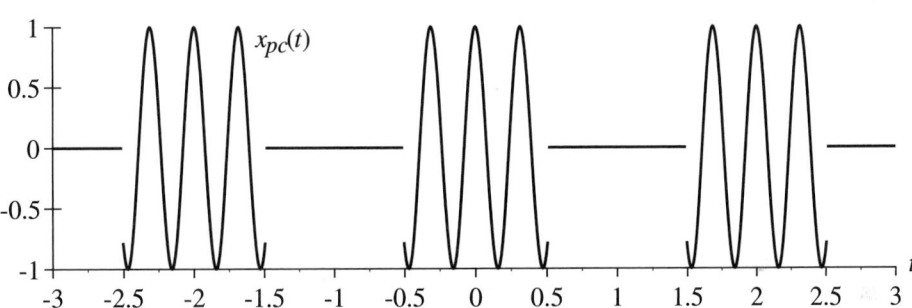

Figure 8.6.5 Signal of Example 8.6.2

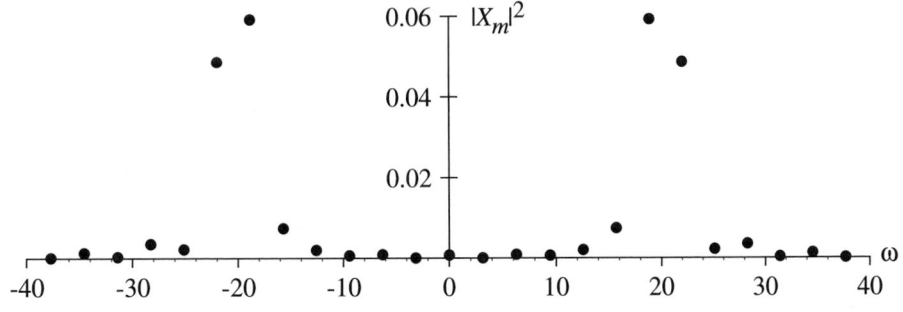

Figure 8.6.6 Two-Sided Power Spectrum of the Signal in (8.6.14)

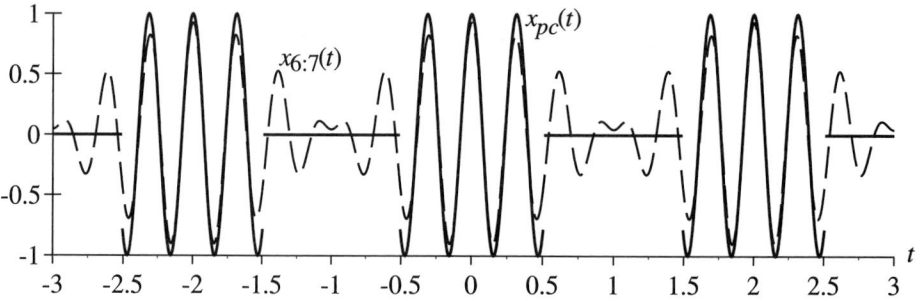

Figure 8.6.7 Approximation to the Signal in (8.6.14)

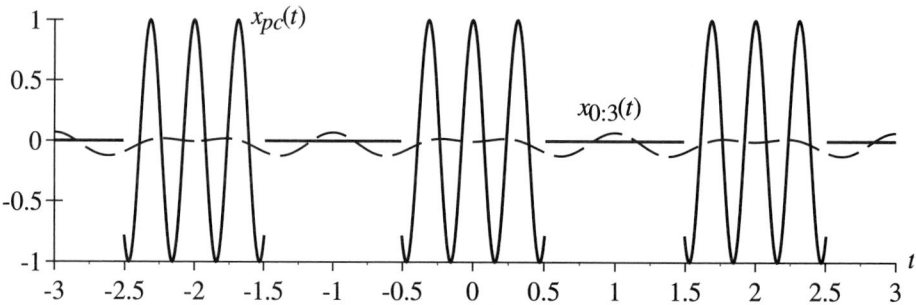

Figure 8.6.8 Second Approximation to Signal in (8.6.14)

The power spectrum is easily derived from the Fourier series (8.6.14). If the parameters of this signal are $A = 1$, $\varepsilon = 1$, $T_0 = 2$, and $\omega_c = 20$, then the two-sided power spectrum of this signal is shown in Figure 8.6.6.

We wish to construct an approximation to the signal $x(t)$ in Figure 8.6.5 by extracting terms from the Fourier series. The analysis in (8.6.2) - (8.6.6) tells us that we should extract those terms with the largest power. These terms are easily identified from the power spectral density in Figure 8.6.6. This power spectrum shows that two terms at approximately 20 rad/sec carry much of the power in this signal. These terms are the 6th and 7th terms. The approximating signal $x_{6:7}(t)$ constructed by using these terms is shown in Figure 8.6.7. To emphasize the importance of the 6th and 7th terms in the Fourier series (8.6.14), suppose that we select the first four terms as our approximating signal. This signal along with $x_c(t)$ is shown in Figure 8.6.8. Clearly, $x_{6:7}(t)$ is a better approximation to $x(t)$ than is $x_{0:3}(t)$. The terms of the Fourier series that contribute most to the signal in Figure 8.6.5 are the terms that carry the most power. This example shows that the terms that carry the most power are not necessarily the first terms of the Fourier series.

Power calculations serve to quantify the intuition drawn from the graphs above. The power of the signal $x(t)$ is calculated using Theorem 8.5.1 for periodic signals. We have

Table 8.6.2 Power in the Signal (8.6.7) and Its Approximating Partial Sums

Terms in the Approximation	Power	Percent of Total Power
all	0.2614	100
0 - 3	0.0039	1.5
5 - 7	0.2307	88.3
0 - 12	0.2538	97.1

$$P = \frac{1}{T_0} \int_{T_0} |x(t)|^2 \, dt = \frac{1}{T_0} \int_{-\frac{\varepsilon}{2}}^{\frac{\varepsilon}{2}} |A \cos \omega_c t|^2 \, dt = \frac{A^2}{2T_0} \left(\varepsilon + \frac{1}{\omega_c} \sin(\omega_c \varepsilon) \right). \tag{8.6.15}$$

The power of the various approximating signals is shown in Table 8.6.2. We see that the approximating signals that include the 6 - 7th terms also carry the most power.

If we compare the power spectra of the signals in Examples 8.6.1 and 8.6.2, we see that the terms with the most power have been shifted from the low frequency band in Figure 8.6.2 to the frequency band of 10 - 30 rad/sec in the power spectrum in Figure 8.6.6. This shift in power is due to the multiplication of each period of the signal in (8.6.7) by $\cos(\omega_c t)$ in (8.6.13). Signals of this type are commonly used in communication systems.

▲▲

8.6.4 Summary

In this section we have illustrated, by example, the concept of the spectral content of a periodic signal. This concept is based on the observation that a periodic signal is composed of sinusoids at specific frequencies. Furthermore, not all of these sinusoids are of equal importance to the periodic signal. When we calculate the power in this signal, Parseval's theorem tells us that we can calculate the power in each component of the Fourier series separately and sum the individual terms. Hence, we can associate with each term in the Fourier series the power that term contributes to the total power of the signal. Finally, we observe that each term in the Fourier series is identified by its frequency. Now the power in the signal is coupled to the frequency of the terms in the Fourier series allowing us to plot the power in the terms of the Fourier series as a function of the frequency of the that term of the Fourier series. This plot is, of course, the power spectrum of the signal. Those terms with the largest power are correlated to the frequency of those terms. This concept is the concept of the spectral content of the signal. This concept linking the power in a signal to certain frequency bands is one of the fundamental concepts in signal analysis.

Two signals are considered in this section. The power spectrum of the first signal is shown in Figure 8.6.2. The power in this signal is concentrated in the lowest frequencies, below 5 rad/sec, with the highest power contained in the zero frequency (DC) term. The power spectrum of the second signal is shown in Figure 8.6.6. Unlike the first signal, the power of the second signal is not concentrated in the low frequency terms. Rather the power is contained in those terms around 20 rad/sec. These examples show that the spectral content of a signal depends on the waveform.

The spectral content of a signal is also of fundamental importance when analyzing the propagation of a signal through a system. We will explore this topic in Chapter 15.

8.7 STATIC NONLINEARITIES

8.7.1 A Nonlinear System

In this section we will study systems that are different than most of the other systems we discuss in this text. We will introduce these systems by way of example. Consider the op amp circuit in Figure 8.7.1.

Note: The first op amp is an inverting op amp. This amplifier will simplify the discussion below. We will consider this op amp to be ideal.

We think of this circuit as a system where the input signal is the voltage $v_i(t)$ and the output signal is the voltage $v_o(t)$. For low voltages, the relationship between the input signal and the output signal of the circuit in Figure 8.7.1 is

$$\frac{V_o(s)}{V_i(s)} = \left(-\frac{R_{in}}{R_{in}}\right)\left(-\frac{R_f}{R_1}\right) = \frac{R_f}{R_1} = K_g. \tag{8.7.1}$$

Now suppose that the input voltage is increased. For some input voltage, say v_{imax}, the electronics will saturate and the output voltage $v_o(t)$ will be a constant voltage v_{max}. We can model this nonideal behavior at large input signal levels using a saturation characteristic of the electronics as

$$S(v_i) = \begin{cases} v_{max}, & v_{imax} < v_i \\ K_g v_i, & -v_{imax} < v_i < v_{imax} \\ -v_{max}, & -v_{imax} < v_i \end{cases} \tag{8.7.2}$$

where

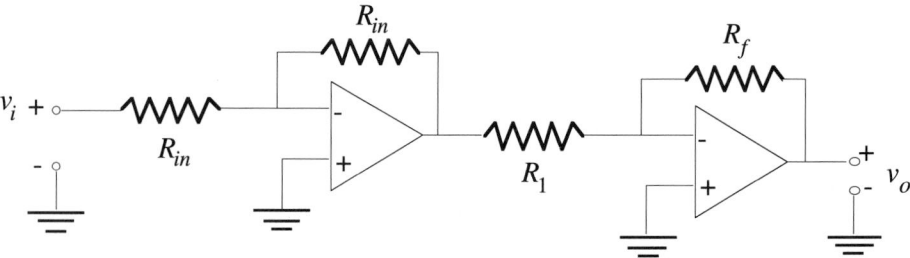

Figure 8.7.1 Operational Amplifier

$$v_{max} = K_g v_{imax}.$$
(8.7.3)

Let $K_g = 1$ for simplicity. Now the function in (8.7.2) is shown in Figure 8.7.2. We think of the saturation nonlinearity as a system, and we use the terminology associated with systems. This system is shown pictorially with an input signal and an output signal. Internally, the input signal appears on the abscissa axis and the output signal appears on the ordinate axis of the nonlinearity.

In this section we will consider a sinusoidal input signal to this system to be

$$v_i(t) = \cos(\omega_0 t).$$
(8.7.4)

The output signal from the nonlinearity is a clipped cosine wave and it is shown in Figure 8.7.3. The output signal can be constructed graphically, and this graphical construction is described in Section 13.7. This section can be read at this time.

8.7.2 Harmonic Distortion

The input signal to the op amp is the cosine (8.7.4) while Figure 8.7.3 shows the output signal from the nonlinearity. If we think of the input signal propagating through the nonlinearity to produce the output signal in Figure 8.7.3, then we see that the nonlinearity has a dramatic effect on the signal. It is difficult, however, to characterize the effects of this nonlinearity on the signal in a general way directly from the waveform. The Fourier series, however, provides one method for characterizing the effects of a nonlinearity that is widely used. We will illustrate this analysis in this subsection on the op amp example above.

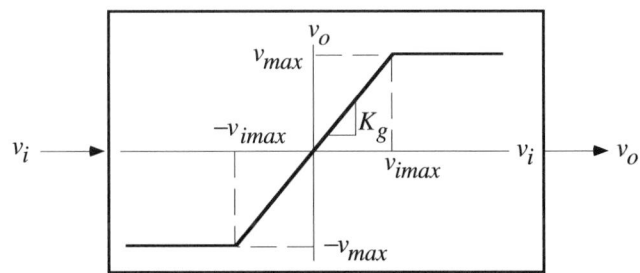

Figure 8.7.2 Saturation Nonlinearity

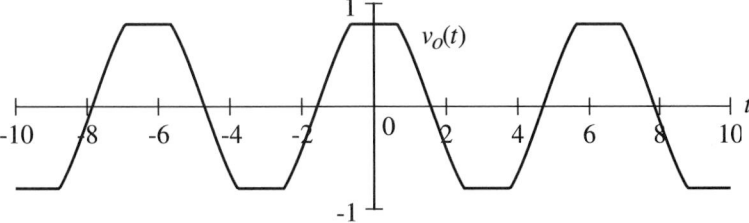

Figure 8.7.3 Clipped Cosine

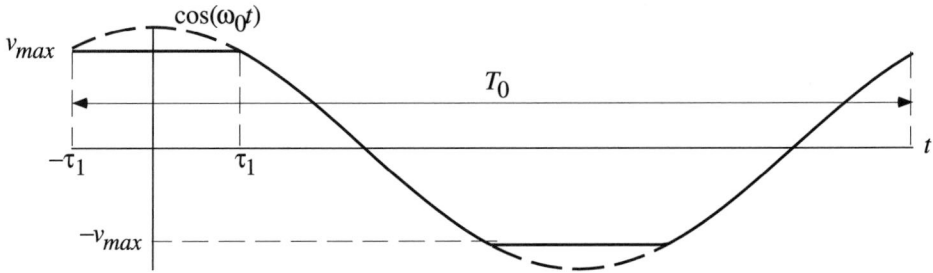

Figure 8.7.4 One Period of a Clipped Cosine

The cosine input signal (8.7.4) to the nonlinearity in Figure 8.7.2 is periodic. The output signal in Figure 8.7.3 is also a periodic signal. Since the output signal is periodic, it has a Fourier series. The Fourier series of the output signal gives us insight into the effect of the nonlinearity. To derive the Fourier series of the output signal, observe that one period of the output signal is shown in Figure 8.7.4. The signal in Figure 8.7.3 is the periodic extension of the function

$$v_{T_0}(t) = \begin{cases} v_{max}, & -\tau_1 < t \leq \tau_1 \\ \cos(\omega_0 t), & \tau_1 < t \leq \dfrac{T_0}{2} - \tau_1 \\ -v_{max}, & \dfrac{T_0}{2} - \tau_1 < t \leq \dfrac{T_0}{2} + \tau_1 \\ \cos(\omega_0 t), & \dfrac{T_0}{2} + \tau_1 < t \leq T_0 - \tau_1 \end{cases} \qquad (8.7.5)$$

The time τ_1 is determined from

$$v_o(\tau_1) = \cos(\omega_0 \tau_1) = v_{max}. \qquad (8.7.6)$$

To derive the Fourier series, we will employ a series of "tricks" to reduce the computation. The idea is to write the signal in Figure 8.7.4 in terms of the signal that has been "clipped" from the cosine by the nonlinearity. That is, we want to write the signal in Figure 8.7.3 as

$$v_o(t) = \cos(\omega_0 t) - v_c(t). \qquad (8.7.7)$$

The signal $v_c(t)$ is shown in Figure 8.7.5. In Figure 8.7.5 we have shown the clipped signal as the sum of two components

$$v_c(t) = v_p(t) + v_n(t). \qquad (8.7.8)$$

We will compute the Fourier series of each of the components and then sum the Fourier series to construct the Fourier series of the clipped sinusoid.

We begin with the signal $v_p(t)$. This signal is given by

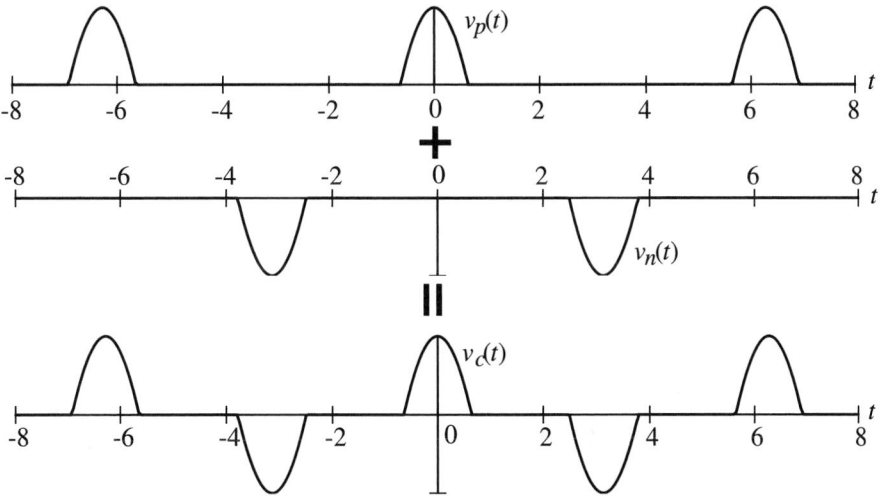

Figure 8.7.5 Component Signals of a Clipped Sinusoid

$$v_p(t) = \begin{cases} \cos(\omega_0 t) - v_{max}, & -\tau_1 \leq t < \tau_1 \\ 0, & \tau_1 < t < T_0 - \tau_1 \end{cases} \tag{8.7.9}$$

We will compute the trigonometric Fourier series next. The signal in (8.7.9) is even. Therefore, the coefficients b_m of the trigonometric Fourier series will be zero. The Fourier series of this signal is

$$v_p(t) = a_0 + \sum_{m=1}^{\infty} a_m \cos(m\omega_0 t). \tag{8.7.10}$$

The constant term will cancel with the constant from the signal $v_n(t)$ so we will not calculate it. The other coefficients of the trigonometric Fourier series are given by

$$a_m = \frac{2}{T_0} \int_{-\tau_1}^{T_0-\tau_1} v_p(t)\cos(m\omega_0 t)\, dt \tag{8.7.11}$$

$$= \frac{2}{T_0} \int_{-\tau_1}^{\tau_1} \cos(\omega_0 t)\cos(m\omega_0 t)\, dt - \frac{2}{T_0} \int_{-\tau_1}^{\tau_1} v_{max}\cos(m\omega_0 t)\, dt.$$

To evaluate the first integral in (8.7.11) we use the cosine identity

$$\cos(\omega_0 t)\cos(m\omega_0 t) = \frac{1}{2}\left[\cos((m-1)\omega_0 t) + \cos((m+1)\omega_0 t)\right]. \tag{8.7.12}$$

Using (8.7.12) in (8.7.11) we obtain

$$\frac{2}{T_0} \int_{-\tau_1}^{\tau_1} \cos(\omega_0 t) \cos(m\omega_0 t) \, dt = \frac{2\tau_1}{T_0} \Big[\mathrm{Sa}\big((m-1)\omega_0\tau_1\big) + \mathrm{Sa}\big((m+1)\omega_0\tau_1\big) \Big]. \tag{8.7.13}$$

The integration in (8.7.13) doesn't hold for $m = 1$ as can be seen from the denominator of (8.7.13). For $m = 1$ we have

$$\frac{2}{T_0} \int_{-\tau_1}^{\tau_1} \cos(\omega_0 t) \cos(\omega_0 t) \, dt = \frac{2\tau_1}{T_0} - \frac{1}{2\pi} \sin(2\omega_0\tau_1). \tag{8.7.14}$$

Evaluating the second integral in (8.7.11) we have

$$\frac{2}{T_0} \int_{-\tau_1}^{\tau_1} v_{max} \cos(m\omega_0 t) \, dt = \frac{4}{mT_0\omega_0} v_{max} \sin(m\omega_0\tau_1). \tag{8.7.15}$$

Now the coefficients of the Fourier series in (8.7.10) are given by

$$a_1 = \frac{2\tau_1}{T_0} - \frac{1}{2\pi} \sin(2\omega_0\tau_1) - \frac{4}{T_0\omega_0} v_{max} \sin(\omega_0\tau_1), \tag{8.7.16}$$

$$a_m = \frac{2\tau_1}{T_0} \Big[\mathrm{Sa}\big((m-1)\omega_0\tau_1\big) + \mathrm{Sa}\big((m+1)\omega_0\tau_1\big) \Big] - \frac{4v_{max}}{mT_0\omega_0} \sin(m\omega_0\tau_1).$$

Next consider the signal $v_n(t)$ in Figure 8.7.5. This signal can be derived from $v_p(t)$ by observing that this signal is shifted in time and inverted. We have

$$v_n(t) = -v_p\left(t - \frac{T_0}{2}\right) = -a_0 + \sum_{m=1}^{\infty} -a_m \cos\left[m\omega_0\left(t - \frac{T_0}{2}\right)\right]. \tag{8.7.17}$$

Using the trigonometric identity we can rewrite the terms in the summation (8.7.17) as

$$-a_m \cos\left[m\omega_0\left(t - \frac{T_0}{2}\right)\right] = -a_m\big[\cos(m\omega_0 t - m\pi)\big] = -(-1^m)a_m \cos(m\omega_0 t). \tag{8.7.18}$$

By combining (8.7.10), (8.7.17), and (8.7.18) we arrive at the Fourier series

$$v_c(t) = \sum_{\substack{m=1 \\ m \text{ odd}}}^{\infty} 2a_m \cos(m\omega_0 t). \tag{8.7.19}$$

Finally, the clipped sinusoid in (8.7.8) is given by

$$(8.7.20)$$

$$v_o(t) = \cos(\omega_0 t) - v_c(t) = (1 - 2a_1)\cos(\omega_0 t) - \sum_{\substack{m=3 \\ m \text{ odd}}}^{\infty} 2a_m \cos(m\omega_0 t).$$

8.7.3 Total Harmonic Distortion

To summarize the calculations above, the input signal to the nonlinearity, (8.7.2), is a cosine at a single frequency. The output signal is periodic; however, it is not a cosine. This signal has an infinite number of cosine terms with frequencies at integer multiples of the frequency of the input signal. The power spectral density of the input and output signals shown in Figure 8.7.6 graphically illustrate this observation.

Terminology: The term in the output signal at the same frequency as the input signal carries the most power. This term is called the <u>fundamental</u> term. The term whose frequency is twice the frequency of the fundamental is called the <u>first harmonic</u>. The first harmonic is not present in (8.7.20). The term whose frequency is three times the frequency of the fundamental is called the <u>second harmonic</u>. And so on. The other terms in the Fourier series are called <u>higher-order harmonics</u>. While the signal that goes into the nonlinearity is a pure cosine, the signal at the output of the nonlinearity looks more like a square wave. We say that the nonlinearity has <u>distorted</u> the input signal. We call this type of distortion <u>harmonic</u> distortion because we can characterize the distortion in terms of the presence of the higher-order harmonics. To quantify this distortion we define the total harmonic distortion of $v_o(t)$. Let P_o be the power in the output signal, and P_1 the power in the fundamental of the Fourier series of $v_o(t)$ in (8.7.20). Then the <u>total harmonic distortion</u> THD is defined as

$$\text{THD\%} = \frac{P_o - P_1}{P_o} \times 100\%. \qquad (8.7.21)$$

To calculate this distortion for the output signal (8.7.20), note that

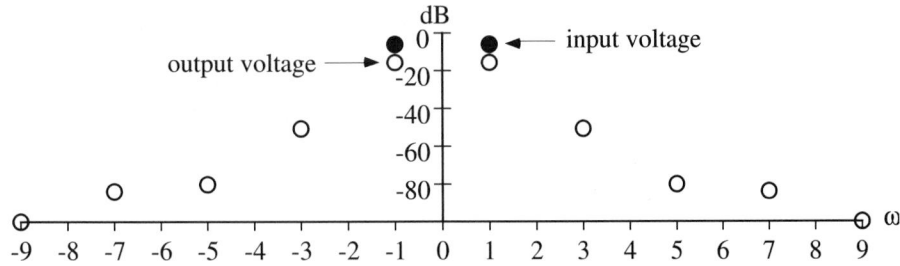

Figure 8.7.6 Power Spectral Density of the Input and Output Signals

$$P_f - P_{har} = P_1 = \frac{(1-2a_1)^2}{2}.$$

(8.7.22)

The power in the output signal can be approximated by summing the power in the remaining terms of the power spectrum.

Remark 8.7.1: In general the power in the output signal is not equal to the power in the input signal. The physical system that is represented by the nonlinearity can add power to or absorb power from the input signal.

▲▲

The total harmonic distortion for the amplifier saturation discussed above obviously depends on the relationship between the amplitude of the input signal and the saturation level of the amplifier. It is clear that as the voltage limit v_{max} is increased, the total harmonic distortion will decrease. As less of the input signal is clipped, the output signal will look more and more like a cosine.

If the amplitude of the input signal is less then the saturation limit, then there is no distortion introduced into the signal. Thus, the distortion of the output signal is dependent on the *absolute* values of the signal amplitude and the saturation limit. In contrast, for linear systems the output signal scales proportionally with the input signal. That is, the properties of the output signal can only be determined relative to the input signal. The dependence of the output signal on absolute values of the input signal is a characteristic of nonlinear systems.

Remark 8.7.2: This example sheds light on the frequency response theorem, Theorem 14.1.1. A static nonlinearity acts on a signal is a different way than a linear system. In the example above, the input signal was a sinusoid at a certain frequency. The output signal, however, is a signal that contained many frequencies; i.e., it is a Fourier series. This type of behavior is excluded from linear systems. Indeed, Theorem 14.1.1 says that if the input signal to a linear system is a sinusoid with a given frequency, the output signal is also a sinusoid *at the same frequency*. Furthermore, there is only one sinusoid in the output signal at the frequency of the input signal. This observation is one of the fundamental differences between linear and nonlinear systems.

▲▲

In general, when a system contains a nonlinear element, it is difficult to analyze. The Fourier series, however, gives us a tools to understand the effects of the nonlinearity on a signal passing through it. Basically, if the input signal is periodic, the nonlinearity will introduce additional harmonics into the output signal. The power in these harmonics is used to characterize the distortion introduced by the nonlinearity, the distortion measured being the THD.

8.8 MATLAB EXPERIMENTS

8.8.1 Introduction

In this section we will give several M-files that simulate the periodic signals that we have discussed in the last two chapters. With these M-files, the examples can be reproduced. Furthermore, some additional exploratory exercises are proposed.

8.8.2 Square Wave

Consider the periodic signal in Figure 8.8.1. The Fourier series for this signal is developed in Example 7.3.4, and it is given by

$$x(t) = \sum_{m=-\infty}^{\infty} \left[\frac{A\varepsilon}{T_0} e^{-jm\omega_0 t_0} \, \mathrm{Sa}\left(\frac{m\pi\varepsilon}{T_0} \right) \right] e^{jm\omega_0 t}. \tag{8.8.1}$$

Since MATLAB doesn't have an "Sa" command, we will rewrite (8.8.1) in terms of a **sinc** command. We have

$$x(t) = \sum_{m=-\infty}^{\infty} \left[\frac{A\varepsilon}{T_0} e^{-jm\omega_0 t_0} \, \mathrm{sinc}\left(\frac{m\varepsilon}{T_0} \right) \right] e^{jm\omega_0 t}. \tag{8.8.2}$$

The following M-file calculates the coefficients of this Fourier series, displays the Fourier series, the two-sided amplitude spectrum, and the two-sided power spectrum. The user can chose the signal parameters shown in Figure 8.8.1 as well as the number of terms of the Fourier series in the calculated by the M-file.

```
% Generic Square Wave
clear
% Define the signal parameters
T0 = 2;          % period of the signal
ep = 1;          % pulse width, less than T0
t0 = 0;          % offset
A = 1;           % amplitude
w0 = 2*pi/T0;    % fundamental frequency
```

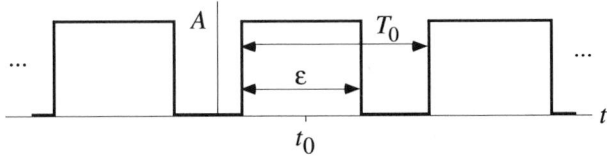

Figure 8.8.1 Generic Square Wave

```
% Number of terms in the Fourier series
nterm = 10;

% Calculate coefficients of exponential Fourier series
X0 = A*ep/T0;                                          % m = 0 term
for kk = 1:nterm
     X(kk) = X0*(exp(-j*kk*w0*t0))*sinc(kk*ep/T0);   % positive coefficients
end

% Calculate coefficients of cosine Fourier series
Am = 2*abs(X);
qm = angle(X);

% Calculate the partial sums of the Fourier series
t = linspace(-2*T0+ep/2,2*T0-ep/2,400);      % time vector
x = X0*ones(size(t));                         % constant term
for jj = 1:nterm
     x = x + Am(jj)*cos(jj*w0*t+qm(jj));
end

% Calculate two-sided amplitude spectrum
Xt = abs(X);
Amp = [fliplr(Xt),X0,Xt];                     % Amplitude spectrum

% Calculate two-sided power spectrum
Pt = abs(X).^2;
Pwr = [fliplr(Pt),X0^2,Pt];                   % Power spectrum

% Calculate frequencies of Fourier series terms
w = [-nterm:nterm]*w0;
```

```
% Plot Fourier series      % Plot amplitude spectrum                  % Plot power spectrum
figure(1)                  figure(2)                                 figure(3)
plot(t,x)                  plot(w,Amp,'o')                           plot(w,Pwr,'o')
title('Fourier Series')    title('Two-Sided Amplitude Spectrum')     title('Power Spectrum')
xlabel('Time')             xlabel('Frequency, rad/sec')              xlabel('Frequency, rad/sec')
```

Exploratory Exercise 8.8.1: Change the signal parameters and note the change in the amplitude and power spectra. First, vary ε between 0.1 and 1.8. What is the effect on the power spectrum? Second, vary T_0 and ε such that their ratio is equal. Again, what is the effect on the power spectrum?

▲▲

Exploratory Exercise 8.8.2: Modify this M-file to plot on one graph several partial sums of the Fourier series. Note how the buildup of the number of terms in the partial sum produces a better approximation of the square wave. Compare these results with the amplitude spectrum and power spectrum for each partial sum.

Also note that the overshoot at the discontinuity doesn't decrease with the number of terms in the partial sum.   This behavior is known as the <u>Gibb's phenomenon</u>.

▲▲

Exploratory Exercise 8.8.3: Modify this M-file so that any combination of terms may be included in the partial sum. Verify the results of Example 8.6.1.

▲▲

8.8.3 Modulated Square Wave

As a second example, consider the signal in Example 7.3.6. Over one period the signal is defined as

$$x_c(t) = \begin{cases} A\cos(\omega_c t), & -\dfrac{\varepsilon}{2} < t \le \dfrac{\varepsilon}{2} \\[2mm] 0, & \dfrac{\varepsilon}{2} < t \le T_0 - \dfrac{\varepsilon}{2} \end{cases} \tag{8.8.3}$$

The modulated square wave is defined by periodic extension of (8.8.3). A graph of this signal is shown in Figure 8.8.2. The Fourier series of this signal is developed in Example 7.3.6. Again, we rewrite its expression in terms of the sinc function. We get

$$x_{pc}(t) = \frac{2A}{\omega_c T_0}\sin\left(\frac{\omega_c \varepsilon}{2}\right) \tag{8.8.4}$$

$$+ \frac{A\varepsilon}{T_0}\sum_{m=1}^{\infty}\left[\operatorname{sinc}\big((f_c - mf_0)\varepsilon\big) + \operatorname{sinc}\big((f_c + mf_0)\varepsilon\big)\right]\cos(m\omega_0 t)$$

where $\omega_c = 2\pi f_c$ and $\omega_0 = 2\pi f_0$.

The following M-file calculates the partial sums of this Fourier series and plots these partial sums over a graph of the modulated square wave. It also plots the two-sided power spectrum. The number of coefficients of the exponential Fourier series

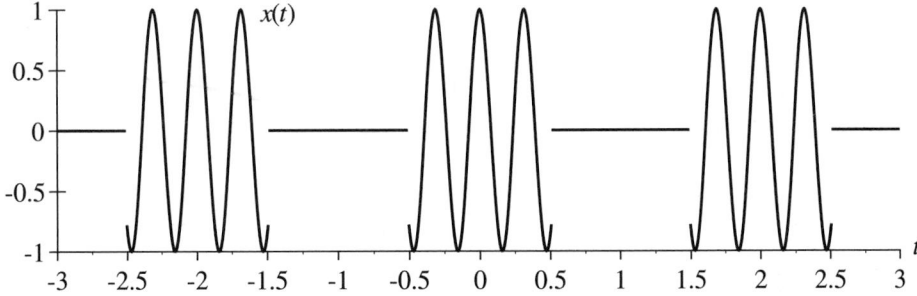

Figure 8.8.2 Modulated Square Wave

that is calculated is specified by the variable nterm. These terms are plotted in the two-sided power spectrum. The user can independently specify the terms that are contained in the partial sum through the variables ni and nf. Of course, the user can specify all of the signal parameters.

% Modulated Square Wave

```
clear
% Define signal parameters
T0 = 2;              % fundamental period
A = 1;               % amplitude
ep = 1;              % pulse width, less than T0      .
wc = 20;             % frequency of the pulse, rad/sec
fc = wc/2/pi;        % frequency of the pulse, Hz
w0 = 2*pi/T0;        % fundamental frequency, rad/sec
f0 = w0/2/pi;        % fundamental frequency, Hz


% Construct periodic waveform
step = 0.005;                        % increment in time vector
t = [-2*T0+T0/2:step:2*T0-T0/2];     % time vector for waveform
wvfm = zeros(size(t));               % signal waveform
for kk = -1:1
    i1 = max(find(t<T0*kk-ep/2));
    i2 = min(find(t>T0*kk+ep/2));
    wvfm(i1:i2) = A*cos(wc*(t(i1:i2)-T0*kk));
end


% Fourier series
% Calculate the coefficients of the exponential Fourier series
nterm = 12;          % number of coefficients calculated
for kk = 0:nterm
    if kk == 0
        X0 = (2*A/(T0*wc))*sin((wc*ep/2));
    else
        X(kk) = (A*ep/T0/2)*(sinc((fc-kk*f0)*ep)+sinc((fc+kk*f0)*ep));
    end
end


% Calculate coefficients of cosine Fourier series
amp = 2*abs(X);      % Amplitude
ang = angle(X);      % Phase


% Calculate partial sums
ni = 3;              % First term in the partial sum
nf = 10;             % Last term in the partial sum
if ni ==0
    x = X0*ones(size(t));
    ns = 1;
else
```

```
    x = zeros(size(t));
    ns = ni;
end

for jj = ns:nf
    x = x + amp(jj)*cos(jj*w0*t+ang(jj));
end

% Power Calculate power spectrum
Xs = [fliplr(X),X0,X];          % Two-sided coefficient vector
Xmag2 = abs(Xs).^2;             % Power spectrum
freq = [-nterm:nterm]*w0;       % Frequency axis

figure(1)                                   figure(2)
plot(t,x,t,wvfm)                            plot(freq,Xmag2,'o')
title('Waveform and Fourier Series')        title('Power Spectrum')
xlabel('Time')                              xlabel('Frequency, rad/sec')
```

Exploratory Exercise 8.8.4: Verify the results of Example 8.6.2.

▲▲

Exploratory Exercise 8.8.5: Change the signal parameters and note the change in the amplitude and power spectrum. First, vary ε between 0.1 and 1.8. What is the effect on the power spectrum? Second, vary ω_c. What is the effect of the on the power spectrum?

▲▲

8.9 CHAPTER SUMMARY

8.9.1 Spectral Content of a Signal

In this chapter we have developed the concept of the spectral content of a signal. The development of this concept begins with the definition of the amplitude and phase spectra of the signal which are summarized in Table 8.9.1. The importance of these concepts lies in part in that they allow for a simple graphical representation of a signal in the frequency domain.

The two key concepts in the development of the spectral content of the signal are energy and power.

Definition 8.2.1: The <u>energy</u> E of the signal $x(t)$ is defined as

$$E = \lim_{\gamma \to \infty} \int_{-\gamma}^{\gamma} |x(t)|^2 dt$$

if the limit exists and it is nonzero.

▲▲

Table 8.9.1 Definitions of Amplitude and Phase Spectra

Signal Representation		Definition
cosine Fourier series	(one-sided) amplitude and phase spectra	8.1.1
exponential Fourier series	(two-sided) amplitude and phase spectra	8.1.2
Fourier transform	amplitude and phase spectra	8.1.7

Table 8.9.2 Definitions of Energy and Power Spectra

Signal Representation		Definition
energy signal	energy spectral density	8.3.2
power signal	power spectral density	8.4.2
cosine Fourier series	(one-sided) power spectrum	8.5.5
exponential Fourier series	(two-sided) power spectrum	8.5.5

Definition 8.2.2: The <u>power</u> P of the signal $x(t)$ is defined as

$$P = \lim_{\gamma \to \infty} \frac{1}{2\gamma} \int_{-\gamma}^{\gamma} |x(t)|^2 \, dt$$

if the limit exists and it is nonzero.

▲▲

   The next step in the development is to relate the energy to the frequency domain representation of the signal.   This relationship is accomplished with Parseval's theorem.

Theorem 8.3.1: **Parseval's Theorem** Let $x(t)$ be an energy signal which has a Fourier transform $X(\omega)$. Then the energy in this signal is given by

$$E = \int_{-\infty}^{\infty} |x(t)|^2 \, dt = \frac{1}{2\pi} \int_{-\infty}^{\infty} |X(\omega)|^2 \, d\omega.$$

▲▲

Theorem 8.3.1 is extended to the Fourier series in Theorem 8.5.3 and Corollary 8.5.4 and to power signals in Theorem 8.4.3. This theorem relates the amplitude spectrum (squared) to the energy in the signal.   Thus, we are able to characterize the distribution of the energy in terms of frequency.   This observation gives rise to the definitions of energy and power spectral density given in Table 8.9.2. Examples of the interpretation of the energy and power spectral density are given in the later sections of the chapter.   This concept leads naturally to the signal bandwidth (Definition 8.3.4). These ideas are also applied to the analysis of the output signal of a static nonlinearity. They are used to define total harmonic distortion (THD) in Section 8.7.

8.9.2 Classification of Signals

In the discussion of signal modeling in Chapters 5, 7, and 8, what have essentially defined several classes of signals will be summarized next.

Periodic/Aperiodic We can divide all signals into two classes: periodic signals and aperiodic signals. For most signals in which we are interested, the periodic signals can be represented using a Fourier series. The Fourier series gives us direct insight into their frequency domain properties through their amplitude and phase spectra and power spectra. The second class of signals is signals that are not periodic: aperiodic signals. For aperiodic signals, the Fourier transform is used to expose their frequency domain properties. The interpretation of the spectral content of both classes of signals is quite similar.

Energy/Power Signals A second way to divide up signals is based on the definitions of energy and power. (Again some signals are not included in these two sets, but these signals are of lesser interest.) For these two sets of signals the energy spectral density and power spectral density give insight into the properties of these signals in the frequency domain.

Deterministic/Random Signals A third way to classify signals is to distinguish between signals that depend on a random variable and those that don't — deterministic signals. In this text we have discussed only those signals that are deterministic although some of the analysis techniques we have introduced can be extended to random processes. We mention random processes because they form a very important class of signal models which, unfortunately, are beyond the scope of this text. Random signals complement all of the deterministic signal models we have introduced here.

8.10 HOMEWORK FOR CHAPTER 8

Homework Problems for Section 8.1

8.1.1   The amplitude spectrum of the signal $x(t)$ is shown in Figure P8.1.1. From this signal we construct the following signals by modulation. Sketch the amplitude spectrum of each signal.
(i) $x_1(t) = x(t)\cos(100t)$ (ii) $x_2(t) = x(t)\cos(10t)$

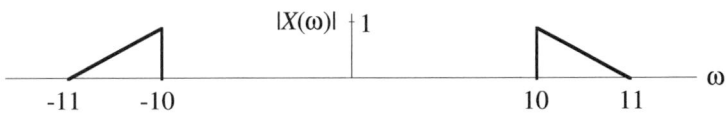

Figure P8.1.1

8.1.2   Consider the following Fourier series. For each signal:
(a) Plot the two-sided amplitude and phase spectra.
(b) Plot the one-sided amplitude and phase spectra.
(c) Plot the signal itself.

(i) $x(t) = \displaystyle\sum_{\substack{m=-\infty \\ m\neq 0}}^{\infty} \left[\frac{2}{jm\pi}\left(1 - \cos\left(\frac{m\pi}{2}\right)\right)\right] e^{j\frac{m\pi}{2}t}$

(ii) $x(t) = 1 + \displaystyle\sum_{\substack{m=-\infty \\ m\neq 0}}^{\infty} \left(\frac{1}{m\pi} e^{j\frac{\pi}{2}}\right) e^{jm\pi t}$

(iii) $x(t) = \displaystyle\sum_{m=-\infty}^{\infty} \frac{0.5}{(1+j\pi m)} e^{j\frac{\pi m t}{3}}$

(iv) $x(t) = \displaystyle\sum_{\substack{m=-\infty \\ m\neq 0}}^{\infty} \frac{1}{j2\pi m}\left(1 + e^{-j\frac{2\pi m}{3}} - 2e^{-j\frac{4\pi m}{3}}\right) e^{j\frac{2\pi m t}{3}}$

(v) $x(t) = \displaystyle\sum_{m=1}^{\infty} \left(\frac{2}{2m^2+1}\right)\cos(2mt - 0.5m)$

(vi) $x(t) = 1 + \displaystyle\sum_{\substack{m=-\infty \\ m\neq 0}}^{\infty} \frac{1}{m} e^{j\left(\frac{m\pi}{4}\right)} e^{j2mt}$

(vii) $x(t) = \displaystyle\sum_{m=1}^{\infty} \frac{3}{4m^2}\cos\left(2mt - \frac{\pi m}{4}\right)$

8.1.3   For each of the following three functions:
(a) Plot each signal.
(b) Find the Fourier transform of each signal for $\varepsilon = 1$.
(c) Plot the amplitude and phase spectra of each signal.

(i) $x_1(t) = A\Pi\left(\dfrac{t}{\varepsilon}\right)$

(ii) $x_2(t) = B\left[1 + \cos\left(\dfrac{2\pi t}{\varepsilon}\right)\right]\Pi\left(\dfrac{t}{\varepsilon}\right)$

(iii) $x_3(t) = C\left[\cos\left(\dfrac{2\pi t}{\varepsilon}\right)\right]\Pi\left(\dfrac{t}{\varepsilon}\right)$

8.1.4   Consider the amplitude and phase spectra shown in Figure P8.1.4.
(a) What is the fundamental period of this signal?
(b) Write the corresponding Fourier series in exponential form.
(c) Write the corresponding Fourier series in cosine form.
(d) Write the corresponding Fourier series in trigonometric form.

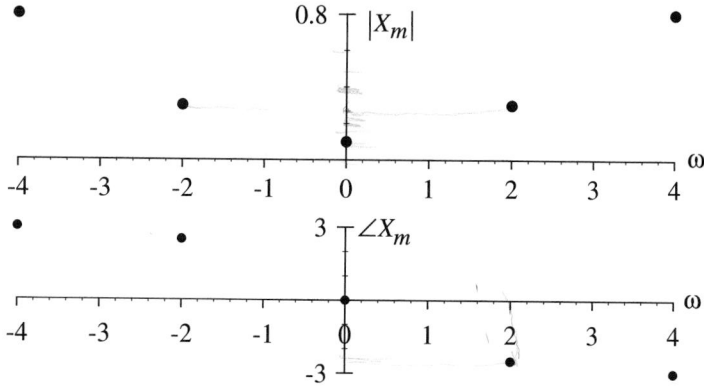

Figure P8.1.4

8.1.5   Plot the amplitude and phase spectra of the following signals.

(i) $x(t) = \text{Sa}(3t)$

(ii) $x(t) = \dfrac{1}{2\pi}\left[\text{Sa}\left(\dfrac{t}{2}\right)\right]^2 \cos(20t)$

(iii) $x(t) = e^{-2|t|}\cos(20t)$

(iv) $x(t) = \left(0.5025e^{-0.2t}\sin 1.99t\right)u_s(t)$

(v) $x(t) = (\cos(t))(\cos(10t))$

(vi) $x(t) = \Lambda(t)\cos(0.1t)$

8.1.6   Consider the amplitude and phase spectra shown in Figure P8.1.6.
(a) What is the fundamental period of $x(t)$?
(b) Find the corresponding Fourier series in exponential form.
(c) Find the corresponding Fourier series in its cosine form.

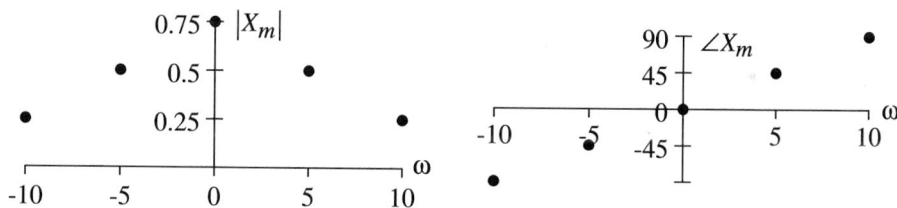

Figure P8.1.6

8.1.7   Prove the following statement: If $x(t)$ is a real signal, then amplitude spectrum is an even function of ω, and the phase spectrum is an odd function of ω.

8.1.8   Find and plot the amplitude and phase spectra of the signal

$$x(t) = e^{-\alpha t}u_s(t), \quad \alpha > 0$$

Homework Problems for Section 8.2

8.2.1   Show that a periodic signal can never be an energy signal.

8.2.2   Define a signal

$$\tilde{b}(t) = \begin{cases} b_k(t), & -\dfrac{T_b}{2} < t \le \dfrac{T_b}{2} \\ 0, & \text{elsewhere} \end{cases}$$

Also define the sequence $\tilde{a} = \begin{bmatrix} 1 & 0 & 1 & 1 \end{bmatrix}$. Now define the signal

$$s(t) = \sum_{m=0}^{3} \tilde{b}(t - mT_b)$$

(a) Plot $s(t)$ for each of the following signals $b_k(t)$ when $T_b = 10^{-6}$.
(b) Compute the energy in each signal $s(t)$.
(c) Repeat parts (a-b) for the sequence $\tilde{a} = \begin{bmatrix} 0 & 1 & 0 & 1 \end{bmatrix}$.
(d) Suppose the sequence $\tilde{a}$ is a long sequence of random 0's and 1's. What can you say about the energy of each signal?

(i) $b_1(t) = \Pi\left(\dfrac{t}{T_b}\right)$

(iii) $b_3(t) = \dfrac{\cos(\pi t)}{1 - (2t)^2}\,\text{sinc}\left(\dfrac{t}{T_b}\right)$

(ii) $b_2(t) = \Pi\left(\dfrac{t - \dfrac{T_b}{4}}{\dfrac{T_b}{2}}\right) - \Pi\left(\dfrac{t + \dfrac{T_b}{4}}{\dfrac{T_b}{2}}\right)$

(iv) $b_4(t) = \cos\left(\dfrac{\pi t}{2T_b}\right)$

8.2.3   Determine which of the following signals are energy signals and which are power signals.

(i) $x(t) = \cos(4t)u_s(t)$

(ii) $x(t) = \sin(4t) + \sin(\pi t)$

(iii) $x(t) = e^{5t}u_s(t)$

(iv) $x(t) = e^{-4|t|}$

(v) $x(t) = \displaystyle\sum_{m=0}^{\infty} \Pi(t - 3m)$

(vi) $x(t) = \displaystyle\sum_{m=0}^{\infty} \dfrac{1}{m^2}\Pi(t - 3m)$

Homework Problems for Section 8.3

8.3.1   For each of the three signals given below:
(a) Plot each signal.
(b) Compute the energy in each signal using the time-domain definition of energy.
(c) For a fixed ε, select the constants A, B, and C such that each signal has the same energy.

(d) Find the energy spectral density of each signal.
(e) Find the 90% bandwidth of this signal.   (Requires numerical integration.)

(i) $x_1(t) = A\Pi\left(\dfrac{t}{\varepsilon}\right)$

(ii) $x_2(t) = B\left[1 + \cos\left(\dfrac{2\pi t}{\varepsilon}\right)\right]\Pi\left(\dfrac{t}{\varepsilon}\right)$

(iii) $x_3(t) = C\left[\cos\left(\dfrac{2\pi t}{\varepsilon}\right)\right]\Pi\left(\dfrac{t}{\varepsilon}\right)$

8.3.2   The Fourier transform of the signal $x(t)$ is shown in Figure P8.3.2.
(a) Find the signal $x(t)$.
(b) What is the energy in this signal?

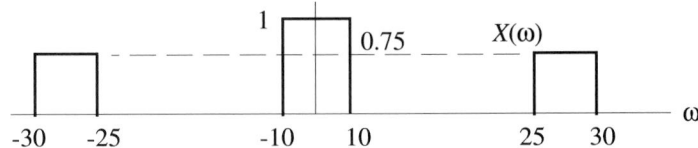

Figure P8.3.2

8.3.3   Consider the signals given below.
(a) Plot the energy spectral density of this signal.
(b) What is the bandwidth of each signal?
(c) What is the bandlimit of each signal?

(i) $x(t) = \text{Sa}(3t)$ (iii) $x(t) = \sin(\omega t)\Pi(t)$

(ii) $x(t) = \dfrac{1}{2\pi}\left[\text{Sa}\left(\dfrac{t}{2}\right)\right]^2 \cos(20t)$

8.3.4   Show that the energy in a signal $x(t)$ is given by

$$E = \int_{-\infty}^{\infty} |x(t)|^2\, dt = \int_{-\infty}^{\infty} |X(f)|^2\, df$$

where the Fourier transform of this signal is $X(f)$.

8.3.5   Consider a signal $x(t)$ whose amplitude spectrum is shown in Figure P8.3.5. Also consider the signal

$$x_1(t) = x(t)\cos(100t)$$

(a) Find the amplitude spectrum $x_1(t)$.
(b) What is the bandwidth and bandlimit of each of these signals?

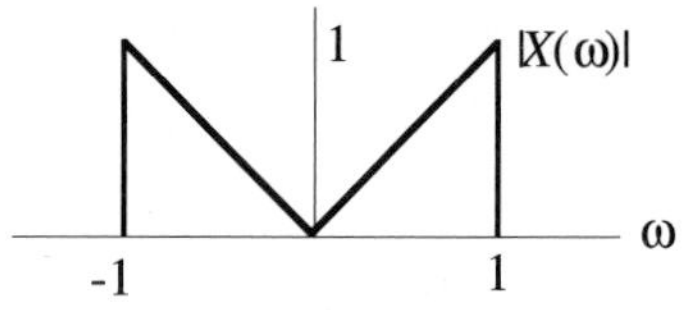

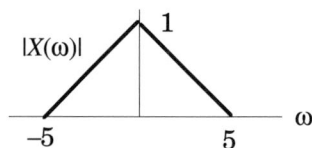

Figure P8.3.5

8.3.6 Consider the following signals. Using MATLAB find the 95% energy bandwidth of each signal.

(i) $x(t) = \text{Sa}(3t)$

(ii) $x(t) = \dfrac{1}{2\pi}\left[\text{Sa}\left(\dfrac{t}{2}\right)\right]^2 \cos(20t)$

(iii) $x_1(t) = e^{-2|t|}$ and $x_2(t) = e^{-2|t|}\cos(20t)$

8.3.7 The Fourier transform of a signal is shown in Figure P8.3.7.

(a) Sketch the amplitude spectrum of the signal $x_1(t) = x(t)\cos(15t)$.

(b) What is the bandwidth and bandlimit of both $x(t)$ and $x_1(t)$.

Figure P8.3.7

Homework Problems for Section 8.4

8.4.1 Compute the power spectral density of the signal $x(t) = A\sin(\omega t)$.

8.4.2 Find the power spectral density of the signal

$$x(t) = \sum_{m=-\infty}^{\infty} \Lambda(t - 2m).$$

8.4.3 Compute the power in the signal whose power spectral density is

$$S(\omega) = \frac{A^2\pi}{2}\left[\delta(\omega - \omega_0) + \delta(\omega + \omega_0)\right].$$

Homework Problems for Section 8.5

8.5.1   The Fourier series of a periodic signal is

$$x(t) = \sum_{\substack{m=-\infty \\ m \neq 0}}^{\infty} \left[\frac{2}{jm\pi} \left(1 - \cos\left(\frac{m\pi}{2}\right) \right) \right] e^{j\frac{m\pi}{2}t}$$

(a) Plot this signal.
(b) Plot the discrete power spectrum of this Fourier series.
(c) Plot the power spectral density of this Fourier series.
(d) Find the power in $x(t)$ using Parseval's theorem (numerical calculation).
(e) Plot the power in the partial sums as a function of the number of terms in the partial sum.

8.5.2   Consider the signal

$$x(t) = \begin{cases} e^{-|t|}, & -1 < t \leq 1 \\ 0, & \text{elsewhere} \end{cases}$$

Define the signal $y(t)$ as the periodic extension of $x(t)$.
(a) Sketch $x(t)$.
(b) Sketch $y(t)$.
(c) Find the power in $y(t)$.

8.5.3   The amplitude spectrum of a signal $x(t)$ is shown in Figure P8.5.3. Assume that the phase spectrum is zero.
(a) Construct $x(t)$.
(b) Is this signal a power signal or an energy signal?   Find the corresponding power or energy.
(c) What is the bandwidth and bandlimit of this signal?

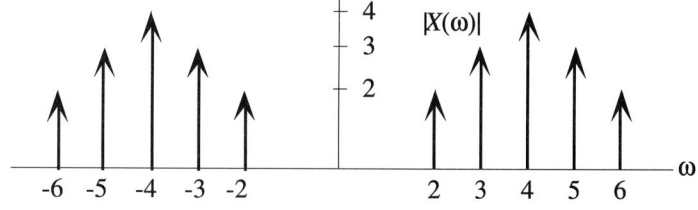

Figure P8.5.3

8.5.4   Prove Theorem 8.5.1.

8.5.5   Using MATLAB determine the 95% power bandwidth of each of the signals in Problems 7.2.2, 7.2.3, and 7.2.4.

Homework Problems for Section 8.6

8.6.1   Consider the signal $x(t)$ whose amplitude spectrum is shown in Figure P8.6.1. Assume the phase spectrum is zero.
(a) What is the fundamental period of this signal?
(b) Write out the Fourier series of this signal in exponential form.
(c) Write out the Fourier series of this signal in cosine form.
(d) What is the power in $x(t)$?
(e) What sinusoidal component of $x(t)$ carries the most power?

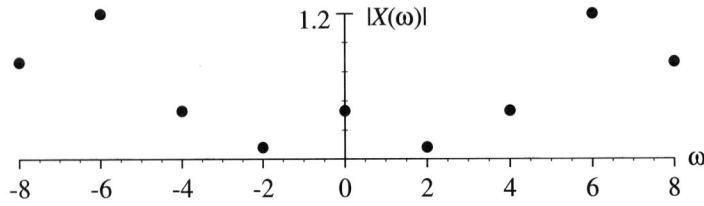

Figure P8.6.1

8.6.2   Consider the signal

$$
x_c(t) = \begin{cases} A\cos(\omega_c t), & -\dfrac{\varepsilon}{2} < t \le \dfrac{\varepsilon}{2} \\ 0, & \dfrac{\varepsilon}{2} < t \le T_0 - \dfrac{\varepsilon}{2} \end{cases}
$$

Let the signal $x(t)$ be defined by periodic extension of $x_c(t)$. Calculate the power in $x(t)$.

8.6.3   Consider the two periodic signals in Examples 8.6.1 and 8.6.2. Let $x_{0:m_1}(t)$ be an approximating signal for the Fourier series. For each periodic signal plot the power in the corresponding approximating signal as a function of the number of terms in the approximating signal. Compare these two plots.

8.6.4   Suppose the signal $x(t)$ is defined as the periodic extension of

$$x_1(t) = t, \quad 0 \le t < 2$$

The exponential Fourier series of this signal is

$$x(t) = 1 + \sum_{\substack{m=-\infty \\ m \ne 0}}^{\infty} \left(\frac{1}{m\pi} e^{j\frac{\pi}{2}} \right) e^{jm\pi t}$$

(a) Sketch this signal.
(b) Find the power in this signal.
(c) Find and plot the power spectrum.

(d) What is the minimum number of terms in this series that carry more than 95% of the power?

(e) Plot the partial sums of this series.

8.6.5   The exponential Fourier series of a periodic signal is

$$x(t) = \sum_{\substack{m=-\infty \\ m \neq 0}}^{\infty} \frac{1}{j2\pi m} \left(1 + e^{-j\frac{2\pi m}{3}} - 2e^{-j\frac{4\pi m}{3}}\right) e^{j\frac{2\pi m t}{3}}$$

(a) Plot the partial sums of this series.

(b) Find the power in this signal.

(c) Find and plot the power spectrum.

(d) What is the minimum number of terms in this series that carry more than 95% of the power?

Homework Problems for Section 8.7

8.7.1   Consider the static nonlinearity shown in Figure P8.7.1. Suppose that the input to this nonlinearity is $x(t) = \cos(\omega_0 t)$.

(a) Find the output signal for this input signal.

(b) Find the Fourier series of the output signal.

Note: This nonlinearity will generate an <u>exact</u> 2nd order harmonic in response to the sinusoidal input signal. This response is sometimes used to generate a carrier signal in a communication channel.

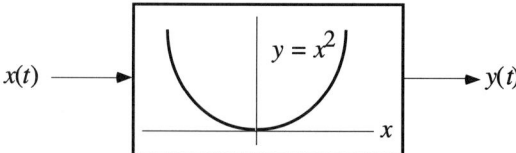

Figure P8.7.1

8.7.2   Consider the nonlinearity in Figure P8.7.2. The input signal to this nonlinearity is a $x(t) = \cos(\pi t)$.

(a) What is the functional description of this nonlinearity?

(b) Find the Fourier series of the output signal.

(c) Plot the power spectrum of the output signal.

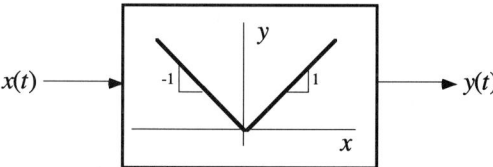

Figure P8.7.2

8.7.3   Consider the nonlinearity in Figure P8.7.3.   The input signal to this nonlinearity is a $x(t) = \cos(\omega_0 t)$.

(a) What is the functional description of this nonlinearity?

(b) Find the Fourier series of the output signal.

(c) Draw a simple circuit with a diode and resistor such that the input and output voltages are related according to the nonlinearity in Figure P8.7.3.

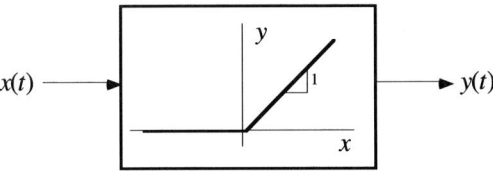

Figure P8.7.3

Chapter 9

The Laplace Transform

Chapter Outline

A signal is a function that represents the time variation of a physical variable. To this point we have introduced two quite distinct representations of a signal. The first class of representations of a signal, developed in Chapter 5, is the class of functions that depend on time. This type of representation is quite obvious. The second class of signal representations, derived in Chapter 7, is based on the Fourier transform of the signal representations in the first class (when the Fourier transform exists). These signal representations depend on a real frequency variable ω. In this chapter we will define a third class of signal representations that depend on a complex variable s. This class of signal representations is the Laplace transform of the first class of signal representations (when the Laplace transform exists). This third class of signal representations, like the second class of signal representations, doesn't represent the signals directly in the time domain. There are many advantages and insights to this type of signal representation, which we will develop in the coming chapters. We can make similar comments about system representations, which we will discuss in detail in Chapters 10 - 15.

In this chapter we introduce the Laplace transform as a mathematical tool for analyzing signals and systems. After the definition of the Laplace transform the properties of the transform are developed. Finally, these properties are applied to the solution of a differential equation. The last section of this chapter discusses the relationship between the definitions of the Laplace and Fourier transforms. In keeping with the philosophy of the text, the primary focus of this chapter is on the mathematical details of the Laplace transform. In the following chapters we will apply this transform to the analysis of signals and systems.

Computer Usage: The Laplace transform has also appeared as a useful representation for signals and systems for computer packages. Hence, knowledge of the transform is required for effective use of many computer packages.

Summary of Sections

Section 9.1:   We define the Laplace transform.

Section 9.2:   We discuss the properties of the Laplace transform.

Section 9.3:   We introduce partial fraction expansions and inversion of the Laplace transform.

Section 9.4:   We discuss the solution of differential equations using the Laplace transform.

Section 9.5:   We discuss the relationship between the Laplace and Fourier transform.

Section 9.6   Chapter summary section.

Coverage of the Text

This chapter requires only the background material on singularity function and complex functions in Chapters 2 and 3. The basic terminology of signals and systems, introduced in Chapters 5 and 6, is also used. Section 9.5 requires knowledge of the Fourier transform from Section 7.4. Section 9.5 can be skipped on first reading without loss of continuity.

9.1 DEFINITION OF THE LAPLACE TRANSFORM

9.1.1 Definitions

Let $x(t)$ be a signal representation. The Laplace transform is a mathematical tool that transforms this signal representation (a function of time) into a completely different signal representation (a function of a complex variable s). There is a one-to-one correspondence between the signal $x(t)$ and its Laplace transform $X(s)$.

Hence, the Laplace transform of a signal is another representation of the signal. The basic concept is in many ways similar to the Fourier transform.[1]

The main purpose of this section is to give the definition of the Laplace transform. This definition gives us two important pieces of information about the Laplace transform. First, it identifies which functions of time can be Laplace transformed, since not all signal representations have a Laplace transform. The definition tells us when we can apply the mathematical tool. Second, this definition gives us a computational formula for computing the Laplace transform of a particular signal. This formula is the starting place for generating the Laplace transform for a specific signal. Hence, this definition gives us information that is routinely, if indirectly, used.

The definition of the Laplace transform is as follows.

Definition 9.1.1: Let $x(t)$, $-\infty < t < \infty$, be a signal that satisfies

(a) $x(t) = 0$ for $t < 0$,

(b) $\int_0^\infty |x(t)| e^{-\sigma t} dt < \infty$ for $0 \le \sigma_0 < \sigma < \infty$

Then the <u>Laplace transform</u> of $x(t)$ is defined as

$$\mathcal{L}\{x(t)\} = \int_0^\infty x(t) e^{-st} dt = X(s).$$

The <u>inverse</u> Laplace transform of $X(s)$ is given by

$$x(t) = \frac{1}{2\pi j} \int_{\sigma - j\omega}^{\sigma + j\omega} X(s) e^{st} ds = \mathcal{L}^{-1}\{X(s)\}.$$

▲▲

Notation: The argument of a Laplace transform is called the <u>Laplace variable,</u> and it is denoted by s. Broken into real and imaginary parts, the Laplace variable is written in standard notation as $s = \sigma + j\omega$.

Definition 9.1.2: The pair

$$\mathcal{L}\{x(t)\} = X(s) \leftrightarrow x(t) = \mathcal{L}^{-1}\{X(s)\}$$

is called a <u>Laplace transform pair.</u>

▲▲

In particular, the notation $x(t) \leftrightarrow X(s)$ means that the Laplace transform of $x(t)$ exists.

---

[1] The relationship between Laplace and Fourier transforms is discussed in Section 9.5.

Computer Usage: The Laplace transform is an alternate representation of a signal $x(t)$. This representation is useful for representing signals in computer packages.

Terminology: The calculation of the inverse Laplace transform is sometimes called inverting $X(s)$.

Terminology: We say that $x(t)$ is in the time domain because it is a function of time, and $X(s)$ is in the frequency domain because it depends on the Laplace variable s. This terminology will be justified over the course of the text.

9.1.2 Existence of the Laplace Transform

There are two conditions in the definition of a Laplace transform that characterize the signals that may be Laplace transformed. The first condition generally dictates when the Laplace transform can be used in practice. In words, it says that the signals must be zero for negative time.

Example 9.1.3: Consider the signals

$$(a) \quad \sin(\omega t), -\infty < t < \infty, \tag{9.1.1}$$

$$(b) \quad [\sin(\omega t)]u_s(t), -\infty < t < \infty.$$

According to condition (a) of Definition 9.1.1, the signal (a) is *not* Laplace transformable because this signal is nonzero for negative time.[2] The signal (b) does satisfy (a) of Definition 9.1.1. This signal is Laplace transformable.

▲▲

The existence of the Laplace transform depends on the existence of the integral in Definition 9.1.1. Since the integral is over an infinite interval, its existence is not guaranteed for all signals that satisfy condition (a) of Definition 9.1.1. The existence of this integral is guaranteed by condition (b) of the definition. To show that the integral exists, we must show that

$$\left| \int_0^\infty x(t)e^{-st} dt \right| < \infty \tag{9.1.2}$$

for $\sigma > \sigma_0$. Using an integral inequality we have

$$\left| \int_0^\infty x(t)e^{-st} dt \right| \le \int_0^\infty \left| x(t)e^{-st} \right| dt \le \int_0^\infty \left| x(t) \right| \left| e^{-st} \right| dt. \tag{9.1.3}$$

If we let $s = \sigma + j\omega$ with $\sigma \ge 0$, then we have

$$\left| e^{-st} \right| = \left| e^{-(\sigma+j\omega)t} \right| = e^{-\sigma t} \left| e^{-j\omega t} \right| = e^{-\sigma t} \tag{9.1.4}$$

---

[2] See also Section 2.1.

since

$$\left|e^{-j\omega t}\right| = 1. \tag{9.1.5}$$

Using (9.1.4) in (9.1.3) we have

$$\left|\int_0^\infty x(t)e^{-st}\,dt\right| \le \int_0^\infty \left|x(t)\right|e^{-\sigma t}\,dt < \infty. \tag{9.1.6}$$

The right-hand side of (9.1.6) is just (b) of Definition 9.1.1. Hence, the defining integral of the Laplace transform will exist when (b) is satisfied.

Example 9.1.4: This example illustrates the role of condition (b) when using the Laplace transform. Consider the signal

$$x(t) = e^{at}u_s(t). \tag{9.1.7}$$

This signal satisfies (a) of Definition 9.1.1. To check that condition (b) is satisfied, we have

$$\lim_{\gamma \to \infty} \int_0^\gamma \left|x(t)\right|e^{-\sigma t}\,dt = \lim_{\gamma \to \infty} \int_0^\gamma e^{at}e^{-\sigma t}\,dt = \lim_{\gamma \to \infty} \left.\frac{e^{(a-\sigma)t}}{a-\sigma}\right|_0^\gamma \tag{9.1.8}$$

$$= \lim_{\gamma \to \infty} \frac{e^{(a-\sigma)\gamma}-1}{a-\sigma} = \frac{-1}{a-\sigma}, \quad \sigma > a.$$

Clearly, the integral (9.1.8) is finite if $\sigma > \sigma_0 \ge a$. Now using Definition 9.1.1 the Laplace transform is given by

$$\int_0^\infty x(t)e^{-st}\,dt = \lim_{\gamma \to \infty} \int_0^\gamma e^{at}e^{-st}\,dt = \lim_{\gamma \to \infty} \left.\frac{e^{(a-s)t}}{a-s}\right|_0^\gamma = \lim_{\gamma \to \infty} \frac{e^{(a-s)\gamma}-1}{a-s}. \tag{9.1.9}$$

To evaluate the limit in (9.1.9) let

$$a - s = a - (\sigma - j\omega). \tag{9.1.10}$$

Substituting (9.1.10) into (9.1.9) we find

$$\lim_{\gamma \to \infty} \frac{1}{a-s}\left(e^{(a-\sigma)\gamma}e^{-j\omega\gamma}-1\right) = \frac{1}{s-a} \quad \text{for} \quad \sigma > a. \tag{9.1.11}$$

So we have

$$L\{e^{at}u_s(t)\} = \frac{1}{s-a}. \qquad (9.1.12)$$

If we let $a \to 0$, the above calculation yields

$$L\{u_s(t)\} = \frac{1}{s}. \qquad (9.1.13)$$

▲▲

As a practical matter, we don't often find signals that don't satisfy condition (b). Usually, condition (a) determines if the Laplace transform is used.

9.1.3 The Impulse Function

Suppose that $x(t) = \delta(t)$. Using Definition 9.1.1 the Laplace transform of this signal would be

$$\int_0^\infty \delta(t)e^{-st}dt. \qquad (9.1.14)$$

The integral in (9.1.14), however, is undefined because the lower limit of (9.1.14) coincides with the zero value of the argument of the impulse function (see Remark 2.2.2). When a signal $x(t)$ satisfies Definition 9.1.1, but this signal contains an impulse function or derivatives of the impulse function, then the definition of the Laplace transform is replaced by

$$L\{x(t)\} = \lim_{\gamma \uparrow 0} \int_\gamma^\infty x(t)e^{-st}dt = \int_{0^-}^\infty x(t)e^{-st}dt. \qquad (9.1.15)$$

The limit in (9.1.15) is computed as γ approaches zero from below. Evaluating (9.1.15) we have

$$L\{\delta(t)\} = \int_{0^-}^\infty \delta(t)e^{-st}dt = 1. \qquad (9.1.16)$$

The impulse function plays a central role in this text.

Note: When the Laplace transform is used to transform an impulse function or its derivatives, it is understood that the integral is evaluated using the limit from below as in (9.1.15).

9.1.4 Region of Convergence[3]

The existence of the Laplace transform depends on the existence of the integral in (b) of Definition 9.1.1. This integral must exist for at least one complex number. It

---

[3] This subsection applies to only a few specialized results in the following chapters.

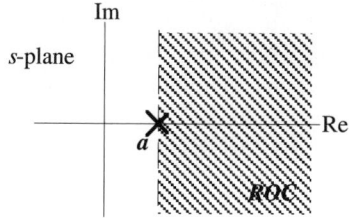

Figure 9.1.1 *ROC* of Example 9.1.6

is usually the case, however, that if the integral exists for one complex number it will exist for a certain set in the complex plane. Occasionally, this set is useful in this text, so we give it a name.

Definition 9.1.5: The set of all complex numbers for which the defining integral of the Laplace transform exists is called the <u>region of convergence</u> (*ROC*).

▲▲

Terminology: Similar terminology is used for discussing z-transforms.

Example 9.1.6: Consider the signal in (9.1.7). The region of convergence for this signal is

$$ROC = \left\{ s = \sigma + j\omega \in C \mid \sigma > \sigma_0 = a \right\}. \tag{9.1.17}$$

This set is shown in Figure 9.1.1.

▲▲

Note: The region of convergence depends on the signal $x(t)$.

This next result will find application in stability theory and Fourier transforms.

Theorem 9.1.7: Suppose that the Laplace transform of $x(t)$ exists for some $\sigma > 0$. Suppose that $X(s)$ is a proper rational function and its poles are given by $p_i = \sigma_i + j\omega_i, \quad i = 1, 2, \ldots, n$. Let

$$\sigma_0 = \max_i |\sigma_i|.$$

Then the *ROC* of $x(t)$ is the half plane defined by

$$ROC = \left\{ s = \sigma + j\omega \in C \mid \sigma > \sigma_0 \right\},$$

and conversely.

▲▲

If $\sigma_0 < 0$, then all of the poles of $X(s)$ are in the open LHP.[4] Since the region of convergence means the region where the integral (b) in Definition 9.1.1 converges we can set $\sigma = 0$. Then we have the following corollary of Theorem 9.1.7.

Theorem 9.1.8: Suppose that the Laplace transform of $x(t)$, $X(s)$ is a proper rational function. Then all of the poles of $X(s)$ are in the open LHP if and only if

$$\int_{-\infty}^{\infty} |x(t)| \, dt < \infty.$$

▲▲

Theorem 9.1.8 can be restated as follows. If $x(t)$ has a Laplace transform which has all of its poles in the open LHP, then $x(t)$ is absolutely integrable, and conversely.

9.2 PROPERTIES OF THE LAPLACE TRANSFORM

The Laplace transform is widely used as an operational transform. That is, it is used as a problem-solving tool. There are two reasons for its widespread use. First, a large number of common signals have simple Laplace transforms. We derived two in the last section, and in this section we will derive several more. A table of Laplace transform pairs is contained in the summary section of this chapter for reference. Second, several properties of the Laplace transform allow this transform to be applied to many problems of practical interest. In this section we present these properties of the Laplace transform. In Sections 9.3 and 9.4 we show how these properties are used.

The first property is the principle on which much of the analysis is based.

Property 9.2.1: (Linearity) Given the Laplace transform pairs $x_1(t) \leftrightarrow X_1(s)$ and $x_2(t) \leftrightarrow X_2(s)$ and any real numbers a_1 and a_2, then

$$L\{a_1 x_1(t) + a_2 x_2(t)\} = a_1 X_1(s) + a_2 X_2(s).$$

▲▲

Terminology: We say that the Laplace transform is linear.

This property of the Laplace transform embodies the idea of superposition. The Laplace transform of each of the signals can be calculated separately and the result summed, or, the signals can be summed and the Laplace transform calculated.

Example 9.2.2: The linearity of the Laplace transform can be used to generate new Laplace transforms from known transforms. To find the Laplace transform of the cosine function we first write it using Euler's formula

$$x(t) = [\cos(\omega t)]u_s(t) = \frac{1}{2}\left(e^{j\omega t} + e^{-j\omega t}\right)u_s(t). \tag{9.2.1}$$

---

[4] Left-half plane. See Definition 3.1.13.

(See Definition 3.1.4.) Define the function

$$x_1(t) = \left(e^{j\omega_0 t}\right)u_s(t). \tag{9.2.2}$$

Following Example 9.1.4, we find that

$$L\{x_1(t)\} = \frac{1}{s - j\omega_0}. \tag{9.2.3}$$

Similarly, we have

$$L\{x_2(t)\} = L\left\{e^{-j\omega_0 t}u_s(t)\right\} = \frac{1}{s + j\omega_0}. \tag{9.2.4}$$

Now take $a_1 = a_2 = 1/2$ in Property 9.2.1. Then we have

$$L\{x(t)\} = L\{[\cos(\omega t)]u_s(t)\} = L\{a_1 x_1(t) + a_2 x_2(t)\} = L\left\{\frac{1}{2}e^{j\omega t} + \frac{1}{2}e^{-j\omega t}\right\} \tag{9.2.5}$$

$$= \frac{1}{2}\left(\frac{1}{s + j\omega}\right) + \frac{1}{2}\left(\frac{1}{s - j\omega}\right) = \frac{s}{s^2 + \omega^2}.$$

▲▲

Property 9.2.3: (Convolution) Given the Laplace transform pairs $x(t) \leftrightarrow X(s)$ and $h(t) \leftrightarrow H(s)$, then

$$L\left\{\int_{0^-}^{\infty} h(t - \lambda)x(\lambda)\, d\lambda\right\} = H(s)X(s).$$

▲▲

The convolution property of the Laplace transform is the fundamental link between transfer functions and time-domain analysis of systems via the convolution integral.

Example 9.2.4: The transfer function of the RC network in Section 6.3 is

$$\frac{Y(s)}{X(s)} = H(s) = \frac{1}{s + \dfrac{1}{RC}}. \tag{9.2.6}$$

The impulse response function is

$$L^{-1}\{H(s)\} = h(t) = e^{-\frac{1}{RC}t}u_s(t). \tag{9.2.7}$$

So the convolution representation of this network is

$$y(t) = \int_0^\infty e^{-(t-\lambda)\frac{1}{RC}} u_s(t-\lambda)x(\lambda)\, d\lambda. \tag{9.2.8}$$

▲▲

The following two properties along with linearity allow the Laplace transform to be used for solving differential equations. These properties are also central to the development of block diagrams, a graphical analysis tool for networks and other systems described by differential equations.

Property 9.2.5: (Integration) Given the Laplace transform pair $x(t) \leftrightarrow X(s)$, we have

$$\mathcal{L}\left\{ \int_0^t x(\lambda)\, d\lambda \right\} = \frac{1}{s} X(s).$$

▲▲

Example 9.2.6: The unit ramp function is the integral of the unit step function. Property 9.2.5 tells us that

$$\mathcal{L}\{r(t)\} = \mathcal{L}\left\{ \int_0^t u_s(\lambda)\, d\lambda \right\} = \frac{1}{s}\mathcal{L}\{u_s(t)\} = \left(\frac{1}{s}\right)\frac{1}{s} = \frac{1}{s^2}. \tag{9.2.9}$$

▲▲

Property 9.2.6 says that to take an integral in the time domain we should multiply by $1/s$ in the Laplace domain. This trick is often useful.

Property 9.2.7: (Differentiation) Given the Laplace transform pair $x(t) \leftrightarrow X(s)$, assume that

$$\dot{x}(t) = \frac{dx}{dt}$$

satisfies (2) of Definition 9.1.1. Then

$$\mathcal{L}\{\dot{x}(t)\} = sX(s) - x(0^-).$$

▲▲

Property 9.2.7 says that to take a derivative in the time domain we should multiply by s in the Laplace domain when $x(0^-) = 0$. This trick is often useful.

Remark 9.2.8: Using Property 9.2.7 the Laplace transform of higher order derivatives can be derived. Consider the second-order derivative

$$\ddot{x}(t) = \frac{d}{dt}[\dot{x}(t)]. \tag{9.2.10}$$

Let

$$\dot{x}(t) = w(t). \tag{9.2.11}$$

Substituting (9.2.11) into (9.2.10) and using Property 9.2.7 we have

$$L\left\{\frac{d}{dt}[w(t)]\right\} = sW(s) - w(0^-). \tag{9.2.12}$$

Again using Property 9.2.7 on (9.2.11) we have

$$W(s) = L\{\dot{x}(t)\} = sX(s) - x(0^-). \tag{9.2.13}$$

Combining (9.2.13) with (9.2.12) results in

$$L\{\ddot{x}(t)\} = s\left[sX(s) - x(0^-)\right] - \dot{x}(0^-) = s^2 X(s) - sx(0^-) - \dot{x}(0^-). \tag{9.2.14}$$

In general the Laplace transform of the nth order derivative is

$$L\{x^{(n)}(t)\} = s^n X(s) - s^{(n-1)}x(0^-) - s^{(n-2)}x^{(1)}(0^-) - \cdots - x^{(n-1)}(0^-). \tag{9.2.15}$$

▲▲

The following property is useful in the analysis of discrete signals and systems.

Property 9.2.9: (Time Delay) Given the Laplace transform pair $x(t) \leftrightarrow X(s)$, for any positive number $T_s > 0$

$$L\{x(t - T_s)\} = e^{-T_s s} X(s).$$

▲▲

Example 9.2.10: Consider a pulse which begins at the origin

$$p(t) = u_s(t) - u_s(t - T_s), \quad T_s \geq 0. \tag{9.2.16}$$

Using the linearity of the Laplace transform and Property 9.2.9, we have

$$L\{u_s(t) - u_s(t - T_s)\} = \frac{1}{s} - \frac{e^{-sT_s}}{s} = \frac{1 - e^{-sT_s}}{s}. \tag{9.2.17}$$

▲▲

The following property is very useful for generating new Laplace transforms.

Property 9.2.11: (Time Scaling) Given the Laplace transform pair $x(t) \leftrightarrow X(s)$, then for any positive number a,

$$L\{x(at)\} = \frac{1}{a} X\left(\frac{s}{a}\right).$$

▲▲

The next two properties of the Laplace transform are useful for control system analysis. There are also useful for checking computational results (by hand and computer), particularly for graphs.

Property 9.2.12: (Initial Value Theorem)   Given the Laplace transform pair $x(t) \leftrightarrow X(s)$, suppose that

$$\lim_{t \downarrow 0} x(t) = x(0^+)$$

exists. Then we have

$$\lim_{s \to \infty} sX(s) = x(0^+).$$

▲▲

Example 9.2.13: This example shows when and how to use the initial value theorem. Consider the signal

$$x_1(t) = (\cos t)u_s(t). \tag{9.2.18}$$

The limit of the cosine as we approach 0 from the right exists. Using the formula in Property 9.2.12 we obtain

$$\lim_{s \to \infty} sX(s) = \lim_{s \to \infty} \frac{s^2}{s^2 + 1} = \lim_{s \to \infty} \frac{1}{1 + \frac{1}{s^2}} = 1 \tag{9.2.19}$$

as we expect.
    Consider the signal

$$x_2(t) = \delta(t) + u_s(t). \tag{9.2.20}$$

This signal contains an impulse function located at the origin. The limit as we approach 0 from the right doesn't exist, and so we can't apply the formula in Property 9.2.13.

▲▲

Property 9.2.14: (Final Value Theorem)   Given the Laplace transform pair $x(t) \leftrightarrow X(s)$, suppose that

$$\lim_{t \to \infty} x(t)$$

exists. Then we have that

$$\lim_{t \to \infty} x(t) = \lim_{s \to 0} sX(s).$$

▲▲

Example 9.2.15: The following two examples show how to use the final value theorem and when we can't apply this theorem.

If

$$x(t) = e^{-at} u_s(t), \quad a > 0 \tag{9.2.21}$$

then the limit exists as $t \to \infty$. The final value theorem yields

$$\lim_{s \to 0} sX(s) = \lim_{s \to 0} \frac{s}{s+a} = 0. \tag{9.2.22}$$

If

$$x(t) = (\sin t) u_s(t), \tag{9.2.23}$$

then the limit as $t \to \infty$ of (9.2.23) does not exist. The Laplace transform of (9.2.23) is

$$X(s) = \frac{1}{s^2 + 1}. \tag{9.2.24}$$

The limit

$$\lim_{s \to 0} sX(s) = \lim_{s \to 0} \frac{s}{s^2 + 1} = 0. \tag{9.2.25}$$

is well defined but does not correspond to the behavior of $x(t)$ as $t \to \infty$. ▲▲

The following result is useful for determining when the conditions of the final value theorem are satisfied.

Theorem 9.2.16: Suppose that $X(s)$ is the Laplace transform of $x(t)$ and that $X(s)$ is a proper rational function. If all of the poles of $sX(s)$ are in the open LHP, then $\lim_{t \to \infty} x(t)$ exists.

▲▲

Example 9.2.17: Consider again the signal in (9.2.23). The Laplace transform in (9.2.24) has poles on the imaginary axis. This signal does not satisfy Theorem 9.2.16 so the final value theorem can't be applied, as noted in Example 9.2.15.

▲▲

9.3 PARTIAL FRACTION EXPANSION

9.3.1 Definition

In this section we present the partial fraction expansion of a rational function. The partial fraction expansion is a practical way to invert the Laplace transform $X(s)$. Given a Laplace transform $X(s)$ we can compute the corresponding time function $x(t)$ by using only a simple computation and a table of Laplace transforms. The method itself, however, also lends insight into the structure of signals and transfer functions.

Computer Usage: Even when computers are widely available to invert Laplace transforms, an understanding of partial fraction expansions is useful conceptually.

Terminology: The terminology in this section is drawn from the discussion of rational functions in Section 3.2.

The following theorem is a formal statement of a partial fraction expansion.

Theorem 9.3.1: Let $X(s)$ be a strictly proper rational function which has n poles.
(a) Suppose that all of the poles are distinct. Then there exist complex numbers c_i such that $X(s)$ can be represented as

$$X(s) = \frac{b(s)}{(s-p_1)(s-p_2)\cdots(s-p_n)} = \frac{c_1}{(s-p_1)} + \frac{c_2}{(s-p_2)} + \cdots + \frac{c_n}{(s-p_n)}.$$

(b) Suppose the rational function has n distinct poles, but pole p_1 is repeated r times. Then there exist complex numbers c_i and d_i such that $X(s)$ can be represented as

$$X(s) = \frac{b(s)}{(s-p_1)^r(s-p_2)\cdots(s-p_n)} = \frac{d_1}{(s-p_1)} + \frac{d_2}{(s-p_1)^2} + \cdots + \frac{d_r}{(s-p_1)^r}$$

$$+ \frac{c_2}{(s-p_2)} + \cdots + \frac{c_n}{(s-p_n)}.$$

▲▲

Definition 9.3.2: The representation of $X(s)$ in Theorem 9.3.1 is called a partial fraction representation. The numbers c_i and d_k are called the residues of the function $X(s)$.

▲▲

Note: The extension of Theorem 9.3.1 to rational functions with multiple repeated poles is immediate.

Next we will prove part (a) of Theorem 9.3.1 as this proof lends insight into the computation of partial fraction expansions.

Proof of Theorem 9.3.1(a): Multiplying each side of the expansion by $(s - p_1)$ we obtain

$$(s - p_1)X(s) = (s - p_1)\frac{c_1}{(s - p_1)} + (s - p_1)\frac{c_2}{(s - p_2)} + \cdots + (s - p_1)\frac{c_n}{(s - p_n)} \tag{9.3.1}$$

$$= c_1 + (s - p_1)\frac{c_2}{(s - p_2)} + \cdots + (s - p_1)\frac{c_n}{(s - p_n)}.$$

There are no cancellations on the right-hand side of (9.3.1) except the first term because all of the poles are distinct. Now evaluate each side of (9.3.1) at $s = p_1$. On the right-hand side of (9.3.1) all of the terms except the first term are zero. We are left with

$$(s - p_1)X(s)\Big|_{s=p_1} = \frac{b(s)}{(s - p_2)\cdots(s - p_n)}\Big|_{s=p_1} = c_1. \tag{9.3.2}$$

Repeat this procedure to compute the rest of the numbers c_i.

▲▲

The following examples illustrates how to compute a partial fraction expansion.

Example 9.3.3: Consider the rational function

$$X(s) = \frac{s+2}{s(s+1)}. \tag{9.3.3}$$

Since the poles of this function are isolated, the partial fraction expansion of (9.3.3) is given by

$$X(s) = \frac{s+2}{s(s+1)} = \frac{c_1}{s} + \frac{c_2}{s+1}. \tag{9.3.4}$$

Using (9.3.2) we compute

$$sX(s)\Big|_{s=0} = (s)\left(\frac{s+2}{s(s+1)}\right) = \frac{s+2}{s+1}\Big|_{s=0} = 2 = c_1, \tag{9.3.5}$$

and

$$(s+1)X(s)\Big|_{s=-1} = (s+1)\left(\frac{s+2}{s(s+1)}\right)\Big|_{s=-1} = \frac{s+2}{s}\Big|_{s=-1} = -1 = c_2. \tag{9.3.6}$$

Now the partial fraction expansion of (9.3.3) is

$$X(s) = \frac{s+2}{s(s+1)} = \frac{2}{s} + \frac{-1}{s+1}. \qquad (9.3.7)$$

▲▲

The calculation of partial fractions holds when the coefficients are complex as well. The next example illustrates this type of calculation.

Example 9.3.4: Consider the rational function

$$X(s) = \frac{s+3}{(s+1)^2 + 4}. \qquad (9.3.8)$$

The partial fraction expansion is given by

$$X(s) = \frac{c}{s-p} + \frac{\bar{c}}{s-\bar{p}} = \frac{c}{s-(-1+j2)} + \frac{\bar{c}}{s-(-1-j2)}. \qquad (9.3.9)$$

The second residue is the conjugate of the first residue. The residue c is computed using (9.3.2). We have

$$c = \left(s-(-1+2j)\right)X(s)\Big|_{s=-1+2j} = \left(s-(-1+2j)\right)\frac{s+3}{(s+1)^2+4}\Big|_{s=-1+2j} \qquad (9.3.10)$$

$$= \frac{s+3}{s-(-1-2j)}\Big|_{s=-1+2j} = \frac{(-1+2j)+3}{(-1+2j)-(-1-2j)} = \frac{2+j2}{4j} = 0.5 - 0.5j.$$

Now the partial fraction expansion is

$$X(s) = \frac{0.5 - j0.5}{s-(-1+j2)} + \frac{0.5 + j0.5}{s-(-1-j2)}. \qquad (9.3.11)$$

▲▲

We will illustrate a method for determining a partial fraction expansion of a Laplace transform with repeated poles by way of example.

Example 9.3.5: Consider the rational function

$$X(s) = \frac{1}{s^2(s+a)}. \qquad (9.3.12)$$

The partial fraction expansion of (9.3.12) is

$$X(s) = \frac{1}{s^2(s+a)} = \frac{d_1}{s} + \frac{d_2}{s^2} + \frac{c_1}{s+a}. \qquad (9.3.13)$$

Since this rational function has a repeated root at $s = 0$, the procedure above can't be used to find the partial fraction expansion. The following procedure is one way of finding the partial fraction expansion of (9.3.12). Cross-multiplying in (9.3.13) we get

$$1 = d_1 s(s + a) + d_2(s + a) + c_1 s^2 = (c_1 + d_1)s^2 + (ad_1 + d_2)s + d_2 a. \tag{9.3.14}$$

Equation (9.3.14) is the equality of two polynomials. These polynomials are equal if their coefficients are equal. Equating coefficients in (9.3.14) we obtain

$$0 = c_1 + d_1, \quad 0 = ad_1 + d_2, \quad 1 = d_2 a. \tag{9.3.15}$$

Solving this system of equations leads to

$$d_2 = \frac{1}{a}, \quad d_1 = -\frac{1}{a}d_2 = -\frac{1}{a^2}, \quad c_1 = -d_1 = \frac{1}{a^2}. \tag{9.3.16}$$

Inserting the results in (9.3.16) into (9.3.13) gives

$$X(s) = \frac{1}{a^2}\left(\frac{a}{s^2} - \frac{1}{s} + \frac{1}{s+a}\right). \tag{9.3.17}$$

Note: The residues c_1 and d_2 could have been calculated using the method of Example 9.3.3.

▲▲

9.3.2 Partial Fraction Inversion

We are given the Laplace transform $X(s)$ which is a rational function. Suppose we want to find the inverse Laplace transform of $X(s)$, $x(t)$. Partial fraction expansions make it easy to accomplish this inversion. We suppose that the poles of $X(s)$ are real and distinct so that $X(s)$ has a partial fraction expansion as in Theorem 9.3.1 with real coefficients. Notice that each term of the partial fraction expansion can be inverted using the Laplace transform pair

$$c_i e^{p_i t} u_s(t) \leftrightarrow \frac{c_i}{s - p_i}. \tag{9.3.18}$$

Then using the *linearity* of the Laplace transform we see that $x(t)$ is given by

$$x(t) = c_1 e^{p_1 t} u_s(t) + c_2 e^{p_2 t} u_s(t) + \cdots + c_n e^{p_n t} u_s(t). \tag{9.3.19}$$

Example 9.3.6: The partial fraction expansion of (9.3.3) is

$$X(s) = \frac{2}{s} + \frac{-1}{s+1}. \tag{9.3.7}$$

The corresponding function of time is

$$x(t) = 2u_s(t) - e^{-t}u_s(t). \tag{9.3.20}$$

▲▲

Remark 9.3.7: We can always write a rational function as a partial fraction expansion of linear factors if the poles are distinct. If the rational function has complex poles, however, we are left with complex residues, which are often inconvenient. Usually, terms with complex conjugate poles are combined to produce a real signal with real coefficients as follows. Suppose that $X(s)$ is a rational function with real coefficients with exactly two complex poles. Then $X(s)$ can be written as

$$X(s) = \frac{c}{s - p} + \frac{\bar{c}}{s - \bar{p}}. \tag{9.3.21}$$

The residues, c and $\bar{c}$, of these poles in the partial fraction expansion will also appear in complex pairs. Let $p = \sigma + j\omega$. Then inverting (9.3.21) and using (9.3.18) we have

$$x(t) = \left(ce^{pt} + \bar{c}e^{\bar{p}t}\right)u_s(t) = \left(ce^{(\sigma+j\omega)t} + \bar{c}e^{(\sigma-j\omega)t}\right)u_s(t). \tag{9.3.22}$$

Note: In (9.3.22) the residue c is associated with the pole that has a positive imaginary part. The conjugate of c is associated with the pole with negative real part.

Write the complex constant c in its polar representation. Using Euler's identity we can rewrite the term in brackets in (9.3.22) as

$$ce^{(\sigma+j\omega)t} + \bar{c}e^{(\sigma-j\omega)t} = |c|e^{j\angle c}e^{(\sigma+j\omega)t} + |c|e^{-j\angle c}e^{(\sigma-j\omega)t}. \tag{9.3.23}$$

Next we rewrite (9.3.23) so we can apply Euler's identity. We obtain

$$|c|e^{j\angle c}e^{(\sigma+j\omega)t} + |c|e^{-j\angle c}e^{(\sigma-j\omega)t} = \left(2|c|e^{\sigma t}\right)\frac{1}{2}\left[e^{j(\omega t + \angle c)} + e^{-j(\omega t + \angle c)}\right]. \tag{9.3.24}$$

Recognizing the bracketed term in (9.3.24) as a cosine (using Euler's identity), $x(t)$ takes the form

$$x(t) = \left[2|c|e^{\sigma t}\cos(\omega t + \angle c)\right]u_s(t). \tag{9.3.25}$$

▲▲

Example 9.3.8: The partial fraction expansion of the rational function

$$X(s) = \frac{s + 3}{(s+1)^2 + 4} = \frac{0.5 - j0.5}{s - (-1 + j2)} + \frac{0.5 + j0.5}{s - (-1 - j2)} \tag{9.3.26}$$

was obtained in Example 9.3.4 above. The inverse Laplace transform from (9.3.25) is

$$x(t) = \left[2|0.5 - j0.5|e^{-1t} \cos(4t + \angle(0.5 - j0.5)) \right] u_s(t) \tag{9.3.27}$$

$$= \left[2 \left(\frac{\sqrt{2}}{2} \right) e^{-1t} \cos\left(4t - \frac{\pi}{4} \right) \right] u_s(t).$$

▲▲

Repeated poles of a partial fraction expansion are handled separately. A table of Laplace transforms is used to invert each term of the expansion. The following example illustrates the procedure.

Remark 9.3.9: Consider partial fraction expansion of the rational function

$$X(s) = \frac{1}{s^2(s+a)} = \frac{1}{a^2} \left(\frac{a}{s^2} - \frac{1}{s} + \frac{1}{s+a} \right). \tag{9.3.28}$$

The inverse Laplace transform of the repeated pole at the origin, derived in Example 9.2.6, is

$$\frac{1}{s^2} \leftrightarrow t u_s(t). \tag{9.3.29}$$

The inverse Laplace transform yields

$$x(t) = \frac{1}{a^2} \left(at u_s(t) - u_s(t) + e^{-at} u_s(t) \right). \tag{9.3.30}$$

In this example the squared factor of the Laplace transform in (9.3.29) leads to a multiplication by t of the signal in (9.3.29). This behavior is often observed of repeated poles.

▲▲

Theorem 9.3.1 required the rational function to be strictly proper. When $X(s)$ is not strictly proper, it is rewritten using long division as

$$X(s) = N(s) + X_{sp}(s) \tag{9.3.31}$$

where $X_{sp}(s)$ is strictly proper and $N(s)$ is a polynomial. Now $X_{sp}(s)$ can be inverted using the techniques above. Each term of $N(s)$ can be inverted using

$$\frac{d^n \delta(t)}{dt^n} \leftrightarrow s^n. \tag{9.3.32}$$

Example 9.3.10: Suppose $X(s)$ is given by

$$X(s) = \frac{s^2 + 6s + 6}{s^2 + 3s + 2}. \tag{9.3.33}$$

Long division in (9.3.33) leads to

$$X(s) = 1 + \frac{3s + 4}{s^2 + 3s + 2}. \tag{9.3.34}$$

Next expand the second term in (9.3.34) by partial fraction expansion. The result is

$$X(s) = 1 + \frac{1}{s+1} + \frac{2}{s+2}. \tag{9.3.35}$$

Now using the tables we find that

$$x(t) = \delta(t) + e^{-t}u_s(t) + 2e^{-2t}u_s(t). \tag{9.3.36}$$

▲▲

9.3.3 MATLAB Experiments

MATLAB can be used to carry out the partial fraction expansion using the **residue** command. The following M-file calculates the residues in Example 9.3.3.

Example 9.3.3: (*continued*)

```
clear
b = [1,2];              % numerator polynomial
a = [1,1,0];            % denominator polynomial
[r,p,k] = residue(b,a)  % calculation of residues
```

▲▲

Exploratory Exercise 9.3.11: Verify the residues in Examples 9.3.4 - 9.3.5.

▲▲

9.4 LAPLACE TRANSFORM SOLUTION TO DIFFERENTIAL EQUATIONS

9.4.1 Solving Differential Equations

In this section we will show how to find the solution of linear, constant coefficient differential equations using the Laplace transform. We consider differential equations of the form

$$y^{(n)}(t) + a_{n-1}y^{(n-1)}(t) + \cdots + a_1\dot{y}(t) + a_0 y(t) \tag{9.4.1}$$

$$= b_m x^{(m)}(t) + b_{m-1}x^{(m-1)}(t) + \cdots + b_0 x(t),$$

$$y^{(n-1)}(0^-) = y_{n-1},\, y^{(n-2)}(0^-) = y_{n-2}, \cdots, y(0^-) = y_0, \quad t \geq 0$$

where the coefficients a_i and b_i are real numbers and $x(t)$ is a given function which is Laplace transformable. A solution for this equation which satisfies the initial conditions can be found using Laplace transforms. We illustrate this procedure using the second-order equation

$$\ddot{y}(t) + a_1 \dot{y}(t) + a_0 y(t) = b_1 \dot{x}(t) + b_0 x(t), \tag{9.4.2}$$

$$\dot{y}(0^-) = y_1, \ y(0^-) = y_0, \quad x(0^-) = x_0 \quad t \geq 0.$$

Taking the Laplace transform of both sides of (9.4.2) we get

$$L\{\ddot{y}(t)\} + a_1 L\{\dot{y}(t)\} + a_0 L\{y(t)\} = b_1 L\{\dot{x}(t)\} + b_0 L\{x(t)\}. \tag{9.4.3}$$

In (9.4.3) we have used the fact that the Laplace transform is linear, Property 9.2.1. Next, applying Property 9.2.7 for differentiation, (9.4.3) becomes

$$\left\{ s^2 Y(s) - s y_0 - y_1 \right\} + a_1 \left\{ s Y(s) - y_0 \right\} + a_0 \left\{ Y(s) \right\} \tag{9.4.4}$$

$$= b_1 \left\{ s X(s) - x_0 \right\} + b_0 \left\{ X(s) \right\}.$$

Solving (9.4.4) for $Y(s)$ we get

$$Y(s) = \frac{(y_0 s + y_1 + a_1 y_0 - b_1 x_0) + (b_1 s + b_0) X(s)}{s^2 + a_1 s + a_0}. \tag{9.4.5}$$

Now (9.4.5) can be inverted to find $y(t)$.

In (9.4.2) a derivative of the forcing function $x(t)$ appears on the right-hand side of the equation. In this situation, first take the Laplace transform of $x(t)$. Then insert this Laplace transform into (9.4.5), effectively taking the derivative in the Laplace transform domain.

Example 9.4.1: Consider the differential equation

$$\ddot{y}(t) + 6\dot{y}(t) + 8y(t) = \dot{x}(t) + x(t), \tag{9.4.6}$$

$$\dot{y}(0) = 3, \quad y(0) = 1, \quad x(t) = u_s(t).$$

We want to find the output signal $y(t)$. First we take the Laplace transform of both sides of (9.4.6). We obtain

$$\left(s^2 Y(s) - s y(0) - \dot{y}(0)\right) + 6\left(s Y(s) - y(0)\right) + 8 Y(s) = \left(s X(s) - x(0^-)\right) + X(s). \tag{9.4.7}$$

Substituting in the initial conditions from (9.4.6) into (9.4.7) and using the Laplace transform of the forcing function we have

$$\left(s^2 Y(s) - 1s - 3\right) + 6\left(sY(s) - 1\right) + 8Y(s) = \left(s\frac{1}{s} - 0\right) + \frac{1}{s}. \tag{9.4.8}$$

The Laplace transform of the differential equation (9.4.7) requires an initial condition for the derivative of $x(t)$ at the origin. Since the forcing function $x(t)$ is a unit step function, the derivative is an impulse function $\delta(t)$. The impulse function is not defined at the origin. So we take the initial condition at $t = 0^-$. Solving (9.4.8) for $Y(s)$ we get

$$Y(s) = \frac{s^2 + 10s + 1}{s(s+4)(s+2)} = \left(\frac{1}{8}\right)\frac{1}{s} + \left(\frac{15}{4}\right)\frac{1}{s+2} + \left(-\frac{23}{8}\right)\frac{1}{s+4}. \tag{9.4.9}$$

We have also calculated the partial fraction expansion of $Y(s)$ in (9.4.9). Now taking the inverse Laplace transform we obtain

$$y(t) = \frac{1}{8}u_s(t) + \frac{15}{4}e^{-2t}u_s(t) - \frac{23}{8}e^{-4t}u_s(t). \tag{9.4.10}$$

▲▲

9.4.2 Implications of the Pole Locations

The use of the partial fraction expansion for the inversion of a Laplace transform lends insight into the relationship between the Laplace transform and the corresponding function of time. Reflection on this process shows that the poles of the Laplace transform determine exponentials that appear in the function of time. Hence, the locations of the poles of the Laplace transform can lend insight into the qualitative behavior of the function of time without explicitly constructing this function. This observation will be exploited repeatedly in this text. The following example amplifies this observation.

Example 9.4.2: In Example 9.4.1 the poles of the solution of the differential equation $Y(s)$ in (9.4.9) are 0, −2, and −4. Each of these poles is isolated. Hence, each corresponds to one term of the partial fraction expansion in (9.4.9). Each term in the partial fraction expansion, in turn, is associated with an exponential in the inverse Laplace transform $y(t)$ in (9.4.10). Furthermore, the exponent of each exponential is determined by the corresponding pole. By simply observing the poles of the Laplace transform, we know the structure of $y(t)$.

We can also deduce something of the behavior of $y(t)$ as time goes to infinity. The two poles in the open LHP[5] correspond to decaying exponentials. These two signals go to zero at infinity. The other pole corresponds to a constant step function which approaches a constant value. Hence, the solution of the differential equation $y(t)$ will go to a constant as time goes to infinity. This observation doesn't require the actual solution to be constructed since we only need the poles of the Laplace transform of the solution.

Furthermore, we notice that the solution of the differential equation doesn't go to infinity as the time goes to infinity but remains finite. Again this observation follows

---

[5] Left-half plane. See Definition 3.1.13.

directly from poles of the Laplace transform. This discussion leads to the important concept of stability in Chapter 13.

We might also notice that the two poles, −2 and −4, are associated with the differential equation (system) while the other pole at 0 is associated with the forcing function $x(t)$ (input signal). Hence, the output signal of this system $y(t)$ is composed of two parts: one part due to the system and one part due to the input signal. This observation is traced directly to partial fraction expansion (which is based on the linearity of the Laplace transform) of the Laplace transform. We will explore this relationship in Chapter 10.

The point of this discussion is that certain properties of the solution of the differential equation can be deduced directly from the Laplace transform of the differential equation with a very few computations. This approach is representative of the method of system analysis which is developed in this text.

▲▲

We can expand on the discussion of pole locations in the last example. This discussion is facilitated by frequent references to the Table of Laplace Transform Pairs, Table 9.6.2. Suppose that $x(t)$ has a Laplace transform $X(s)$ that is a rational function. Each pole corresponds to one term in the partial fraction expansion of $X(s)$. First, suppose that each pole is isolated. If the pole is in the RHP, it corresponds to a growing exponential. If the pole is in the LHP, it corresponds to a decaying exponential.

A similar relationship exists between a Laplace transform with a pair of complex poles and the corresponding function of time. If the real part of the poles is in the LHP, then the function of time is a sinusoid with a decaying amplitude. If the real part of the poles is in the RHP, then the signal is a sinusoid with a growing amplitude. If the real part of the poles is on the imaginary axis, then $x(t)$ is a sinusoid. This discussion is summarized in Figure 9.4.1.

We have similar observations for Laplace transforms with repeated poles. Recall that the Laplace transform of the unit step function is

$$u_s(t) \leftrightarrow \frac{1}{s}. \tag{9.4.11}$$

If we multiply the unit step by $(1/s)$ then $x(t)$ is

$$tu_s(t) \leftrightarrow \frac{1}{s^2}. \tag{9.4.12}$$

If the Laplace transform contains a double pole at the origin, then $x(t)$ has a factor t in it. This trend is followed in higher-order repeated poles. It can be shown that

$$\frac{t^{r-1}e^{-at}}{(r-1)!} \leftrightarrow \frac{1}{(s+a)^r}. \tag{9.4.13}$$

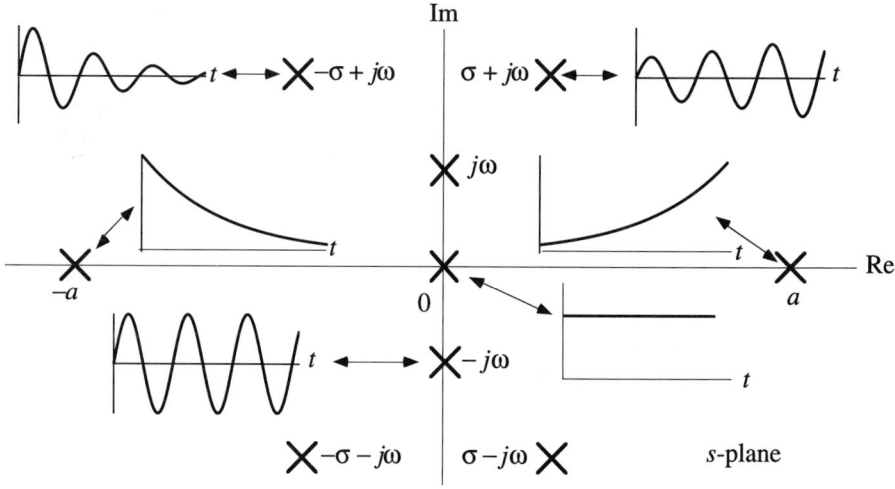

Figure 9.4.1 Relationship Between Pole Locations of the Laplace Transform and the Corresponding Inverse Laplace Transform for Transforms with No Repeated Poles

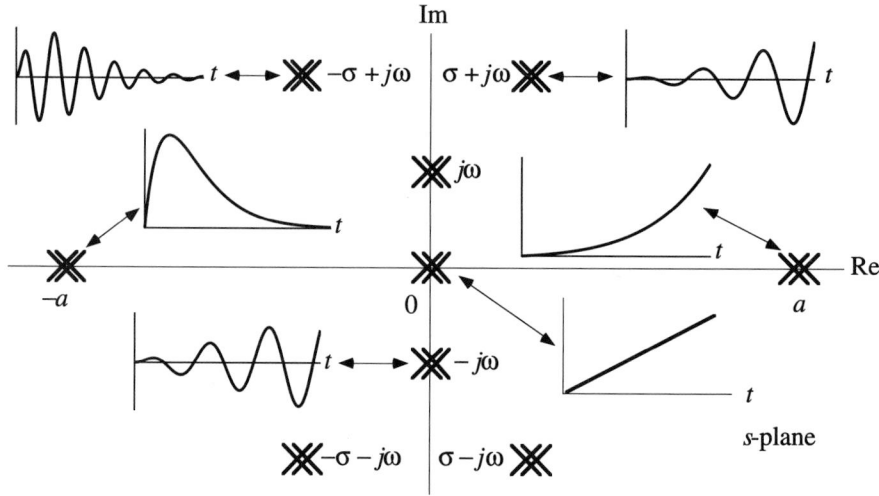

Figure 9.4.2 Relationship Between Pole Locations of the Laplace Transform and the Corresponding Inverse Laplace Transform for Transforms with Repeated Poles

Complex poles follow a similar trend. We have that

$$x(t) = t\cos\omega t \leftrightarrow \frac{s^2 - \omega^2}{(s^2 + \omega^2)^2}.$$

(9.4.14)

The graph of a function with a Laplace transform with a repeated pole is slightly different from the graph of a function with a Laplace transform having a single pole. Graphs of several of the functions are shown in Figure 9.4.2.

9.5 RELATIONSHIP TO FOURIER TRANSFORMS

In Chapter 7 we introduced the Fourier transform. There is a close relationship between Laplace and Fourier transforms which we discuss next. The definitions of these two transforms form the basis of the comparison. For convenience, we repeat these definitions.

Definition 9.1.1: Let $x(t)$, $-\infty < t < \infty$, be a signal that satisfies

(a) $x(t) = 0$ for $t < 0$,

(b) $\int_0^\infty |x(t)| e^{-\sigma t} \, dt < \infty$ for $0 \le \sigma_0 < \sigma < \infty$

Then the <u>Laplace transform</u> of $x(t)$ is defined as

$$L\{x(t)\} = \int_0^\infty x(t)e^{-st} \, dt = X(s).$$

▲▲

Definition 7.4.1: Let $x(t)$ be a signal that satisfies:

(a) $x(t)$, $-\infty < t < \infty$, and

(b) $\int_{-\infty}^\infty |x(t)| \, dt \le M < \infty$,   for $0 < M < \infty$

Then the <u>Fourier transform</u> of $x(t)$ is defined as

$$X(\omega) = \int_{-\infty}^\infty x(t)e^{-j\omega t} \, dt = \mathcal{F}\{x(t)\}.$$

▲▲

In comparing the defining integrals of the Laplace and Fourier transforms, it would appear that

$$L\{x(t)\}\big|_{s=j\omega} = \left[\int_0^\infty x(t)e^{-st} \, dt = \right]_{s=j\omega} = \int_{-\infty}^\infty x(t)e^{-j\omega t} \, dt = \mathcal{F}\{x(t)\}. \qquad (9.5.1)$$

That is, the Fourier transform is computed from the Laplace transform by substituting $s = j\omega$. It is the purpose of this section to identify those signals for which (9.5.1) is true.

The Laplace and Fourier transforms of a signal $x(t)$ are related as in (9.5.1) when this signal satisfies both definitions of the Laplace and the Fourier transform. Define the set

$$L\mathcal{F} = \{x(t) \mid x(t) \text{ is Laplace and Fourier transformable}\}. \tag{9.5.2}$$

First, if $x(t) \in L\mathcal{F}$, then according to the definition of the Laplace transform, condition (a), $x(t) = 0$ for $t < 0$. Condition (a) for the Fourier transforms does not impose this restriction so here the Laplace transform restricts the members of the set $L\mathcal{F}$. Second, for $x(t) \in L\mathcal{F}$, condition (b) of the Fourier transform is

$$\int_{-\infty}^{\infty} |x(t)| \, dt = \int_{0^-}^{\infty} |x(t)| \, dt \leq M < \infty \tag{9.5.3}$$

because the Laplace transform requires that $x(t) = 0$ for $t < 0$. Comparing (9.5.3) to condition (b) for the Laplace transform, we see that

$$\int_{0^-}^{\infty} |x(t)| \, dt = \left[\int_{0^-}^{\infty} |x(t)| e^{-\sigma t} \, dt \right]_{\sigma=0}. \tag{9.5.4}$$

It follows that if $x(t) \in L\mathcal{F}$, then $x(t)$ must satisfy condition (b) of the Fourier transform. This restriction excludes some signals that are Laplace transformable. These results are summarized by the following theorem.

Theorem 9.5.1: Let $x(t)$ be a signal which satisfies

(a) $x(t) = 0$ for $t < 0$,

(b) $\displaystyle\int_{0}^{\infty} |x(t)| \, dt \leq M < \infty$ for $0 < M < \infty$

Then the Laplace transform of $x(t)$, $X(s)$ and the Fourier transform of $x(t)$, $X(\omega)$ are related by

$$\mathcal{F}\{x(t)\} = X(\omega) = X(s)\big|_{s=j\omega} = L\{x(t)\}\big|_{s=j\omega}.$$

▲▲

Remark 9.5.2: Theorem 9.5.1 requires that the Fourier transform for $x(t)$ exist. This theorem explicitly excludes signals that have a generalized Fourier transform.

▲▲

The following theorem gives a simple test for condition (b) of Theorem 9.5.1.

Theorem 9.5.3: Let $x(t)$ be a signal such that

$$\mathcal{L}\{x(t)\} = X(s)$$

where $X(s)$ is a rational, strictly proper transfer function.[6] Then Theorem 9.5.1 holds if all the poles of $X(s)$ are in the open LHP.

▲▲

Example 9.5.4: Consider the signal

$$x(t) = e^{-at}u_s(t), \quad a > 0. \tag{9.5.5}$$

This signal is zero for negative time, so it satisfies condition (a) from both transforms. Substituting (9.5.5) into condition (b) in the definition of the Laplace transform, Definition 9.1.1, we obtain

$$\int_0^\infty \left| e^{-at}u_s(t) \right| e^{-\sigma t} dt < \infty \quad \text{for} \quad -a < \sigma. \tag{9.5.6}$$

as shown in Example 9.1.4. For $a > 0$, we can chose $\sigma = 0$ in (9.5.4). So this signal does belong to the set $\mathcal{LF}$. The Laplace and Fourier transforms of this signal are, respectively,

$$\mathcal{L}\{e^{-at}u_s(t)\} = \frac{1}{s+a} \quad \text{and} \quad \mathcal{F}\{e^{-at}u_s(t)\} = \frac{1}{j\omega + a}. \tag{9.5.7}$$

Then by Theorem 9.5.1

$$X(s)\big|_{s=j\omega} = \frac{1}{s+a}\bigg|_{s=j\omega} = \frac{1}{j\omega + a} = X(\omega). \tag{9.5.8}$$

Suppose we set $a = 0$ in (9.5.5) so that $x(t)$ becomes a step function. Then (9.5.6) is changed to

$$\int_0^\infty |u_s(t)| e^{-\sigma t} \, dt < \infty \quad \text{for} \quad 0 < \sigma. \tag{9.5.9}$$

Now we can't set $\sigma = 0$ in (9.5.9) because the integral won't exist. Hence, we can't compute the Fourier transform of the unit step function from the Laplace transform of that function.

The step function does have a generalized Fourier transform

$$\mathcal{F}\{u_s(t)\} = \pi\delta(\omega) + \frac{1}{j\omega}. \tag{9.5.10}$$

---

[6] See Definitions 3.2.6 and Theorem 3.2.7.

It's just that this generalized Fourier transform can't be computed from the Laplace transform by substitution of $j\omega$. That is,

$$X(s)\big|_{s=j\omega} = \frac{1}{s}\bigg|_{s=j\omega} = \frac{1}{j\omega} \neq X(\omega) = \pi\delta(\omega) + \frac{1}{j\omega}.$$

(9.5.11)

▲▲

Example 9.5.5: Consider the signal

$$x(t) = e^{-a|t|}, \quad a > 0, \quad -\infty < t < \infty.$$

(9.5.12)

This signal is Fourier transformable, but it is not Laplace transformable because it is not zero for negative time.

▲▲

Example 9.5.6: Consider the signal

$$x(t) = [\sin(\omega t)]u_s(t).$$

(9.5.13)

This signal does meet condition (a) for both transforms. This signal is Laplace transformable, but it is not Fourier transformable as we have defined it in Definition 7.4.1 because it is not absolutely integrable. That is, this signal does not satisfy condition (b) of Definition 7.4.1. The generalized Fourier transform of (9.5.13) can't be calculated from the Laplace transform.

▲▲

9.6 CHAPTER SUMMARY

In this chapter we introduced the Laplace transform.

Definition 9.1.1: Let $x(t)$, $-\infty < t < \infty$, be a signal that satisfies

    (a) $x(t) = 0$ for $t < 0$,

    (b) $\displaystyle\int_0^\infty |x(t)|e^{-\sigma t}\,dt < \infty$ for $0 \le \sigma_0 < \sigma < \infty$

Then the <u>Laplace transform</u> of $x(t)$ is defined as

$$\mathcal{L}\{x(t)\} = \int_0^\infty x(t)e^{-st}\,dt = X(s).$$

The <u>inverse</u> Laplace transform of $X(s)$ is given by

$$x(t) = \frac{1}{2\pi j}\int_{\sigma-j\omega}^{\sigma+j\omega} X(s)e^{st}\,ds = \mathcal{L}^{-1}\{X(s)\}.$$

▲▲

Table 9.6.1 Laplace Transform Properties

Property	Description	Mathematical Description
9.2.1	Linearity	$\mathcal{L}\{a_1x_1(t)+a_2x_2(t)\}=a_1X_1(s)+a_2X_2(s)$
9.2.3	Convolution	$\mathcal{L}\left\{\int_{0^-}^{\infty}h(t-\lambda)x(\lambda)\,d\lambda\right\}=H(s)X(s)$
9.2.5	Integration	$\mathcal{L}\left\{\int_0^t x(\lambda)\,d\lambda\right\}=\dfrac{1}{s}X(s)$
9.2.7	Differentiation	$\mathcal{L}\{\dot{x}(t)\}=sX(s)-x(0^-)$
9.2.9	Time Delay	$\mathcal{L}\{x(t-T_s)\}=e^{-sT_s}X(s).$
9.2.11	Time Scaling	$\mathcal{L}\{x(at)\}=\dfrac{1}{a}X\left(\dfrac{s}{a}\right),\quad a>0$
9.2.12	Initial Value Theorem	$\lim_{s\to\infty}sX(s)=x(0^+)$
9.2.14	Final Value Theorem	$\lim_{t\to\infty}x(t)=\lim_{s\to0}sX(s)$

The Laplace transform associates each signal as a function of time with a function of the complex variable s, $x(t)\leftrightarrow X(s)$.

We introduced several properties of the Laplace transform. These properties are summarized in Table 9.6.1.

Laplace transforms are inverted by using a partial fraction expansion which is given in the following theorem.

Theorem 9.3.1: Let $X(s)$ be a strictly proper rational function which has n poles.
(a) Suppose that all the poles are distinct. Then there exist complex numbers c_i such that $X(s)$ can be represented as

$$X(s)=\frac{c_1}{(s-p_1)}+\frac{c_2}{(s-p_2)}+\cdots+\frac{c_n}{(s-p_n)}.$$

(b) Suppose the rational function has n distinct poles, but pole p_1 is repeated r times. Then there exist complex numbers c_i and d_i such that $X(s)$ can be represented as

$$X(s)=\frac{d_1}{(s-p_1)}+\frac{d_2}{(s-p_1)^2}+\cdots+\frac{d_r}{(s-p_1)^r}+\frac{c_2}{(s-p_2)}+\cdots+\frac{c_n}{(s-p_n)}.$$

▲▲

Definition 9.3.2: The representation of $X(s)$ in Theorem 9.3.1 is called a partial fraction representation. The numbers c_i and d_i are called the residues of the function $X(s)$.

▲▲

The inverse Laplace transform is constructed by inverting each term of the partial fraction expansion. This inversion is accomplished with the use of a table of Laplace transform pairs, Table 9.6.2.

There is a close relationship between the Laplace and Fourier transform which is given in the following theorem.

Theorem 9.5.1: Let $x(t)$ be a signal which satisfies

(a) $x(t) = 0$ for $t < 0$,

(b) $\int_0^\infty |x(t)|\, dt \le M < \infty$ for $0 < M < \infty$

Then the Laplace transform of $x(t)$, $X(s)$ and the Fourier transform of $x(t)$, $X(\omega)$ are related by

$$\mathcal{F}\{x(t)\} = X(\omega) = X(s)\big|_{s=j\omega} = \mathcal{L}\{x(t)\}\big|_{s=j\omega}.$$

▲▲

If the Laplace transform is given by a rational function, then the assumptions of Theorem 9.5.1 are satisfied when all of the poles of the Laplace transform are in the open LHP.

Table 9.6.2 Table of Laplace Transform Pairs

$X(s)$	$x(t),\ t \ge 0$
1	$\delta(t)$
$\dfrac{1}{s}$	$u_s(t)$
$\dfrac{1}{s^2}$	$t u_s(t)$
$\dfrac{1}{s^n}$	$\left[\dfrac{1}{(n-1)!}t^{n-1}\right]u_s(t),\quad n$ positive integer
s^n	$\dfrac{d^n \delta(t)}{dt^n}$
$\dfrac{1}{s+a}$	$e^{-at}u_s(t)$
$\dfrac{1}{(s+a)^n}$	$\left[\dfrac{1}{(n-1)!}t^{n-1}e^{-at}\right]u_s(t)$

Table 9.6.2 (*continued*) Table of Laplace Transform Pairs

$X(s)$	$x(t), \ t \geq 0$
$\dfrac{1}{(s+a)(s+b)}$	$\left[\dfrac{1}{b-a}\left(e^{-at} - e^{-bt}\right)\right]u_s(t)$
$\dfrac{1}{s(s+a)(s+b)}$	$\left[\dfrac{1}{ab}\left(1 - \dfrac{b}{b-a}e^{-at} + \dfrac{a}{b-a}e^{-bt}\right)\right]u_s(t)$
$\dfrac{(s+c)}{s(s+a)(s+b)}$	$\left[\dfrac{1}{ab}\left(c - \dfrac{b(c-a)}{b-a}e^{-at} - \dfrac{a(b-c)}{b-a}e^{-bt}\right)\right]u_s(t)$
$\dfrac{\omega}{s^2 + \omega^2}$	$[\sin(\omega t)]u_s(t)$
$\dfrac{s}{s^2 + \omega^2}$	$[\cos(\omega t)]u_s(t)$
$\dfrac{s+c}{s^2 + \omega^2}$	$\left[\left(\dfrac{\sqrt{c^2 + \omega^2}}{\omega}\right)\sin(\omega t + \phi)\right]u_s(t), \quad \phi = \tan^{-1}\left(\dfrac{\omega}{c}\right)$
$\dfrac{1}{s(s^2 + \omega^2)}$	$\left[\dfrac{1}{\omega^2}(1 - \cos(\omega t))\right]u_s(t)$
$\dfrac{(s+\alpha)}{(s+\alpha)^2 + \omega_c^2}$	$\left[e^{-\alpha t}\cos(\omega_c t)\right]u_s(t)$
$\dfrac{1}{(s+\alpha)^2 + \omega_c^2}$	$\left[\dfrac{1}{\omega_c}e^{-\alpha t}\sin(\omega_c t)\right]u_s(t)$
$\dfrac{\omega_n^2}{s^2 + 2\zeta\omega_n s + \omega_n^2}$	$\left[\dfrac{\omega_n}{\sqrt{1-\zeta^2}}e^{-\zeta\omega_n t}\sin(\omega_c t)\right]u_s(t), \qquad \omega_c = \omega_n\sqrt{1-\zeta^2}$
$\dfrac{s+a}{(s+\alpha)^2 + \omega_c^2}$	$\left[\dfrac{\sqrt{(a-\alpha)^2 + \omega_c^2}}{\omega_c}e^{-\alpha t}\sin(\omega_c t + \phi)\right]u_s(t), \quad \phi = \tan^{-1}\left(\dfrac{\omega_c}{a-\alpha}\right)$
$\dfrac{\omega_n^2}{s\left(s^2 + 2\zeta\omega_n s + \omega_n^2\right)}$	$\left[1 - \dfrac{1}{\sqrt{1-\zeta^2}}e^{-\zeta\omega_n t}\sin(\omega_c t + \phi)\right]u_s(t)$ $\phi = \cos^{-1}\zeta, \quad \omega_c = \omega_n\sqrt{1-\zeta^2}$
$\dfrac{s^2 - \omega^2}{(s^2 + \omega^2)^2}$	$[t\cos(\omega t)]u_s(t)$

9.7 HOMEWORK FOR CHAPTER 9

Homework Problems for Section 9.1

9.1.1   Find an example of a signal that satisfies condition (a) of the Definition 9.1.1, but doesn't satisfy condition (b).

9.2.1   For what values of $\lambda > 0$ is the following function Laplace transformable? (Do not find the Laplace transform.)

$$x(t) = 5\Pi\left(\frac{t-10}{\lambda}\right).$$

9.1.3   Which of the following signals can be Laplace transformed? Why?

(i) $x(t) = \displaystyle\sum_{m=-\infty}^{\infty} \frac{3}{\pi m^2} e^{-j\frac{m\pi}{3}}$

(ii) $x(t) = (\sin(t))\,\text{sgn}(t)$

(iii) $x(t) = \delta(t+1)$

(iv) $x(t) = \delta(t-1)$

(v) $x(t) = \delta^{(n)}(t)$

(vi) $x(t) = \log(t)$

9.1.4   Find the *ROC* of each of the following Laplace transforms.

(i) $X(s) = \dfrac{s+1}{(s+2)(s+4)}$

(ii) $X(s) = \dfrac{4}{(s+1)(s-3)}$

(iii) $X(s) = \dfrac{4}{s^2+s+4}$

(iv) $X(s) = \dfrac{s+3}{(s-2)(s-8)}$

Homework Problems for Section 9.2

9.2.1   Suppose that the Laplace transform of the signal $x(t)$ is $X(s) = \dfrac{1}{s+2}$. Find the Laplace transform of each of the following signals.

(i) $\displaystyle\int_0^{\infty} x(\lambda)x(t-\lambda)\,d\lambda$

(ii) $\dfrac{dx(t)}{dt}$

(iii) $x(t-3)u_s(t-3)$

(iv) $x(5t)$

9.2.2   Consider the signal $x(t) = e^{-t}u_s(t)$ with the Laplace transform $X(s)$. Determine the inverse Laplace transform of the following signals.

(i) $e^{-3s}X(s)$

(ii) $\dfrac{X(s)}{s}$

(iii) $sX(s)$

(iv) $X(2s)$

9.2.3   Find the Laplace transform of the following signal when it exists.

$$x(t) = 5\Pi\left(\frac{t-10}{\lambda}\right).$$

9.2.4   Consider the Laplace transform

$$X(s) = \frac{b_m s^m + \cdots + b_0}{s^n + a_{n-1}s^{n-1} + \cdots + a_0}\left(\frac{1}{s}\right)$$

Using the final value theorem show that

$$\lim_{t\to\infty} x(t) = \frac{b_0}{a_0}.$$

9.2.5   The Laplace transforms of several signals are shown below
   (a) Find $\lim_{t\to\infty} x(t)$ using the final value theorem, if the limit exists. If the
       limit doesn't exist, state why not.
   (b) Find $\lim_{t\to 0} x(t)$ using the initial value theorem, if the limit exists. If the
       limit doesn't exist, state why not.

   (i) $\dfrac{1}{s^3 + s^2 + 16s + 16}$ (iv) $\dfrac{3}{s^2 - 1}$

   (ii) $\dfrac{s+3}{s^2 + 2s + 1}$ (v) $\dfrac{s^2 + s}{s+2}$

   (iii) $\dfrac{s}{s+2}$ (vi) $\dfrac{1}{s(s+1)}$

Homework Problems for Section 9.3.

9.3.1   Consider the partial fraction expansion of a second-order rational function
with two complex poles. Show that the residues are complex conjugates of
each other.

9.3.2   Find the Laplace transform of the following functions.
   (i) $x(t) = e^{2t}u_s(t) - \left(e^{-3t}\sin(30t)\right)u_s(t)$ (iii) $x(t) = e^{at}\left[u_s(t) - u_s(t-T)\right]$
   (ii) $x(t) = \left(e^{4t}\cos(2t)\right)u_s(t)$ (iv) $x(t) = \left(e^{3t} - e^{-4t}\right)u_s(t)$

9.3.3   Find the inverse Laplace transform of the following functions.

(i) $X(s) = \dfrac{(s+1)}{(s+3)(s+12)}$

(iii) $X(s) = \dfrac{s+1.2}{(s+1)(s+4)}$

(ii) $X(s) = \dfrac{s+3}{(s+1)(s+10)}$

(iv) $X(s) = \dfrac{s-2}{s^2 + s + 1}$

9.3.4   Suppose that the signal $x(t)$ has a Laplace transform $X(s)$. The poles of $X(s)$ are shown in Figure P9.3.4.
(a) Find ζ and ω_n for these poles.
(b) Find and sketch the signal.

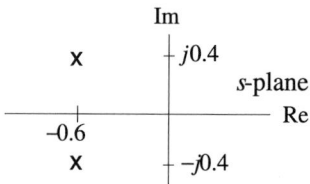

Figure P9.3.4

Homework Problems for Section 9.4

9.4.1   Solve the following differential equations using the Laplace transform.

(i) $\ddot{y}(t) - \dot{y}(t) - 6y(t) = e^{-t}u_s(t)$

$\dot{y}(0) = 5, \quad y(0) = 0$

(ii) $\ddot{y}(t) + 7\dot{y}(t) + 10y(t) = x(t)$

$y(0) = 3, \quad \dot{y}(0) = 2, \quad x(t) = (1 - e^{-3t})u_s(t)$

(iii) $\dot{y}(t) + 6y(t) = x(t)$

$y(0) = 2, \quad x(t) = e^{-3t}u_s(t)$

(iv) $\ddot{y} + 11\dot{y} + 10y = e^{-2t}u_s(t),$

$y(0) = 2, \quad \dot{y}(0) = -3$

(v) $\ddot{y}(t) + 0.4\dot{y}(t) + 0.03y(t) = e^{-0.2t}u_s(t)$

$y(0) = -2, \quad \dot{y}(0) = -1$

(vi) $\ddot{y}(t) + 4\dot{y}(t) + 3y(t) = 2\ddot{x}(t) - 4\dot{x}(t) - x(t)$

$y(0) = -2, \quad \dot{y}(0) = 1, \quad x(t) = u_s(t)$

(vii) $\ddot{y} + 8\dot{y} + 12y = 2\dot{x} + x,$

$\dot{y}(0) = 1, \quad y(0) = 4, \quad x(t) = u_s(t)$

9.4.2   Consider the network in Figure P9.4.2.
   (a) Find the differential equation that corresponds to this network from the transfer function.
   (b) Solve this differential equation for the initial current and voltage given when the input voltage is zero $v_s(t) = 0$ (short circuit).

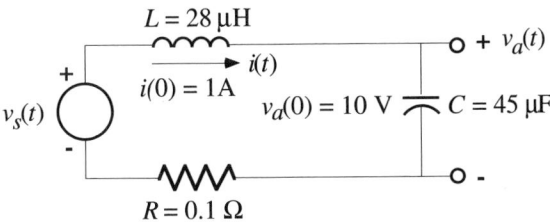

Figure P9.4.2

9.4.3   Consider the systems that are given below.
   (a) Find the step response of each of the following systems.
   (b) Plot all of these step responses on one graph in MATLAB.
   (c) Determine the initial and final value of each step response.

   (i) $\dfrac{Y(s)}{X(s)} = \dfrac{1}{s+1}$

   (ii) $\dfrac{Y(s)}{X(s)} = \dfrac{10}{(s+1)(s+10)}$

   (iii) $\dfrac{Y(s)}{X(s)} = \dfrac{9-s}{9(s+1)}$

   (iv) $\dfrac{Y(s)}{X(s)} = \dfrac{s+3.05}{(s+1)(s+3)}$

9.4.4   Prove the following statement: Suppose that $X(s)$ is the Laplace transform of $x(t)$ and that $X(s)$ is a proper rational function. If all of the poles of $sX(s)$ are in the open LHP, then $\lim\limits_{t \to \infty} x(t)$ exists.

9.4.5   Prove the following statement: Suppose the signal $x(t)$ has a Laplace transform $X(s)$ which is strictly proper. Further suppose that the poles of $sX(s)$ are in the open LHP. Then this signal is bounded.

Homework Problems for Section 9.5

9.5.1   Each signal below, find the Fourier transform from the Laplace transform, if possible. If it is not possible to find the Fourier transfom from the Laplace transform, give the reason.

   (i) $X(s) = \dfrac{s^2 + 2}{(s-1)(s^2 + s + 2)}$

   (ii) $X(s) = \dfrac{s+2}{(s+1)(s+5)}$

   (iii) $x(t) = te^{-3t}u_s(t)$

   (iv) $x(t) = e^{3t}u_s(t)$

   (v) $x(t) = (\sin(t))\,\mathrm{sgn}(t)$

   (vi) $x(t) = \displaystyle\sum_{k=-\infty}^{\infty} \dfrac{3}{\pi k^2} e^{j2\pi kt}$

   (vii) $x(t) = r_p(t)$

   (viii) $x(t) = e^{-6t}\sin(6\pi t)u_s(t)$

9.5.2   Consider the signal whose Laplace transform is given by

$$X(s) = \frac{\omega_n^2}{s^2 + 2\zeta\omega_n s + \omega_n^2}$$

(a) For what values of ζ, $-1 \leq \zeta \leq 1$, can the Fourier transform be computed from the Laplace transform?

(b) For these values of ζ, what is the Fourier transform?

9.5.3   Explain all of the differences between the Laplace and Fourier transform versions of the convolution properties.

Chapter 10

Transfer Functions and State Space Representations

Chapter Outline

When we introduced the definition of a system in Chapter 6, we also introduced the concept of a system representation. A system representation is an explicit mathematical expression of a system. In Chapter 6 we identified four system representations which we will investigate extensively: differential equations, Laplace transfer functions, Fourier transfer functions, and the convolution integral. It is the system representation that we use to analyze a system. The kind of analysis tools that we can bring to bear on the system depend heavily on the particular system representation. The study of system representations is a major component of system analysis. We shall devote a good deal of this text to the study of system representations, their properties, and their interrelationships.

A given system can have several, basically equivalent, system representations. A major goal of this text is to introduce four major system representations, and explain how they are interrelated. In addition, we will introduce several minor variations of each system representation. We will also show how to transform one system representation into another representation.

The need for different but equivalent system representations is derived from the way these system representations are used. As discussed in Chapter 1, a system representation must yield an acceptable approximation to the experimentally observed physical process. The derivation of the system representation from experimental data may lead quite naturally to a particular system representation. Second, a system representation must lend itself to analysis and design. The analysis could be based on theoretical results or the analysis could be carried out using a computer simulation tool. Almost all theoretical results apply to a particular type of system representation. If the system representation derived from the experimental data is not of the appropriate form, then the system representation must be changed to analyze the system. Similar comments apply to computer analysis of systems. To begin the computer analysis it is necessary to enter the system representation into the computer package. Not all system representations can be entered directly into a computer package. In this case the system representation must be changed to fit the data structure of the computer package. In other cases, one system representation may be preferable to another because of the computer interface with the user or because of numerical computation reasons. All of these observations testify for the need for multiple system representations and an understanding of the relationship between these representations.

In this chapter we introduce the study of system representations with the differential equation. Differential equations occur naturally in the study of physical processes, occurring frequently in physics, dynamics, and network analysis. We use the differential equation to introduce the first major system representation, the transfer function. Quite simply it is shown that the transfer function is natural part of the analysis of a differential equation using the Laplace transform.

In this chapter we also introduce a pictorial representation of a transfer function called a block diagram. We define block diagrams, and show how they can be derived from differential equations. Block diagrams are widely used because they provide a picture of a complex system which exposes the structure of the model of the system. We also show how a block diagram can be manipulated to simplify the system or change it to suit a particular analysis tool.

There is a very close relationship between transfer functions, block diagrams, and differential equations. To expose this relationship we introduce a special form of the block diagram, called an all-integrator block diagram, and a special form of the differential equation, called a state space representation. State space representations are the second major system representation, being the preferred form of differential equations. The thrust of the development is to explain how a state space representation is derived from an all-integrator block diagram and conversely.

Both of these representations play a central role in system theory. State space representations are further developed in the next chapter.

Computer Usage: Both the transfer function and the state space representation are widely used as a data structure for computer packages. State space representations, which are based on matrices, are often preferred for the internal computations because the matrices can be conditioned for optimal performance of the numerical algorithms. We will show how to use MATLAB in conjunction with both of these system representations.

Summary of Sections

Section 10.1:   We develop the transfer function from the differential equation.

Section 10.2:   We introduce block diagrams as a graphical method for displaying transfer functions and modeling systems.

Section 10.3:   We construct the block diagrams of several specific systems to illustrate their role in the development of a system model.

Section 10.4:   We present a method for the manipulation and reduction of block diagrams.

Section 10.5:   We introduce a special class of block diagrams called all-integrator block diagrams. Using all-integrator block diagrams we define state space representations.

Section 10.6:   Chapter summary section.

Coverage of the Text

This chapter requires the definition of a signal and a system discussed in Chapters 5 and 6. Laplace transforms, discussed in Chapter 9, are also used extensively.

10.1 THE TRANSFER FUNCTION

10.1.1 Introduction

Consider the system

$$y(t) = \mathcal{H}[x(t)] \tag{10.1.1}$$

shown in Figure 10.1.1. It is the purpose of the following four chapters to introduce and discuss the properties of several system representations. In this chapter we will discuss two system representations for (10.1.1): the differential equation and the transfer function. The differential equation is widely used in many branches of

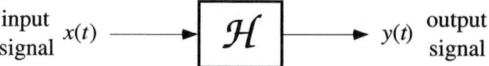

Figure 10.1.1 Cartoon of a System

engineering and physics. Because of the familiarity of this representation, we will start with it.

When we solve differential equations using the Laplace transform we obtain quite naturally the transfer function. This system representation turns out to be extremely useful as we shall demonstrate over the course of the text. In this section we will introduce the definition of the transfer function and define its basic properties.

10.1.2 Differential Equations

In this text we will be primarily concerned with differential equations of the form

$$y^{(n)}(t) + a_{n-1}y^{(n-1)}(t) + \cdots + a_1\dot{y}(t) + a_0 y(t) \tag{10.1.2}$$

$$= b_m x^{(m)}(t) + b_{m-1}x^{(m-1)}(t) + \cdots + b_0 x(t),$$

$$y^{(n-1)}(0) = y_{n-1}, \; y^{(n-2)}(0) = y_{n-2}, \ldots, \dot{y}(0) = y_1, \; y(0) = y_0.$$

Notation: In (10.1.2) we assume that the coefficients a_k and b_k are real numbers; i.e., they are constants.

The input to this system is the signal $x(t)$ and the output of this system is $y(t)$. The "system" is the relationship between $x(t)$ and $y(t)$ implied by the differential equation. The order of the highest derivative in (10.1.4) is n.

Terminology: We say that the differential equation in (10.1.2) is an _nth order_ differential equation.

For purposes of this discussion we will consider the second-order equation

$$\ddot{y}(t) + a_1\dot{y}(t) + a_0 y(t) = b_1\dot{x}(t) + b_0 x(t), \quad t \geq 0 \tag{10.1.3}$$

$$\dot{y}(0^-) = y_1, \; y(0^-) = y_0, \quad x(0^-) = x_0$$

without loss of generality. Next we will develop explicit dependence of the output signal $y(t)$ on the input signal $x(t)$ and on the initial conditions. To that end, when we take the Laplace transform of both sides of (10.1.3) we get

$$L\{\ddot{y}(t)\} + a_1 L\{\dot{y}(t)\} + a_0 L\{y(t)\} = b_1 L\{\dot{x}(t)\} + b_0 L\{x(t)\}. \tag{10.1.4}$$

In (10.1.4) we have used the fact that the Laplace transform is linear, Property 9.2.1. Next, applying Property 9.2.7 for differentiation, (10.1.4) becomes

$$\left\{s^2 Y(s) - s y_0 - y_1\right\} + a_1\left\{s Y(s) - y_0\right\} + a_0\left\{Y(s)\right\} = b_1\left\{s X(s) - x_0\right\} + b_0\left\{X(s)\right\}. \tag{10.1.5}$$

When taking the Laplace transform of the derivative in (10.1.4), we use the initial condition at $t = 0^-$ if appropriate. Solving (10.1.5) for $Y(s)$ we get

$$Y(s) = \frac{y_0 s + y_1 + a_1 y_0 - b_1 x_0}{s^2 + a_1 s + a_0} + \frac{b_1 s + b_0}{s^2 + a_1 s + a_0} X(s). \tag{10.1.6}$$

In (10.1.6) we have separated the initial conditions from the input signal. Each term in (10.1.6) can now be inverted to yield

$$y(t) = y_{zi}(t) + y_{zs}(t), \quad t \geq 0. \tag{10.1.7}$$

The form of the Laplace transform in (10.1.7) shows that the function $y(t)$ contains two separate components. The first term in (10.1.7)

$$y_{zi}(t) = \mathcal{L}^{-1}\left\{\frac{y_0 s + y_1 + a_1 y_0 - b_1 x_0}{s^2 + a_1 s + a_0}\right\} \tag{10.1.8}$$

is that part of solution of the differential equation that depends on the initial conditions. The second term of (10.1.7)

$$y_{zs}(t) = \mathcal{L}^{-1}\left\{\frac{b_1 s + b_0}{s^2 + a_1 s + a_0} X(s)\right\} \tag{10.1.9}$$

is due to the forcing function $x(t)$. In general, the solution to the differential equation (10.1.3) will always be decomposed in this way.

Definition 10.1.1: Let the solution of the differential equation (10.1.3) be given by (10.1.7). Then the component of the output signal $y_{zi}(t)$ is called the <u>zero input response</u>. The component of the output signal $y_{zs}(t)$ is called the <u>zero state response</u>.

▲▲

Terminology: The term *zero input response* is obvious: it is the solution to the differential equation when the input signal is zero. The term *zero state response* is derived from the theory of state space representations, which will be introduced later in the chapter. It implies that the initial conditions are zero.

The zero state response isolates the relationship between the input signal and the output signal. This relationship is the basis of the definition of a transfer function that is introduced below.

   With Definition 10.1.1 we have given names to familiar notions. The next example ties these names to network analysis.

Example 10.1.2: *RC network* Consider the network shown in Figure 10.1.2.

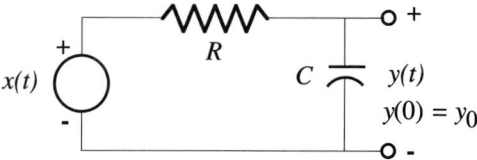

Figure 10.1.2 RC Network

Suppose that the input voltage is a unit step function $x(t) = u_s(t)$, and we want to find the output signal $y(t)$ across the capacitor with the given initial voltage. First, we derive the differential equation of this circuit. Applying Kirchhoff's law, the current in the network is given by

$$C\dot{y}(t) = i(t) = \frac{x(t) - y(t)}{R}. \tag{10.1.10}$$

From (10.1.10) the voltage across the capacitor is given by the differential equation

$$\dot{y}(t) + \frac{y(t)}{RC} = \frac{x(t)}{RC}, \quad y(0) = y_0. \tag{10.1.11}$$

Taking the Laplace transform of (10.1.11) using $x(t) = u_s(t)$, we obtain

$$sY(s) - y_0 + \frac{Y(s)}{RC} = \frac{X(s)}{RC} = \left(\frac{1}{s}\right)\frac{1}{RC}. \tag{10.1.12}$$

Solving (10.1.12) for $Y(s)$ and separating out the initial condition yields the zero input and zero state response

$$Y(s) = \underbrace{\frac{y_0}{s + \dfrac{1}{RC}}}_{Y_{zi}(s)} + \underbrace{\frac{1}{s\left(s + \dfrac{1}{RC}\right)RC}}_{Y_{zs}(s)}. \tag{10.1.13}$$

The partial fraction expansion of the zero state response in (10.1.13) is

$$Y_{zs}(s) = \frac{1}{s\left(s + \dfrac{1}{RC}\right)RC} = \frac{1}{s} + \frac{-1}{\left(s + \dfrac{1}{RC}\right)}. \tag{10.1.14}$$

Combining (10.1.14) with (10.1.13) and taking the inverse Laplace transform we get

$$y(t) = \left(e^{-\frac{t}{RC}}\right)y_0 + \left(1 - e^{-\frac{t}{RC}}\right)u_s(t). \tag{10.1.15}$$

In (10.1.15) the zero input response is the part of $y(t)$ due to the initial voltage on the capacitor

$$y_{zi}(t) = y_0 e^{-\frac{t}{RC}}, \tag{10.1.16}$$

and the zero state response is the part of $y(t)$ due to the forcing function

$$y_{zs}(t) = \left(1 - e^{-\frac{t}{RC}}\right) u_s(t). \tag{10.1.17}$$

▲▲

10.1.3 Definition of the Transfer Function

The above example shows that when a differential equation is used to represent a system, the output signal depends on the initial conditions as well as the input signal. The initial conditions represent internal energy storage or other physical phenomena which affect the output signal of the system, but which are independent of the input signal. In many situations the initial conditions are not important to the analysis of the system, and they can be excluded by setting them to zero. For these systems we are primarily interested in the relationship between the output signal and the input signal. The idea is to abstract this idea and define a new system representation. To express this relationship concisely, we can use the Laplace transform. This development leads to the first important system representation: the transfer function. The basic concept of a transfer function is to relate the input signal to the output signal of the system. Therefore the definition of a transfer function is based on this concept.

Definition 10.1.3: The <u>transfer function</u> $H(s)$ of a system (10.1.1) is the ratio of the Laplace transform of the output signal over the Laplace transform of the input signal

$$\frac{\mathcal{L}\{y(t)\}}{\mathcal{L}\{x(t)\}} = \frac{Y(s)}{X(s)} = H(s).$$

▲▲

The transfer function is one of the most important system representations we shall study in this text.

Usually in this text the transfer function is a rational function; a ratio of polynomials

$$\frac{Y(s)}{X(s)} = \frac{b_m s^m + b_{m-1} s^{m-1} + \cdots + b_0}{s^n + a_{n-1} s^{n-1} + \cdots + a_0} = \frac{b(s)}{a(s)} = H(s). \tag{10.1.18}$$

Notation: The indexing of the coefficients of polynomials plays an important role in some of the results below. The index of the coefficient is the same as the order of s of that term in the polynomial.

As suggested by the analysis above the transfer function can be easily calculated from the differential equation.

Theorem 10.1.4:   Relationship Between Differential Equations and Transfer Functions   Suppose a system is represented by a differential equation

$$y^{(n)}(t) + a_{n-1}y^{(n-1)}(t) + \cdots + a_1\dot{y}(t) + a_0 y(t)$$

$$= b_m x^{(m)}(t) + b_{m-1}x^{(m-1)}(t) + \cdots + b_0 x(t),$$

$$y^{(n-1)}(0) = 0, \ y^{(n-2)}(0) = 0, \ldots, y(0) = 0, \quad t \geq 0$$

where the *initial conditions are all zero*.   Then the transfer function of this system is given by

$$\frac{Y(s)}{X(s)} = \frac{b_m s^m + b_{m-1}s^{m-1} + \cdots + b_0}{s^n + a_{n-1}s^{n-1} + \cdots + a_0} = \frac{b(s)}{a(s)} = H(s).$$

▲▲

Terminology: Theorem 10.1.4 relates two system representations: the differential equation and the transfer function.   All of the mathematical quantities of the differential equation are related to real functions in the time variable t.   So we say that this representation exists in the <u>time domain</u>.   All of the mathematical quantities related to the transfer function are complex functions of the Laplace transform variable s.   So we say this system representation exists in the <u>frequency domain</u>. The motivation of this terminology will be more clear after the discussion of frequency response in Chapters 14 and 15.

To construct the transfer function from the differential equation requires that we know the coefficients and the order of the corresponding derivative.   Conversely, we can easily reconstruct the differential equation by taking the inverse Laplace transform of the transfer function.   Hence, there is a one-to-one correspondence between the transfer function and the differential equation.   This relationship plays a central role in the use of computers for the analysis of differential equations.   We will explore this idea in more detail in the next chapter.

It is worth emphasizing again that the concept of a transfer function is independent of the concept of initial conditions.   *When the transfer function is calculated from a differential equation, the initial conditions are always set to zero.*   If the initial conditions play an important role in the study of a particular system, then the differential equation system representation must be chosen instead of the transfer function.   (Usually state space equations are chosen to study systems where the initial conditions must be accounted for.   This system representation is discussed later in this chapter and in the next chapter.)

Because transfer functions are usually rational functions, the terminology of rational functions, introduced in Section 3.2, is associated with transfer functions. We have the following definition.

Definition 10.1.5: The order of the transfer function in (10.1.18) is the order of the denominator polynomial; the order of the transfer function in (10.1.18) is n. If the order of the numerator polynomial is strictly less than the order of the denominator polynomial $m < n$, we say the transfer function is strictly proper. If the order of the numerator polynomial is less than or equal to the order of the denominator polynomial $m \leq n$, we say the transfer function is proper. If the order of the numerator polynomial is greater than the order of the denominator polynomial $m > n$, we say the transfer function is not proper.

▲▲

Note: The order of the transfer function coincides with the order of the differential equation.

The transfer function can be displayed in terms of its polynomials as in (10.1.18), or the polynomials can be factored into their roots as

$$\frac{Y(s)}{X(s)} = \frac{K(s-z_1)(s-z_2)\cdots(s-z_m)}{(s-p_1)(s-p_2)\cdots(s-p_m)} = \frac{b(s)}{a(s)} = H(s). \tag{10.1.19}$$

Notation: Whenever we write a transfer function as in (10.1.19) or (10.1.18) we *always* assume that there are no common roots between the numerator and denominator. That is, the numerator and denominator are coprime. If common roots occur, they are canceled out.

We shall often find it convenient to work with the transfer function factored as in (10.1.19). Therefore, we give this form a name.

Definition 10.1.6: The roots of the numerator polynomial $b(s)$ in (10.1.18) are called the zeros of the transfer function. The roots of the denominator polynomial $a(s)$ in (10.1.18) are called the poles of the transfer function. If the poles and zeros are graphically presented as location in the s-plane, then we call this diagram a pole-zero diagram.

▲▲

Notation: The transfer function in (10.1.18) is completely defined by the coefficients of the numerator and denominator polynomials. The same transfer function is equivalently defined by the poles, zeros, and gain in (10.1.19). Both of these representations are used by MATLAB.

Notation: Suppose the transfer function is given by (10.1.18). In (10.1.18) there are three rational functions. Two of these rational functions, $X(s)$ and $Y(s)$, represent signals and the third rational function $H(s)$ represents the system. Thus, these rational functions have very different physical interpretations.

Obviously, a transfer function can't be defined unless the input and output signals are Laplace transformable. Hence, this assumption is made implicitly when we use the transfer function. Less obviously, the ratio of the signals must make sense for all signals. This assumption is examined in detail in Chapter 13. In this chapter we will restrict ourselves to systems for which the transfer function exists.

The system includes the input and output signals $X(s)$ and $Y(s)$ as well as the transfer function $H(s)$. The transfer function is defined as the ratio of the transform of the output signal over the transform of the input signal. This definition is a re-iteration of the definition of a system in terms of the input and output signals.

In this text we will be mainly concerned with transfer functions that are derived from differential equations. The transfer function, however, can be calculated directly from experimental data as well. Hence, it is important to keep in mind the definition of a transfer function.

10.1.4 Examples

The next three examples illustrate the concepts introduced in this section.

Example 10.1.7:   Consider again the RC network in Example 10.1.2.   The differential equation of this system with the initial condition set to zero is

$$\dot{y}(t) + \frac{y(t)}{RC} = \frac{x(t)}{RC}, \quad y(0) = 0. \tag{10.1.20}$$

Taking the Laplace transform of (10.1.20) and solving for the transfer function yields

$$\frac{Y(s)}{X(s)} = \frac{\dfrac{1}{RC}}{s + \dfrac{1}{RC}}. \tag{10.1.21}$$

The pole-zero diagram of this system is shown Figure 10.1.3. This transfer function always has one pole in the open LHP and no zeros.   ▲▲

Example 10.1.8: Mass-spring-damper system   Consider a mass moving on a frictionless plane attached to a rigid support by a spring and damper as shown in Figure 10.1.4.   The mass is m_{st}, the spring constant is k_{st}, and the damping coefficient is c_{st}. The force applied to the mass is $f_{st}(t)$, and the displacement of the mass due to this force is $y_{st}(t)$. The differential equations for this system are derived in Section 6.4 using Newton's second law.   We have

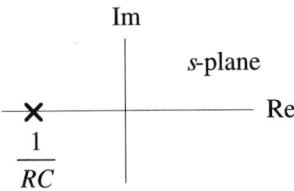

Figure 10.1.3 Pole-Zero Diagram of an RC Network

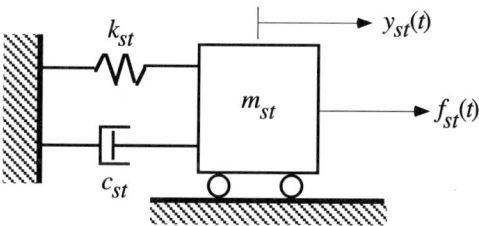

Figure 10.1.4 Mass-Spring-Damper System

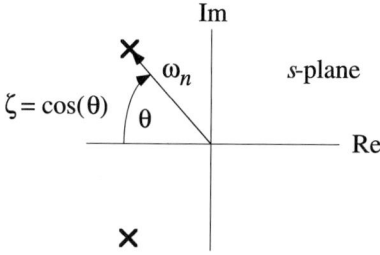

Figure 10.1.5 Pole-Zero Diagram of a Mass-Spring-Damper System

$$m_{st}\ddot{y}_{st}(t) + c_{st}\dot{y}_{st}(t) + k_{st}y_{st}(t) = f_{st}(t),$$ (10.1.22)
$$\dot{y}_{st}(0) = y_{st}(0) = 0.$$

Taking the Laplace transform of (10.1.22) and solving for the transfer function yields

$$\frac{Y_{st}(s)}{F_{st}(s)} = \frac{\dfrac{1}{m_{st}}}{s^2 + \dfrac{c_{st}}{m_{st}}s + \dfrac{k_{st}}{m_{st}}}.$$ (10.1.23)

The pole-zero diagram for the transfer function (10.1.23) is shown in Figure 10.1.5. The location of the poles depends on the coefficients of the transfer function as derived in Section 6.4. The geometric interpretation of these pole locations is given in Section 3.2. In this text we will be interested in those parameters values that lead to a pair of complex poles as shown in Figure 10.1.5.

▲▲

Example 10.1.9: Proof-Mass Actuator Consider the proof-mass actuator shown in Figure 10.1.6. In Section 6.5 the differential equation that describes the motion of the proof-mass is shown to be

$$m_{pm}\ddot{y}_{pm}(t) = K_{ef}v_{pm}(t),$$ (10.1.24)
$$\dot{y}_{pm}(0) = y_{pm}(0) = 0.$$

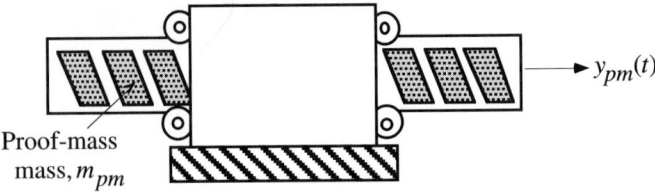

Figure 10.1.6 Proof-Mass Actuator

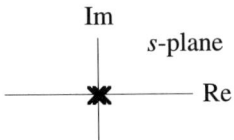

Figure 10.1.7 Pole-Zero Diagram of a Proof-Mass Actuator

Taking the Laplace transform of the system in (10.1.24) and solving for the transfer function yields

$$\frac{Y_{pm}(s)}{V_{pm}(s)} = \frac{K_{ef}}{m_{pm}s^2}. \tag{10.1.25}$$

The pole-zero diagram of this transfer function is shown in Figure 10.1.7. The transfer function (10.1.25) has two poles at the origin.

▲▲

10.1.5 MATLAB Experiments

A computer simulation tool for signals and systems must contain several essential ingredients. With respect to systems, two important components are the ability to enter system representations into the computer package and the capability to simulate the output signal of the system for a given input signal. Over the course of this text we will discuss how MATLAB implements each of these aspects of system analysis.

MATLAB has several provisions for storing a system representation. In this section we have discussed transfer functions, so we begin with the way MATLAB stores transfer functions. MATLAB allows a transfer function to be stored in two formats: as a ratio of polynomials and as poles, zeros, and a gain. We begin with the polynomial representation of a transfer function. (See (10.1.18).) To enter this form of the transfer function into MATLAB it is necessary to enter the two polynomials.

Next, we briefly review polynomial manipulation in MATLAB. The manual should be consulted for a more thorough discussion. A polynomial is stored as a vector. For example, a second-order polynomial is represented as

$$a_2 s^2 + a_1 s + a_0 \ \leftrightarrow \ \begin{bmatrix} a_2 & a_1 & a_0 \end{bmatrix}. \tag{10.1.26}$$

Polynomial addition is the same as vector addition. Polynomial multiplication is accomplished by the **conv** command.

The Control System Toolbox in MATLAB allows a system to be represented as an object in MATLAB. For a complete discussion of the properties of these objects, the reader is referred to the manual. We give only a brief introduction here. In the MATLAB manual, the type of systems we are discussing here are referred to as "LTI Models." This terminology will be explained in Chapter 13.

To enter a transfer function as a system, first the numerator and denominator polynomials are entered into MATLAB. Then the system is defined with the command **tf**. The input arguments are the numerator and denominator polynomials. The output argument is the name of the system (which refers to an object). Henceforth, MATLAB treats this variable name as a system, and performs calculations accordingly.

An alternative representation of a transfer function is to factor the numerator and denominator into linear factors. Then the transfer function is represented by its poles and zeros along with a constant gain factor. This system representation can also be entered into MATLAB with the command **zpk**. This command functions like the command **tf**. The input arguments are the two vectors and a scalar containing the zeros and poles along with the constant gain factor. (See (10.1.19).) The output argument is the name of the system.

The commands **tf** and **zpk** can also be used to translate a system representation from one form to another. If a system is entered as a transfer function using the **tf** command, the system is stored as two polynomials. If the system is used as the input argument to the **zpk** command, the system representation will be translated into zeros and poles with a constant gain.

As with all computer simulation packages, the numerical accuracy of the computations is a concern. While these numerical computation issues are beyond the scope of this text, the reader should be aware of these issues. For example, not all system representations are equal when numerical computations are concerned. The system representation that is used in the computations can affect the accuracy of the answer. Therefore, it is important to understand the relationships between the system representations so that the appropriate representation can be selected.

A second important tool for system analysis is the capability to simulate the output signal of a system for a given input signal. MATLAB has several ways to simulate a system, but the most convenient command for the type of systems in which we are interested is **lsim**. The complete syntax associated with this command is explained in MATLAB help. In its simplest usage, the input arguments are: (1) a system representation, (2) the time vector, and (3) the input signal. The output argument of **lsim** is the output signal. When this command is used without output arguments, a plot is generated.

A system can also be simulated using the command **ltiview**. A description of this advanced simulation tool is left to the MATLAB manual.

MATLAB Version 4 The previous version of MATLAB didn't have the capability to represent systems as object. In this version of MATLAB a transfer function is simply entered as two polynomials which must be tracked independently by the user.

The M-files below can be modified for Version 4 by replacing the system name by the two polynomials which define the transfer function. For example to simulate the system below, the **lsim** command is used

```
y = lsim(h,x,t);        % simulate system
```

where h is the name of the system transfer function. The transfer function is defined by the numerator and denominator polynomials b and a respectively. Therefore, the system simulation is also accomplished by

```
y = lsim(b,a,x,t);      % simulate system
```

This modification can be made throughout the text to make the M-file compatible with Version 4.

In Version 4 there are a separate set of commands to change a system representation from one form into another. The command which changes a transfer function into zero pole form is **tf2zp**. The input arguments to this command are the two vectors containing the numerator and denominator of the transfer function. The output arguments are the vectors containing the zeros and poles and the constant gain. Again the three vectors which define this system representation must be tracked independently by the user. There exist a number of similar commands for changing the system representation into other forms.   ▲▲

The following examples illustrate the use of MATLAB to simulate a system.

Example 10.1.7: (*continued*) The following M-file simulates the *RC* network in Example 10.1.7 with the input signal used in Example 5.1.5. The input signal is

$$x(t) = u_s(t) - u_s(t - t_0).$$ (10.1.27)

```
clear
% Define system
RC = 1;                 % define RC time constant
b = [1/RC];             % define numerator
a = [1,1/RC];           % define denominator
h = tf(b,a);            % define system
% Define signal
ic = 0.05;              % time increment
tf = 10;                % final time
t = [0:ic:tf];          % define time vector
pl = 100;               % pulse length of input signal
x = zeros(size(t));     % define input signal
x(1:pl) = ones(1,pl);
y = lsim(h,x,t);        % simulate system
plot(t,x,t,y,'--')
xlabel('Time, sec')
title('Input and Output Signal Of a RC Network')                    ▲▲
```

Exploratory Exercise 10.1.10: In the above M-file vary the pulse width between 2 and 8 ($20 < pl < 180$), and observe the effect on the output signal. Next fix the pulse width and vary the RC time constant between $0.4 < RC < 4$. What conclusions can you draw about the shape of the output signal and the relationship between the signal parameters (pulse width) and the system parameters (RC constant)?

▲▲

Example 10.1.8: (*continued*) The following M-file simulates the mass-spring-damper system in Example 10.1.8. (See also the discussion in Section 6.4.) The M-file calculates the poles and zeros of the transfer function by using the command **zpk**. This M-file also plots the poles and zeros in the *s*-plane and calculates the damping ratio and natural frequency of the poles.

```
clear
% enter system
mst = 400;                          % mass
cst = 80;                           % damping
kst = 600;                          % stiffness
bst = [1/mst];                      % numerator of the transfer function
ast = [1,cst/mst,kst/mst];          % denominator of the transfer function
hst = tf(bst,ast);                  % define system
zpk(hst)                            % calculate zeros and poles
figure(1)
pzmap(hst)                          % plot poles and zeros
axis([-0.15,0,-1.5,1.5])
title('zero-pole map')
grid
% calculate natural frequency and damping ratio of the poles
damp(hst)
% plot the step response of this system
wn = sqrt(kst/mst);                 % natural frequency
t = linspace(0,5*(2*pi/wn),300);    % generate time vector
fst = ones(size(t));                % generate step input signal
yst = lsim(hst,fst,t);              % simulate system
figure(2)
plot(t,yst)
title('Step Response')
xlabel('Time')
```

▲▲

Exploratory Exercise 10.1.11: In Example 10.1.8 vary the stiffness as $10 < kst < 1500$. Plot the change in the natural frequency ω_n and damping ratio ζ. Note the change in the poles and the step response. Repeat this exercise for a change in the damping, $10 < cst < 1000$, and a change in the mass, $100 < mst < 2000$.

▲▲

10.2 BLOCK DIAGRAMS

10.2.1 Definition of a Block Diagram

In this section we introduce a pictorial representation of a transfer function called a block diagram. This pictorial representation of transfer functions is extremely useful for visualizing the structure of a system, particularly if it is composed of many interconnected subsystems. It is also useful for analyzing a system. We will explain this interconnection structure in this section. In the coming sections in this chapter we will show how to use block diagrams to represent complex systems.

We assume that a system can be represented by a transfer function

$$\frac{Y(s)}{X(s)} = H(s), \text{ or } Y(s) = H(s)X(s). \tag{10.2.1}$$

In (10.2.1) we see that by using the Laplace transform, a system is modeled by an algebraic relationship of polynomials in s. We can graphically represent this relationship as shown in Figure 10.2.1.

Definition 10.2.1:   The pictorial representation in Figure 10.2.1 of the transfer function in (10.2.1) is called a <u>block diagram</u>.   ▲▲

Always bear in mind that $X(s)$ is the input *signal* which causes the *system $H(s)$* to produce the output *signal* $Y(s)$. That is, there is a definite flow implied by Figure 10.2.1. The direction of the arrows from input signal to output signal is intended to emphasize this flow. The reverse direction is not defined by Figure 10.2.1. If a system can be modeled by reversing the arrows and inverting the transfer function, then this system must be modeled separately.

The block diagram in Figure 10.2.1 is an equivalent mathematical expression for the algebraic equation in (10.2.1). It should always be remembered that block diagram manipulation is equivalent to precisely defined mathematical operations. Thus, block diagrams stand in stark contrast to the cartoons used previously as a visual aid for remembering the definition of a system.

Often a complex system is formed by joining together many simpler systems. The usefulness of block diagrams is that they can be used to represent this interconnection of simple systems to build a more complex system. Block diagrams serve to visualize the structure of the total system, that is, the interconnection of the components of the system. Block diagrams also have the flexibility to represent the total system at various levels of complexity. Furthermore, block diagrams can be used as a computational tool. First, the block diagram is assembled from the block diagrams of the component parts. Then the transfer function of the whole system is

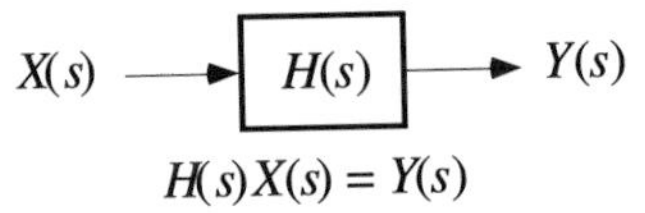

$$H(s)X(s) = Y(s)$$

Figure 10.2.1 Definition of a Block Diagram

computed from the block diagram. We will develop these ideas in detail in this and the next section.

Computer Usage: Computer programs have evolved to include a graphical interface based on block diagrams. These programs have a built-in interface that generates a system representation (transfer function or state space equations) based on the block diagram. Thus, block diagrams should be viewed as one of the interface languages between human and computer.

10.2.2 Simple Interconnections of Systems

Block diagrams are particularly useful for representing complex systems that are formed by joining together simpler systems. Next we discuss three fundamental interconnections of systems that form the basis of more complicated block diagrams. In the following discussion we make reference to the two systems shown in Figure 10.2.2. To join together several elementary blocks requires two additional ideas to establish the interconnection structure between the blocks.

Definition 10.2.2: At a <u>branch point</u>, shown in Figure 10.2.3, a signal is duplicated as two or more signals. All signals leaving the branch point are the same as the signal entering the branch point.

▲▲

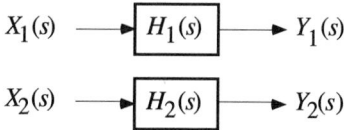

Figure 10.2.2 Two Systems

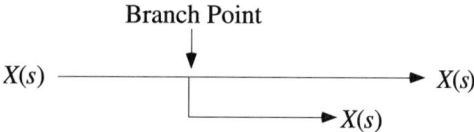

Figure 10.2.3 Branch Point

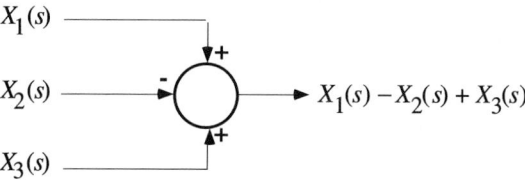

Figure 10.2.4 Summing Point

Terminology: Branch points are sometimes called <u>pickoff points</u>.

Definition 10.2.3: At a <u>summing point</u>, shown in Figure 10.2.4, several signals are added or subtracted to create the output signal. The signal leaving the summing point is the sum of all signals entering the summing point. If the signal enters the summing point with a + sign, then that signal is added to the output signal. If the signal enters the summing point with a − sign, then that signal is subtracted from the output signal.

▲▲

Notation: Convention allows only one signal to leave a summing point.

Using branch points and summing nodes, we can create new systems from existing systems in Figure 10.2.2. There are three basic ways we can join the systems in Figure 10.2.2. We begin with the simplest interconnection structure.

System Structure 10.2.4: **Cascade Interconnection**   Given the two systems in Figure 10.2.2, we connect the output signal of the first system to the input signal of the second system. The input signal of the new system is $x_1(t)$ and the output signal is $y_2(t)$. The block diagram is shown in Figure 10.2.5a.

Terminology: The System Structure 10.2.4 is also called a <u>series interconnection</u>.

The equations governing the composite system in Figure 10.2.5a are

$$H_1(s)X_1(s) = Y_1(s), \quad Y_1(s) = X_2(s), \quad H_2(s)X_2(s) = Y_2(s). \tag{10.2.2}$$

From the equations in (10.2.2) the transfer function of the new system is

$$\frac{Y_2(s)}{X_1(s)} = H_1(s)H_2(s). \tag{10.2.3}$$

The transfer function of the composite system is obtained by performing the (simple) algebraic operations in (10.2.3). The block diagram of the composite system is shown in Figure 10.2.5b.

▲▲

In the last example we proposed a system structure in terms of a block diagram shown in Figure 10.2.5. Then we used algebraic manipulations to effectively reduce the total system into a simple transfer function in (10.2.3). Henceforth, the reduction of the system in Figure 10.2.5a to the system in Figure 10.2.5b can be carried out graphically without resort to the underlying algebraic equations. We will repeat this procedure in the following two examples as well. Then we will expand this idea to more complex block diagrams. The relationship between the picture, the block diagram, and the underlying equations is of prime importance.

Using the two systems in Figure 10.2.2 we can interconnect them in a second way.

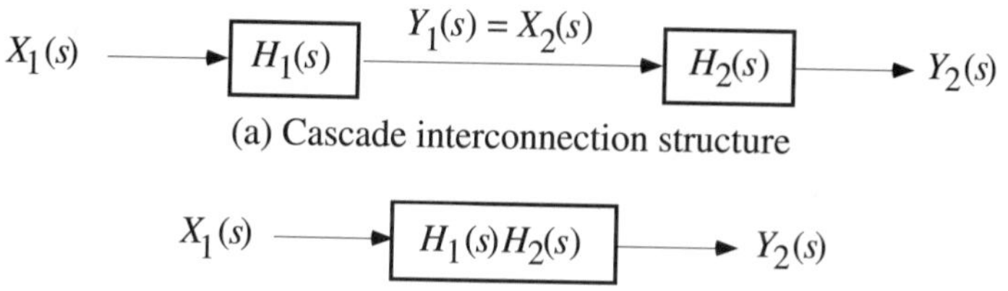

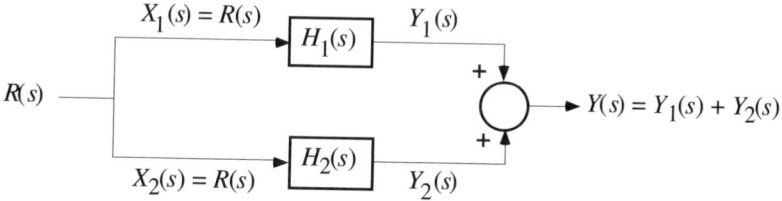

(a) Cascade interconnection structure

$$X_1(s) \longrightarrow \boxed{H_1(s)H_2(s)} \longrightarrow Y_2(s)$$

(b) Composite system

Figure 10.2.5 Cascade Interconnection of Two Systems

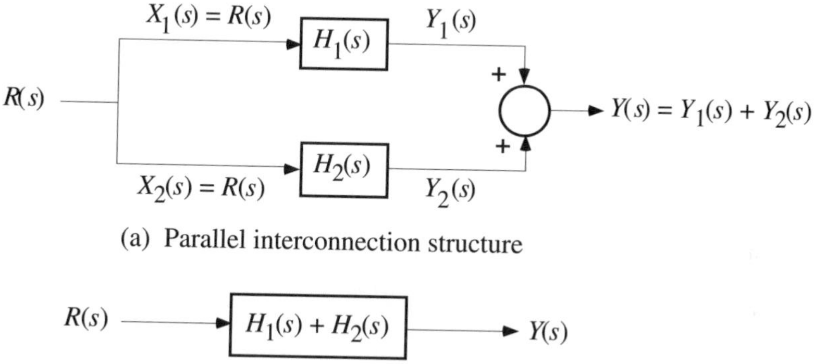

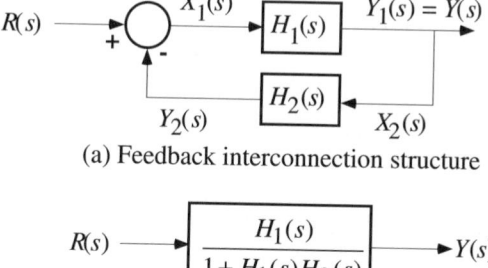

(a) Parallel interconnection structure

$$R(s) \longrightarrow \boxed{H_1(s) + H_2(s)} \longrightarrow Y(s)$$

(b) Composite system

Figure 10.2.6 Parallel Interconnection of Two Systems

(a) Feedback interconnection structure

$$R(s) \longrightarrow \boxed{\dfrac{H_1(s)}{1 + H_1(s)H_2(s)}} \longrightarrow Y(s)$$

(b) Composite system

Figure 10.2.7 A Feedback Loop

System Structure 10.2.5: Parallel Interconnection Another connection between the systems in Figure 10.2.2 is obtained by giving each system in Figure 10.2.2 the same input signal $r(t)$ and summing the output signals. The input signal to this composite system is $r(t)$, and the output signal is $y(t)$. The corresponding block diagram is shown in Figure 10.2.6a. The governing equations are

$$X_1(s) = R(s) = X_2(s) \qquad \begin{array}{l} H_1(s)R(s) = Y_1(s) = Y(s) \\[4pt] H_2(s)R(s) = Y_2(s) \end{array} \qquad Y(s) = Y_1(s) + Y_2(s) \qquad (10.2.4)$$

From the equations in (10.2.4), the transfer function of the composite system in Figure 10.2.6b is

$$\frac{Y(s)}{R(s)} = H_1(s) + H_2(s). \qquad (10.2.5)$$

▲▲

The third system structure is more complicated than the last two system interconnection structures.

System Structure 10.2.6:   Feedback Loop   Figure 10.2.7 shows another interconnection of the two systems in Figure 10.2.2. The output of the first system is connected to the input of the second system.   Now an external signal $r(t)$ is introduced and the output signal of the second system is subtracted from $r(t)$.   The resultant signal is used as an input to the first system.   In the composite system $r(t)$ is considered the input signal and $y_1(t)$ is considered the output signal.

The equations from the block diagram of the composite system are

$$R(s) - Y_2(s) = X_1(s) \qquad \begin{array}{l} Y_1(s) = H_1(s)X_1(s) \\[4pt] Y_2(s) = H_2(s)Y_1(s) \end{array} \qquad Y_1(s) = X_2(s) = Y(s) \qquad (10.2.6)$$

The transfer function of the composite system, shown in Figure 10.2.7b, is found by first combining the first and second equations in (10.2.6) to yield

$$Y(s) = Y_1(s) = H_1(s)\big[R(s) - Y_2(s) \big]. \qquad (10.2.7)$$

Next write $Y_2(s)$ in terms of $Y_1(s)$. Using this expression in (10.2.7) we get

$$Y(s) = H_1(s)\big[R(s) - H_2(s)Y(s) \big]. \qquad (10.2.8)$$

Now we can solve (10.2.8) for the transfer function

$$\frac{Y(s)}{R(s)} = \frac{H_1(s)}{1 + H_1(s)H_2(s)}. \qquad (10.2.9)$$

Any system which is configured as the block diagram in Figure 10.2.7a is called a feedback loop for obvious reasons.

Note: The + sign in the denominator of (10.2.9) can be traced directly to the − sign at the summing junction in Figure 10.2.7.

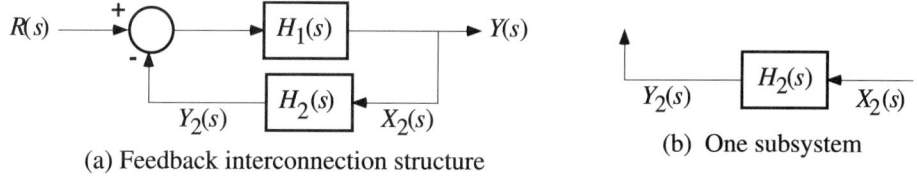

(a) Feedback interconnection structure   (b) One subsystem

Figure 10.2.8 Definition of a Subsystem

The formula in (10.2.9) can be remembered using the following terminology. Starting from the input signal $r(t)$ follow the arrows taking the shortest path to the output signal. The product of all blocks along this path is called the <u>forward path gain</u>. Next follow the arrows around the loop. The product of all blocks around the feedback loop is called the <u>loop gain</u>. The transfer function in (10.2.9) is (forward path gain)/(one plus the loop gain).

▲▲

Next we introduce some terminology for special kinds of feedback loops.

Definition 10.2.7: The transfer function $H_1(s)$ is called the <u>open loop</u> system. The transfer function (10.2.9) is called the <u>closed loop</u> system. If there is a negative sign in the summing junction of the feedback loop, the system is said to be a <u>negative feedback loop</u>. If the transfer function $H_2(s)$ is a unity gain $H_2(s) = 1$, then we say this system is a <u>negative unity feedback</u> system.

▲▲

Control theory is devoted to the study of feedback loops.

Much of the power of block diagrams is that they express the composite system as an interconnection of the components of the system. Because we routinely think of systems in this context, we attach a name to these "smaller systems."

Definition 10.2.8: Let a block diagram be composed of a number of elements interconnected together. If we extract one of these components from the block diagram, then we say that this component is a <u>subsystem</u>, if this component is itself a system.

▲▲

Example 10.2.9: A subsystem of Figure 10.2.8a is shown in Figure 10.2.8b.

▲▲

10.3 EXAMPLES OF BLOCK DIAGRAMS

10.3.1 Introduction

In this section we present two examples of the use of block diagrams for system representation. These examples are intended to illustrate the flexibility of a block diagram to represent systems that are composed of several subsystems. These examples will show how to construct a block diagram from complex impedances of a network and how to construct a block diagram from the differential equations of a physical process.

10.3.2 *LRC* Network

In this subsection we consider the *LRC* network shown in Figure 10.3.1. Here we will analyze this network by block diagram analysis. Through this example we connect the methods of analysis of this text with the methods of analysis discussed in a networks course.

Note: If we define $x(t)$ as the input signal and $y(t)$ as the output signal, then this circuit becomes a system. We propose to develop a block diagram of this system. This block diagram will give us a relationship between the input signal $x(t)$ and the output signal $y(t)$. Hence, the choice of the input signal and the output signal determines the analysis that follows. If we define another output signal we will obtain another system.

Next we develop the block diagram for this system using the tools of network analysis. First, we label the current, voltages, and complex impedances as shown in Figure 10.3.2. Starting with the output voltage $Y(s)$ we see that $I_3(s)$ is related to $Y(s)$ as

$$Y(s) = \left(\frac{1}{sC_2}\right)I_3(s).$$

(10.3.1)

The corresponding block diagram is shown in Figure 10.3.3. Summing currents into

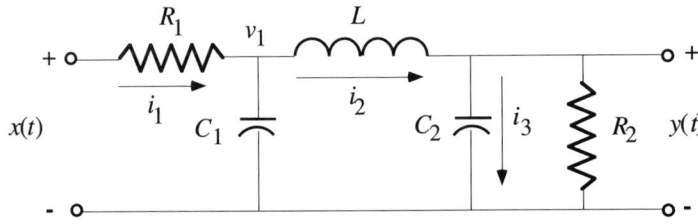

Figure 10.3.1 *LRC* Network

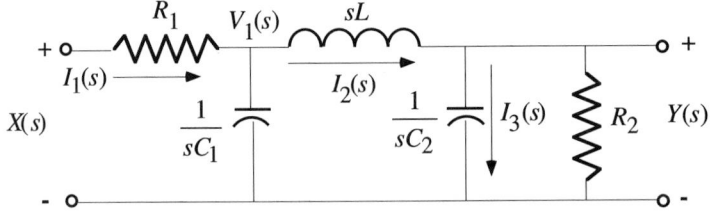

Figure 10.3.2 Network of Figure 10.3.1 with Complex Impedances

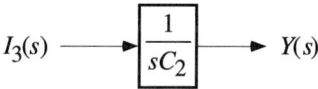

Figure 10.3.3 First Step in the Construction of the Block Diagram

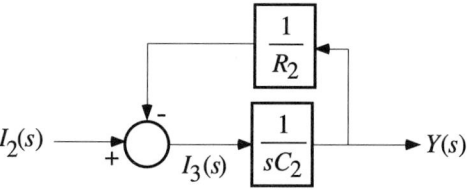

Figure 10.3.4 Second Step in Construction of the Block Diagram

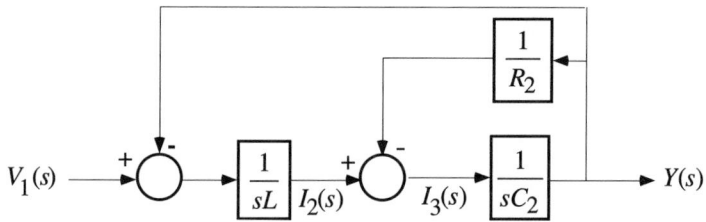

Figure 10.3.5 Third Step in Construction of the Block Diagram

the capacitor C_2, we get

$$I_3(s) = I_2(s) - \left(\frac{1}{R_2}\right) Y(s). \tag{10.3.2}$$

Combining (10.3.1) and (10.3.2), we have the following block diagram shown in Figure 10.3.4. Now the current $I_2(s)$ is determined from the voltages as

$$I_2(s) = \frac{V_1(s) - Y(s)}{sL}. \tag{10.3.3}$$

Including (10.3.3) with the block diagram in Figure 10.3.4 we have the diagram in Figure 10.3.5. Next the voltage $V_1(s)$ is given by

$$V_1(s) = \frac{I_1(s) - I_2(s)}{sC_1}. \tag{10.3.4}$$

The updated block diagram is shown in Figure 10.3.6. Finally, the current $I_1(s)$ is the voltage drop across the input resistor

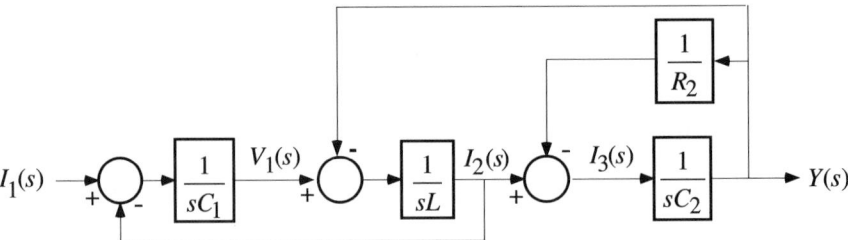

Figure 10.3.6 Fourth Step in Construction of the Block Diagram

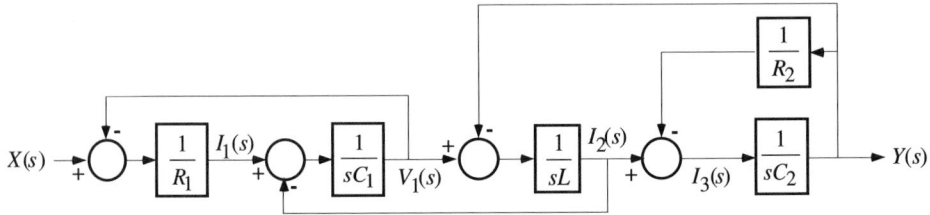

Figure 10.3.7 Block Diagram of the Network in Figure 10.3.1

$$I_1(s) = \frac{X(s) - V_1(s)}{R_1}. \qquad\qquad (10.3.5)$$

Combining (10.3.5) with Figure 10.3.6 we obtain the block diagram of the whole network as shown in Figure 10.3.7.

10.3.3 Proof-Mass Actuator/Mass-Spring-Damper System

Introduction In Section 6.5 we introduced a proof-mass actuator as a device that can apply forces to a structure to suppress vibrations in that structure. A simple model of vibrations in a structure is provided by the mass-spring-damper system in Section 6.4.   For this reason we will call the mass-spring-damper system a "structure," and we will use the subscript st to indicate quantities related to this system.   In this subsection we attach the proof-mass actuator to a mass-spring-damper system and develop a block diagram model that can be used to study the interaction of these two systems.   This discussion will illustrate the process of constructing a block diagram model of a complex system from the block diagrams of the subsystems.

Equations of Motion of the Two Masses The proof-mass actuator attached to the structure is shown in Figure 10.3.8. A simple model of the proof-mass actuator was given in Section 6.5 and a model of the structure in Section 6.4 which should be reviewed at this time. If a voltage signal is applied to the proof-mass actuator, the electromagnetic coupling between the coil windings and the proof-mass will exert a force on the proof-mass causing it to accelerate. The force applied to the proof-mass will cause an equal and opposite force to be applied to the mass of the structure.

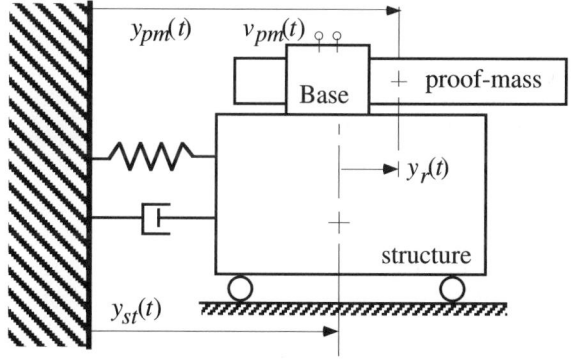

Figure 10.3.8 Structure with Proof-Mass Actuator

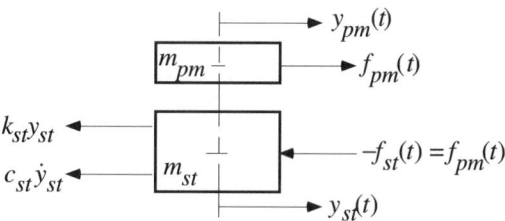

Figure 10.3.9 Free-Body Diagram of the Proof-Mass Actuator and Structure

Hence, oscillations (vibrations) will result in the structure. We want to model the relationship between the voltage applied to the proof-mass, $v_{pm}(t)$, the position of the proof-mass, $y_{pm}(t)$, and the position of the structure, $y_{st}(t)$.

The free-body diagram of the system in Figure 10.3.8 is shown in Figure 10.3.9. The motion of the proof-mass is determined by the force exerted on it. This force is related to the input voltage by

$$m_{pm}\ddot{y}_{pm} = f_{pm}(t) = K_{ef}v_{pm}(t). \tag{10.3.6}$$

The motion of the structure is given by

$$m_{st}\ddot{y}_{st}(t) + c_{st}\dot{y}_{st}(t) + k_{st}y_{st}(t) = f_{st}(t). \tag{10.3.7}$$

The force applied to the proof-mass is applied equal and opposite to the mass of the structure. We have

$$f_{pm}(t) = -f_{st}(t). \tag{10.3.8}$$

The coupled set of equations (10.3.6) - (10.3.8) describe the system in Figure 10.3.8. Next, we construct a block diagram of this system. The coupling between the proof-mass actuator and the structure is the signal in (10.3.8). This signal should appear explicitly in the block diagram. Accordingly, we split the equations in (10.3.6) into two equations as shown in (10.3.6). Taking the Laplace transform of each equation in (10.3.6) we have

$$\frac{F_{pm}(s)}{V_{pm}(s)} = K_{ef} \quad \text{and} \quad \frac{Y_{pm}(s)}{F_{pm}(s)} = \frac{1}{m_{pm}s^2}. \tag{10.3.9}$$

This block diagram of each of the transfer functions in (10.3.9) is shown in Figure 10.3.10 along with the block diagram of the transfer function of (10.3.7). Next the two transfer functions in Figure 10.3.10 are connected using (10.3.8). The result is shown in Figure 10.3.11.

Instrumentation of the Actuator   A proof-mass actuator is used to suppress vibrations in a structure. This vibration suppression is implemented by sensing motion of the structure and applying a restoring force via the actuator. Hence, a sensor is required to sense the motion of the structure. We must be able to sense where the proof-mass is relative to the base. If this measurement is not available, then the proof-mass may run against its stops, rendering it useless. The sensor

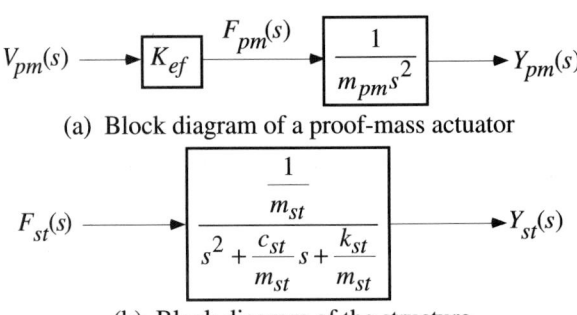

(a) Block diagram of a proof-mass actuator

(b) Block diagram of the structure

Figure 10.3.10 Block Diagram of a Proof-Mass Actuator on a Structure

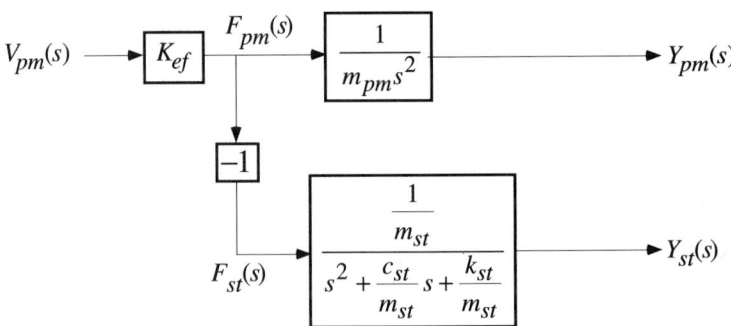

Figure 10.3.11 Block Diagram of the System in Figure 10.3.8

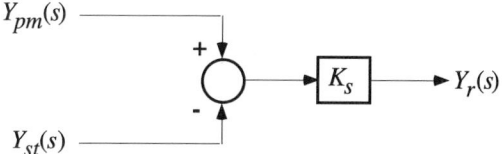

Figure 10.3.12 Block Diagram of the Position Sensor

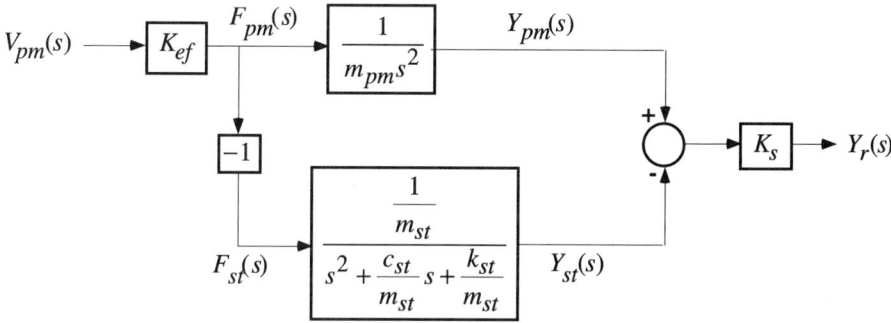

Figure 10.3.13 Block Diagram of the Proof-Mass-Structure System

measurement of the position of the proof-mass will allow us to keep the proof-mass centered. Taking into account these observations, we attach a sensor to the base of the actuator. This sensor measures the displacement of the proof-mass with respect to the base of the actuator. If we denote the output signal as $y_r(t)$ then the output of the sensor is related to the displacements of the proof-mass and the structure by

$$y_r(t) = K_s\big(y_{pm}(t) - y_{st}(t)\big). \tag{10.3.10}$$

The constant K_s is called a <u>calibration</u> constant with units of meters/volt. The block diagram of the model of the sensor is shown in Figure 10.3.12.

Complete System Model If we include this sensor in the block diagram of the actuator and structure in Figure 10.3.11 we obtain the block diagram in Figure 10.3.13. The block diagram in Figure 10.3.13 clearly indicates the coupling between the proof-mass actuator and the structure, and the coupling of the displacements of the proof-mass and the structure through the sensor.

10.4 BLOCK DIAGRAM REDUCTION

10.4.1 Introduction

In the previous sections we have shown how a block diagram can be used to represent systems that are composed of several subsystems. The basic idea is that a

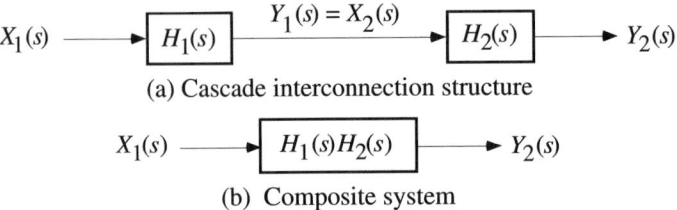

(a) Cascade interconnection structure

(b) Composite system

Figure 10.4.1 Cascade Connection of Two Systems

picture, a block diagram, can be used to represent a transfer function, an equation. Hence, the picture is mathematically equivalent to the equation. This observation suggests that the equations can be manipulated by manipulating the pictures. It turns out that this idea is a very profitable approach to working with transfer functions as we will show in this section.

To introduce this idea of block diagram manipulation we will start with the simple interconnection structures in Section 10.2 and show how they can be reduced. Then we will progress to more complicated examples.

System Structure 10.2.6: Cascade Interconnection   Figure 10.4.1a shows two systems in cascade. The composite system in Figure 10.4.1a can be reduced to a single transfer function as shown in Figure 10.4.1b. This example illustrates the basic idea of block diagram reduction. We will extend this concept to more complicated block diagrams in this section.

▲▲

The process of manipulating a block diagram carries a name.

Definition 10.4.1: The process of manipulating the block diagram into another form is called <u>block diagram reduction</u>.

▲▲

Transfer functions are *algebraic* equations in the complex variable s. Block diagrams are a pictorial representation of these algebraic equations. Block diagram manipulations can be used to simplify algebraic equations by manipulating the block diagram, rather than working directly on the equations themselves. Henceforth, the reduction of the block diagram can be carried out graphically without resort to the underlying equations. Note that block diagram reduction is a well-defined mathematical process.

We are developing this concept because it turns out that the manipulation of the block diagram is simpler than calculating the transfer function from the equations directly. The process of manipulating a block diagram into another form is called block diagram <u>reduction</u> because the purpose of this reduction is usually (but not always) to simplify the diagram. In this section we will present several examples of block diagram reduction.

Block diagram reduction consists of a few basic manipulations applied repeatedly to different parts of the block diagram. These basic manipulation rules are given in Figure 10.6.1 in the Chapter Summary Section. The table entries can be

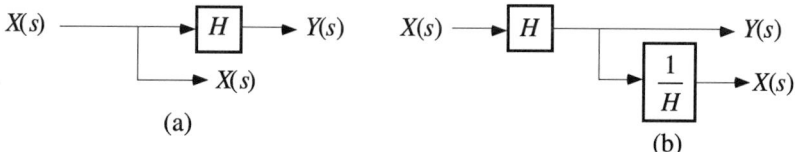

Figure 10.4.2 One Entry in the Table of Block Diagram Manipulations

verified from the corresponding algebraic equations. The following example illustrates these calculations.

Example 10.4.2: Consider the two equivalent block diagrams in Figure 10.4.2. The input signal in Figure 10.4.2a is $X(s)$. This system has two output signals. The output signal of the lower branch is $X(s)$ while the output signal for the upper branch is

$$Y(s) = H(s)X(s). \tag{10.4.1}$$

The input signal in Figure 10.4.2b is also $X(s)$. This system also has two output signals. The upper branch is still $Y(s)$ in (10.4.1), while the output signal of the lower branch is

$$X(s) = \frac{1}{H(s)}\big(H(s)X(s)\big) = X(s). \tag{10.4.2}$$

Since both of the systems in Figure 10.4.2a and Figure 10.4.2b have the same input and output signals, they must be equivalent.

▲▲

10.4.2 Examples of Block Diagram Reduction

The following examples illustrate the use of block diagram reduction for the calculation of the transfer function for a system from the block diagram.

Example 10.4.3: In the last section we developed the block diagram of a proof-mass actuator attached to a structure shown in Figure 10.3.8. This block diagram is shown in Figure 10.3.13. We can use the block diagram in Figure 10.3.13 to find the transfer function from the electrical input signal to the actuator, $V_{pm}(s)$, to the electrical output signal of the sensor, $Y_r(s)$. Now we see that this system is of a parallel interconnection structure, System Structure 10.2.5, between the dynamics of the proof-mass and the dynamics of the structure. To find the transfer function for this system we can use the results of the parallel interconnection structure and the cascade interconnection structure to write this transfer function directly from Figure 10.3.13. The result is

$$\frac{Y_r(s)}{V_{pm}(s)} = K_s \left(\frac{1}{m_{pm}s^2} - \frac{-\dfrac{1}{m_{st}}}{s^2 + \dfrac{c}{m_{st}}s + \dfrac{k}{m_{st}}} \right) K_{ef}. \tag{10.4.3}$$

▲▲

The next example illustrates the use of the feedback loop to reduce a block diagram.

Example 10.4.4: Feedback Control of the Proof-Mass Actuator The proof-mass actuator as discussed in Section 6.5 must be modified for use in a vibration suppression loop. The form of the modification is shown in Figure 10.4.3. From the block diagram we can see the physical changes that have been made to the proof-mass actuator. The modified proof-mass actuator has two additional feedback[1] loops. The pick-off point of the outer loop is from the position of the proof-mass. This pick-off point indicates the presence of a sensor that has an electronic output signal that is proportional to the position of the proof-mass. This electronic signal is subtracted from an external reference signal using an op amp, say. The inner feedback loop has a pick-off point from the velocity of the proof-mass. This pick-off point indicates the presence of a sensor that has an electronic output signal that is proportional to the velocity of the proof-mass. This electronic signal is amplified with a gain of K_{va} and electronically subtracted from the signal from the outer feedback loop. This signal is amplified with a gain of K_{pa} and used as an input signal to the motor.

To find the transfer function from the input signal $R_{pm}(s)$ to the displacement of the proof-mass, $Y_{pm}(s)$, the block diagram reduction is performed in two steps. First, the inner loop is reduced through the use of the feedback interconnection structure. The transfer function of the inner loop is calculated as

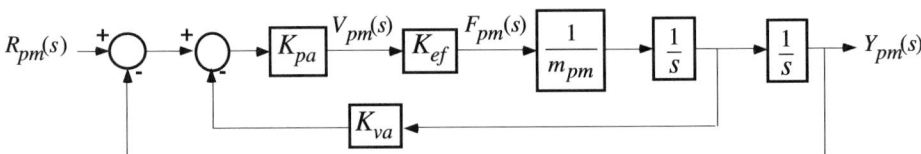

Figure 10.4.3 Modified Proof-Mass Actuator

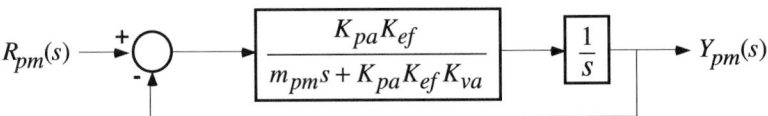

Figure 10.4.4 Block Diagram Reduction of the Modified Proof-Mass Actuator

---

[1] See System Structure 10.2.6.

$$\frac{\dfrac{K_{pa}K_{ef}}{m_{pm}s}}{1+(K_{va})\left(\dfrac{K_{pa}K_{ef}}{m_{pm}s}\right)} = \frac{K_{pa}K_{ef}}{m_{pm}s + K_{pa}K_{ef}K_{va}}.$$ (10.4.4)

The reduced block diagram is shown in Figure 10.4.4. Reducing the outer loop in Figure 10.4.4, the transfer function of this system is

$$\frac{Y_{pm}(s)}{R_{pm}(s)} = \frac{\dfrac{K_{pa}K_{ef}}{\left(m_{pm}s + K_{pa}K_{ef}K_{va}\right)(s)}}{1+\dfrac{K_{pa}K_{ef}K_{va}}{\left(m_{pm}s + K_{pa}K_{ef}K_{va}\right)(s)}} = \frac{\dfrac{K_{pa}K_{ef}}{m_{pm}}}{s^2 + \dfrac{K_{pa}K_{ef}}{m_{pm}}s + \dfrac{K_{pa}K_{ef}}{m_{pm}}}.$$ (10.4.5)

▲▲

Example 10.4.5: LRC Network In the last section we discussed the network shown in Figure 10.3.1 and Figure 10.3.2. The block diagram of the network is shown in Figure 10.3.7. Next we will show how block diagram reduction can be used to find the transfer function of this system. To be concrete we assume that the parameter values of the network to be: $R_1 = 1\ \Omega$, $R_2 = 1\ \Omega$, $C_1 = 1$ F, $C_2 = 1$ F, and $L = 1$ H. Using these parameter values the transfer functions in the individual blocks in Figure 10.3.7 are shown in Table 10.4.1. The block diagram in Figure 10.3.7 is shown in Figure 10.4.5 using the transfer functions in Table 10.4.1.

The block diagram reduction begins by reducing the feedback loop at the output voltage of the diagram. The reduced block diagram is shown in Figure 10.4.6 where

$$G_{45} = \frac{G_4}{1+G_4G_5} = \frac{1}{s+1}.$$ (10.4.6)

Block diagram reduction proceeds by first moving the pickoff point at $I_2(s)$ to $Y(s)$ as shown in Figure 10.4.7. Then the feedback loop in Figure 10.4.7 is reduced as shown in Figure 10.4.8 where

$$G_{345} = \frac{G_3G_{45}}{1+G_3G_{45}} = \frac{\left(\dfrac{1}{s}\right)\left(\dfrac{1}{s+1}\right)}{1+\left(\dfrac{1}{s}\right)\left(\dfrac{1}{s+1}\right)} = \frac{1}{s^2+s+1}.$$ (10.4.7)

Table 10.4.1 Parameter Values for the LRC Network

$G_1(s)$	$G_2(s)$	$G_3(s)$	$G_4(s)$	$G_5(s)$
$\dfrac{1}{R_1}$	$\dfrac{1}{sC_1}$	$\dfrac{1}{sL}$	$\dfrac{1}{sC_2}$	$\dfrac{1}{R_2}$
1	$\dfrac{1}{s}$	$\dfrac{1}{s}$	$\dfrac{1}{s}$	1

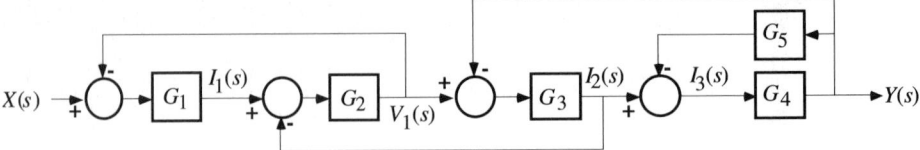

Figure 10.4.5 Block Diagram of Figure 10.3.7 Using Table 10.4.1

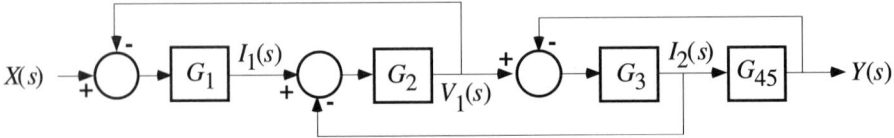

Figure 10.4.6 First Step in the Reduction of Block Diagram of Figure 10.4.5

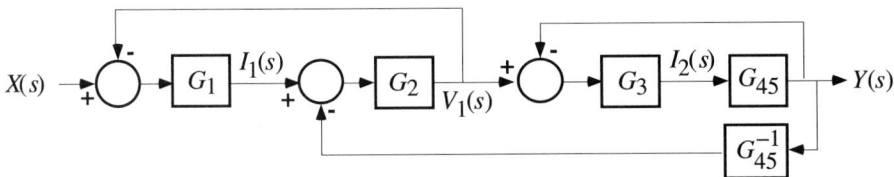

Figure 10.4.7 Second Step in the Reduction of the Block Diagram

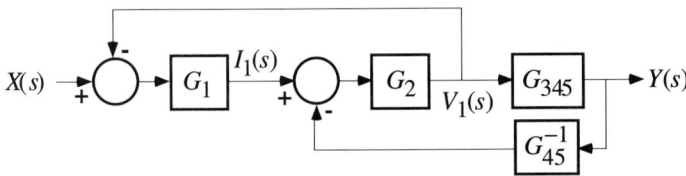

Figure 10.4.8 Third Step in the Reduction of the Block Diagram

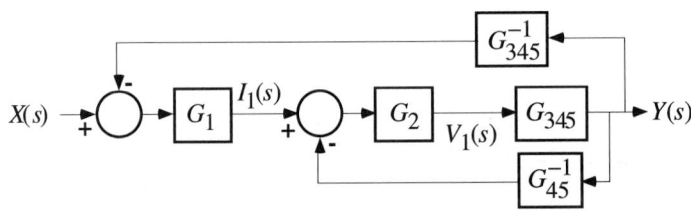

Figure 10.4.9 Fourth Step in the Reduction of the Block Diagram

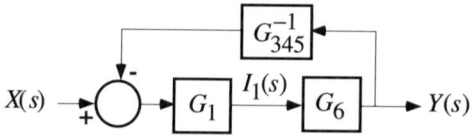

Figure 10.4.10 Final Step in the Reduction of the Block Diagram

Next, the pickoff point at $V_1(s)$ is moved to $Y(s)$ as shown in Figure 10.4.9. Again, the feedback loop is reduced so that the block diagram is as shown in Figure 10.4.10 where

$$G_6 = \frac{G_2 G_{345}}{1 + G_2 G_{345} G_{45}^{-1}} = \frac{1}{(s)(s^2 + s + 1) + (s + 1)}. \tag{10.4.8}$$

Now the transfer function is obtained by reducing the feedback loop in Figure 10.4.10. We get

$$\frac{Y(s)}{X(s)} = \frac{G_1 G_6}{1 + G_1 G_6 G_{345}^{-1}}. \tag{10.4.9}$$

For the parameter values given above, the transfer function is

$$\frac{Y(s)}{X(s)} = \frac{1}{s^3 + 2s^2 + 3s + 2}. \tag{10.4.10}$$

▲▲

10.4.3 MATLAB Experiments

MATLAB has the capability to perform the algebraic calculations of systems that accompany the block diagram reduction above. These calculations are very straightforward. First, each of the subsystems of the block diagram is defined in MATLAB. Then the algebra that follows from the block diagram reduction can be carried out in MATLAB. The following M-file performs the calculations in the analysis of the *LRC* network in Example 10.4.5.

When MATLAB performs algebraic calculation with systems it doesn't necessarily carry out pole-zero calculations. It is necessary to perform these cancellations with the command **minreal**.

Example 10.4.5: (*continued*)
```
clear
% define parameters
R1 = 1; R2 = 1; C1 = 1; C2 = 1; L = 1;

% define transfer functions
G1 = tf(1/R1);
G2 = tf(1,[C1,0]);
```

```
G3 = tf(1,[L,0]);
G4 = tf(1,[C2,0]);
G5 = tf(1/R2);

% step 1
G45 = G4/(1+G4*G5);
G45 = minreal(G45);
display('G45 = ')
display(G45)

% step 2
G345 = G3*G45/(1+G3*G45);
G345 = minreal(G345);
display('G345 = ')
display(G345)

% step 3
G6 = G2*G345/(1+G2*G345*inv(G45));
G6 = minreal(G6);
display('G6 = ')
display(G6)

% step 4
H = G1*G6/(1+G1*G6*inv(G345));
H = minreal(H);
display('H = ')
display(H)
```
▲▲

Exploratory Exercise 10.4.6: Write a MATLAB M-file to perform the block diagram reductions in Example 10.4.4.
▲▲

10.5 ALL-INTEGRATOR BLOCK DIAGRAMS AND STATE SPACE REPRESENTATIONS

10.5.1 Introduction

This section has two purposes. First, we introduce a special form of a block diagram called an all-integrator block diagram. This type of block diagram finds application in the synthesis of circuits from block diagrams and in computer applications. Second, we introduce a second system representation called a state space representation. This representation is an alternative to a differential equation. A state space representation is derived from an all-integrator block diagram establishing the continuity between transfer functions and differential equations. In the next chapter we will discuss systematic procedures for the construction of all-integrator block diagrams from transfer functions.

Transfer functions and state space equations are two complementary representations for a system.   As far as system analysis is concerned, these two representations expose very different properties of the system, the transfer function being in the frequency domain and the differential equations being in the time domain.   Together, they allow a deeper understanding of the system than either of them separately.   In the analysis and design of systems, both of these representations are routinely used.

10.5.2 All-Integrator Block Diagrams

Static and Dynamic Subsystems   First, we develop a special form of block diagrams, called all-integrator block diagrams, that play an important role in development of state space equations.   These diagrams are also important in their own right.   To define these special block diagrams we must introduce a distinction between the functional form of the blocks in a block diagram.   Suppose the transfer function of a system is

$$\frac{Y(s)}{X(s)} = \frac{b_2 s^2 + b_1 s + b_0}{s^3 + a_2 s^2 + a_1 s + a_0} = H(s). \tag{10.5.1}$$

Cross-multiplying we have

$$\left(s^3 + a_2 s^2 + a_1 s + a_0\right)Y(s) = \left(b_2 s^2 + b_1 s + b_0\right)X(s). \tag{10.5.2}$$

Using the inverse Laplace transform (10.5.2) the underlying differential equation is

$$\dddot{y}(t) + a_2 \ddot{y}(t) + a_1 \dot{y}(t) + a_0 y(t) = b_2 \ddot{x}(t) + b_1 \dot{x}(t) + b_0 x(t). \tag{10.5.3}$$

We assume that the initial conditions of the differential equation (10.5.3) are zero. Equations (10.5.1) - (10.5.3) show that there is a one-to-one relationship between a transfer function that is a function of s and the underlying differential equation.

Suppose we wanted to solve (10.5.3) for $y(t)$ with zero initial conditions and for a given input signal $x(\tau), 0 \le \tau \le t$.   We could construct a solution in several ways, but note that we need the input signal $x(\tau)$ over the time interval $0 \le \tau \le t$.   Because we need the input signal over a finite (nonzero) interval of time, we say that the physical process is dynamic.   Consider, on the other hand, the system

$$y(t) = Kx(t). \tag{10.5.4}$$

To compute the output signal for this system we need the input signal at time t only. That is, we need the input signal only at a single point in time.

Definition 10.5.1:   In a block diagram, a subsystem is said to be <u>dynamic</u> if the corresponding transfer function is a function of the Laplace variable s.   If the subsystem is just a real scalar gain, the subsystem is said to be <u>static</u>.

▲▲

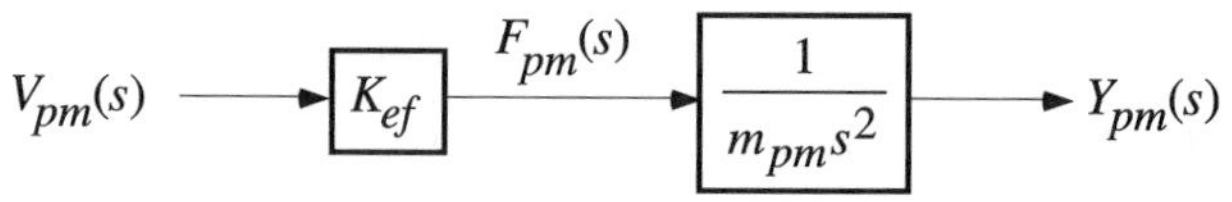

Figure 10.5.1 Block Diagram of a Proof-Mass Actuator

Example 10.5.2: Consider the block diagram of the proof-mass actuator shown in Figure 10.5.1. The subsystem

$$\frac{Y_{pm}(s)}{F_{pm}(s)} = \frac{1}{m_{pm}s^2} \tag{10.5.5}$$

is a dynamic subsystem, and the subsystem

$$\frac{F_{pm}(s)}{V_{pm}(s)} = K_{ef} \tag{10.5.6}$$

is a static subsystem.

▲▲

Integrators and All-Integrator Block Diagrams   The system shown in Figure 10.5.2 is an important subclass of blocks in a block diagram. The block diagram in Figure 10.5.2 corresponds to the Laplace transform equations

$$Y(s) = \frac{1}{s} X(s). \tag{10.5.7}$$

The corresponding signals in the time domain are

$$y(t) = \int_0^t x(\lambda)\, d\lambda \tag{10.5.8}$$

where Property 9.2.5 of the Laplace transform was used. The block in Figure 10.5.2 in important enough to carry a name.

Definition 10.5.3: The block diagram in Figure 10.5.2 is called an <u>integrator</u>.

▲▲

Using only integrators and static blocks we can build up a block diagram.

Definition 10.5.4: A block diagram is an <u>all-integrator block diagram</u> if the only dynamic blocks are integrators.

▲▲

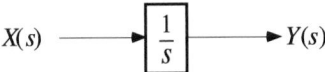

Figure 10.5.2 An Integrator

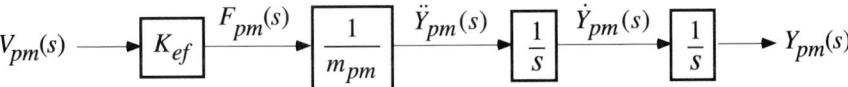

Figure 10.5.3 All-Integrator Block Diagram of a Proof-Mass Actuator

Example 10.5.5:   Consider the proof-mass actuator in Figure 10.5.1.   An all-integrator block diagram of this system can be constructed by factoring the transfer function. This all-integrator block diagram is shown in Figure 10.5.3.

▲▲

10.5.3 State Space Equations

We know that the transfer function is extracted from the differential equation representation of the system as stated in Theorem 10.1.4.   There is an alternative representation of the differential equation that is more commonly used, however. Next we will develop this system representation from an all-integrator block diagram.

There is another way to label the signals of an integrator in Figure 10.5.2. Taking the time derivative of (10.5.8) yields

$$\frac{dy}{dt} = x(t). \tag{10.5.9}$$

Equation (10.5.9) shows that the input signal is the derivative of the output signal. Hence, the block diagram in Figure 10.5.2 can be labeled as in Figure 10.5.4. In fact the labeling of the integrator in Figure 10.5.4 suggests the relationship (10.5.9) because the derivative of a signal is explicitly shown.

Notation: It has been common practice to label the Laplace transform of the input signal and the output signal in a block diagram by the corresponding signal in the time domain as shown in Figure 10.5.4. While this practice is technically incorrect, we shall frequently use it.

To start the development of the second system representation, we consider the proof-mass actuator in Example 10.5.5.

Example 10.5.6: In this example we will construct the state space equations for the proof-mass actuator. The block diagram of this system is shown in Figure 10.5.5. Using Figure 10.5.4 we have labeled the output signals of each integrator with the variables q_i, $i = 1,2$. We have also labeled the input signals to the integrators with the derivatives of the output signals in Figure 10.5.5. When we write the equations that express the relationship between the signals from the block diagram we obtain

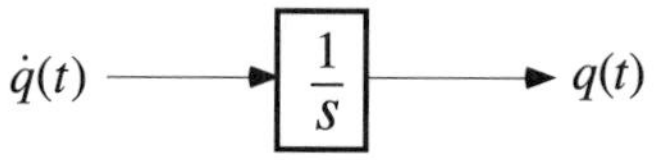

$$\dot{q}(t) \longrightarrow \boxed{\dfrac{1}{s}} \longrightarrow q(t)$$

Figure 10.5.4 An Integrator with the Input and Output Signals Relabeled

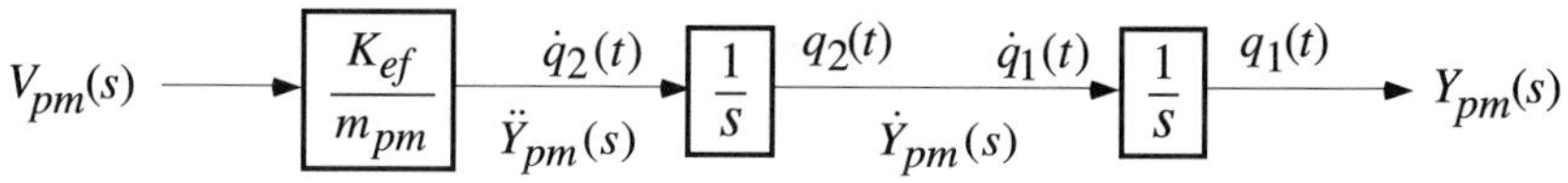

$$V_{pm}(s) \longrightarrow \boxed{\dfrac{K_{ef}}{m_{pm}}} \xrightarrow[\ddot{Y}_{pm}(s)]{\dot{q}_2(t)} \boxed{\dfrac{1}{s}} \xrightarrow[\dot{Y}_{pm}(s)]{q_2(t)} \xrightarrow{\dot{q}_1(t)} \boxed{\dfrac{1}{s}} \xrightarrow{q_1(t)} Y_{pm}(s)$$

Figure 10.5.5 All-Integrator Block Diagram of a Proof-Mass Actuator
with State Variables Assigned

$$\dot{q}_2(t) = \frac{1}{m_{pm}} f_{pm}(t), \tag{10.5.10}$$

$$\dot{q}_1(t) = q_2(t),$$

$$y_{pm}(t) = q_1(t).$$

Combining the equations in (10.5.10) we obtain

$$\ddot{y}_{pm} = \ddot{q}_1 = \dot{q}_2 = \frac{1}{m_{pm}} f_{pm}(t) = \frac{1}{m_{pm}} K_{ef} v_{pm}(t), \tag{10.5.11}$$

and so we recover the original differential equation (6.5.3).

Next we introduce another representation of the differential equation in (10.5.11). To that end we rewrite (10.5.10) by inserting each of the signals $q_i(t)$, $i = 1,2$, in each differential equation in (10.5.10) with zero coefficients. We obtain

$$\dot{q}_1(t) = 0q_1(t) + q_2(t) + 0v_{pm}(t), \tag{10.5.12}$$

$$\dot{q}_2(t) = 0q_1(t) + 0q_2(t) + \frac{K_{ef}}{m_{pm}} v_{pm}(t),$$

$$y_{pm}(t) = q_1(t).$$

The only signals that appear in (10.5.12) are the q_i, $i = 1,2$, the input signal, and the output signal. Next we define the vector

$$\vec{q}(t) = \begin{bmatrix} q_1(t) \\ q_2(t) \end{bmatrix}. \tag{10.5.13}$$

Using (10.5.13), (10.5.12) is rewritten into the matrix form. We have

$$\dot{\vec{q}}(t) = \begin{bmatrix} 0 & 1 \\ 0 & 0 \end{bmatrix} \vec{q}(t) + \begin{bmatrix} 0 \\ \dfrac{K_{ef}}{m_{pm}} \end{bmatrix} v_{pm}(t), \tag{10.5.14}$$

$$y_{pm}(t) = \begin{bmatrix} 1 & 0 \end{bmatrix} \vec{q}(t) + 0 v_{pm}(t).$$

Note: As a check, when the matrix multiplications in (10.5.14) are carried out, the resulting scalar equations should be the same as the equations in (10.5.12).   ▲▲

The matrix equations in (10.5.14) form the system representation in which we are interested.   The next example illustrates how to construct such a differential equation from any all-integrator block diagram.

Example 10.5.7: Consider the all-integrator block diagram in Figure 10.5.6.
Step 1 We start with an all-integrator block diagram in Figure 10.5.6.
Step 2 In Figure 10.5.6 we have labeled the output signal of all of the integrators with $q_i(t)$ and input signal with the derivatives $\dot{q}_i(t)$ for i = 1,2.
Step 3 Next we express the derivatives $\dot{q}_i(t)$ in terms of the signals in the block diagram. From Figure 10.5.6 we have

$$\dot{q}_1(t) = a_{11}q_1(t) + a_{12}q_2(t) + b_1 x(t), \tag{10.5.15}$$

$$\dot{q}_2(t) = a_{21}q_1(t) + a_{22}q_2(t) + b_2 x(t),$$

$$y(t) = c_1 q_1(t) + c_2 q_2(t) + dx(t).$$

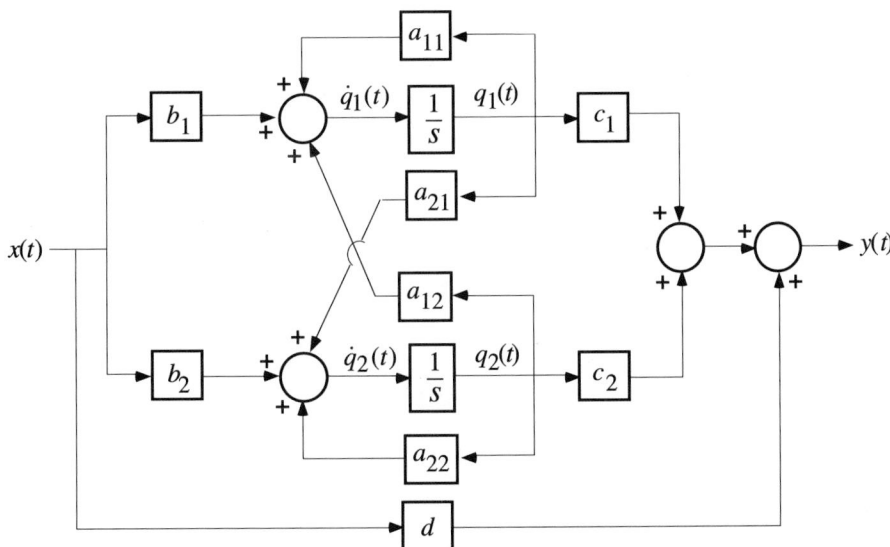

Figure 10.5.6 All-Integrator Block Diagram

The coupled differential equations in (10.5.15) represent the system shown in Figure 10.5.6. These equations are expressed in matrix form.

Step 4 Define the vector $\vec{q}(t)$ as

$$\vec{q}(t) = \begin{bmatrix} q_1(t) \\ q_2(t) \end{bmatrix}. \tag{10.5.16}$$

Now using (10.5.16) we rewrite the equations in (10.5.15) in matrix form. It follows that

$$\dot{\vec{q}}(t) = \begin{bmatrix} \dot{q}_1(t) \\ \dot{q}_2(t) \end{bmatrix} = \begin{bmatrix} a_{11} & a_{12} \\ a_{21} & a_{22} \end{bmatrix} \vec{q}(t) + \begin{bmatrix} b_1 \\ b_2 \end{bmatrix} x(t) \tag{10.5.17}$$

$$y(t) = \begin{bmatrix} c_1 & c_2 \end{bmatrix} \vec{q}(t) + [d]x(t).$$

The equations in (10.5.17) can be written in the more compact form

$$\dot{\vec{q}}(t) = A\vec{q}(t) + Bx(t), \tag{10.5.18}$$

$$y(t) = C\vec{q}(t) + Dx(t),$$

where

$$A = \begin{bmatrix} a_{11} & a_{12} \\ a_{21} & a_{22} \end{bmatrix}, \quad B = \begin{bmatrix} b_1 \\ b_2 \end{bmatrix}, \tag{10.5.19}$$

$$C = \begin{bmatrix} c_1 & c_2 \end{bmatrix}, \quad D = [d].$$

▲▲

The set of equations in (10.5.18) is the system representation in which we are interested. The system representation in (10.5.18) is completely defined by the matrices in (10.5.19).

Definition 10.5.8: The set of equations

$$\dot{\vec{q}}(t) = A\vec{q}(t) + Bx(t),$$

$$y(t) = C\vec{q}(t) + Dx(t),$$

is called a state space representation of a system. The vector $\vec{q}(t)$ is called the state vector, or the state of the system (10.5.18). The components of the state vector $q_i(t)$, $i = 1,\ldots,n$ are called the state variables. We say the system is of order n. We call A the state matrix. We call B the input matrix. We call C the output matrix. We call D the direct feedthrough matrix.

▲▲

Terminology: State space representations are also called <u>state (space) equations</u>, <u>state space models</u>, and <u>state variable equations</u>.

In Example 10.5.7 we constructed a state space representation from an all-integrator block diagram.   This procedure will translate any all-integrator block diagram into a state space representation.

Procedure 10.5.9:   Construction of State Space Representation from an All-Integrator Block Diagram

Step 1 Develop an all-integrator block diagram.[2]

Step 2 To the output of the ith integrator, assign the state variable $q_i(t)$. It follows that the input signal to the integrator is $\dot{q}_i(t)$. The input signal is so labeled.

Step 3 From the block diagram express the derivatives of the state variables $\dot{q}_i(t)$ in terms of the other state variables $q_i(t)$ and the input signal $x(t)$. Also express the output signal $y(t)$ in terms of the state variables and the input signal.

Step 4 Assemble the scalar state equations into a matrix representation, (10.5.18). ▲▲

In constructing state equations use only the signals $\dot{q}_i(t)$, $q_i(t)$, $x(t)$, and $y(t)$. All other signals are excluded. Procedure 10.5.9 is completely general and it will translate any all-integrator block diagram into differential equations of the form (10.5.18).   The next example illustrates the construction of a state space representation from a block diagram using Procedure 10.5.9.

Example 10.5.10: Consider the system in Figure 10.5.7. Following Procedure 10.5.9 we will construct the state space representation.

Step 1   The block diagram in Figure 10.5.7 is already an all-integrator block diagram.

Step 2 The state variables have been assigned to each integrator in Figure 10.5.7.

Step 3 From Figure 10.5.7 the scalar equations are

$$\dot{q}_1(t) = q_2(t), \qquad\qquad\qquad (10.5.20)$$
$$\dot{q}_2(t) = -\omega^2 q_1(t) + b_2 x(t),$$
$$\dot{q}_3(t) = -a q_3(t) + b_3 x(t),$$
$$y(t) = c_1 q_1(t) + c_3 q_3(t) + d x(t).$$

Step 4 Define the vector

$$\vec{q}(t) = \begin{bmatrix} q_1(t) \\ q_2(t) \\ q_3(t) \end{bmatrix}. \qquad\qquad\qquad (10.5.21)$$

---

[2] In the next chapter we will describe how to translate any block diagram into an all-integrator block diagram.

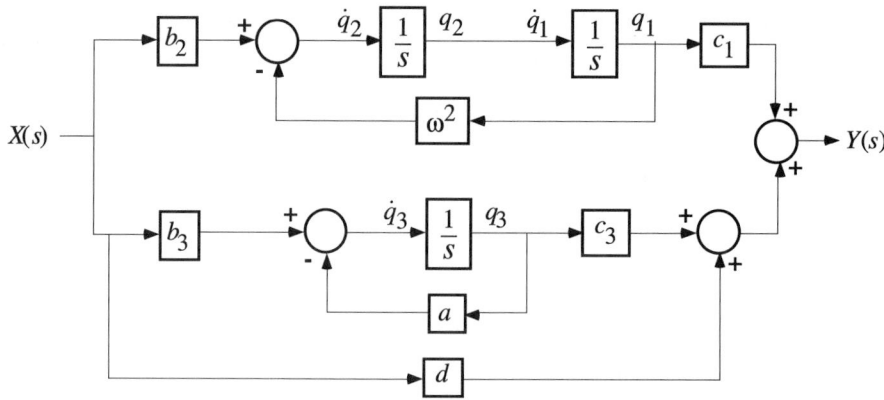

Figure 10.5.7 Block Diagram of Example 10.5.10

Rearranging (10.5.20) into matrix form, we get the state space representation

$$\begin{bmatrix} \dot{q}_1(t) \\ \dot{q}_2(t) \\ \dot{q}_3(t) \end{bmatrix} = \begin{bmatrix} 0 & 1 & 0 \\ -\omega^2 & 0 & 0 \\ 0 & 0 & -a \end{bmatrix} \begin{bmatrix} q_1(t) \\ q_2(t) \\ q_3(t) \end{bmatrix} + \begin{bmatrix} 0 \\ b_2 \\ b_3 \end{bmatrix} x(t),$$

(10.5.22)

$$y(t) = \begin{bmatrix} c_1 & 0 & c_3 \end{bmatrix} \begin{bmatrix} q_1(t) \\ q_2(t) \\ q_3(t) \end{bmatrix} + [d] x(t).$$

▲▲

Note: As a check, when the matrix multiplications in (10.5.22) are carried out, the resulting scalar equations should be the same as the equations in (10.5.20).

In the examples above we have arbitrarily assigned state variables to the output signals of the integrators. A second state space representation could be constructed by permuting the assignment of the state variables. It will be shown in Section 11.3 that any state variable assignment will lead to an equivalent state space representation of the block diagram. For our purposes in this text, any assignment of the state variables will suffice.

Consider the state space representations in (10.5.14), (10.5.17), and (10.5.22). All of these state space equations have similar matrix dimensions. An understanding of these matrix dimensions is essential to the use of the state space representation, particularly when the computer is employed.

Note: When constructing the state equations, the rules of matrix multiplication apply. These rules require that the matrix dimensions must be compatible.

Suppose that an all-integrator block diagram has n integrators. If we assign a state variable $q_i(t)$ to the output signal of each integrator, there will be n state

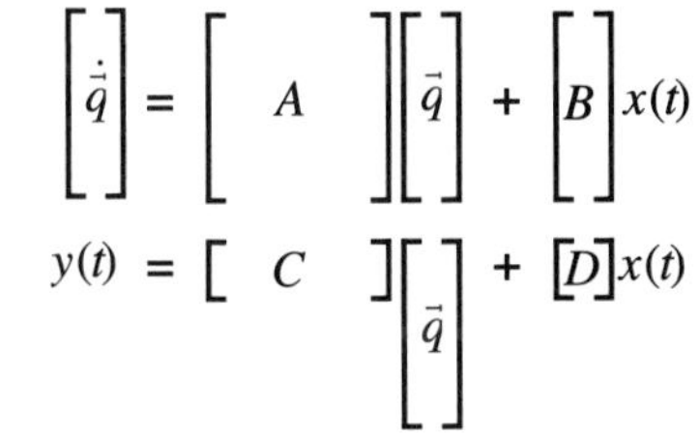

$$\left[\,\dot{\vec{q}}\,\right] = \left[\quad A \quad\right]\left[\,\vec{q}\,\right] + \left[B\right]x(t)$$

$$y(t) = \left[\quad C \quad\right]\left[\begin{array}{c} \\ \vec{q} \\ \\ \end{array}\right] + \left[D\right]x(t)$$

Figure 10.5.8 Dimensions of State Space Equations

variables. If we assign these state variables into a vector $\vec{q}(t)$ the vector will be of length n. From (10.5.18) this n-vector satisfies

$$\dot{\vec{q}}(t) = A\vec{q}(t) \tag{10.5.23}$$

if we set $x(t) = 0$. The matrix product $A\vec{q}(t)$ must be an n-vector. It follows that A will be an $n \times n$ matrix. We will usually consider systems that have one input signal $x(t)$. From

$$\dot{\vec{q}}(t) = A\vec{q}(t) + Bx(t) \tag{10.5.24}$$

we see that the product of $Bx(t)$ must be an n-vector. It follows that B must be an $n \times 1$ matrix. Similarly, we will usually focus on systems having one output signal $y(t)$. From the output equation

$$y(t) = C\vec{q}(t) + Dx(t) \tag{10.5.25}$$

we see that the matrix product $C\vec{q}(t)$ must be a scalar. So C is a $1 \times n$ matrix. For a system that has one input signal and one output signal the product of D and $x(t)$ is a scalar. It follows that D a 1×1 matrix. A visualization of the matrix dimensions is shown in Figure 10.5.8.

10.5.4 All-Integrator Block Diagrams from State Equations

Procedure 10.5.9 is used to construct state equations from block diagrams. This procedure can be reversed to construct all-integrator block diagrams from state equations.

Procedure 10.5.11: Construction of All-Integrator Block Diagrams from State Equations
Step 1 Given the state equations, multiply out the matrix equations to obtain the scalar equations.
Step 2 Draw as many integrators as there are state variables. Label the output signal of the integrator with a state variable. Label the input signal of each integrator as the derivative of the output signal of the integrator.

Step 3 Connect the input and output signals of the integrators according to the relationship expressed by the scalar equations.

▲▲

Example 10.5.12: Consider the system

$$\dot{\vec{q}}(t) = \begin{bmatrix} 0 & 1 & 0 \\ 0 & 0 & 1 \\ -a_0 & -a_1 & -a_2 \end{bmatrix} \begin{bmatrix} q_1(t) \\ q_2(t) \\ q_3(t) \end{bmatrix} + \begin{bmatrix} 0 \\ 0 \\ 1 \end{bmatrix} x(t) \tag{10.5.26}$$

$$y(t) = \begin{bmatrix} b_0 & b_1 & b_2 \end{bmatrix} \vec{q}(t).$$

We will draw the all-integrator block diagram of this system using Procedure 10.5.11.

Step 1 First we expand the matrix equations in (10.5.26) into scalar equations. We obtain

$$\begin{aligned} \dot{q}_1(t) &= q_2(t), \\ \dot{q}_2(t) &= q_3(t), \\ \dot{q}_3(t) &= -a_0 q_1(t) - a_1 q_2(t) - a_2 q_3(t) + x(t), \\ y(t) &= b_0 q_1(t) + b_1 q_2(t) + b_2 q_3(t). \end{aligned} \tag{10.5.27}$$

Step 2 Next we draw an integrator for each state variable as shown in Figure 10.5.9. The output signal to each integrator is labeled with a state variable, and the input signal to the integrator is labeled with its derivative.

$\dot{q}_1(t)$
$\frac{1}{s}$
$q_1(t)$
$\dot{q}_2(t)$
$\frac{1}{s}$
$q_2(t)$
$\dot{q}_3(t)$
$\frac{1}{s}$
$q_3(t)$

Figure 10.5.9 Second Step in Constructing an All-Integrator Block Diagram from State Equations

$\dot{q}_3(t)$
$\frac{1}{s}$
$q_3(t)$
$\dot{q}_2(t)$
$\frac{1}{s}$
$q_2(t)$
$\dot{q}_1(t)$
$\frac{1}{s}$
$q_1(t)$

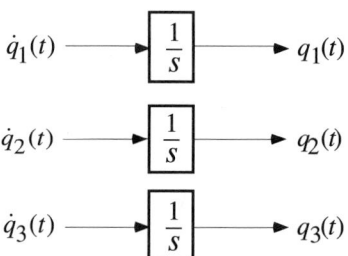

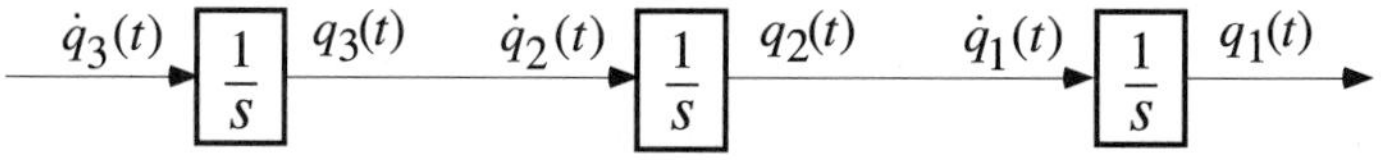

Figure 10.5.10 Interconnection Structure as Determined by the First Two Equations in (10.5.27)

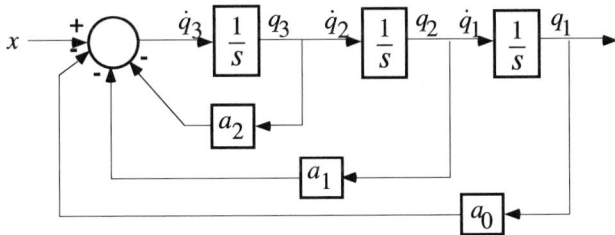

Figure 10.5.11 Interconnection Structure as Determined by the
First Three Equations in (10.5.27)

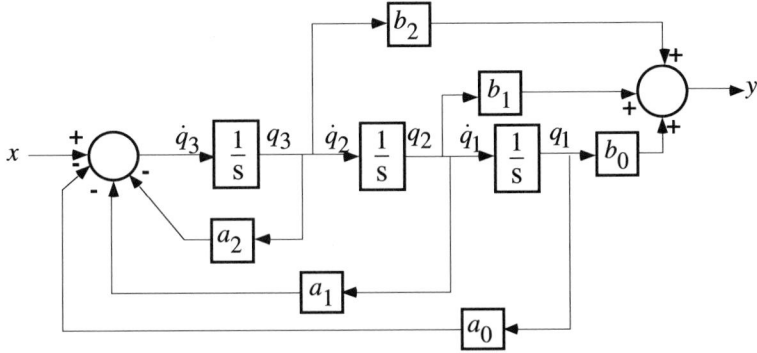

Figure 10.5.12 Block Diagram of the State Equations in (10.5.26)

Step 3. Now the block diagram is constructed by interconnecting the integrators according to the scalar equations in (10.5.27). The first two equations lead to the interconnections shown in Figure 10.5.10. The interconnections implied by the third equation in (10.5.27) is shown in Figure 10.5.11. Incorporating the output signal into Figure 10.5.11 leads to the block diagram in Figure 10.5.12. The transfer function of this system is

$$\frac{Y(s)}{X(s)} = \frac{b_2 s^2 + b_1 s + b_0}{s^3 + a_2 s^2 + a_1 s + a_0}. \tag{10.5.28}$$

This transfer function can be verified by block diagram reduction.

▲▲

10.5.5 MATLAB Experiments

Because MATLAB is based on matrix calculations, the state space representation of a system is a very natural representation in MATLAB. In many ways, it is the most convenient representation. Furthermore, the state space representation is the preferred representation for numerical calculations.

The state space representation is defined by the four matrices (A,B,C,D) in MATLAB. Like the transfer function, this system representation can be given a name with the command **ss** (state space). In all subsequent calculations, this variable is treated as a system object. The following M-file illustrates the use of the state space representation to compute the response of a system to a sinusoidal input signal.

MATLAB Version 4   In Version 4, a state space representation is simply represented as four matrices. In the commands that call for a system, the four matrices can be substituted for the system name.

Example 10.5.6: (*continued*) M-file

```
clear
% define parameters
mpm = 10;                        % proof-mass mass
Kef = 1;                         % gain
% define state space system
Apm = [0,1;0,0];                 % state matrix
Bpm = [0;Kef/mpm];               % input matrix
Cpm = [1,0];                     % output matrix
Dpm = 0;                         % direct feedthrough matrix
Hpm = ss(Apm,Bpm,Cpm,Dpm);  % define system
% simulate system
t = linspace(0,20,400);          % time vector
vpm = sin(t);                    % sinusoidal input signal
ypm = lsim(Hpm,vpm,t);           % simulate system
plot(t,ypm)
xlabel('time')
ylabel('displacement')
```

▲▲

Exploratory Exercise 10.5.13: Modify the M-file above to calculate the response of the proof-mass actuator to a square wave input signal with the same frequency as the sinusoid. Qualitatively compare the two output signals.

▲▲

Exploratory Exercise 10.5.14: Write an M-file to simulate the output signal of the system in Example 10.5.10. Assume that the input signal is a sinusoid with a frequency of 1 rad/sec. Let $\omega = 2$ and $a = 1$. How does the output signal change if $\omega = 1$?

▲▲

10.6 CHAPTER SUMMARY

10.6.1 System Representations

In this chapter we initiated the study of systems

$$y(t) = \mathcal{H}\big[x(t)\big]. \tag{10.6.1}$$

We introduced two of the major system representations we will study in this text: transfer functions and state space representations. The <u>transfer function</u> is defined (Definition 10.1.3) as the ratio of the Laplace transform of the output signal over the Laplace transform of the input signal

$$\frac{\mathcal{L}\{y(t)\}}{\mathcal{L}\{x(t)\}} = \frac{Y(s)}{X(s)} = H(s). \tag{10.6.2}$$

We also introduced a <u>state space representation</u> (Definition 10.5.8)

$$\dot{\vec{q}}(t) = A\vec{q}(t) + Bx(t), \tag{10.6.3}$$
$$y(t) = C\vec{q}(t) + Dx(t),$$

where the vector $\vec{q}(t)$ is called the <u>state vector</u>, or the <u>state</u> of the system. The components of the state vector $q_i(t)$ are called the <u>state variables</u>.

In addition to this algebraic representation of a transfer function, we also introduced a pictorial representation of the transfer function called a <u>block diagram</u> (Definition 10.2.1). Block diagrams are useful for representing pictorially the structure of a system composed of several subsystems. The three basic interconnection structures between two subsystems are: <u>cascade or series interconnection</u> (System Structure 10.2.4), <u>parallel interconnection</u> (System Structure 10.2.5), and <u>feedback loop</u> (System Structure 10.2.6). A block diagram is built up from these interconnection structures.

In this chapter we also discussed the manipulation of block diagrams using certain rules so that the underlying algebraic equations are preserved. This manipulation of block diagrams is called <u>block diagram reduction</u> (Definition 10.4.1). The rules for these transformations are summarized in Figure 10.6.1. The construction of block diagrams is described in Section 10.3 and the reduction of block diagrams is described in Section 10.4.

10.6.2 Relationship Between System Representations

One of the main themes of this text is the relationship between the various system representations. We have begun the development of these relationships in this chapter. The first result is the relationship between a differential equation and a transfer function (Theorem 10.1.4). The transfer function is obtained from the differential equation by taking the Laplace transform of the differential equation when the initial conditions are zero. The differential equation is obtained from the transfer function by cross-multiplying the transfer function and taking the inverse Laplace transform.

We also investigated the relationship between a special form of block diagrams and state space representations. This relationship depends on the block diagram representation of an <u>integrator</u> (Definition 10.5.3). In some block diagrams the only dynamic blocks are integrators. These block diagrams are called <u>all-integrator block</u>

diagrams (Definition 10.5.4). A state space representation can be derived directly from an all-integrator block diagram (Procedure 10.5.9). Briefly, the idea is to label the output signal of each integrator with a state variable, label the input signal to the integrator with the derivative of the state variable, and then write the state space equations directly from the block diagram. This procedure can be reversed to draw a block diagram from a state space representation (Procedure 10.5.11). In the next chapter we will show how to transform an arbitrary block diagram into an all-integrator block diagram, thereby completing the development describing the relationship between block diagrams and state space representations. We will also clarify the relationship between state space representations and transfer functions.

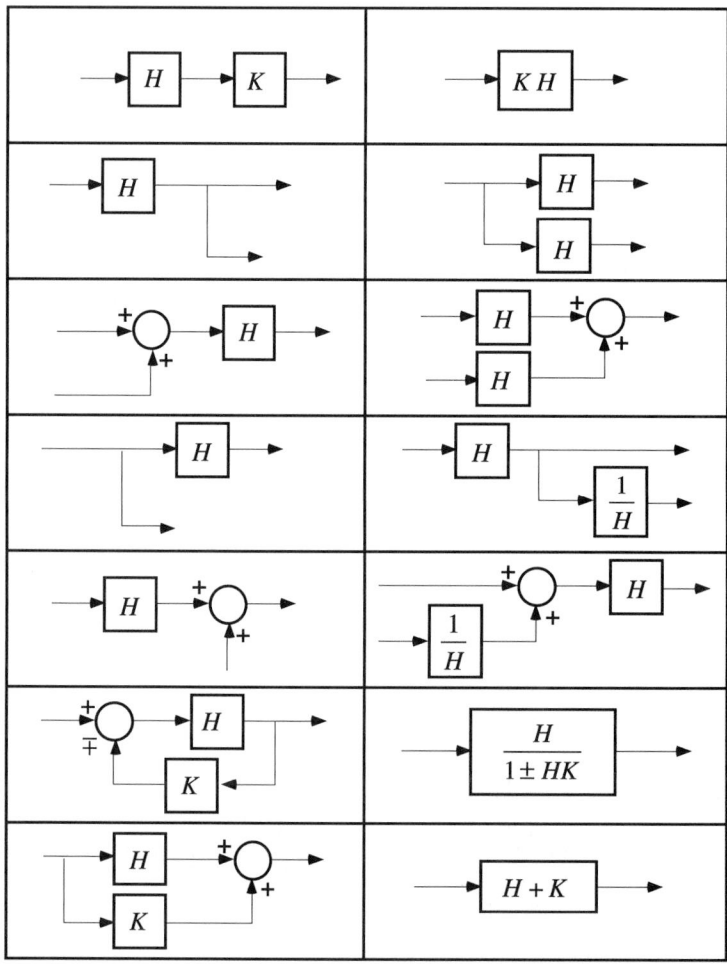

Figure 10.6.1 Block Diagram Manipulation Rules

10.7 HOMEWORK FOR CHAPTER 10

Homework Problems for Section 10.1

10.1.1 Draw the pole-zero diagrams of the following transfer functions. Label ζ and ω_n of the complex poles.

(i) $\dfrac{Y(s)}{X(s)} = \dfrac{s+1}{(s+10)(s^2+s+9)}$

(iii) $\dfrac{Y(s)}{X(s)} = \dfrac{(s+0.8)}{(s+0.5)(s^2+2.4s+4)}$

(ii) $\dfrac{Y(s)}{X(s)} = \dfrac{s(s-2)}{(s-3)(s^2+3s+15)}$

(iv) $\dfrac{Y(s)}{X(s)} = \dfrac{s}{s^2+0.4s+4}$

10.1.2 Find the transfer function of the following systems.

(i) $\ddot{y}(t) + 2\zeta\omega_n\dot{y}(t) + \omega_n^2 y(t) = b_0 x(t)$

(ii) $\ddot{y}(t) + 2\zeta\omega_n\dot{y}(t) + \omega_n^2 y(t) = b_1 \dot{x}(t)$

(iii) $\ddot{y}(t) + 2\zeta\omega_n\dot{y}(t) + \omega_n^2 y(t) = b_2 \ddot{x}(t)$

10.1.3 Find the differential equation associated with the following transfer functions.

(i) $\dfrac{Y(s)}{X(s)} = \dfrac{s+1}{(s+10)(s^2+s+9)}$

(iv) $\dfrac{Y(s)}{X(s)} = \dfrac{s}{s^2+0.4s+4}$

(ii) $\dfrac{Y(s)}{X(s)} = \dfrac{s(s-2)}{(s-3)(s^2+3s+15)}$

(v) $\dfrac{Y(s)}{X(s)} = \dfrac{s+4}{s(s+1)}$

(iii) $\dfrac{Y(s)}{X(s)} = \dfrac{s+10}{s^2+3s+2}$

(vi) $\dfrac{Y(s)}{X(s)} = \dfrac{e^{-s}}{s+1}$

10.1.4 The poles and zeros of two transfer functions are shown in Figure P10.1.4.
(a) What are the transfer functions? (Assume an arbitrary gain constant.)
(b) Express the complex poles in terms of ζ and ω_n.
(c) Find the differential equations associated with the pole-zero diagrams.

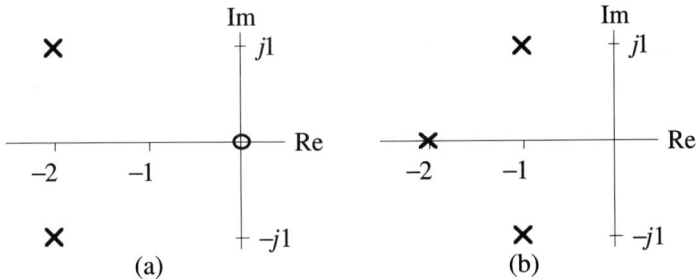

Figure P10.1.4

10.1.5 Consider the following systems.
(a) Plot the pole-zero plot of each of these systems.
(b) Plot the impulse and step response of each of the following systems.
(c) Plot the output signal of the systems when the input signal is $x(t) = (\sin t)u_s(t)$.
(d) Explain the qualitative difference between the output signals of (i) and (ii).
(e) Why does the step response of (v) approach zero?
(f) Why does the step response of (iv) go negative before it reaches it steady state value?
(g) Why do the impulse responses of (iii) - (v) oscillate while the impulse response of (i) doesn't?
(h) Explain the qualitative difference between the output signals of (v) and (vi).

(i) $\dfrac{Y(s)}{X(s)} = \dfrac{1}{(s+1)(s+2)}$ (iv) $\dfrac{Y(s)}{X(s)} = \dfrac{s-2}{s^2+s+2}$

(ii) $\dfrac{Y(s)}{X(s)} = \dfrac{1}{(s+1)(s-2)}$ (v) $\dfrac{Y(s)}{X(s)} = \dfrac{s}{s^2+s+2}$

(iii) $\dfrac{Y(s)}{X(s)} = \dfrac{1}{s^2+s+2}$ (vi) $\dfrac{Y(s)}{X(s)} = \dfrac{s-2}{s^2-s+2}$

10.1.6 Consider the system

$$\frac{Y(s)}{X(s)} = \frac{s+2}{(s+1)(s+3)}$$

Suppose that the input signal is $x(t) = e^{\alpha t}u_s(t)$.
(a) Show that the output signal contains a term of the form

$$H(\alpha)e^{j\alpha t}u_s(t).$$

(b) What happens if $\alpha = -2$?
(c) What happens if $\alpha = -1$?

10.1.7 Consider the following systems.
(a) Find the step and impulse response of each of these systems.
(b) Plot the step and impulse responses using MATLAB.
   (i) $\ddot{y}(t) + 2\dot{y}(t) + y(t) = x(t)$
   (ii) $\dddot{y}(t) + 4\ddot{y}(t) + 2\dot{y}(t) + y(t) = 3\dot{x}(t) - x(t)$
   (iii) $\dddot{y} + 3\ddot{y} + \dot{y} + 3y = \dddot{x} + 4\ddot{x} + 2\dot{x} + 7x$

10.1.8 Consider the following systems.
(a) Plot the pole-zero plot.
(b) Find and plot the step response of each system.
(c) Find and plot the impulse response of each system.

(i) $\dfrac{Y(s)}{X(s)} = \dfrac{3}{(s+10)(s+100)}$

(v) $\dfrac{Y(s)}{X(s)} = \dfrac{s+10}{s^2+3s+2}$

(ii) $\dfrac{Y(s)}{X(s)} = \dfrac{s+1}{s-3}$

(vi) $\dfrac{Y(s)}{X(s)} = \dfrac{1}{s-4}$

(iii) $\dfrac{Y(s)}{X(s)} = \dfrac{1}{s+4}$

(vii) $\dfrac{Y(s)}{X(s)} = \dfrac{s+2}{(s+6)(s+9)}$

(iv) $\dfrac{Y(s)}{X(s)} = \dfrac{100}{s^2+6s+100}$

(viii) $\dfrac{Y(s)}{X(s)} = \dfrac{1}{s(s+4)}$

10.1.9 The device shown in Figure P10.1.9 is an accelerometer.

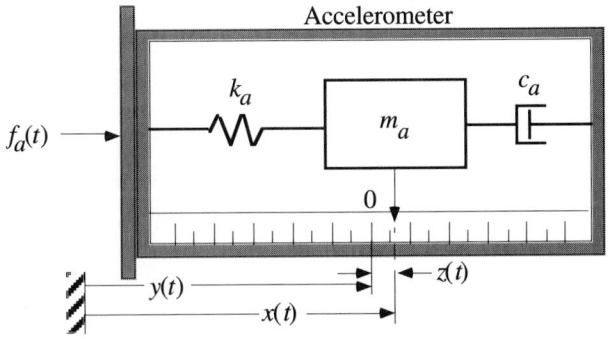

Figure P10.1.9 Accelerometer

When this device is attached to a structure that is accelerating, the output signal is a voltage which is proportional the acceleration. Let $y(t)$ be the inertial position of the structure; $x(t)$ be the inertial position of the mass m_a; $z(t) = x(t) - y(t)$ be the relative displacement between the structure and the mass; $a(t) = -K_a z(t)$ be the voltage produced by the electronics that is proportional to the acceleration of the structure; $f_a(t)$ be the force applied to the mass of the accelerometer. Summing the forces on the mass we obtain

$$m_a \ddot{x}(t) + c_a(\dot{x}(t) - \dot{y}(t)) + k_a(x(t) - y(t)) = 0.$$

Rewriting this equation in terms of $z(t)$ yields

$$m_a \ddot{z}(t) + c_a \dot{z}(t) + k_a z(t) = -m_a \ddot{y}(t) = -f_a(t).$$

For this system, take $f_a(t)$ to be the input signal and $a(t)$ to be the output signal.

(a) Starting with (i) derive (ii).

(b) Find the transfer function $\dfrac{A(s)}{F(s)}$.

(c) The parameters for this accelerometer are given in the table below. For each of the parameter sets A - C, answer the following questions.
Note: This calculation measures the accelerometer's capability to measure rapidly changing accelerations.

(i)   Find the poles of the transfer function. Express them in terms of ζ and ω_n.

(ii)   Assume that the input signal to the accelerometer is a unit step function. Plot the output signal of the accelerometer.

(iii)   How many seconds does it take the accelerometer output signal to reach the true value of the input signal?

(iv)   How many seconds does it take the accelerometer output signal to stay within 2% of the true value of the input signal?

Parameters of the Accelerometer

Parameter Set	m_a	c_a	k_a	K_a
A	1	10	100	100
B	1	5	100	100
C	1	10	1000	1000

10.1.10 Consider a system represented by the transfer function

$$\frac{Y(s)}{X(s)} = \frac{\omega_n^2}{s^2 + 2\zeta\omega_n s + \omega_n^2}$$

(a) Find the impulse and step response of this system.
(b) Write a MATLAB M-file to plot the step and impulse response based on the formulas for (a).
(c) Let $\omega_n = 1$. Plot the step response for $\zeta = 0.1, 0.3, 0.7, 0.9$. What is the difference between these step responses?
(d) Repeat (c) for the impulse response.
(e) Let $\zeta = 0.5$. Plot the step response for $\omega_n = 0.5, 1, 5$. What is the difference between these step responses?
(f) Repeat (e) for the impulse response.

10.1.11 Suppose that a system is known to be represented by a transfer function with two poles, no zeros, and a constant gain of 1. For each of the parameter values given below:
(a) Find the transfer function for each system.
(b) Plot the poles in the s-plane.
(c) Compare and contrast the response of each system.

(i) $(\zeta, \omega_n) = (.7, 3)$ (iv) $(\zeta, \omega_n) = (0.8, 10)$
(ii) $(\zeta, \omega_n) = (0.707, 1)$ (v) $(\zeta, \omega_n) = (0.15, 20)$
(iii) $(\zeta, \omega_n) = (0.01, 1)$ (vi) $(\zeta, \omega_n) = (0.5, 20)$

10.1.12 Consider the system in Figure P10.1.12.
   (a) Find the transfer function in terms of L, C, and R.
   (b) Find (ζ, ω_n) for the poles in terms of L, R, and C. Draw (by hand) the pole-zero plot of the transfer function using parameter values given. Label (ζ, ω_n).
   (c) Write a MATLAB M-file to calculate the output signal when the input signal is

$$x(t) = \Pi\left(\frac{t}{T_b}\right) + \Pi\left(\frac{t - 2T_b}{T_b}\right), \quad T_b = 10^{-2}$$

   (d) Change the bit width to $T_b = 10^{-3}$. Plot the output signal of the system. How does the output signal change?
   (e) Change the value of the resistance to $R = 100$. Calculate (ζ, ω_n) and compare to (b). Plot the output signal of the system using the input signal when $T_b = 10^{-2}$.

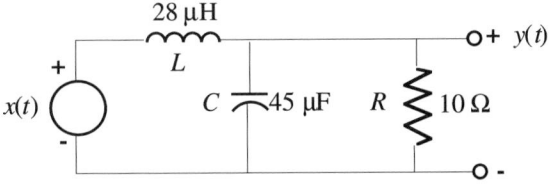

Figure P10.1.12

Homework Problems for Section 10.2

10.2.1 Consider the transfer functions

$$\frac{Y_1(s)}{X_2(s)} = H_1(s) = \frac{100}{s^2 + 6s + 100} \quad \text{and} \quad \frac{Y_2(s)}{X_2(s)} = H_2(s) = \frac{s}{s + 20}.$$

   (a) Suppose these two systems are to be connected in a cascade interconnection where $X_1(s)$ is the input signal to the composite system, and $Y_2(s)$ is the output signal. Draw the appropriate block diagram, and find the transfer function of the composite system.
   (b) Suppose these two systems are to be connected in a parallel interconnection structure. Draw the appropriate block diagram and find the transfer function of the composite system.
   (c) Suppose these two systems are to be connected in a feedback configuration where $Y(s)$ is to be the output signal from the composite system. Draw the appropriate block diagram and find the transfer function of the composite system.

10.2.2   Consider the transfer functions

$$\frac{Y_1(s)}{X_1(s)} = \frac{b(s)}{a(s)}, \quad \frac{Y_2(s)}{X_2(s)} = \frac{n(s)}{p(s)}.$$

(Remember that we always assume that the numerator and denominator polynomials will not have any factors in common.)

(a) Suppose these transfer functions are connected in a cascade interconnection? What are the zeros of the resulting composite system? When will pole-zero cancellations occur?

(b) Suppose these transfer functions are connected in a parallel interconnection? What are the zeros of the resulting composite system? When will pole-zero cancellations occur?

(c) Suppose these transfer functions are connected in a feedback interconnection? What are the zeros of the resulting composite system? When will pole-zero cancellations occur?

10.2.3   Consider two systems in a feedback configuration shown in Figure P10.2.3. The closed loop transfer function is

(i) $\dfrac{Y(s)}{R(s)} = \dfrac{H_1(s)}{1 + H_1(s)H_2(s)}$ if   (ii) $1 + H_1(s)H_2(s) \neq 0$.

If the quantity in (ii) is zero, then the mathematical expression in (i) is not defined, and the system in Figure P10.2.3 doesn't exist. We say the system is not well-defined. So before we can write the closed loop system we must verify the (ii). If one of the transfer functions is strictly proper, then (ii) is always satisfied. Find two proper transfer functions such that (ii) is not true; find two proper transfer functions such that the system is not well-defined.

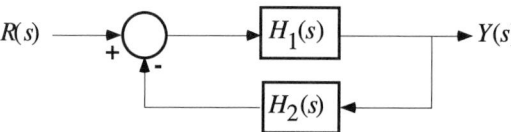

Figure P10.2.3

10.2.4   Consider the feedback system in Figure P10.2.3. Let the transfer function of this system be $T(s)$. Consider again the feedback system in Figure P10.2.3, but with a different signal identified as the output signal. This system is shown in Figure P10.2.4.

(a) Find the transfer function

$$\frac{Z(s)}{X(s)} = S(s)$$

(b) Show that $T(s) + S(s) = 1$.

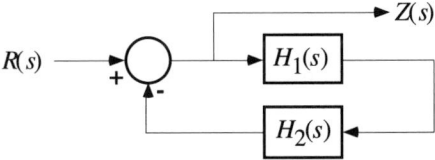

Figure P10.2.4

10.2.5 Consider the transfer functions

$$\frac{Y_1(s)}{X_1(s)} = H_1(s) = \frac{1}{s+1}, \quad \frac{Y_2(s)}{X_2(s)} = H_2(s) = \frac{1}{s}, \quad \frac{Y_3(s)}{X_3(s)} = H_3(s) = \frac{1}{s-1}.$$

Suppose the first two transfer functions are to be interconnected in a cascade connection. Then this cascade interconnection is to be joined with the third transfer function in a feedback interconnection. The output signal of the complete system in $Y_2(s)$.

(a) Draw the appropriate block diagram.
(b) Find the transfer function of the complete system.

10.2.6 Consider the system in Figure P10.2.6.
(a) Find the transfer function $Y(s)/R(s) = G(s)$.
(b) Draw the pole-zero plot that describes this system. Label (ζ, ω_n).
(c) Suppose that the input signal is $r(t) = e^{-4t}u_s(t)$. Find the output signal.
The signal $e^{-4t}u_s(t)$ doesn't appear in the output signal. Why?

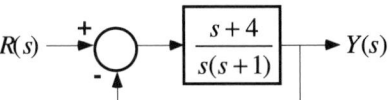

Figure P10.2.6

Homework Problems for Section 10.3

10.3.1 Consider the capacitor in Figure P10.3.1.
(a) Find the transfer function $V(s)/I(s)$.
(b) Draw the block diagram of this transfer function.
Note: Use this example to compare the block diagram to the circuit element diagram.

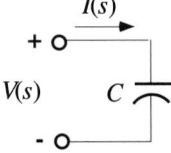

Figure P10.3.1

10.3.2   Consider the network in Figure P10.3.2.

(a) Show that the following equations describe this network.   The charge delivered to the capacitor is $Q_C(s)$.

$$I(s) = \frac{1}{sL}[V_s(s) - V_a(s)]$$

$$I(t) = I_r(s) + I_c(s)$$

$$I_r(s) = \frac{V_a(s)}{R}$$

$$V_a(s) = \frac{Q_c(s)}{C} = \frac{I(s)}{sC}$$

(b) Draw a block diagram of this system using these equations.   Take the output signal to be $V_a(s)$.

(c) Find the transfer function from the voltage $v_s(t)$ to the output voltage $v_a(t)$.

(d) Find the transfer function from the voltage $v_s(t)$ to the current $i(t)$.

(e) Find the transfer function from the voltage $v_s(t)$ to the current $i_c(t)$.

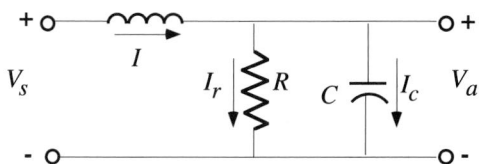

Figure P10.3.2

10.3.3   Consider the network in Figure P10.3.3.

(a) Find the transfer function from the input voltage $x(t)$ to the output voltage $y(t)$.

(b) Find the transfer function from the voltage $x(t)$ to the current $i_2(t)$.

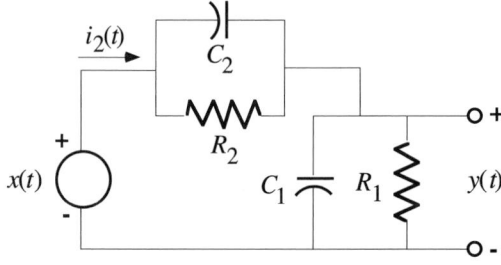

Figure P10.3.3

10.3.4 Consider the network in Figure P10.3.4. Find the transfer function from the input voltage $x(t)$ to the output voltage $y(t)$.

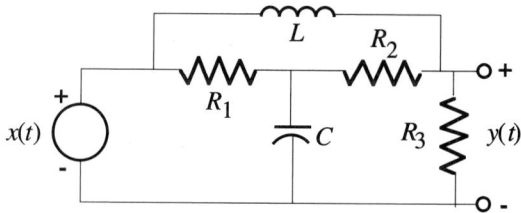

Figure P10.3.4

10.3.5 Consider the network in Figure P10.3.5. Find the transfer function from the input voltage $x(t)$ to the output voltage $y(t)$.

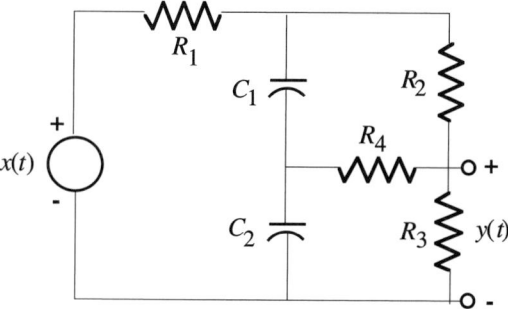

Figure P10.3.5

10.3.6 Consider the network in Figure P10.3.6. Find the transfer function from the input current $x(t)$ to the output voltage $y(t)$.

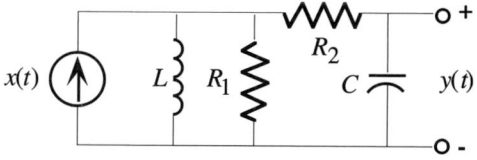

Figure P10.3.6

10.3.7 Consider the network in Figure P10.3.7.
   (a) Find the transfer function from the current $x(t)$ to the voltage $y(t)$.
   (b) Find the transfer function from the input current to the current through the resistor, $i_r(t)$.
   (c) Find the transfer function from the input current to the current through the resistor, $i_C(t)$.

(d) Find the transfer function from the input current to the current through the inductor, $i_L(t)$.

Let $R = 1$ kΩ, $C = 2$ μF, and $L = 5$ mH.

(e) Plot the impulse and step response of each of the systems in parts (a) - (d).

(f) Suppose the input signal is $x(t) = [\sin(\omega_0 t)]u_s(t)$. If $\omega_0 = 10^4$ rad/sec, plot the output signal for each of the transfer functions in parts (a) - (d).

(g) Notice that the output signals become sinusoidal after an initial transient. What is the amplitude of this oscillation?

(h) What is the relative phase between the output signals (a) - (d)?

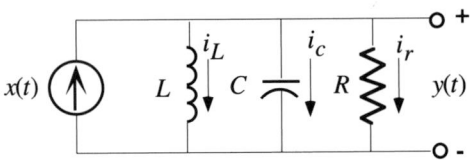

Figure P10.3.7

10.3.8 Consider the network in Figure P10.3.8.

(a) Find the transfer function from the input voltage $x(t)$ to the output voltage $y(t)$.

(b) Find the transfer function from the input voltage to the current through the resistor, $i_r(t)$.

(c) Find the transfer function from the input voltage to the voltage across the resistor, $v_c(t)$.

(d) Find the transfer function from the input voltage to the voltage across the inductor, $v_L(t)$.

Let $R = 0.1$ Ω, $C = 30$ μF, and $L = 16$ μH.

(e) Plot the impulse and step response of each of the systems in (a) - (d).

(f) Suppose the input signal is $x(t) = [\sin(\omega_0 t)]u_s(t)$. If $\omega_0 = 5 \times 10^4$ rad/sec, plot the output signal for each of the transfer functions in parts (a) - (d).

(g) Notice that the output signals become sinusoidal after an initial transient. What is the amplitude of this oscillation?

(h) What is the relative phase between the output signals (a) - (d)?

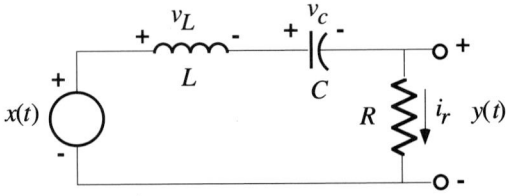

Figure P10.3.8

10.3.9 Consider the mass-spring-damper system in Figure P10.3.9.
(a) Show that the equations that govern the motion of this system are

$$m_1\ddot{y}_1(t) = -c_1\dot{y}_1(t) - c_2(\dot{y}_1(t) - \dot{y}_2(t)) - k_1y_1(t) - k_2(y_1(t) - y_2(t)) + f(t)$$
$$m_2\ddot{y}_2(t) = -c_2(\dot{y}_2(t) - \dot{y}_3(t)) - c_3(\dot{y}_2(t) - \dot{y}_3(t)) - k_2(y_1(t) - y_2(t))$$
$$-k_3(y_2(t) - y_3(t))$$
$$m_3\ddot{y}_3(t) = -c_3(\dot{y}_3(t) - \dot{y}_2(t)) - k_3(y_3(t) - y_2(t))$$

(b) Draw a block diagram of this system.

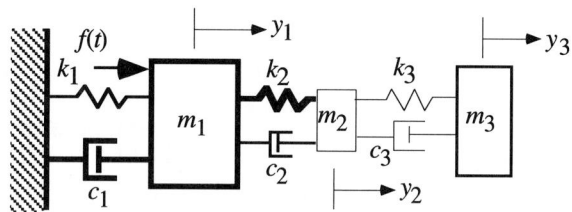

Figure P10.3.9

Homework Problems for Section 10.4

10.4.1 Find the transfer function in Figure P10.4.1 by block diagram manipulation.

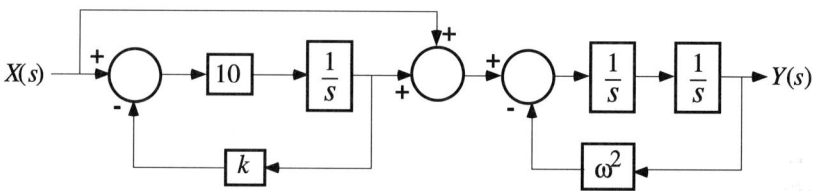

Figure P10.4.1

10.4.2 Find the transfer function in Figure P10.4.2 by block diagram reduction.

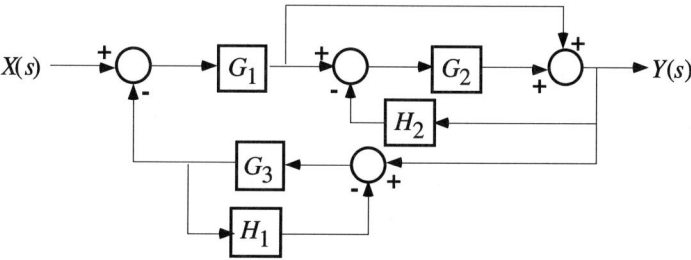

Figure P10.4.2

10.4.3 Find the transfer function in Figure P10.4.3 by block diagram reduction.

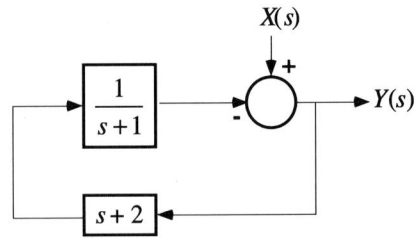

Figure P10.4.3

10.4.4 Find the transfer function in Figure P10.4.4 by block diagram reduction.

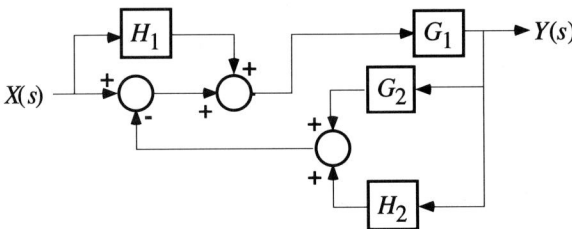

Figure P10.4.4

10.4.5 Consider the system in Figure P10.4.5.

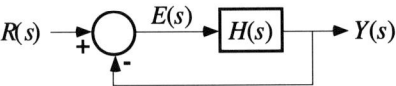

Figure P10.4.5

Assume that the closed loop transfer function has all of its poles in the open LHP.

(a) Suppose the transfer function is

$$\frac{Y(s)}{E(s)} = H(s) = \left(\frac{1}{s}\right)H_1(s)$$

where all possible pole-zero cancellations have been carried out. If $r(t) = u_s(t)$, show that $e(t) \to 0$ as $t \to \infty$.

(b) Suppose the transfer function is

$$\frac{Y(s)}{E(s)} = H(s) = \left(\frac{1}{s^2}\right)H_1(s)$$

where all possible pole-zero cancellations have been carried out. If the input signal is the unit ramp function, $r(t) = r_p(t)$, show that $e(t) \to 0$ as $t \to \infty$.

10.4.6 Find the transfer function in Figure P10.4.6 by block diagram reduction.

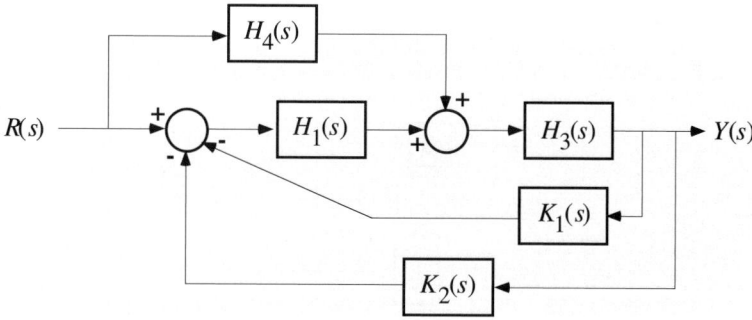

Figure P10.4.6

10.4.7 Find the transfer function in Figure P10.4.7 by block diagram reduction.

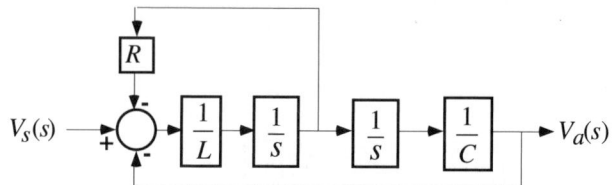

Figure P10.4.7

10.4.8 Find the transfer function in Figure P10.4.8 by block diagram reduction.

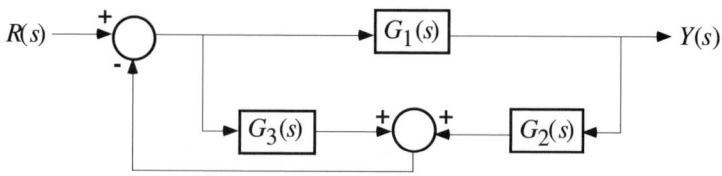

Figure P10.4.8

10.4.9 Verify all of the block diagram manipulations in Figure 10.6.1.

Homework Problems for Section 10.5

10.5.1   Consider the block diagram in Figure P10.5.1.
   (a) Find the transfer function by block diagram reduction.
   (b) Assign state variables to the block diagram and write the state equations.

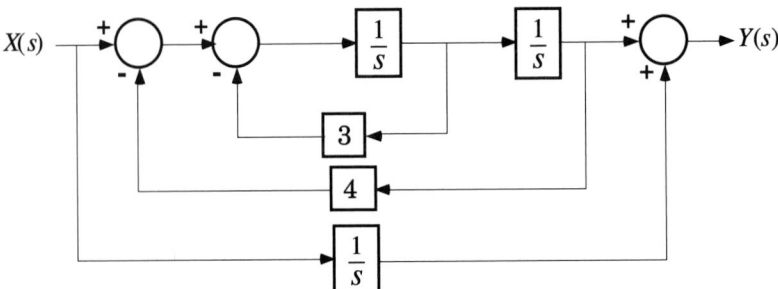

Figure P10.5.1

10.5.2   Consider the block diagram in Figure P10.5.2.
   (a) Find the transfer function by block diagram reduction.
   (b) Assign state variables to the block diagram and write the state equations.

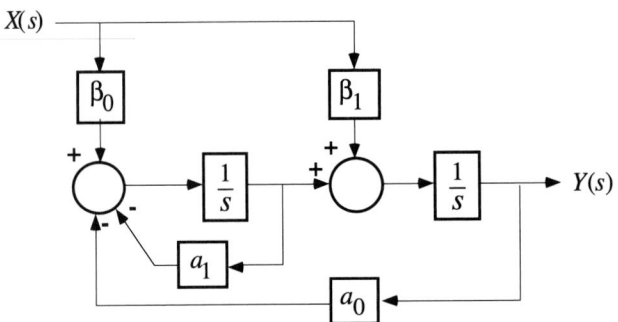

Figure P10.5.2

10.5.3   Consider the systems below represented by the state equations.
   (a) For each system draw an all-integrator block diagram.
   (b) Find the transfer function using block diagram reduction.
   (c) For the systems in (i) - (v), enter each system into MATLAB and plot the step response.
   (d) For the system in (vi), let $\sigma = 0.5$ and $\omega = 1$. Plot the step response. Then let $\sigma = 0.05$ and plot the step response. Compare these two graphs and explain the differences based on the poles.
   (e) For the systems in (vii) and (viii), let $\zeta = 0.3$ and $\omega_n = 1$. Plot the step response for each system and compare the results. Can you explain the differences?

(i) $\dot{\vec{q}} = \begin{bmatrix} 1 & 0 \\ 0 & -1 \end{bmatrix} q + \begin{bmatrix} 3 \\ 5 \end{bmatrix} x,$

$y = \begin{bmatrix} 2 & 1 \end{bmatrix} q.$

(v) $\dot{\vec{q}} = \begin{bmatrix} 3 & 5 \\ 0 & 7 \end{bmatrix} \vec{q} + \begin{bmatrix} 0 \\ 2 \end{bmatrix} x$

$y = \begin{bmatrix} 1 & 4 \end{bmatrix} \vec{q} + [8] x$

(ii) $\dot{\vec{q}} = \begin{bmatrix} 2 & 1 & 0 \\ 7 & 3 & 5 \\ 0 & 6 & 4 \end{bmatrix} \vec{q} + \begin{bmatrix} 8 \\ 0 \\ 0 \end{bmatrix} x,$

$y = \begin{bmatrix} 9 & 0 & 0 \end{bmatrix} \vec{q} + 10x$

(vi) $\dot{\vec{q}} = \begin{bmatrix} -\sigma & \omega \\ -\omega & -\sigma \end{bmatrix} \vec{q} + \begin{bmatrix} 1 \\ 1 \end{bmatrix} x$

$y = \begin{bmatrix} 1 & 1 \end{bmatrix} \vec{q}$

(iii) $\dot{\vec{q}} = \begin{bmatrix} -1 & 1 \\ 0 & 2 \end{bmatrix} \vec{q} + \begin{bmatrix} 0 \\ 3 \end{bmatrix} x$

$y = \begin{bmatrix} 1 & 0 \end{bmatrix} \vec{q} + 4x$

(vii) $\dot{\vec{q}} = \begin{bmatrix} 0 & 1 \\ -\omega_n^2 & -2\zeta\omega_n \end{bmatrix} \vec{q} + \begin{bmatrix} 0 \\ 1 \end{bmatrix} x$

$y = \begin{bmatrix} 1 & 0 \end{bmatrix} \vec{q}$

(iv) $\dot{\vec{q}} = \begin{bmatrix} 0 & 0 & 2 \\ 5 & 0 & -1 \\ 3 & 0 & 0 \end{bmatrix} \vec{q} + \begin{bmatrix} 0 \\ 1 \\ 0 \end{bmatrix} x$

$y = \begin{bmatrix} 6 & 0 & 0 \end{bmatrix} \vec{q} + 4x$

(viii) $\dot{\vec{q}} = \begin{bmatrix} 0 & 1 \\ -\omega_n^2 & -2\zeta\omega_n \end{bmatrix} \vec{q} + \begin{bmatrix} 0 \\ 1 \end{bmatrix} x$

$y = \begin{bmatrix} 0 & 1 \end{bmatrix} \vec{q}$

10.5.4 Consider the block diagram in Figure P10.5.4.
(a) Find the transfer function by block diagram reduction.
(b) Assign state variables to the block diagram and write the state equations.

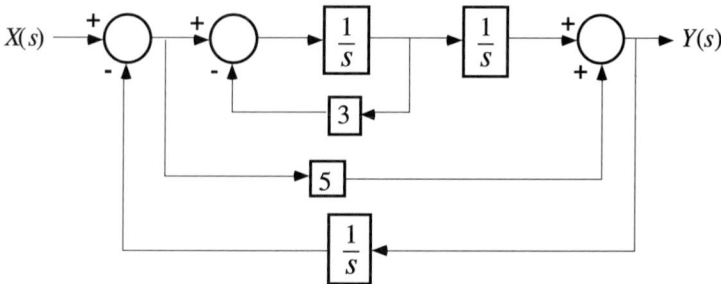

Figure P10.5.4

10.5.5 Consider the block diagram in Figure P10.5.5.
(a) Find the transfer function by block diagram reduction.
(b) Assign state variables to the block diagram and write the state equations.

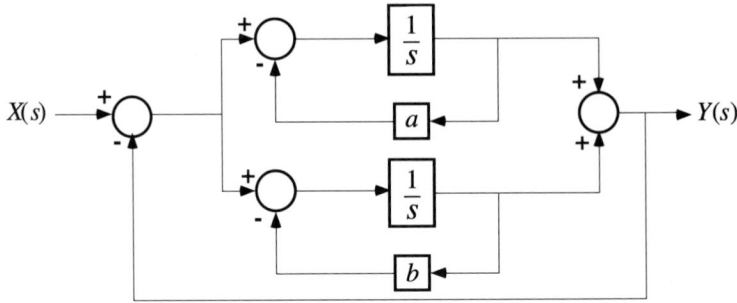

Figure P10.5.5

10.5.6 Consider the block diagram in Figure P10.5.6.
(a) Find the transfer function by block diagram reduction.
(b) Assign state variables to the block diagram and write the state equations.

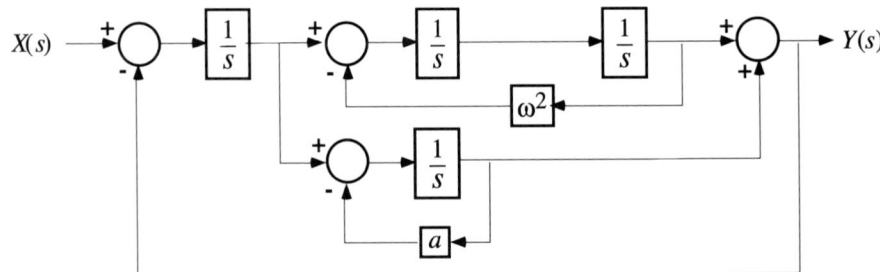

Figure P10.5.6

10.5.7 Consider the following systems represented by a set of coupled differential equations. Here $x(t)$ is considered the input signal and $y(t)$ is considered the output signal.
(a) Write the state space representation for each system.
(b) Draw the all-integrator block diagram of each system. Label the state variables.

(i) $\dot{q}_1(t) = q_2(t),$

$\dot{q}_2(t) = q_3(t),$

$\dot{q}_3(t) = -3q_1(t) + 2q_2(t) - q_3(t) + 4x(t),$

$y(t) = -2q_1(t).$

(ii) $\dot{q}_1(t) = -3q_1(t) + 3x(t),$

$\dot{q}_2(t) = 4q_2(t) + 6x(t),$

$y(t) = -2q_1(t) + 7q_2(t) + 2x(t).$

Chapter 11

Introduction to Realization Theory

Chapter Outline

In the last chapter we introduced the state space representation as an outgrowth from an all-integrator block diagram. It is obvious that there is a close relationship between the transfer functions, block diagrams, and the state space representations. In this chapter we will develop many of these relationships. In general, the theory that describes this relationship between transfer functions and state space representations is called realization theory; hence, the chapter title.

There are three themes that are developed in this chapter. The first theme is the relationship between a state space representation of a system and its transfer function. First, we give the formula for computing the transfer function from a state space representation. This formula is straightforward. We also note several key relationships between the state space system and its transfer function. Second, we develop a method for translating a transfer function into a state space representation. This reverse direction, developing a state space representation from a transfer function, is difficult. Most of the theory is beyond the scope of this text. We present two simple ways of translating a transfer function or block diagram into a state space representation, which are sufficient for our purposes. Third, we develop the relationship between state space representations that have the same transfer function. This result indicates the flexibility of the state space representation for modeling and analyzing a system.

The second theme that is discussed in this chapter is the development of a state space representation directly from the differential equations of the system that are derived from physical laws. We also discuss how to translate a differential equation into a state space representation for systems which are governed by higher order differential equations. When these insights are combined with the previous results in this chapter, the state space representation emerges as an extremely powerful tool for modeling and analysis of systems.

Computer Usage: The third theme that is developed in this chapter is the use of a state space system to simplify the entry of a system model into the computer. We want to have the tools to enter a system into the computer in such a way that we are able to carry out an analysis of this system. The construction of a state space realization from a transfer function is a step in this direction. The state space representation is the appropriate tool for this task. We show how to use the structure of the block diagram to simplify the data entry. The results for single input single output systems are extended to systems with multiple input and output signals.

Summary of Sections

Section 11.1:   We develop the relationship between the transfer function and a state space representation.

Section 11.2:   We introduce a method to translate a transfer function to a state space representation.

Section 11.3:   We develop the relationship between state space representations that have the same transfer function.

Section 11.4:   We discuss the derivation of a state space representation directly from physical laws.

Section 11.5:   We discuss the construction of a system representation for a system with more than one input and/or one output signal.

Section 11.6:   Chapter summary section.

Coverage of the Text

This chapter requires all of the results of Chapter 10.

11.1 CALCULATION OF A TRANSFER FUNCTION FROM A STATE SPACE REPRESENTATION

11.1.1 Introduction

In the last chapter we have shown how to a derive state space representation from an all-integrator block diagram. We know that the block diagram is equivalent to a transfer function because of block diagram reduction. It follows that the state space representation should be equivalent to a transfer function. In this section we will show how to compute a transfer function from a state space representation. We will also discuss the relationship between the poles of the transfer function and the eigenvalues of the state matrix. These results form the basis of the relationship between these two system representations.

11.1.2 State Space Representation to Transfer Functions

Scalar Equations Next we will show how to compute the transfer function from the state space equations. In order to motivate the matrix calculations, we will start with the first-order state space equation

$$\dot{q}(t) = aq(t) + bx(t), \quad q(0) = 0, \tag{11.1.1}$$

$$y(t) = cq(t) + dx(t),$$

where $q(t)$ is a scalar signal and a, b, c, and d are all scalar constants. The notation for the matrix equations is also motivated by this example.

Taking the Laplace transform of the equations in (11.1.1) assuming that the initial condition is zero, we get

$$sQ(s) = aQ(s) + bX(s), \tag{11.1.2a}$$
$$Y(s) = cQ(s) + dX(s). \tag{11.1.2b}$$

Note: To compute the transfer function we always set the initial conditions to zero. See Theorem 10.1.4.

Now we solve (11.1.2) for the transfer function $Y(s)/X(s)$. Solving for $Q(s)$ in (11.1.2a) we get

$$(s - a)Q(s) = bX(s), \quad \text{or} \tag{11.1.3a}$$
$$Q(s) = (s - a)^{-1} bX(s). \tag{11.1.3b}$$

Inserting (11.1.3b) into (11.1.2b), we find that the transfer function is

$$\frac{Y(s)}{X(s)} = c(s-a)^{-1}b + d. \tag{11.1.4}$$

Example 11.1.1:   Consider the RC network in Figure 11.1.1.   The state space representation for the RC circuit is

$$\dot{q}(t) = -\frac{1}{RC}q(t) + \frac{1}{RC}x(t), \quad q(0) = 0 \tag{11.1.5}$$

$$y(t) = q(t).$$

Comparing (11.1.5) with (11.1.1) we see that

$$a = -\frac{1}{RC}, \quad b = \frac{1}{RC}, \quad c = 1, \quad d = 0. \tag{11.1.6}$$

From (11.1.4) the transfer function is

$$\frac{Y(s)}{X(s)} = \frac{1}{RC}\left(s + \frac{1}{RC}\right)^{-1}(1) + 0 = \frac{\frac{1}{RC}}{s + \frac{1}{RC}}. \tag{11.1.7}$$

▲▲

Matrix Equations   Motivated by the development above, consider the state space representation

$$\dot{\vec{q}}(t) = A\vec{q}(t) + Bx(t), \tag{11.1.8}$$

$$y(t) = C\vec{q}(t) + Dx(t).$$

Taking the Laplace transform of (11.1.5) while assuming zero initial conditions we get

$$s\vec{Q}(s) = A\vec{Q}(s) + BX(s), \tag{11.1.9a}$$

$$Y(s) = C\vec{Q}(s) + DX(s). \tag{11.1.9b}$$

In (11.1.9a) we replace $s\vec{Q}(s)$ by $sI\vec{Q}(s)$, where I is an n x n identity matrix.   Solving the first equation for $\vec{Q}(s)$ we get

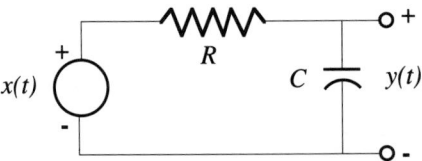

Figure 11.1.1 RC Network

$$(sI - A)\vec{Q}(s) = BX(s).$$ \hfill (11.1.10)

To solve for $\vec{Q}(s)$ we multiply (11.1.10) on the *left* by $(sI - A)^{-1}$. We obtain

$$\vec{Q}(s) = (sI - A)^{-1}BX(s).$$ \hfill (11.1.11)

Note: When the dimension of the state vector is greater than one, all operations become matrix operations. The rules of matrix algebra must be observed. We must multiply on the left by the inverse.

Now inserting (11.1.11) into (11.1.9b) we obtain

$$Y(s) = \left[C(sI - A)^{-1}B + D\right]X(s).$$ \hfill (11.1.12)

To find the transfer function in (11.1.12) we simply solve for $Y(s)/X(s)$. These results are summarized in the following theorem.

Theorem 11.1.2: **Reduction of State Space Equations to a Transfer Function**
Consider the state space representation

$$\dot{\vec{q}}(t) = A\vec{q}(t) + Bx(t), \quad \vec{q}_0 = 0,$$
$$y(t) = C\vec{q}(t) + Dx(t),$$

of a system where the initial conditions are set to zero. Then the transfer function of this system is given by

$$\frac{Y(s)}{X(s)} = C(sI - A)^{-1}B + D.$$

▲▲

Theorem 11.1.2 is well known and widely used.

The formula (11.1.12) requires the existence of the inverse of the matrix $(sI - A)$. The inverse exists if the determinant is nonzero. The leading term in the determinant of $(sI - A)$ is s^n, which is never zero. Hence, this inverse always exists.

The mathematical derivation from the state equations (11.1.8) to the transfer function (11.1.12) requires that: (1) the state equations in (11.1.8) are decomposed into scalar equations, (2) the Laplace transform is applied to each scalar equation, and (3) the equations are reassembled into matrix form in (11.1.12). The details are left to the reader.

Computer Usage: All computer packages contain a function to perform the calculation in Theorem 11.1.2. If we want to find the transfer function of a complicated system, often the best way is to develop a state space representation of the system, enter this system into the computer, and then allow the computer to calculate the transfer function via Theorem 11.1.2. The results in this chapter are presented for just such a scenario. The results in this chapter have many other uses as well.

The following examples illustrate the use of Theorem 11.1.2.

Example 11.1.3: The state variable model of the proof-mass actuator is

$$\dot{\vec{q}}(t) = \begin{bmatrix} 0 & 1 \\ 0 & 0 \end{bmatrix} \vec{q}(t) + \begin{bmatrix} 0 \\ \dfrac{K_{ef}}{m_{pm}} \end{bmatrix} v_{pm}(t),$$ (11.1.13)

$$y_{pm}(t) = \begin{bmatrix} 1 & 0 \end{bmatrix} \vec{q}(t) + 0 v_{pm}(t).$$

This model was derived in Example 10.5.6. To compute the transfer function from this model, first we form the matrix

$$(sI - A) = s \begin{bmatrix} 1 & 0 \\ 0 & 1 \end{bmatrix} - \begin{bmatrix} 0 & 1 \\ 0 & 0 \end{bmatrix} = \begin{bmatrix} s & -1 \\ 0 & s \end{bmatrix}.$$ (11.1.14)

The inverse of this matrix can be computed using a standard formula for 2 x 2 matrices. The result is

$$(sI - A)^{-1} = \frac{1}{s^2} \begin{bmatrix} s & 1 \\ 0 & s \end{bmatrix} = \begin{bmatrix} \dfrac{1}{s} & \dfrac{1}{s^2} \\ 0 & \dfrac{1}{s} \end{bmatrix}.$$ (11.1.15)

Now the transfer function is computed using Theorem 11.1.2. We have

$$\frac{Y_{pm}(s)}{V_{pm}(s)} = C(sI - A)^{-1}B + D = \begin{bmatrix} 1 & 0 \end{bmatrix} \begin{bmatrix} \dfrac{1}{s} & \dfrac{1}{s^2} \\ 0 & \dfrac{1}{s} \end{bmatrix} \begin{bmatrix} 0 \\ \dfrac{K_{ef}}{m_{pm}} \end{bmatrix} + 0 = \frac{K_{ef}}{m_{pm}s^2}.$$ (11.1.16)

Of course, this result agrees with Section 6.5.

▲▲

Example 11.1.4: Consider the system discussed in Example 10.5.10. The state space representation of that system is

$$\begin{bmatrix} \dot{q}_1(t) \\ \dot{q}_2(t) \\ \dot{q}_3(t) \end{bmatrix} = \begin{bmatrix} 0 & 1 & 0 \\ -\omega^2 & 0 & 0 \\ 0 & 0 & -a \end{bmatrix} \begin{bmatrix} q_1(t) \\ q_2(t) \\ q_3(t) \end{bmatrix} + \begin{bmatrix} 0 \\ b_2 \\ b_3 \end{bmatrix} x(t),$$ (11.1.17)

$$y(t) = \begin{bmatrix} c_1 & 0 & c_3 \end{bmatrix} \vec{q}(t) + \begin{bmatrix} d \end{bmatrix} x(t).$$

First, we compute the inverse of

$$(sI - A) = \begin{bmatrix} s & -1 & 0 \\ \omega^2 & s & 0 \\ 0 & 0 & s+a \end{bmatrix}. \tag{11.1.18}$$

Since this matrix is a block diagonal matrix, the inverse can be computed from the inverse of the diagonal blocks. We have

$$(sI - A)^{-1} = \frac{1}{(s^2+\omega^2)(s+a)} \begin{bmatrix} s(s+a) & (s+a) & 0 \\ -\omega^2(s+a) & s(s+a) & 0 \\ 0 & 0 & s^2+\omega^2 \end{bmatrix}. \tag{11.1.19}$$

Now using Theorem 11.1.2 we have

$$C(sI - A)^{-1}B + D \tag{11.1.20}$$

$$= \frac{1}{(s^2+\omega^2)(s+a)} \begin{bmatrix} c_1 & 0 & c_3 \end{bmatrix} \begin{bmatrix} s(s+a) & (s+a) & 0 \\ -\omega^2(s+a) & s(s+a) & 0 \\ 0 & 0 & s^2+\omega^2 \end{bmatrix} \begin{bmatrix} 0 \\ b_2 \\ b_3 \end{bmatrix} + d$$

$$= \frac{1}{(s^2+\omega^2)(s+a)} \begin{bmatrix} c_1 s(s+a) & c_1(s+a) & c_3(s^2+\omega^2) \end{bmatrix} \begin{bmatrix} 0 \\ b_2 \\ b_3 \end{bmatrix}$$

$$= \frac{c_1 b_2}{(s^2+\omega^2)} + \frac{c_3 b_3}{(s+a)} + d.$$

▲▲

Example 11.1.4 shows that if the direct feed-through matrix is nonzero, $D \neq 0$, the transfer function will be proper. Otherwise, the transfer function will be strictly proper.

11.1.3 Basic Relationships Between the Transfer Function and State Space Representations

Theorem 11.1.2 provides the basic relationship between a state space representation and the corresponding transfer function. Often we would like to understand this relationship in a deeper way. We will next develop a relationship between the poles of the transfer function and the eigenvalues of the state matrix A. In the following discussion we will refer to Example 11.1.4.

Let us start with an all-integrator block diagram with n integrators. We assign a state variable to the output of each integrator. Then we construct a state space representation as in Procedure 10.5.9. This results in an nth-order state space representation. That is, the order of the state matrix A is n.

Next we consider the computation of the computation of $(sI - A)^{-1}$ in the calculation of the transfer function in Theorem 11.1.2. From matrix theory, this inverse is given by

$$(sI - A)^{-1} = \frac{\text{adj}(sI - A)}{\det(sI - A)}. \tag{11.1.21}$$

The adjoint matrix $\text{adj}(sI - A)$ of the matrix $(sI - A)$ is

$$\text{adj}(sI - A) = G^T = \left[g_{ij} \right]^T. \tag{11.1.22}$$

where the ijth cofactor g_{ij} of this matrix is

$$g_{ij} = (-1)^{i+j} \det(M_{ij}). \tag{11.1.23}$$

The ijth minor M_{ij} of a matrix $(sI - A)$ is obtained by striking the ith row and jth column of $(sI - A)$. This discussion shows that the adjoint matrix is a polynomial matrix (a matrix whose elements are polynomials).

To illustrate this last comment, consider the construction of the inverse of the matrix in (11.1.18). For example, we calculate the cofactor of the $(2,1)$ element of $(sI - A)$. The appropriate minor is obtained by striking the second row and first column in (11.1.18). We have

$$M_{21} = \begin{bmatrix} -1 & 0 \\ 0 & s+a \end{bmatrix}. \tag{11.1.24}$$

Then the cofactor is

$$g_{21} = (-1)^{2+1} \left[-1(s+a) \right] = s + a. \tag{11.1.25}$$

Then the adjoint matrix is

$$\text{adj}(sI - A) = G^T = \begin{bmatrix} * & * & * \\ s+a & * & * \\ * & * & * \end{bmatrix}^T = \begin{bmatrix} * & s+a & * \\ * & * & * \\ * & * & * \end{bmatrix}. \tag{11.1.26}$$

This process is continued until all elements of the adjoint matrix are computed. Then the $(sI - A)^{-1}$ is computed as

$$(sI - A)^{-1} = \frac{1}{\det(sI - A)} \text{adj}(sI - A) = \frac{1}{\det(sI - A)} \begin{bmatrix} * & s+a & * \\ * & * & * \\ * & * & * \end{bmatrix}. \tag{11.1.27}$$

The purpose of this discussion is to show that the adjoint matrix $\text{adj}(sI - A)$ contains only polynomials. This fact is obvious from the calculation above. First, $(sI - A)$ contains only polynomials. Second, calculation of the cofactor involves the calculation of a determinant of a submatrix of $(sI - A)$. Hence, the minor (determinant) must be a polynomial because it is a product of the elements of the matrix. It follows that the adjoint matrix is a matrix whose entries are polynomials. Furthermore, we conclude that $C[\text{adj}(sI - A)]B$ is a polynomial because B and C are constant matrices.

Next we consider the $\det(sI - A)$. This determinant is a polynomial. In fact, it is the characteristic equation of A, as discussed in Section 4.2. Hence, the roots of this polynomial are the eigenvalues of A.

Now consider the transfer function

$$\frac{Y(s)}{X(s)} = C(sI - A)^{-1}B + D = \frac{1}{\det(sI - A)} C[\text{adj}(sI - A)]B + D = \frac{b(s)}{a(s)}. \qquad (11.1.28)$$

When the ratio of these two polynomials is formed, the ratio of $C[\text{adj}(sI - A)]B$ and $\det(sI - A)$, only cancellations can occur between these polynomials. After these cancellations have been carried out, the roots of $\det(sI - A)$ that are left will be the poles of the transfer function. Hence, we see that the poles of the transfer function in (11.1.28) must be contained within the set of eigenvalues of A. We summarize these observations in the following theorem.

Theorem 11.1.5: Let a system be represented by an all-integrator block diagram with n integrators. Then:

1. The corresponding state space representation will be of order n.
2. The transfer function will be at most order n.
3. The poles of the transfer function will be among the eigenvalues of the state matrix A.

▲▲

According to Theorem 11.1.5, when the transfer function is formed, two cases can occur. In the first case, there are no cancellations between the numerator and denominator so all of the eigenvalues of A appear as poles of the transfer function. In Example 11.1.4 the poles of the transfer function, (11.1.20), are exactly the roots of the determinant. In the second case, some cancellations occur between the numerator and denominator in (11.1.17), so that some of the roots of $\det(sI - A)$ don't appear in the transfer function. The next example illustrates this case.

Example 11.1.6: Consider again the system in Example 11.1.4, except that we set $c_3 = 0$. The new system is shown in Figure 11.1.2. The transfer function for this system, as derived from (11.1.20) is

$$C(sI - A)^{-1}B + D = \frac{c_1(s + a)b_2 + c_3(s^2 + \omega^2)b_3}{(s^2 + \omega^2)(s + a)} + d. \qquad (11.1.29)$$

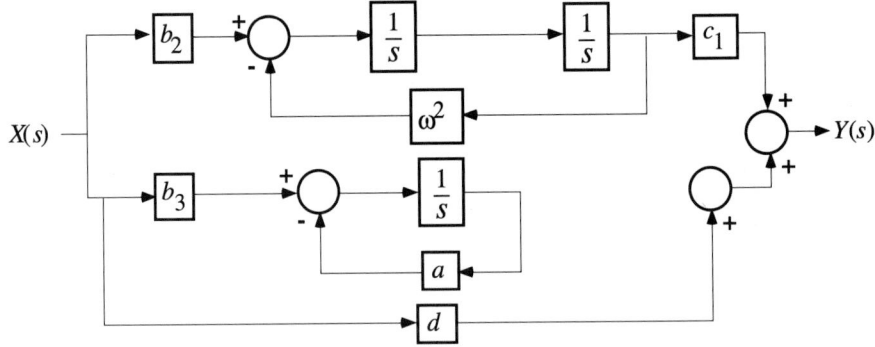

Figure 11.1.2 Block Diagram of Example 11.1.6

If we set $c_3 = 0$ in (11.1.29) the last line becomes

$$\frac{c_1(s+a)b_2 + (0)\left(s^2 + \omega^2\right)b_3}{\left(s^2 + \omega^2\right)(s+a)} + d = \frac{c_1(s+a)b_2}{\left(s^2 + \omega^2\right)(s+a)} + d = \frac{c_1 b_2}{\left(s^2 + \omega^2\right)} + d. \qquad (11.1.30)$$

The eigenvalues of the state matrix A are $\lambda_{1/2} = \pm j\omega$ and $\lambda_3 = -a$. In this example, however, the eigenvalue $\lambda_3 = a$ cancels with a zero in the numerator in (11.1.30), and so it doesn't appear in the transfer function. Hence, one of the eigenvalues of the state matrix A doesn't appear in the transfer function. The transfer function in (11.1.30) should be compared with the transfer function in (11.1.20). ▲▲

There is a complex, but well understood, relationship between the state space representation and transfer function of a system. The complete development of this relationship is beyond the scope of this text.

11.1.4 MATLAB Experiments

As discussed in the last chapter, MATLAB has the capability to translate a system from one representation into another representation. This capability is contained in the commands that create the system objects: **tf**, **zpk**, and **ss**. When the input argument is a system object, the output argument is a system object with the specified representation. The following M-file illustrates this calculation.

MATLAB Version 4 Version 4 of MATLAB contains a special series of commands to convert one representation into another. One such command is

[b,a] = **ss2tf**(A,B,C,D)

The input arguments define one system representation. The output arguments define another system representation. A table of these commands is given in the MATLAB manual.

Example 11.1.3: (*continued*) M-file

```
clear
% define state space system
mpm = 10;                    % proof-mass mass
Kef = 1;                     % gain
Apm = [0,1;0,0];             % state matrix
Bpm = [0;Kef/mpm];           % input matrix
Cpm = [1,0];                 % output matrix
Dpm = 0;                     % direct feedthrough matrix
Hpm = ss(Apm,Bpm,Cpm,Dpm); % define system
% calculate transfer function
hpm = tf(Hpm)
```

▲▲

Exploratory Exercise 11.1.7: Write an M-file to verify the results of Examples 11.1.4 and 11.1.6.

▲▲

11.2 TWO REALIZATIONS

11.2.1 Introduction

In the last section we showed how a transfer function can be computed from a state space representation. In this section we discuss the reverse process of constructing a state space representation from a transfer function. While the calculation of the transfer function from the state equations is straightforward, the construction of state space representation from the transfer function is not so easy. A complete treatment of this subject is beyond the scope of this text. We will only present a few simple, but useful results.

The central concept of this chapter is embodied in the following definition.

Definition 11.2.1: Given the transfer function of a system

$$\frac{Y(s)}{X(s)} = H(s),$$

if there is a state space representation

$$\dot{\vec{q}}(t) = A\vec{q}(t) + Bx(t),$$
$$y(t) = C\vec{q}(t) + Dx(t),$$

such that

$$C(sI - A)^{-1}B + D = H(s)$$

the state space equations are called a <u>realization</u> of the transfer function.

▲▲

Terminology: Realization theory is the mathematical theory that describes the relationship between the transfer function and state space equations. Realization theory includes the algorithms for constructing a state space representation from the transfer function, which is the main topic of this chapter.

In general there exist many state space representations for a given transfer function. Because all of these state space representations have the same transfer function, they are all considered equivalent from the input-output viewpoint. Among all these state space representations, there exist several that are particularly easy to construct from the transfer function. We will develop three of these representations in this chapter.

11.2.2 First Realization

In this section we consider systems that are represented by a transfer function

$$\frac{Y(s)}{X(s)} = \frac{b_2 s^2 + b_1 s + b_0}{s^3 + a_2 s^2 + a_1 s + a_0} = \frac{b(s)}{a(s)} = H(s). \tag{11.2.1}$$

The extension of the results below to transfer functions of other orders is immediate. To develop the first realization for (11.2.1), we cross-multiply in (11.2.1) to get

$$s^3 Y(s) + a_2 s^2 Y(s) + a_1 Y(s) + a_0 Y(s) = b_2 s^2 X(s) + b_1 s X(s) + b_0 X(s). \tag{11.2.2}$$

Multiplying (11.2.2) by $1/s^3$ and rearranging terms results in

$$Y(s) = \frac{1}{s}\left[b_2 X(s) - a_2 Y(s)\right] + \frac{1}{s^2}\left[b_1 X(s) - a_1 Y(s)\right] + \frac{1}{s^3}\left[b_0 X(s) - a_0 Y(s)\right]. \tag{11.2.3}$$

Regrouping terms yields

$$Y(s) = \frac{1}{s}\left\{b_2 X(s) - a_2 Y(s) + \frac{1}{s}\left[b_1 X(s) - a_1 Y(s) + \frac{1}{s}\left(b_0 X(s) - a_0 Y(s)\right)\right]\right\}. \tag{11.2.4}$$

From the equation (11.2.4) we will construct a block diagram. First we write (11.2.4) as

$$Y(s) = \frac{1}{s}\left\{b_2 X(s) - a_2 Y(s) + X_1(s)\right\}. \tag{11.2.5}$$

From (11.2.5) the output signal of the system is the output signal from an integrator. The input signals into the integrator are the output signal multiplied by a_2, the input signal multiplied by b_2, and the signal $X_1(s)$. The block diagram of (11.2.5) is shown in Figure 11.2.1. Using (11.2.4) we write the signal $X_1(s)$ as

$$X_1(s) = \frac{1}{s}\left[b_1 X(s) - a_1 Y(s) + X_2(s)\right]. \tag{11.2.6}$$

Thus, we see that the signal $X_1(s)$ is also the output signal to an integrator. The input signals into the integrator are the output signal multiplied by a_1, the input signal multiplied by b_1, and the signal $X_2(s)$. We add this integrator to the block diagram in Figure 11.2.1 to get the block diagram in Figure 11.2.2. The signal $X_2(s)$ in Figure 11.2.2 is obtained from (11.2.4). We have

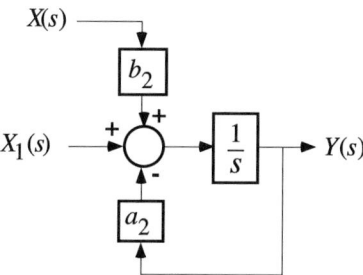

Figure 11.2.1 First Step in the Construction of a Block Diagram for (11.2.4)

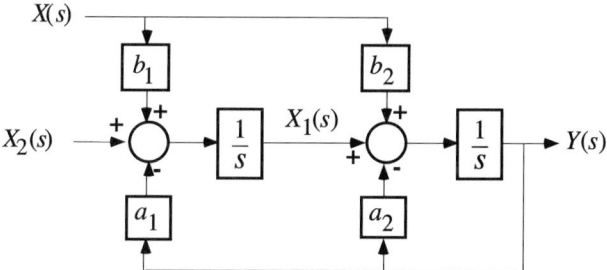

Figure 11.2.2 Second Step in the Construction of a Block Diagram for (11.2.4)

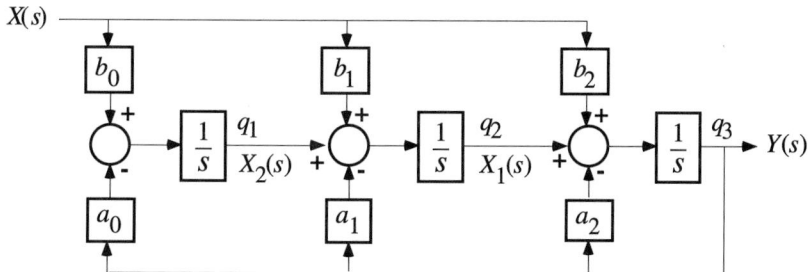

Figure 11.2.3 The Block Diagram of the First Canonical Realization

$$X_2(s) = \frac{1}{s}\big(b_0 X(s) - a_0 Y(s)\big).$$

(11.2.7)

Again the signal $X_2(s)$ is the output signal of an integrator. The input signals into the integrator are the output signal multiplied by a_0, and the input signal multiplied by b_0. Now this integrator is added to the block diagram in Figure 11.2.2. The resulting block diagram is shown in Figure 11.2.3. This block diagram accounts for all of the signals in (11.2.4). It can be verified by block diagram reduction that the transfer function of this block diagram is indeed (11.2.1).

The regular structure of this block diagram indicates how a block diagram of a different order transfer function would be constructed.

Notation: Note that the indexing of the coefficients of the transfer function are crucial to the construction of the block diagram in Figure 11.2.3.

The state space equations corresponding to the all-integrator block diagram in Figure 11.2.3 can be constructed according to Procedure 10.5.9. Using the state variables labeled in Figure 11.2.3 the corresponding equations are

$$\dot{q}_1(t) = -a_0 q_3(t) + b_0 x(t),$$

(11.2.8)

$$\dot{q}_2(t) = -a_1 q_3(t) + q_1(t) + b_1 x(t),$$

$$\dot{q}_3(t) = -a_2 q_3(t) + q_2(t) + b_2 x(t),$$

$$y(t) = q_3(t).$$

These equations are then assembled into matrix equations. This discussion is summarized as follows.

Realization 11.2.2: Given the differential equation

$$\dddot{y}(t) + a_2 \ddot{y}(t) + a_1 \dot{y}(t) + a_0 y(t) = b_2 \ddot{x}(t) + b_1 \dot{x}(t) + b_0 x(t)$$

with the transfer function

$$\frac{Y(s)}{X(s)} = \frac{b_2 s^2 + b_1 s + b_0}{s^3 + a_2 s^2 + a_1 s + a_0} = \frac{b(s)}{a(s)} = H(s),$$

a realization of this transfer function is

$$\dot{\vec{q}}(t) = \begin{bmatrix} \dot{q}_1(t) \\ \dot{q}_2(t) \\ \dot{q}_3(t) \end{bmatrix} = \begin{bmatrix} 0 & 0 & -a_0 \\ 1 & 0 & -a_1 \\ 0 & 1 & -a_2 \end{bmatrix} \vec{q}(t) + \begin{bmatrix} b_0 \\ b_1 \\ b_2 \end{bmatrix} x(t),$$

$$y(t) = \begin{bmatrix} 0 & 0 & 1 \end{bmatrix} \vec{q}(t) + \begin{bmatrix} 0 \end{bmatrix} x(t).$$

The block diagram corresponding to this state space representation is shown in Figure 11.2.3.

▲▲

The block diagram in Figure 11.2.3 depends only on the parameters of the transfer function in (11.2.1). Hence, from the transfer function we can immediately draw the block diagram in Figure 11.2.3. Indeed, no calculations are required.

Having derived the state equations from the block diagram, we see that the intermediate step of constructing a block diagram is not required to write the state equations from the transfer function. We need only to substitute the coefficients from the transfer function into the corresponding entries of the state space equations. Note that the matrices in Realization 11.2.2 have a fixed structure. First, there appears only 0's, 1's, and the coefficients of the transfer function. The numerator coefficients of the transfer function define the elements of the input matrix. The denominator coefficients appear in the last column of the state matrix. The rest of the elements of the A matrix are 0's and 1's. The 1's appear on the subdiagonal of the state matrix with 0's everywhere else. (The structure of the 1's and 0's is determined by the all-integrator block diagram. Verify!) Output matrix consists of zeros with a 1 in the last position.

Realization 11.2.2 is defined only for strictly proper transfer functions. Hence, the direct feedthrough term is always zero. A proper transfer function should be reduced to a strictly proper transfer function plus a constant term by long division. The constant term then becomes the direct feedthrough term in the state space realization. The all-integrator block diagram is modified accordingly.

Notation: The indexing of the coefficients in the transfer function and the corresponding entries of the state equations in Realization 11.2.2 is crucial.

Now we see that the third-order differential equation is equivalent to the state space representation (a first-order differential equation). The state space representation is preferred because, among other reasons, a large number of results from matrix theory can be used to study the behavior of the state space representation both theoretically and numerically. The special form of the state space (matrix) equations carries a name.

Definition 11.2.3:   The block diagram in Figure 11.2.3 is called a <u>canonical</u> block diagram. The state space equations in Realization 11.2.2 are called <u>canonical</u> state space equations.

▲▲

Note: All canonical block diagrams are all-integrator block diagrams, but not all all-integrator block diagrams are canonical block diagrams.

There are several other canonical forms that are not discussed explicitly in this text.

Example 11.2.4:   Consider the mass-spring-damper system in Section 6.4. The transfer function is

$$\frac{Y_{st}(s)}{F_{st}(s)} = \frac{\dfrac{1}{m_{st}}}{s^2 + \dfrac{c_{st}}{m_{st}}s + \dfrac{k_{st}}{m_{st}}}. \tag{11.2.9}$$

The block diagram of Realization 11.2.2 of this system is shown in Figure 11.2.4. The state space equations for the transfer function (11.2.9) are

$$\dot{\bar{q}}(t) = \begin{bmatrix} 0 & -\dfrac{k_{st}}{m_{st}} \\ 1 & -\dfrac{c_{st}}{m_{st}} \end{bmatrix} \bar{q}(t) + \begin{bmatrix} \dfrac{1}{m_{st}} \\ 0 \end{bmatrix} f_{st}(t), \tag{11.2.10}$$

$$y(t) = \begin{bmatrix} 0 & 1 \end{bmatrix} \bar{q}(t).$$

▲▲

Example 11.2.5: Consider a proof-mass actuator attached to a mass-spring-damper system as discussed in Subsection 10.3.3. The transfer function of that system is given by

$$\begin{aligned}
\frac{Y_r(s)}{V_{pm}(s)} &= K_s \left[\frac{\dfrac{1}{m_{pm}}}{s^2} - \frac{\dfrac{1}{m_{st}}}{s^2 + \dfrac{c_{st}}{m_{st}}s + \dfrac{k_{st}}{m_{st}}} \right] K_{ef} \tag{11.2.11} \\[2em]
&= \frac{K_s K_{ef} \left[\left(\dfrac{1}{m_{st}} - \dfrac{1}{m_{pm}} \right) s^2 + \dfrac{c_{st}}{m_{pm}m_{st}}s + \dfrac{k_{st}}{m_{pm}m_{st}} \right]}{s^4 + \dfrac{c_{st}}{m_{st}}s^3 + \dfrac{k_{st}}{m_{st}}s^2}.
\end{aligned}$$

Using Realization 11.2.2 a state space representation of this system is given by

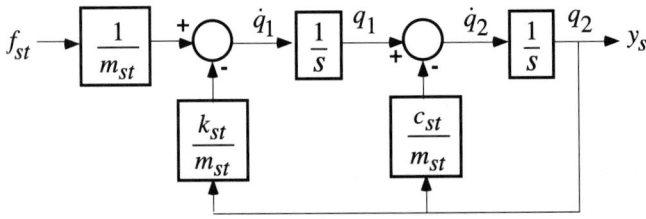

Figure 11.2.4 Realization 11.2.2 for the Mass-Spring-Damper System

$$
\dot{\vec{q}}(t) =
\begin{bmatrix}
0 & 0 & 0 & 0 \\
1 & 0 & 0 & 0 \\
0 & 1 & 0 & -\dfrac{k_{st}}{m_{st}} \\
0 & 0 & 1 & -\dfrac{c_{st}}{m_{st}}
\end{bmatrix}
\vec{q}(t) +
\begin{bmatrix}
\dfrac{K_s K_{ef} k_{st}}{m_{pm} m_{st}} \\[4pt]
\dfrac{K_s K_{ef} c_{st}}{m_{pm} m_{st}} \\[4pt]
K_s K_{ef}\left(\dfrac{1}{m_{st}} - \dfrac{1}{m_{pm}}\right) \\[4pt]
0
\end{bmatrix}
v_{pm}(t),
$$

(11.2.12)

$$
y_r(t) = [0 \ \ 0 \ \ 0 \ \ 1]\vec{q}(t) + [0]x(t).
$$

Note that the 1's in the state matrix appear on the subdiagonal.

▲▲

11.2.3 Second Realization

There is another variation on the realizations given above. Consider the transfer function

$$
\frac{Y(s)}{X(s)} = H(s) = [H(s)]^T.
$$

(11.2.13)

Suppose we take the transpose of this function as shown in (11.2.13). Since the transfer function is just a scalar, the transpose is equal to itself. If we substitute in the state space realization via the formula for the transfer function in Theorem 11.1.2, and carry out the transpose on the matrices we obtain

$$
\frac{Y(s)}{X(s)} = H(s) = [H(s)]^T = \left[C(sI - A)^{-1}B\right]^T = B^T\left(sI - A^T\right)^{-1}C^T.
$$

(11.2.14)

The last equality shows that if we: (1) interchange the input and output matrices and take their transpose, and (2) replace the state matrix with its transpose, then we haven't changed the transfer function. If we apply this idea to Realization 11.2.2 we obtain a second realization. Thus we have

$$
A^T =
\begin{bmatrix}
0 & 0 & -a_0 \\
1 & 0 & -a_1 \\
0 & 1 & -a_2
\end{bmatrix}^T
=
\begin{bmatrix}
0 & 1 & 0 \\
0 & 0 & 1 \\
-a_0 & -a_1 & -a_2
\end{bmatrix},
\quad
B^T = [0 \ \ 0 \ \ 1]^T =
\begin{bmatrix}
0 \\
0 \\
1
\end{bmatrix}
$$

(11.2.15)

$$
C^T =
\begin{bmatrix}
b_0 \\
b_1 \\
b_2
\end{bmatrix}^T
= [b_0 \ \ b_1 \ \ b_2],
\quad
D^T = 0^T = 0.
$$

These calculations give rise to the following realization.

Realization 11.2.6: Given the differential equation

$$\dddot{y}(t) + a_2 \ddot{y}(t) + a_1 \dot{y}(t) + a_0 y(t) = b_2 \ddot{x}(t) + b_1 \dot{x}(t) + b_0 x(t)$$

with the transfer function

$$\frac{Y(s)}{X(s)} = \frac{b_2 s^2 + b_1 s + b_0}{s^3 + a_2 s^2 + a_1 s + a_0},$$

a realization of this transfer function is

$$\dot{\vec{q}}(t) = \begin{bmatrix} \dot{q}_1(t) \\ \dot{q}_2(t) \\ \dot{q}_3(t) \end{bmatrix} = \begin{bmatrix} 0 & 1 & 0 \\ 0 & 0 & 1 \\ -a_0 & -a_1 & -a_2 \end{bmatrix} \vec{q}(t) + \begin{bmatrix} 0 \\ 0 \\ 1 \end{bmatrix} x(t),$$

$$y(t) = \begin{bmatrix} b_0 & b_1 & b_2 \end{bmatrix} \vec{q}(t) + \begin{bmatrix} 0 \end{bmatrix} x(t).$$

The corresponding all-integrator block diagram is shown in Figure 11.2.5. This block diagram is derived using Procedure 10.5.11 in Example 10.5.12. ▲▲

11.2.4 MATLAB Experiments

The MATLAB command **ss** will translate a system in a transfer function representation into a state space representation. The input argument is the name of the system represented as a transfer function. The output argument is the name of the system represented as a state space system. The following M-file illustrates this calculation.

MATLAB Version 4 In Version there are a separate set of commands that translate a different system representation into a state space representation. For example, the command **tf2ss** translates a transfer function into a state space representation.

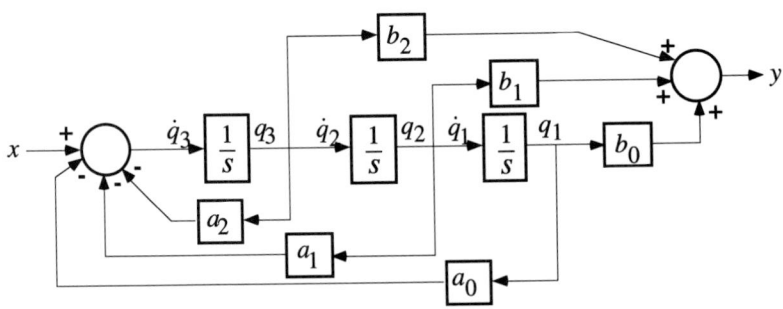

Figure 11.2.5 Block Diagram of Realization 11.2.6

Example 11.2.4: (*continued*) M-file

```
clear
% enter system
mst = 400;                        % mass
cst = 80;                         % damping
kst = 600;                        % stiffness
hst = tf([1/mst],[1,cst/mst,kst/mst]);    % define system

% calculate state space realization
Hst = ss(hst)
```

▲▲

Exploratory Exercise 11.2.7: Write an M-file to calculate the state space model of the system in Example 11.2.5. Enter the transfer function of the proof-mass actuator and the mass-spring-damper system separately. Compute the state space representation of the composite system in two ways. First, compute the state space models of each subsystem separately, and then add the systems. Second, add the two transfer functions, and then compute the state space representation. Compare your two answers. Verify that both state space representations have the same transfer function.

▲▲

11.3 EQUIVALENT DYNAMICAL SYSTEMS

11.3.1 Introduction

Consider the state space representation of a system

$$\dot{\vec{q}}(t) = A\vec{q}(t) + Bx(t),$$ (11.3.1)

$$y(t) = C\vec{q}(t) + Dx(t).$$

In Theorem 11.1.2 we showed that this system representation is related to its transfer function by the formula

$$\frac{Y(s)}{X(s)} = H(s) = C(sI - A)^{-1}B + D.$$ (11.3.2)

We observe that the state of the system (11.3.1) doesn't appear in the transfer function (11.3.2). Hence, the question arises "Do other state space representations have the same transfer function?" A complete answer to this question is beyond the scope of this text; however, we will give a partial answer. This answer involves the concept of equivalent dynamical systems. This concept is central to the study of state space representations. We will develop this concept next.

11.3.2 Transformations of States

Consider the state space representation (11.3.1). Recall that for each time t the state vector $\vec{q}(t)$ is an n-dimensional vector. Let T be a $n \times n$ nonsingular matrix (the inverse of T exists). Define a new state vector $\vec{\tilde{q}}(t)$ by

$$\vec{\tilde{q}}(t) = T\vec{q}(t). \tag{11.3.3}$$

Definition 11.3.1: The matrix T in (11.3.3) is called a <u>state transformation matrix</u>.
▲▲

Because T is a nonsingular matrix, we can write

$$\vec{q}(t) = T^{-1}\vec{\tilde{q}}(t). \tag{11.3.4}$$

Therefore, at each point in time, the state vectors are uniquely related to each other. In fact, the new state variables are linear combinations of the old state variables.

Example 11.3.2: Suppose the state vector is given by

$$\vec{q}(t) = \begin{bmatrix} q_1(t) \\ q_2(t) \end{bmatrix}. \tag{11.3.5}$$

Let the state transformation matrix be given by

$$T = \begin{bmatrix} 1 & 1 \\ -1 & 1 \end{bmatrix}. \tag{11.3.6}$$

It can be verified that this transformation matrix is nonsingular. The new state vector is given by

$$\vec{\tilde{q}}(t) = \begin{bmatrix} \tilde{q}_1(t) \\ \tilde{q}_2(t) \end{bmatrix} = T\vec{q}(t) = \begin{bmatrix} 1 & 1 \\ -1 & 1 \end{bmatrix} \begin{bmatrix} q_1(t) \\ q_2(t) \end{bmatrix} = \begin{bmatrix} q_1(t) + q_2(t) \\ q_2(t) - q_1(t) \end{bmatrix}. \tag{11.3.7}$$

Here the new state variables are just the sum and difference of the old state variables.

As a second example, let the state transformation matrix be given by

$$T = \begin{bmatrix} 0 & 1 \\ 1 & 0 \end{bmatrix}. \tag{11.3.8}$$

It can be verified that this transformation matrix is nonsingular. The new state vector is given by

$$\vec{\tilde{q}}(t) = \begin{bmatrix} \tilde{q}_1(t) \\ \tilde{q}_2(t) \end{bmatrix} = T\vec{q}(t) = \begin{bmatrix} 0 & 1 \\ 1 & 0 \end{bmatrix}\begin{bmatrix} q_1(t) \\ q_2(t) \end{bmatrix} = \begin{bmatrix} q_2(t) \\ q_1(t) \end{bmatrix}. \tag{11.3.9}$$

For this state transformation matrix, the new state vector is just a reordering of the states of the original state vector. Of course, there are an infinite number of other choices for the state transformation matrix.

▲▲

If we do introduce a transformation of the states, how does this transformation affect the state space representation of the system? To answer this question, we take the derivative of the new state vector $\vec{\tilde{q}}(t)$ in (11.3.3). We obtain

$$\dot{\vec{\tilde{q}}}(t) = T\dot{\vec{q}}(t). \tag{11.3.10}$$

Substituting (11.3.1) into (11.3.10) we obtain

$$\dot{\vec{\tilde{q}}}(t) = T\dot{\vec{q}}(t) = T\big(A\vec{q}(t) + Bx(t)\big) = TA\vec{q}(t) + TBx(t). \tag{11.3.11}$$

Using (11.3.4) in (11.3.11) we arrive at

$$\dot{\vec{\tilde{q}}}(t) = TA\vec{q}(t) + TBx(t) = TAT^{-1}\vec{\tilde{q}}(t) + TBx(t). \tag{11.3.12}$$

We also use (11.3.4) in the output equation of (11.3.1). Hence, we obtain the new state space representation

$$\dot{\vec{\tilde{q}}}(t) = TAT^{-1}\vec{\tilde{q}}(t) + TBx(t), \tag{11.3.13}$$

$$y(t) = CT^{-1}\vec{\tilde{q}}(t) + Dx(t).$$

Multiplying out the matrices in (11.3.13) we obtain

$$\begin{aligned} \tilde{A} &= TAT^{-1}, & n \times n \text{ matrix} \\ \tilde{B} &= TB, & n \times 1 \text{ matrix} \\ \tilde{C} &= CT^{-1}, & 1 \times n \text{ matrix} \\ \tilde{D} &= D, & 1 \times 1 \text{ matrix} \end{aligned} \tag{11.3.14}$$

That is, (11.3.13) can be rewritten as

$$\dot{\vec{\tilde{q}}}(t) = \tilde{A}\vec{\tilde{q}}(t) + \tilde{B}x(t), \tag{11.3.15}$$

$$y(t) = \tilde{C}\vec{\tilde{q}}(t) + \tilde{D}x(t).$$

These calculations suggest the following definition.

Definition 11.3.3: If two state vectors are related by a state space transformation matrix as in (11.3.3), then the two state space representations (11.3.1) and (11.3.15) are said to be (dynamically) equivalent.

▲▲

Note that two systems can be dynamically equivalent only if they have the same order.

11.3.3 Input-Output Relationships

The input and output signals of two equivalent state space representations (11.3.1) and (11.3.15) are the same. Next we will show that these two equivalent representations also have the same transfer function. If we compute the transfer function of (11.3.15) using the formula in Theorem 11.1.2 we get

$$\tilde{H}(s) = \tilde{C}(sI - \tilde{A})^{-1}\tilde{B} + \tilde{D} = CT^{-1}(sTT^{-1} - TAT^{-1})^{-1}TB + D \qquad (11.3.16)$$

$$= (CT^{-1})[T(sI - A)T^{-1}]^{-1}(TB) + D$$

where we have used (11.3.14). Next we use the fact that

$$(VWZ)^{-1} = Z^{-1}W^{-1}V^{-1} \qquad (11.3.17)$$

if V, W, and Z are three nonsingular matrices. Applying this rule to the inverse in (11.3.16) we obtain

$$\tilde{H}(s) = (CT^{-1})[T(sI - A)T^{-1}]^{-1}(TB) + D \qquad (11.3.18)$$

$$= (CT^{-1})[T(sI - A)^{-1}T^{-1}](TB) + D = C(sI - A)^{-1}B + D = H(s).$$

Hence, the transfer functions of the two systems (11.3.1) and (11.3.15) are the same.

Theorem 11.3.4: The transfer functions of two dynamically equivalent state space representations are the same. If two state space representations of the same order are not dynamically equivalent, then they have different transfer functions.

▲▲

This theorem divides all state space representations of a given order into classes. The representations in one class are all dynamically equivalent. And the state space representations in this class all have the same transfer function. By choosing different state transformation matrices, we can investigate the structure of the system without changing the transfer function. Transformation of the states is a powerful tool for investigating the structure of state space representations.

This theorem doesn't answer the question of whether state space representations of a different order can have the same transfer function. We leave this question to advanced treatments of state space systems.

Example 11.3.5: This example illustrates the use of state space transformations to construct a dynamically equivalent system. Consider the state space representation in Realization 11.2.2

$$\dot{\vec{q}}(t) = \begin{bmatrix} \dot{q}_1(t) \\ \dot{q}_2(t) \\ \dot{q}_3(t) \end{bmatrix} = \begin{bmatrix} 0 & 0 & -a_0 \\ 1 & 0 & -a_1 \\ 0 & 1 & -a_2 \end{bmatrix} \vec{q}(t) + \begin{bmatrix} b_0 \\ b_1 \\ b_2 \end{bmatrix} x(t), \tag{11.3.19}$$

$$y(t) = \begin{bmatrix} 0 & 0 & 1 \end{bmatrix} \vec{q}(t) + [0]x(t).$$

Define the state space transformation

$$\begin{bmatrix} \tilde{q}_1(t) \\ \tilde{q}_2(t) \\ \tilde{q}_3(t) \end{bmatrix} = T\vec{q}(t) = \begin{bmatrix} 0 & 0 & 1 \\ 0 & 1 & 0 \\ 1 & 0 & 0 \end{bmatrix} \begin{bmatrix} q_1(t) \\ q_2(t) \\ q_3(t) \end{bmatrix} = \begin{bmatrix} q_3(t) \\ q_2(t) \\ q_1(t) \end{bmatrix}. \tag{11.3.20}$$

This transformation corresponds to a renumbering of the state variables in the all-integrator block diagram in Figure 11.2.3. It can be verified that $T = T^{-1}$. Applying this transformation to (11.3.19) using the matrix relationships in (11.3.14) we obtain

$$\dot{\tilde{\vec{q}}}(t) = \begin{bmatrix} -a_2 & 1 & 0 \\ -a_1 & 0 & 1 \\ -a_0 & 0 & 0 \end{bmatrix} \tilde{\vec{q}}(t) + \begin{bmatrix} b_2 \\ b_1 \\ b_0 \end{bmatrix} x(t), \tag{11.3.21}$$

$$y(t) = \begin{bmatrix} 1 & 0 & 0 \end{bmatrix} \tilde{\vec{q}}(t) + [0]x(t).$$

The state space representations in (11.3.19) and (11.3.21) have the same transfer function. This example shows that it doesn't matter how the states are labeled in the all-integrator block diagrams that were used to construct the realizations in the last section. Note, however, that the position of the coefficients of the transfer function in the matrices of the state space representation depend on the choice of states. ▲▲

11.3.4 MATLAB Experiments

The MATLAB command **ss2ss** automates the calculations (11.3.13) - (11.3.14) for the transformation of a state space representation using a state transformation matrix. The input arguments are the name of the system along with the state transformation matrix. The output argument is the name of the transformed system.

Example 11.3.5: (*continued*) M-file

```
clear
% define system
A = [0,0,-1;1,0,-2;0,1,-3];  % state matrix
B = [4;5;6];                 % input matrix
C = [0,0,1];                 % output matrix
D = 0;                       % direct feedthrough matrix
```

H = ss(A,B,C,D)   % define transformation
T = [0,0,1;0,1,0;1,0,0];   % perform transformation
HT = ss2ss(H,T)

▲▲

Exploratory Exercise 11.3.6: Verify that the system and the transformed system have the same transfer function. What does the theory say this transfer function should be?

▲▲

11.4 STATE EQUATIONS FROM PHYSICAL LAWS

11.4.1 A Network Example

Up to this point we have emphasized the relationship between state space representations, block diagrams, and transfer functions. This treatment has been rather abstract. State variables often have a natural physical interpretation, however, and arise quite naturally in the analysis of physical systems. In this section we will illustrate this connection by way of examples.

Example 11.4.1: Consider the *LRC* network of Subsection 10.3.2 and shown in Figure 11.4.1. In Figure 11.4.1 we have assigned voltages and currents. We will develop a state space representation directly from this network. This development begins by writing the differential equations that describe the relationship between the voltage and current in the inductor and capacitors based on the physical laws. We have

$$C_2 \frac{dv_0(t)}{dt} = i_3(t), \tag{11.4.1}$$

$$L \frac{di_2(t)}{dt} = v_1(t) - v_0(t),$$

$$C_1 \frac{dv_1(t)}{dt} = i_1(t) - i_2(t).$$

Based on our knowledge of the form of the state variable equations, (11.4.1) suggests how we should assign state variables to the voltages and currents. We chose the following state variables as shown in Table 11.4.1. In terms of the state variables, (11.4.1) becomes

$$C_2 \frac{dq_1(t)}{dt} = C_2 \frac{dv_0(t)}{dt} = i_3(t), \tag{11.4.2}$$

$$L \frac{dq_2(t)}{dt} = L \frac{di_2(t)}{dt} = v_1(t) - v_0(t),$$

$$C_1 \frac{dq_3(t)}{dt} = C_1 \frac{dv_1(t)}{dt} = i_1(t) - i_2(t).$$

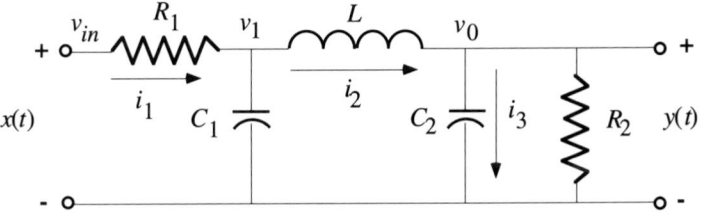

Figure 11.4.1 *LRC* Network

Table 11.4.1 State Variables of the *LRC* Network[1]

State Variable	Signal
$q_1(t)$	voltage across the capacitor C_2, $v_o(t)$
$q_2(t)$	current through the inductor L, $i_2(t)$
$q_3(t)$	voltage across the capacitor C_1, $v_1(t)$

Next we couple these equations together using Kirchhoff's node and loop equations. The goal is to use only the state variables along with the input and output signals. In terms of the state variables we obtain

$$C_2 \frac{dq_1(t)}{dt} = i_3(t) = i_2(t) - \frac{v_0(t)}{R_2} = q_2(t) - \frac{q_1(t)}{R_2} \tag{11.4.3}$$

$$L \frac{dq_2(t)}{dt} = v_1(t) - v_0(t) = q_3(t) - q_1(t),$$

$$C_1 \frac{dq_3(t)}{dt} = i_1(t) - i_2(t) = \frac{v_{in}(t) - v_1(t)}{R_1} - i_2(t) = \frac{x(t) - q_3(t)}{R_1} - q_2(t).$$

Now we isolate the derivatives of the state variables and introduce the output equation for the state space representation. We have

$$\frac{dq_1(t)}{dt} = -\frac{q_1(t)}{C_2 R_2} + \frac{q_2(t)}{C_2} \tag{11.4.4}$$

$$\frac{dq_2(t)}{dt} = -\frac{q_1(t)}{L} + \frac{q_3(t)}{L},$$

$$\frac{dq_3(t)}{dt} = -\frac{q_2(t)}{C_1} - \frac{q_3(t)}{C_1 R_1} + \frac{x(t)}{C_1 R_1},$$

$$y(t) = q_1(t).$$

Finally, these equations are rewritten into state space form. Then (11.4.4) becomes

---

[1] Of course, we could make other choices of the state variables.

$$(11.4.5)$$

$$\dot{\vec{q}}(t) = \begin{bmatrix} -\dfrac{1}{C_2 R_2} & \dfrac{1}{C_2} & 0 \\ -\dfrac{1}{L} & 0 & \dfrac{1}{L} \\ 0 & -\dfrac{1}{C_1} & -\dfrac{1}{C_1 R_1} \end{bmatrix} \vec{q}(t) + \begin{bmatrix} 0 \\ 0 \\ \dfrac{1}{C_1 R_1} \end{bmatrix} x(t),$$

$$y(t) = \begin{bmatrix} 1 & 0 & 0 \end{bmatrix} \vec{q}(t).$$

As this example demonstrates, it is straightforward to rewrite the equations that describe a network into a state space representation.

▲▲

11.4.2 Phase Variables

It is easy to write the differential equations that describe a network into a state space representation because the underlying differential equations (11.4.2) are in first-order form. For other physical systems, the differential equations don't naturally arise in first-order form. For example, the differential equations for a mass-spring-damper in Section 6.4 are naturally second-order. For these systems we can use the following procedure to reduce the differential equation to first-order form. We will illustrate this procedure with the third-order differential equation

$$\dddot{y}(t) + a_2 \ddot{y}(t) + a_1 \dot{y}(t) + a_0 y(t) = b_0 x(t).$$ (11.4.6)

We chose the state variables as

$$q_1(t) = y(t), \quad q_2(t) = \dot{y}(t), \quad q_3(t) = \ddot{y}(t).$$ (11.4.7)

Differentiating the first two state variables we obtain

$$\dot{q}_1(t) = \dot{y}(t) = q_2(t),$$ (11.4.8)
$$\dot{q}_2(t) = \ddot{y}(t) = q_3(t).$$

Differentiating the third state variable and using the system differential equation in (11.4.6) we obtain

$$\dot{q}_3(t) = \dddot{y}(t) = -a_2 \ddot{y}(t) - a_1 \dot{y}(t) - a_0 y(t) + b_0 x(t)$$ (11.4.9)
$$= -a_2 q_3(t) - a_1 q_2(t) - a_0 q_1(t) + b_0 x(t).$$

Also note that the output signal is given in (11.4.7). The scalar equations in (11.4.8) and (11.4.9) are rewritten into matrix form to obtain the following realization.

Realization 11.4.2: Given the differential equation

$$\dddot{y}(t) + a_2 \ddot{y}(t) + a_1 \dot{y}(t) + a_0 y(t) = b_0 x(t)$$

with the transfer function

$$\frac{Y(s)}{X(s)} = \frac{b_0}{s^3 + a_2 s^2 + a_1 s + a_0},$$

a realization of this transfer function is

$$\dot{\bar{q}}(t) = \begin{bmatrix} \dot{q}_1(t) \\ \dot{q}_2(t) \\ \dot{q}_3(t) \end{bmatrix} = \begin{bmatrix} 0 & 1 & 0 \\ 0 & 0 & 1 \\ -a_0 & -a_1 & -a_2 \end{bmatrix} \bar{q}(t) + \begin{bmatrix} 0 \\ 0 \\ b_0 \end{bmatrix} x(t),$$

$$y(t) = \begin{bmatrix} 1 & 0 & 0 \end{bmatrix} \bar{q}(t) + [0]x(t).$$

A block diagram of this realization is shown in Figure 11.4.2.

▲▲

Terminology: The state variables of Realization 11.4.2 are called <u>phase variables</u>.

The differential equation in (11.4.6) doesn't have any derivatives of the input signal on the right-hand side of the equations. Hence, the transfer function has only a constant, not a polynomial, in the numerator. This realization is restricted to equations of this form. If the differential equation (11.4.6) has derivatives of the input signal, then one of the realizations in Section 11.2 must be used.

Example 11.4.3: This example illustrates the use of Realization 11.4.2. Consider the mass-spring-damper system in Example 11.2.4. The corresponding differential equation is

$$\ddot{y}_{st}(t) + \frac{c_{st}}{m_{st}} \dot{y}_{st}(t) + \frac{k_{st}}{m_{st}} y_{st}(t) = \frac{1}{m_{st}} f_{st}(t). \tag{11.4.10}$$

This equation fits the form required by Realization 11.4.2. We assign the state variables

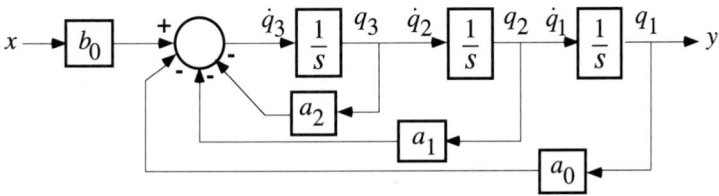

Figure 11.4.2 Block Diagram of Realization 11.4.2

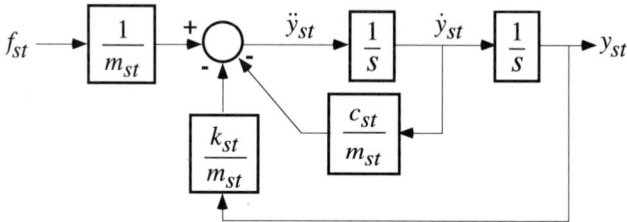

Figure 11.4.3 Realization 11.4.2 of the Mass-Spring-Damper System

$$q_1(t) = y_{st}(t),$$ (11.4.11)

$$q_2(t) = \dot{y}_{st}(t).$$

Note that the state variables are the position and velocity of the mass. The state space equations are

$$\dot{\vec{q}}(t) = \begin{bmatrix} 0 & 1 \\ -\dfrac{k_{st}}{m_{st}} & -\dfrac{c_{st}}{m_{st}} \end{bmatrix} \begin{bmatrix} q_1(t) \\ q_2(t) \end{bmatrix} + \begin{bmatrix} 0 \\ \dfrac{1}{m_{st}} \end{bmatrix} f_{st}(t),$$ (11.4.12)

$$y_{st}(t) = \begin{bmatrix} 1 & 0 \end{bmatrix} \vec{q}(t).$$

Note: The state space equations in (11.4.12) should be compared to the state space equations in (11.2.10).

The block diagram for this system is shown in Figure 11.4.3.

▲▲

11.4.3 Incorporation of Initial Conditions into State Space Equations

Suppose we are given the following differential equation

$$\ddot{y}(t) + a_2 \ddot{y}(t) + a_1 \dot{y}(t) + a_0 y(t) = b_2 \ddot{x}(t) + b_1 \dot{x}(t) + b_0 x(t)$$ (11.4.13)

$$\ddot{y}(0) = y_2, \quad \dot{y}(0) = y_1, \quad y(0) = y_0$$

We would like to translate this differential equation into a state space representation, and then simulate the system for a given input signal $x(t)$. We can translate the differential equation into a state space representation

$$\dot{\vec{q}}(t) = A\vec{q}(t) + Bx(t), \quad \vec{q}(0) = \vec{q}_0.$$ (11.4.14)

$$y(t) = C\vec{q}(t) + Dx(t)$$

using any of the canonical forms introduced in Section 11.2 or this section. The question is how to translate the initial conditions on the differential equation in (11.4.13) into appropriate initial conditions for the state space representation in

(11.4.14). It turns out that this calculation is straightforward. Since the initial conditions in (11.4.13) are independent of the input signal, assume that $x(t) \equiv 0$. Then (11.4.14) is reduced to

$$\dot{\vec{q}}(t) = A\vec{q}(t), \quad \vec{q}(0) = \vec{q}_0. \tag{11.4.15}$$
$$y(t) = C\vec{q}(t).$$

Next note that the value of $y(t)$ at $t = 0$ is

$$y(0) = C\vec{q}(0). \tag{11.4.16}$$

This equation doesn't fully specify the initial condition for the state space system, $\vec{q}(0)$. We haven't used the information on the derivatives of the output signal yet. Taking a derivative of the output signal in (11.4.15) we obtain

$$\dot{y}(t) = C\dot{\vec{q}}(t) = C\big(A\vec{q}(t)\big) = CA\vec{q}(t), \tag{11.4.17}$$

or

$$\dot{y}(0) = CA\vec{q}(0).$$

Combining (11.4.16) and (11.4.17) we get

$$\begin{bmatrix} y(0) \\ \dot{y}(0) \end{bmatrix} = \begin{bmatrix} C \\ CA \end{bmatrix} \vec{q}(0). \tag{11.4.18}$$

Repeating this calculation for the second derivative of $y(t)$ (11.4.17) becomes

$$\ddot{y}(t) = CA\dot{\vec{q}}(t) = CA\big(A\vec{q}(t)\big), \tag{11.4.19}$$

or

$$\ddot{y}(0) = CA^2\vec{q}(0).$$

Using (11.4.19) in (11.4.18) we obtain

$$\begin{bmatrix} y(0) \\ \dot{y}(0) \\ \ddot{y}(0) \end{bmatrix} = \begin{bmatrix} C \\ CA \\ CA^2 \end{bmatrix} \vec{q}(0) = O\vec{q}_0. \tag{11.4.20}$$

Now consider the matrix equation in (11.4.20). Since the differential equation in (11.4.13) is third-order, the state space representation in (11.4.14) will have three states. Furthermore, the output matrix C will be a 1 x 3 matrix. Hence, the matrix O in (11.4.20) is a square matrix. If the inverse of O exists, then we can solve (11.4.20) by multiplying on the left by its inverse. We obtain

$$\vec{q}_0 = O^{-1} \begin{bmatrix} y(0) \\ \dot{y}(0) \\ \ddot{y}(0) \end{bmatrix} \tag{11.4.21}$$

which expresses the initial conditions of the state space representation (11.4.14) in terms of the initial conditions on the differential equation in (11.4.13). The initial condition in (11.4.21) can then be used to solve for the states and output signal in (11.4.14) given the input signal $x(t)$.

In general, the differential equation in (11.4.13) will be nth order. For an nth-order differential equation the matrix O will be given by

$$O = \begin{bmatrix} C \\ CA \\ \vdots \\ CA^{n-1} \end{bmatrix}. \tag{11.4.22}$$

Terminology: The matrix O is called the <u>observability matrix</u> for reasons explained in advanced treatments of state space representations.

An nth-order differential equation (11.4.13) will correspond to a state space representation (11.4.14) with n states, and the output matrix C will be a 1 x n matrix. Therefore, the matrix O will be an n x n matrix.

In order to solve for the initial conditions in (11.4.20), the inverse of the matrix O must exist. This condition must be explicitly checked for a given state space representation, for there exist state space representations for which this matrix is singular. In this case, the initial conditions of the state space representation are not uniquely related to the differential equation. The investigation of the properties of this matrix is a gateway into the advanced study of state space representations.

Example 11.4.4:   In this example we will illustrate the use of the observability matrix to translate the initial conditions from a differential equation to the state space model. Consider the differential equation

$$\ddot{y}(t) + 6\dot{y}(t) + 8y(t) = \dot{x}(t) + x(t), \tag{11.4.23}$$
$$\dot{y}(0) = 3, \quad y(0) = 1.$$

This differential equation is discussed in Example 9.4.1.   Suppose we wish to translate this differential equation into a state space representation.   Using Realization 11.2.2, a state space representation of (11.4.23) is given by

$$\dot{\vec{q}}(t) = \begin{bmatrix} 0 & -8 \\ 1 & -6 \end{bmatrix} \vec{q}(t) + \begin{bmatrix} 1 \\ 1 \end{bmatrix} x(t), \tag{11.4.24}$$
$$y(t) = \begin{bmatrix} 0 & 1 \end{bmatrix} \vec{q}(t).$$

For this example, $n = 2$.   The observability matrix is

$$O = \begin{bmatrix} C \\ CA \end{bmatrix} = \begin{bmatrix} 0 & 1 \\ \hline 1 & -6 \end{bmatrix}. \qquad (11.4.25)$$

Therefore, the initial conditions for the state space representation are given by

$$\vec{q}_0 = \begin{bmatrix} 9 \\ 1 \end{bmatrix} = O^{-1} \begin{bmatrix} y(0) \\ \dot{y}(0) \end{bmatrix} = \begin{bmatrix} 0 & 1 \\ 1 & -6 \end{bmatrix}^{-1} \begin{bmatrix} 1 \\ 3 \end{bmatrix}. \qquad (11.4.26)$$

Using these initial conditions with a given input signal, the output signal is determined.

▲▲

11.4.4 MATLAB Experiments

The MATLAB command **lsim** is used to simulate the output signal of a state space system with nonzero initial conditions. The input arguments include the system, the input signal vector, the time vector, and the vector of initial conditions. The following M-file simulates the output signal for Example 11.4.4.

Example 11.4.4: (*continued*) Consider again the system in Example 11.4.4. Suppose the input signal is a unit step function. The following MATLAB M-file plots the corresponding output signal.

```
clear
% define the system
A = [0,-8;1,-6];
B = [1;1];
C = [0,1];
D = 0;
H = ss(A,B,C,D);
y0 = [1;3];
% calculate initial states
Obs = [C;C*A];          % observability matrix
q0 = inv(Obs)*y0;       % initial states
% define input signal
t = linspace(0,4,400);
x = ones(size(t));
% simulate output signal
y = lsim(H,x,t,q0);
plot(t,y)
xlabel('time')
ylabel('output signal')
```

▲▲

Exploratory Exercise 11.4.5: Write a MATLAB M-file to simulate the response of the mass-spring-damper system in Example 11.4.3 when the applied force is a sinusoid and the mass is initially displaced by 1 unit.

▲▲

11.5 MULTIVARIABLE SYSTEMS

11.5.1 Introduction

The systems we have considered up to this point have had one input signal and one output signal. Obviously, a real system can have more than one input or output signal. An automobile is an example of a system with more than one input signal (steering, brake, and accelerator pedal) and more than one output signal (position and velocity in two directions). We distinguish these systems with many input and/or output signals from the systems we have considered previously.

Definition 6.1.6: A system with one input signal and one output signal is called an single-input-single-output (SISO) system. A system with more than one input signal and/or more than one output signal is called an multivariable (MIMO) system.

▲▲

Terminology: A multivariable system is defined by the number of input and output signals. In particular, it is not defined by the number of states in a state space representation. That is, a state space representation with 20 state variables and one input signal and one output signal is a SISO system.

In this text we will focus on the tools that allow us to analyze and synthesize a SISO system. Some of these tools can be extended to MIMO systems. Therefore, in this section we will discuss the modeling of MIMO systems. There are two ways to represent a multivariable system: using a transfer function matrix or using state space equations. We discuss each in turn.

Computer Usage: The matrix format of state space multivariable systems makes them the model of choice for many CAD packages. These representations are quite useful when analyzing complex (or simple) systems, particularly for simulating these systems. Usually, we wish to look at several signals at several points in the block diagram. By defining and entering a multivariable system into the computer we save ourselves the effort of entering several SISO systems for the several signals of interest. Here again we see that the state space representation plays an important role in the entry of a system representation into the computer.

11.5.2 State Space Representations

The state space representation of a MIMO system is only a minor variation of a state space representation of a SISO system. The matrices only need to be enlarged to accommodate the extra input and output signals. The following example illustrates the concept.

Example 11.5.1: We have discussed the proof-mass actuator in Section 6.5. This device is not useful as we have described it because almost any input signal will cause the proof-mass to drift toward its stops as indicated by its impulse and step response in Section 6.5. Therefore, we must modify this device. The way we

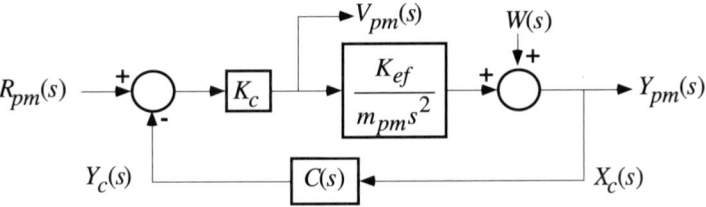

Figure 11.5.1 Compensated Proof-Mass Actuator

propose to modify the device is to add an additional system, called a <u>compensator</u>, in a feedback loop as shown in Figure 11.5.1. The compensator would be an electronic system with the transfer function

$$\frac{Y_c(s)}{X_c(s)} = \frac{s+z}{s+p} = 1 + \frac{z-p}{s+p}.$$ (11.5.1)

The input signal to the compensator is the output signal from the sensor on the base of the proof-mass that senses the displacement of the proof-mass. (See Subsection 10.3.3.) The output signal of this system would be an input signal into an op amp which would implement the summing junction. An external <u>reference</u> signal[2] $r_{pm}(t)$ (a voltage signal) is also an input signal into the summing junction. The output signal of the summing op amp would be an input signal to a gain op amp which would in turn feed the input signal to the power electronics of the motor.[3]

We have also added an external force input signal $w(t)$ acting on the proof-mass in Figure 11.5.1. This signal could represent an external load or a nonconservative force due to the mechanical construction of the actuator. Generally, this signal is beyond our control; it is present whether we want it or not. Hence, this signal is called a <u>disturbance</u> signal.

Finally we have identified the output signal $v_{pm}(t)$. This signal is the input signal to the power electronics of the actuator. Given a reference input signal, we want to monitor this signal to see if this signal exceeds the maximum allowable voltage of the motor.

In this example we will develop a state space model of the compensated proof-mass actuator in Figure 11.5.1. This development begins with the translation of the block diagram in Figure 11.5.1 into an all-integrator block diagram in Figure 11.5.2. This all-integrator block diagram uses the decomposition of the compensator in (11.5.1).

To construct the MIMO state space representation, we proceed as with SISO systems as given in Procedure 10.5.9. First, we assign state variables to the outputs of the integrators as shown in Figure 11.5.2. From the block diagram we write the equations for each derivative of the state variables. For the first state we have

---

[2] The reference signal would be selected to control the displacement of the proof-mass. We would like the proof-mass to follow the reference signal.

[3] In this description we assume that all of the signals are voltages.

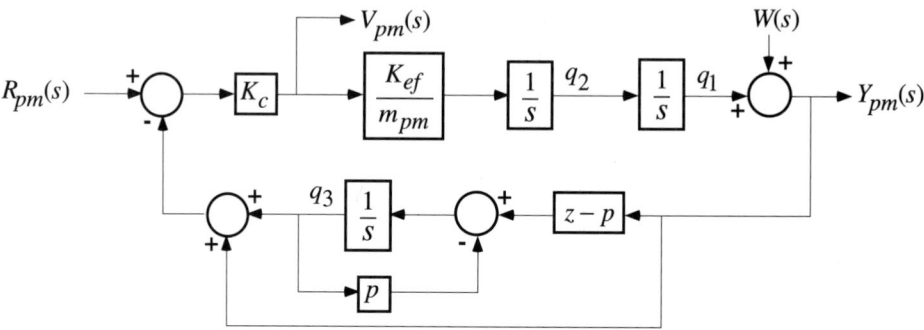

Figure 11.5.2 All-Integrator Block Diagram of the Modified Proof-Mass Actuator

$$\dot{q}_1(t) = q_2(t).\tag{11.5.2}$$

The second state yields

$$\dot{q}_2(t) = \frac{K_c K_{ef}}{m_{pm}} \left[r_{pm}(t) - \left(y_{pm}(t) + q_3(t) \right) \right]\tag{11.5.3}$$

$$= \frac{K_c K_{ef}}{m_{pm}} r_{pm}(t) - \frac{K_c K_{ef}}{m_{pm}} q_1(t) - \frac{K_c K_{ef}}{m_{pm}} w(t) - \frac{K_c K_{ef}}{m_{pm}} q_3(t).$$

The third state equation is

$$\dot{q}_3(t) = -p q_3(t) + (z - p) y_{pm}(t) = -p q_3(t) + (z - p) q_1(t) + (z - p) w(t).\tag{11.5.4}$$

There are two output signals for this system, so there are two equations for these signals. We have

$$y_{pm}(t) = q_1(t) + w(t),\tag{11.5.5}$$

and

$$v_{pm}(t) = K_c \left[r_{pm}(t) - \left(y_{pm}(t) + q_3(t) \right) \right]\tag{11.5.6}$$

$$= K_c r_{pm}(t) - K_c q_1(t) - K_c w(t) - K_c q_3(t).$$

In the final expressions (11.5.2) - (11.5.6), only the state variables and the input and output signals may appear. We define the state vector as

$$\vec{q}(t) = \begin{bmatrix} q_1(t) \\ q_2(t) \\ q_3(t) \end{bmatrix}.\tag{11.5.7}$$

Then the state equations in (11.5.2) - (11.5.6) in matrix form become

$$\dot{\vec{q}}(t) = \begin{bmatrix} 0 & 1 & 0 \\ \dfrac{K_c K_{ef}}{m_{pm}} & 0 & -\dfrac{K_c K_{ef}}{m_{pm}} \\ (z-p) & 0 & -p \end{bmatrix} \vec{q}(t) + \begin{bmatrix} 0 \\ \dfrac{K_c K_{ef}}{m_{pm}} \\ 0 \end{bmatrix} r_{pm}(t) + \begin{bmatrix} 0 \\ -\dfrac{K_c K_{ef}}{m_{pm}} \\ (z-p) \end{bmatrix} w(t),$$ (11.5.8)

$$y_{pm}(t) = \begin{bmatrix} 1 & 0 & 0 \end{bmatrix} \vec{q}(t) + w(t),$$

$$v_{pm}(t) = \begin{bmatrix} -K_c & 0 & -K_c \end{bmatrix} \vec{q}(t) + K_c r_{pm}(t) - K_c w(t).$$

Since there are more than one input signal and output signal, we define the input and output signal vectors as

$$\vec{x}(t) = \begin{bmatrix} r_{pm}(t) \\ w(t) \end{bmatrix}, \quad \text{and} \quad \vec{y}(t) = \begin{bmatrix} y_{pm}(t) \\ v_{pm}(t) \end{bmatrix}.$$ (11.5.9)

Using these vectors and the equations in (11.5.8), we obtain

$$\dot{\vec{q}}(t) = \begin{bmatrix} 0 & 1 & 0 \\ \dfrac{K_c K_{ef}}{m_{pm}} & 0 & -\dfrac{K_c K_{ef}}{m_{pm}} \\ (z-p) & 0 & -p \end{bmatrix} \vec{q}(t) + \begin{bmatrix} 0 & 0 \\ \dfrac{K_c K_{ef}}{m_{pm}} & -\dfrac{K_c K_{ef}}{m_{pm}} \\ 0 & (z-p) \end{bmatrix} \vec{x}(t),$$ (11.5.10)

$$\vec{y}(t) = \begin{bmatrix} 1 & 0 & 0 \\ -K_c & 0 & -K_c \end{bmatrix} \vec{q}(t) + \begin{bmatrix} 0 & 1 \\ K_c & -K_c \end{bmatrix} \vec{x}(t).$$

▲▲

In general, a MIMO state space representation has the form

$$\dot{\vec{q}}(t) = A\vec{q}(t) + B\vec{x}(t),$$ (11.5.11)
$$\vec{y}(t) = C\vec{q}(t) + D\vec{x}(t).$$

Notation: We denote the number of input signals by n_i. We denote the number of output signals by n_o.

We make the following observations.

1. All of the remarks on SISO state variable equations apply to multivariable state space representations.
2. Since there is more than one input signal, we create a vector of input signals. Then the input matrix B has more than one column. The number of input signals is the same as the number of columns of B. That is, B is an $n \times n_i$ matrix.

3. Since there is more than one output signal, we create a vector of output signals. Then the output matrix C has more than one row. The number of output signals is the same as the number of rows of C. That is, C is an n_o x n matrix.
4. The direct feedthrough term represents a coupling between each input signal and each output signal. These coupling terms are arranged into a direct feedthrough matrix D. This matrix has the same number of rows as there are rows of C. It has the same number of columns as B. That is, D is an n_o x n_i matrix.
5. The number of input signals need not equal the number of output signals.

The above example shows that constructing a state space representation for a MIMO system is the same as constructing a state space representation for a SISO except that we must define the input and output signals to be vectors.

11.5.3 Transfer Functions

An alternative way of representing a MIMO system is based on transfer functions. Since there are multiple input and output signals, there are multiple transfer functions. We arrange these transfer functions into a transfer function matrix $H(s)$.

Notation: We will use a bold font, say $H(s)$, to indicate a matrix of rational functions.

For example, in the modified proof-mass actuator above, we group the input and output signals into vectors

$$\begin{bmatrix} Y_{pm}(s) \\ V_{pm}(s) \end{bmatrix} \quad \text{and} \quad \begin{bmatrix} R_{pm}(s) \\ W(s) \end{bmatrix}. \tag{11.5.12}$$

Since this system has two input and output signals, these signals are related by a 2 x 2 matrix

$$\begin{bmatrix} Y_{pm}(s) \\ V_{pm}(s) \end{bmatrix} = \begin{bmatrix} H_{11}(s) & H_{12}(s) \\ H_{21}(s) & H_{22}(s) \end{bmatrix} \begin{bmatrix} R_{pm}(s) \\ W(s) \end{bmatrix} = H(s) \begin{bmatrix} R_{pm}(s) \\ W(s) \end{bmatrix}. \tag{11.5.13}$$

When we consider systems with multiple input or output signals we must consider a matrix of transfer functions. A complete treatment transfer function matrices is beyond the scope of this text.

Suppose we set $W(s) = 0$ in (11.5.13). Then the transfer function between $R_{pm}(s)$ and $Y_{pm}(s)$ is $H_{11}(s)$. This observation shows how to compute the transfer function matrix of a MIMO system. All input signals except one are set to zero. Then the individual transfer functions are computed from the remaining input signal to all output signals. This process is repeated for each input signal until all entries of the transfer function matrix are known.

In Theorem 11.1.2 we developed the formula for calculating the transfer function from a state space representation. This formula also holds for MIMO systems. That is, given a MIMO system we have

$$\vec{Y}(s) = \left[C(sI - A)^{-1}B + D\right]\vec{X}(s) = \boldsymbol{H}(s)\vec{X}(s). \tag{11.5.14}$$

Of course, the ratio of the output signal vector over the input signal vector no longer makes sense.

11.5.4 MATLAB Experiments

MATLAB naturally accommodates multivariable systems. In the state space representations, it is only necessary to expand the dimension of the input and output matrices. MATLAB also handles transfer function matrices. The reader is referred to the MATLAB manual for details.

There are frequent references in the MATLAB manual to "indices of the input vector." This terminology refers to MIMO systems where the ith input index refers to the ith input signal.

The following M-file simulates the multivariable system in Example 11.5.1. To construct the system model this M-file begins by constructing the state space model for the SISO system from the reference input signal to the displacement of the proof-mass. This SISO model is built using the **feedback** command which automatically forms the interconnection of two subsystems. In order to generate the MIMO state space model in Example 11.5.1, it is necessary to augment the input matrix, the output matrix, and the direct feedthrough matrix of the composite system model in Example 11.5.1 according to (11.5.10). To change each of these matrices, it is necessary to: (1) retrieve the input or output matrix of the composite system with the **get** command, (2) create the new input or output matrix, and (3) insert the new input, output and direct feedthrough matrices into the state space model using the **set** command. The reader is referred to the MATLAB manual for a complete discussion of defining, accessing, and changing the properties of LTI models.

Example 11.5.1: (*continued*) M-file

```
clear
% define state space system for proof-mass
mpm = 10;                    % proof-mass mass
Kef = 1;                     % gain
Apm = [0,1;0,0];             % state matrix
Bpm = [0;Kef/mpm];           % input matrix
Cpm = [1,0];                 % output matrix
Dpm = 0;                     % direct feedthrough matrix
Hpm = ss(Apm,Bpm,Cpm,Dpm);   % create system

% define compensator transfer function
p = -2;                 % pole of compensator
z = -0.5;               % zero of compensator
Kc = 20;                % loop gain
hc = zpk(z,p,Kc);       % create compensator
Hc = ss(hc);            % transform to state space
```

```
% calculate the closed loop system
Hcloop = feedback(Hpm,hc);

% create MIMO system
% augment input matrix
Bc = get(Hc,'b');              % retrieve input matrix of compensator
B1 = [Kc*Bpm;Bc];             % create addition input
B = get(Hcloop,'b');           % retrieve input matrix of composite system
Bt = [B,B1];                   % augment input matrix
% augment output matrix
C1 = [-Kc,0,-Kc];              % create additional output
C = get(Hcloop,'c');           % retrieve output matrix of composite system
Ct = [C;C1];                   % augment output matrix
% augment direct feedthrough matrix
Dt = [0,1;Kc,-Kc];            % create direct feedthrough matrix
% redefine composite system
set(Hcloop,'b',Bt,'c',Ct,'d',Dt);
display(Hcloop)

% simulate system
t = linspace(0,30,400);       % time vector
rpm = 0.15*sin(0.5*t);        % reference signal
w = .01*sin(10*t);            % disturbance signal
u = [rpm;w]';                 % input signal vector
y = lsim(Hcloop,u,t);         % simulate system
figure(1)
plot(t,y(:,1))
xlabel('time')
ylabel('displacement')
title('Original Output')
figure(2)
plot(t,y(:,2))
xlabel('time')
ylabel('motor voltage')
title('Augmented Output')
```

▲▲

Exploratory Exercise 11.5.2: Change the frequency of the disturbance signal to 2 rad/sec. What is the change in the output signals?

▲▲

Exploratory Exercise 11.5.3: Define the velocity of the proof-mass to be a third output signal. Modify the M-file above to include this output signal in the composite system model.

▲▲

Exploratory Exercise 11.5.4: Modify the above M-file to extract the subsystem defined by the disturbance input signal and the position of the proof-mass. Calculate the transfer function.

▲▲

11.6 CHAPTER SUMMARY

In this chapter we have studied the relationship between transfer functions, all-integrator block diagrams, and state space representations completing the discussion begun in the last chapter. The transfer function is calculated from the state space representation by a formula (Theorem 11.1.2). Given the state space representation

$$\dot{\vec{q}}(t) = A\vec{q}(t) + Bx(t), \quad \vec{q}_0 = 0, \tag{11.6.1}$$

$$y(t) = C\vec{q}(t) + Dx(t),$$

of a system where the initial conditions are set to zero. Then the transfer function is given by

$$\frac{Y(s)}{X(s)} = C(sI - A)^{-1}B + D. \tag{11.6.2}$$

This formula shows that if the system is represented by an all-integrator block diagram with n integrators, the state space representation will be of order n, the transfer function will be at most order n, and the poles of the transfer function will be among the eigenvalues of the state matrix A (Theorem 11.1.5).

In this chapter we have also developed the relationship between a differential equation, an all-integrator block diagram, a state space representation, and a transfer function. This relationship was expressed in two different forms: Realization 11.2.2 and Realization 11.2.6. When no derivatives of the input signal appear on the right-hand side of the differential equation, then Realization 11.4.2 can also be used. These realizations allow one representation to be easily translated into another representation.

The choice of state variables is not unique. Two state vectors contain the same information, however, if they are related by a state transformation matrix $\tilde{q}(t) = Tq(t)$. Then the two state space representations are said to be (dynamically) equivalent (Definition 11.3.3). Dynamically equivalent systems have different (internal) representations, but they have the same input/output relationship (Theorem 11.3.4).

State variable models are often used because the states have physical meaning. Here the state space equations are written directly from the physics of the process. This idea is illustrated by example in Section 11.4.

Many systems have more than one input and/or output signal; they are multivariable systems. It is straightforward to extend state space representations to these systems. The input and output matrices are simply expanded as described in Section 11.5. It is also possible to define transfer function matrices for multivariable systems as described in this section.

11.7 HOMEWORK FOR CHAPTER 11

Homework Problems for Section 11.1

11.1.1   The state space equations of the *LRC* network in Section 10.3 are

$$\dot{\vec{q}}(t) = \begin{bmatrix} -\dfrac{1}{R_2 C_2} & \dfrac{1}{C_2} & 0 \\ -\dfrac{1}{L} & 0 & \dfrac{1}{L} \\ 0 & -\dfrac{1}{C_1} & -\dfrac{1}{R_1 C_1} \end{bmatrix} \vec{q}(t) + \begin{bmatrix} 0 \\ 0 \\ \dfrac{1}{C_1 R_1} \end{bmatrix} v_{in}(t),$$

$$v_0(t) = \begin{bmatrix} 1 & 0 & 0 \end{bmatrix} \vec{q}(t).$$

Find the transfer function by hand calculation using the formula for calculation of transfer function from state equations.   Use the following parameter values: $R_1 = 1\ \Omega$, $R_2 = 1\ \Omega$, $C_1 = 1$ F, $C_2 = 1$ F, and $L = 1$ H.

11.1.2   Consider the systems represented by state space equations.   Find the transfer function using the matrix formula.

(i) $\dot{\vec{q}} = \begin{bmatrix} 0 & 0 & 2 \\ 5 & 0 & -1 \\ 3 & 0 & 0 \end{bmatrix} \vec{q} + \begin{bmatrix} 0 \\ 1 \\ 0 \end{bmatrix} x$

$y = \begin{bmatrix} 6 & 0 & 0 \end{bmatrix} \vec{q} + 4x$

(ii) $\dot{\vec{q}} = \begin{bmatrix} 1 & 0 \\ 0 & -1 \end{bmatrix} \dot{\vec{q}} + \begin{bmatrix} 3 \\ 5 \end{bmatrix} x$

$y = \begin{bmatrix} 2 & 1 \end{bmatrix} \dot{\vec{q}}$

(iii) $\dot{\vec{q}} = \begin{bmatrix} -1 & 1 \\ 0 & 2 \end{bmatrix} \vec{q} + \begin{bmatrix} 0 \\ 3 \end{bmatrix} x$

$y = \begin{bmatrix} 1 & 0 \end{bmatrix} \vec{q} + 4x$

(iv) $\dot{\vec{q}} = \begin{bmatrix} -\sigma & \omega \\ -\omega & -\sigma \end{bmatrix} \vec{q} + \begin{bmatrix} 1 \\ 1 \end{bmatrix} x$

$y = \begin{bmatrix} 1 & 1 \end{bmatrix} \vec{q}$

(v) $\dot{\vec{q}} = \begin{bmatrix} 0 & 1 \\ -\omega_n^2 & -2\zeta\omega_n \end{bmatrix} \vec{q} + \begin{bmatrix} 1 \\ 0 \end{bmatrix} x$

$y = \begin{bmatrix} 1 & 0 \end{bmatrix} \vec{q}$

(vi) $\dot{\vec{q}} = \begin{bmatrix} 0 & 1 \\ -\omega_n^2 & -2\zeta\omega_n \end{bmatrix} \vec{q} + \begin{bmatrix} 0 \\ 1 \end{bmatrix} x$

$y = \begin{bmatrix} 1 & 0 \end{bmatrix} \vec{q}$

11.1.3   Consider the system

$$\dot{\vec{q}}(t) = \begin{bmatrix} \lambda_1 & 0 \\ 0 & \lambda_2 \end{bmatrix} \vec{q}(t) + \begin{bmatrix} b_1 \\ b_2 \end{bmatrix} x(t),$$

$$y(t) = \begin{bmatrix} c_1 & c_2 \end{bmatrix} \vec{q}(t) + [d] x(t).$$

(a) Compute the transfer function of this system.
(b) What is the order of the transfer function when $\lambda_1 = \lambda_2$?

11.1.4 Consider the system

$$\begin{bmatrix} \dot{q}_1(t) \\ \dot{q}_2(t) \end{bmatrix} = \begin{bmatrix} 0 & 1 \\ -\omega_n^2 & -2\zeta\omega_n \end{bmatrix}\begin{bmatrix} q_1(t) \\ q_2(t) \end{bmatrix} + \begin{bmatrix} 0 \\ b \end{bmatrix} f_{st}(t),$$

$$y_{st}(t) = \begin{bmatrix} 1 & 0 \end{bmatrix}\bar{q}(t).$$

(a) Draw the all-integrator block diagram of this system.
(b) Suppose we modify this system by the addition of the feedback

$$f_{st}(t) = -k_p y_{st}(t) + r(t).$$

Add this feedback to the block diagram of (a).   Now we call the composite system the closed loop system.
(c) Find the state space representation of the closed loop system.
(d) Find the transfer function of the closed loop system.
(e) How do the poles of the closed loop system vary as a function of the feedback gain k_p?

11.1.5 Consider the system

$$\begin{bmatrix} \dot{q}_1(t) \\ \dot{q}_2(t) \end{bmatrix} = \begin{bmatrix} 0 & 1 \\ -\omega_n^2 & -2\zeta\omega_n \end{bmatrix}\begin{bmatrix} q_1(t) \\ q_2(t) \end{bmatrix} + \begin{bmatrix} 0 \\ b \end{bmatrix} f_{st}(t),$$

$$y_v(t) = \begin{bmatrix} 0 & 1 \end{bmatrix}\bar{q}(t)$$

(a) Draw the all-integrator block diagram of this system.
(b) Suppose we modify this system by the addition of the feedback

$$f_{st}(t) = -k_v y_v(t) + r(t).$$

Add this feedback to the block diagram of (a).   Now we call the composite system the closed loop system.
(c) Find the state space representation of the closed loop system.
(d) Find the transfer function of the closed loop system.
(e) How do the poles of the closed loop system vary as a function of the feedback gain k_v?

11.1.6 Consider the system

$$\begin{bmatrix} \dot{q}_1(t) \\ \dot{q}_2(t) \end{bmatrix} = \begin{bmatrix} 0 & 1 \\ -\omega_n^2 & -2\zeta\omega_n \end{bmatrix}\begin{bmatrix} q_1(t) \\ q_2(t) \end{bmatrix} + \begin{bmatrix} 0 \\ b \end{bmatrix} f_{st}(t),$$

$$y_{st}(t) = \begin{bmatrix} 1 & 0 \end{bmatrix}\bar{q}(t)$$

(a) Draw the all-integrator block diagram of this system.

(b) Suppose we modify this system by the addition of the feedback

$$f_{st}(t) = -\begin{bmatrix} k_p & k_v \end{bmatrix}\begin{bmatrix} q_1(t) \\ q_2(t) \end{bmatrix} + r(t)$$

Add this feedback to the block diagram of (a).   Now we call the composite system the closed loop system.

(c) Find the state space representation of the closed loop system.
(d) Find the transfer function of the closed loop system.
(e) How do the poles of the closed loop system vary as a function of the feedback gains $[k_p, k_v]$?

11.1.7 Consider the system

$$\dot{\vec{q}}(t) = \begin{bmatrix} a_{11} & A_{12} \\ A_{21} & A_{22} \end{bmatrix}\vec{q}(t) + \begin{bmatrix} 1 \\ 0 \end{bmatrix}x(t),$$

$$y(t) = \begin{bmatrix} 1 & 0 \end{bmatrix}\vec{q}(t).$$

Show that the zeros of the transfer function of this system are given by $b(s) = \det(sI - A_{22})$.

11.1.8 Consider the two mass-spring-damper system in Figure P11.1.8. A set of state space equations for this system is

$$\dot{\vec{q}}(t) = \begin{bmatrix} 0 & 1 & 0 & 0 \\ -\dfrac{k_a+k_c}{m_a} & -\dfrac{c_a}{m_a} & \dfrac{k_c}{m_a} & 0 \\ 0 & 0 & 0 & 1 \\ \dfrac{k_c}{m_{st}} & 0 & -\dfrac{k_c}{m_{st}} & 0 \end{bmatrix}\vec{q}(t) + \begin{bmatrix} 0 \\ \dfrac{1}{m_a} \\ 0 \\ 0 \end{bmatrix}f_a(t)$$

$$y_{st}(t) = \begin{bmatrix} 0 & 0 & 10 & 0 \end{bmatrix}\vec{q}(t)$$

(a) Write an M-file to compute the transfer function of this system. Use the following parameter values: $k_a = 10$ N/m, $c_a = 8$ N/m/sec, $m_a = 1$ kg, $k_c = 100$ N/m, and $m_{st} = 0.1$ kg.
(b) Change the transfer function into pole-zero form.

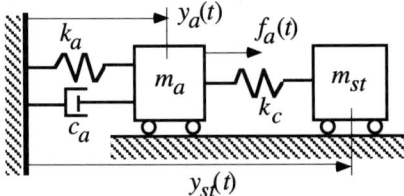

Figure P11.1.8

11.1.9 Consider the two mass-spring-damper system shown in Figure P11.1.9. A set of state space equations for this system is

$$\dot{\vec{q}}(t) = \begin{bmatrix} 0 & 1 & 0 & 0 \\ -\dfrac{k_a+k_c}{m_a} & -\dfrac{c_a}{m_a} & \dfrac{k_c}{m_a} & 0 \\ 0 & 0 & 0 & 1 \\ \dfrac{k_c}{m_{st}} & 0 & -\dfrac{k_c}{m_{st}} & -\dfrac{c_{st}}{m_{st}} \end{bmatrix} \vec{q}(t) + \begin{bmatrix} 0 \\ \dfrac{1}{m_a} \\ 0 \\ 0 \end{bmatrix} f_a(t)$$

$$y_{st}(t) = \begin{bmatrix} 0 & 0 & 10 & 0 \end{bmatrix} \vec{q}(t)$$

(a) Write an M-file to compute the transfer function of this system. Use the following parameter values: $k_a = 10$ N/m, $c_a = 8$ N/m/sec, $m_a = 1$ kg, $k_c = 100$ N/m, $c_{st} = 20$ N/m/sec, and $m_{st} = 100$ kg.

(b) Change the transfer function into pole-zero form.

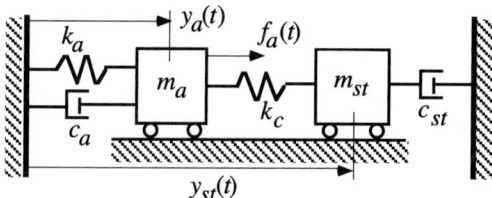

Figure P11.1.9

Homework Problems for Section 11.2

11.2.1 Consider the following systems.
(a) Find a realization of each system using Realization 11.2.2. Draw the corresponding all-integrator block diagram.
(b) Find a realization of each system using Realization 11.2.6. Draw the corresponding all-integrator block diagram.

(i) $\dfrac{Y(s)}{X(s)} = \dfrac{s^2 + 3s + 4}{s^3 + 12s^2 + 2s + 5}$

(iv) $\dfrac{Y(s)}{X(s)} = \dfrac{1}{3s^2 + 2s + 4}$

(ii) $\dfrac{Y(s)}{X(s)} = \dfrac{3s + 6}{s^4 + 3s^2 + 2s + 4}$

(v) $\dfrac{Y(s)}{X(s)} = \dfrac{s^3}{s^3 + 2s^2 + 3}$

(iii) $\dfrac{Y(s)}{X(s)} = \dfrac{5s + 20}{s^4 + 10s^3 + 32s^2 + 80s + 10}$

(vi) $\dfrac{Y(s)}{X(s)} = \dfrac{s + 4}{s(s + 1)}$

11.2.2 For the following transfer function find a realization for this transfer function of the form in Realization 11.2.6. Draw the all-integrator block diagram, and find the state equations.

$$\frac{Y(s)}{X(s)} = H(s) = \frac{\beta_2 s^2 + \beta_1 s + \beta_0}{s^3 + a_2 s^2 + a_1 s + a_0} + b_3$$

11.2.3   Consider the system

$$\dot{\vec{q}}(t) = \begin{bmatrix} \dot{q}_1(t) \\ \dot{q}_2(t) \\ \dot{q}_3(t) \end{bmatrix} = \begin{bmatrix} 0 & 1 & 0 \\ 0 & 0 & 1 \\ -6 & -3 & -2 \end{bmatrix} \vec{q}(t) + \begin{bmatrix} 0 \\ 0 \\ 1 \end{bmatrix} x(t),$$

$$y(t) = \begin{bmatrix} 10 & 5 & 0 \end{bmatrix} \vec{q}(t) + \begin{bmatrix} 0 \end{bmatrix} x(t)$$

(a) Find the transfer function of this system.
(b) Suppose the output signal is fedback to the input signal according to

$$x(t) = -y(t) + r(t)$$

Draw a block diagram of the closed loop system using the transfer function of the system.
(c) Find the state space equations of the closed loop system.
(d) Find the transfer function of the closed loop system.
(e) Find the poles of the original system and the closed loop system.

11.2.4   Consider the system

$$y^{(4)}(t) + 7\dddot{y}(t) + 4\ddot{y}(t) + 3\dot{y}(t) + 6y(t) = 5\ddot{x}(t) + x(t)$$

(a) Find a state space representation of this system.
(b) Draw an all-integrator block diagram of this system.

11.2.5   Consider the system

$$\dot{\vec{q}}(t) = \begin{bmatrix} 0 & 0 & -3 \\ 1 & 0 & -4 \\ 0 & 1 & -2 \end{bmatrix} \vec{q}(t) + \begin{bmatrix} 1 \\ 7 \\ 8 \end{bmatrix} x(t)$$

$$y(t) = \begin{bmatrix} 0 & 0 & 1 \end{bmatrix} \vec{q}(t) + 5x(t)$$

Find the transfer function of this system.

Homework Problems for Section 11.3

11.3.1   (a) Verify the new state space representation in Example 11.3.5 by re-ordering the state variables in the block diagram Figure 11.2.3 and deriving the state space representation from the block diagram.
(b) Verify that the transformation in (11.3.20) will transform the system in (11.3.19) into the system (11.3.21).

11.3.2 Consider the state transformation matrix

$$T = \begin{bmatrix} \alpha & 0 \\ 0 & \beta \end{bmatrix}$$

which is defined for second-order systems. Show that this transformation matrix results in a simple scaling of the state variables.

Homework Problems for Section 11.4

11.4.1 Consider the network shown in Figure P11.4.1. Let the current through the inductor be one state variable, $i_L(t) = q_1(t)$, and the voltage across the resistor be the second state variable, $y(t) = q_2(t)$.
(a) Write a set of state space equations for this system.
(b) Draw the corresponding all-integrator block diagram. Label the state variables.

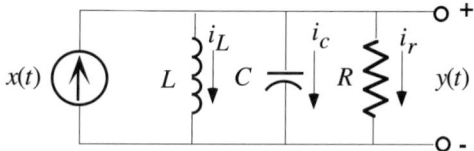

Figure P11.4.1

11.4.2 Consider the network shown in Figures P11.4.2a-c.
(a) Write a set of state space equations for this system.
(b) Draw the corresponding all-integrator block diagram. Label the state variables.

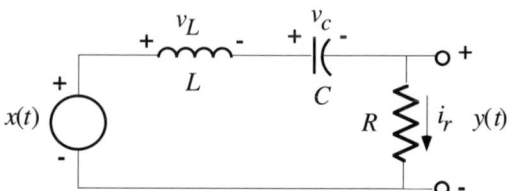

Figure P11.4.2a

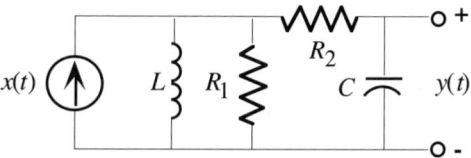

Figure P11.4.2b

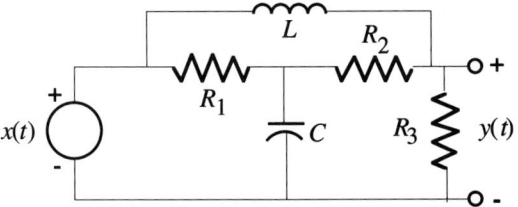

Figure P11.4.2c

11.4.3   Consider the block diagram shown in Figure P11.4.3.   This is a block diagram of an *LRC* network.   The various signals are voltages and currents. The signal $Q(s)$ is the charge delivered to the capacitor.   Let the state variables be $q_1(t) = i(t)$ and $q_2(t) = q(t)$.

(a) Find a set of state space equations for this system.
(b) Reconstruct the network from the block diagram.

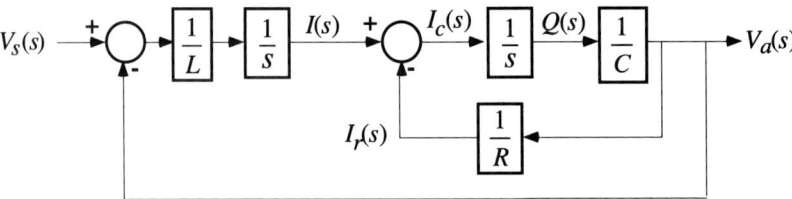

Figure P11.4.3

11.4.4   Consider the network shown in Figure P11.4.4.   Suppose that $v_s(t) = u_s(t)$.

(a) Find a state space representation for this system using the current $i(t)$ and the voltage $v_a(t)$ as state variables.
(b) Write a MATLAB M-file to simulate the output signal $v_a(t)$ of this system.   Plot the output signal.

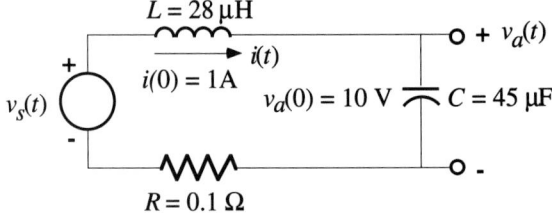

Figure P11.4.4

11.4.5   Show that the observability matrix for the third-order example in Realization 11.2.2 is always nonsingular.   Conclude that the observability matrix for any system in this canonical form is nonsingular.

Homework Problems for Section 11.5

11.5.1 Consider the system shown in Figure P11.5.1.
(a) Find the transfer function matrix of this system.
(b) Find a state space model of this system.

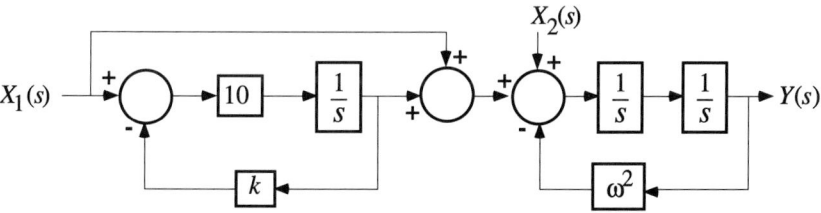

Figure P11.5.1

11.5.2 Consider the system shown in Figure P11.5.2. Find the transfer function matrix of this system.

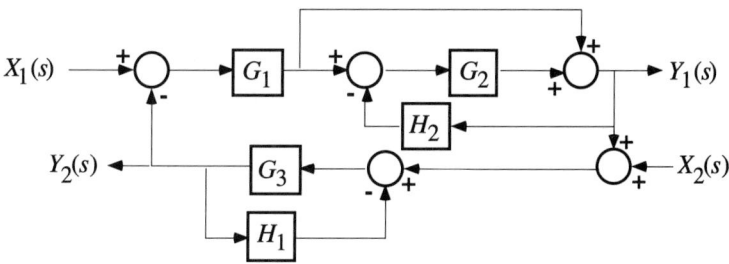

Figure P11.5.2

11.5.3 Consider the system shown in Figure P11.5.3. Find the transfer function matrix of this system.

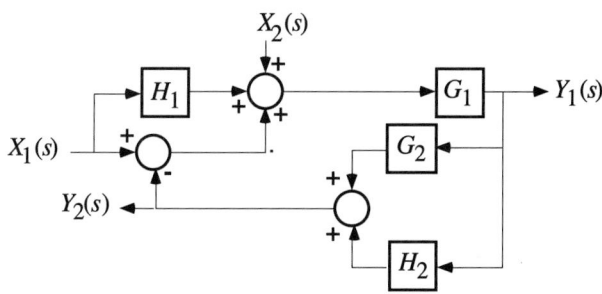

Figure P11.5.3

11.5.4 Consider the system shown in Figure P11.5.4.
(a) Find the transfer function matrix of this system.
(b) Find a state space model of this system.

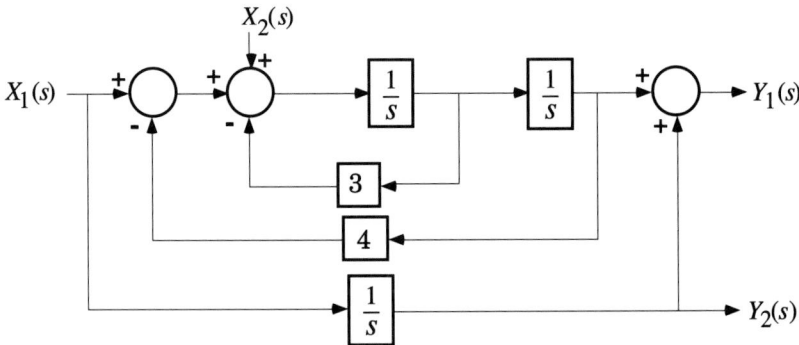

Figure P11.5.4

11.5.5 Consider the mass-spring-damper system in Figure P11.5.5. The parameters of this system are $k_1 = 20$ N/m, $k_2 = 50$ N/m, $k_3 = 10$ N/m, $c_1 = 4$ N/m/sec, $c_2 = 0.2$ N/m/sec, $c_3 = 0.1$ N/m/sec, $m_1 = 2$ kg, $m_2 = 0.15$ kg, and $m_3 = 0.8$ kg. The equations that govern the motion of this system are

$$m_1 \ddot{y}_1(t) = -c_1 \dot{y}_1(t) - c_2(\dot{y}_1(t) - \dot{y}_2(t)) - k_1 y_1(t) - k_2(y_1(t) - y_2(t)) + f(t)$$
$$m_2 \ddot{y}_2(t) = -c_2(\dot{y}_2(t) - \dot{y}_3(t)) - c_3(\dot{y}_2(t) - \dot{y}_3(t)) - k_2(y_1(t) - y_2(t)) - k_3(y_2(t) - y_3(t))$$
$$m_3 \ddot{y}_3(t) = -c_3(\dot{y}_3(t) - \dot{y}_2(t)) - k_3(y_3(t) - y_2(t))$$

(a) Show that these equations can be written in the form

$$\begin{bmatrix} m_1 & 0 & 0 \\ 0 & m_2 & 0 \\ 0 & 0 & m_3 \end{bmatrix} \ddot{\vec{y}} + \begin{bmatrix} c_1 + c_2 & -c_2 & 0 \\ -c_2 & c_2 + c_3 & -c_3 \\ 0 & -c_3 & c_3 \end{bmatrix} \dot{\vec{y}}$$
$$+ \begin{bmatrix} k_1 + k_2 & -k_2 & 0 \\ -k_2 & k_2 + k_3 & -k_3 \\ 0 & -k_3 & k_3 \end{bmatrix} \vec{y} = \begin{bmatrix} 1 \\ 0 \\ 0 \end{bmatrix} f$$

or

$$M\ddot{\vec{y}} + C\dot{\vec{y}} + K\vec{y} = Bf(t).$$

(b) Define the state vector

$$\vec{q}(t) = \begin{bmatrix} y(t) \\ \dot{y}(t) \end{bmatrix}.$$

Show that the equations in (a) can be written in state space form as

$$\dot{q}(t) = \begin{bmatrix} \ddot{y}(t) \\ \dot{y}(t) \end{bmatrix} = \begin{bmatrix} \mathbf{0}_3 & I_3 \\ -M^{-1}K & -M^{-1}C \end{bmatrix} \begin{bmatrix} \dot{y}(t) \\ y(t) \end{bmatrix} + \begin{bmatrix} \mathbf{0}_{13} \\ M^{-1}B \end{bmatrix} x(t)$$

$$y(t) = \begin{bmatrix} I_3 & \mathbf{0}_{31} \end{bmatrix} \begin{bmatrix} \dot{y}(t) \\ y(t) \end{bmatrix}$$

(c) Write a MATLAB M-file to enter this system into the computer. Compute the transfer function matrix.

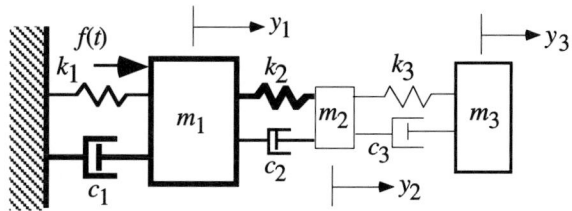

Figure P11.5.5

11.5.6 Consider the *LRC* network discussed in Section 11.4.
  (a) Develop an all-integrator block diagram from the state space representation of this circuit.
  (b) Define the voltage across the inductor is

  $$Y_2(s) = V_1(s) - V_0(s).$$

  Add this relation to the block diagram of (a). Also define the output signal

  $$Y_3(s) = V_1(s).$$

  Add this output signal to the block diagram.
  (c) Modify the output equation of the state space representation to include the two new output signals.
  (d) Write a MATLAB M-file to simulate this system. Use the following parameter values: $R_1 = 1\ k\Omega$, $R_2 = 1\ k\Omega$, $C_1 = 1\ \mu F$, $C_2 = 1\ \mu F$, and $L = 1\ H$. Suppose the input signal to this network is as follows. For each input signal plot the output signal for this input signal. (Three plots for each input signal.)
  (i) $x(t) = [\cos(500t)]u_s(t)$ (ii) $x(t) = [\cos(2000t)]u_s(t)$

11.5.7   Consider the two mass-spring-damper system shown in Figure P11.5.7. The equations of motion for the two masses are

$$m_a \frac{d^2 y_a(t)}{dt^2} = -k_a y_a(t) - k_c \left[y_a(t) - y_{st}(t) \right] - c_a \dot{y}_a(t) + f_a(t)$$

$$m_{st} \frac{d^2 y_{st}(t)}{dt^2} = -k_c \left[y_{st}(t) - y_a(t) \right] - f_{ext}(t)$$

The input signals to this system are the two forces $f_a(t)$ and $f_{ext}(t)$. The output signals are the two displacements $y_a(t)$ and $y_{st}(t)$.

(a) Find a set of state space equations (multivariable) of the full system.
(b) Draw an all-integrator block diagram of this system.
(c) Consider the system defined by the input signal $f_a(t)$ and the output signal $y_{st}(t)$. Find the transfer function from the state space equations using the matrix formula. (By hand.)
(d) Develop a MATLAB M-file to enter the state space model of this system.   Use the parameter values: $k_a = 10$ N/m, $k_c = 100$ N/m, $c_a = 8$ N/m/s, $m_a = 1$ kg, and $m_{st} = 0.1$ kg.
(e) Compute the transfer function defined by $\dfrac{Y_{st}(s)}{F_a(s)}$ using MATLAB. Also compute the poles and zeros of this system.
(f) Compute the transfer function defined by $\dfrac{Y_a(s)}{F_{ext}(s)}$ using MATLAB. Also compute the poles and zeros of this system.

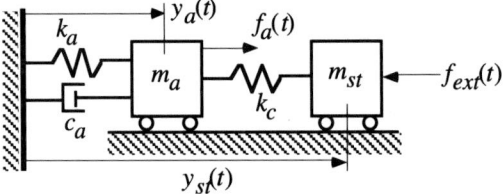

Figure P11.5.7

Chapter 12

The Convolution Representation and the Fourier Transfer Function

Chapter Outline

In Chapters 10 and 11 we have studied the Laplace transfer function and state space equations as system representations. In this chapter we introduce the third and fourth system representations: the convolution integral and the Fourier transfer function. These two system representations complete the set of major system representations we will study in this text.

The convolution representation is somewhat different from the previous system representations in that it expresses the relationship between the input and output signal in terms of an integral, not a differential equation or a transform of a differential equation. This representation has certain advantages in the representation of systems which are not conveniently expressed as one of the two previous representations. Hence, this representation is very useful as an analysis tool in advanced system analysis. We also present a graphical method for evaluating the convolution integral.

The second system representation we introduce in this chapter is the Fourier transfer function. The Fourier transfer function is very similar in concept to the Laplace transfer function except that the Fourier transform is used instead of the Laplace transform. The differences between the transforms leads to differences in the types of problems these two transfer functions are used to solve. The Fourier transfer function plays a central role in the analysis of the frequency response of systems studied in Chapters 14 and 15.

The convolution integral is closely related to the Laplace transfer function and (not so obviously) to the state space representations. These relationships are developed in this chapter. In the course of this development we introduce a general formula for the solution of the state equations which is of interest in its own right. The relationship between the Fourier transfer function and the other three system representations is also developed.

Summary of Sections

Section 12.1:   We introduce the convolution integral and develop its basic properties.

Section 12.2:   We present a graphical evaluation of the convolution integral.

Section 12.3:   We develop the relationship between the convolution integral, the Laplace transfer function, and state space equations.

Section 12.4:   We discuss the Fourier transform representation of a system.

Section 12.5:   Chapter summary section.

Coverage of the Text

Sections 12.1 and 12.2 require only results from Chapter 6. Section 12.4 requires knowledge of the Fourier transform from Chapter 7. Section 12.3 requires results from Chapters 10 and 11.

12.1 THE CONVOLUTION REPRESENTATION

12.1.1 Introduction

Consider the system

$$y(t) = \mathcal{H}[x(t)].$$

(12.1.1)

We have already discussed in detail the transfer function and state space representations for (12.1.1). In this section we will discuss the third major system representation: the convolution integral. We will introduce the notation of this system representation and give several examples of its use.

The origin of the convolution integral is obvious if we start from the transfer function. Suppose that the system (12.1.1) can be represented by a transfer function

$$\frac{Y(s)}{X(s)} = H(s). \tag{12.1.2}$$

Cross-multiplying and inverting back to the time domain using the convolution property (Property 9.2.3) of the Laplace transform we obtain

$$y(t) = \int_0^\infty h(t - \lambda)x(\lambda)\, d\lambda. \tag{12.1.3}$$

Taking liberties with the lower limit of integration in (12.1.3), we introduce the following definition.

Definition 12.1.1: A system representation of (12.1.1) in the form of

$$y(t) = \int_{-\infty}^\infty h(t - \lambda)x(\lambda)\, d\lambda$$

is called a <u>convolution representation</u> of the system.

▲▲

The integral in the convolution representation always takes the form shown in Definition 12.1.1. A particular system is characterized by the function $h(t)$.

Definition 12.1.2: The function $h(t)$ is called the <u>impulse response function</u>.

▲▲

The definition of the function $h(t)$ comes from the following observation. If the input signal is an impulse function, $x(t) = \delta(t)$, then the output signal $y_h(t)$ is

$$y_h(t) = \int_{-\infty}^\infty h(t - \lambda)x(\lambda)\, d\lambda = \int_{-\infty}^\infty h(t - \lambda)\delta(\lambda)\, d\lambda = h(t). \tag{12.1.4}$$

Hence, the response of the system (the output signal) to an impulsive input signal is the impulse response function $h(t)$. (See also the discussion in Section 6.2.)

Note: In the definition of the convolution representation it should be remembered that $x(t)$ and $y(t)$ are the input and output *signals* while the impulse response function $h(t)$ along with the convolution integral define the *system*.

In defining the convolution integral we have taken liberties with the expression (12.1.3). We have changed the lower limit to negative infinity. This change allows us to use input signals and impulse response functions that are nonzero for negative time, signals that are not Laplace transformable. (Recall that the Laplace transform, Definition 9.1.1, requires the signals to be zero for negative time.) Hence, the convolution representation applies to a larger class of systems than the transfer function representation. The relationship between the convolution integral, transfer functions, and state space representations is investigated in Section 12.3.

12.1.2 Examples

The next two examples illustrate the calculation of the impulse response function and the use of the convolution integral for computing the output signal for a given input signal.

Example 12.1.3: Consider the *RC* network in Example 10.1.7 where it was shown that the transfer function is

$$\frac{Y(s)}{X(s)} = H(s) = \frac{\dfrac{1}{RC}}{s + \dfrac{1}{RC}}. \tag{12.1.5}$$

It follows that the impulse response function is

$$h(t) = \mathcal{L}^{-1}\left\{\frac{\dfrac{1}{RC}}{s + RC}\right\} = \frac{1}{RC} e^{-\frac{t}{RC}} u_s(t). \tag{12.1.6}$$

Now the convolution representation of this system is obtained by inserting (12.1.6) into the convolution integral and replacing t by $t - \lambda$. We get

$$y(t) = \int_{-\infty}^{\infty} h(t - \lambda)x(\lambda)\, d\lambda = \int_{-\infty}^{\infty} \left(\frac{1}{RC} e^{-\frac{(t-\lambda)}{RC}} u_s(t - \lambda)\right) x(\lambda)\, d\lambda. \tag{12.1.7}$$

Next we will illustrate the evaluation of the convolution integral. Suppose that the input signal to the network is a unit step voltage. Using the convolution representation we have

$$y(t) = \int_{-\infty}^{\infty} \left(\frac{1}{RC} e^{-\frac{(t-\lambda)}{RC}} u_s(t - \lambda)\right) u_s(\lambda)\, d\lambda. \tag{12.1.8}$$

Since the input signal is zero when its argument is negative, the lower limit of integration becomes

$$y(t) = \frac{1}{RC} \int_{0}^{\infty} \left(e^{-\frac{(t-\lambda)}{RC}} u_s(t - \lambda)\right)(1)\, d\lambda. \tag{12.1.9}$$

Next the unit step function inside the brackets is zero for $t < \lambda$. That is, when $t < \lambda$, the impulse response function is zero. So the upper limit of integration in (12.1.9) is

$$y(t) = \frac{1}{RC} \int_{0}^{t} \left[e^{-\frac{(t-\lambda)}{RC}} (1)\right] d\lambda. \tag{12.1.10}$$

The variable of integration is λ so the variable t can be taken outside of the integral. Then (12.1.10) is evaluated as

$$y(t) = \frac{e^{-\frac{t}{RC}}}{RC} \int_0^t e^{\frac{\lambda}{RC}} d\lambda = e^{-\frac{t}{RC}} \left[e^{\frac{\lambda}{RC}} \right]_{\lambda=0}^t = 1 - e^{-\frac{t}{RC}}, \quad t \geq 0. \tag{12.1.11}$$

This answer agrees with the solution obtained using the Laplace transform in Section 6.3.

▲▲

Example 12.1.4: Consider the proof-mass actuator. In Example 10.1.9 it was shown that the transfer function is

$$\frac{Y_{pm}(s)}{V_{pm}(s)} = H_{pm}(s) = \frac{K_{ef}}{m_{pm}s^2}. \tag{12.1.12}$$

It follows that the impulse response function is

$$\mathcal{L}^{-1}\left\{ \frac{K_{ef}}{m_{pm}s^2} \right\} = \frac{K_{ef}}{m_{pm}}(t)u_s(t). \tag{12.1.13}$$

Now the convolution representation is

$$y_{pm}(t) = \frac{K_{ef}}{m_{pm}} \int_{-\infty}^{\infty} (t-\lambda)u_s(t-\lambda)v_{pm}(\lambda)\, d\lambda. \tag{12.1.14}$$

Suppose that $v_{pm}(t)$ is given by

$$v_{pm}(t) = \left(\sin(\omega_i t)\right)u_s(t). \tag{12.1.15}$$

We want to find the displacement of the proof-mass in response to this input signal using the convolution integral. When we substitute (12.1.15) into (12.1.14), we get

$$y_{pm}(t) = \frac{K_{ef}}{m_{pm}} \int_{-\infty}^{\infty} (t-\lambda)u_s(t-\lambda)\left[\sin(\omega_i\lambda)u_s(\lambda)\right] d\lambda \tag{12.1.16}$$

$$= \frac{K_{ef}}{m_{pm}} \left\{ \int_0^t (t)\left[(\sin\omega_i\lambda)\right] d\lambda - \int_0^t (\lambda)\left[\sin(\omega_i\lambda)\right] d\lambda \right\}$$

where the limits of integration in the second line were discussed in the previous example. The first integral evaluates to

$$\int_0^t (t)\left[\sin(\omega_i\lambda)\right] d\lambda = \frac{t}{\omega_i}\left[-\cos(\omega_i\lambda)\right]_{\lambda=0}^t = \frac{1}{\omega_i}\left[t - t\cos(\omega_i t)\right]. \tag{12.1.17}$$

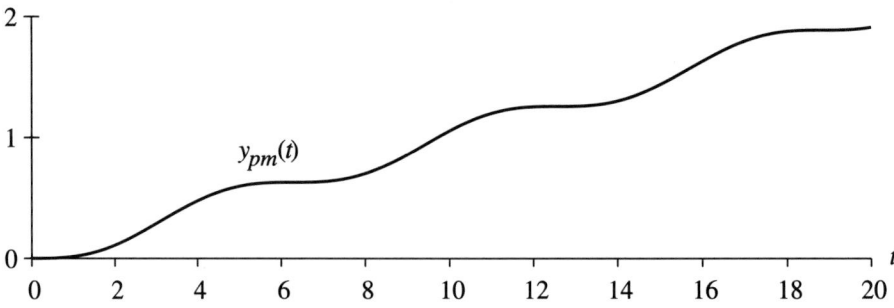

Figure 12.1.1 Displacement of the Proof-Mass for a Sinusoidal Input Signal

The second integral evaluates to

$$\int_0^t \lambda \sin(\omega_i \lambda) \, d\lambda = \frac{1}{\omega_i^2} \left[\sin(\omega_i t) - t \cos(\omega_i t) \right]. \tag{12.1.18}$$

Subtracting (12.1.18) from (12.1.17) we find that the output signal in response to the input signal is

$$y_{pm}(t) = \frac{K_{ef}}{m_{pm} \omega_i^2} \left[\omega_i t - \sin(\omega_i t) \right]. \tag{12.1.19}$$

A graph of $y_{pm}(t)$ is shown in Figure 12.1.1. This example shows that if the proof-mass actuator is driven by a sinusoidal input signal the displacement of the proof-mass will drift until the proof-mass reaches the end of its linear motion. To reproduce the graph in Figure 12.1.1, see MATLAB experiments in Section 10.5.

▲▲

12.1.3 Forms of the Convolution Integral

There are several variations on the exact expression of the convolution integral in Definition 12.1.1 in which the functions $h(t)$ and $x(t)$ have special properties. For many systems,[1] the impulse response function satisfies

$$h(t) = 0, \quad t \leq 0. \tag{12.1.20}$$

In light of (12.1.20) we observe that

$$h(t - \lambda) = 0 \quad \text{for} \quad t < \lambda. \tag{12.1.21}$$

Using (12.1.21) in the convolution integral the upper limit of integration becomes

---

[1] Systems that satisfy (12.1.20) are causal. See Theorem 13.5.1.

$$y(t) = \int_{-\infty}^{t} h(t-\lambda)x(\lambda)\, d\lambda. \tag{12.1.22}$$

This property is illustrated in the two previous examples. If, in addition, the input signal is zero for negative time, $x(t) = 0$ for $t < 0$, then (12.1.22) becomes

$$y(t) = \int_{0}^{t} h(t-\lambda)x(\lambda)\, d\lambda. \tag{12.1.23}$$

Note: The expression for the convolution representation in (12.1.23) is widely used.

There is a certain symmetry in the convolution integral in Definition 12.1.1. To see this symmetry, let $\gamma = t - \lambda$ be a change of variables of integration. Then the convolution integral can be written as

$$y(t) = \int_{-\infty}^{\infty} h(t-\lambda)x(\lambda)\, d\lambda = \int_{-\infty}^{\infty} h(\gamma)x(t-\gamma)\, d\gamma. \tag{12.1.24}$$

There is a simplified mathematical expression for the convolution representation based on the equivalence of the expressions in (12.1.24).

Notation: Since the functions $x(t)$ and $h(t)$ enter into the integral in (12.1.24) symmetrically, we write

$$y(t) = \int_{-\infty}^{\infty} h(t-\lambda)x(\lambda)\, d\lambda = \int_{-\infty}^{\infty} h(\gamma)x(t-\gamma)\, d\lambda = h(t) * x(t). \tag{12.1.25}$$

The usefulness of this notation is illustrated with a cascade interconnection of two systems shown in Figure 12.1.2. The cascade system in Figure 12.1.2 has a convolution representation obtained by composing the convolution representation of the two systems. This expression is rather messy, however. Using the notation in (12.1.25) the composite system in Figure 12.1.2 can be represented as

$$y(t) = h_2(t) * h_1(t) * x_1(t). \tag{12.1.26}$$

Note that the order of the systems in Figure 12.1.2 and (12.1.26) can be reversed, but the relationship between the input and output signals will remain the same. This notation is useful for systems that don't have transfer functions, or for systems with input signals that are nonzero in negative time.

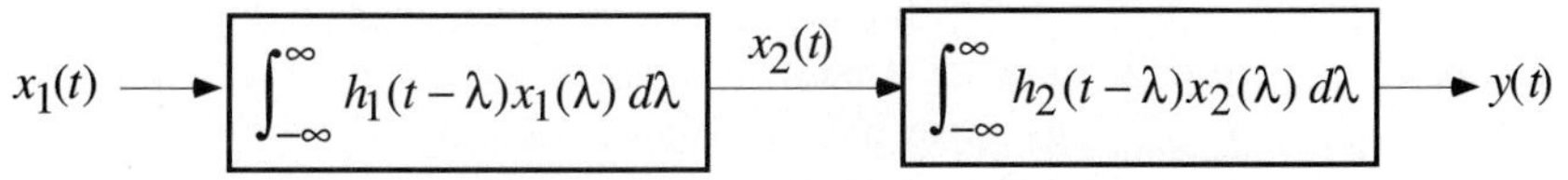

Figure 12.1.2 Cascade System Interconnection

12.2 GRAPHICAL CONVOLUTION

In the last two sections we have seen that the convolution integral

$$y(t) = \int_{-\infty}^{\infty} h(t - \lambda)x(\lambda)\, d\lambda \tag{12.2.1}$$

naturally arises from the discussion of transfer functions. Indeed, it plays an important role in system theory. Unfortunately, the convolution integral is typically hard to evaluate, even for simple functions. In this section we present a graphical method for evaluating the convolution integral. Once mastered, this method provides a quick way to estimate the value of the convolution integral for simple functions. This method also provides insight into the structure of the integration that is useful in understanding the action of some systems.

This method will be presented by way of example. Let the impulse response function $h(t)$ be given by

$$h(t) = 2\Pi\left(\frac{t-1}{2}\right), \tag{12.2.2}$$

and let the input signal be given by

$$x(t) = \Pi\left(\frac{t-2}{4}\right). \tag{12.2.3}$$

These two functions are shown in Figure 12.2.1.

Consider again the convolution integral (12.2.1). In the integrand, t appears as a constant with respect to the variable of integration λ. For each t, $-\infty < t < \infty$, we evaluate the integral to find $y(t)$. We start with $t = 0$.

$t = 0$ For this value of t the convolution integral is given by

$$y(0) = \int_{-\infty}^{\infty} h(0 - \lambda)x(\lambda)\, d\lambda. \tag{12.2.4}$$

To graphically construct the convolution of $h(\lambda)$ and $x(\lambda)$ at $t = 0$, we follow the steps described next.

Step 1 For each t we graph the functions $x(\lambda)$ and $h(t - \lambda)$ on the same graph and construct graphs of the product $x(\lambda)h(t - \lambda)$ as well as the graph $y(t)$. These graphs are shown in Figure 12.2.2 for $t = 0$.

In this graph there are four coordinate axes. The lower axis in Figure 12.2.2c is the axis for $y(t)$ as we will construct it below. This axis is for t. The functions $x(\lambda)$ and $h(0 - \lambda)$ are shown in the upper axis in Figure 12.2.2a. On the middle axis in Figure 12.2.2b is the product of $x(\lambda)$ and $h(0 - \lambda)$, shown as a dashed line. These two axes are for λ. The origins of all of these axes are aligned.

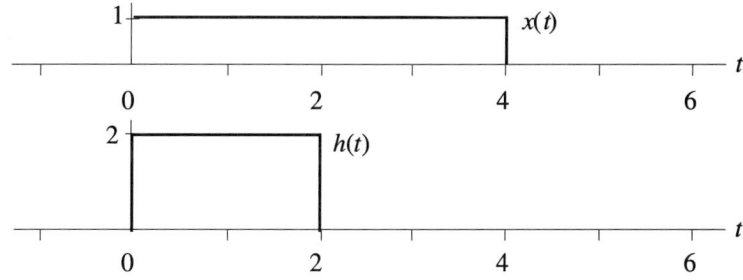

Figure 12.2.1 The Impulse Response Function $h(t)$ and the Input Signal $x(t)$

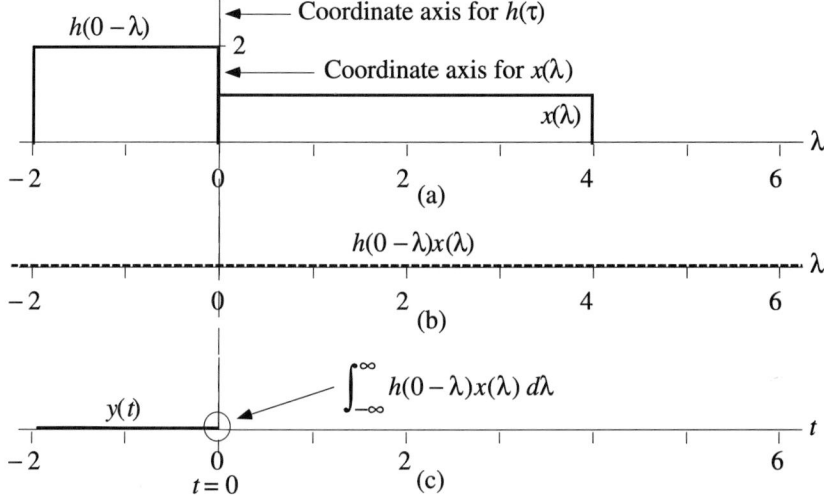

Figure 12.2.2 Graphical Evaluation of the Convolution Integral at $t = 0$

There is a fourth axis in Figure 12.2.2a for $h(\tau)$ where $\tau = 0 - \lambda$. The origin of this axis is aligned with $t = 0$, the value of t for which we are evaluating the integral. On this axis, $h(t)$ in Figure 12.2.1 has simply been flipped over. When the τ axis is thus aligned with the λ axis, the functions $h(\tau)$ and $h(0 - \lambda)$ overlay each other. The τ axis is not shown explicitly.

In the procedure that follows the function $x(\lambda)$ will not change. We will have to construct a new function $h(t_0 - \lambda)$ for each t_0. There are two ways to construct this function. The first way is to use the function $h(\tau)$, which is the function $h(t)$ flipped about the ordinate axis. Then the origin of this axis is aligned with t_0 on the t axis and overlaid on the λ axis. This method is purely graphical.

To construct the signal $h(t_0 - \lambda)$ analytically on the λ axis, recall the signal modeling from Section 5.2. At $\lambda = t_0$, the argument of this function is zero. Hence, the origin of $h(\lambda)$ has been shifted to $\lambda = t_0$. Next if we substitute $\lambda = t_0 + 1$, say, we have that

$$h\big(t_0 - (t_0 + 1)\big) = h(-1).$$ (12.2.5)

The time scale of $h(t)$ has been reversed. That is, $h(t)$ has been flipped about the ordinate axis. Substituting $t_0 = 0$, we obtain the graph of $h(0 - \lambda)$ in Figure 12.2.2.

Step 2 Next, the integral (12.2.1) says that the two functions $x(\lambda)$ and $h(0 - \lambda)$ should be multiplied together. This product is shown in Figure 12.2.2b as a dashed line on the middle graph. In this case, the product is identically zero.

Step 3 Finally, the integral in (12.2.1) says that the product of the two functions, $h(0 - \lambda)x(\lambda)$, is integrated over the entire real line, $-\infty < \lambda < \infty$. (Integration is, of course, equivalent to determining the area under the curve in Figure 12.2.2b.) This area is shown as a shaded area in Figure 12.2.2b and the figures below. In Figure 12.2.2, $h(0 - \lambda)x(\lambda) = 0$, so that the integral is zero. On the lower graph in Figure 12.2.2c, we set the value of $y(t)$ at $t = 0$ to zero; i.e., $y(0) = 0$.

In fact, this analysis shows that the product $h(t - \lambda)x(\lambda) = 0$ for any $t \le 0$. These values of the integral $y(t)$ are shown in the lower coordinate system as zero in Figure 12.2.2c for $t \le 0$.

The steps described above are repeated at each value of t to construct $y(t)$. These steps are followed in the discussion below.

$t = 1$ For $t = 1$, the two functions $x(\lambda)$ and $h(1 - \lambda)$ are shown in Figure 12.2.3a along with their product in Figure 12.2.3b. In addition, the area under the product $h(1 - \lambda)x(\lambda)$ is also shown as the cross-hatched area in Figure 12.2.3c. Since the area under $h(1 - \lambda)x(\lambda)$ is $(1)(2) = 2$, we have that $y(1) = 2$. This value is marked in Figure 12.2.3c. We also conclude that the values of $y(t)$ for $0 \le t \le 2$ are as shown in Figure 12.2.3c.

Consider the construction of $h(1 - \lambda)$ using $h(\tau)$. In comparing Figure 12.2.2a to Figure 12.2.3a, it appears that the origin of the τ axis has been shifted to the left to align with $t = 1$. In doing so the effect on the graph of $h(\tau)$ has been to slide across the graph of $x(\lambda)$. At each intervening value of t, the product $h(t - \lambda)x(\lambda)$ is formed and the integral is evaluated. The result is the graph of $y(t)$ for $0 < t < 1$ in Figure 12.2.3c. The reader is urged to consider this interpretation in the remaining graphs.

$t = 3$ For $t = 3$, the two functions $x(\lambda)$ and $h(3 - \lambda)$ are shown in Figure 12.2.4a. Their product and the area under the product are shown in Figure 12.2.4b. Since the area under $h(3 - \lambda)x(\lambda)$ is $(2)(2) = 4$, we have that $y(3) = 4$. This value is marked in Figure 12.2.4c on the axis for $y(t)$. We also conclude that the values of $y(t)$ for $2 \le t \le 4$ are as shown in Figure 12.2.4c.

$t = 5$ For $t = 5$, the two functions $x(\lambda)$ and $h(5 - \lambda)$ are shown in Figure 12.2.5a. Their product is shown in Figure 12.2.5b. In addition, the area under the product $h(5 - \lambda)x(\lambda)$ is shown. The corresponding value of the convolution integral is also shown in Figure 12.2.5c. The construction of the value of the convolution integral at this value of t is similar to the construction at $t = 1$ in Figure 12.2.3.

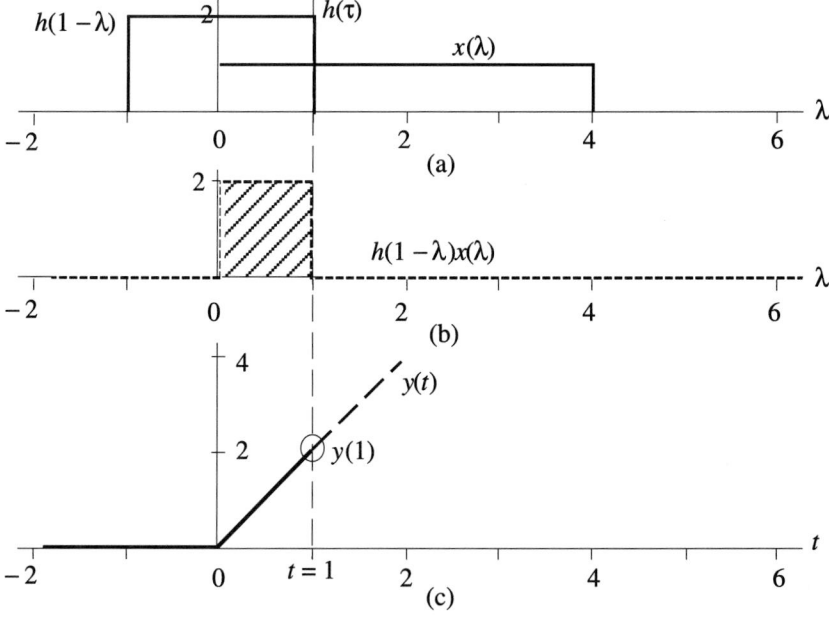

Figure 12.2.3 Graphical Convolution of $h(t)$ and $x(t)$ for $t \le 1$

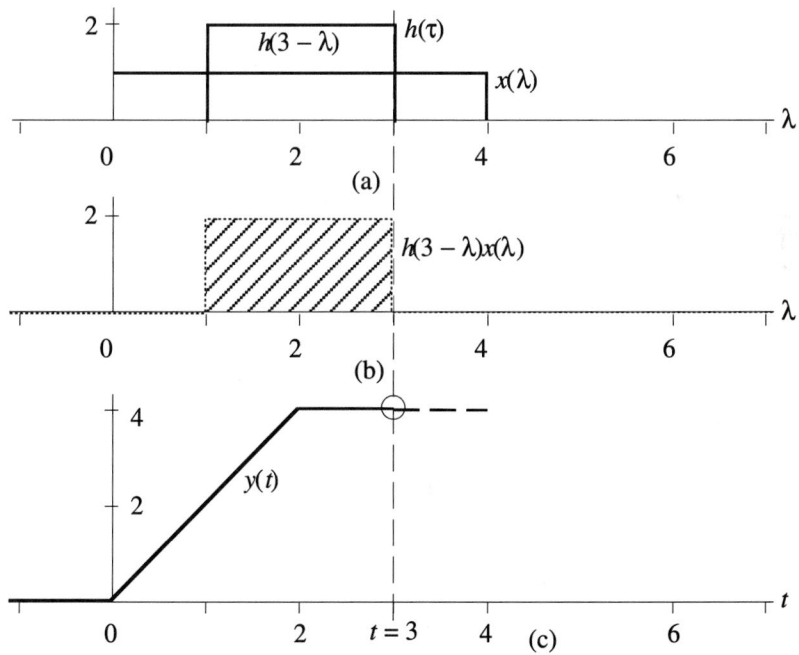

Figure 12.2.4 Graphical Convolution of $h(t)$ and $x(t)$ for $t \le 3$

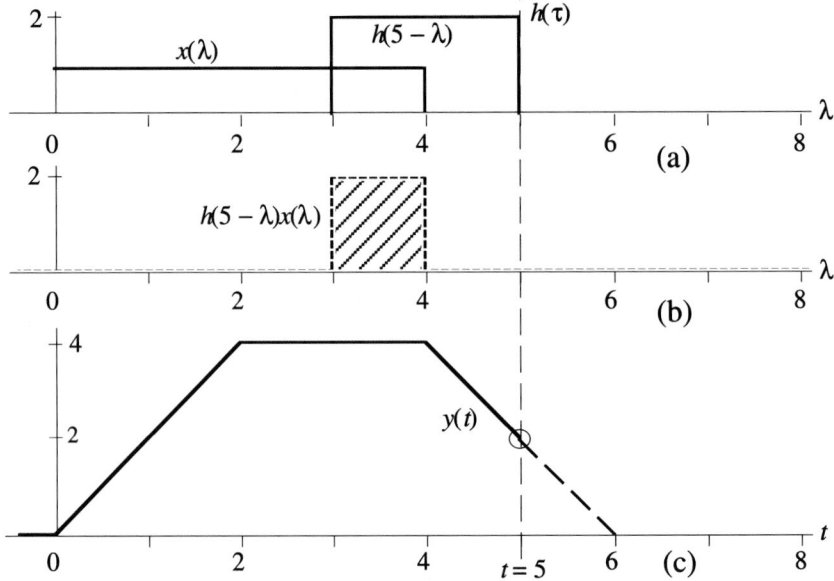

Figure 12.2.5 Graphical Convolution of $h(t)$ and $x(t)$ for $t \le 5$

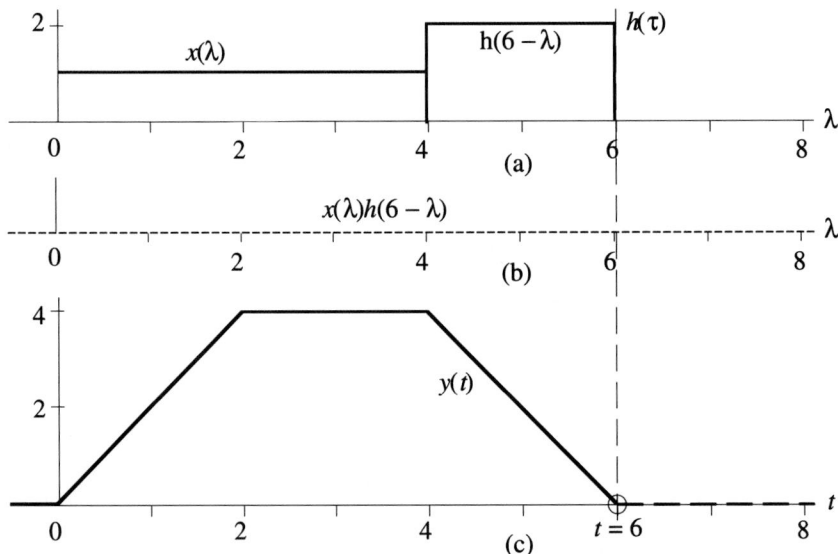

Figure 12.2.6 Graphical Convolution of $h(t)$ and $x(t)$ for $t \le 6$

t = 6 For $t = 6$, the two functions $x(\lambda)$ and $h(6 - \lambda)$ are shown in Figure 12.2.6a along with their product in Figure 12.2.6b. The area under the product $h(6 - \lambda)x(\lambda)$ is also shown in Figure 12.2.6b; it is zero. It is clear from the graph that $h(t - \lambda)x(\lambda) = 0$ for all $t \geq 6$. The convolution of the two functions $h(\lambda)$ and $x(\lambda)$ is shown in its entirety in Figure 12.2.6.

This example illustrates the basic idea behind graphical convolution. With a little practice, simple convolution integrals can be evaluated very rapidly with this graphical technique.

12.3   THE RELATIONSHIP BETWEEN THE CONVOLUTION INTEGRAL AND OTHER SYSTEM REPRESENTATIONS

12.3.1 Introduction

In this chapter we introduced the convolution representation

$$y(t) = \int_{-\infty}^{\infty} h(t - \lambda)x(\lambda)\, d\lambda \qquad \text{(12.3.1)}$$

for this system. The convolution representation is the third system representation we have introduced after the transfer function and the state space representation. In this section we will develop the relationship between these three system representations.

12.3.2 Relationship Between the Transfer Function and Convolution Integral

We begin with the representation between the convolution integral and the transfer function. This relationship is derived by taking the Laplace transform of the convolution integral in (12.3.1). The only restriction is that the signals and the impulse response function must be Laplace transformable according to Definition 9.1.1. This definition requires that the impulse function and the input signal must satisfy

$$h(t) = 0, \quad x(t) = 0, \quad t < 0. \qquad \text{(12.3.2)}$$

In this case the convolution integral takes on the form

$$y(t) = \int_{0}^{\infty} h(t - \lambda)x(\lambda)\, d\lambda \qquad \text{(12.3.3)}$$

as we derived in Subsection 12.1.3. We also assume that the impulse response function and the input signal satisfy the second condition in Definition 9.1.1 (which is usually the case). Now we take the Laplace transform of the modified convolution integral (12.3.3) to obtain the transfer function. These observations are summarized in the following theorem.

Theorem 12.3.1: Given the system represented by the convolution integral in (12.3.1), suppose that:
(a)   the input signal is zero for negative time, $x(t) = 0$, $t < 0$
(b)   the impulse response function is zero for negative time, $h(t) = 0$, $t < 0$
(c)   both $x(t)$ and $h(t)$ satisfy (b) of Definition 9.1.1
Then the system has a transfer function representation

$$\frac{Y(s)}{X(s)} = H(s).$$

If the system has a transfer function representation, then the system can be represented by a convolution representation of the form

$$y(t) = \int_0^\infty h(t - \lambda)x(\lambda)\, d\lambda.$$

▲▲

In general, the convolution integral can be used to model systems that can't be modeled with a transfer function.   The convolution representation can also accommodate input signals that are defined for

$$x(t), \quad -\infty < t < \infty. \tag{12.3.4}$$

Such input signals are not Laplace transformable, and so the transfer function can't be used to represent the system.

12.3.3 Relationship with the State Space Representation

Next we consider the relationship between the convolution integral representation in (12.3.1) and the state space representation

$$\dot{\vec{q}}(t) = A\vec{q}(t) + Bx(t), \quad \vec{q}(0) = 0 \tag{12.3.5}$$

$$y(t) = C\vec{q}(t) + Dx(t).$$

We assume the initial conditions for the state space equations in (12.3.5) are zero. This assumption is required because the convolution integral doesn't model any internal behavior of the system.   In order to discuss this relationship, we must develop a general solution to the state equations in (12.3.5) using the Laplace transform.
      To motivate the following notation and development, first we consider the scalar state space equations

$$\dot{q}(t) = aq(t) + bx(t), \tag{12.3.6}$$

$$y(t) = cq(t) + dx(t).$$

Solving for the transfer function we obtain

$$Y(s) = \left(c(s-a)^{-1}b + d\right)X(s).$$ (12.3.7)

To get $y(t)$ we take the inverse Laplace transform of (12.3.7) keeping in mind that the product of two Laplace transforms is a convolution integral in the time domain. We obtain

$$y(t) = \int_0^\infty [ce^{a(t-\lambda)}b + d\delta(t-\lambda)]x(\lambda)\, d\lambda.$$ (12.3.8)

In particular, we see that

$$(s-a)^{-1} \leftrightarrow e^{at}.$$ (12.3.9)

Now we return to the matrix equations in (12.3.5). From Theorem 11.1.2 the transfer function of (12.3.5) is

$$H(s) = C(sI - A)^{-1}B + D.$$ (12.3.10)

We want to take the inverse Laplace transform of (12.3.10) to obtain the impulse response function. So we must find an inverse Laplace transform for the matrix $(sI - A)^{-1}$. Note that the matrix

$$(sI - A)^{-1} = \left[\frac{b_{ij}(s)}{a_{ij}(s)}\right]$$ (12.3.11)

is a matrix of rational functions. Each entry in this matrix has an inverse Laplace transform

$$\eta_{ij}(t) = \mathcal{L}^{-1}\left\{\frac{b_{ij}(s)}{a_{ij}(s)}\right\}.$$ (12.3.12)

By reassembling these Laplace transforms of the individual elements into a matrix, we can define an inverse Laplace transform for (12.3.11). Guided by the scalar example above, we introduce the following definition and notation.

Definition 12.3.2: Let A be an $n \times n$ matrix of real numbers. Then the inverse Laplace transform of $(sI - A)^{-1}$ is defined as

$$e^{At} = \mathcal{L}^{-1}\left\{(sI - A)^{-1}\right\} = \left[\eta_{ij}(t)\right].$$

The function e^{At} is called the <u>state transition matrix</u>.

▲▲

Notation: The function e^{At} is called a <u>matrix function</u>. It has different properties from the scalar function e^{at}. A complete treatment of matrix functions is left for a more advanced text.

Terminology: The name *state transition matrix* is taken from an advanced treatment of state space representations.

Example 12.3.3: Consider the proof-mass actuator discussed in Section 6.5. From Example 10.5.6 we have that the state space model of the proof-mass actuator is

$$\dot{\vec{q}}(t) = \begin{bmatrix} 0 & 1 \\ 0 & 0 \end{bmatrix} \vec{q}(t) + \begin{bmatrix} 0 \\ \dfrac{K_{ef}}{m_{pm}} \end{bmatrix} v_{pm}(t), \tag{12.3.13}$$

$$y_{pm}(t) = \begin{bmatrix} 1 & 0 \end{bmatrix} \vec{q}(t).$$

The matrix in (12.3.11) for this example is

$$(sI - A)^{-1} = \left(\begin{bmatrix} s & 0 \\ 0 & s \end{bmatrix} - \begin{bmatrix} 0 & 1 \\ 0 & 0 \end{bmatrix} \right)^{-1} = \begin{bmatrix} s & -1 \\ 0 & s \end{bmatrix}^{-1} = \begin{bmatrix} \dfrac{1}{s} & \dfrac{1}{s^2} \\ 0 & \dfrac{1}{s} \end{bmatrix}. \tag{12.3.14}$$

Now the state transition matrix is

$$e^{At} = \mathcal{L}^{-1} \left\{ \begin{bmatrix} \dfrac{1}{s} & \dfrac{1}{s^2} \\ 0 & \dfrac{1}{s} \end{bmatrix} \right\} = \begin{bmatrix} u_s(t) & t u_s(t) \\ 0 & u_s(t) \end{bmatrix}. \tag{12.3.15}$$

▲▲

Using the state transition matrix, Definition 12.3.2, with the formula for the transfer function (12.3.10) we find that the impulse response function is

$$h(t) = \mathcal{L}^{-1}\left\{ C(sI - A)^{-1}B + D \right\} = C\left(\mathcal{L}^{-1}\left\{ (sI - A)^{-1} \right\} \right)B + D\delta(t) \tag{12.3.16}$$

$$= Ce^{At}B + D\delta(t).$$

Hence, Definition 12.3.2 leads to an explicit representation of the impulse response function in terms of the defining matrices of the state space representation. The impulse response function, in turn, defines the convolution representation. These results are summarized in the following theorem.

Theorem 12.3.4: **Relationship of the Convolution Integral to the State Space Representation** Given the state space representation

$$\dot{\vec{q}}(t) = A\vec{q}(t) + Bx(t), \quad \vec{q}(0) = 0,$$
$$y(t) = C\vec{q}(t) + Dx(t),$$

convolution representation of this system is given by

$$y(t) = \int_0^\infty \left[Ce^{A(t-\lambda)}B + D\delta(t-\lambda) \right] x(\lambda)\, d\lambda$$

where the impulse response function of this system is

$$h(t) = Ce^{At}B + D\delta(t). \qquad \blacktriangle\blacktriangle$$

The convolution representation of a state space representation is essentially obtained from the transfer function of the state space representation. Hence, the remarks regarding the relationship between the transfer function and the convolution integral given above also apply here.

Example 12.3.5: Consider again Example 12.3.3. Using the state transition matrix in (12.3.15) the impulse response function of the proof-mass actuator is

$$h(t) = \begin{bmatrix} 1 & 0 \end{bmatrix} \begin{bmatrix} u_s(t) & tu_s(t) \\ 0 & u_s(t) \end{bmatrix} \begin{bmatrix} 0 \\ \dfrac{K_{ef}}{m_{pm}} \end{bmatrix} = \dfrac{K_{ef}}{m_{pm}} tu_s(t). \qquad (12.3.17)$$

$$\blacktriangle\blacktriangle$$

12.3.4 Nonzero Initial Conditions

The state space representation allows for a nonzero initial condition. So we can include initial conditions in the solution of (12.3.5). When we take the Laplace transform of the first line in (12.3.5) with a nonzero initial condition we obtain

$$s\vec{Q}(s) - \vec{q}_0 = A\vec{Q}(s) + BX(s), \qquad (12.3.18)$$

Solving (12.3.18) for $\vec{Q}(s)$ we obtain

$$\vec{Q}(s) = (sI - A)^{-1}\vec{q}_0 + (sI - A)^{-1}BX(s). \qquad (12.3.19)$$

Now $Y(s)$ is given by

$$Y(s) = C(sI - A)^{-1}\vec{q}_0 + \left[C(sI - A)^{-1}B + D \right]X(s). \qquad (12.3.20)$$

The inverse Laplace transform of (12.3.20) leads to the solution in Theorem 12.3.4 with an additional term.

Corollary 12.3.6: Given the state space representation

$$\dot{\vec{q}}(t) = A\vec{q}(t) + Bx(t), \quad \vec{q}(0) = \vec{q}_0,$$
$$y(t) = C\vec{q}(t) + Dx(t),$$

the solution to this differential equation is given by

$$y(t) = Ce^{At}\vec{q}_0 + \int_0^\infty \left[Ce^{A(t-\lambda)}B + D\delta(t-\lambda) \right] x(\lambda)\, d\lambda.$$

▲▲

The output signal in Corollary 12.3.6 is composed of two components: (1) The response due to the initial condition, and (2) the response due to the input signal $x(t)$ respectively.   These two components were identified in Definition 10.1.1. Specifically, the zero state response is

$$y_{zs}(t) = \int_0^\infty \left[Ce^{A(t-\lambda)}B + D\delta(t-\lambda) \right] x(\lambda)\, d\lambda \tag{12.3.21}$$

The zero input response is

$$y_{zi}(t) = Ce^{At}\vec{q}_0. \tag{12.3.22}$$

The zero state response can also be calculated using the transfer function or the convolution representation. The response due the initial condition, however, can't be calculated from either of these representations because these representations are based on an input-output model. The initial condition models the internal dynamics of the system that give rise to an output signal that are not excited by an input signal. As noted above, the convolution integral requires that initial condition to be zero.   Hence, the state space representation can be used to model certain phenomenon which the convolution integral, or transfer function cannot.

Example 12.3.7: Consider again the proof-mass actuator discussed in Example 12.3.3. Suppose the initial condition for this system is

$$\vec{q}_0 = \begin{bmatrix} 1 \\ 1 \end{bmatrix}, \tag{12.3.23}$$

and that the input signal is zero. Then the output signal, the displacement of the proof-mass, is given by

$$y_{pm}(t) = Ce^{At}\vec{q}_0 = \begin{bmatrix} 1 & 0 \end{bmatrix} \begin{bmatrix} u_s(t) & tu_s(t) \\ 0 & u_s(t) \end{bmatrix} \begin{bmatrix} 1 \\ 1 \end{bmatrix} = \begin{bmatrix} u_s(t) & tu_s(t) \end{bmatrix} \begin{bmatrix} 1 \\ 1 \end{bmatrix} \qquad (12.3.24)$$

$$= u_s(t) + tu_s(t).$$

This example shows that any initial displacement or velocity of the proof-mass will cause the proof-mass to drift to its stops.

▲▲

12.4 THE FOURIER TRANSFER FUNCTION

12.4.1 Introduction

Thus far we have introduced three major system representations: the Laplace transfer function, the state space representation, and the convolution representation. In this section we introduce the fourth, and last, system representation, the Fourier transfer function. This system representation is very similar to the Laplace transfer function in concept, but differs in certain details. The relationship of the Fourier transfer function with other system representations is given in this section.

12.4.2 Definition

Suppose that the system

$$y(t) = \mathcal{H}[x(t)] \qquad (12.4.1)$$

can be represented as a convolution integral

$$y(t) = \int_{-\infty}^{\infty} h(t - \lambda)x(\lambda)\, d\lambda. \qquad (12.4.2)$$

Applying the Convolution Property 7.5.7 of Fourier transforms to (12.4.2) we obtain

$$Y(\omega) = H(\omega)X(\omega). \qquad (12.4.3)$$

This Fourier transform suggests a fourth system representation.

Definition 12.4.1: The <u>Fourier (transform) transfer function</u> of a system is the ratio of the Fourier transform of the output signal divided by the Fourier transform of the input signal

$$\frac{\mathcal{F}\{y(t)\}}{\mathcal{F}\{x(t)\}} = \frac{Y(\omega)}{X(\omega)} = H(\omega).$$

▲▲

There are three functions of frequency in the expression of the Fourier transfer function. The functions $X(\omega)$ and $Y(\omega)$ are the Fourier transforms of *signals*. The function $H(\omega)$ is the transfer function of the *system*.

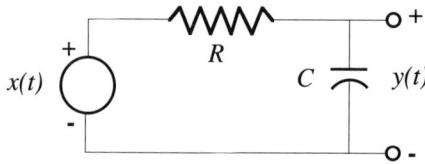

Figure 12.4.1 *RC* Network

Example 12.4.2:   The Fourier transform representation is well-known in network analysis. Consider the *RC* network shown in Figure 12.4.1. The transfer function of this system is

$$\frac{Y(s)}{X(s)} = \frac{1}{s + \dfrac{1}{RC}}.$$

(12.4.4)

Inverting (12.4.4) the impulse response function of this system is

$$h(t) = e^{-\frac{t}{RC}} u_s(t).$$

(12.4.5)

The Fourier transfer function, the Fourier transform of the impulse response function in (12.4.5), is

$$\mathcal{F}\left\{ e^{-\frac{t}{RC}} u_s(t) \right\} = H(\omega) = \frac{1}{\dfrac{1}{RC} + j\omega}.$$

(12.4.6)

The transfer function can also be derived directly by assigning complex impedances to circuit components in Figure 12.4.1.

▲▲

12.4.3 Relationship to the Convolution Representation

The definition of the Fourier transfer function, Definition 12.4.1, can only be used if all of the Fourier transforms exist. Insight into this restriction can be gained from the convolution representation in (12.4.2). At a minimum, the input and output signals along with the impulse response function must have Fourier transforms. Suppose that the input signal is bounded

$$|x(t)| < M.$$

(12.4.7)

Then we can bound the output signal as

$$|y(t)| = \left| \int_{-\infty}^{\infty} h(\lambda) x(t - \lambda)\, d\lambda \right| < \int_{-\infty}^{\infty} |h(\lambda)| |x(t - \lambda)|\, d\lambda < M \int_{-\infty}^{\infty} |h(\lambda)|\, d\lambda.$$

(12.4.8)

Hence, the output signal will be bounded if the impulse response function satisfies

$$\int_{-\infty}^{\infty} |h(t)|\, dt < \infty. \tag{12.4.9}$$

Note: The restriction in (12.4.9) on the impulse response function is significant. We will return to it again after we have introduced properties of systems in Chapter 13.

The existence of the Fourier transform of the impulse response function is guaranteed by (12.4.9) which equivalent to condition (b) of Definition 7.4.1. Now we can apply the convolution property of Fourier transform to obtain the Fourier transform of the output signal. These results are summarized in the following theorem.

Theorem 12.4.3: Given the system (12.4.2) suppose the impulse response function satisfies (12.4.9). Also assume the input signal $x(t)$ has a generalized Fourier transform. Then the output signal also has a generalized Fourier transform

$$\mathcal{F}\{y(t)\} = Y(\omega).$$

Furthermore, this output signal can be represented by

$$Y(\omega) = H(\omega)X(\omega)$$

where

$$\mathcal{F}\{h(t)\} = H(\omega) \quad \text{and} \quad \mathcal{F}\{x(t)\} = X(\omega).$$

▲▲

Theorem 12.4.3 requires that the Fourier transform of the impulse response $h(t)$ exist. In particular, impulse response functions with *generalized* Fourier transforms are excluded.

12.4.4 Examples

The following examples show when the Fourier transfer function exists for a system.

Example 12.4.4: Consider the proof-mass actuator discussed in Example 10.1.9. The impulse response of the proof-mass actuator is

$$h_{pm}(t) = \frac{K_{ef}}{m_{pm}}\, t u_s(t). \tag{12.4.10}$$

This impulse function does not have a Fourier transform, so this system doesn't have a Fourier transform representation.

▲▲

Example 12.4.5: Consider the mass-spring-damper system in Example 10.1.8. The impulse response function of this system is

$$h_{st}(t) = \left[\left(\frac{1}{\omega_c m_{st}} \right) e^{-\zeta \omega_n t} \sin \omega_c t \right] u_s(t) \tag{12.4.11}$$

where

$$\omega_c = \omega_n^2 \sqrt{1 - \zeta^2}, \quad \omega_n^2 = \frac{k_{st}}{m_{st}}, \quad \zeta = \frac{c_{st}}{2\sqrt{k_{st} m_{st}}}. \tag{12.4.12}$$

When $\zeta > 0$ the poles of the system are in the open LHP. So the Fourier transfer function exists, and it can be computed from the Laplace transform of the impulse response function in (12.4.11) as discussed in Section 9.5. The transfer function[2] for the mass-spring-damper system is

$$H_{st}(s)\Big|_{s=j\omega} = H_{st}(\omega) = \frac{\dfrac{1}{m_{st}}}{\left(\dfrac{k_{st}}{m_{st}} - \omega^2 \right) + j\omega \left(\dfrac{c_{st}}{m_{st}} \right)} = \frac{Y_{st}(\omega)}{F_{st}(\omega)}. \tag{12.4.13}$$

If we set the damping of this system to zero, $\zeta = 0$, in (12.4.11) the impulse response function is

$$h_{st}(t) = \left[\frac{1}{\sqrt{m_{st} k_{st}}} \sin(\omega_n t) \right] u_s(t). \tag{12.4.14}$$

This system has a generalized Fourier transform, but not an ordinary Fourier transform. In particular, this impulse response function doesn't satisfy (12.4.9). The system with the impulse response function in (12.4.14) doesn't have a Fourier representation.

▲▲

12.4.5 Relationship to the Laplace Transfer Function

The Fourier transfer function is closely related to the Laplace transfer function. This relationship is straightforward when we recognize that both transfer functions are calculated from the impulse response function of the convolution representation (12.4.2). These two transfer functions are essentially the same when both the Laplace and Fourier transform exist for the impulse response function. The conditions under which a function is both Laplace and Fourier transformable are given in Section 9.5.

There are essentially two conditions that must be satisfied. First, both the Laplace and Fourier transform exist for both the impulse response function and the input signal. In practice, this means that the signals must satisfy

$$h(t) = 0, \quad x(t) = 0, \quad t < 0. \tag{12.4.15}$$

---

[2] This calculation is discussed in detail in Example 10.1.8.

Second, the Fourier transform must exist for the impulse response function and the generalized Fourier transform must exist for the input signal.   In this case the Laplace transfer function will exist for the impulse response function as discussed in Section 9.5. These observations are summarized in the following theorem.

Theorem 12.4.6: Consider a system represented by the convolution integral (12.4.2). Suppose the impulse response function has a Laplace transfer function $H(s)$. Suppose further that the poles of this transfer function are in the open LHP. Then the Fourier transfer function can be computed from the Laplace transfer function according to

$$H(s)\big|_{s=j\omega} = H(\omega).$$

If in addition the input signal has both a Laplace and Fourier transform, then the Fourier transfer function representation is equivalent to the Laplace transfer function representation.

▲▲

While the Laplace and Fourier transfer functions are closely related for many systems, these two transforms are generally used to solve different types of problems. The transform that is chosen for a particular problem is determined by the system properties and the characteristics of the input signal. If the system doesn't satisfy (12.4.9), then the Laplace transform must be used.   For systems which satisfy (12.4.9), then the input signal must be examined.   If the input signal satisfies (12.4.15), then, potentially, either the Laplace or Fourier transform can be applied. Because the system is at rest until $t = 0$, the output signal will exhibit transient behavior when the input signal is applied. This type of problem is often addressed using the Laplace transform. On the other hand, if the input signal is nonzero at negative infinity, then the Fourier transform must be used to analyze the problem because the Laplace transform can't be applied to the input signal. In this situation, the system is said to be operating in steady state. The analysis of the steady state behavior of a system is commonly accomplished using the Fourier transform. If we think back over network analysis, we recall that sinusoidal steady state behavior is generally tackled using the Fourier transform, while problem with switches are generally addressed with the Laplace transform.

12.5 CHAPTER SUMMARY

In this chapter we introduced two new system representations.   The first representation is the convolution representation (Definition 12.1.1)

$$y(t) = \int_{-\infty}^{\infty} h(t - \lambda)x(\lambda)\, d\lambda = h(t) * x(t) \tag{12.5.1}$$

where $h(t)$ is the <u>impulse response function</u> (Definition 12.1.2). In Section 12.2 we show how to graphically evaluate this integral.

If the impulse response function satisfies $h(t) = 0$ for $t < 0$, and, in addition, the input signal is zero for negative time, $x(t) = 0$ for $t < 0$, then the convolution representation can be written as

$$y(t) = \int_0^t h(t - \lambda)x(\lambda)\, d\lambda. \tag{12.5.2}$$

Under these conditions the system also has a transfer function representation (Theorem 12.3.1) which is given by

$$\frac{Y(s)}{X(s)} = H(s) = \mathcal{L}\{h(t)\}. \tag{12.5.3}$$

A state space representation

$$\dot{\vec{q}}(t) = A\vec{q}(t) + Bx(t), \quad \vec{q}(0) = 0, \tag{12.5.4}$$
$$y(t) = C\vec{q}(t) + Dx(t),$$

also has a representation as a convolution integral. Define the state transition matrix (Definition 12.3.2) as

$$e^{At} = \mathcal{L}^{-1}\{(sI - A)^{-1}\}. \tag{12.5.5}$$

Then the convolution representation of this system is given by (Theorem 12.3.5/Corollary 12.3.6)

$$y(t) = Ce^{At}\vec{q}_0 + \int_0^\infty \left[Ce^{A(t-\lambda)}B + D\delta(t - \lambda) \right]x(\lambda)\, d\lambda. \tag{12.5.6}$$

The second system representation we introduced in this chapter is the <u>Fourier (transform) transfer function</u> (Definition 12.4.1)

$$\frac{\mathcal{F}\{y(t)\}}{\mathcal{F}\{x(t)\}} = \frac{Y(\omega)}{X(\omega)} = H(\omega). \tag{12.5.7}$$

▲▲

The Fourier transfer function is closely related to the convolution representation and the Laplace transfer function.   Given a convolution representation, suppose the impulse response function satisfies

$$\int_{-\infty}^{\infty} |h(t)|\, dt < \infty. \tag{12.5.8}$$

Also assume the input signal $x(t)$ has a generalized Fourier transform.   Then the output signal also has a generalized Fourier transform (Theorem 12.4.3)

$$\mathcal{F}\{y(t)\} = Y(\omega) = H(\omega)X(\omega). \tag{12.5.9}$$

The relationship between the Fourier transfer function and the Laplace transfer function is derived from the convolution representation in Theorem 12.4.7. Suppose the impulse response function of a convolution representation has a Laplace transfer function $H(s)$. Suppose further that the poles of this transfer function are in the open LHP. Then the Fourier transfer function can be computed from the Laplace transfer function according to

$$H(s)\big|_{s=j\omega} = H(\omega). \tag{12.5.10}$$

12.6 HOMEWORK FOR CHAPTER 12

Homework Problems for Section 12.1

12.1.1 Suppose the impulse response function of the proof-mass actuator is given by $h(t) = (0.1t)u_s(t)$. Find the output if the input is $v_{pm}(t) = r(t)\Pi(t - 1/2)$.

12.1.2 What is the impulse response function of the following system?

$$y(t) = \int_{-\infty}^{\infty} \left[(t - \lambda)e^{-2(t-\lambda)} \cos(3(t - \lambda))\right]u_s(t - \lambda)x(\lambda)\, d\lambda$$

12.1.3 Suppose that the impulse response function of a system is

$$h(t) = \Pi\left(\frac{t - 4}{2}\right).$$

Find the step response of this system.

12.1.4 Show that

(i) $\quad \Pi\left(\dfrac{t}{\varepsilon}\right) * \Pi\left(\dfrac{t}{\varepsilon}\right) = \dfrac{\varepsilon}{2}\Lambda\left(\dfrac{t}{\varepsilon}\right).$ (ii) $\quad \Pi\left(\dfrac{t}{\varepsilon}\right) * \delta(t - t_0) = \Pi\left(\dfrac{t - t_0}{\varepsilon}\right)$

12.1.5 Find a convolution representation for each of the following systems.
(i) $\ddot{y}(t) + 3\dot{y}(t) + 2y(t) = 2x(t)$ (ii) $\ddot{y} + 8\dot{y} + 12y = 2\dot{x} + x$

12.1.6 Consider the two mass-spring-damper system in Figure P12.1.6. A set of state space equations for this system is

$$\dot{\vec{q}}(t) = \begin{bmatrix} 0 & 1 & 0 & 0 \\ -\dfrac{k_a + k_c}{m_a} & -\dfrac{c_a}{m_a} & \dfrac{k_c}{m_a} & 0 \\ 0 & 0 & 0 & 1 \\ \dfrac{k_c}{m_{st}} & 0 & -\dfrac{k_c}{m_{st}} & 0 \end{bmatrix} \vec{q}(t) + \begin{bmatrix} 0 \\ \dfrac{1}{m_a} \\ 0 \\ 0 \end{bmatrix} f_a(t)$$

$$y_{st}(t) = \begin{bmatrix} 0 & 0 & 10 & 0 \end{bmatrix} \vec{q}(t)$$

Use the parameter values: $k_a = 10$ N/m, $k_c = 100$ N/m, $c_a = 8$ N/m/sec, $m_a = 1$ kg, and $m_{st} = 0.1$ kg.   Using MATLAB find the impulse response function of this system.

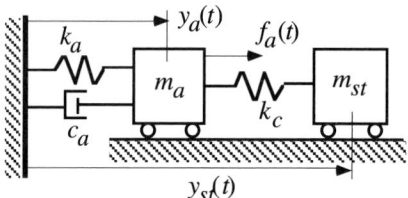

Figure P12.1.6

12.1.7   Find the convolution integral that describes each of the following systems.

(i) $\dfrac{Y(s)}{X(s)} = \dfrac{s}{s^2 + 0.4s + 4}$ (ii) $\dfrac{Y(s)}{X(s)} = \dfrac{s+4}{s(s+1)}$

12.1.8   Suppose the impulse response function of a system is

$$h(t) = u_s(t) - u_s(t - T).$$

(a) Find the convolution representation in its simplest form.   Why is this system called an averaging filter?
(b) Suppose the input signal is given by $x(t) = \cos(\omega t)$.   Find the output signal of this system.

Homework Problems for Section 12.2

12.2.1   Graphically evaluate the convolution integral

$$y_i(t) = \int_{-\infty}^{\infty} h(t - \lambda) x_i(\lambda)\, d\lambda, \quad i = 1, 2$$

for the functions shown in Figure P12.2.1.

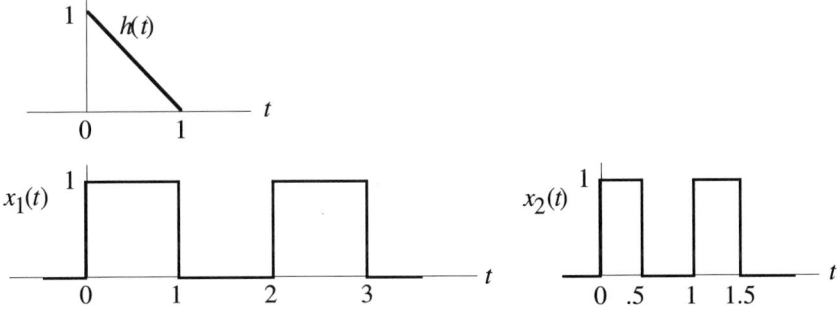

Figure P12.2.1

12.2.2 Graphically convolve the two signals in Figure P12.2.2.

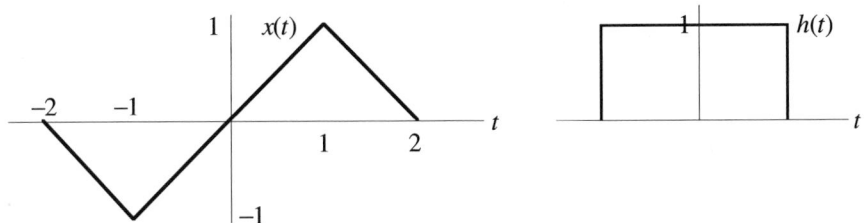

Figure P12.2.2

12.2.3 Graphically evaluate $x(t)*x(t)$ for $x(t)$ shown in Figure P12.2.3.

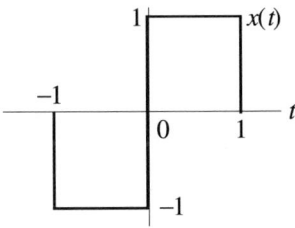

Figure P12.2.3

12.2.4 Graphically convolve the signals in Figure P12.2.4.

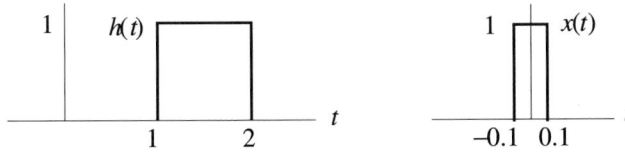

Figure P12.2.4

12.2.5   Graphically evaluate $h(t)*x(t)$ in Figure P12.2.5.

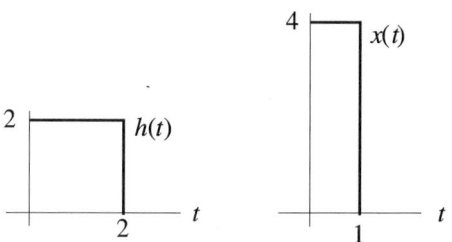

Figure P12.2.5

12.2.6   Suppose that the impulse response of a system is given by

$$h(t) = u_s(t) - u_s(t-3).$$

Graphically find $y(t) = h(t) * x(t)$ if the input signal is

$$x(t) = \left[u_s(t+1) - u_s(t)\right] - \left[u_s(t) - u_s(t-1)\right]$$

12.2.7   Suppose the impulse response function of a system is $h(t) = \Pi(t)$.   If the input signal $x(t)$ is shown in Figure P12.2.7 find the output signal by graphical convolution.

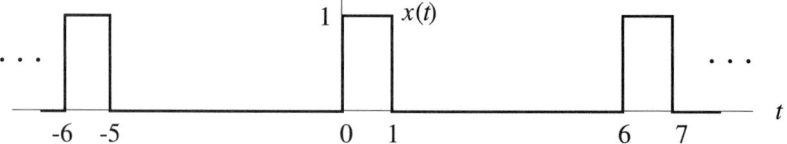

Figure P12.2.7

12.2.8   Suppose that the impulse response function of a system is $h(t) = u_s(t)$. Suppose the input signal is $x(t) = \Pi(t - 2) + \Pi(t + 1)$. Find $y(t) = h(t) * x(t)$ using graphical convolution.

12.2.9   Graphically convolve the two signals in Figure P12.2.9.

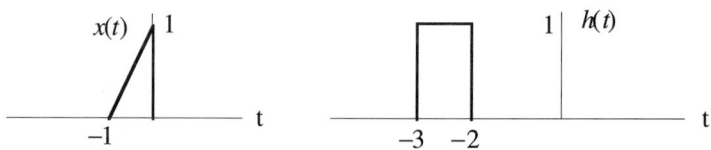

Figure P12.2.9

12.2.10 Graphically convolve the following two signals in Figure P12.2.10.

Figure P12.2.10

Homework Problems for Section 12.3

12.3.1 For a state space model, find $y(t)$ when $x(t) = \delta(t)$. Show that this input signal is equivalent to a certain initial condition.

12.3.2 Find a transfer function for each of the following systems. Also find a state space representation for each systems.

(i) $\quad y(t) = \int_{-\infty}^{\infty} \left[\left(e^{-3(t-\lambda)} + e^{5(t-\lambda)} \right) u_s (t-\lambda) \right] x(\lambda)\, d\lambda$

(ii) $\quad y(t) = \int_{-\infty}^{\infty} \left[\sin(\omega(t-\lambda)) \right] u_s (t-\lambda) x(\lambda)\, d\lambda$

(iii) $\quad y(t) = \int_{-\infty}^{\infty} e^{-3(t-\lambda)} \left[\sin(t-\lambda) \right] u_s (t-\lambda) x(\lambda)\, d\lambda$

(iv) $\quad y(t) = \int_{-\infty}^{\infty} \left(e^{-\lambda} + 0.1 e^{-2\lambda} \sin(10\lambda) \right) u_s (\lambda) x(t-\lambda)\, d\lambda$

(v) $\quad y(t) = \int_{-\infty}^{\infty} [e^{-2(t-\lambda)} \sin(4(t-\lambda)] x(\lambda)\, d\lambda$

12.3.3 (a) Find the state transition matrix for each of the following equations.
(b) Find the impulse response function of each system.

(i) $\quad \dot{\vec{q}} = \begin{bmatrix} 0 & 0 & 2 \\ 5 & 0 & -1 \\ 3 & 0 & 0 \end{bmatrix} \vec{q} + \begin{bmatrix} 0 \\ 1 \\ 0 \end{bmatrix} x$

$\quad y = \begin{bmatrix} 6 & 0 & 0 \end{bmatrix} \vec{q} + 4x$

(ii) $\quad \dot{\vec{q}} = \begin{bmatrix} 1 & 0 \\ 0 & -1 \end{bmatrix} \vec{q} + \begin{bmatrix} 3 \\ 5 \end{bmatrix} x$

$\quad y = \begin{bmatrix} 2 & 1 \end{bmatrix} \vec{q}$

(iii) $\quad \dot{\vec{q}} = \begin{bmatrix} -1 & 1 \\ 0 & 2 \end{bmatrix} \vec{q} + \begin{bmatrix} 0 \\ 3 \end{bmatrix} x$

$\quad y = \begin{bmatrix} 1 & 0 \end{bmatrix} \vec{q} + 4x$

(iv) $\quad \dot{\vec{q}} = \begin{bmatrix} -\sigma & \omega \\ -\omega & -\sigma \end{bmatrix} \vec{q} + \begin{bmatrix} 1 \\ 1 \end{bmatrix} x$

$\quad y = \begin{bmatrix} 1 & 1 \end{bmatrix} \vec{q}$

(v) $\quad \dot{\vec{q}} = \begin{bmatrix} 0 & 1 \\ -\omega_n^2 & -2\zeta\omega_n \end{bmatrix} \vec{q} + \begin{bmatrix} 1 \\ 0 \end{bmatrix} x$

$\quad y = \begin{bmatrix} 1 & 0 \end{bmatrix} \vec{q}$

(vi) $\quad \dot{\vec{q}} = \begin{bmatrix} 0 & 1 \\ -\omega_n^2 & -2\zeta\omega_n \end{bmatrix} \vec{q} + \begin{bmatrix} 0 \\ 1 \end{bmatrix} x$

$\quad y = \begin{bmatrix} 1 & 0 \end{bmatrix} \vec{q}$

12.3.4 Consider the equation

$$\dot{\vec{q}}(t) = \begin{bmatrix} -\sigma & \omega \\ -\omega & -\sigma \end{bmatrix} \vec{q}(t)$$

Let $\sigma = 1$ and $\omega = 2$.
(a) What are the eigenvalues of this state matrix?
(b) Plot the states of this equation on one graph for the following initial conditions.

(i) $\vec{q}_0 = \begin{bmatrix} 1 \\ 0 \end{bmatrix}$ (ii) $\vec{q}_0 = \begin{bmatrix} 0 \\ 1 \end{bmatrix}$ (iii) $\vec{q}_0 = \begin{bmatrix} 1 \\ 1 \end{bmatrix}$

12.3.5 Repeat Problem 12.3.4 using $\sigma = 0.1$ and $\omega = 2$.

12.3.6 Consider the system

$$\dot{\vec{q}}(t) = \begin{bmatrix} -1 & 0 \\ 0 & 1 \end{bmatrix} \vec{q}(t) + \begin{bmatrix} 1 \\ 2 \end{bmatrix} x(t), \qquad \vec{q}_0 = \begin{bmatrix} -1 \\ 2 \end{bmatrix},$$

$$y(t) = \begin{bmatrix} 2 & 3 \end{bmatrix} \vec{q}(t).$$

Suppose that the input signal is a unit step function. Solve for the state variables and the output signal.

12.3.7 Consider a system whose impulse response function is

$$h(t) = \delta(t) + e^{-t} u_s(t).$$

(a) What is the transfer function of this system?
(b) If the input signal for this system is $x(t) = [1 + e^{-t}] u_s(t)$, find the limit of the output signal as $t \to \infty$.

Homework Problems for Section 12.4

12.4.1 Find a Fourier transfer function for each of the following systems if it exists.

(i) $y(t) = \displaystyle\int_{-\infty}^{\infty} \left[\left(e^{-3(t-\lambda)} + e^{5(t-\lambda)} \right) u_s(t-\lambda) \right] x(\lambda)\, d\lambda$

(ii) $y(t) = \displaystyle\int_{-\infty}^{\infty} \left[\sin(\omega(t-\lambda)) \right] u_s(t-\lambda) x(\lambda)\, d\lambda$

(iii) $y(t) = \displaystyle\int_{-\infty}^{\infty} e^{-3(t-\lambda)} \left(\sin(t-\lambda) \right) u_s(t-\lambda) x(\lambda)\, d\lambda$

(iv) $y(t) = \displaystyle\int_{-\infty}^{\infty} \left[e^{-\lambda} + 0.1 e^{-2\lambda} \sin(10\lambda) \right] u_s(\lambda) x(t-\lambda)\, d\lambda$

(v) $y(t) = \displaystyle\int_{-\infty}^{\infty} [e^{-2(t-\lambda)} \sin(4(t-\lambda))] x(\lambda)\, d\lambda$

12.4.2 For each of the following networks, find the Fourier transfer function.

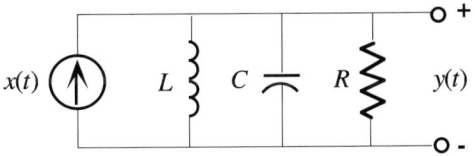

Figure P12.4.2a

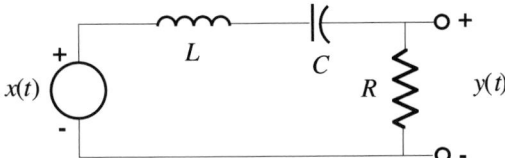

Figure P12.4.2b

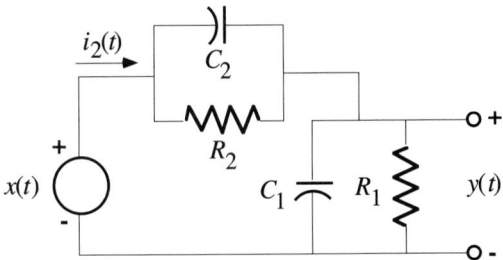

Figure P12.4.2c

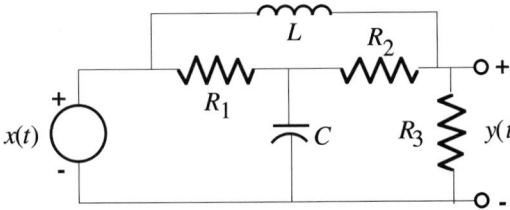

Figure P12.4.2d

12.4.3   (a) Consider the circuit in Figure P12.4.3a.

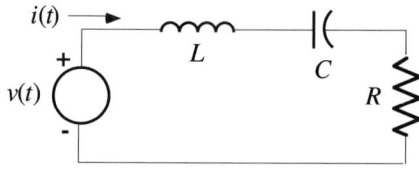

Figure P12.4.3a

The relationship between the voltage and current in this circuit is given by

$$L\frac{di(t)}{dt} + Ri(t) + \frac{1}{C}\int_{-\infty}^{t} i(\lambda)\, d\lambda = v(t)$$

The impedance of this circuit is defined as the Fourier transfer function $\frac{V(\omega)}{I(\omega)}$. Find this impedance.

   (b) Consider the mass-spring-damper system in Figure P12.4.3b.

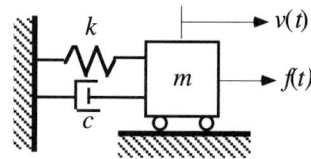

Figure 12.4.3b

Define the velocity of the mass to be $v(t)$. Designate this signal as the output signal. The relationship between the force on the mass and the velocity of the mass is given by

$$m\frac{dv(t)}{dt} + cv(t) + k\int_{-\infty}^{t} v(\lambda)\, d\lambda = f(t)$$

The impedance of this system is defined as the Fourier transfer function $\frac{F(\omega)}{V(\omega)}$. Find this impedance.

12.4.4   Suppose the impulse response function of a system is

$$h(t) = u_s(t) - u_s(t - T).$$

Find the Fourier transfer function of this system if it exists.

Chapter 13

Properties of Systems

Chapter Outline

In this chapter we take a more abstract view of a system. Starting with the basic definition of a system, we introduce four properties of this system. These four properties are: causality, linearity, time invariance, and relaxed. It is then shown that these four properties can be used to derive the convolution representation. We have already established that the convolution representation is closely related to the Laplace transfer function, the Fourier transfer function, and state space representations. So from a mathematical viewpoint, these four properties essentially define all of these system representations.

This abstract characterization of system representations has several practical implications. In order for system analysis to be useful, the system representation must match the observed physical process. The system properties introduced above are a basic link between the experimentally observed data and the system representation used to model the process. Often we can directly verify if the physical process satisfies (or doesn't satisfy) the four properties above. Hence, these properties can be used to determine if the system representations are appropriate models for the observed physical process.

The four system representations we are focusing on in this text are in many ways essentially equivalent. There are some differences between them, however. The four system properties we will introduce in this chapter will allow us to distinguish between these system representations. By carefully characterizing each system representation, we gain insight into selecting the proper system representation to model a certain system. It is the purpose of this chapter to explore the similarities and differences between these system representations.

A fifth system property that is singled out for special treatment is stability. Roughly, a system is stable if a small input signal leads to a small output signal. If a system is not stable, then a small input signal will cause an arbitrarily large output signal. As the output signal grows large the system will cease to function properly. Unstable systems behave quite differently from stable systems so we generally want to avoid them. We develop several tests for stability that can be applied to each of the system representations.

In this text we concentrate primarily on the four primary system representations. These system representations all share the property of being linear, allowing us to use the Laplace and Fourier transforms, for example. In the last section of this chapter, however, we discuss a simple type of nonlinear system, a static nonlinearity. Almost all physical systems contain some nonlinearity of this type. The graphical representation of this type of nonlinear system is introduced. We show how to construct the output signal for a given input signal. Furthermore, the effects of this type of nonlinearity can be analyzed using some of the techniques in Chapter 8. See Section 8.7 for the analysis of a static nonlinearity using Fourier series.

Summary of Sections

Section 13.1:   We define the four fundamental system properties and show how these properties lead to the convolution integral.

Section 13.2:   We expand on the implications of the system properties defined in the first section using physical examples.

Section 13.3:   We define stability and give a characterization of the stability of a convolution integral in terms of the impulse response function.

Section 13.4:   We discuss the stability characterizations of transfer function and state space representations.

Section 13.5:   We discuss the system properties defined in Section 13.1 for each of the system representations introduced in the previous chapters.

Section 13.6:   We discuss static nonlinearities.

Section 13.7:   Chapter summary section.

Coverage of the Text

Sections 13.1 and 13.3 require Chapter 12. The rest of the sections also require Chapters 10 and 11. Section 13.6 is independent of the rest of the chapter

13.1 DEFINITION OF THE SYSTEM PROPERTIES

13.1.1 Introduction

Consider the system

$$y(t) = \mathcal{H}[x(t)]. \tag{13.1.1}$$

In the previous chapters we began by assuming that this system could be represented by a Laplace or Fourier transfer function, state space equations, or a convolution integral. In this section we adopt a more abstract view. We start by defining several properties for (13.1.1).   Then we show that the previously defined system representations can be derived from these properties.   The four properties we will discuss here are: causal, linear, time invariant, and relaxed.

13.1.2 System Properties

Causal The first property is possessed by all physical processes.

Definition 13.1.1: A system is <u>causal</u> if for all input signals $x_1(t)$ and $x_2(t)$ that satisfy

$$x_1(t) = x_2(t) \quad \text{for} \quad t \le t_0,$$

the corresponding output signals satisfy

$$y_1(t) = y_2(t) \quad \text{for} \quad t \le t_0$$

for all t_0.

▲▲

The above definition is rather abstract (but elegant).   It is simply a clever mathematical expression for a property which coincides with the usual English usage. The following example shows how to apply this definition.

Example 13.1.2: Consider the system

$$y(t-1) = x(t).\tag{13.1.2}$$

To check the causality of this system, we must choose two input signals that are identical for $t \le t_0$ but that differ for $t > t_0$ for some t_0. Let $x_1(t) = 0$ shown in Figure 13.1.1a and let $x_2(t) = u_s(t-a)$ also shown in Figure 13.1.1a where $t_0 < a < t_0 + 1$. It is clear that

$$x_1(t) = x_2(t) \quad \text{for} \quad t \le t_0.\tag{13.1.3}$$

The output signals for each of these two input signals are also shown in Figure 13.1.1. The output signal for the first input signal is just identically zero as shown in Figure 13.1.1b. The second input signal goes from 0 to 1 at $t = a$. The corresponding output signal goes from 0 to 1 at time $t = a - 1$ according to the system in (13.1.2). This output signal is shown in Figure 13.1.1b.

    By (13.1.3) these two signal satisfy the first condition of Definition 13.1.1 but

$$y_1(t_0) = 0 \ne 1 = y_2(t_0) \quad \text{for} \quad t = t_0\tag{13.1.4}$$

violating Definition 13.1.1. The difference in the output signals at $t = t_0$ could only be attributed to variations in the input signals for times "in the future," that is, for values of $t > t_0$. Hence, this system is not causal.

▲▲

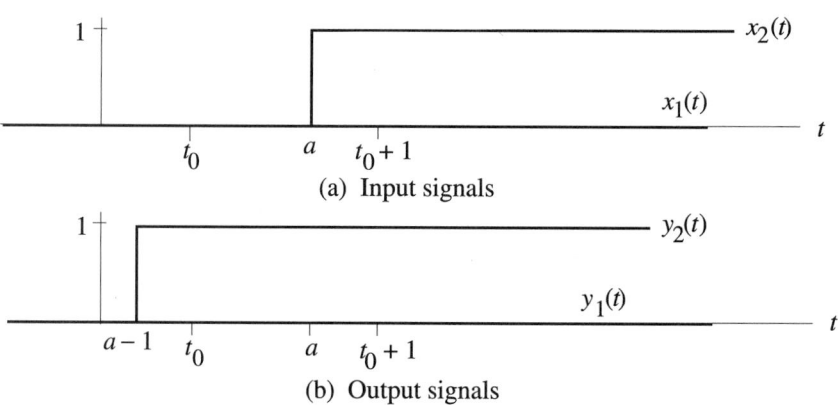

Figure 13.1.1 The Input and Output Signals for Example 13.1.2

Example 13.1.3: Any system modeled with an ordinary differential equation is a causal system.

▲▲

Linear The following property is required of most of the systems we will study in this text.

Definition 13.1.4: Let $x_1(t)$ and $x_2(t)$ be input signals such that

$$y_i(t) = \mathcal{H}\big[x_i(t)\big], \quad i = 1, 2.$$

A system is <u>linear</u> if for any two real numbers a_1 and a_2

$$\mathcal{H}\big[a_1 x_1(t) + a_2 x_2(t)\big] = a_1 y_1(t) + a_2 y_2(t).$$

Otherwise, we say the system is <u>nonlinear</u>.

ss

If a system is linear it satisfies two familiar properties.   First, this system satisfies the <u>superposition</u> principle

$$\mathcal{H}\big[x_1(t) + x_2(t)\big] = y_1(t) + y_2(t). \tag{13.1.5}$$

It is also <u>homogenous</u>

$$\mathcal{H}\big[a_1 x_1(t)\big] = a_1 y_1(t). \tag{13.1.6}$$

Both of these properties are frequently used to solve network problems.
If a system is linear, then its response to a complicated signal

$$x(t) = a_1 x_1(t) + \cdots + a_n x_n(t) \tag{13.1.7}$$

can be predicted by knowing the response of the system to the signals

$$\mathcal{H}\big[x_i(t)\big] = y_i(t), \quad i = 1, \ldots, n. \tag{13.1.8}$$

This ability to predict the response of the system to a large class of input signals based on the output signal of the system for a relatively few input signals is the essence of the property of linearity.

Example 13.1.5: Any differential equation of the form

$$y^{(n)}(t) + a_{n-1}(t) y^{(n-1)}(t) + \cdots + a_0(t) y(t) \tag{13.1.9}$$
$$= b_m(t) x^{(m)}(t) + b_{m-1}(t) x^{(m-1)}(t) + \cdots + b_0 x(t)$$

is a linear system. The following systems are linear.

(1) $\ddot{y}(t) + 2\zeta\omega_n\dot{y}(t) + \omega_n^2 y(t) = x(t)$ (13.1.10)

(2) $\dot{y}(t) + ty(t) = x(t)$

The following systems are nonlinear.

(1) $\ddot{y}(t) + \alpha y(t) + [y(t)]^3 = x(t)$ (13.1.11)

(2) $\dot{y}(t) + y(t)x(t) = x(t)$

▲▲

A class of nonlinear systems, called static nonlinearities, is discussed in Section 13.7. An example of the analysis of the output signal of a static nonlinearity is given in Section 8.7.

Time Invariant   A system is defined in terms of two signals: the input signal and the output signal. A signal is defined as a *time* variation of a physical variable. Hence, a system implicitly depends on time. The next definition draws a distinction between systems that depend explicitly on time, and those that only depend implicitly on time.

Definition 13.1.6: A system is <u>time invariant</u> if for all t_0

$$\mathcal{H}[x(t - t_0)] = y(t - t_0).$$

Otherwise, the system is <u>time-varying</u>.

▲▲

If a system is time invariant then we can chose the origin of the time axis wherever we want.

Example 13.1.7:   If the coefficients in (13.1.9) vary as functions of time, then (13.1.9) is time-varying. A system of the form

$$y^{(n)}(t) + a_{n-1}y^{(n-1)}(t) + \cdots + a_0 y(t)$$ (13.1.12)

$$= b_m x^{(m)}(t) + b_{m-1}x^{(m-1)}(t) + \cdots + b_0 x(t)$$

where the a_i and b_i are constant is a time invariant system.

▲▲

Relaxed   A system generates an output signal because it is stimulated. The way in which the system is stimulated will determine how the system is analyzed. The next definition distinguishes among the input signals and initial conditions that lead to a system output signal.

Definition 13.1.8: A system is <u>relaxed at time t_0</u> if the output signal $y(t)$ on the interval $t_0 \le t$ is solely and uniquely excited by the input signal $x(t)$ on the interval $t_0 \le t$ for $-\infty < t_0$.

▲▲

Typically, a relaxed system will have a zero output signal up until some time t_0. At time t_0 an input signal is applied to the system, after which the output signal will be nonzero.

There are two ways that a system is not relaxed. First, the system is not relaxed if the input signal is zero, but the system contains stored energy. Usually, this type of system is modeled using differential equations with nonzero initial conditions. Second, a system is not relaxed if it has an input signal present for all time. Here the system is operating in steady state.

13.1.3 Derivation of the Convolution Integral

Next we will give a heuristic argument that a system which is *linear, time invariant, and relaxed* can be represented by a convolution representation of the system. We make no assumptions about the physics of the underlying process. The intention is to develop the convolution integral as a modeling tool that applies to many different systems. Because of the relationship between the convolution integral and the other system representations, these results extend to the other system representations as well. This discussion is undertaken in Section 13.5.

To begin we assume that we are given a system

$$y(t) = \mathcal{H}\big[x(t)\big]. \tag{13.1.13}$$

We will show that by using only the properties of the systems given above and the impulse response function, we can construct a convolution representation of the system. This construction will proceed by: (1) determining the preliminary data that characterizes the system, (2) constructing an approximation to an arbitrary input signal, (3) computing the output signal due to each input approximating signal, (4) applying the properties introduced above, and (5) reassembling the output signal.

We assume that the impulse response function $h(t)$ of the system is known over the time interval $t_0 \le t \le t_f$. The impulse response function is obtained by applying an impulse input signal at time $t = t_0$ and then determining the output signal $y(t) = h(t)$. This impulse response function could have been derived using physical principles, or it could have been measured. In either case, we require the system to be *relaxed* because we assume that the impulse response, as the output signal, is solely due to the impulse function as the input signal.

Let $x(t)$, $t_0 < t < t_f$, ($t_0 < 0$, say) be an arbitrary input signal. To compute the corresponding output signal, we first decompose the input signal. We divide the interval into several smaller intervals $\Delta\lambda$ wide. Then we approximate $x(t)$ by

$$\hat{x}(t) = \sum_{m=-M}^{N} x(m\,\Delta\lambda)\Pi\!\left(\frac{t - m\,\Delta\lambda}{\Delta\lambda}\right), \qquad t_0 = -M\,\Delta\lambda, \qquad t_f = N\,\Delta\lambda. \tag{13.1.14}$$

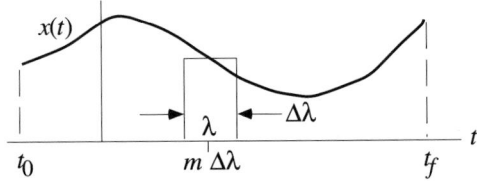

Figure 13.1.2 Approximation of $x(t)$

Thus, $x(t)$ is approximated by a series of pulse functions one of which is shown in Figure 13.1.2. Through rearrangement and multiplication by a unity factor we obtain

$$\hat{x}(t) = \frac{\Delta\lambda}{\Delta\lambda} \sum_{m=-M}^{N} x(m\,\Delta\lambda)\,\Pi\!\left(\frac{t-m\,\Delta\lambda}{\Delta\lambda}\right) \tag{13.1.15}$$

$$= \sum_{m=-M}^{N} x(m\,\Delta\lambda)\left[\frac{1}{\Delta\lambda}\Pi\!\left(\frac{t-m\,\Delta\lambda}{\Delta\lambda}\right)\right]\Delta\lambda.$$

We would like to take a limit $\Delta\lambda \to 0$ in (13.1.15). When we take that limit, we assume that intervals $\Delta\lambda$ are re-indexed for each $\Delta\lambda$ so that

$$\lim_{\substack{\Delta\lambda\to0\\M,N\to\infty}} m\,\Delta\lambda = \lambda, \qquad \lim_{\substack{M\to\infty\\\Delta\lambda\to0}} -M\,\Delta\lambda = t_0, \qquad \text{and} \qquad \lim_{\substack{N\to\infty\\\Delta\lambda\to0}} N\,\Delta\lambda = t_f. \tag{13.1.16}$$

Using the approximation of the input signal in (13.1.15) we approximate output of the system (13.1.13) by

$$\hat{y}(t) = \mathcal{H}[\hat{x}(t)] = \mathcal{H}\left[\sum_{m=-M}^{N} x(m\,\Delta\lambda)\left[\frac{1}{\Delta\lambda}\Pi\!\left(\frac{t-m\,\Delta\lambda}{\Delta\lambda}\right)\right]\Delta\lambda\right]. \tag{13.1.17}$$

Assuming the system is *linear* we can write the output signal of the system as a sum of the contributions of each of the piecewise approximation of the input signal. We have

$$\hat{y}(t) = \sum_{m=-M}^{N} x(m\,\Delta\lambda)\mathcal{H}\left[\frac{1}{\Delta\lambda}\Pi\!\left(\frac{t-m\,\Delta\lambda}{\Delta\lambda}\right)\right]\Delta\lambda. \tag{13.1.18}$$

Recall from Example 5.2.8 that

$$\lim_{\substack{\Delta\lambda\to0\\M,N\to\infty}} \frac{1}{\Delta\lambda}\Pi\!\left(\frac{t-m\,\Delta\lambda}{\Delta\lambda}\right) = \delta(t-\lambda). \tag{13.1.19}$$

Hence, the term in the brackets in (13.1.18) is essentially the impulse function in (13.1.19). Assuming the system is *time invariant* we have

$$\mathcal{H}\left[\frac{1}{\Delta\lambda}\Pi\left(\frac{t-m\,\Delta\lambda}{\Delta\lambda}\right)\right] = \hat{h}(t-m\,\Delta\lambda) \tag{13.1.20}$$

where $\hat{h}(t-m\,\Delta\lambda)$ is an approximation to the impulse response function over the time interval $t_0 \le t \le t_f$. Using (13.1.20), (13.1.18) becomes

$$\hat{y}(t) = \sum_{m=-M}^{N} x(m\,\Delta\lambda)\hat{h}(t-m\,\Delta\lambda)\Delta\lambda. \tag{13.1.21}$$

Equation (13.1.21) says that the system output is approximated by a weighted sum of impulse response functions. In the limit the output signal of the system is

$$y(t) = \lim_{\substack{\Delta\lambda\to 0 \\ M,N\to\infty}} \hat{y}(t) \tag{13.1.22}$$

$$= \lim_{\substack{\Delta\lambda\to 0 \\ M,N\to\infty}} \sum_{m=-M}^{N} x(m\,\Delta\lambda)\hat{h}(t-m\,\Delta\lambda)\,\Delta\lambda = \int_{t_0}^{t_f} h(t-\lambda)x(\lambda)\,d\lambda.$$

Hence, we recover the convolution representation of the system.

We have shown that given a system (13.1.13) that is linear and time invariant, it can be modeled by the convolution integral in (13.1.11) if we know its impulse response function. In the previous chapters we have shown that under certain conditions the convolution integral is equivalent to a Laplace and Fourier transfer function and to a state space representation. We immediately conclude that if a system is linear and time invariant it can also be modeled with any of these other three system representations. (There are certain restrictions which are discussed fully in Section 13.5.)

13.1.4 Summary

The analysis in this section lends important insight into the use of a system representation to model observed physical processes. Suppose that we are studying a physical process and we wish to construct a representation of this process. How do we know that one of the representations we have introduced can be used to model the observed data? This question is answered by the analysis above. We must verify if the process is linear and time invariant. Quite often these properties can be verified from independent information. Or these properties can be tested by applying the appropriate input signals. Once we have verified these two properties, we know that it can be modeled by a convolution integral, or an equivalent representation. Or if the system is nonlinear or time-varying, we know not to use one of these models. this information is extremely important at the outset of the modeling process.

Moreover, this analysis indicates that any method of arriving at this representation is valid. One method for obtaining this model is to use the laws of physics to derive the relevant differential equation. From the differential equation the transfer function is calculated and then the impulse response function is calculated from the transfer function. Another approach, however, is to measure the impulse response function directly from experimental input-output data. We are directly computing a model from measured signals. This calculated impulse response function can then be transformed into an appropriate representation. In general, this process of model building is called *system identification*. System identification is justified by the results in this section and it is often used.

Terminology: The system representations we have introduced, the Laplace and Fourier transfer function, state space representations, and the convolution integral, are sometimes referred to as <u>linear, time invariant</u> (LTI) systems. (MATLAB uses this terminology, for example.) The analysis in this section explains this terminology. All these system representations have the properties of linearity and time invariance.

13.2 DISCUSSION OF PROPERTIES OF SYSTEMS

13.2.1 Introduction

Consider the system

$$y(t) = \mathcal{H}[x(t)]. \qquad (13.2.1)$$

In the last section we defined four properties for such a system: causal, linear, time invariant, and relaxed. We then showed that the last three properties lead to convolution representation of the system. Clearly, these four properties are important to the systems which we are studying in this text. In this section we will give several examples which indicate the role these properties play in system analysis.

13.2.2 Causality

Causality is a property possessed of all physical processes. Hence, any representation which accurately models a physical process will be causal. If a representation turns out not to be causal, typically it is discarded. We won't discuss causality further.

13.2.3 Linearity and Time Invariance

The premise of system theory is that a physical process can be studied by analyzing the representation of that physical process using the tools of mathematics (e.g., the Laplace or Fourier transform). The representation, however, must have certain properties before these tools can be applied. So there is a close relationship between the representations with which we model a process and the methodology we use to analyze those representations.

Linearity and time invariance are two system properties that establish a connection between system representations and the mathematical tools that can be used to analyze them. For example, both the Laplace and Fourier transform can only be used for linear, time invariant system analysis. If a system is not linear, particularly if it is modeled by a nonlinear differential equation, the analysis can be very difficult. This difficulty stems, in part, from the fact that many of the powerful techniques developed in this text apply only to linear systems.

Example 13.2.1: Consider the simple differential equation

$$a\dot{q}(t) + q(t) = x(t), \quad q(0) = 0. \tag{13.2.2}$$

Taking the Laplace transform of this equation we obtain

$$\mathcal{L}\{a\dot{q}(t) + q(t) = x(t)\} \quad \Rightarrow \quad a(sQ(s) - 0) + Q(s) = X(s). \tag{13.2.3}$$

Now we can easily solve (13.2.3) for $Q(s)$ and through the inverse Laplace transform obtain $q(t)$. Suppose, however, that the coefficient is time-varying as in the equation

$$a(t)\dot{q}(t) + q(t) = x(t), \quad q(0) = 0. \tag{13.2.4}$$

Taking the Laplace transform of the first term we obtain

$$\mathcal{L}\{a(t)\dot{q}(t)\} = \int_{-\infty}^{\infty} A(s - \lambda)[\lambda Q(\lambda)] \, d\lambda \tag{13.2.5}$$

because of the convolution property of Laplace transforms, Property 9.2.3. This expression is not simple, and we can't easily solve (13.2.4) using (13.2.5). This example shows why time-varying systems are generally not solvable using the Laplace transform.

▲▲

The next example shows how linearity and time invariance are used in the routine analysis of networks.

Example 13.2.2: Consider again the simple RC circuit in Figure 13.2.1. Suppose the input signal is the pulse

$$x(t) = u_s(t) - u_s(t - t_0). \tag{13.2.6}$$

We would like to determine the output signal for this input signal. We will use this simple example to show where all the system properties come into play.
The model for this system is

$$\dot{y}(t) + \frac{y(t)}{RC} = \frac{x(t)}{RC}. \tag{13.2.7}$$

Since this system is linear we can compute the time response of each component of the signal in (13.2.6) and add the output signals. Since the system is time invariant we can compute the output signal due to the second input signal from a time delayed

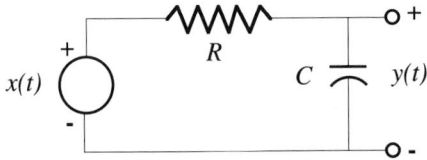

Figure 13.2.1 *RC* Network

version of the first input signal.

Considering the first input signal

$$x_1(t) = u_s(t),$$
(13.2.8)

we see that the system is relaxed because the input signal is zero for negative time, and the initial condition is zero. The linearity of the system allows us to use the Laplace transform to solve the differential equation. Hence, we can solve for the output signal using the transfer function via the Laplace transform. The transfer function of this system is

$$\frac{Y(s)}{X(S)} = H(s) = \frac{\frac{1}{RC}}{s + \frac{1}{RC}}.$$
(13.2.9)

Following the usual procedure to solve for the output signal given the input signal (13.2.8) we obtain

$$y_1(t) = \left(1 - e^{-\frac{t}{RC}}\right) u_s(t).$$
(13.2.10)

The second step function in (13.2.6) is a time delayed version of the first signal. Since the system is time invariant, we can calculate the output signal of the second step function in (13.2.6) from the first output signal in (13.2.10). The output signal of the system due to the second input signal is

$$y_2(t) = \left(1 - e^{-\frac{(t-t_0)}{RC}}\right) u_s(t - t_0).$$
(13.2.11)

Subtracting (13.2.11) from (13.2.10) (because the system is linear) and simplifying the expression we arrive at the output signal

$$y(t) = \begin{cases} 1 - e^{-\frac{t}{RC}}, & 0 < t < t_0 \\ e^{-\frac{(t-t_0)}{RC}} - e^{-\frac{t}{RC}}, & t > t_0 \end{cases}$$
(13.2.12)

▲▲

13.2.4 Relaxed

The property of a system being relaxed is used primarily to determine which system representation is appropriate, as we will see in the later sections of this chapter. The next example shows how a system may not be relaxed.

Example 13.2.3:   Consider the *RC* network in Figure 13.2.2. Suppose the input signal is a step function. From Section 6.3, the output signal is

$$y(t) = y_0 e^{-\frac{t}{RC}} + \left(1 - e^{-\frac{t}{RC}}\right) u_s(t). \tag{13.2.13}$$

If $y_0 \neq 0$, the output signal contains a term that can't be traced to the input signal. Hence, this system is not relaxed. In this example a nonzero initial condition corresponds to stored energy in the capacitor. If there is no stored energy, then the initial condition is zero and the system is relaxed.

Next, suppose that the initial condition is zero, but that the input signal is

$$x(t) = \cos(\omega_i t), \quad -\infty < t < \infty. \tag{13.2.14}$$

This system is not relaxed because for any time t_0 the output signal from the system will be determined by the input signal for $t < t_0$.

▲▲

13.2.5 A Mechanical System

The four properties defined in this last section are independent from one another. That means a system can possess any three of the properties, but not the fourth. The following example illustrates this point.

Example 13.2.4:   Consider a mass-spring-damper system in which the mass is suspended in a gravity field as shown in Figure 13.2.3. We wish to derive a model for this system. We have two choices for coordinate systems for the displacement. The first reference position $y_g(t)$ of the mass is when the spring is undeformed. The second reference $y_{st}(t)$ of the mass when it is in equilibrium. If we use the first coordinate system referenced to the undeformed position of the mass, then the differential equation derived from the free-body diagram in Figure 13.2.4 using Newton's second law is

$$m_{st}\ddot{y}_g(t) = -k_{st}y_g(t) - c_{st}\dot{y}_g(t) + f_{st}(t) - m_{st}g. \tag{13.2.15}$$

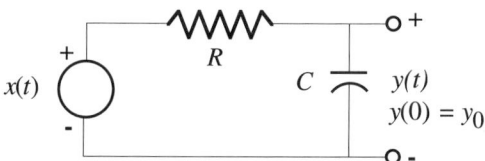

Figure 13.2.2 *RC* Network

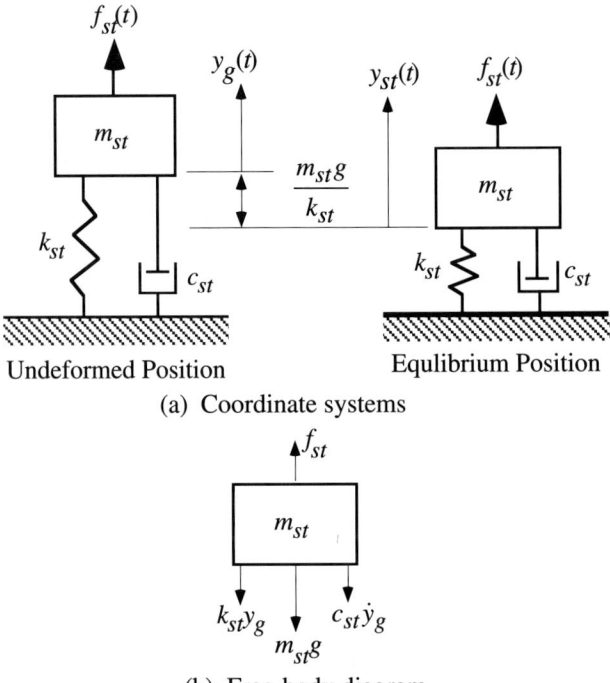

(a) Coordinate systems

(b) Free-body diagram

Figure 13.2.3 Mass-Spring-Damper System in a Gravity Field

This system is not linear because of the constant term due the effect of gravity. If this system were linear, when we applied a zero input force $f_{st}(t) = 0$, we would obtain zero displacement. For the model in (13.2.15), however, a zero input force will lead to a displacement due to the gravity term. Hence, this system is not linear.

   We can derive the second coordinate system from (13.2.15) by noting that the offset of the equilibrium position from the undeformed position is $m_{st}g/k_{st}$. Now define a new displacement variable

$$y_{st}(t) = y_g(t) - \frac{m_{st}g}{k_{st}}.$$ (13.2.16)

Substituting (13.2.16) into (13.2.15) we get

$$m_{st}\ddot{y}_{st}(t) = -k_{st}y_{st}(t) - c_{st}\dot{y}_{st}(t) + f_{st}(t).$$ (13.2.17)

Here we have derived a model of the motion of the mass around the equilibrium point of the mass-spring-damper system.

   If we want to model the motion of the mass after it is released from the undeformed position, we can attached the initial condition to the differential equation (13.2.17)

$$m_{st}\ddot{y}_{st}(t) = -k_{st}y_{st}(t) - c_{st}\dot{y}_{st}(t) + f_{st}(t), \quad y_{st}(0) = \frac{m_{st}g}{k_{st}}. \tag{13.2.18}$$

This system is not relaxed because of the stored potential energy modeled by the initial condition.

   If we are interested in the motion of the mass from the equilibrium position due to applied forces, then the system model is given by (13.2.17). This system is causal, linear, time invariant, and relaxed. Now we can apply the methods of modeling and analysis we introduced in the previous chapters. These methods generally don't apply to (13.2.18).

   Consider again the system defined in (13.2.17). Suppose that the mass is a water tank that has developed a leak at time $t = 0$. As a result, the mass is given by

$$m_{st}(t) = m_{st0}e^{-at} + 1, \quad a > 0, \quad m_{st0} > 1. \tag{13.2.19}$$

Now the differential equation for this system is

$$m_{st}(t)\ddot{y}_{st}(t) = -k_{st}y_{st}(t) - c_{st}\dot{y}_{st}(t) + f_{st}(t). \tag{13.2.20}$$

This system is causal, linear, relaxed, but time-varying because the mass coefficient is time-varying.

   Finally, consider again the system defined in (13.2.17). In deriving this equation, we assumed that Hooke's law was used to model the restoring force of the spring to the mass. Hooke's law is a linear model of a spring. This model is represented by a (dashed) straight line in Figure 13.2.4. The restoring force may be nonlinear, however. For example, if the spring is a hard spring, the load displacement curve is shown in Figure 13.2.4. The curve in Figure 13.2.4 shows that the restoring force due to the spring resulting from a displacement of the mass is not linear, because it applies an increasing larger force with increasing displacement. A hard spring can be modeled by the addition of a cubic term as

$$-\gamma[y_{st}(t)]^3. \tag{13.2.21}$$

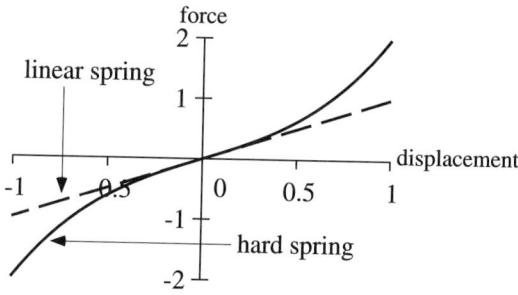

Figure 13.2.4 A Hard Spring

If we use this spring model in the system (13.2.17), we obtain the model

$$m_{st}\ddot{y}_{st}(t) = -k_{st}y_{st}(t) - \gamma\left[y_{st}(t)\right]^3 - c_{st}\dot{y}_{st}(t) + f_{st}(t).$$
(13.2.22)

This model is causal, time invariant, relaxed, but nonlinear due to the cubic term in (13.2.22).

▲▲

13.3 BIBO STABILITY

13.3.1 Introduction

In this section we continue the discussion of properties of the system

$$y(t) = \mathcal{H}[x(t)]$$
(13.3.1)

with a discussion of stability. We have singled out stability for special treatment, because it plays an important role in the synthesis of systems.

To introduce stability, we begin by asking a very basic question about the system (13.3.1). If the input $x(t)$ is "small," will the output $y(t)$ also be "small"? Clearly, for many systems this property is desirable. If you press the accelerator pedal on your car just a "little bit" you expect the car to go only a "little bit" faster. If you turn up the volume on your stereo just a "little," you expect the music to be just a "little" louder. If a small signal causes an arbitrarily large output signal, the system could damage itself and possibly other systems in the near vicinity. It is this second type of behavior we wish to avoid (usually).

Example 13.3.1: The proof-mass actuator, as described in Section 6.5, has the type of behavior we would like to avoid. The transfer function for this device is

$$\frac{Y_{pm}(s)}{V_{pm}(s)} = \frac{K_{ef}}{m_{pm}s^2}.$$
(13.3.2)

Let the input signal to this device be

$$v_{pm}(t) = \left[A\sin(\omega_i t)\right]u_s(t).$$
(13.3.3)

The corresponding output signal is

$$y_{pm}(t) = \frac{AK_{ef}}{m_{pm}\omega_i^2}\left[\omega_i t - \sin(\omega_i t)\right]u_s(t).$$
(13.3.4)

(See Example 12.1.4.) The input and output signals are shown in Figure 13.3.1. Note that while the input signal is bounded, the output signal is increasing without bound.

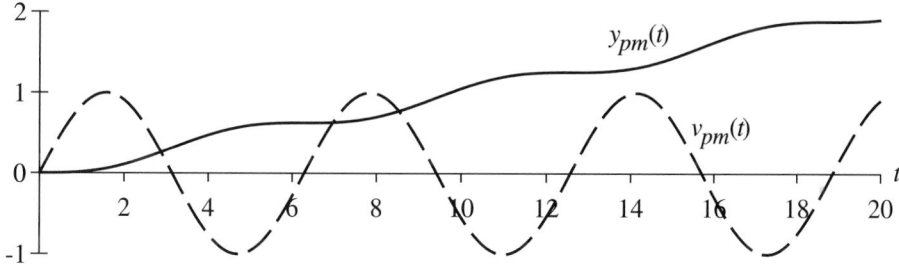

Figure 13.3.1 Position of the Proof-Mass for a Sinusoidal Input Voltage

The proof-mass actuator is an example of what we will define as an unstable system.
    Mathematically, the output signal of an unstable system will increase without bound. This model of the physical process usually breaks down once the signals pass certain limits, and the model of the system ceases to be valid. Recall that a proof-mass actuator has a proof-mass with a finite stroke length. Hence, if the proof-mass displacement exceeds the stroke length, then the proof-mass will run against its stops. Obviously, the actuator is ineffectual when the proof-mass is pinned to its stops. The contact of the proof-mass against the stops also will impart shocks to the actuator possibly causing damage. It is clear from Figure 13.3.1 the position of the proof-mass will eventually exceed its stroke length no matter how small the amplitude A of the input sinusoidal voltage. Here a "small" input signal leads to unbounded growth in the output signal. Eventually, the actuator is rendered useless. Therefore, this device must be modified before it can be used. Typically, unstable systems are worthless in practice.

▲▲

13.3.2 Definition of BIBO Stability

We want to give a precise mathematical description of the behavior illustrated by the proof-mass actuator above.   This description begins with a mathematical definition of the measure of how "small" signal is.

Definition 13.3.2: A signal $x(t)$ is <u>bounded</u> if there exists a constant M, $0 < M < \infty$, such that

$$|x(t)| \leq M \text{ for all } t.$$

▲▲

In particular, a bounded signal does *not* go to infinity as $t \to \infty$.

Example 13.3.3: The following signals are bounded.

$$x(t) = \sin(\omega t), \quad x(t) = e^{-at}u_s(t). \tag{13.3.5}$$

The following signals are unbounded.

$$x(t) = t, \quad x(t) = \frac{1}{t}, \quad x(t) = \tan(t), \quad x(t) = e^{-at}. \tag{13.3.6}$$

▲▲

Remark 13.3.4: In this discussion of stability, we do not allow the impulse function $\delta(t)$ to be an input signal. The reason is that in terms of simple function theory we can't decide on a bound for the impulse function because it is not a function in the ordinary sense. This restriction doesn't interfere with the applications of the concept of stability in this book. A rigorous treatment of BIBO stability which includes a discussion of impulse functions as input signals can be found in advanced texts on system theory.

▲▲

The next definition captures the idea that a "small" input signal results in a "small" output signal.

Definition 13.3.5: The system (13.3.1) is <u>bounded-input-bounded-output (BIBO) stable</u> if every bounded input signal results in a bounded output signal. Otherwise, the system is BIBO <u>unstable</u>.

▲▲

Terminology: This definition of stability is stated in terms of the input signal and the output signal of the system. Therefore, this kind of stability is called <u>input-output</u> stability.

There are several other definitions of stability that are commonly used. These definitions of stability (not introduced here) are stated only in terms of the state of the system (implying a state space model). It is important to distinguish which concept of stability is being discussed. In this book we will only use BIBO stability.

The definition of BIBO stability applies to any system described by (13.3.1) including nonlinear, time-varying systems as well as the systems represented by transfer functions, convolution integrals, and state space equations. In this text, however, we will mostly consider linear, time invariant systems.

The definition of BIBO stability requires that *every* bounded input signal cause a bounded output signal. The following examples illustrates this point.

Example 13.3.6: Consider the system

$$y(t) = \int_0^\infty h(t - \lambda)x(\lambda)\, d\lambda = \int_0^\infty u_s(t - \lambda)x(\lambda)\, d\lambda = \int_0^t x(\lambda)\, d\lambda \tag{13.3.7}$$

whose impulse response function is a step function. Clearly, this system is an integrator. If the input signal into this system is

$$x_1(t) = e^{-at}u_s(t), \tag{13.3.8}$$

then the output signal of this system is

$$y_1(t) = \int_0^t e^{-a\lambda} \, d\lambda = \frac{1}{a}\left(1 - e^{-at}\right)u_s(t). \tag{13.3.9}$$

For the bounded input signal in (13.3.8), the output signal in (13.3.9) is bounded. If the input signal to this system is

$$x_2(t) = u_s(t), \tag{13.3.10}$$

then the output signal is

$$y_2(t) = \int_0^t 1 \, d\lambda = t u_s(t) \tag{13.3.11}$$

which is unbounded. This example shows that for a given system there are many bounded input signals that result in bounded output signals. This system is BIBO unstable, however, because we have found *one* bounded input signal, (13.3.10), that causes the output signal, (13.3.11), to be unbounded. Hence, a key word in Definition 13.3.5 is *every*.

▲▲

In the last example the output signal (13.3.11) goes to infinity even though the input signal (13.3.10) is bounded. If we examine the output signal on any finite interval, however, the output signal would be bounded. The output signal goes to infinity only over an infinite time interval. Implicit in the (mathematical) definition of BIBO stability is the consideration of the input and output signals over an infinite time interval. This concept is essential to the definition of stability.

From a practical point of view, we need not wait until infinity to determine if the output signal has gone to infinity. In practice, "very large" is a good substitute for "infinity." Indeed, in practice output signals can become "very large" in a very short time. Stability plays a central role in the practice of system theory.

This definition is motivated by practical considerations. Because we will not always know what the input signal to the system will be we must consider all possible input signals. For example, if we consider a car to be a system, we won't know in advance what input signal the driver will give the car. Hence, we want the car to behave in a reasonable way for all input signals. So the definition of stability requires that the system respond with bounded output signal for *every* bounded input signal.

Unfortunately, it is not possible to check system stability by calculating the output signal for every input signal. Rather, we would like to have a simple, one-time test that could be applied to the system representation. Therefore, we will develop several tests for BIBO stability based on the representation of the system discussed in the previous chapters.

13.3.3 Convolution Integral

Next we will develop a test for BIBO stability for the convolution representation

$$y(t) = \int_{-\infty}^{\infty} h(t - \lambda) x(\lambda) \, d\lambda \qquad\qquad (13.3.12)$$

based on the impulse response function.

Theorem 13.3.7: A system modeled by the convolution integral (13.3.12) is BIBO stable if and only if the impulse response function satisfies

$$\int_{-\infty}^{\infty} |h(t)| \, dt \le M_h < \infty.$$

▲▲

Proof: We assume that the integral of the impulse response function is bounded, and we assume that $x(t)$ is bounded by M_x. Taking the absolute value of both sides of (13.3.12) we have

$$|y(t)| = \left| \int_{-\infty}^{\infty} h(t - \lambda) x(\lambda) \, d\lambda \right| \le \int_{-\infty}^{\infty} |h(t - \lambda)| |x(\lambda)| \, d\lambda. \qquad (13.3.13)$$

Now using the bound on $x(t)$ we have

$$|y(t)| \le \int_{-\infty}^{\infty} |h(t - \lambda)| |x(\lambda)| \, d\lambda \le M_x \int_{-\infty}^{\infty} |h(t - \lambda)| \, d\lambda \le M_x M_h. \qquad (13.3.14)$$

From (13.3.14) we see that if the integral of $h(t)$ is bounded, then the output signal is also bounded. Thus, a bounded input signal gives a bounded output signal. This proves one direction of the implication in Theorem 13.3.7. The rest of the proof of Theorem 13.3.7 is omitted.

▲▲

Example 13.3.8: The convolution representation of an *RC* network was derived in Section 6.3. Applying Theorem 13.3.7 to this system we find that

$$\int_{-\infty}^{\infty} |h(t)| \, dt = \lim_{\gamma \to \infty} \int_{-\gamma}^{\gamma} \left| \frac{1}{RC} \exp\left(\frac{-t}{RC}\right) u_s(t) \right| dt = \lim_{\gamma \to \infty} \left[1 - \exp\left(\frac{-\gamma}{RC}\right) \right] = 1. \qquad (13.3.15)$$

This system is BIBO stable for all values of R and C such that $RC > 0$. Hence, this system is BIBO stable for all choices of resistors and capacitors.

▲▲

Example 13.3.9: Consider the proof-mass actuator of Example 13.3.1. The impulse response function for this device is

$$h(t) = t u_s(t). \qquad\qquad (13.3.16)$$

Using Theorem 13.3.7 to check the stability of this system, we have

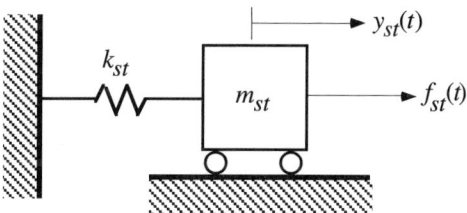

Figure 13.3.2 Mass-Spring System

$$\lim_{\gamma \to \infty} \int_{-\gamma}^{\gamma} |h(t)|\, dt = \lim_{\gamma \to \infty} \int_{0}^{\gamma} t\, dt = \lim_{\gamma \to \infty} \frac{\gamma^2}{2} = \infty. \qquad (13.3.17)$$

This system is unstable. This result is confirmed in Example 13.3.1 where it is shown that a bounded input signal gives an unbounded output signal.

▲▲

Example 13.3.10: Consider the mass-spring-damper system in Section 6.4. Suppose that the damping coefficient is zero, $c_{st} = 0$. This system is shown in Figure 13.3.2. The impulse response function is

$$h(t) = \frac{1}{\omega_n}\left[\sin(\omega_n t)\right] u_s(t). \qquad (13.3.18)$$

The impulse response in (13.3.18) was obtained setting $c_{st} = 0$ in (6.4.4). Applying Theorem 13.3.7 to (13.3.18) we have

$$\lim_{\gamma \to \infty} \int_{-\gamma}^{\gamma} |h(t)|\, dt = \lim_{\gamma \to \infty} \int_{0}^{\gamma} |\sin(\omega_n t)|\, dt = \infty. \qquad (13.3.19)$$

By Theorem 13.3.7 this system is BIBO unstable.

▲▲

13.4 BIBO STABILITY OF TRANSFER FUNCTIONS AND STATE SPACE REPRESENTATIONS

13.4.1 Introduction

In this section we will continue our discussion of the BIBO stability of the system

$$y(t) = \mathcal{H}\left[x(t)\right]. \qquad (13.4.1)$$

Here we will consider systems represented by transfer functions and state space representations. We will derive tests for the stability of these systems that can be determined from the system representations themselves rather than checking the input and output signals as per the definition.

13.4.2 BIBO Stability of Transfer Function Representations

Theorems We begin our discussion with transfer functions. The following theorem is the main result in this area.

Theorem 13.4.1: Suppose that a system (13.4.1) is represented by a proper rational transfer function

$$\frac{Y(s)}{X(s)} = H(s).$$

Then this system is BIBO stable if and only if all of the poles of the transfer function are in the open LHP.[1]

▲▲

To check the stability of a system using Theorem 13.4.1, we must determine the poles of the transfer function. This result sparked a lot of research into how to determine the poles of a polynomial from its coefficients in the days before computers, yielding the famous Routh stability criteria. Today all CAD packages devoted to system analysis include a command to compute the roots of a polynomial. The poles can also be displayed in a pole-zero plot, making this plot an easy way to determine the stability of the system.

The proof of this theorem is facilitated by the following characterization of bounded signals using Laplace transforms.

Lemma 13.4.2: Suppose that the signal $x(t)$ has a rational, proper Laplace transform $X(s)$. Then this signal is bounded if and only if all of the poles of $X(s)$ are in the open LHP, or any poles on the imaginary axis are not repeated.

▲▲

Note: Lemma 13.4.2 applies to *signals,* whereas Theorem 13.4.1 applies to *systems.* The concept of stability applies only to systems, not signals.

The proof of Theorem 13.4.1 is obvious from the discussion of partial fractions in Section 9.3. The following discussion outlines some of the details.

Proof of Theorem 13.4.1: Given a bounded input signal $x(t)$ with a proper rational Laplace transform $X(s)$ the Laplace transform of the output signal is

$$Y(s) = H(s)X(s). \tag{13.4.2}$$

---

[1] The LHP is the left half plane in the s-plane. "Open" means that the LHP doesn't contain the imaginary axis. See Definition 3.1.12. Equivalently, the poles must have strictly negative real parts.

Case 1 Suppose that the poles of the input signal are distinct from the poles of the transfer function. Then the partial fraction expansion of the output signal is given by

$$Y(s) = \underbrace{\frac{c_1}{s - p_1} + \cdots + \frac{c_n}{s - p_n}}_{\text{poles of transfer function}} + \underbrace{\frac{c_{n+1}}{s - p_{n+1}} + \cdots + \frac{c_{n+m}}{s - p_{n+m}}}_{\text{poles of input signal}}. \tag{13.4.3}$$

From (13.4.3) we see that $y(t)$ is derived from two sets of poles. First, there are poles that come from the input signal. The terms in $y(t)$ that come from these poles must be bounded because the input signal is bounded. The second set of poles come from the transfer function. If $y(t)$ is bounded, then the terms in $y(t)$ due to these poles must be bounded. We conclude that these poles can't be in the RHP. So the transfer function is prohibited from having poles in the open[2] RHP, because these poles lead to unbounded behavior of the corresponding time function.

Case 2 We are left with the possibility that the poles of the system can be on the imaginary axis. If the transfer function has a pair of poles on the imaginary axis, the transfer function will have the form

$$\frac{Y(s)}{X(s)} = H_1(s)\left(\frac{\omega_0}{s^2 + \omega_0^2}\right) = H(s). \tag{13.4.4}$$

We take the input signal to be

$$x(t) = [\sin(\omega_0 t)]u_s(t). \tag{13.4.5}$$

Using this input signal, the partial fraction expansion of the Laplace transform of the output signal is

$$Y(s) = \left(\frac{c_1}{s - p_1} + \cdots + \frac{c_{n-2}}{s - p_{n-2}}\right) + \left[\frac{c_{n-1}s + c_n}{\left(s^2 + \omega_0^2\right)^2}\right]. \tag{13.4.6}$$

The last term in the partial fraction expansion in (13.4.6) will invert to a term of the form

$$\tilde{y}(t) = At\sin(\omega_0 t + \theta) \tag{13.4.7}$$

where the factor t in (13.4.7) is due to the squared denominator factor in (13.4.6). Hence, the output signal will have a term in it which is unbounded. We have just shown that if a transfer function has a pair of poles on the imaginary axis, then there is a bounded input signal (13.4.5) that leads to an unbounded output signal. So the transfer function can't have any poles on the imaginary axis.

---

[2] The poles can't have positive real parts.

A corresponding analysis for poles of the transfer function in the open LHP shows that the output signal is unbounded for all bounded input signals.

We have shown that the transfer function can't have any poles in the RHP or on the imaginary axis. Hence, we are left with the conclusion of the theorem.

▲▲

We know that the transfer function is related to the impulse response function by

$$h(t) = \mathcal{L}^{-1}\{H(s)\}. \tag{13.4.8}$$

Therefore, the question comes up as to whether a system is stable when represented by one system representation, but not stable when represented by another system representation. The answer to this question is no; it must be no if this theory is to be practical. We have the following result.

Theorem 13.4.3: Suppose that the impulse response function of a system is related to the transfer function according to (13.4.8). Then

$$\int_{-\infty}^{\infty} |h(\lambda)| \, d\lambda < M < \infty$$

if and only if all poles of the transfer function are in the open LHP.

▲▲

The proof of this theorem is beyond the scope of this text.

This theorem says that if a system represented by a convolution representation is BIBO stable by Theorem 13.3.7, then the system represented by a transfer function is also BIBO stable by Theorem 13.4.1 and conversely.

Examples The following examples illustrate the use of Theorem 13.4.1.

Example 13.4.4: The transfer function of the RC network is

$$\frac{Y(s)}{X(s)} = \frac{\dfrac{1}{RC}}{s + \dfrac{1}{RC}}. \tag{13.4.9}$$

This transfer function has a single pole at $s = -1/RC$ as shown in Figure 13.4.1. This pole is always in the open LHP. Therefore, this system is always stable for all values of R and C such that $RC > 0$. It can be shown that any network with only resistors, capacitors, and inductors is BIBO stable.

▲▲

Example 13.4.5: The transfer function of the proof-mass actuator is

$$\frac{Y_{pm}(s)}{V_{pm}(s)} = \frac{K_{ef}}{m_{pm}s^2}. \tag{13.4.10}$$

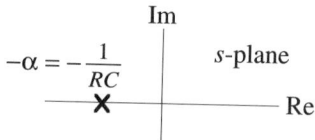

Figure 13.4.1 Pole-Zero Diagram of an RC Network

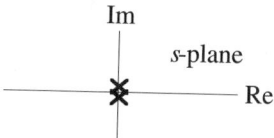

Figure 13.4.2 Poles-Zero Diagram of a Proof-Mass Actuator

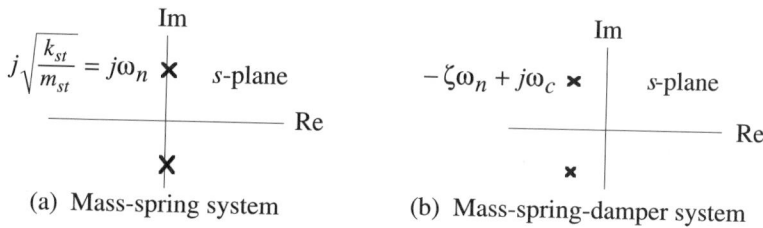

(a) Mass-spring system   (b) Mass-spring-damper system

Figure 13.4.3 Poles-Zero Diagram of the Mass-Spring System

This system has a pair of poles at the origin as shown in Figure 13.4.2. This system is BIBO unstable. See also Example 13.3.1.

▲▲

Example 13.4.6: Now consider the mass-spring-damper system with the transfer function

$$\frac{Y_{st}(s)}{F_{st}(s)} = \frac{\dfrac{1}{m_{st}}}{s^2 + \dfrac{c_{st}}{m_{st}}s + \dfrac{k_{st}}{m_{st}}}.$$

(13.4.11)

First, suppose the damping is zero, $c_{st} = 0$. This situation corresponds to a mass-spring system (no damper). The poles of this transfer function (13.4.11) are a pair of complex poles located on the imaginary axis as shown in Figure 13.4.3a. Since these poles are not in the open LHP, this system is BIBO unstable for all positive values of m_{st} and k_{st}. The instability of this system can be demonstrated by selecting an input signal of the form in (13.4.5). This sinusoidal input signal drives the system at resonance, resulting in arbitrarily large output signals.

Next suppose the damping is positive, $c_{st} > 0$, so that the system includes a

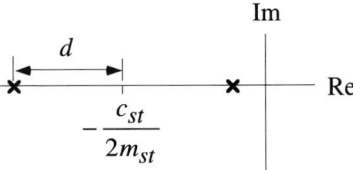

Figure 13.4.4 Poles of a Spring-Mass-Damper System with High Damping

damper. Positive damping implies that $\zeta > 0$. The poles of (13.4.11) are

$$-\frac{c_{st}}{2m_{st}} \pm \sqrt{\frac{c_{st}^2 - 4m_{st}k_{st}}{4m_{st}^2}}. \tag{13.4.12}$$

If c_{st} is such that $c_{st}^2 - 4m_{st}k_{st} < 0$, then $0 < \zeta < 1$ and the poles are a complex pair shown in Figure 13.4.3b. This system is BIBO stable by Theorem 13.4.1.

If c_{st} is such that $c_{st}^2 - 4m_{st}k_{st} \geq 0$, then $\zeta \geq 1$ and the poles are both real. In this situation, (13.4.12) shows that

$$\frac{c_{st}}{2m_{st}} \geq \sqrt{\frac{c_{st}^2}{4m_{st}^2}} \geq \sqrt{\frac{c_{st}^2}{4m_{st}^2} - \frac{k_{st}}{m_{st}}} = d. \tag{13.4.13}$$

Figure 13.4.4 shows the location of the poles based on (13.4.13). This system is always BIBO stable when the damping is nonzero.   ▲▲

Example 13.4.7: Systems with a Nonproper Rational Function Differentiater Consider the system

$$\frac{Y(s)}{X(s)} = s. \tag{13.4.14}$$

This transfer function is not a proper rational function. Therefore, Theorem 13.4.1 can't be used to check the stability of this system. This system is unstable, because the bounded input signal

$$x(t) = \sin(e^{at}), \quad a > 0 \tag{13.4.15}$$

leads to an unbounded output signal

$$y(t) = ae^{at}\cos(e^{at}). \tag{13.4.16}$$

This system is a pure differentiater. It is difficult to build a device with this transfer function because the small amplitude noise leads to large output signals. Now we say this system is BIBO unstable.

Time Delay Consider the system

$$y(t) = x(t - T), \quad T > 0. \tag{13.4.17}$$

This system is a pure delay. To determine if this system is stable, let $x(t)$ be bounded. Then

$$|x(t)| = |y(t + T)| \le M < \infty. \tag{13.4.18}$$

Hence, this system is BIBO stable. Using Property 9.2.11 of Laplace transforms, the transfer function is

$$\frac{Y(s)}{X(s)} = e^{-sT}. \tag{13.4.19}$$

This transfer function has no poles anywhere. Theorem 13.4.1 does not apply because the transfer function in (13.4.19) is not rational.

▲▲

13.4.3 BIBO Stability of State Space Representations

The stability of a system can also be determined directly from the state space representation

$$\dot{\vec{q}}(t) = A\vec{q}(t) + Bx(t), \tag{13.4.20}$$
$$y(t) = C\vec{q}(t) + Dx(t).$$

From Theorem 11.1.2 the transfer function of this system is

$$\frac{Y(s)}{X(s)} = C(sI - A)^{-1}B + D. \tag{13.4.21}$$

Recall from Theorem 11.1.5 that the poles of the transfer function are contained in the set of eigenvalues of A. Now applying Theorem 13.4.1, we have the following result.

Theorem 13.4.8: Suppose that a system is represented by state space equations. If all of the eigenvalues of A are in the open LHP, then the system (13.4.20) is BIBO stable.

▲▲

The next example illustrates the use of Theorem 13.4.8.

Example 13.4.9: The state space representation of the mass-spring-damper system was derived in Example 11.4.3, and it is repeated here

$$
\dot{\vec{q}}(t) = \begin{bmatrix} 0 & 1 \\ -\dfrac{k_{st}}{m_{st}} & -\dfrac{c_{st}}{m_{st}} \end{bmatrix} \vec{q}(t) + \begin{bmatrix} 0 \\ -\dfrac{1}{m_{st}} \end{bmatrix} f_{st}(t),
\tag{13.4.22}
$$

$$
y(t) = \begin{bmatrix} 1 & 0 \end{bmatrix} \vec{q}(t).
$$

The eigenvalues of A in (13.4.22) are computed from the characteristic polynomial

$$
a(s) = \det(sI - A) = \det \begin{bmatrix} s & -1 \\ \dfrac{k_{st}}{m_{st}} & s + \dfrac{c_{st}}{m_{st}} \end{bmatrix} = s^2 + \dfrac{c_{st}}{m_{st}} s + \dfrac{k_{st}}{m_{st}}.
\tag{13.4.23}
$$

For this system the $\det(sI - A)$ matches the denominator polynomial of the transfer function in (13.4.11). Hence, the eigenvalues of (13.4.23) match the poles of the transfer function.

(i)   If $c_{st} > 0$, then both the poles are in the open LHP, and the system is BIBO stable.

(ii)   If $c_{st} = 0$, then both the poles are on the imaginary axis, and the system is BIBO unstable.   ▲▲

Example 13.4.10: Consider the system

$$
\dot{\vec{q}}(t) = \begin{bmatrix} -1 & 0 \\ 0 & 1 \end{bmatrix} \vec{q}(t) + \begin{bmatrix} b_1 \\ b_2 \end{bmatrix} x(t),
\tag{13.4.24}
$$

$$
y = \begin{bmatrix} c_1 & 0 \end{bmatrix} \vec{q}(t).
$$

The characteristic polynomial for this system is

$$
a(s) = \det \begin{bmatrix} s+1 & 0 \\ 0 & s-1 \end{bmatrix} = (s+1)(s-1).
\tag{13.4.25}
$$

From (13.4.25) the eigenvalues of A are 1 and -1. The transfer function for this system is

$$
H(s) = C(sI - A)^{-1} B + D = \dfrac{c_1 b_1}{s+1} + \dfrac{0 \cdot b_2}{s-1} = \dfrac{c_1 b_1}{s+1}.
\tag{13.4.26}
$$

Applying Theorem 13.4.1 to the transfer function in (13.4.26) shows that this system is BIBO stable even though the A matrix has an eigenvalue in the RHP.

This example shows that some state space representations are BIBO stable even if they have an eigenvalue in the RHP. This eigenvalue doesn't appear in the transfer function as a pole, however. The problem here is that the state space representation models part of the system that is not captured by the transfer function. A full treatment of this subject can be found in an advanced treatment of linear system theory.

▲▲

13.5 PROPERTIES OF SYSTEM REPRESENTATIONS

13.5.1 Introduction

In the previous chapters we have introduced four system representations: (1) the convolution representation, (2) the Laplace transfer function, (3) state space representations, and (4) the Fourier transfer function.   In this chapter we have defined five system properties: (1) causal, (2) linear, (3) time invariant, (4) relaxed, and (5) BIBO stable. While these system properties are defined for any system, they certainly apply to systems which have one or more of the above representations. In this section we will discuss the system properties of the system representations we have introduced in the last three chapters. This discussion serves two purposes. First, it will clarify the relationships between these system representations. The relationships between these system representations have been developed in the previous three chapters. By integrating this discussion of system properties with the previous results, we are able to give system theoretic interpretations to certain mathematical conditions. Second, this discussion is helpful when deciding which system representation to use when analyzing a system. In many situations we can use several of the system representations. In other situations, only one or two of the representations can be used. Often, the properties of the underlying physical process determine which system representation must be chosen.

13.5.2 The Convolution Integral

We start with the convolution integral.   The derivation of the convolution representation in Section 13.1 shows that the convolution representation

$$y(t) = \int_{t_o}^{t_f} h(t-\lambda)x(\lambda)\,d\lambda \qquad (13.5.1)$$

is linear and time invariant. A careful analysis of integration shows that a system must be linear to have an integral representation of the form in (13.5.1).

The time invariance of the convolution representation is due to the fact that the argument of the impulse response function in (13.5.1) depends on the difference $t - \lambda$. For a time-varying system, the convolution representation becomes

$$y(t) = \int_{t_0}^{t_f} h(t,\lambda)x(\lambda)\,d\lambda. \qquad (13.5.2)$$

The analysis of time-varying systems is not easy, and will not be pursued here.

The convolution representation can also be used to represent noncausal systems, since causality is not required for the derivation of (13.5.1). Causality can be characterized in terms of the impulse response function as follows. Let $h(t - \lambda)$ be the output signal of a system to an impulse applied at time $t = \lambda$. If

$$h(t - \lambda) \neq 0 \quad \text{for} \quad t < \lambda, \tag{13.5.3}$$

then the system output signal $y(t)$ is nonzero before the input signal is applied at time λ. Such a system must be noncausal. Hence, the convolution system representation (13.5.1) is causal if and only if $h(t) = 0$ for $t < 0$.

If we assume that the system represented by the convolution integral in (13.5.1) is causal, then this integral can be written as

$$y(t) = \int_{t_o}^{t} h(t - \lambda)x(\lambda)\, d\lambda \tag{13.5.4}$$

where we have used (13.5.3) to replace the upper limit in the integral in (13.5.1). This change follows from the fact that for $t < \lambda < t_f$, the argument of the impulse response function $(t - \lambda)$ is negative. Hence, impulse response function is zero.

If there is an input signal present from $t = -\infty$ the convolution integral takes on the form

$$y(t) = \int_{-\infty}^{t_f} h(t - \lambda)x(\lambda)\, d\lambda. \tag{13.5.5}$$

In (13.5.5) the lower limit of the integration is $-\infty$. We must verify that the integral is finite over this infinite interval of integration. This integral will exist if the input signal is bounded and

$$\left| \int_{-\infty}^{t_f} h(t - \lambda)x(\lambda)\, d\lambda \right| \leq \int_{-\infty}^{\infty} |h(t - \lambda)||x(\lambda)|\, d\lambda \leq M \int_{-\infty}^{\infty} |h(\lambda)|\, d\lambda. \tag{13.5.6}$$

If the impulse response function is absolutely integrable, then the system is BIBO stable by Theorem 13.3.7, and conversely. Hence, this infinite integral exists if the system is BIBO stable. So when the system is BIBO stable, the convolution integral can be used to model systems that are not relaxed by virtue of a input signal that is nonzero at negative infinity.

Finally, the system is relaxed if the input signal is zero for time less than some finite constant. This property was discussed in Example 13.2.4.

We summarize these observations with the following theorem.

Theorem 13.5.1: Consider the convolution representation (13.5.2). The convolution representation is linear and time invariant. If the impulse response satisfies $h(t) = 0$ for $t < 0$, the convolution representation is causal and conversely. In this case the system can be represented as in (13.5.4). If the system is BIBO stable the convolution representation can be used to model systems with a bounded input signal that is nonzero on a (semi-)infinite interval.   In this case the convolution representation becomes

$$y(t) = \int_{-\infty}^{\infty} h(t-\lambda)x(\lambda)\, d\lambda.$$

If the input signal $x(t)$ satisfies $x(t) = 0$, for $t < t_0$, then the convolution representation is relaxed. If the input signal is nonzero over $-\infty < t < t_f$, then the system is not relaxed.

▲▲

13.5.3 Transfer Functions

We can find the Laplace transfer function from the convolution integral (13.5.1) simply by taking the Laplace transform of that equation. We obtain

$$\frac{Y(s)}{X(s)} = H(s). \tag{13.5.7}$$

The Laplace transform imposes several restrictions on the system and the input and output signals, however.

First, the fact that the Laplace transform can be used to transform the convolution integral into a rational function relies on the fact that this system is linear, a condition that is also required for the convolution integral to exist.

Second, the system must be time invariant.   Given the convolution representation of a time invariant system, the transfer function can be calculated using the convolution property of the Laplace transform, Property 9.2.3.   If the convolution integral is time-varying, the transfer function, in general, will not exist.

Third, the impulse response function has to be Laplace transformable.   The definition of the Laplace transform, Definition 9.1.1, implies that the impulse response must satisfy $h(t) = 0$ for $t < 0$. That is, the system must be causal. See the discussion above for convolution integrals.

Fourth, the input signal must be Laplace transformable. Hence, the input signal must also satisfy $x(t) = 0$ for $t < 0$.   This condition with the convolution representation implies that the system is relaxed. We have also noted that if a transfer function is calculated from a differential equation, the initial conditions must be set to zero.[3] Hence, the system must be relaxed to use the transfer function as a representation of a system.

Fifth, the transfer function representation doesn't require the system to be BIBO stable.

We summarize the discussion above with the following theorem.

Theorem 13.5.2: If the system (13.5.1) can be represented by a transfer function (13.5.7), then the system is causal, linear, time invariant, and relaxed.

▲▲

---

[3] Theorem 10.1.4.

13.5.4 State Space Equations

Next we consider state space representations. State space equations are essentially equivalent differential equations of the form

$$\dddot{y}(t) + a_2 \ddot{y}(t) + a_1 \dot{y}(t) + a_0 y(t) = b_2 \ddot{x}(t) + b_1 \dot{x}(t) + b_0 x(t) \qquad (13.5.8)$$

$$\ddot{y}(0) = y_2, \, \dot{y}(0) = y_1, \, y(0) = y_0$$

where a third-order equation was selected for convenience. State space equations are only a rewriting of the differential equation (13.5.8) as discussed in Section 11.2. These equations take the form

$$\dot{\vec{q}}(t) = A\vec{q}(t) + Bx(t), \quad \vec{q}(0) = q_0, \qquad (13.5.9)$$

$$y(t) = C\vec{q}(t) + Dx(t).$$

In Section 13.1 it was established that the differential equation in (13.5.8) is causal, linear, and time invariant. The state space representation in (13.5.9) inherits these properties from the differential equation in (13.5.8).

State space representations (differential equations) naturally include initial conditions. Initial condition represent internal energy storage, and as such they occur in systems that are not relaxed. If the initial conditions are zero, then the system is relaxed.

Finally, differential equations don't require the system to be BIBO stable. Hence, state space representations can be used for unstable systems.

Theorem 13.5.3: The state space representation in (13.5.9) is causal, linear, and time invariant. If $\vec{q}_0 = 0$, and $x(t) = 0$ for $t < 0$, then the system is also relaxed.
▲▲

13.5.5 Fourier Transfer Function

The Fourier transfer function

$$\frac{Y(\omega)}{X(\omega)} = H(\omega) \qquad (13.5.10)$$

is obtained by taking the Fourier transform of the convolution integral

$$y(t) = \int_{-\infty}^{\infty} h(t - \lambda) x(\lambda) \, d\lambda. \qquad (13.5.11)$$

Hence, the Fourier transfer function inherits linearity and time invariance from the convolution representation.

Suppose that the system (13.5.11) is noncausal; $h(t) \neq 0$, $t < 0$. The Fourier transform is defined for such functions so the Fourier transfer function can be used for noncausal systems.

If the input signal is zero as time approaches negative infinity, then the Fourier transfer function is relaxed.

If the input signal is nonzero at $t = -\infty$, then the system is not relaxed. Furthermore, we must ensure that output signal in (13.5.11) exists for all time so that we can take the Fourier transform. The existence of the output signal is discussed in Section 12.4. There it is shown that if the input signal is bounded and the impulse response function satisfies

$$\int_{-\infty}^{\infty} |h(t)| \, dt < \infty, \tag{13.5.12}$$

then the output signal does exist. Using Theorem 13.3.7 we see that (13.5.12) is equivalent to the system being BIBO stable. Hence, the Fourier transfer function requires the system to be BIBO stable.

The observation are summarized in the following theorem.

Theorem 13.5.4: The Fourier transfer function (13.5.9) is linear, time invariant, and BIBO stable. The system is relaxed if the input signal is zero at $t = -\infty$.

▲▲

13.5.6 Summary

All four system representations discussed above can be used to represent causal, linear, time invariant, relaxed, BIBO stable systems. If the system is noncausal, the convolution representation or Fourier transfer function can be used. If the system is not relaxed by virtue of a persistent input signal and the system is BIBO stable, the convolution representation or the Fourier transfer function can be used. If the system is not relaxed because of internal energy storage, the state space representation (differential equation) must be used. If the system is not BIBO stable, then the Fourier transfer function can't be used. These observations are useful when deciding which system representation to chose for analyzing a particular system.

13.6 STATIC NONLINEARITIES

13.6.1 A Nonlinear System

In this section we will study systems that are different than most of the other systems we discuss in this text. We will introduce these systems by way of example. Consider the op amp circuit in Figure 13.6.1.

Note: The first op amp is an inverting op amp. This amplifier will simplify the discussion below. We will consider this op amp to be ideal.

We think of this circuit as a system where the input signal is the voltage $v_i(t)$ and the output signal is the voltage $v_o(t)$. The relationship between the input signal and the output signal of the circuit in Figure 13.6.1 for small voltage levels is

$$\frac{V_o(s)}{V_i(s)} = \left(-\frac{R_{in}}{R_{in}} \right)\left(-\frac{R_f}{R_1} \right) = \frac{R_f}{R_1} = K_g. \tag{13.6.1}$$

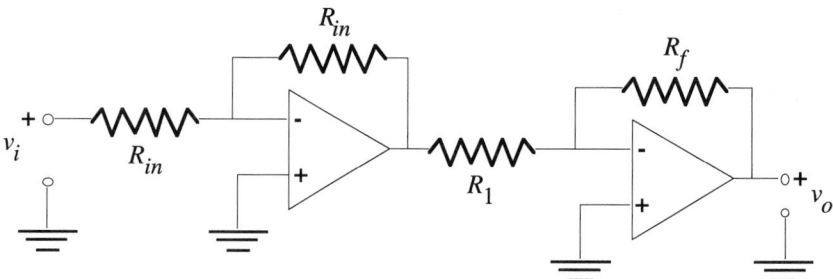

Figure 13.6.1 Operational Amplifier

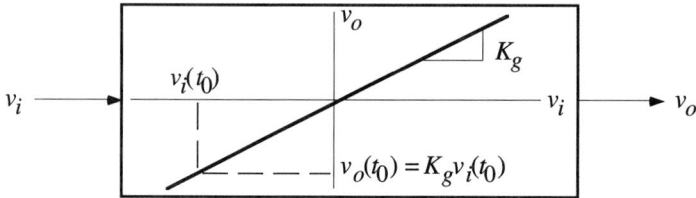

Figure 13.6.2 Static System in (13.6.1)

This input-output relationship can be pictorially displayed as shown in Figure 13.6.2. We interpret the "block diagram" in Figure 13.6.2 as follows. For a given time t_0, the value of the input signal is $v_i(t_0)$. This value is shown in Figure 13.6.2 on the abscissa axis. From the function in the block in Figure 13.6.2 we read off the corresponding value of the voltage

$$v_o(t_0) = K_g v_i(t_0) \tag{13.6.2}$$

also marked in Figure 13.6.2 on the ordinate axis. The value $v_o(t_0)$ is the value of the output signal for the input signal $v_i(t_0)$ at time $t = t_0$. The transfer function of this particular system (13.6.1) is a gain K_g which is also the slope of the line in Figure 13.6.2.

The model of the op amp in (13.6.1) is for input signals with low voltages. If the input signal is increased, however, the model in (13.6.1) and Figure 13.6.2 is no longer correct. Next we will investigate a more realistic model of the second op amp in Figure 13.6.1. For some input voltage, say v_{imax}, the electronics will saturate and the output voltage $v_o(t)$ will be a constant v_{max}. We can model this nonideal behavior at large input signal levels using a diagram similar to the one in Figure 13.6.2. First, we model the saturation characteristic of the electronics as

$$S(v_i) = \begin{cases} v_{max}, & v_{imax} < v_i \\ K_g v_i, & -v_{imax} < v_i < v_{imax} \\ -v_{max}, & -v_{imax} < v_i \end{cases} \tag{13.6.3}$$

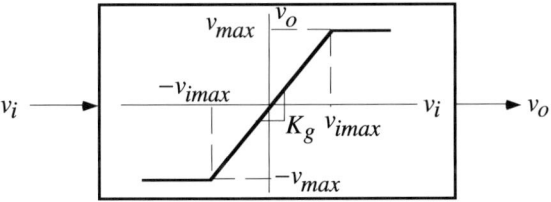

Figure 13.6.3 Saturation Nonlinearity

where

$$v_{max} = K_g v_{imax}. \tag{13.6.4}$$

Now the function in (13.6.3) is shown in Figure 13.6.3. We think of the saturation nonlinearity as a system, and we use the terminology associated with systems.

13.6.2 Graphical Construction of the Output Signal from a Nonlinearity

Suppose that we are given the input signal to a nonlinearity and we want to find the output signal. Next we will give a graphical technique for finding this output signal. We again consider the saturation nonlinearity in Figure 13.6.3. Let $K_g = 1$, for simplicity. Suppose that the input signal is the cosine

$$v_i(t) = \cos(\omega_0 t) \tag{13.6.5}$$

as shown in Figure 13.6.4a. The output signal from this static nonlinearity is shown in Figure 13.6.4c. The methodology for constructing this signal is given next.

This output signal is constructed for each time instant. We will demonstrate the method for several points in time. Beginning at $t = t_1$ in Figure 13.6.4a we see that $v_i(t_1) = 0$. From the nonlinearity in Figure 13.6.4b we read off the output signal at t_1, $v_o(t_1) = 0$. This value of the output signal is labeled in Figure 13.6.4c. So for every time t_k such that $v_i(t_k) = 0$, then $v_o(t_k) = 0$.

Next consider the time point $t = t_2$ in Figure 13.6.4a. The signal at $t = t_2$, $v_i(t_2)$, is positive, and it is labeled on the abscissa of the nonlinearity in Figure 13.6.4b. For the nonlinearity in Figure 13.6.4b we read the corresponding value of the signal $v_o(t_2)$ on the ordinate axis. Because the nonlinearity is a straight line, we have that $v_o(t_2) = v_i(t_2)$. Now in Figure 13.6.4c at the time point $t = t_2$ we enter $v_o(t_2)$ for the output signal .

Next consider the time point $t = t_3$ in Figure 13.6.4a. The value $v_i(t_3)$ is labeled on the abscissa of the nonlinearity in Figure 13.6.4b. For the nonlinearity in Figure 13.6.4b we read the corresponding value of the signal $v_o(t_3)$ on the ordinate axis. This value of the input signal $v_i(t_3)$ is greater than the saturation limit, the value of the output signal is $v_o(t_3) = v_{max}$. Now at the time point $t = t_3$ in Figure 13.6.4c for the output signal we enter $v_o(t_3)$.

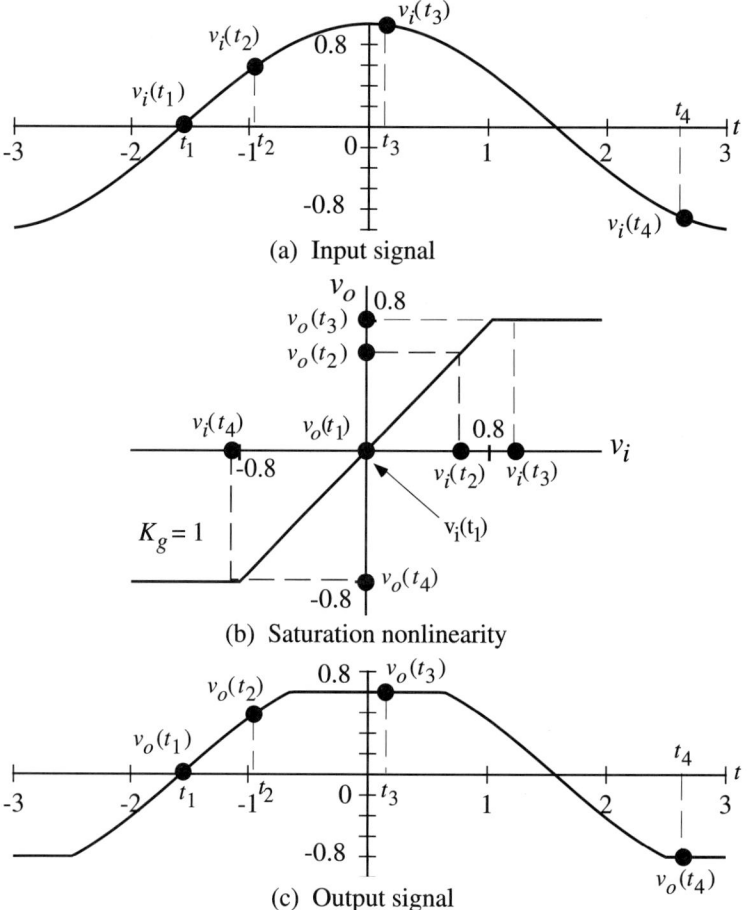

Figure 13.6.4 Input and Output Signal for a Saturation Nonlinearity

Finally, consider the point $t = t_4$ in Figure 13.6.4a. The value of the input signal at this point is $v_i(t_4)$, and it is labeled on the nonlinearity in Figure 13.6.4a. Here again the op amp is saturated. The corresponding value of the output signal $v_o(t_4)$ is also labeled on the nonlinearity in Figure 13.6.4b, and the output signal in Figure 13.6.4c.

We proceed in this way, point by point, until all points on the output signal are labeled. We arrive at the clipped cosine wave in Figure 13.6.5.

13.6.3 Summary

In this section we have considered special systems of the type

$$y = g(x). \tag{13.6.6}$$

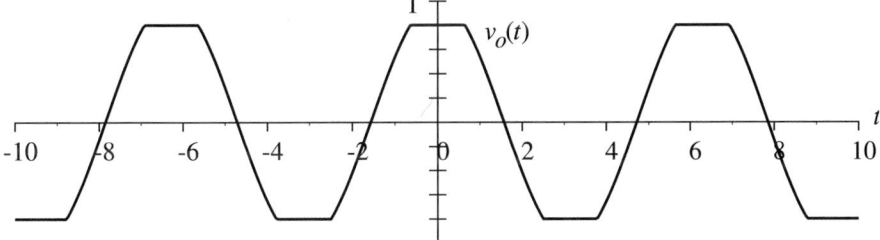

Figure 13.6.5 Clipped Cosine

If $g(x)$ is

$$y = ax, \quad a \text{ constant} \tag{13.6.7}$$

then the system in (13.6.6) is linear (see Definition 13.1.4).

Definition 13.6.1: We call the system in (13.6.6) a <u>static nonlinearity</u> if is not a linear system as in (13.6.7).

▲▲

Static nonlinearities are commonly represented graphical as shown in Figure 13.6.3.

Note: The pictorial representation of the input-output relationship in Figure 13.6.3 is reminiscent of a "block diagram" (introduced in Section 10.2). In general, these block diagrams don't behave like standard block diagrams, however. For example, they can't be manipulated using block diagram reduction techniques.

13.7 CHAPTER SUMMARY

In this chapter we introduced five system properties, shown in Table 13.7.1. The first four properties must be verified from the definition. The fifth property, BIBO stability, means that every bounded input signal results in a bounded output signal. Generally, this property must be checked using one of the three tests based on the system representation introduced in Sections 13.3-4. According to Theorem 13.3.7, a

Table 13.7.1 Definitions of System Properties

Property	Definition
Causal	Definition 13.1.1
Linear	Definition 13.1.4
Time Invariant	Definition 13.1.6
Relaxed	Definition 13.1.8
BIBO Stable	Definition 13.3.5

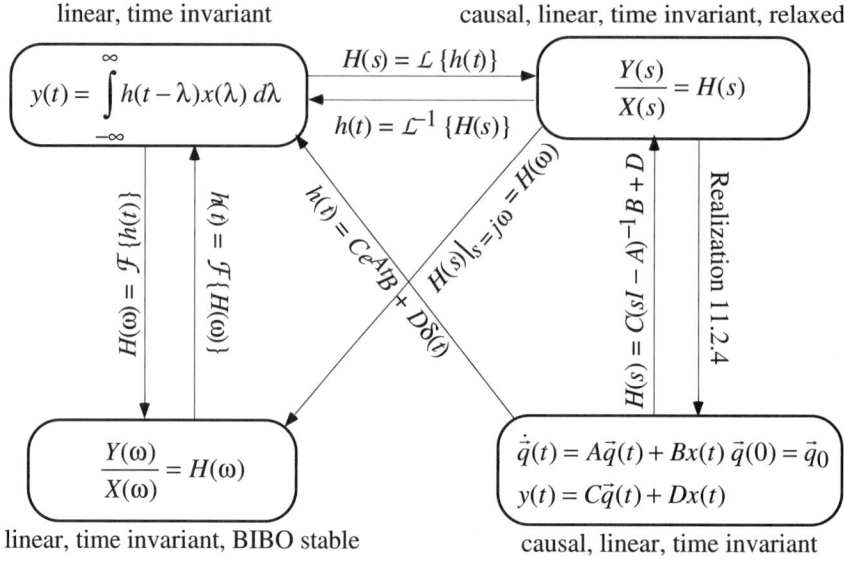

Figure 13.7.1 Relationships Between System Representations

system modeled by the convolution integral is BIBO stable if and only if the impulse response function satisfies

$$\int_{-\infty}^{\infty} |h(t)|\, dt \le M_h < \infty. \tag{13.7.1}$$

If a system is represented by a proper, rational Laplace transfer function, then it is BIBO stable if and only if all the poles of the transfer function are in the open LHP (Theorem 13.4.1). Finally, if a system is represented by state space equations and all the eigenvalues of the state matrix A are in the open LHP, then this system is BIBO stable (Theorem 13.4.8).

In Section 13.1 we showed that if a system is linear and time invariant, then it can be represented by a convolution representation provided we know the impulse response function. It follows that all of the system representations we have been discussing in this text are linear and time invariant. These system representations differ in their other properties, however. The properties inherent in each of the system representations are summarized in Table 13.7.2. The relationships between these four system representations are summarized in Figure 13.7.1.

In Section 13.6 we discussed systems that can be represented by a static nonlinearity. We showed how to graphically construct the output signal given an input signal. The analysis of the output signal of these systems is given in Section 8.7.

Table 13.7.2 Properties of Systems

Representation	causal	linear	time invariant	relaxed	BIBO stable
			System Property		
convolution integral		X	X		
Laplace transfer function	X	X	X	X	
state space equations	X	X	X		
Fourier transfer function		X	X		X

13.8 HOMEWORK FOR CHAPTER 13

Homework Problems for Section 13.1

13.1.1 Consider the system shown in Figure P13.1.1. Is this system causal? Linear? Time Invariant? Relaxed?

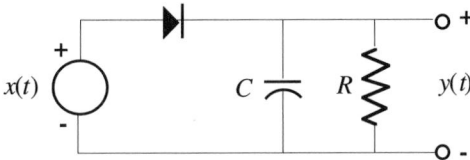

Figure P13.1.1

13.1.2 Consider the following systems. Are these systems causal? Linear? Time invariant? Relaxed?
(i) $\ddot{y}(t) + \sin(y(t)) = x(t)$, $t \geq 0$, $y(0) = 1$.
(ii) $y(t) = x(t)\cos(\omega_0 t)$, $-\infty < t < \infty$.
(iii) $y(t) = x(t)\cos(\omega_0 t)$, $0 \leq t < \infty$.

13.1.3 Show that the saturation characteristic in Figure P13.1.3 is a nonlinear system. Is this system causal? Time invariant? Relaxed? Why?

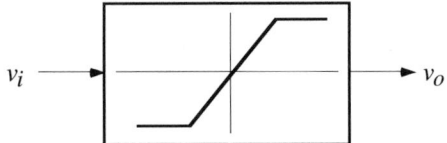

Figure P13.1.3

13.1.4 Consider the network in Figure P13.1.4. Suppose that $v_s(t) = u_s(t)$. Is this system causal? Linear? Time invariant? Relaxed? Why?

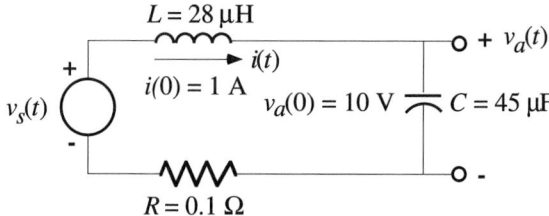

Figure P13.1.4

Homework Problems for Section 13.2

13.2.1 Consider the two mass-spring-damper system shown in Figure P13.2.1. Suppose the damping in this system satisfies a square law

$$force = c\left[\text{sgn}(\dot{y})|\dot{y}|^2\right]$$

Is this system causal? Linear? Time invariant? Relaxed?

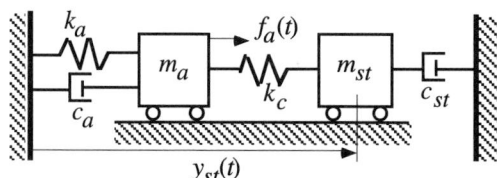

Figure P13.2.1

13.2.2 Consider the *RC* network in Figure P13.2.2. Suppose that the input signal is

$$x(t) = \cos(\omega_i t), \quad -\infty < t < \infty.$$

Compute the output signal for this network. Is this system relaxed? Why?

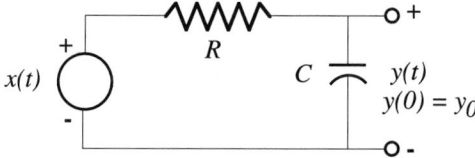

Figure P13.2.2

Homework Problems for Section 13.3

13.3.1 The impulse response function of a system is shown in Figure P13.3.1. Is this system BIBO stable? Why?

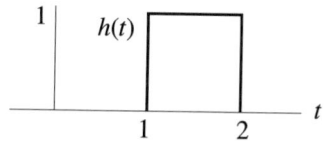

Figure P13.3.1

13.3.2 The impulse response functions for several systems are given below. Determine the stability of each system.

(i) $h(t) = (e^{-t} + e^{2t})u_s(t)$

(ii) $h(t) = \left(e^{4t}\sin(12t)\right)u_s(t)$

(iii) $h(t) = \Pi\left(\dfrac{t+4}{2}\right)$

(iv) $h(t) = \left(\dfrac{1}{t}\right)u_s(t)$

(v) $h(t) = [\ln(t)]u_s(t)$

(vi) $h(t) = e^{-3t}u_s(t) + 4\delta(t)$

13.3.3 Suppose a system is represented by a convolution integral where the impulse response function is shown in Figure P13.3.3. Is this system BIBO stable? Justify your answer.

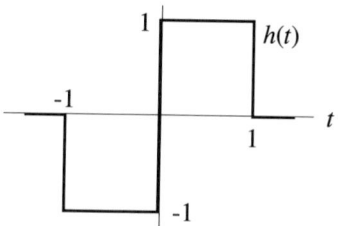

Figure P13.3.3

13.3.4 Consider the static, nonlinear system shown in Figure P13.3.4.
(a) Is this system BIBO stable? Why?
(b) This system can be represented as $y(t) = g(x(t))$. Under what conditions on $g(x)$ will systems in this class be BIBO stable?

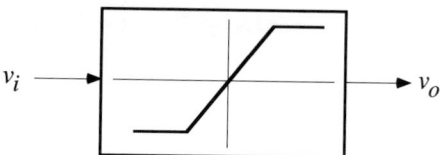

Figure P13.3.4

Homework Problems for Section 13.4

13.4.1   Consider the block diagram in Figure P13.4.1.
(a) Find the transfer function of this system. Is it BIBO stable?
(b) Find a state space representation of this system. Is it BIBO stable?

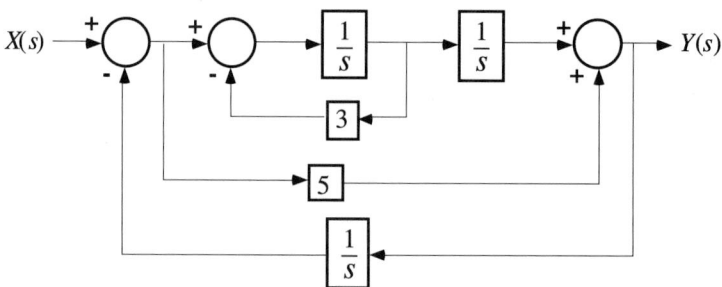

Figure P13.4.1

13.4.2   Consider the following transfer function

$$\frac{Y(s)}{X(s)} = \frac{25(s-1)}{s^2 + 2as + 25}$$

where a is a parameter.
(a) Find ω_n and ζ of the poles as a function of the parameter a.
(b) If $a = 3$, draw the pole-zero plot. Label ω_n and ζ of the poles. Is this transfer function BIBO stable for this value of a?
(c) Repeat (b) if the value of $a = -4$.

13.4.3   Is each of the following systems BIBO stable? Why?
    (i) $\dfrac{Y(s)}{X(s)} = \dfrac{s+4}{s^3 + 2s^2 + 3}$ (iii) $\dfrac{Y(s)}{X(s)} = \dfrac{s}{s^2 + 0.4s + 4}$
    (ii) $\dfrac{Y(s)}{X(s)} = \dfrac{s+4}{s(s+1)}$ (iv) $\dfrac{Y(s)}{X(s)} = \dfrac{s-3}{s^4 - 0.4s + 4}$

13.4.4   Suppose that a system is defined by the convolution integral

$$y(t) = \int_0^t \left(e^{-(t-\lambda)} + e^{2(t-\lambda)} \right) x(\lambda) \, d\lambda.$$

(a) What is the transfer function of this system?
(b) Is this system BIBO stable? Why?

13.4.5   Consider the system in Figure P13.4.5. Is this system stable?

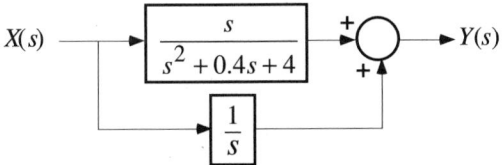

Figure P13.4.5

13.4.6 Consider the system represented by the block diagram in Figure P13.4.6. For what values of k is this system BIBO stable?

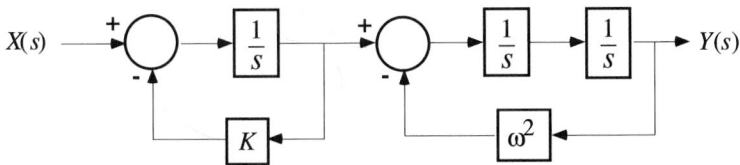

Figure P13.4.6

13.4.7 (a) For each of the following systems determine if the system is BIBO stable.

$$\frac{Y(s)}{X(s)} = H_1(s) = \frac{100}{s^2 + 6s + 100} \qquad \frac{V(s)}{R(s)} = H_2(s) = \frac{s}{s + 20}$$

(b) Suppose that these two systems were connected as shown in Figure P13.4.7. Is the system in Figure P13.4.7 BIBO stable?

(c) For each of the following systems determine if the system is BIBO stable. Is the system in Figure P13.4.7 BIBO stable?

$$\frac{Y(s)}{X(s)} = H_1(s) = \frac{1}{(s-1)(s+2)} \qquad \frac{V(s)}{R(s)} = H_2(s) = \frac{s-1}{s+20}$$

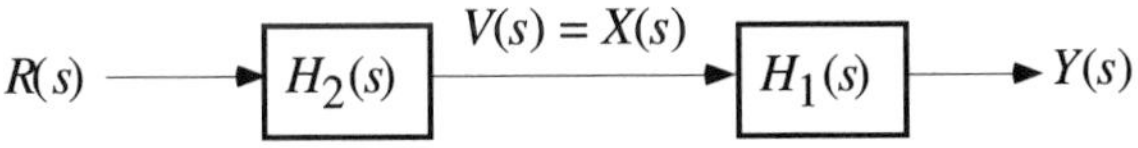

$$R(s) \longrightarrow \boxed{H_2(s)} \xrightarrow{\; V(s) = X(s) \;} \boxed{H_1(s)} \longrightarrow Y(s)$$

Figure P13.4.7

13.4.8 For what values of K is the following system BIBO stable?

$$\frac{Y(s)}{X(s)} = \frac{s+3}{s^2 + 10Ks + 25}, \qquad -1 \le K \le 1.$$

13.4.9 Consider the system in Figure P13.4.9. Find the range of values of K such that this system is BIBO stable.

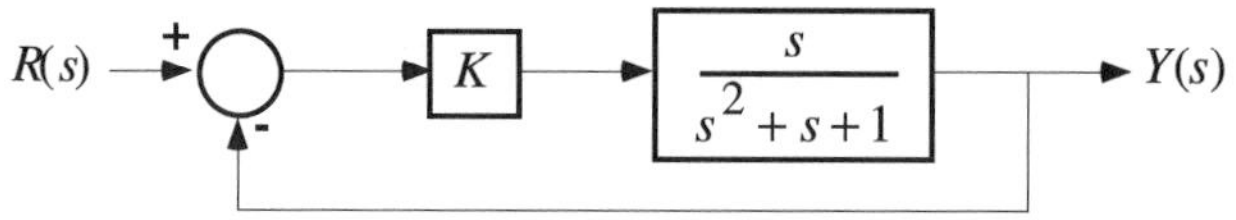

$$R(s) \xrightarrow{+} \bigcirc \xrightarrow{} \boxed{K} \xrightarrow{} \boxed{\dfrac{s}{s^2+s+1}} \xrightarrow{} Y(s)$$

Figure P13.4.9

13.4.10 Consider the two mass-spring-damper system in Figure P13.4.10.
(a) Show that a set of state space equations for this system is

$$\dot{\vec{q}}(t) = \begin{bmatrix} 0 & 1 & 0 & 0 \\ -\dfrac{k_a+k_c}{m_a} & -\dfrac{c_a}{m_a} & \dfrac{k_c}{m_a} & 0 \\ 0 & 0 & 0 & 1 \\ \dfrac{k_c}{m_{st}} & 0 & -\dfrac{k_c}{m_{st}} & 0 \end{bmatrix} \vec{q}(t) + \begin{bmatrix} 0 \\ \dfrac{1}{m_a} \\ 0 \\ 0 \end{bmatrix} f_a(t)$$

$$y_{st}(t) = \begin{bmatrix} 0 & 0 & 10 & 0 \end{bmatrix} \vec{q}(t)$$

Suppose the parameter values of this system are: $k_a = 10$ N/m, $c_a = 8$ N/m/sec, $m_a = 1$ kg, $k_c = 100$ N/m, and $m_{st} = 0.1$ kg.
(b) Develop a MATLAB state space model of this system.
(c) Compute the transfer function of this system using MATLAB.
(d) Based on the transfer function, is this system BIBO stable? Why?
(e) Based on the state space system, is this system BIBO stable? Why?

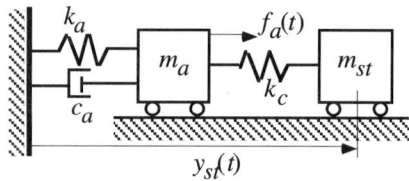

Figure P13.4.10

13.4.11 Consider the following systems. Is each system BIBO stable? Why?

(i) $\dot{\vec{q}} = \begin{bmatrix} -1 & 0 \\ 0 & 4 \end{bmatrix} \vec{q} + \begin{bmatrix} 3 \\ 5 \end{bmatrix} x$

$y = \begin{bmatrix} 2 & 8 \end{bmatrix} \vec{q} + 6x$

(ii) $\dot{\vec{q}} = \begin{bmatrix} 0 & 0 & 2 \\ 5 & 0 & -1 \\ 3 & 0 & 0 \end{bmatrix} \vec{q} + \begin{bmatrix} 0 \\ 1 \\ 0 \end{bmatrix} x$

$y = \begin{bmatrix} 6 & 0 & 0 \end{bmatrix} \vec{q} + 4x$

(iv) $\dot{\vec{q}} = \begin{bmatrix} 2 & 1 & 0 \\ 7 & 3 & 5 \\ 0 & 6 & 4 \end{bmatrix} \vec{q} + \begin{bmatrix} 8 \\ 0 \\ 0 \end{bmatrix} x$

$y = \begin{bmatrix} 9 & 0 & 0 \end{bmatrix} \vec{q} + 10x$

(v) $\dot{\vec{q}} = \begin{bmatrix} -1 & 1 \\ 0 & 2 \end{bmatrix} \vec{q} + \begin{bmatrix} 0 \\ 3 \end{bmatrix} x$

$y = \begin{bmatrix} 1 & 0 \end{bmatrix} \vec{q} + 4x$

(iii) $\dot{\vec{q}} = \begin{bmatrix} 1 & 0 \\ 0 & -1 \end{bmatrix} \vec{q} + \begin{bmatrix} 3 \\ 5 \end{bmatrix} x$

$y = \begin{bmatrix} 2 & 1 \end{bmatrix} \vec{q}$

(vi) $\dot{\vec{q}} = \begin{bmatrix} 0 & 1 & 0 \\ 0 & 0 & 1 \\ -6 & -3 & -2 \end{bmatrix} \vec{q} + \begin{bmatrix} 0 \\ 0 \\ 1 \end{bmatrix} x$

$y = \begin{bmatrix} 10 & 5 & 0 \end{bmatrix} \vec{q}$

13.4.12 Is the following system BIBO stable?

$$\ddot{y}(t) - \dot{y}(t) - 6y(t) = 4x(t)$$

13.4.13 Consider the two mass-spring-damper system in Figure P13.4.13.
(a) Show that a set of state space equations for this system is

$$\dot{\vec{q}}(t) = \begin{bmatrix} 0 & 1 & 0 & 0 \\ -\dfrac{k_a+k_c}{m_a} & -\dfrac{c_a}{m_a} & \dfrac{k_c}{m_a} & 0 \\ 0 & 0 & 0 & 1 \\ \dfrac{k_c}{m_{st}} & 0 & -\dfrac{k_c}{m_{st}} & -\dfrac{c_{st}}{m_{st}} \end{bmatrix} \vec{q}(t) + \begin{bmatrix} 0 \\ \dfrac{1}{m_a} \\ 0 \\ 0 \end{bmatrix} f_a(t)$$

$$y_{st}(t) = \begin{bmatrix} 0 & 0 & 10 & 0 \end{bmatrix} \vec{q}(t)$$

The parameter values for this system are: $k_a = 10$ N/m, $c_a = 8$ N/m/sec, $m_a = 1$ kg, $k_c = 100$ N/m, $c_{st} = 20$ N/m/sec, and $m_{st} = 100$ kg.
(b) Develop a MATLAB state space model of this system.
(c) Compute the transfer function of this system using MATLAB.
(d) Based on the transfer function, is this system BIBO stable? Why?
(e) Based on the state space system, is this system BIBO stable? Why?

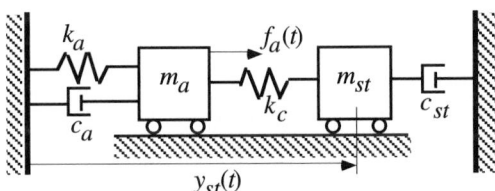

Figure P13.4.13

Homework Problem for Section 13.5

13.5.1 The impulse response function of a system is shown in Figure P13.5.1.
(a) Is this system causal? Why?
(b) What other system representations can be used for this system? Find these representations.

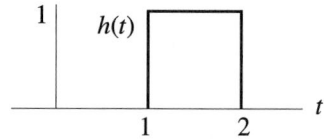

Figure P13.5.1

13.5.2 Which of the following systems has a Fourier transform representation?

(i) $\dfrac{Y(s)}{X(s)} = \dfrac{1}{s^2 - 1}$

(iii) $\dot{y}(t) = y^2(t) + x(t)$

(ii) $y(t) = \displaystyle\int_0^\infty \left(\dfrac{1}{t - \lambda}\right) u_s(t - \lambda - t_0) x(\lambda)\, d\lambda$ (iv) $\dot{\vec{q}}(t) = \begin{bmatrix} -1 & 0 \\ 0 & -2 \end{bmatrix} \vec{q}(t) + \begin{bmatrix} 1 \\ 1 \end{bmatrix} x(t)$

$t_0 > 0$

$y(t) = \begin{bmatrix} 2 & 1 \end{bmatrix} \vec{q}(t) + x(t)$

13.5.3 The impulse functions of several systems are given below. Determine whether each system is causal.

(i) $h(t) = \sin(t)\Pi\left(\dfrac{t - 4}{2}\right)$

(iii) $h(t) = \cos(t)u_s(t) + \delta(t)$

(ii) $h(t) = \mathrm{sinc}(t)$

(iv) $h(t) = \cos(t)u_s(t) + \delta(t + 1)$

13.5.4 Suppose a system is represented by a convolution integral where the impulse response function is shown in Figure P13.5.4. Suppose the input signal for this system is

$$x(t) = \sum_{k=-\infty}^{\infty} \Pi\left(\dfrac{t - k}{0.1}\right)$$

(a) Is this system causal? Linear? Time invariant? Relaxed?
(b) What other representations can be used for this system?

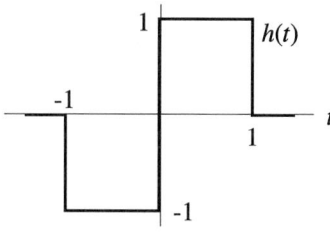

Figure P13.5.4

13.5.5 Consider the circuit in Figure P13.5.5.
   (a) Suppose the input signal is $x(t) = \cos(\omega_0 t)$, $-\infty < t < \infty$.
      (i)   Is this system causal? linear? time invariant? relaxed?
      (ii) Which system representations can be used to represent this system?
   (b) Suppose the input signal is $x(t) = \left(1 - e^{-\alpha t}\right) u_S(t)$.
      (i)   Is this system causal? linear? time invariant? relaxed?
      (ii) Which system representations can be used to represent this system?

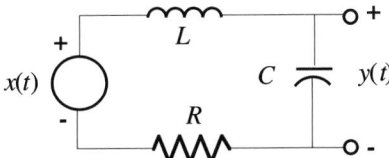

Figure P13.5.5

13.5.6 Consider the mass-spring system in Figure P13.5.6.   What system representations can be used to describe this system?

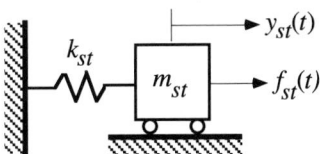

Figure P13.5.6

Homework Problems for Section 13.6

13.6.1 Consider the static nonlinearity shown in Figure P13.6.1.
   (a) Sketch the output signal when the input signal is $x(t) = A\sin(2t)$.
   (b) Determine the Fourier series of the output signal.
   (c) Plot the power spectrum of the output signal for $A = 0.6, 2.5$.

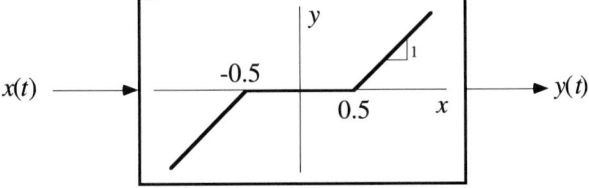

Figure P13.6.1

13.6.2 Consider the static nonlinearity in Figure P13.6.2. Suppose the input signal is

$$x(t) = \Pi(t - 0.5) - \Pi(t + 0.5)$$

Find the output signal.

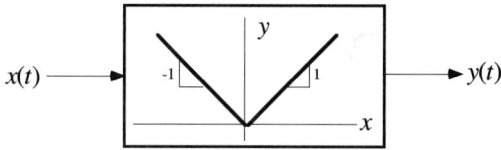

Figure P13.6.2

13.6.3 Consider the static nonlinearity in Figure P13.6.3. Suppose the input signal is

$$x(t) = \Pi(t - 0.5) - \Pi(t + 0.5)$$

Find the output signal. Compare the result to Problem 13.6.2.

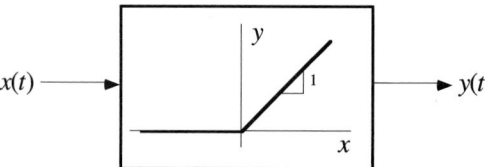

Figure P13.6.3

Chapter 14

The Frequency Response Theorem

Chapter Outline

This text is devoted to the study of signals and systems. By definition, an input signal to a system produces an output signal. In this chapter we begin to study how the characteristics of the input signal and the characteristics of the system interact to produce the output signal. To frame this discussion, we will adopt the following terminology. We say the input signal <u>propagates</u> through the system to generate the output signal. This terminology is consistent with the definition of a system, which defines the system in terms of the input and output signal. This terminology suggests that the relationship between the input and output signals and the system can be studied by comparing the input signal to the output signal. We can attribute change between these two signals to the system. We shall pursue this idea in the next two chapters.

This analysis will require the tools of signal analysis and system analysis. The signal analysis will use the signal representations developed in Chapter 8. The system analysis will use the system representations and their properties developed in Chapters 10, 11, 12, and 13. Through an analysis that combines these powerful results, we will obtain deep insight into the propagation of a signal through a system.

Our method of analysis will start with simple input signals, sinusoids, to establish the basic link between the input signal, the system, and the output signal. From these simple signals we will proceed to periodic signals, infinite sums of sinusoids, and then to aperiodic signals in Chapter 15.

In this chapter we study the response of a system to a sinusoidal input signal. The output signal is characterized in terms of a complex quantity calculated from the transfer function of the system. So the output *signal* of a system is directly related to the *system* representation. This result is precisely stated in the frequency response theorem which is discussed in this chapter. The linkage between the input and output signals and the system in the frequency response theorem is of immense importance in many aspects of system theory. The frequency response theorem can be stated using either the Laplace transfer function or the Fourier transfer function. Both versions offer insight into the fundamental concept embodied in this theorem.

The frequency response theorem identifies the frequency response function as the system property that links the system to the signal propagating through it. This function is identical to the Fourier transfer function, and it can be easily calculated from the Laplace transfer function. It is really the frequency response function that is of central interest in this chapter. Graphical interpretations of the frequency response function are studied in detail. Several important properties of the frequency response function are identified including the bandwidth of the system.

It turns out that the shape of the frequency response function tells us which sinusoidal signals are passed by the system and which signals are blocked. This idea is called <u>filtering</u>. Some sinusoidal signals are allowed to pass through the system unattenuated while other signals are blocked, or filtered out. This terminology suggests that we call a system a filter. A system as a filter can be classified depending on the frequencies of which sinusoids are transmitted by the system and which sinusoids are blocked. The notion of the bandwidth of the system is introduced to describe the shape of the frequency response function. This concept plays a fundamental role in system theory. The concept of filtering is taken one step further in the filter design process. Here we work the problem backwards. We start with a desired frequency response function. We want to build a system that has that particular frequency response function. In this chapter we briefly introduce filter design.

In Chapter 16, we study the structure of the frequency response function via Bode plots. Bode plots, which are constructed from the poles and zeros of the transfer function, lend insight into the shape of the frequency response function. The shape of the frequency response theorem, in turn, determines how a signal will propagate through a system. Hence, Bode plots offer insight into the way a system affects the signals propagating through it by linking the poles and zeros of the (Laplace) transfer function to the shape of the frequency response function.

Summary of Sections

Section 14.1   The frequency response theorem is stated using Laplace transforms.

Section 14.2   The frequency response theorem is stated using Fourier transforms.

Section 14.3   The frequency response function is defined.

Section 14.4   A graphical interpretation of the frequency response function is given.

Section 14.5   The notion of the bandwidth of a system is defined and discussed.

Section 14.6   The concept of filtering is introduced and ideal filters are defined.

Section 14.7   The process of filter design is introduced.

Section 14.8   Chapter summary section.

Coverage of the Text

This chapter can be read without significant loss of content while covering only one of Sections 14.1 or 14.2. That is, this chapter can be covered using only the Laplace or Fourier transfer function. Bode plots in Chapter 16 are naturally related to the frequency response function discussed in Sections 14.3 and 14.4.

14.1 THE FREQUENCY RESPONSE THEOREM USING LAPLACE TRANSFORMS

In this section we will consider the linear, time invariant BIBO stable system

$$y(t) = \mathcal{H}[x(t)]. \tag{14.1.1}$$

We will assume that the input signal is a sinusoid for $t \geq 0$ as shown in Figure 14.1.1. The output signal is also a sinusoid, after a transient response, as shown in Figure

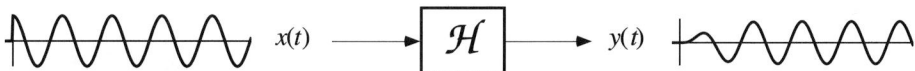

Figure 14.1.1 System and Input Signals Considered in This Section

14.1.1. We will characterize this output signal in terms of the frequency of the input signal and two parameters calculated from the transfer function of the system. While it is not immediately obvious, it turns out that this relationship between the input *signal* and the *system* is one of the cornerstones of system theory. The following theorem is the main result. (It is restated in Section 14.2 in terms of Fourier transforms.)

Theorem 14.1.1: Frequency Response Theorem Using the Laplace Transform
Consider a system that is represented by its Laplace transfer function[1]

$$\frac{Y(s)}{X(s)} = H(s).$$

Also assume that this system is BIBO stable. Let the input signal be

$$x(t) = \left[\cos(\omega_i t)\right] u_s(t).$$

Then the output signal of the system $y(t)$ is given by

$$y(t) = y_{tr}(t) + y_{ss}(t)$$

where

$$y_{ss}(t) = \left|H(\omega_i)\right| \cos\left(\omega_i t + \angle H(\omega_i)\right), \quad t \geq 0,$$

with

$$H(s)\big|_{s=j\omega_i} = H(\omega_i) = \left|H(\omega_i)\right| e^{j\angle H(\omega_i)}$$

and where

$$\lim_{t \to \infty} y_{tr}(t) = 0.$$

▲▲

Proof: The Laplace transform of the input signal is

$$X(s) = \frac{s}{s^2 + \omega_i^2}.$$

(14.1.2)

If the transfer function is

---

[1] Definition 10.1.3.

$$\frac{Y(s)}{X(s)} = H(s) = \frac{b(s)}{a(s)} = \frac{b(s)}{(s - p_1) \cdots (s - p_n)}, \tag{14.1.3}$$

then the Laplace transform of the output signal is given by

$$Y(s) = H(s)X(s) = \frac{b(s)}{a(s)} \frac{s}{s^2 + \omega_i^2}. \tag{14.1.4}$$

Since the system is BIBO stable, all of the poles will be in the open LHP. In particular, the system will not have any poles on the imaginary axis. (This observation is key to the proof.) Hence, the poles of the last factor in (14.1.4), which come from the Laplace transform of the input signal, are isolated poles. Expanding (14.1.4) in partial fractions[2] we find that

$$Y(s) = \frac{c_1}{s - p_1} + \cdots + \frac{c_n}{s - p_n} + \frac{d}{s - j\omega_i} + \frac{\bar{d}}{s + j\omega_i}. \tag{14.1.5}$$

We collect the first n terms in the partial fraction expansion in (14.1.5), and we rewrite them as

$$\frac{g(s)}{a(s)} = \sum_{m=1}^{n} \frac{c_m}{s - p_m}. \tag{14.1.6}$$

The last two coefficients in (14.1.5) are computed using Remark 9.3.7. For the first coefficient we have

$$d = \left[(s - j\omega_i)Y(s) \right]\Big|_{s = j\omega_i} = H(\omega_i) \frac{j\omega_i}{(j\omega_i + j\omega_i)} = \frac{H(\omega_i)}{2}. \tag{14.1.7}$$

It follows that

$$\bar{d} = \left[(s + j\omega_i)Y(s) \right]\Big|_{s = -j\omega_i} = \frac{\overline{H(\omega_i)}}{2}. \tag{14.1.8}$$

Combining (14.1.7) and (14.1.8) and taking the inverse Laplace transform we get

$$y_{ss}(t) = \frac{1}{2}\left[H(\omega_i)e^{j\omega_i t} + \overline{H(\omega_i)}e^{-j\omega_i t} \right] = |H(\omega_i)|\cos(\omega_i t + \angle H(\omega_i)), \quad t \geq 0. \tag{14.1.9}$$

Hence, we have established the first part of the theorem.

---

[2] In the expansion (14.1.5) we could allow some or all of the poles of the transfer function to be repeated without affecting the following discussion. For simplicity we assume that the transfer function has only isolated poles.

The other terms of the partial fraction expansion are obtained by inverting (14.1.6) into the time domain, i.e.,

$$y_{tr}(t) = \mathcal{L}^{-1}\left\{\frac{g(s)}{a(s)}\right\}. \tag{14.1.10}$$

Applying the final value theorem of Laplace transforms (Property 9.2.14) to (14.1.10) we find that

$$\lim_{t \to \infty} y_{tr}(t) = \lim_{s \to 0} \frac{sg(s)}{a(s)} = 0. \tag{14.1.11}$$

Because the system is BIBO stable, all of the roots of $a(s)$ are in the open LHP. Then the factor s will not cancel with any pole, and the limit in (14.1.11) will hold. Thus we have established the last part of the theorem.

▲▲

Next we introduce some terminology that allows us to describe the results of Theorem 14.1.1.

Definition 14.1.2:   The term $y_{ss}(t)$ in Theorem 14.1.1 is called the (sinusoidal) steady state response of the system.   The term $y_{tr}(t) = y(t) - y_{ss}(t)$ is called the transient response of the system.

▲▲

Terminology: The term *response* in the phrase *steady state response* refers to a function of time, i.e., the output signal.

The essence of the frequency response theorem is that it gives us a complete description of the steady state response of the system in terms of the function $H(\omega)$ (the transfer function) without directly inverting the Laplace transform of the output signal.   The next example illustrates this idea.

Example 14.1.3:   First we will investigate the response of the RC network in Figure 14.1.2 to sinusoidal inputs using Theorem 14.1.1.   The transfer function of this system is

$$\frac{Y(s)}{X(s)} = H(s) = \frac{\dfrac{1}{RC}}{s + \dfrac{1}{RC}} = \frac{\alpha}{s + \alpha}, \quad \alpha = \frac{1}{RC}. \tag{14.1.12}$$

The input signal is

$$x(t) = \left[\cos(\omega_i t)\right] u_s(t). \tag{14.1.13}$$

Since the frequency of the input signal is ω_i, then

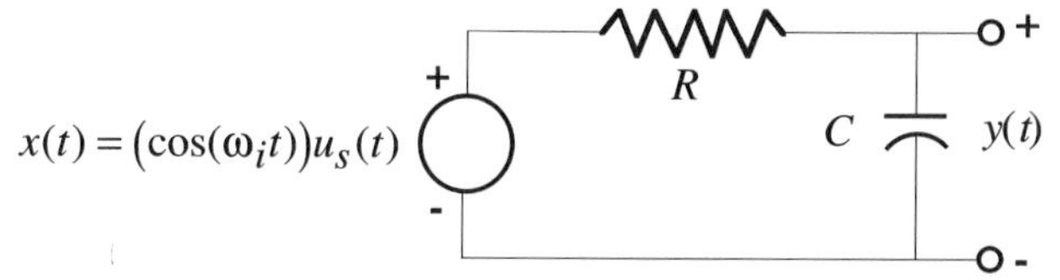

Figure 14.1.2 *RC* Network

$$H(s)\big|_{s=j\omega_i} = \frac{\alpha}{s+\alpha}\bigg|_{s=j\omega_i} = \frac{\alpha}{\alpha+j\omega_i} = \frac{\alpha}{\sqrt{\alpha^2+\omega_i^2}}\, e^{-j\tan^{-1}\frac{\omega_i}{\alpha}} \qquad (14.1.14)$$

$$= \frac{\alpha}{\sqrt{\alpha^2+\omega_i^2}}\, e^{-j\phi}.$$

Suppose that the input signal is

$$x(t) = [\cos(1t)]u_s(t), \qquad (14.1.15)$$

and $RC = 1$. Then $\omega_i = 1$ and the sinusoidal steady state response of this network is

$$y_{ss}(t) = |H(\omega_i)|\cos(\omega_i t + \angle H(\omega_i))u_s(t) = \left[\frac{\alpha}{\sqrt{\alpha^2+\omega_i^2}}\right]\cos(\omega_i t - \phi)u_s(t) \qquad (14.1.16)$$

$$= (0.707)\cos(1t - 0.785)u_s(t).$$

The transient response can be computed using the inverse Laplace transforms. It can be verified that the transient response is

$$y_{tr}(t) = \frac{-\alpha^2}{\alpha^2+\omega_i^2}\exp\left(-\frac{t}{\alpha}\right)u_s(t) = -\frac{1}{1+1}e^{-t}u_s(t) = -0.5e^{-t}u_s(t). \qquad (14.1.17)$$

The input and output signals are shown in Figure 14.1.3. The output signal contains both the transient and steady state response according to Theorem 14.1.1. The transient response dies out after a short time, and the output signal consists only of the steady state output signal. This situation is depicted in Figure 14.1.3.

Computer Usage: The Laplace transform version of the frequency response theorem gives us a way to compute the steady state output signal. The input signal begins at $t = 0$. Hence, any simulation package can be used to generate the output signal from a suitable representation of the system and the input signal. Figure 14.1.3 was generated in this way.

▲▲

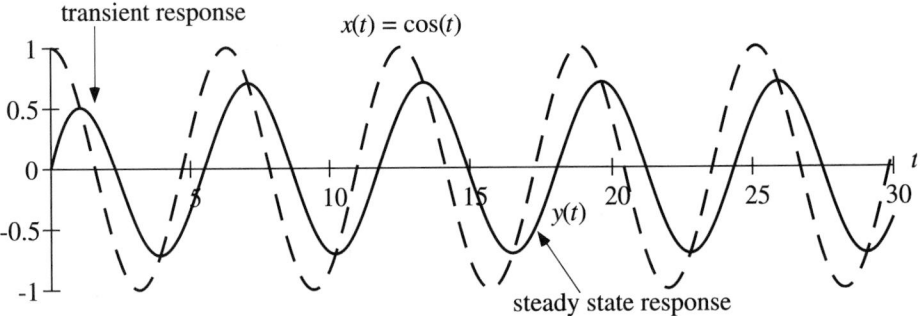

Figure 14.1.3 Input and Output Signals for Example 14.1.3

14.2 THE FREQUENCY RESPONSE THEOREM USING FOURIER TRANSFORMS

Consider the linear, time invariant BIBO stable system in Figure 14.2.1. In the last section we analyzed this system assuming that the system representation is a Laplace transfer function. In this section we will assume that the system representation is a Fourier transfer function, and that the input signal is a sinusoid defined on the entire real line as shown in Figure 14.2.2 Hence, the output signal is also a sinusoid defined for all time. The main theorem of this section, the frequency response theorem, gives the relationship between the input and output signals in terms of the Fourier transfer function. This result is the analog of Theorem 14.1.1 stated in terms of Fourier transforms instead of Laplace transforms.

Theorem 14.2.1: Frequency Response Theorem Using Fourier Transforms
Consider a system represented by its Fourier transfer function

$$\frac{Y(\omega)}{X(\omega)} = H(\omega).$$

If the input signal is

$$x(t) = \cos(\omega_i t), \qquad -\infty < t < \infty,$$

then the output signal is

$$y_{ss}(t) = |H(\omega_i)|\cos(\omega_i t + \angle H(\omega_i)), \qquad -\infty < t < \infty.$$

▲▲

Proof: Suppose that the input signal to the system is

$$x(t) = \cos(\omega_i t) \qquad -\infty < t < \infty. \tag{14.2.1}$$

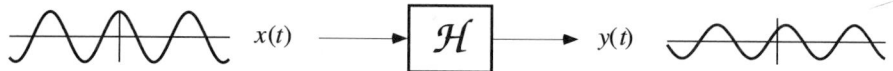

Figure 14.2.1 System and Input Signal Considered in This Section

Using Euler's formula[3] we can express $x(t)$ as

$$x(t) = \cos(\omega_i t) = \frac{1}{2}\left(e^{j\omega_i t} + e^{-j\omega_i t}\right).$$ (14.2.2)

We will compute the output signal $y(t)$ by computing the output signal to each exponential in (14.2.2) and then summing these two output signals. (This system is linear.)

We will analyze this system using its convolution representation

$$y(t) = \int_{-\infty}^{\infty} h(\lambda)x(t-\lambda)\,d\lambda$$ (14.2.3)

where the impulse response function $h(t)$ is related to the Fourier transfer function by

$$H(\omega) = \int_{-\infty}^{\infty} h(\lambda)e^{-j\omega t}\,dt,$$ (14.2.4)

First, consider the input signal

$$x_1(t) = e^{j\omega_i t}.$$ (14.2.5)

Using (14.2.3) the corresponding output signal is

$$y_1(t) = \int_{-\infty}^{\infty} h(\lambda)e^{j\omega_i(t-\lambda)}\,d\lambda = e^{j\omega_i t}\int_{-\infty}^{\infty} h(\lambda)e^{-j\omega_i\lambda}\,d\lambda.$$ (14.2.6)

The exponential has been taken outside the integral because t is a constant with respect to the variable of integration λ. Comparing (14.2.6) to (14.2.4), we see that the last integral in (14.2.6) is just the Fourier transform $H(\omega)$ evaluated at $\omega = \omega_i$. Using this observation (14.2.6) reduces to

$$y_1(t) = e^{j\omega_i t} H(\omega_i).$$ (14.2.7)

Next consider the input signal

$$x_2(t) = e^{-j\omega_i t}.$$ (14.2.8)

---

[3] Definition 3.1.6.

Since the input $x_2(t)$ in (14.2.8) is the complex conjugate of $x_1(t)$ in (14.2.5), the output signal will be the conjugate of (14.2.7). We have

$$y_2(t) = e^{-j\omega_i t} \int_{-\infty}^{\infty} h(\lambda)e^{j\omega_i \lambda} d\lambda = e^{-j\omega_i t} \overline{\int_{-\infty}^{\infty} h(\lambda)e^{-j\omega_i \lambda} d\lambda} = e^{-j\omega_i t} \overline{H(\omega_i)}. \tag{14.2.9}$$

The total system response is now computed by combining the outputs in (14.2.7) and (14.2.9) and multiplying by 1/2. The result is

$$y_{ss}(t) = \frac{1}{2}\left[H(\omega_i)e^{j\omega_i t} + \overline{H(\omega_i)}e^{-j\omega_i t}\right]. \tag{14.2.10}$$

The complex number $H(\omega_i)$ can be written as

$$H(\omega_i) = |H(\omega_i)|e^{j\angle H(\omega_i)}. \tag{14.2.11}$$

Using (14.2.11), the symmetry properties of $H(\omega)$, and Euler's formula we can rewrite (14.2.10) as

$$y_{ss}(t) = \frac{1}{2}|H(\omega_i)|\left[e^{j(\omega_i t + \angle H(\omega_i))} + e^{-j(\omega_i t + \angle H(\omega_i))}\right] \tag{14.2.12}$$

$$= |H(\omega_i)|\cos(\omega_i t + \angle H(\omega_i)) \qquad -\infty < t < \infty.$$

▲▲

Theorem 14.2.1 says that the output signal of the system is a sinusoid at the frequency of the input signal with constant amplitude and phase *for all time*. This observation gives rise to the following definition.

Definition 14.2.2: Given a system with a Fourier transfer function and an input signal which is a sinusoid, the output signal is called the (sinusoidal) steady state response of the system.

▲▲

Terminology: The term *response* in the phrase *steady state response* refers to a function of time, i.e., the output signal.

In Section 14.1 we stated the frequency response theorem in terms of Laplace transforms. In that theorem we also identify an output signal $y_{ss}(t)$, also called the steady state response. Although these two output signals are computed in different ways, they are the same signal. This fact is established in Section 14.3.

The next example illustrates how to calculate the output signal for a given sinusoidal input signal and Fourier transfer function using Theorem 14.2.1.

Example 14.2.3: Consider the *RC* circuit shown in Figure 14.2.2. From Example 12.4.2 the Fourier transfer function for this system is

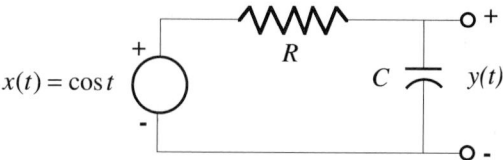

Figure 14.2.2 *RC* Network

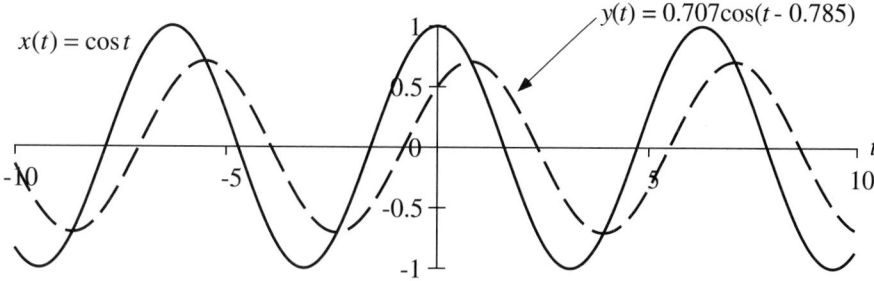

Figure 14.2.3 Input and Output Signals in Example 14.2.2

$$H(\omega) = \frac{1}{1 + j\omega} = \frac{1}{\sqrt{1 + \omega^2}} e^{-j\tan^{-1}\left(\frac{\omega}{1}\right)}. \tag{14.2.13}$$

Suppose that the input signal to this system is

$$x(t) = \cos(t). \tag{14.2.14}$$

The frequency of the input signal is $\omega_i = 1$. The magnitude and phase of the output signal is calculated by evaluating the frequency response function at this frequency. We have

$$H(\omega)\big|_{\omega=1} = H(1) = \frac{1}{1 + j1} = \frac{1}{\sqrt{1 + 1^2}} e^{-j\tan^{-1}(1)} = \frac{1}{\sqrt{2}} e^{-j\frac{\pi}{4}}. \tag{14.2.15}$$

According to the frequency response theorem, the corresponding output signal is

$$y_{ss}(t) = \frac{1}{\sqrt{2}} \cos\left(1t - \tan^{-1}(1)\right) = 0.707 \cos\left(t - \frac{\pi}{4}\right). \tag{14.2.16}$$

The input signal in (14.2.14) and the output signal (14.2.16) are plotted in Figure 14.2.3.

Note: Compare Figure 14.2.3 with Figure 14.2.1.

Terminology: In Figure 14.2.3 we see that the input and output signals have the same frequency, but the output signal has suffered a time delay (negative phase shift) relative to the input signal. This time delay can be traced to the negative phase shift between the input and output signal in (14.2.16) as predicted in Theorem 14.2.1. We say the output signal <u>lags</u> the input signal. If the output signal is advanced in time with respect to the input signal, we say the output signal <u>leads</u> the input signal. This terminology is widely used.

Computer Usage: The Fourier version of the frequency response theorem gives us a way to compute the *steady state* output signal when the input signal is a sinusoid. Because the input signal begins at negative infinity, we can't directly simulate the output signal using a representation of the system. Indeed, we require an initial condition at a finite time to begin the simulation. Hence, a direct simulation of the system is not possible. We can compute the two complex numbers

$$H(\omega_i) = |H(\omega_i)| e^{\angle H(\omega_i)} \qquad (14.2.17)$$

from the Fourier transfer function. Then the numbers in (14.2.17) can be substituted into the steady state frequency response in Theorem 14.2.1 and the signal can be plotted directly. This method was used in Figure 14.2.3. This method should be compared to the calculation of the output signal when the Laplace transform version of the frequency response theorem.

▲▲

The circuit analysis in Example 14.2.3 can also be carried out using phasors as discussed in books on network analysis. The frequency response theorem places this analysis tool in a broader context, allowing it to be applied to a broader class of systems.

14.3 THE FREQUENCY RESPONSE FUNCTION

14.3.1 Introduction

In this section we will continue our discussion of the frequency response theorem. The essence of this theorem is as follows. If we compare the steady state response of the system to a sinusoidal input signal, the output signal is a sinusoid at the same frequency, but the amplitude and phase have been changed. This change is determined by the function $H(\omega)$, which is calculated from the transfer function (Laplace or Fourier) of the system. In this section and the next we will investigate in more detail the relationship between the function $H(\omega)$, the sinusoidal input signal, and the steady state response of the system. In this section we will show that the Laplace and Fourier versions of the frequency response theorem are the same. We will give a name to the function $H(\omega)$ and we will interpret the input and output signals in terms of their power.

14.3.2 System Representations and the Frequency Response Theorem[4]

We begin our discussion of the frequency response theorem by establishing that the two versions of the frequency response theorem given in each of the first two sections of this chapter are the same. To compare the two results we must select a system to which both theorems can be applied. We represent this system with a convolution integral

$$y(t) = \int_{-\infty}^{\infty} h(t-\lambda)x(\lambda)\,d\lambda. \tag{14.3.1}$$

Assume that the system in (14.3.1) is causal, linear, time invariant, and BIBO stable. Also assume that this system has a Fourier transfer function

$$\frac{Y(\omega)}{X(\omega)} = H_F(\omega) = \mathcal{F}\{h(t)\}, \tag{14.3.2}$$

and a Laplace transfer function

$$\frac{Y(s)}{X(s)} = H_L(s) = \mathcal{L}\{h(t)\}. \tag{14.3.3}$$

(Note the special notation for the transfer functions that is used in this subsection only.) The Fourier transfer function doesn't require the system to be causal, but the Laplace transfer function can only be used for a causal system. The Laplace transfer function can be used for unstable systems, but the frequency response theorem requires the system to be stable. Because the system is BIBO stable, all the poles of the Laplace transfer function are in the open LHP. Now using Theorem 9.5.1 we have that the Fourier transfer function is related to the Laplace transfer function by

$$H_L(s)\big|_{s=j\omega} = H(\omega) = H_F(\omega). \tag{14.3.4}$$

Consider now the steady state response of the system as computed by each of the two versions of the frequency response theorem. Since each of these two output signals are computed from the same function, they must be the same. These observations establish the relationship between the Laplace version of the frequency response theorem and the Fourier version of the frequency response theorem. We state this result formally.

Corollary 14.3.1: Consider the system in (14.3.1). Assume that the system in (14.3.1) is causal, linear, time invariant, and BIBO stable. Assume that the input signal to the system (14.3.1) is

$$x_L(t) = \big[\cos(\omega_i t)\big]u_s(t).$$

---

[4] This subsection should be skipped if only Section 14.1 or 14.2 has been covered.

The steady state response of the system is

$$\hat{y}_{ss}(t) = \left[\left\|H_L(\omega_i)\right|\cos(\omega_i t + \angle H_L(\omega_i))\right]u_s(t).$$

Next, suppose that the input signal to the system (14.3.1) is

$$x_F(t) = \cos(\omega_i t), \quad -\infty < t < \infty.$$

The steady state response of the system is

$$\tilde{y}_{ss}(t) = \left|H_F(\omega_i)\right|\cos(\omega_i t + \angle H_F(\omega_i)).$$

Then

$$\hat{y}_{ss}(t) = \tilde{y}_{ss}(t), \quad t \geq 0.$$

▲▲

Corollary 14.3.1 implies that the frequency response theorem holds independently of the system representation. The system must be linear, time invariant, and BIBO stable. For example, the steady state response can be computed from a state space representation.

Terminology: Based on Corollary 14.3.1, we will refer to the *frequency response theorem*. The appropriate version of this theorem can be determined from context.

14.3.3 Definition of the Frequency Response Function

Now we can turn our attention to the main topic of this section. The following corollary is a straightforward extension of the frequency response theorem, but it serves to highlight the focus of the discussion below.

Corollary 14.3.2: Consider the system in (14.3.1). Assume that this system satisfies the assumptions of the frequency response theorem. Let the input signal be

$$x(t) = A\cos(\omega_i t + \phi).$$

Then the steady state response of this system is

$$y_{ss}(t) = A\left|H(\omega_i)\right|\cos(\omega_i t + \angle H(\omega_i) + \phi)$$

where $H(\omega)$ is given in (14.3.4).

▲▲

If the input signal is scaled by A, the output signal is also scaled by A because the system is linear. The phase shift in the input signal is also transmitted to the output signal because the system is time invariant.

In the rest of this section we focus on the steady state output signal

$$y_{ss}(t) = A|H(\omega_i)|\cos(\omega_i t + \angle H(\omega_i) + \phi). \tag{14.3.5}$$

The frequency response theorem specifies that the input signal to the system is a single sinusoid at a given frequency. No other signals appear in the output signal, and the frequency of (14.3.5) is identical to the frequency of the input signal. The implications of this fact, which are many, are explored over this chapter and the next. This observation is not true for time-varying or nonlinear systems. For example, Section 8.7 shows that the output signal of a static nonlinearity contains an infinite number of cosines with frequencies that are multiples of the frequency of the input signal. *The single most important fact provided by the frequency response theorem is that the steady state response is a pure sinusoid at the same frequency as the input signal.*

In order to fully understand the frequency response theorem, it is important to recognize that the steady state response can only be understood by comparing it to the input signal. Corollary 14.3.2 isolates the effect of the system on the input signal as it propagates through the system. The measure of change in the output signal is referenced to the input signal. The ratio of the amplitudes is

$$\frac{\text{peak amp } y_{ss}(t)}{\text{peak amp } x(t)} = \frac{A|H(\omega_i)|}{A} = |H(\omega_i)|. \tag{14.3.6}$$

That is, the absolute amplitude of the input signal is not the determining factor in the output signal; it is the relative change between the input signal as compared to the output signal. Furthermore, the phase shift between the input and output signals is

$$\text{phase of } y_{ss}(t) - \text{phase of } x(t) = (\omega_i t + \angle H(\omega_i) + \phi) - (\omega_i t + \phi) = \angle H(\omega_i). \tag{14.3.7}$$

Again the absolute phase of the input signal is not important; it is the relative shift in the output signal as compared to the input signal that determines the steady state response of the system. *Hence, the concept of frequency response is tied directly to the definition of a system as a relationship between the input and the output signals.*

Equations (14.3.6) and (14.3.7) show that the relative change of the output signal as compared to the input signal is determined by one complex number calculated from the transfer function of the system. This complex number depends on the frequency of the input signal. Suppose we consider input signals of all frequencies. As the frequency of the input signal varies from 0 to ∞ we obtain a complex function.

Definition 14.3.3: Consider a linear, time invariant, BIBO stable system with a Laplace and/or Fourier transfer function. Then the function

$$H(\omega) = H_F(\omega) = H_L(s)\big|_{s=j\omega}$$

is called the <u>frequency response (function)</u> of the system.[5]

▲▲

---

[5] If only one transfer function exists, then the other transfer function is ignored.

Terminology: The frequency response function of the system is a function of the *frequency* variable ω (in rad/sec). This terminology is a change from the "impulse response," "step response," and "steady state response" of a system which refer to functions of *time*.

Terminology: The name "frequency response function" can be motivated by observing that if the input signal is an impulse function $x(t) = \delta(t)$, then the Fourier transform of the output signal is given by

$$Y(\omega) = H(\omega)X(\omega) = H(\omega).$$ (14.3.8)

The Fourier transform of the output signal, the response of the system in the frequency domain, is the frequency response function when the input signal is an impulse function.

Notation: Often the frequency variable is f with units of Hz. Using this notation, the frequency response function is written as $H(f)$. This transfer function can be computed from

$$H(f) = H(\omega)\big|_{\omega = 2\pi f}.$$ (14.3.9)

All of the interpretations given for $H(\omega)$ apply equally well to $H(f)$, however. We will not use this frequency variable in this text.

Example 14.3.4: Consider again the *RC* network discussed in Examples 14.1.3 and 14.2.3. The frequency response function for this system is

$$H(\omega) = \frac{1}{1 + j\omega} = \frac{1}{\sqrt{1 + \omega^2}} e^{-j\tan^{-1}\left(\frac{\omega}{1}\right)}.$$ (14.3.10)

▲▲

14.3.4 Power Considerations

There is a close relationship between the power in the input signal and the power in the steady state output signal and the frequency response function. Recall that the power[6] in the sinusoidal input signal and the output signal is, respectively,

$$P_x = \frac{A^2}{2} \quad \text{and} \quad P_y = \frac{A^2|H(\omega_i)|^2}{2}.$$ (14.3.11)

Comparing the two quantities in (14.3.11) we observe that

$$P_y = |H(\omega_i)|^2 P_x.$$ (14.3.12)

---

[6] The power of a signal is defined in Definition 8.2.2, and it is discussed in Sections 8.2 - 4. These sections should be reviewed at this time.

We will frequently use this relationship between the power in the input signal and the output signal.

The ratio of the power in the steady state output signal, (14.3.5), to the power in the input signal is

$$\frac{\text{power of } y_{ss}(t)}{\text{power of } x(t)} = \frac{\dfrac{A^2 |H(\omega_i)|^2}{2}}{\dfrac{A^2}{2}} = |H(\omega_i)|^2. \qquad (14.3.13)$$

Hence, we see that the frequency response function determines the power distribution among the steady state components of the output signal over frequency.

A common unit used to measure the relative change in power is the decibel.

Definition 14.3.5: The decibel (dB) is defined as

$$x \text{ dB} = 10 \log\left(\frac{\text{output power}}{\text{input power}}\right)$$

where the logarithm is log base 10.

▲▲

In terms of decibels, the steady state output of a system is given by

$$10 \log\left[\frac{\text{power in } y_{ss}(t)}{\text{power in } x(t)}\right] = 10 \log\left[\frac{\dfrac{A^2 |H(\omega_i)|^2}{2}}{\dfrac{A^2}{2}}\right] = 20 \log|H(\omega_i)|. \qquad (14.3.14)$$

Hence, the power of the output signal can be expressed in terms of the power of the input signal and the frequency response function. (This concept again underscores the definition of a system as a ratio of the input signal to the output signal.) For passive networks, the power in the output signal is always less than the power in the input signal.

Definition 14.3.6: The quantity $20 \log|H(\omega)|$ is called the attenuation of the system.

▲▲

Terminology: If the attenuation is less than one (0 dB), we say the output signal has been attenuated.

The quantity $20 \log|H(\omega)|$ is used in one of the most common graphical representations of the frequency response function, Bode plots, which are discussed in Chapter 16.

14.4 GRAPHICAL INTERPRETATIONS OF THE FREQUENCY RESPONSE FUNCTION

14.4.1 Introduction

Consider the linear, time invariant, BIBO stable system

$$y(t) = \mathcal{H}[x(t)]. \tag{14.4.1}$$

Assume that the system in (14.4.1) has a Fourier transfer function

$$\frac{Y(\omega)}{X(\omega)} = H_F(\omega), \tag{14.4.2}$$

and/or a Laplace transfer function

$$\frac{Y(s)}{X(s)} = H_L(s). \tag{14.4.3}$$

Then the frequency response function (Definition 14.3.3) of this system is

$$H(\omega) = H_F(\omega) = H_L(s)\big|_{s=j\omega} \tag{14.4.4}$$

The frequency response function gives us a wealth of information that is not immediately obvious from any of the system representations. We will discuss the frequency response function in depth in this and the next chapter. In this section we will point out the fundamental properties of the frequency response function. In particular, we will present graphical interpretation of the frequency response function.

14.4.2 Magnitude and Phase Graphs

The frequency response function is a complex function of a real variable ω. For each frequency ω_i, the value of the frequency response function is a complex number $H(\omega_i)$. This complex number is completely defined by its magnitude and phase; i.e.,

$$H(\omega_i) = |H(\omega_i)|e^{\angle H(\omega_i)}. \tag{14.4.5}$$

Hence, the frequency response function is determined by specifying the magnitude and phase, separately, at each frequency. This information is usually displayed in graphical form in terms of two plots.

Definition 14.4.1: Let $H(\omega)$ be a frequency response function. The graph

$|H(\omega)|$ vs. frequency

is called the <u>magnitude graph</u> of the frequency response function. The graph

$\angle H(\omega)$ vs. frequency

is called the <u>phase graph</u> of the frequency response function.

▲▲

Notation: The scale on $|H(\omega)|$ can be in decibels, or log magnitude, or linear as suits the application. The frequency scale can be linear or logarithmic; the units of the frequency scale can be in hertz or rad/sec. The frequency range can be $0 < \omega < \infty$ or, if the scale is linear, $-\infty < \omega < \infty$. Of particular importance are Bode plots, described in Chapter 16, where the magnitude graph is dB vs. $\log(\omega)$ and the phase graph is given by phase vs. $\log(\omega)$.

Recall that $|H(\omega)|$ is an even function of ω and $\angle H(\omega)$ is an odd function of ω. Thus, the information in the frequency range $0 < \omega < \infty$ is repeated in the frequency range $-\infty < \omega < 0$. Nonetheless, the frequency response function is often plotted over the entire frequency axis.

There is one other quantity used to characterize the phase of the frequency response function. To define this quantity let the phase be given by $\Theta(\omega) = \angle H(\omega)$.

Definition 14.4.2: The <u>group delay</u> $\tau(\omega)$ of the frequency response function is defined as

$$\tau(\omega) = -\frac{d\Theta(\omega)}{d\omega}.$$

▲▲

The group delay measures how much the phase changes as a function of frequency.

The following example illustrates the frequency response function.

Example 14.4.3: Consider again the *RC* network discussed in Examples 14.1.3 and 14.2.3. The frequency response function for this system is

$$H(\omega) = \frac{1}{1+j\omega} = \frac{1}{\sqrt{1+\omega^2}}\, e^{-j\tan^{-1}\left(\frac{\omega}{1}\right)}. \tag{14.4.6}$$

The magnitude of this function is

$$|H(\omega)| = \left|\frac{1}{1+j\omega}\right| = \frac{1}{\sqrt{1+\omega^2}}. \tag{14.4.7}$$

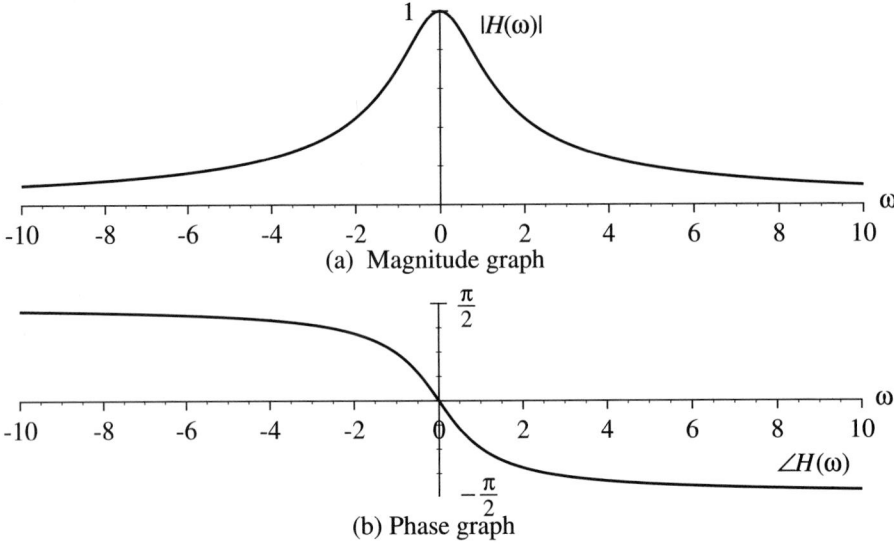

Figure 14.4.1 Magnitude and Phase Graphs of the *RC* Network

The phase of this function is

$$\angle H(\omega) = -\tan^{-1}(\omega). \tag{14.4.8}$$

The magnitude and phase graphs are shown in Figure 14.4.1.

▲▲

The following example will aid in the interpretation of the frequency response function.

Example 14.4.4: (A Butterworth Filter) Consider the system

$$\frac{Y(s)}{X(s)} = \frac{1}{s^4 + 2.613s^3 + 3.414s^2 + 2.613s + 1}. \tag{14.4.9}$$

This system is called a 4th-order Butterworth filter. The magnitude of the frequency response function is shown in Figure 14.4.2 and the phase is shown in Figure 14.4.3. The frequency response function for this system could be analytically calculated from the transfer function in (14.4.9). This function, however, would be very complicated. The analytical representation would not lend much insight into the implications of this function. The two graphs in Figure 14.4.2 are simple. This situation is common when using frequency response functions. The analytical representation, when available, is complicated while the magnitude and phase graphs are simple. Hence, the frequency response function is almost always presented in graphical form.

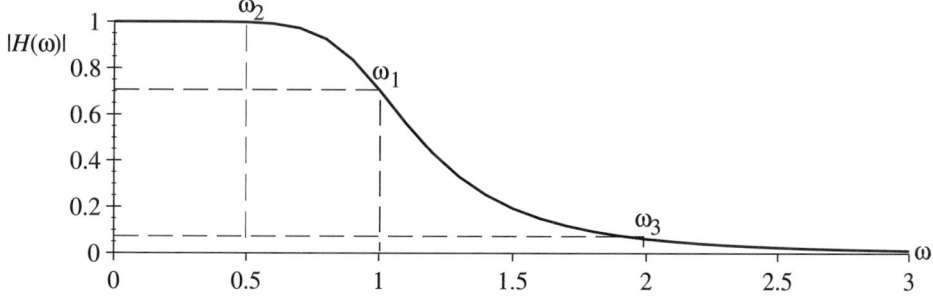

Figure 14.4.2 Magnitude Graph of the Frequency Response Function for a Butterworth Filter in (14.4.9)

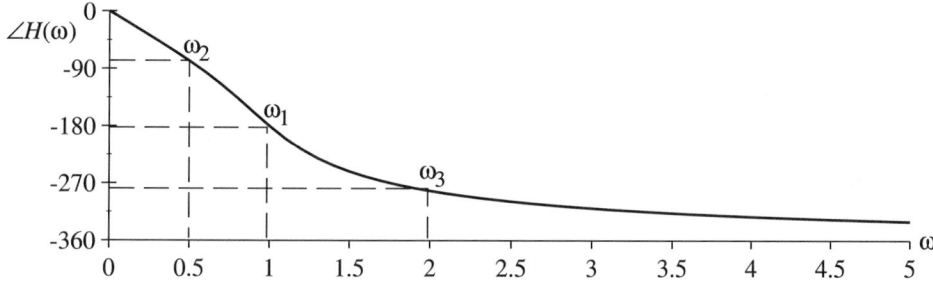

Figure 14.4.3 Phase Graph of the Frequency Response Function for a Butterworth Filter in (14.4.9)

Chapter 16 is devoted to the graphical construction of the frequency response function from the poles and zeros of the Laplace transfer function. This development lends insight into the structure of the frequency response function, particularly the asymptotic properties of the graphs.

▲▲

14.4.3 Graphical Interpretations

Next we will show how to determine the steady state response from the frequency of the input signal and the magnitude and phase plots of the frequency response function of the system. If the input signal is

$$x(t) = \cos(\omega_i t), \tag{14.4.10}$$

the steady state response in given by

$$y_{ss}(t) = |H(\omega_i)| \cos(\omega_i t + \angle H(\omega_i)). \tag{14.4.11}$$

The amplitude of the steady state component of the output signal can be determined by evaluating the magnitude graph of the frequency response function at the frequency of the input signal, i.e., by evaluating $|H(\omega_i)|$. Similarly, the phase shift of the output signal is determined by evaluating the phase graph of the frequency response function at the frequency of the input signal, i.e., by evaluating $\angle H(\omega_i)$.

Example 14.4.5:   Consider again the Butterworth filter in (14.4.9). Suppose the input signal is

$$x(t) = [\cos(t)]u_s(t) = [\cos(\omega_1 t)]u_s(t). \tag{14.4.12}$$

For the input signal (14.4.12) we can read the value of the magnitude of the frequency response function at the frequency of the input signal, $\omega_1 = 1$ from the graph in Figure 14.4.2. We see that

$$|H(\omega_1)| = |H(1)| = 0.707. \tag{14.4.13}$$

Similarly, we can read the phase from the graph in Figure 14.4.3. Again the input frequency is marked at $\omega_1 = 1$. We have

$$\angle H(\omega_1) = \angle H(1) = -180° = -\pi. \tag{14.4.14}$$

Substituting these two numbers into the steady state output signal in (14.4.11) yields

$$y_{ss}(t) = 0.707\cos(t - \pi). \tag{14.4.15}$$

Then the output signal in (14.4.15) is shown in Figure 14.4.4 along with the input signal in (14.4.12). Clearly (14.4.15) describes the steady state response of the system shown in Figure 14.4.4. Since the amplitude of the input signal (14.4.12) is $A = 1$, the amplitude of the output signal should be

$$(A)|H(\omega_1)| = (1)|H(1)| = (1)(0.707). \tag{14.4.16}$$

The amplitude of the output signal (14.4.15) can be read off from the graph and it agrees with (14.4.16). The phase of the input signal (14.4.12) is zero (0). Using (14.4.14) and the frequency response theorem, the phase of the output signal should be

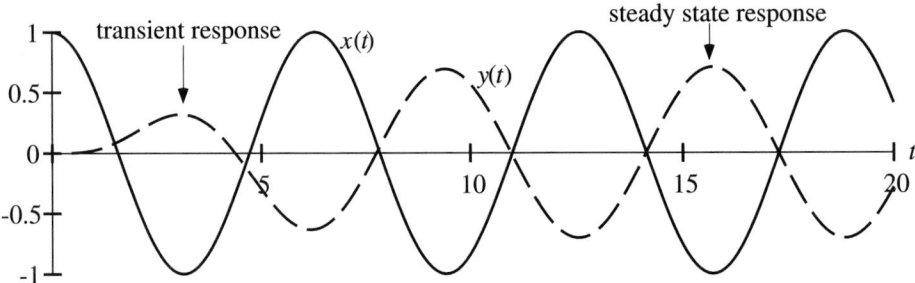

Figure 14.4.4 Input Signal (14.4.12) and the Corresponding Output Signal

$$\phi + \angle H(\omega_1) = 0 + \angle H(1) = 0 + (-\pi) = -\pi. \tag{14.4.17}$$

Figure 14.4.4 shows that the steady state response of the system is 180° output of phase with the input signal as predicted from (14.4.17). This example clearly shows that the steady state response of the system to a signal at a fixed frequency is simple to determine from the graphs of the frequency response function.

▲▲

The previous example showed how to interpret the graph of the frequency response function for a sinusoidal input signal of a given frequency. From this basic interpretation we move on to a broader interpretation of this function. We ask ourselves, "How does the system's response to a sinusoid at a given frequency compare to the system's response to a sinusoid at another frequency?" The following example lends insight into the answer to this question.

Example 14.4.6: We again consider the Butterworth filter in (14.4.9). Suppose we choose three input signals

$$x_2(t) = \cos(0.5t) = \cos(\omega_2 t), \tag{14.4.18}$$
$$x_1(t) = \cos(1t) = \cos(\omega_1 t),$$
$$x_3(t) = \cos(2t) = \cos(\omega_3 t).$$

The output signal for each of these input signals is shown in Figure 14.4.5. Consider the steady state responses of the Butterworth filter to the input signals in (14.4.18) as shown in Figure 14.4.5. We are interested in the relative amplitude of these sinusoidal signals on the interval $t > 15$, say. This figure clearly shows the dependence of the amplitude of the steady state response on the frequency of the input signal. For the input signal with a frequency of $\omega_2 = 0.5$ rad/sec, the amplitude of the output signal shows only a very small attenuation. For the input signal with a frequency of $\omega_1 = 1$ rad/sec, the amplitude of the output signal shows some attenuation. The amplitude of the input signal with a frequency of 2 rad/sec has been reduced by more than an order of magnitude, however.

This system behavior is clearly evident in the frequency response function. From the magnitude graph of the frequency response function in Figure 14.4.2 (where the frequencies of the input signals in (14.4.18) are marked), we see that as the frequency of the input signal increases, the amplitude of the output signal is reduced. In Figure 14.4.2 the magnitude of the frequency response function is near unity for low frequencies. Input sinusoids with frequencies in this frequency band ($\omega < 0.5$, say) will pass through the system with little attenuation. The magnitude of the frequency response function is near zero for high frequencies. Input sinusoids with frequencies in this frequency band ($\omega > 3$, say) will pass through the system with large attenuation. Figure 14.4.5 shows that the *shape* of the frequency response function across all frequencies determines the system response to sinusoidal input signals of all frequencies. *This concept is central to the discussion of the propagation of more complex signals through a system.*

It is important to know the shape of the frequency response function. The shape is determined by fluctuations over certain frequency bands and asymptotic limits as ω goes to zero and infinity. Therefore, when we graph a magnitude graph, say, it is

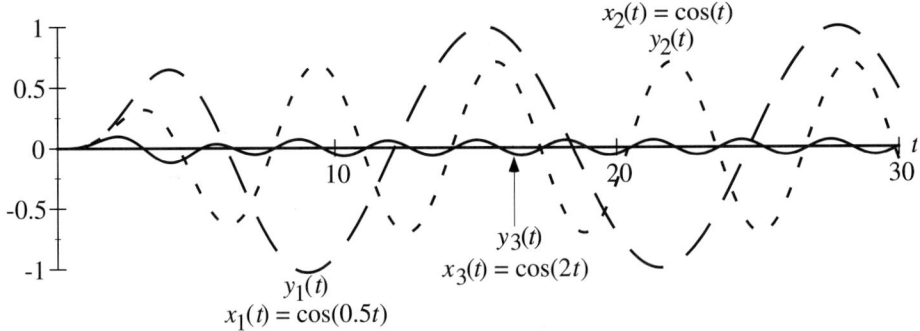

Figure 14.4.5 Output Signals from the Butterworth Filter in Example 14.4.12

important to know on which frequency bands the fluctuations occur. This information can be determined from the poles and zeros of the transfer function as described in Chapter 16. See also the discussion of the filters in Chapter 15.

▲▲

Example 14.4.6 shows that a quick glance at the shape of the frequency response function can show how the system responds to input signals whose frequencies vary from 0 to ∞. The next example expands on this idea.

Example 14.4.7: Consider again the Butterworth filter in (14.4.9). This system's behavior can be illustrated further by considering the input signal

$$x(t) = \cos(0.5t) + (0.2)\cos(10t). \tag{14.4.19}$$

This input signal is shown in Figure 14.4.6 along with the output signal. In the input signal the sinusoidal component at 10 rad/sec is clearly visible, while it has been essentially removed in the output signal. The output signal can be predicted from the frequency response function and the frequencies of the input signal.

Note: We can consider each input sinusoid in (14.4.19) separately and sum the results to characterize the output signal *because the system is linear*. The importance of this fact should not be underemphasized.

The output signal for the input signal in (14.4.19) with the frequency of $\omega_2 = 0.5$ rad/sec is shown in Figure 14.4.5. From the magnitude graph of the frequency response function in Figure 14.4.2 we see that its amplitude is attenuated by a small amount. Consider next the second sinusoid in the input signal (14.4.19). The magnitude graph of the frequency response function in Figure 14.4.2 is going to zero as the frequency increases. Hence, the value of the magnitude graph corresponding to the frequency $\omega = 10$ rad/sec of the second sinusoid in the input signal is essentially zero. It follows that the amplitude of the output signal corresponding to the second sinusoid in (14.4.19) is also essentially zero.

We can think of the input signal (14.4.19) as propagating through the

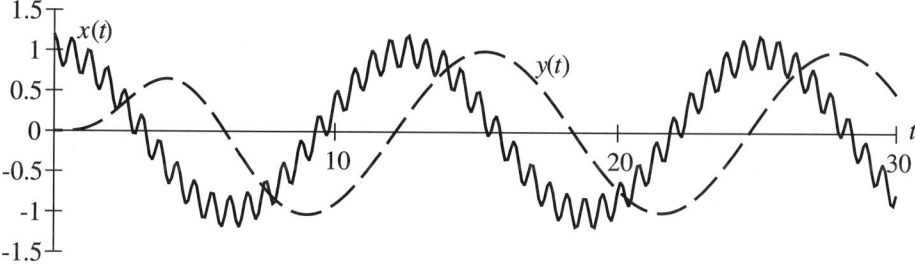

Figure 14.4.6 The Input Signal (14.4.19) and the Output Signal from the Butterworth Filter

Butterworth filter in (14.4.9). (We are interested in the long time behavior of the input and output signals, not the transient response.) Compare the input signal to the output signal. With this interpretation we see that the first component of the output signal propagates through the system with little attenuation while the second component has been essentially removed from the output signal.

Terminology: It is from this property that we derived the name <u>filter</u> for this system. Some input signals are passed by the system unchanged while other signals are filtered out. This concept is pursued in Chapter 15.

▲▲

Terminology: The frequency response theorem and the discussion of this theorem in the last two section have described in detail the relationship between the steady state response

$$y_{ss}(t) = A\left|H(\omega_i)\right|\cos\left(\omega_i t + \angle H(\omega_i) + \phi\right) \tag{14.4.20}$$

of a system and the frequency response function $H(\omega)$. The steady state response in (14.4.20) is a function of t; it exists in the <u>time domain</u>. The amplitude and phase of the steady state response are determined by the frequency response function $H(\omega)$, which is evaluated at the frequency ω_i. That is, the frequency response function is a function of frequency; it exists in the <u>frequency domain</u>. Hence, the frequency response theorem represents the basic coupling between *signals in the time domain*, and the frequency response function of the *system in the frequency domain*. This linkage between the time domain and the frequency domain is of fundamental importance in system theory with far-reaching ramifications for communication, control, and signal processing.

14.4.4   MATLAB Experiments

The following MATLAB M-file was used to generate the Butterworth filter example discussed in this section. The Butterworth filter is discussed in Section 14.7. For the purposes of this example it simply generates a transfer function. The frequency response function is generated by the **bode** command. (See Section 16.4.5 for a discussion of the **bode** command.) Used without left-hand arguments, this command generates Bode plots of the transfer function (which are discussed in Chapter 16). When used with left-hand arguments, it simply computes the frequency response function (not in dB). It is generally preferable to generate a frequency vector independently of the **bode** command. In this M-file this vector is linearly spaced (rather than logarithmic spacing) because the frequency response plots are on linear scales. To reproduce all of the examples in this section, frequency and the time vector of the input signal will have to be modified. This modification is straightforward.

Example 14.4.5: (*continued*) **Butterworth Filter**

```
clear
% model of the system
beta = 1;                          % bandwidth of system
[b,a] = butter(4,beta,'s');        % generate system - 4th order Butterworth filter


% simulation
w1 = 1;                            % frequency of input signal
tmax = 8*(2*pi/w1);
t = linspace(0,tmax,400);          % plot input signal for 8 cycles
x = sin(w1*t);                     % input signal
y = lsim(b,a,x,t);                 % calculate output signal


% Generate frequency response functions
wmin = 0;                          % determine frequency scales
wmax = 3*beta;
w = linspace(wmin,wmax,400);       % determine frequency vector
[mag,phs,w] =bode(b,a,w);          % calculate frequency response function


% plot results
figure(1)
axis([0,tmax,-1,1]);
plot(t,x,t,y);
title('Input and Output Signals')
xlabel('time')
grid


figure(2)
axis([wmin,wmax,-20,0]);
plot(w,mag)
title('Magnitude of the Frequency Response Function')
xlabel('frequency, rad/sec')
grid


figure(3)
axis([wmin,wmax,0,90]);
plot(w,phs)
title('Phase of the Frequency Response Function')
xlabel('frequency, rad/sec')
grid
```

14.5 THE BANDWIDTH OF A SYSTEM

14.5.1 Introduction

In this chapter and the next we are concerned with the propagation of a signals through a system. Our basic method of analysis is to use the frequency response theorem to compute the output signal for a given sinusoidal input signal. Then we compare the input signal with the output signal to determine how much the output signal has changed. The essence of the frequency response theorem is that the change in the output signal relative to the input signal can be determined by the frequency response function of the system. In the last section we showed how the shape of this function determined which sinusoidal signals propagated through the system with little distortion and which sinusoidal input signals were essentially blocked by the system. In this section we will use this concept to classify systems based on the shape of their frequency response function.

Let the system be represented by its Fourier transfer function

$$\frac{Y(\omega)}{X(\omega)} = H(\omega). \tag{14.5.1}$$

We will assume that the input signal is

$$x(t) = A\cos(\omega_i t + \theta). \tag{14.5.2}$$

Then the frequency response theorem, Theorem 14.1.1 or 14.2.1, says that the steady state response is

$$y_{ss}(t) = A|H(\omega_i)|\cos(\omega_i t + \theta + \angle H(\omega_i)). \tag{14.5.3}$$

We observe that both the input signal and the steady state response are sinusoids. Recall that the relationship between the power in the input signal P_x and the power in the output signal P_y is given by

$$P_y = |H(\omega_i)|^2 P_x. \tag{14.5.4}$$

We will exploit this relationship between the power in the input signal and the output signal below.

The relationship between the power of the input and output signals in (14.5.4) shows that it is dependent on the frequency of the input signal. In this section we will restrict ourselves to the simple question: "For what frequencies ω_i is the input signal (14.5.2) transmitted through the system with little loss of power, and for what frequencies is the input signal (14.5.2) transmitted through the system with great loss of power?" The answer to this question will lead to the notion of bandwidth of a system, one of the fundamental properties of a system. The following example illustrates the basic idea.

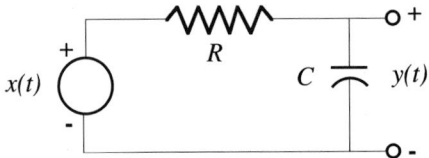

Figure 14.5.1 An *RC* Network with $RC = 1$

Example 14.5.1: Consider the *RC* network in Figure 14.5.1. From Example 14.4.3 the frequency response function of this system is

$$\frac{Y(\omega)}{X(\omega)} = H(\omega) = \frac{1}{j\omega + 1} = \frac{1}{\sqrt{\omega^2 + 1}} e^{-j\tan^{-1}(\omega)}. \qquad (14.5.5)$$

Consider the signals

$$x_1(t) = \cos(0.5t), \quad \text{and} \quad x_2(t) = \cos(10t). \qquad (14.5.6)$$

For the input signal $x_1(t)$ with frequency $\omega_1 = 0.5$ rad/sec the corresponding output signal is

$$y_1(t) = |H(\omega_i)|\cos(0.5t + \angle H(\omega_i)) = \frac{1}{\sqrt{1 + (0.5)^2}} \cos\left(0.5t - \tan^{-1}(0.5)\right) \qquad (14.5.7)$$

$$= (0.995)\cos(0.5t - 0.46).$$

For the input signal $x_2(t)$ with frequency $\omega_2 = 10$ rad/sec the corresponding output signal is

$$y_2(t) = |H(\omega_i)|\cos(10t + \angle H(\omega_i)) = \frac{1}{\sqrt{1 + 10^2}} \cos\left(10t - \tan^{-1}(10)\right) \qquad (14.5.8)$$

$$= (0.1)\cos(10t - 1.47).$$

The power[7] in both input signals is

$$P_{x_1} = P_{x_2} = \frac{A^2}{2} = \frac{1}{2}. \qquad (14.5.9)$$

The power in the output signal $y_1(t)$ is

---

[7] See the discussion of power spectrums for Fourier series in Section 8.5.

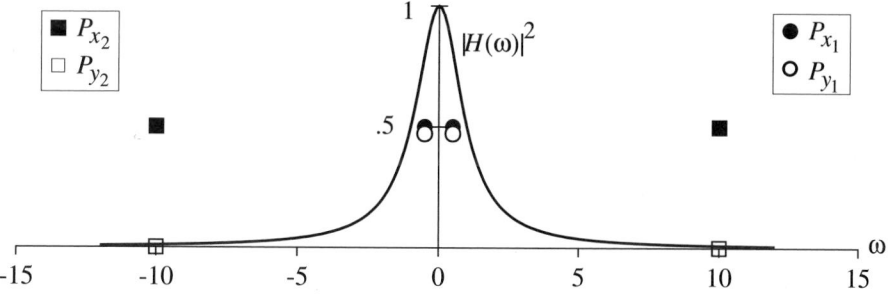

Figure 14.5.2 Magnitude Squared of the Frequency Response Function
and the Power Spectrum of the Input and Output Signals

$$P_{y_1} = \frac{|H(\omega_i)|^2}{2} = \frac{|H(0.5)|^2}{2} = \frac{(0.997)^2}{2} = 0.472, \tag{14.5.10}$$

and the power in the output signal $y_2(t)$ is

$$P_{y_2} = \frac{|H(10)|^2}{2} = \frac{(0.1)^2}{2} = 0.005. \tag{14.5.11}$$

Both of the input signals in (14.5.6) have the same power. The power of the output signal $y_2(t)$, however, has been dramatically reduced compared to the power in the output signal $y_1(t)$.

These results can be presented graphically. The frequency response function squared for the *RC* network and the power of the input and output signals are shown in Figure 14.5.2. The input signals in (14.5.6) are (trivial) Fourier series. The power spectrum of the input and output signals are to be interpreted in this sense. Consider first the component of the input signal at $\omega_1 = 0.5$ rad/sec. The power of the output signal, the circle at $\omega_1 = 0.5$ rad/sec, is obtained by *multiplying* the power of the input signal, the black dot at $\omega_1 = 0.5$ rad/sec, with the magnitude squared of the frequency response function at that frequency. A similar analysis can be applied to the second frequency component at $\omega_2 = 10$ rad/sec.

For the signal $x_1(t)$ the power in the output signal has not been significantly changed over the power in the input signal, because the frequency response function is close to unity at $\omega_1 = 0.5$ rad/sec. For the signal $x_2(t)$, the output signal has been significantly attenuated and the power reduced because the frequency response function is small at $\omega_2 = 10$ rad/sec.

▲▲

14.5.2 Bandwidth of a System

The difference in the way the two signals in Example 14.5.1 propagate through this system can be explained by the frequency response function of the system. From (14.5.4) the ratio of the power of the output signal to the power of the input signal is

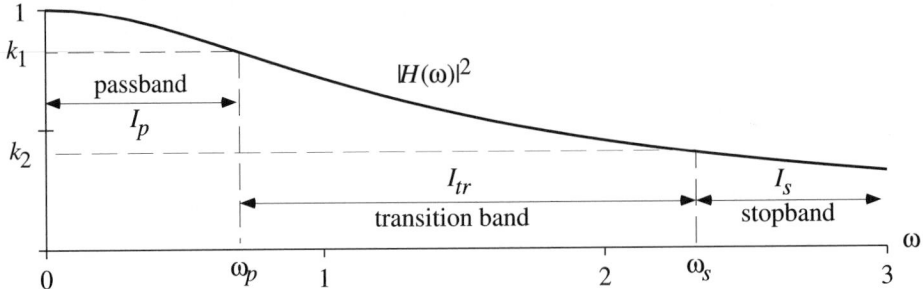

Figure 14.5.3 Passband, Stopband, and Transition Band of a RC Network

$$\frac{P_y}{P_x} = |H(\omega_i)|^2. \tag{14.5.12}$$

First, (14.5.12) tells us that the effect of the system on each of these signals is determined by the *frequency* of the input signal. Furthermore, the relative change in the power of these two signals propagating through the system can be predicted from the *shape* of the magnitude squared of the frequency response function $|H(\omega)|^2$. This observation suggests that we decompose the frequency response function into frequency bands according to the systems effect on the input signal at that frequency.

The frequency response function of an RC network is shown in Figure 14.5.3. Considering Example 14.5.1 we identify three frequency bands based on the frequency response function. These frequency intervals are labeled in Figure 14.5.3.[8] These bands are:

Passband: In the passband the frequency response function is close to unity. If the frequency of the input sinusoid, ω_i, is in frequency interval I_p, the corresponding output signal has approximately the same power as the input signal. The frequency of the input signal $x_1(t)$ is in this band of the RC filter.

Stopband: In the stopband the frequency response function is close to zero. If the frequency of the input sinusoid, ω_i, is in frequency interval I_s, the corresponding output signal experiences large attenuation and the power of the output signal is significantly reduced over the power of the input signal. The frequency of the input signal $x_2(t)$ is in this band.

Transition Band: In the transition band the frequency response function is neither close to unity nor close to zero. If the frequency of the input sinusoid, ω_i, is in frequency interval I_{tr}, the corresponding output signal experiences some attenuation and the power of the output signal is "somewhat" reduced over the power of the input signal.

---

[8] This figure is not drawn to scale. Frequently, the frequency response function is much closer to zero in the stopband.

Notation: The edge frequency of the passband is denoted by ω_p. The edge frequency of the stopband is denoted ω_s.

In Figure 14.5.3 we have shown the passband to be at low frequency followed by the stopband. However, the passband can be located anywhere on the frequency axis. Also it can also be split into several disconnected intervals. A more formal definition of these bands is as follows. Let $H(\omega)$ be the frequency response function a system such that

$$\max_{\omega \geq 0} |H(\omega)| = H_{max}. \tag{14.5.13}$$

Definition 14.5.2: Let k_1 and k_2 be two positive real numbers that satisfy $0 \leq k_2 \leq k_1 < H_{max}^2$.

(a) The <u>passband</u> I_p for the system is the set of frequencies that satisfy

$$I_p = \left\{ \omega > 0 \mid k_1 \leq |H(\omega)|^2 \leq H_{max}^2 \right\}.$$

(b) The <u>stopband</u> I_s for the system is the set of frequencies that satisfy

$$I_s = \left\{ \omega > 0 \mid 0 \leq |H(\omega)|^2 \leq k_2 \right\}.$$

(c) The <u>transition band</u> I_{tr} for the system is the set of frequencies that satisfy

$$I_{tr} = \left\{ \omega > 0 \mid k_2 < |H(\omega)|^2 < k_1 \right\}.$$

▲▲

Suppose that $k_1 = k_2$, so that the length of the transition band is zero.[9] Then the length and location of the passband defines the stopband. That is, the passband, in essence, defines the system. This concept is captured in the following definition.

Definition 14.5.3: Let the passband be a single connected interval

$$I_p = \left\{ \omega \geq 0 \mid \omega_0 \leq \omega \leq \omega_1 \right\}.$$

The <u>bandwidth</u> β of a system is defined as the length of the passband

$$\beta = \omega_1 - \omega_0.$$

Since the bandwidth is defined using k_1, this bandwidth is sometimes called the <u>k_1 power</u> bandwidth.

▲▲

---

[9] Often the transition band is ignored.

Terminology: The bandwidth of a *system*, β, should not be confused with the bandwidth of a *signal*, *B*, in Definition 8.3.5.   To equate the two definitions is to compare apples with oranges.   Unfortunately, these two definitions are entrenched in the literature.   It is left to the practitioner to distinguish the two concepts.   Usually, the right definition can be determined from context.

We make the following observations about the bandwidth of a system.

1.   The bandwidth of a system is defined by comparing an input signal with the output signal.   Here again a system characteristic is derived from the system's effect on the input and output signals.   This observation is consistent with our method of analysis outlined in the introduction to this section.
2.   The passband, stopband, and transition band are defined for *positive* frequencies only.
3.   These bands are defined based on shape of the magnitude of the frequency response function.   That is, they are relative to the system at hand; they are not absolute.
4.   The dividing line between the frequency intervals in Definition 14.5.2 is determined by the numbers k_1 and k_2.   There are no fixed rules for choosing these numbers.   There are a variety criteria depending on the situation.   One criterion is discussed in the next subsection.
5.   The bandwidth of a system is frequently defined in a variety of ways for a variety of reasons.   We have given only one way to define it.   If the definition of this term is important, the exact definition in the context of the application should be verified.
6.   Usually, the passband and/or stopband are the intervals of interest while the transition band is of interest primarily in filtering applications.   So if a frequency is specified as a boundary between the passband and the stopband without mentioning the transition band, this usage is with the understanding that the transition band straddles the dividing frequency.

14.5.3 Half-Power Bandwidth

One common way to define the bandwidth is to use the half-power frequency.   Let the maximum value of the frequency response function be H_{max} as in (14.5.13).   Let ω_{max} be any frequency that corresponds to this peak.   For the input signal

$$x_{max}(t) = \sqrt{2}\,\cos(\omega_{max}t), \tag{14.5.14}$$

the output signal is

$$y_{max}(t) = \left|\sqrt{2}H(\omega_{max})\right|\cos(\omega_{max}t + \angle H(\omega_{max})). \tag{14.5.15}$$

The power in the output signal is

$$P_{max} = \frac{1}{2}\left|\sqrt{2}H(\omega_{max})\right|^2 = \left|H(\omega_{max})\right|^2 = H_{max}^2. \tag{14.5.16}$$

Consider a second input signal to this system

$$x_{hp}(t) = \sqrt{2}\cos(\omega_{hp}t). \tag{14.5.17}$$

Note that the power in the two input signals (14.5.14) and (14.5.17) is the same. The corresponding output signal is

$$y_{hp}(t) = \left|\sqrt{2}H(\omega_{hp})\right|\cos\left(\omega_{hp}t + \angle H(\omega_{hp})\right). \tag{14.5.18}$$

Suppose further that the power in the output signal (14.5.18) is 1/2 of the power of the output signal in (14.5.15). That is,

$$P_{y_{hp}} = \frac{1}{2}\left|\sqrt{2}H(\omega_{hp})\right|^2 = \frac{1}{2}P_{max}. \tag{14.5.19}$$

From (14.5.19) we see that at the frequency ω_{hp} the frequency response function must satisfy

$$\left|H(\omega_{hp})\right|^2 = \frac{1}{2}H_{max}^2 \quad \text{or} \quad \left|H(\omega_{hp})\right| = \frac{H_{max}}{\sqrt{2}}. \tag{14.5.20}$$

If (14.5.20) holds, the output signal in (14.5.18) has one-half the power of the output signal in (14.5.15).

Definition 14.5.4: Given a frequency response function $H(\omega)$ for the system (14.5.1) assume that H_{max} satisfies (14.5.13). Any frequency ω_{hp} that satisfies

$$\left|H(\omega_{hp})\right| = \frac{H_{max}}{\sqrt{2}}$$

is called a <u>half-power</u> frequency.
▲▲

Definition 14.5.5: Define the constants k_1 and k_2 in Definition 14.5.3 using the half-power frequency

$$\left|H(\omega_{hp})\right|^2 = \frac{H_{max}^2}{2} = k_1 = k_2.$$

Then the bandwidth of the system, β_{hp}, is called the <u>half-power</u> bandwidth.
▲▲

Example 14.5.6: Consider the magnitude of the frequency response function RC network in Example 14.5.1 and shown in Figure 14.5.4.

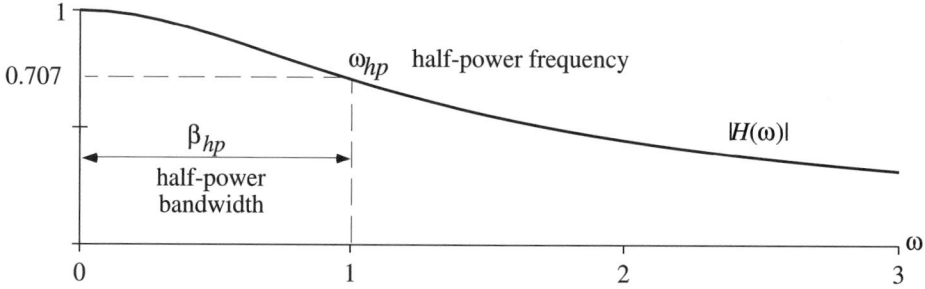

Figure 14.5.4 Half-Power Bandwidth of an *RC* Network

The maximum of the magnitude of the frequency response is $H_{max} = 1$. The half-power frequency is determined from

$$\left[\frac{1}{\sqrt{\omega_{hp}^2 + 1}} \right]^2 = \frac{H_{max}^2}{2} = k_1 = k_2 = \frac{1}{2}, \quad \Rightarrow \quad \omega_{hp} = 1. \tag{14.5.21}$$

The half-power bandwidth of the *RC* network is $\beta_{hp} = 1 - 0 = 1$ rad/sec.

▲▲

Expressed in decibels the definition of the half power frequency in (14.5.20) becomes

$$10 \log |H(\omega_{hp})|^2 = 10 \log \frac{H_{max}^2}{2} = 20 \log(H_{max}) - 20 \log\left(\frac{1}{2}\right) \tag{14.5.22}$$

$$= 20 \log(H_{max}) - 3.01 \; dB.$$

Hence, the power at the half-power frequency is 3 dB below the maximum of magnitude of the frequency response function, expressed in decibels.

Terminology: Because of (14.5.22), the half-power frequency ω_{hp} is sometimes called the −3 dB frequency.

The expression of the half power frequency in terms of decibels makes determination of the bandwidth of a system from its magnitude Bode plot convenient. First, determine the peak value of the Bode plot. Then determine the frequencies which are 3 dB down from the peak. From these frequencies determine the passband and so the bandwidth.

14.6 IDEAL FILTERS

14.6.1 Introduction

In the last section we developed the notion of the bandwidth of a system. The bandwidth depends on the width of the passband. In this section we further refine these notions by naming a system based on the location of the passband. We accomplish this task by introducing the concept of an ideal system. An ideal system precisely locates the passband of the system. Ideal systems are not encountered in the physical world, but they are a useful for explaining filtering and sampling.

14.6.2 Distortionless Filters

We begin with the simplest form of an idealized system. Consider a system defined by the frequency response function (Fourier transfer function)

$$\frac{Y(\omega)}{X(\omega)} = H(\omega) = Ke^{-j\omega t_d} . \tag{14.6.1}$$

If the input signal to this system is

$$x(t) = \cos(0.5t) + \cos(10t), \tag{14.6.2}$$

then by the frequency response theorem (Theorem 14.1.1/14.2.1) the output signal is

$$y_I(t) = K\cos\big(0.5(t - t_d)\big) + K\cos\big(10(t - t_d)\big). \tag{14.6.3}$$

If we compare the two components of the input signal, they have the same amplitude and the relative phase between the two components is zero. Hence, the output signal is just the input signal shifted in time.

Now suppose that the signal in (14.6.2) is the input signal into the RC network in Example 14.4.1. The frequency response function of the RC network is

$$H(\omega) = \frac{1}{\sqrt{1+\omega^2}}\, e^{-j\tan^{-1}(\omega)}. \tag{14.6.4}$$

The output signal of the RC network is

$$y_{rc}(t) = (0.995)\cos(0.5t - 0.46) + (0.01)\cos(10t - 1.47). \tag{14.6.5}$$

If we compare the two components of the output signal (14.6.5), we see that the ratio of the amplitude of the two components is not unity. Furthermore, the relative phase of the two components is nonzero. In contrast to (14.6.3) the RC filter has treated the two components of the input signal differently based on the frequency of each component. The system in (14.6.1) transmits the signal with only a time delay, while the RC network changes, or distorts, the input signal as it propagates through the system. Based on this observation, we introduce the following definition of the system in (14.6.1).

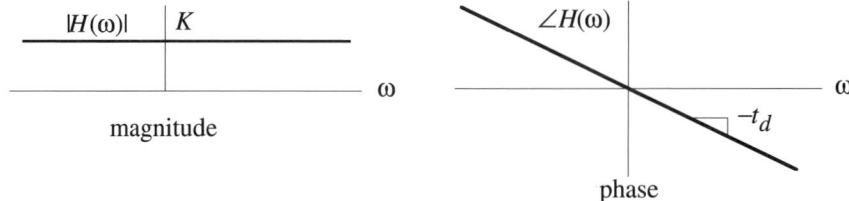

Figure 14.6.1 Frequency Response Function of a Distortionless Filter

Definition 14.6.1: A system is <u>distortionless</u> if its frequency response function is

$$\frac{Y(\omega)}{X(\omega)} = H(\omega) = Ke^{-j\omega t_d}.$$

▲▲

The frequency response function of a distortionless filter is shown in Figure 14.6.1. Suppose that the input signal into a distortionless filter is $x(t)$. Then using Property 7.5.12, the time shift property of the Fourier transform, the output signal is

$$y(t) = Kx(t - t_d). \tag{14.6.6}$$

Hence, the slope of the phase in Figure 14.6.1, the group delay, is equivalent to a pure time delay between the input and output signal of the distortionless filter. The definition of a distortionless filter includes a time delay in the passband because all real filters have some group delay associated with them. See, for example, Figure 14.4.1. A distortionless filter is an ideal filter against which the performance of physically realizable filters are often compared.

14.6.3 Ideal Filters

A distortionless filter (or system) attenuates a signal and introduces a phase delay. No signal is prevented from passing through the system. If we compare this distortionless filter with the frequency response function of an *RC* filter, we see that

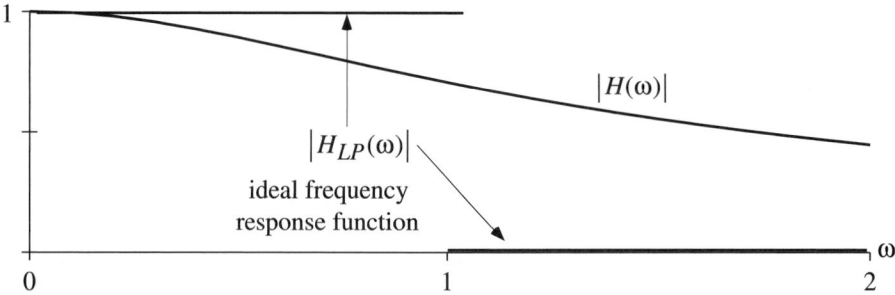

Figure 14.6.2 Magnitude of the Frequency Response Function of the *RC* Network and an Ideal Lowpass Filter

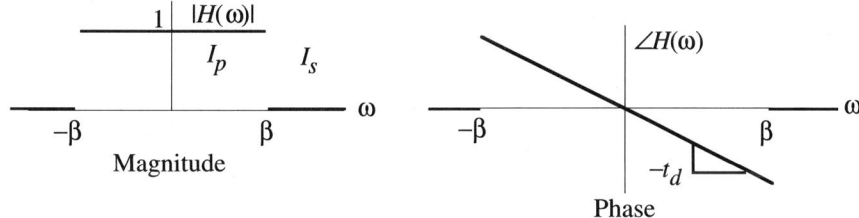

Figure 14.6.3 Frequency Response Function of an Ideal Lowpass Filter

they differ significantly at high frequencies. To better approximate the frequency response function of the *RC* filter, we modify the frequency response function of the distortionless filter by introducing a stopband at high frequencies. The frequency response function of the modified filter is shown in Figure 14.6.2 along with the frequency response function of the *RC* network. Based on Figure 14.6.2, we give the system with the modified frequency response function a name.

Definition 14.6.2: The frequency response function for an <u>ideal lowpass filter</u>, $H_{LP}(\omega)$, is given by

$$|H_{LP}(\omega)| = \begin{cases} 1, & 0 \le |\omega| \le \beta \\ 0, & \beta < |\omega| \end{cases}$$

$$\angle H_{LP}(\omega) = \begin{cases} -\omega t_d, & 0 \le |\omega| \le \beta \\ 0, & \beta < |\omega| \end{cases}$$

▲▲

An ideal lowpass filter is shown in Figure 14.6.3. Suppose that the input signal to an ideal lowpass filter is

$$x(t) = A\cos(\omega_i t), \quad 0 \le \omega_i < \beta. \tag{14.6.7}$$

Then the output signal from the filter can be derived using the frequency response theorem, Theorem 14.1.1 or 14.2.1. The output signal is

$$y(t) = A|H(\omega_i)|\cos(\omega_i t + \angle H_{LP}(\omega_i)) = A\cos(\omega_i t - \omega_i t_d) = A\cos(\omega_i(t - t_d)). \tag{14.6.8}$$

We see that the input signal (14.6.1) propagates through the system unattenuated, but it incurs a time delay of t_d. For signals that satisfy (14.6.7), the system appears to be a distortionless filter. The passband of an ideal lowpass filter is

$$I_p = \{\omega|\ 0 \le |\omega| \le \beta\}. \tag{14.6.9}$$

Terminology: Since the passband is bounded below by zero, the filter is called a *lowpass* filter.

If the frequency of the input signal satisfies $\omega_i > \beta$, then the output signal from the filter will be identically zero. Hence, the stopband of an ideal lowpass filter is

$$I_s = \left\{ \omega \mid \ \beta \geq |\omega| \right\}. \tag{14.6.10}$$

Terminology: This filter has no transition band. Furthermore, this filter transmits signals in its passband perfectly, and it blocks signals in its stopband entirely. Hence, this filter is called a *ideal* lowpass filter.

The classification of the lowpass filter is based on the location of the passband and stopband of the frequency response function. Similarly, other ideal filters can be defined by the rearranging the passband and the stopband. All of the remarks on ideal lowpass filters also pertain to the following filters with obvious modifications.

Definition 14.6.3: The frequency response function for an <u>ideal bandpass filter</u> $H_{BP}(\omega)$ is given by

$$|H_{BP}(\omega)| = \begin{cases} 1, & 0 < \beta_1 \leq |\omega| \leq \beta_2 < \infty \\ 0, & \text{otherwise} \end{cases}$$

$$\angle H_{BP}(\omega) = \begin{cases} -\omega t_d, & 0 < \beta_1 \leq |\omega| \leq \beta_2 < \infty \\ 0, & \text{otherwise} \end{cases}$$

▲▲

The frequency response function of an ideal bandpass filter is shown in Figure 14.6.4. The bandwidth of an ideal bandpass filter is β, the length of the passband

$$\beta = \beta_2 - \beta_1.$$

Definition 14.6.4: The frequency response function for an <u>ideal highpass filter</u> $H_{HP}(\omega)$ is given by

$$|H_{HP}(\omega)| = \begin{cases} 0, & 0 \leq |\omega| \leq \beta \\ 1, & \beta < |\omega| \end{cases}$$

$$\angle H_{HP}(\omega) = \begin{cases} 0, & 0 \leq |\omega| \leq \beta \\ -\omega t_d, & \beta < |\omega| \end{cases}$$

▲▲

The frequency response function of an ideal highpass filter is shown in Figure 14.6.5. Obviously, the bandwidth of this ideal filter is infinite. Usually, a highpass filter is specified by the frequency that separates the stopband from the passband.

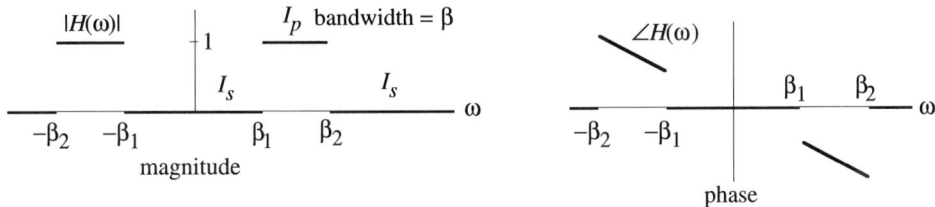

Figure 14.6.4 Frequency Response Function of an Ideal Bandpass Filter

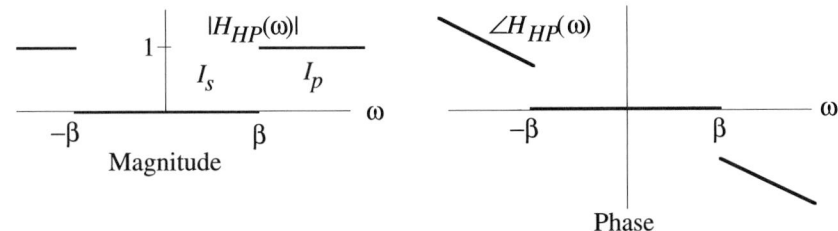

Figure 14.6.5 Frequency Response Function of an Ideal Highpass Filter

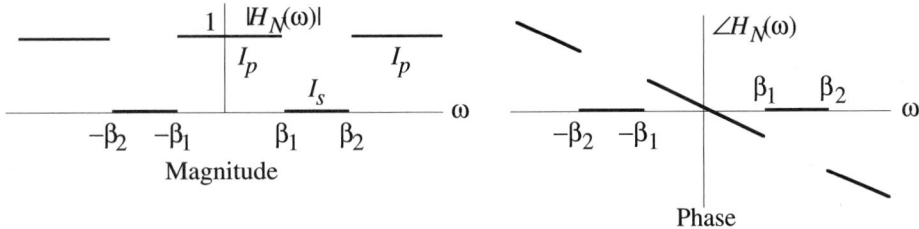

Figure 14.6.6 Frequency Response Function of an Ideal Notch Filter

Definition 14.6.5: The frequency response function for an <u>ideal notch filter</u> $H_N(\omega)$ is given by

$$\left|H_N(\omega)\right| = \begin{cases} 0, & 0 < \beta_1 \le |\omega| \le \beta_2 < \infty \\ 1, & \text{otherwise} \end{cases}$$

$$\angle H_N(\omega) = \begin{cases} 0, & 0 < \beta_1 \le |\omega| \le \beta_2 < \infty \\ -\omega t_d, & \text{otherwise} \end{cases}$$

▲▲

The frequency response function of an ideal notch filter is shown in Figure 14.6.6. A notch filter also has an infinite bandwidth. This type of filter is usually specified by the width and location of the notch (stopband).

Terminology: It is common practice to classify physical systems by the characteristics of their frequency response function. That is, the system is classified based on the best match between the actual frequency response function and one of the ideal filters introduced above.

14.7 INTRODUCTION TO FILTERING

14.7.1 Introduction

Up to this point in the text we have discussed system theory from the point of view of analysis.   That is, given an observed physical process, we try to develop a representation of the signal or system such that the model matches the observed data.   An equally important problem, however, is the reverse process.   We start with a representation of a system, and we try to construct a physical process such that the resulting system matches the representation with which we started.   We call this reverse process system synthesis.   A typical example of the synthesis of a system is the fabrication of an electronic circuit to process an electric signal in a certain way.

The synthesis of a system is, of course, the essence of engineering design.   In this section we will introduce filter design, which is one class of system synthesis techniques.   Filter design is characterized by the way we describe how the system is to perform.   (In order to build a system we must determine what we want the system to do.)   The following steps loosely describe the design process which would lead to the desired filter.

Procedure 14.7.1: **Filter Design**
Step 1   From the problem statement determine the specifications on the desired shape of the magnitude (or phase, or both) of the frequency response function.
Step 2   Find a transfer function that has a frequency response function which meets the specifications of Step 1.
Step 3   Synthesize a system (i.e., hardware or software) such that the system has the given transfer function of Step 2.

▲▲

From Procedure 14.7.1 we see that in filter design the system's performance specifications are given in terms of the frequency response function.   Roughly, the filter design process translates specifications on the frequency response function into a piece of hardware or software that has a frequency response function which meets the specifications.   Step 3, the translation of the transfer function into hardware is, of course, specific to the type of hardware that is being used to construct the system. Step 3 is beyond the scope of this text, and we won't discuss it further.   The part of the filter design process which we will concentrate on in this section is Step 1 and 2 of Procedure 14.7.1.   To give the flavor of the filter design process, next we will discuss a Butterworth filter.

14.7.2 An Example

The following filter is one of the most common and widely used filters.

Definition 14.7.2:   A <u>Butterworth filter</u> is a system where the magnitude of the frequency response function satisfies

$$|H_{bw}(\omega)| = \frac{1}{\sqrt{1+\left(\dfrac{\omega}{\omega_s}\right)^{2n}}}.$$

▲▲

The frequency response function of a Butterworth corresponds to a known transfer function (although it is not trivial to derive that transfer function). For example, if we take $n = 3$ and $\omega_c = 1$, the transfer function of this Butterworth filter is

$$\frac{Y(s)}{X(s)} = \frac{1}{s^3 + 2s^2 + 2s + 1}. \qquad (14.7.1)$$

The magnitude of the frequency response function is shown in Figure 14.7.1.

First, from Figure 14.7.1 we see that this system is a lowpass filter. The frequency response function in Definition 14.7.2 will always give rise to a lowpass filter. At the end of this section we will discuss the other types of filters.

The frequency response function in Definition 14.7.2 has a fixed structure with two parameters: n and ω_c. The parameter n controls the order of the transfer function corresponding to the frequency response function. Taking $n = 3$ yields the third-order transfer function in (14.7.1). Figure 14.7.2 shows the frequency response function of several filters with different orders. The parameter ω_c controls the bandwidth of the filter. Taking $\omega_c = 1$ we get the frequency response function in Figure 14.7.1. Figure 14.7.3 shows the frequency response function of several filters with different bandwidths.

The following example shows how we would use a Butterworth filter in a filter design problem.

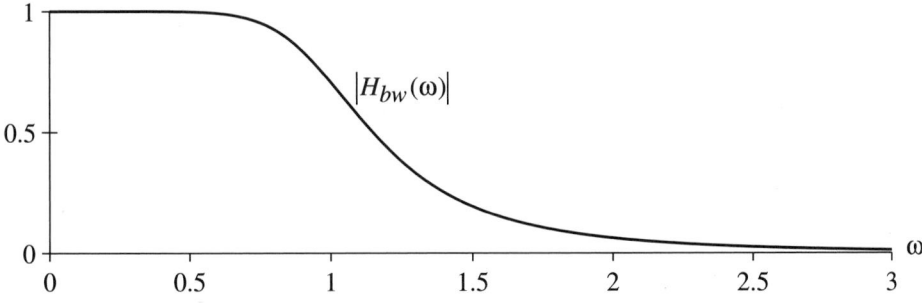

Figure 14.7.1 Magnitude of the Frequency Response Function
of a Butterworth Filter

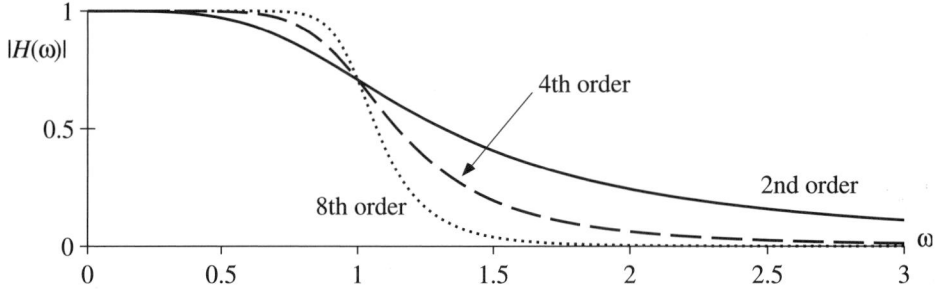

Figure 14.7.2 Magnitude of the Frequency Response Function for Several Butterworth Filters of Different Orders

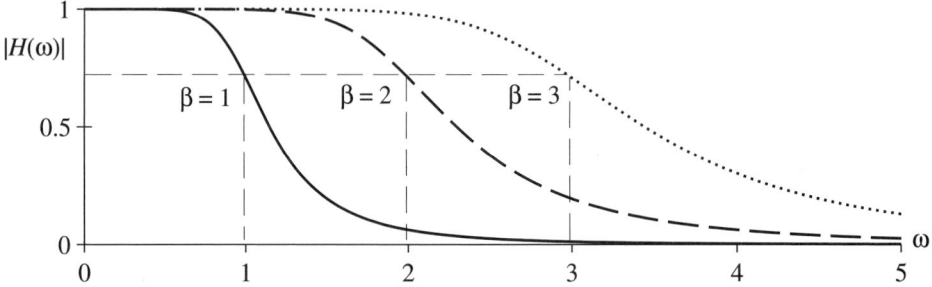

Figure 14.7.3 Magnitude of the Frequency Response Function for Several Butterworth Filters of Different Bandwidths

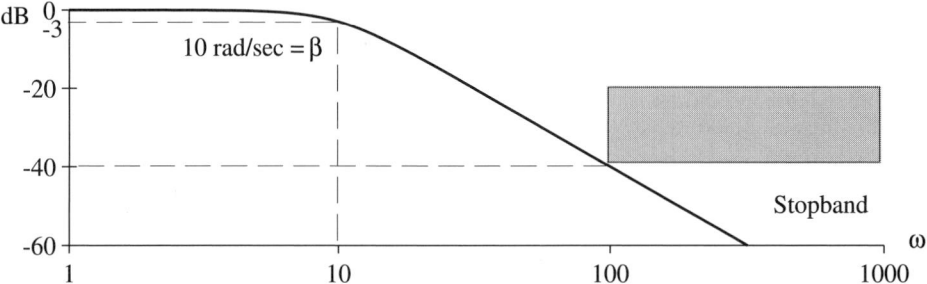

Figure 14.7.4 Bandwidth and Stopband of a Butterworth Filter

Example 14.7.3: Suppose we are to design a Butterworth filter whose bandwidth is 10 rad/sec and whose frequency response function is at least -40 dB for frequencies above 100 rad/sec. These specifications on the frequency response function (Step 1 of Procedure 14.7.1) are shown in Figure 14.7.4. Since the role of the bandwidth is clearly specified in the definition of the Butterworth frequency response function, it only remains to select the order of the filter. From the stopband specifications we have that

$$|H(\omega)| = \frac{1}{\sqrt{1 + \left(\dfrac{\omega}{10}\right)^{2n}}} < 0.01, \quad \omega \geq 100. \tag{14.7.2}$$

Clearing the fraction in (14.7.2) we obtain

$$10^4 - 1 \approx 10^4 \leq \left(\frac{\omega}{10}\right)^{2n}. \tag{14.7.3}$$

Inserting $\omega = 100$ into (14.7.3) and taking the log of both sides we obtain

$$4\log(10) < 2n(\log(100) - \log(10)), \quad \Rightarrow \quad n = 2. \tag{14.7.4}$$

The frequency response function is shown in Figure 14.7.4. The transfer function[10] is given by

$$\frac{Y(s)}{X(s)} = \frac{100}{s^2 + 14.14s + 100}. \tag{14.7.5}$$

▲▲

This example illustrates some of the essentials of the filter design process.

Step 1 The performance specifications for the system are given in terms of the bandwidth and the stopband attenuation. In order to meet these specifications we use a frequency response function for a catalog of such functions. (We will give other frequency response functions below.) The shape of the frequency response function is controlled by a few parameters. These parameters are adjusted until the shape of the magnitude of the frequency response function meets the requirements of the system.

Step 2 Each frequency response function corresponds to a known transfer function. From the parameters of the frequency response function the transfer function is calculated. This parameterization of the frequency response function makes the filter design process very efficient. The mathematical relationship between the frequency response function and the transfer function is in the domain of filtering theory.

Once the transfer function in Step 2 is determined, the synthesis of the system depends entirely on the physics of the system. Filter design is a well-developed area of analog electronics; it is called network synthesis. Over the last thirty years digital electronics have been developed that perform like analog filters (but better in many cases). Step 3 of filter design is mostly outside the scope of this text, and we will not pursue it any further.

---

[10] MATLAB was used to generate the transfer function.

Next we will discuss the specifications by which the frequency response function of the desired filter is prescribed. The form of these specifications is driven by the way the frequency response functions are parameterized as suggested by Example 14.7.3.

14.7.3 Determination of the Frequency Response Function

Specifications The simplest specification on the frequency response function is of the following form. We want the filter to pass sinusoids of all frequencies below a certain frequency ω_c and we want the filter to block all sinusoids with frequencies above this frequency. Mathematically, we want the magnitude of the frequency response function to satisfy

$$H_d(\omega) = \begin{cases} 1, & |\omega| < \omega_c \\ 0, & |\omega| > \omega_c \end{cases} \qquad (14.7.6)$$

This frequency response function defines an ideal lowpass filter with zero phase. (See Definition 14.5.2.)

Definition 14.7.4: The frequency ω_c that determines the dividing line between the passband and the stopband is called the <u>cutoff</u> frequency.

▲▲

Terminology: This filter will pass sinusoids with certain frequencies and block sinusoids with other frequencies. We say this filter is a <u>frequency selective</u> filter.

To begin, we might take the naive approach, and try to find a circuit for the ideal filter in (14.7.6). The impulse response function of this filter is

$$h_d(t) = \frac{\omega_c}{\pi} \, \mathrm{Sa}(\omega_c t). \qquad (14.7.7)$$

Unfortunately, this impulse response function is nonzero for negative time, and so it is noncausal (See Theorem 13.5.1.) Hence, a circuit with the frequency response function in (14.7.6) corresponding to an ideal lowpass filter doesn't exist. Mathematically, the reason the frequency response function in (14.7.6) leads to a noncausal impulse response function is that the frequency response function (14.7.6) has a discontinuity in it.[11] So we must start with a continuous frequency response function if we want the filter design process to lead to a circuit (causal system).

The ideal frequency response function captures exactly the specifications we would like our circuit to have, but it doesn't correspond to a causal system. Therefore, we will relax our specifications on the frequency response function to accommodate reality. The following discussion of frequency response specifications uses the concepts of the passband, stopband, and transition band introduced in Section 14.5 in connection with an *RC* filter.

---

[11] This observation comes from the theory of complex functions.

The ideal frequency response has three components: (1) the edge frequency of the passband is specified by the cutoff frequency ω_c, (2) the magnitude of the frequency response function is a constant one in the passband, and (3) the magnitude of the frequency response function is zero in the stopband. The specifications for the physically realizable filter must come as a relaxation of these three specifications. First, the transition in the frequency response function from the passband to the stopband can't be discontinuous. Therefore, we introduce a transition band with a finite bandwidth to allow the frequency response function to change smoothly from the passband to the stopband. The cutoff frequency falls in the transition band. Frequently, the cutoff frequency is chosen as the bandwidth. Second, we allow the frequency response function to deviate from unity (by a small amount) in the passband. Third, we also allow variations in the stopband because we can't force the frequency response function to be identically zero. These relaxed requirements on the magnitude of the frequency response function are generally shown as constraints on the frequency response function as shown in Figure 14.7.5.

Terminology: We call the variations of the magnitude in the passband and stopband <u>ripples</u>. The rate of decrease of the magnitude function in the transition band is called the <u>rolloff</u> of the frequency response function.

Notation: The maximum allowable ripple in the passband is denoted by ε_p. The maximum allowable ripple in the passband is denoted by ε_s. The edge frequency of the passband is denoted ω_p. The edge frequency of the stopband is denoted ω_s.

In practice, not all specifications are applicable to every filter. Generally, there is a hierarchy of specifications with some specifications being more important than others. This hierarchy is dictated by the physical problem at hand.

We have discussed specifications on the magnitude of the frequency response function. On some filters the specifications are given on the phase. Generally, these specifications call for a constant group delay in the passband. These specifications can be formalized in the same way we have formalized the specifications on the magnitude of the frequency response function.

Selection of the Frequency Response Function   After developing a set of specifications, the next step is to translate the frequency response function into a

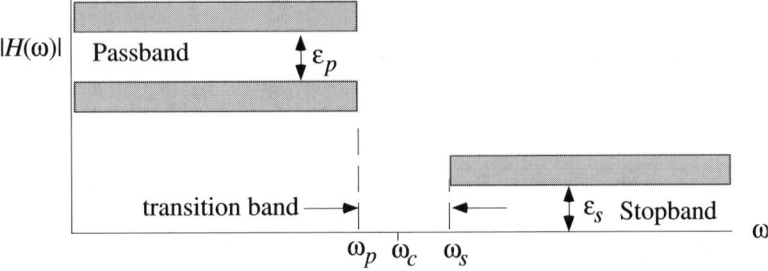

Figure 14.7.5 Specifications for a Filter

transfer function. This step is accomplished by selecting a frequency response function whose transfer function is known. There is a whole catalog of these frequency response functions for different filters, all with different characteristics. We will discuss several of these filters next.

The first filter we will discuss is the Butterworth filter. This filter is defined in Definition 14.7.2. We make the following observations about the Butterworth filter.

1. The magnitude of the frequency response function is maximally flat in the passband.
2. The 3 dB bandwidth is usually adjusted by changing the cutoff frequency ω_c.
3. The width of the transition band is determined by the order of the system, n. As the order of the system increases the transition band becomes narrower, approaching the ideal filter. As the order of the filter increases, the complexity of the system increases.
4. The Butterworth filter is completely specified by the order n and the cutoff frequency ω_c. So when we design a Butterworth filter we need only specify these two parameters.

The Butterworth filter closely approximates the flat magnitude response in the passband of the ideal lowpass filter. Because of this passband characteristic, the Butterworth filter exhibits a rather slow rolloff so that the transition band can be quite large. For some filter designs this specification is of less importance than the stopband. The second type of filter we will consider has a faster rolloff in the transition band at the expense of ripples in the passband or stopband. This type of filter is called a Chebyshev filter.

Definition 14.7.5: The frequency response function of a <u>Chebyshev Type I</u> filter is

$$\left|H_{cb1}(\omega)\right|^2 = \frac{1}{1 + \varepsilon_p C_n^2\left(\dfrac{\omega}{\omega_s}\right)}.$$

The frequency response function of a <u>Chebyshev Type II</u> filter is

$$\left|H_{cb2}(\omega)\right|^2 = \frac{1}{1 + \varepsilon_s\left[\dfrac{C_n^2(\omega_s/\omega_p)}{C_n^2(\omega_s/\omega)}\right]}.$$

▲▲

Terminology: The polynomials $C_n(\omega)$ are called an <u>nth-order Chebyshev</u> polynomial. These polynomials are defined by the recursion

$$C_0(\omega) = 1, \quad C_1(\omega) = \omega C_0(\omega), \quad C_{n+1}(\omega) = 2\omega C_n(\omega) - C_{n-1}(\omega) \tag{14.7.8}$$

The magnitude of the frequency response function of a Type I and Type II Chebyshev filter are shown in Figure 14.7.6.

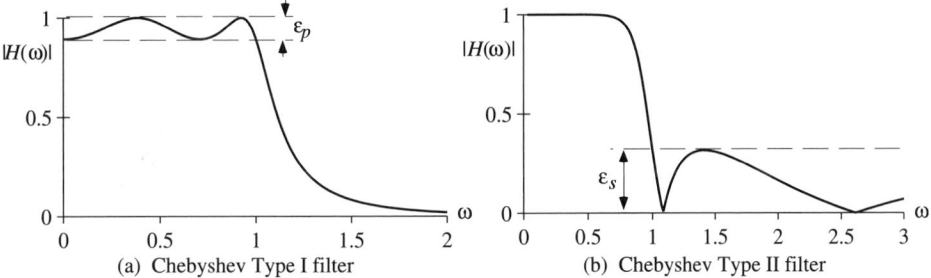

Figure 14.7.6 Magnitude of the Frequency Response Function
of a Chebyshev Filter

Chebyshev filters have the following properties.

1. Chebyshev filters have a shorter transition bandwidth than Butterworth filters with a steeper rolloff of the magnitude of the frequency response function. The rolloff can be adjusted to a certain extent by using the parameter ε_p which controls the ripple in the passband. The more ripple that is allowed, the faster the rolloff. The order of the filter also affects the rolloff, as with the Butterworth filter.
2. The ripple in the stopband of a Chebyshev II filter is controlled by the parameter ε_s. As the stopband ripple is decreased, the transition bandwidth is decreased.
3. Each of the two types of Chebyshev filters is flat in either the passband or stopband. The flat passband introduces less distortion into the signal, and the flat stopband gives better attenuation in the stopband. The type of filter that is selected depends on which band is more important.
4. The bandwidth is determined by the cutoff frequency ω_c.

The third class of filters that are widely used is elliptic filters. The mathematical definition of an elliptic filter is too complicated to be given here. A typical lowpass frequency response function of an elliptic filter is shown in Figure 14.7.7. This class of filters allows ripple in both the passband and the stopband. In return, the elliptic filters have a very narrow transition band.

Example 14.7.6: A Comparison of the Filters The magnitude of the frequency response function of each of the three different filter types is shown in Figure 14.7.8. Each filter is 4th order and each has a bandwidth of 1 rad/sec. The three filter types discussed above differ in three respects. First, their transition bandwidth is different as can be seen in Figure 14.7.8. The Butterworth filters have the largest transition band while elliptic filters have the smallest transition bandwidth. The short transition bandwidth of the elliptic filter, however, is paid for in ripples in the passband and stopband. In the stopband attenuation the Chebyshev I filter is better than both the elliptic and the Butterworth filter. The Butterworth filter introduces the least passband distortion to the signals because it is flat in the passband. ▲▲

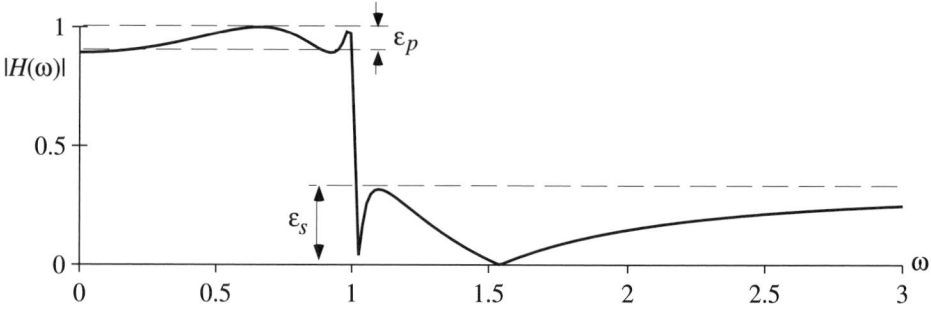

Figure 14.7.7 Magnitude of an Elliptic Filter

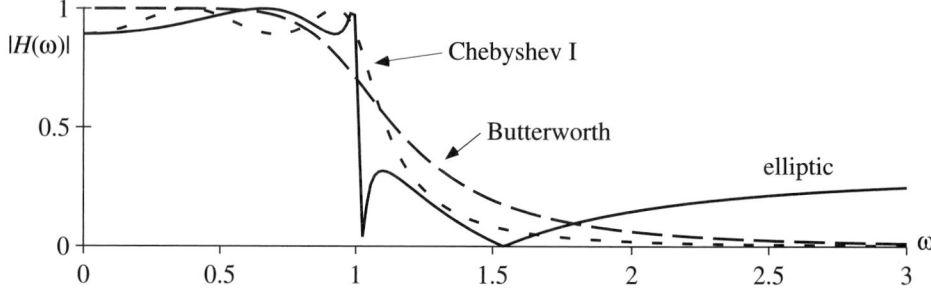

Figure 14.7.8 Frequency Response Function of a Butterworth and Chebyshev Filter

Calculation of the Transfer Function The second step in the filter design process is to calculate the transfer function from the frequency response function. If the frequency response function is one of the standard filters above, then this calculation is nontrivial. In fact, today these calculations are automated in a CAD package. The computer package will automatically generate the transfer function of the filter when the parameters of the filter are specified. In fact, many CAD packages automatically calculate the parameters given the specifications. CAD is having a significant impact on the design of filters in this respect. The filters themselves (Butterworth, Chebyshev, etc.), however, are still widely used. The last subsection of this section describes how to use MATLAB to generate these filters.

14.7.4 Frequency Transformations

Thus far we have discussed only lowpass filters. We are also interested in bandpass, highpass, and notch filters. The design of a filter transfer function is generally carried out in two steps. First, a lowpass filter with a bandwidth of $\beta = 1$ rad/sec is designed. This transfer function is called a <u>lowpass prototype</u> filter. Then this filter is transformed into the desired filter, possibly of a different type. Next we will describe how to transform lowpass prototype filters into bandpass, notch, and highpass filters. The next example will show how to convert a lowpass filter to a bandpass filter.

Example 14.7.7: Consider a second-order lowpass prototype Butterworth filter with a bandwidth of 1 rad/sec

$$\frac{Y(s)}{X(s)} = \frac{1}{s^2 + 1.414s + 1}. \tag{14.7.9}$$

We will construct a bandpass filter by replacing s in (14.7.9) with a suitable complex function. Let ω_{cf} be the center frequency of the bandpass filter and let β be the 3-dB bandwidth. Define the transformation

$$s = \frac{\tilde{s}^2 + \omega_{cf}^2}{\beta\tilde{s}}. \tag{14.7.10}$$

Substituting (14.7.10) into (14.7.9) we obtain

$$\frac{Y(\mathbf{s})}{X(\mathbf{s})} = H(\mathbf{s}) = \frac{1}{\left(\dfrac{\tilde{s}^2 + \omega_{cf}^2}{\beta\tilde{s}}\right)^2 + 1.414\left(\dfrac{\tilde{s}^2 + \omega_{cf}^2}{\beta\tilde{s}}\right) + 1} \tag{14.7.11}$$

$$= \frac{\beta^2\tilde{s}^2}{\tilde{s}^4 + 1.414\beta\tilde{s}^3 + (2\omega_{cf}^2 + \beta^2)\tilde{s}^2 + 1.414\beta\omega_{cf}^2\tilde{s} + \omega_{cf}^4}.$$

Note that when transforming from a lowpass to bandpass or notch filter, the order of the transfer function will double. The magnitude of the frequency response function for a bandwidth of $\beta = 2$ rad/sec and a center frequency of $\omega_{cf} = 5$ rad/sec is shown in Figure 14.7.9.

▲▲

 This example illustrates the process of converting a lowpass prototype filter into a filter of another type. The transformations required to translate a normalized lowpass filter to a bandpass, notch, or highpass filter are shown in Table 14.7.1.

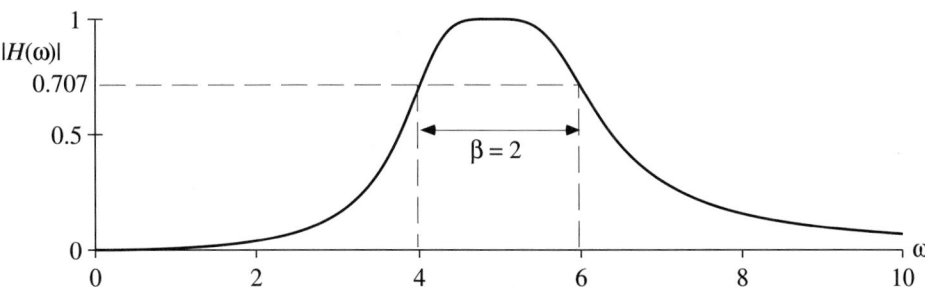

Figure 14.7.9 Magnitude of the Frequency Response Function
of the Bandpass Filter (14.7.11)

Table 14.7.1 Frequency Transformations

Type of Transformation	Formula
Lowpass to Bandpass*	$s = \dfrac{\tilde{s}^2 + \omega_{cf}^2}{\beta \tilde{s}}$
Lowpass to Notch**	$s = \dfrac{\beta \tilde{s}}{\tilde{s}^2 + \omega_{cf}^2}$
Lowpass to Highpass	$s = \dfrac{\omega_c}{\tilde{s}}$

* ω_{cf} is the center frequency of the passband
** ω_{cf} is the center frequency of the stopband

Computer Usage: These transformations are simply a straightforward computation. Most CAD packages embed these transforms directly into their filter design function. Hence, it is possible to design a bandpass filter, say, directly without explicitly doing the transformations in Table 14.7.1.

14.7.5 MATLAB Experiments

The MATLAB commands **butter**, **cheby1**, **cheby2**, and **ellip** automate completely the design process described above for Butterworth, Chebyshev, and elliptic filters. The input arguments to these commands are the order of the filter, the cutoff frequencies, and the passband and/or stopband ripple, if appropriate. To create a bandpass filter, the cutoff frequencies are entered as a vector. An additional optional parameter is required when the filter is a highpass or notch filter. Because these commands generate digital filters it is necessary to add the string 's' to the end of the list of input arguments so that a continuous-time filter is created. The output arguments of these command are the filter representation in one of the three standard representations. The MATLAB help files should be consulted for specifics.

Example 14.7.8: M-file. The following M-file generates a 4th-order Chebyshev Type 1 lowpass filter with 3 dB of ripple in the passband. Then the magnitude and phase of the filter are plotted.

```
clear
order = 4;                      % order of the filter
ripple = 3;                     % passband ripple
cutoff = 1;                     % cutoff frequency
% generate filter
[b,a] = cheby1(order,ripple,cutoff,'s');
% calculate frequency response
w = linspace(0,3,600);          % generate frequency vector
h = freqs(b,a,w);               % calculate frequency response function
mag = abs(h);                   % calculate magnitude
phs = (180/pi)*angle(h);        % calculate phase
```

```
% plot frequency response function
subplot(2,1,1), plot(w,mag)
xlabel('frequency, rad/sec')
title('magnitude')
subplot(2,1,2), plot(w,phs)
xlabel('frequency, rad/sec')
ylabel('deg')
title('phase')
```

▲▲

Exploratory Exercise 14.7.9:   Generate Butterworth, Chebyshev Type 2, and elliptic filters with the same order and cutoff frequencies. Plot the magnitudes on the same graph. Compare your graph with Figure 14.7.8. Plot the phases on the same graph. Also plot the step response of each filter.   What are the differences between these filters?

▲▲

Exploratory Exercise 14.7.10:   Modify the M-file above to generate a bandpass filter with cutoff frequencies $[\omega_1, \omega_2] = [1,2]$.   Then modify this bandpass filter into a notch filter.

▲▲

Exploratory Exercise 14.7.11:   Write an M-file to generate the bandpass filter shown in Figure 14.7.9.

▲▲

14.8 CHAPTER SUMMARY

In this chapter we have studied the response of a linear, time invariant, BIBO stable system to a sinusoidal input signal. We have shown that the steady state response (Definition 14.1.2) of such a system can be characterized by the frequency response function (Definition 14.3.3) $H(\omega)$ where

$$H(\omega) = H_F(\omega) = H_L(s)\big|_{s=j\omega} \tag{14.8.1}$$

and $H_F(\omega)$ is the Fourier transfer function or $H_L(s)$ is the Laplace transfer function of the system. This result is contained in the frequency response theorem which says that if the input signal is

$$x(t) = A\cos(\omega_i t + \phi), \tag{14.8.2}$$

then the steady state response of this system is

$$y_{ss}(t) = A|H(\omega_i)|\cos(\omega_i t + \angle H(\omega_i) + \phi). \tag{14.8.3}$$

This result is given in three different forms in Theorems 14.1.1, 14.2.1 and Corollary 14.3.2. When this theorem is stated for Laplace transfer functions, the output signal of the system contains a transient term as well as the steady state response.

The importance of the frequency response theorem is that the steady state response of the system is completely characterized by the magnitude and phase of the frequency response function. These two functions are usually presented as graphs called, respectively, the magnitude graph and the phase graph (Definition 14.4.1). An explanation of the interpretation of these two graphs is given in Section 14.4.

The graphical representation of the frequency response function lends insight into how a system responds to sinusoidal input signals across the frequency spectrum. Then the magnitude of the frequency response function can be interpreted in terms of power of the input and output signals as discussed in Section 14.3. If P_x is the power of the input signal and P_y is the power of the output signal then

$$P_y = |H(\omega_i)|^2 P_x. \tag{14.8.4}$$

This interpretation of the magnitude graph leads to the notion of the passband, the stopband, the transition band, and the bandwidth of a system (Definition 14.5.2). The passband is the set of frequencies corresponding to input signals which pass through the system with little loss in power. The stopband is the set of frequencies corresponding to input signals that are blocked by the system. Neglecting the transition band, the length and location of the passband defines the stopband. That is, the passband, in essence, defines the system. This observation motivates the definition of the bandwidth of the system as the length of the passband (Definition 14.5.3). It is a matter of engineering judgment to define the dividing line between the passband and stopband. One common criteria, the half-power bandwidth (Definition 14.5.5), is discussed in Section 14.5.

The relative location of the passband and stopband can be used to classify systems. This classification is assisted by the concept of an ideal filter, which is a system that only introduces a time delay to a signal propagating through the system (Definition 14.6.1). When the passband precedes the stopband, this filter is a lowpass filter (Definition 14.6.2). Similar definitions are given for bandpass filters (Definition 14.6.3), highpass filters (Definition 14.6.4), and notch filters (Definition 14.6.5).

In practice, we often want to build a system with a specified frequency response function. The process of synthesizing such a system is called filter design. An introduction to filter design is given in Section 14.7. In this section we discuss how to develop specifications for a frequency response function in terms of the passband, stopband, and transition band. Butterworth (Definition 14.7.2), Chebyshev (Definition 14.7.5), and elliptic filters are introduced as standard filters that are often used to meet these design specifications.

14.9 HOMEWORK FOR CHAPTER 14

Homework Problems for Section 14.1

14.1.1 Consider the mass spring systems in Figures P14.1.1a-c. Here $f_{st}(t)$, the input signal, is the force applied to the mass. The output signal is the displacement of the mass, $y_{st}(t)$. In each case, can the frequency response theorem be applied? If not, why not? If so, find $H(\omega)$.

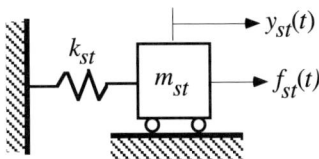

Figure P14.1.1a

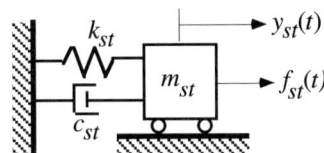

Figure P14.1.1b

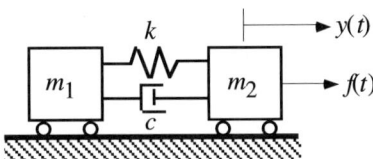

Figure P14.1.1c

14.1.2 For each of the systems given below, suppose the input signal is $x(t) = [\cos(2t)]u_s(t)$. Does the steady state response exist for these systems? If so, find it. If not, why not?

(i) $\dfrac{Y(s)}{X(s)} = \dfrac{1}{s(s+1)}$ (ii) $\dfrac{Y(s)}{X(s)} = \dfrac{(s-2)}{(s+3)(s+1)}$

14.1.3 Consider the op amp in Figure P14.1.3a. The relationship between the input voltage $v_b(t)$ and the output voltage $v_o(t)$ is given by a saturation characteristic as shown in Figure P14.1.3b. Under what conditions can we compute a frequency response function for this system?

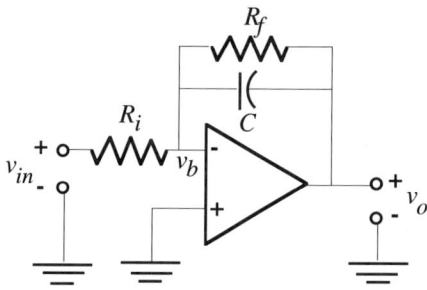

Figure P14.1.3a

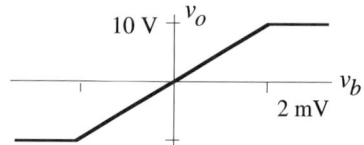

Figure P14.1.3b

14.1.4   Consider the two mass-spring-damper system in Figure P14.1.4.   A set of state space equations for this system is

$$
\dot{\vec{q}}(t) = \begin{bmatrix} 0 & 1 & 0 & 0 \\ -\dfrac{k_a+k_c}{m_a} & -\dfrac{c_a}{m_a} & \dfrac{k_c}{m_a} & 0 \\ 0 & 0 & 0 & 1 \\ \dfrac{k_c}{m_{st}} & 0 & -\dfrac{k_c}{m_{st}} & 0 \end{bmatrix} \vec{q}(t) + \begin{bmatrix} 0 \\ \dfrac{1}{m_a} \\ 0 \\ 0 \end{bmatrix} f_a(t)
$$

$$
y_{st}(t) = \begin{bmatrix} 0 & 0 & 10 & 0 \end{bmatrix} \vec{q}(t)
$$

Suppose the parameter values of this system are: $k_a = 10$ N/m, $c_a = 8$ N/m/sec, $m_a = 1$ kg, $k_c = 100$ N/m, and $m_{st} = 0.1$ kg.
(a) Compute the transfer function of this system using MATLAB.
(b) Suppose the input signal is $f_a(t) = \sin(1t)$.   Plot the output signal.   Also plot the input signal on the same graph.
(c) Identify the transient and steady state components of the output signal.
(d) Compare the amplitude and phase of the sinusoidal part of the signals. What is the change in amplitude between the input and output signal? What is the phase shift between the input and output signals?   Correlate the readings taken from the graph with a calculation from the transfer function.

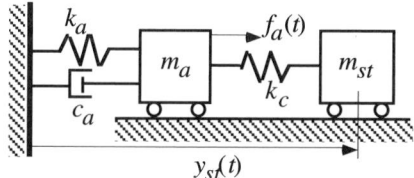

Figure P14.1.4

14.1.5 Consider the *RC* network in Figure P14.1.5. Let $1/RC = 10$.
(a) Find the transfer function of this system.
(b) Suppose $x(t) = \sin(2t)$ is the input signal to the network above. Plot the output signal and the input signal on the same graph.
(c) Suppose $x(t) = \sin(200t)$ is the input signal to the network above. Plot the output signal and the input signal on the same graph.
(d) Find the magnitude and phase of the steady state response in parts (b-c) from the transfer function.

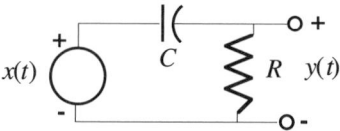

Figure P14.1.5

14.1.6 Consider the system in Figure P14.1.6. Suppose that the input signal is $r(t) = \cos(4t)$.
(a) Find the steady state response using the frequency response theorem.
(b) Confirm this answer through simulation.

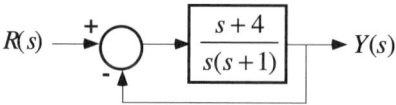

Figure P14.1.6

14.1.7 Consider the system

$$\frac{Y(s)}{X(s)} = H(s) = \frac{s}{s^2 + 0.4s + 4}$$

Suppose that the input signal is $x(t) = u_s(t)$.
(a) Find the output signal using Laplace transforms.
(b) Identify the amplitude of the steady state response using the frequency response theorem. (This input signal is a zero frequency sinusoid.)
(c) Identify the transient response.

Homework Problems for Section 14.2

14.2.1 Consider the network in Figure P14.2.1.
(a) Find the Fourier transfer function of this network.
(b) Let $R/L = 10$. Calculate $|H(\omega)|$ and $\angle H(\omega)$.
(c) Suppose that $x(t) = \cos(10t)$. Calculate the output signal using the frequency response theorem.
(d) Plot the input and output signal on the same graph.

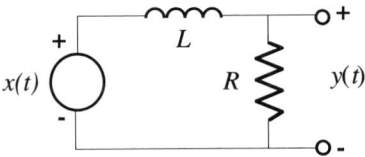

Figure P14.2.1

14.2.2 Consider the network in Figure P14.2.2.
(a) Find the Fourier transfer function of this network.
(b) Let $R/L = 10$. Calculate $|H(\omega)|$ and $\angle H(\omega)$.
(c) Suppose that $x(t) = \cos(10t)$. Calculate the output signal.
(d) Plot the input and output signal on the same graph.
(e) Let $x(t) = \cos(2t)$. Repeat parts (c-d).
(f) Let $x(t) = \cos(200t)$. Repeat parts (c-d).

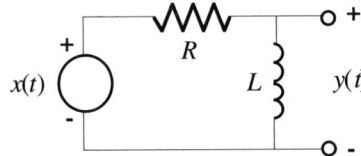

Figure P14.2.2

14.2.3 Consider the network shown in Figure P14.2.3.
(a) Find the Fourier transfer function in terms of L, R, and C.
(b) Calculate $|H(\omega)|$ and $\angle H(\omega)$.
(c) Suppose that $x(t) = \cos(2000\pi t)$. Calculate the output signal.
(d) Plot the input and output signal on the same graph.

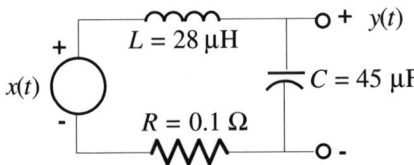

Figure P14.2.3

14.2.4 Consider the system

$$y(t) = [x(t)]^2.$$

Find the output signal is the input signal is $x(t) = \cos(\omega t)$. Conclude that this system doesn't satisfy the frequency response theorem.   What assumptions of the frequency response theorem are violated?

Homework Problems for Section 14.3

14.3.1 Consider the system

$$\frac{Y(s)}{X(s)} = H(s) = \frac{s}{s^2 + 0.4s + 4}$$

Suppose that the input signal is $x(t) = \sin(t)$. Find the steady state response.

14.3.2 Consider the system

$$\frac{Y(s)}{X(s)} = \frac{s+2}{s^2 + s + 4}$$

Suppose the input signal to this system is $x(t) = [3\cos(2t - 0.25)]u_s(t)$. What is the steady state output signal?

14.3.3 Consider the circuit shown in Figure P14.3.3.
(a) Find the frequency response function of this system.
(b) If the input signal is $x(t) = \cos 5t$, $t \geq 0$, plot on one graph the input signal and output signal of this system for $RC = 0.2$ and $RC = 2$.
(c) Determine the value of the frequency response function at $\omega_i = 5$ rad/sec for $RC = 0.2$ and $RC = 2$. Use this calculation to explain the plots in (b).

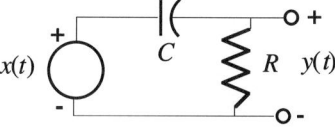

Figure P14.3.3

14.3.4 Consider the parallel resonant circuit shown in Figure P14.3.4. If this circuit is lumped element model of a device, then the input signal is the current into the device and the output signal is the voltage across the device.
(a) Find the Laplace transfer function of this circuit.
(b) Find the poles of this transfer function. Plot them in the s-plane. Label them in terms of (ζ, ω_n) and (α, ω_c).

(c) Assume L and C are fixed. Plot the poles in the s-plane as a function of R.

(d) Find the frequency response function of this system.

(e) Let $C = 28\ \mu\text{F}$ and $L = 28\ \mu\text{H}$. Plot the magnitude of frequency response function of this system for several values of R between $10\ \text{k}\Omega$ and $100\ \text{k}\Omega$. For each value of R, determine the frequency where the magnitude of the frequency response function is a maximum. Compare this frequency to the natural frequency and the critical frequency of the poles.

(f) Show that the half-power bandwidth is given by

$$\beta_{hp} = \frac{1}{RC}.$$

Hint: Show that the maximum of the magnitude of the frequency response function occurs at the natural frequency of the poles.

(g) Suppose the input signal is

$$i(t) = I_m \cos(\omega_n t).$$

Find the steady state response of this system. Also find the current through the inductor.

(h) The instantaneous energy in the capacitor and the energy in the inductor are

$$E_c(t) = \frac{1}{2}C(v(t))^2 \quad \text{and} \quad E_v(t) = \frac{1}{2}L(i(t))^2,$$

respectively. Show that the total energy in the circuit is

$$E_T = E_c(t) + E_L(t).$$

(The total energy in the circuit is not a function of time, although the energy in the capacitor and inductor are time-varying.)

(i) Show that the energy lost in the resistor over one period is

$$E_r = R\int_0^{2\pi} \left(I_m \cos(\omega_n t)\right)^2 dt = \frac{\pi R I_m^2}{\omega_n}.$$

(j) The Q of this circuit is defined as

$$Q = 2\pi\,\frac{\text{maximun energy stored}}{\text{total energy lost per period}}.$$

Now show that

$$Q = 2\pi\omega_n RC = \frac{\omega_n}{\beta_{hp}}.$$

Label the Q factor of each circuit on the corresponding frequency response curve in (e).

Note: The Q factor is inversely proportional to the bandwidth of the system. For high Q factors, the bandwidth of the system is roughly centered on the resonant frequency.

(k) Express ζ in (2) as well as (α, ω_c) in part (a) in terms of the Q factor.

(l) For the input current in (g) show that the amplitude of the current in both the inductor and capacitor is QI_m.

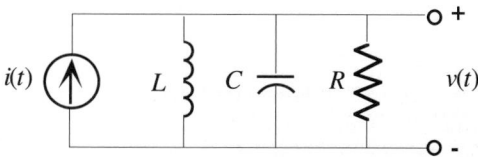

Figure P14.3.4

Homework Problems for Section 14.4

14.4.1 Consider a system whose frequency response function is shown in Figure P14.4.1.

(a) Suppose that the input signal to this system is

$$x_1(t) = \cos(\omega_1 t) = \cos(1.2t).$$

Find the steady state response to this input signal.

(b) Suppose that the input signal to this system is

$$x_2(t) = \cos(\omega_2 t) = \cos(0.2t).$$

Find the steady state response to this input signal.

(c) Suppose that the input signal to this system is

$$x_3(t) = 3\sin(\omega_3 t + \theta_3) = 3\sin(2t - 0.5).$$

Find the steady state response to this input signal.

(d) Suppose the input signal to this system is the sum of the input signals in parts (a) - (c). What is the corresponding output signal?

(e) Plot the input and output signals in (d).

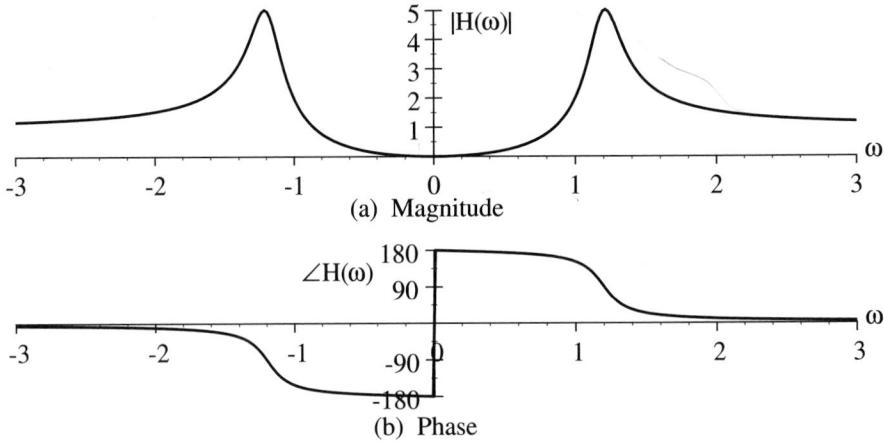

Figure P14.4.1

14.4.2 The frequency response function of a system is shown in Figure P14.4.2.
(a) Suppose the input signal to this system is

$$x(t) = \cos(3t) + \cos(10t) + \cos(40t).$$

Find the steady state output signal for this system. Plot the input and output signals.
(b) Suppose the input signal is $x(t) = 4\sin(9t)$. Find the steady state response of the system. Plot the input and output signals. Is the output signal leading or lagging the output signal?
(c) Suppose the input signal is $x(t) = u_s(t)$. Find the steady state response.

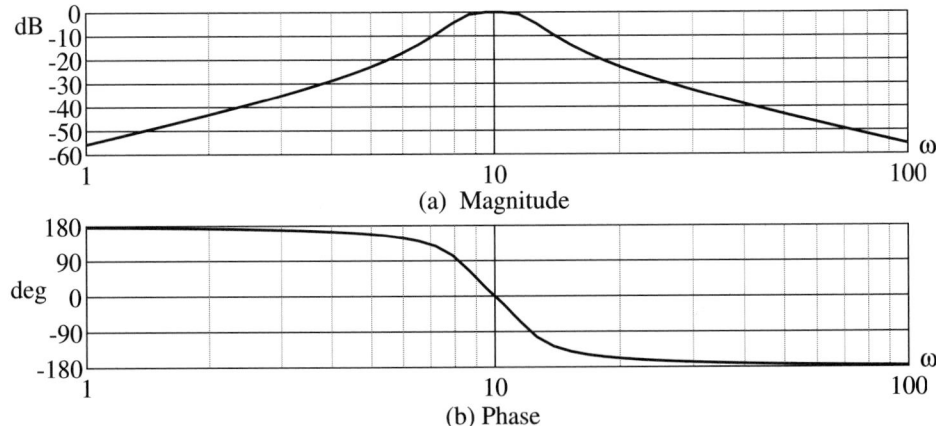

Figure P14.4.2

14.4.3 Consider the Butterworth filter discussed in Section 10.3. A block diagram of this circuit is given in Figure 10.3.7. A state space representation of this system is

$$\dot{\vec{q}}(t) = \begin{bmatrix} -\dfrac{1}{R_2 C_2} & \dfrac{1}{C_2} & 0 \\ -\dfrac{1}{L} & 0 & \dfrac{1}{L} \\ 0 & -\dfrac{1}{C_1} & -\dfrac{1}{R_1 C_1} \end{bmatrix} \vec{q}(t) + \begin{bmatrix} 0 \\ 0 \\ \dfrac{1}{C_1 R_1} \end{bmatrix} v_{in}(t),$$

$$y(t) = \begin{bmatrix} v_0(t) \\ v_1(t) - v_0(t) \\ v_1(t) \end{bmatrix} = \begin{bmatrix} 1 & 0 & 0 \\ -1 & 0 & 1 \\ 0 & 0 & 1 \end{bmatrix} \vec{q}(t).$$

The parameter values are $R_1 = R_2 = 1$ kΩ, $C_1 = C_2 = 1$ μF, and $L = 1$ H.

(a) Identify each of the output signals in the circuit diagram.

(b) Suppose the input signal to this network is $x(t) = [\cos(500t)]u_s(t)$. Plot the output signals for this input signal using MATLAB.

(c) Plot the frequency response function for the transfer function from the input signal to each output signal.

(d) Find the relative phase difference between $v_0(t)$ and $(v_1(t) - v_0(t))$.

(e) Fill in the following table.

Table P14.4.3

Transfer Function				
	dB	Gain	Input Signal Amplitude	Output Signal Amplitude
1				
2				
3				

14.4.4 The magnitude Bode plots of a system is shown Figure P14.4.4.

(a) If the input signal to this system is $x(t) = \cos(\omega_i t)$, at what frequency ω_i is the amplitude of the steady state output signal the smallest?

(b) If the input signal to this system is $x(t) = \cos(6t)$, what is the amplitude of the steady state output signal in decibels and linear units?

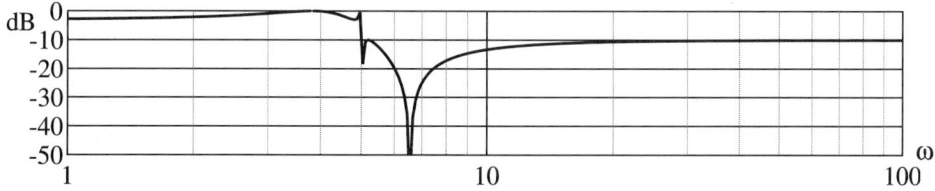

Figure P14.4.4

14.4.5 The magnitude Bode plot of a system is shown in Figure P14.4.5. Suppose the input signal is $x(t) = \cos(\omega_i t)$.

(a) Identify the frequency band where the steady state output signal has an amplitude less than 0.01.

(b) Identify the frequency band where the steady state output signal has an amplitude greater than 0.707.

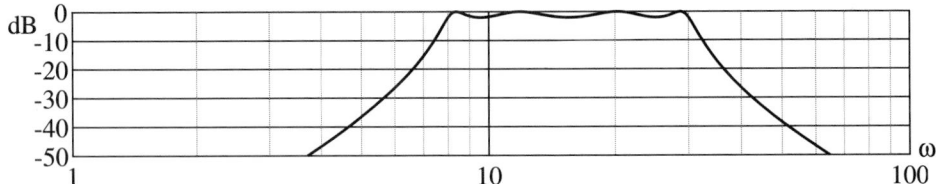

Figure P14.4.5

14.4.6 Consider the magnitude Bode plot of the frequency response function of a system shown in Figure P14.4.6.

(a) Identify the frequency band where the steady state output signal has an amplitude less than 0.1.

(b) Identify the frequency band where the steady state output signal has an amplitude greater than 0.707.

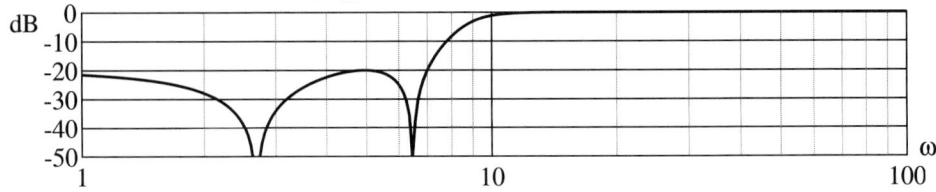

Figure P14.4.6

14.4.7 Consider the R/L network in Figure P14.4.7. Let $R/L = 10$.

(a) Find the transfer function of this system.

(b) Plot the frequency response function of this system.

(c) Graphically determine the amplitude and phase of the steady state response to the input signals $x(t) = \sin(2t)$ and $x(t) = \sin(200t)$.

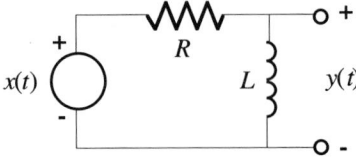

Figure P14.4.7

14.4.8 Consider the *RL* network in Figure P14.4.8. Let $R/L = 10$.
   (a) Find the transfer function of this system.
   (b) Plot the frequency response function of this system.
   (c) Graphically determine the amplitude and phase of the steady state response to the input signals $x(t) = \sin(2t)$ and $x(t) = \sin(20t)$.

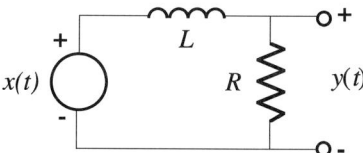

Figure P14.4.8

14.4.9 Consider the mass-spring-damper system in Figure P14.4.9. This system is discussed in detail in Problem 11.5.5. Suppose the input signal is

$$x(t) = [100 \cos(20t)]u_s(t).$$

   (a) Plot the output signals on one graph.
   (b) Plot the magnitude and phase of the frequency response function for the three systems defined by the transfer functions from the force input signal to each separate output signal. Use these plots to explain the amplitudes of the signals part (a).
   (c) At what frequency does a sinusoidal input signal cause a much larger motion in mass 3 relative to the other two masses?
   (d) Over what frequency band does mass 2 oscillate out of phase with the other two masses?
   (e) Use the frequency response functions to explain the relative motions of the three masses for sinusoidal input signals of all frequencies.

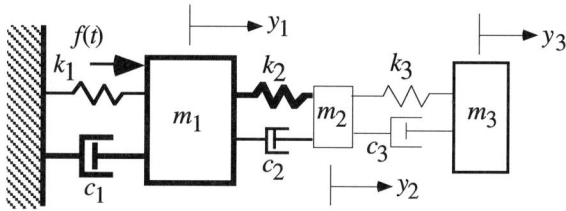

Figure P14.4.9

14.4.10 Consider the system in Figure P14.4.10.
   (a) Sketch the frequency response function for $K = 1, 2\,5$.
   (b) Suppose the input signal is $r(t) = \cos(\omega_i t)$. For what frequency is the amplitude of the steady state response maximum? Consider each K.
   (c) For the frequency in (b), what is the value of the maximum amplitude of the steady state response?

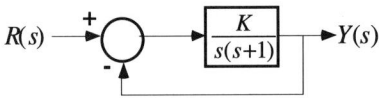

Figure P14.4.10

Homework Problems for Section 14.5

14.5.1 Consider the op amp circuit shown in Figure P14.1.3a. Let $C = 3$ μF, $R_f = 1$ kΩ, and $R_f = 10$ kΩ. What is the bandwidth of this system?

14.5.2 Determine the bandwidth of the system shown in:
(i)   Problem 14.4.2   (ii)   Problem 14.4.4   (iii) Problem 14.4.5
(iv) Problem 14.4.6   (v)   Problem 14.4.7   (vi)   Problem 14.4.8
(vii) Problem 14.4.9

14.5.3 Consider the mass-spring-damper system in Figure P14.5.3. Assume that this system has two complex poles.
(a) Determine the bandwidth of this system as a function of the spring stiffness k_{st}.
(b) Determine the bandwidth of this system as a function of the mass m_{st}.
(c) Determine the bandwidth as a function of the damping c_{st}.

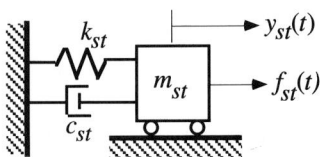

Figure P14.5.3

Homework Problems for Section 14.6

14.6.1 Suppose an ideal lowpass filter has bandwidth of 1000 rad/sec and a group delay of 0.2 sec.
(a) Sketch the frequency response function of this filter.
(b) Find an analytical expression for the frequency response function.

14.6.2 Suppose that the stopband of an ideal highpass filter begins at 1 rad/sec and a group delay of 2 sec.
(a) Sketch the frequency response function of this filter.
(b) Find an analytical expression for the frequency response function.

14.6.3 Suppose an ideal notch filter has a notch with a width of 20 kHz centered on 300 kHz. Suppose that the group delay of this filter is 3 μsec.
(a) Sketch the frequency response function of this filter.
(b) Find an analytical expression for the frequency response function.

14.6.4   Suppose an ideal bandpass filter has a bandwidth of 300 Hz and a group delay of 4 msec.
(a) Sketch the frequency response function of this filter.
(b) Find an analytical expression for the frequency response function.

Homework Problems for Section 14.7

14.7.1   Suppose we want to design a bandpass filter with the following specifications:

passband bandwidth	800 - 1200 Hz
passband ripple	< 0.5 dB
stopband	$f < 700$ Hz and $f > 1300$ Hz
stopband ripple	> 20 dB

(a) Design a minimal order Butterworth filter to meet these specifications.
(b) Design a minimal order Chebyshev Type I filter to meet these specifications.
(c) Design a minimal order Chebyshev Type II filter to meet these specifications.
(d) Design a minimal order elliptic filter to meet these specifications.
(e) What is the order of each of these filters?
(f) In each case plot the magnitude and phase of the frequency response function on the same graph.

14.7.2   Suppose we want to design a bandpass filter with the following specifications:

passband bandwidth	4 - 8 kHz
passband ripple	0.1 dB
order	4

Design a filter to meet these specifications. If you are not able to meet the specifications, explain which specification is violated.
(a) Design a Butterworth filter to meet these specifications.
(b) Design a Chebyshev Type I filter to meet these specifications.
(c) Design a Chebyshev Type II filter to meet these specifications.
(d) Design an elliptic filter to meet these specifications.
(e) In each case plot the magnitude and phase of the frequency response function.

14.7.3   Consider the system

$$\frac{Y(s)}{X(s)} = H(s) = \frac{40,000}{(s^2 + 5.6s + 2^2)(s^2 + 40s + 100^2)}.$$

(a) Plot the step response of this system.

(b) Plot the magnitude and phase frequency response function of this system.

(c) From parts (a) and (b), it can be seen that the output signal will contain a sinusoidal component with a frequency of 100 rad/sec. We wish to filter out this component from the output signal while the components in the output signal due to the other two poles should not be changed. This filter has the transfer function

$$\frac{Y_f(s)}{Y(s)} = H_f(s).$$

(i)   What type of filter is required?
(ii)   What is the bandwidth of this filter?
(iii)   What type of filter would you recommend?
(iv)   Repeat (a) and (b) for the system

$$\frac{Y_f(s)}{X(s)} = H(s)H_f(s).$$

(v)   For the selected filter, determine the attenuation of the unwanted signal components of the output signal for various orders of your filter.
(vi)   How much does the phase of the transfer function in (1) differ from the phase of the transfer function in (2) over the passband of the filter?
(vii)   For the filter designed above, suppose we add the requirement that the phase lag of the filter should be minimized in the stopband. Plot the phase lag of the filter at $\omega_n = 100$ rad/sec as a function of the filter order.

14.7.4 Use MATLAB to generate 4th-order lowpass Chebyshev I and II analog filters with a bandwidth of 1 rad/sec. Assume a 3 dB ripple in the passband of the Chebyshev I filter, and assume the ripple in the stopband of the Chebyshev II filter is 10 dB down.

(a) Plot the magnitude of the frequency response function of both filters on one plot. Use a linear scale for the magnitude, but a log scale for the frequency.

(b) Assume that the input signal to each filter is $x_1(t) = \cos(0.8t)$. Plot the output signal of each filter on the same graph. Choose the time interval sufficiently long to show the steady state behavior.

(c) Explain the amplitude of the steady state response in terms of the magnitude of the frequency response function.

(d) Repeat (b) using the input signal $x_2(t) = \cos(1.5t)$.

Chapter 15

Signal and System Analysis in the Frequency Domain

Chapter Outline

Up to this point we have studied: signals in Chapter 8, and systems in Chapters 10 - 12. Then in Chapter 14 we introduced the frequency response theorem which gives us an explicit link between the system and the signal propagating through it. The frequency response theorem as stated in the last chapter allows for only a pure sinusoidal input signal. In this chapter we will extend the interpretations of the frequency response theorem in two ways. First, we will allow the input signal to be an infinite sum of sinusoids (a Fourier series). This straightforward application of the frequency response theorem extends the theorem to all periodic signals. Second, we extend the frequency response theorem to accommodate energy and power signals as input and output signals through the use of Fourier transforms.

In Chapter 8 we have developed several signal representations: the Fourier series, the Fourier transform, amplitude and phase spectrums, energy spectral density, and the power spectral density. All of these signal representations have graphical representations. By combining these concepts with the frequency response function from the frequency response theorem, we obtain a graphical interpretation of how a signal propagates through a system.

This chapter contains an extended example, a pulse train and a signal pulse propagating through a *RC* network, to illustrate fully the various graphical interpretations of the frequency response theorem using concepts from signal representations. The graphical interpretations are straightforward for Fourier series, but when extended to energy signals, they become a little blurred. By comparing the energy signal analysis to the Fourier series analysis, we gain insight into the extension of this theorem to energy signals. These interpretations lend deep insight into how a signal propagates through a system. The examples in this chapter are necessarily elementary to illustrate the concepts. Nonetheless, the analysis techniques in this chapter extend to more complicated signals and systems. These results also form the basis of the theory of random processes propagating through a system. They form the basis of digital filters. They form the basis for the performance evaluation of control systems. They form the basis for detectors in communication systems. The list goes on. These concepts make the analysis and synthesis of complicated systems tractable.

In many ways this chapter is the logical conclusion to the entire conceptual development in this text (we recognize that the results also extend to discrete signals and systems).

Summary of Sections

Section 15.1:   We develop the description of an energy or power signal propagating through a system in terms of its energy spectral density, or power spectral density.

Section 15.2:   We illustrate the analysis techniques used in the subsequent sections using a single sinusoidal input signal.

Section 15.3:   We describe the propagation of a periodic signal through a system in terms of the relationship between the power spectrum of the signal and the passband of the system.

Section 15.4:   We extend the analysis in the last section to an energy signal propagating through a system.

Section 15.5:   We analyze the performance of a control system on a linear motor.

Section 15.6:   We discuss amplitude modulation and frequency division multiplexing.

Section 15.7:   Chapter summary section.

Coverage of the Text

The system analysis is based on the frequency response theorem and its implications discussed in Chapter 14. The signal analysis is drawn from Chapter 8.

15.1 INTRODUCTION TO SIGNAL AND SYSTEM INTERACTION

15.1.1 An Example

To set the framework of the discussion in this chapter, we will present a simple example. Consider the simple *RC* circuit in Figure 15.1.1. The convolution representation of this system is

$$y(t) = \int_{-\infty}^{\infty} e^{-\frac{(t-\lambda)}{RC}} u_s(t-\lambda)(x(\lambda))\, d\lambda. \tag{15.1.1}$$

Suppose the input signal is the pulse

$$x(t) = \Pi\left(\frac{t}{\varepsilon}\right). \tag{15.1.2}$$

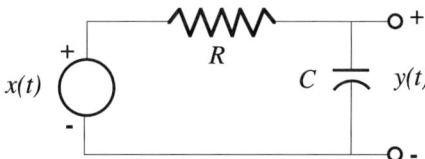

Figure 15.1.1 *RC* Network

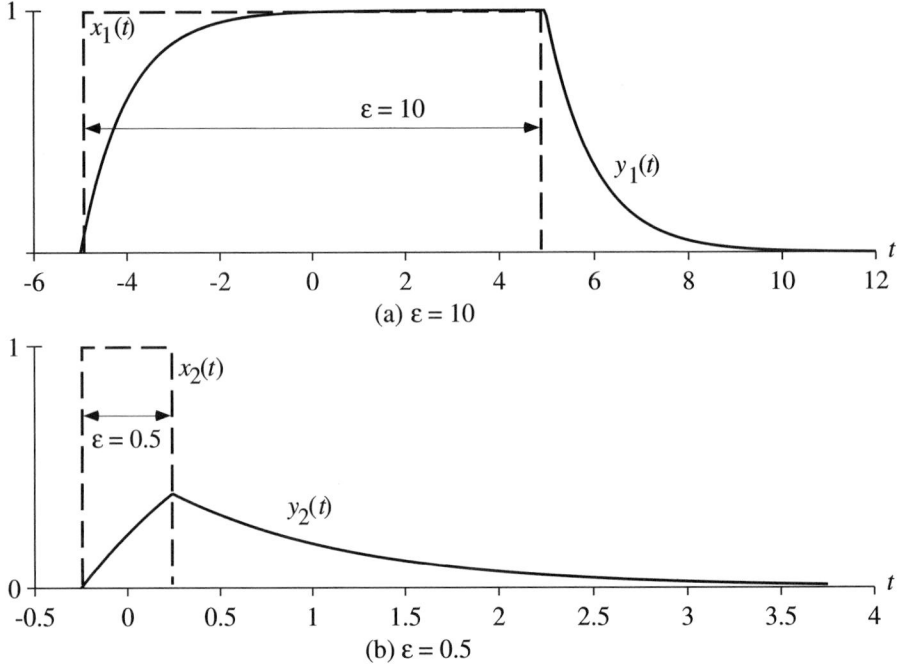

Figure 15.1.2 Two Different Input and Output Signals of the *RC* Network, $RC = 1$

The output signal for the input signal (15.1.2) is

$$
y(t) = \begin{cases} \left(1 - e^{-\frac{1}{RC}\left(t + \frac{\varepsilon}{2}\right)}\right), & -\frac{\varepsilon}{2} < t < \frac{\varepsilon}{2} \\ e^{-\frac{1}{RC}\left(t - \frac{\varepsilon}{2}\right)} - e^{-\frac{1}{RC}\left(t + \frac{\varepsilon}{2}\right)}, & t > \frac{\varepsilon}{2} \end{cases}
$$

(15.1.3)

This output signal is derived in Example 13.2.2. Figure 15.1.2 shows the input and output signals for two different values of ε. Figure 15.1.2 shows that the system responses in a very different way to each of these two input signals. The pulse at the output of the system has suffered only small distortion, while the pulse in Figure 15.1.2b at the output of the system is so badly distorted its relationship to the input signal is barely recognizable. The purpose of this chapter is to explain this difference.

The results in this chapter are presented from the following conceptual viewpoint. The input signal to the system (*RC* network) produces the output signal. To emphasize this concept, the output signals are overlaid on the input signals in Figure 15.1.2. *We say that the input signal (15.1.2) has propagated through the system to yield the output signal (15.1.3).*[1]

---

[1] See also the discussion in the Chapter Introduction of Chapter 14.

It is clear from Figure 15.1.2 that the input is changed as it propagates through the system. To measure this change, we can compare the output signal to the input signal. To distinguish the effect of the system on these signals, we will assume that we want the square pulse in (15.1.2) to be transmitted through the system *unchanged*. If the system doesn't transmit the input signal unchanged, we say the system has *distorted* the input signal. (Almost all systems discussed in this text will distort the signal passing through it.) We are really interested in the distortion in the signal as a result of its transmission through a system.[2]

In the example above we have computed directly the output signals of the *RC* network for two different input signals. While the input signals are similar, the output signals don't resemble each other. Furthermore, the output signal in Figure 15.1.2b is quite different from the input pulse. The amount of the distortion in the output signals can be traced to the two parameters in (15.1.3). One of the parameters is the time constant of the network, *RC*, in (15.1.1), the system parameter. The other parameter is the pulse width ε of the input signal in (15.1.2); the signal parameter. The pulse width of the first signal is wide enough to allow the capacitor to fully charge so that the output signal becomes equal to the input voltage. For the second input signal the capacitor is never able to fully charge before the input signal returns to zero. This example illustrates that the form of the output signal is determined by the *relationship between* the system parameters and the signal characteristics.

This analysis of this very simple *RC* network driven by a pulse input signal explains the behavior of the output signal. It is important to recognize, however, that this type of analysis is confined to simple signals and systems. Here the input signal is described by one parameter, the pulse width. How many parameters would it take to describe an audio signal? The system is also described by only one parameter, the *RC* time constant. Most systems would be described by many parameters. Not even with the help of sophisticated computer packages would it be possible extend this type of analysis to realistic problems. Therefore, to understand the way this input signal propagates through this system, we will shift to the frequency domain. At first glance it is not apparent how frequency domain ideas can be used to characterize time domain behavior of signals and system. It turns out, however, that these ideas lead to very general results that are very powerful in depth and scope for signal and system analysis.

15.1.2 Analysis in the Frequency Domain

In this chapter we will show how frequency domain results can be used to analyze how a signal propagates through a system. The three key results that we will use are: (1) spectral content of a signal, (2) frequency response of a system, and (3) the frequency response theorem.

Spectral Content of a Signal The concept of the spectral content of a signal is developed in Chapter 8. This concept leads to three types of signals. If the signal $x(t)$ is periodic, then it can be represented by a Fourier series

---

[2] In this chapter we adopt this viewpoint, which is basically a system viewed as a filter. This viewpoint is applicable to many real problems. There are many other real world problems for which this viewpoint is not appropriate. Nonetheless, the results in the chapter are still useful in almost all system analysis problems. We chose this viewpoint for this chapter because it seems to readily explain the results.

$$x(t) = \sum_{m=0}^{\infty} A_m \cos(m\omega_0 t + \theta_m), \tag{15.1.4}$$

or

$$x(t) = \sum_{m=-\infty}^{\infty} X_m e^{jm\omega_0 t}. \tag{15.1.5}$$

The power spectral density of this Fourier series is

$$S_x(\omega) = \begin{cases} |X_m|^2, & \omega = m\omega_0 \\ 0, & \omega \neq m\omega_0 \end{cases} \tag{15.1.6}$$

If the signal is an energy signal $x(t)$, then the spectral content is described by the energy spectral density (Definition 8.3.2)

$$D_x(\omega) = |X(\omega)|^2. \tag{15.1.7}$$

If the signal is a power signal $x(t)$, then the spectral content is described by the power spectral density (Definition 8.4.2)

$$S_x(\omega) = |X(\omega)|^2. \tag{15.1.8}$$

Frequency Response of a System We assume that the system is linear, time invariant, and BIBO stable. Then the system that can be represented by its Fourier transfer function

$$\frac{Y(\omega)}{X(\omega)} = H(\omega), \tag{15.1.9}$$

or by its Laplace transfer function

$$\frac{Y(s)}{X(s)} = H(s). \tag{15.1.10}$$

The frequency response function (Definition 14.3.3) of the system is

$$H(s)\big|_{s=j\omega} = H(\omega) \tag{15.1.11}$$

where $H(\omega)$ in (15.1.11) is identical to the Fourier transfer function in (15.1.9).[3]

---

[3] See Corollary 14.3.1.

Frequency Response Theorem   The fundamental result that links signals to systems is the frequency response theorem.[4] If the input signal is

$$x(t) = A\cos(\omega_i t + \theta),$$ (15.1.12)

then the frequency response theorem says that the steady state response is

$$y(t) = A|H(\omega_i)|\cos(\omega_i t + \theta + \angle H(\omega_i)).$$ (15.1.13)

If we compare the input signal (15.1.12) to the output signal (15.1.13), we see that the parameters of the input signal are A, θ, and ω_i. If we vary A, then the amplitude of the output signal will be scaled by the same amount because the system is linear. If we change the phase angle θ, then the phase of the output signal will also change by the same amount because the system is time invariant. If we vary ω_i, however, then both the magnitude and the phase of the output signal (15.1.13) will change because $H(\omega_i)$ is a nontrival function of ω_i. Hence, the frequency of the input signal, ω_i, is the only parameter of the input signal that will give us any information about the output signal. The frequency response function $H(\omega)$ will play a central role in this analysis.

The output signal inherits the parameters A, θ, and ω_i from the input signal. It also inherits the parameters $|H(\omega_i)|$ and $\angle H(\omega_i)$ from the system. Hence, the output signal (15.1.13) represents the interaction of the system (15.1.9) with the input signal (15.1.12). The frequency response theorem is the first and most fundamental result concerning the interaction of a signal and the system through which it propagates. In this chapter we will use this relationship between signal and system to study arbitrary signals passing through a system.

Terminology: In the following sections we will analyze the propagation of a signal through a system using the frequency response function $H(\omega)$ of the system and the energy spectral density $D_x(\omega)$ or power spectral density $S_x(\omega)$ of the signal. All of these functions depend on the frequency variable ω. Hence, we say this analysis is conducted in the *frequency domain*.

The frequency response theorem relates a sinusoidal input signal to the steady state response of the system. Next we will extend this result to periodic, energy, and power signals.

15.1.3 Periodic Input Signals

We consider a linear, time invariant, and BIBO stable system with a periodic input signal as shown in Figure 15.1.3. If the input signal has a period of T_0, the input signal can be represented by the cosine Fourier series representation

---

[4] Since we are concerned with only the steady state behavior, either Theorem 14.1.1 or 14.2.1 can be used for the calculation.

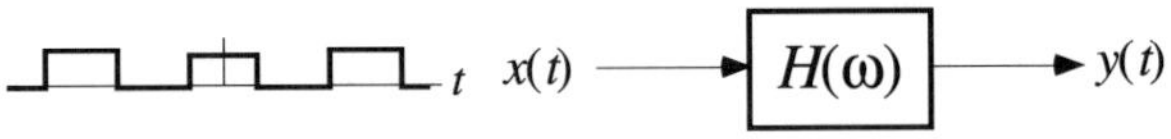

$$x(t) \longrightarrow \boxed{H(\omega)} \longrightarrow y(t)$$

Figure 15.1.3 System with a Periodic Input Signal

$$x(t) = \sum_{m=0}^{\infty} A_m \cos(m\omega_0 t + \theta_m). \tag{15.1.14}$$

To calculate the output signal, first we extract a single component from the Fourier series in (15.1.14)

$$A_m \cos(m\omega_0 t + \theta_m), \tag{15.1.15}$$

and use it as a input signal to the system (15.1.9). Using the frequency response theorem the steady state response is

$$y_m(t) = |H(m\omega_0)|A_m \cos(m\omega_0 t + \theta_m + \angle H(m\omega_0)) = Y_m \cos(m\omega_0 t + \phi_m). \tag{15.1.16}$$

This calculation is repeated for each term in the infinite sum (15.1.14). Next we sum all of the output signals (15.1.16). The result is the Fourier series

$$y(t) = \sum_{m=0}^{\infty} |H(m\omega_0)|A_m \cos(m\omega_0 t + \theta_m + \angle H(m\omega_0)) \tag{15.1.17}$$

$$= \sum_{m=0}^{\infty} Y_m \cos(m\omega_0 t + \phi_m).$$

The discussion above is summarized in Figure 15.1.4.

So we see that if the input signal can be expressed as a Fourier series then the

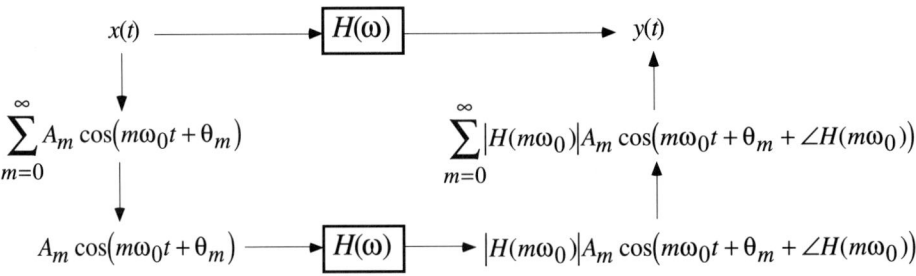

Figure 15.1.4 Fourier Series Representation of a Signal Propagating
Through a System

output signal is also expressible as a Fourier series.[5] Let the power spectral density of the input signal be $S_x(\omega)$ and the power spectral density of the output signal be $S_x(\omega)$. Comparing the Fourier series of the output signal in (15.1.17) with the Fourier series of the input signal in (15.1.14), we find that

$$S_y(\omega) = |H(\omega)|^2 S_x(\omega). \tag{15.1.18}$$

Our analysis will now proceed by comparing the power spectral density of the input signal with the power spectral density of the output signal and the frequency response function of the system. This discussion is taken up in the next section.

15.1.4 Energy Input Signals

If the input signal isn't periodic, but it is an energy signal, we can use the Fourier transform to study the propagation of the signal through the system. Again we consider a linear, time invariant, and BIBO stable system as shown in Figure 15.1.5. The system can be modeled by the Fourier transfer function in (15.1.9). Then the relationship between the Fourier transform of the input signal and the Fourier transform of the output signal is

$$Y(\omega) = H(\omega)X(\omega). \tag{15.1.19}$$

The energy spectral density of the output signal is

$$D_y(\omega) = |Y(\omega)|^2 = |H(\omega)X(\omega)|^2 = |H(\omega)|^2 |X(\omega)|^2 = |H(\omega)|^2 D_x(\omega). \tag{15.1.20}$$

Hence, the energy spectral density of the output signal can be computed from the energy spectral density of the input signal and the magnitude of the frequency response function. The relationship in (15.1.19) can be viewed as a generalization of the relationship in (15.1.17). See the discussion introducing the Fourier transform in Section 7.4.

15.1.5 Aperiodic Power Input Signals

We would like to have a result similar to (15.1.20) for power signals that are not periodic. Consider a power signal $x(t)$ with a power spectral density $S_x(\omega)$. Assume this signal is the input signal to the system in (15.1.9). Define the truncated signal (See Definition 8.4.1) as

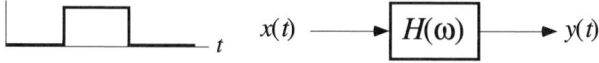

Figure 15.1.5 System with an Energy Input Signal

---

[5] Technically, we must check to see if the infinite sum for the output signal converges. For the system considered in this text, this series does converge.

$$x_\gamma(t) = \begin{cases} x(t), & |t| \le \gamma \\ 0, & |t| > \gamma \end{cases} \tag{15.1.21}$$

Let

$$X_\gamma(\omega) = \mathcal{F}\{x_\gamma(t)\} \tag{15.1.22}$$

be the Fourier transform of the truncated signal in (15.1.21). This energy spectral density in (15.1.22) will exist if the signal $x_\gamma(t)$ is bounded. The energy spectral density of the output signal is

$$D_{y_\gamma}(\omega) = |Y_\gamma(\omega)|^2 = |H(\omega)|^2 |X_\gamma(\omega)|^2 = |H(\omega)|^2 D_{x_\gamma}(\omega). \tag{15.1.23}$$

Letting $\gamma \to \infty$ the power spectral density of the output signal is

$$S_y(\omega) = \lim_{\gamma \to \infty} \frac{D_{y_\gamma}(\omega)}{\gamma} = |H(\omega)|^2 \left[\lim_{\gamma \to \infty} \frac{D_{x_\gamma}(\omega)}{\gamma} \right] = |H(\omega)|^2 S_x(\omega) \tag{15.1.24}$$

provided the limit exists. For the systems considered in this text, the limit in (15.1.24) will exist if the input signal has a well-defined power spectral density.

15.1.6 Summary

All of the results above are summarized in the following theorem. This theorem is a major result describing how signals propagate through systems.

Theorem 15.1.1: Consider a linear, time invariant, BIBO stable system with a frequency response function $H(\omega)$. Suppose that $x(t)$ is an input signal to this system.
(1) If input signal $x(t)$ is an energy signal with a energy spectral density $D_x(\omega)$, then the energy spectral density of the output signal, $D_y(\omega)$, is

$$D_y(\omega) = |H(\omega)|^2 D_x(\omega).$$

(2) If the input signal $x(t)$ is a power signal with a power spectral density $S_x(\omega)$, then the power spectral density of the output signal, $S_y(\omega)$, is

$$S_y(\omega) = |H(\omega)|^2 S_x(\omega).$$

▲▲

This theorem is the primary tool we will use to study the propagation of a signal through a system. We know that a signal is changed as it propagates though a system. This change can be characterized by comparing the input signal to the output signal and the system. In the following sections we will compare the input signal to the output signal using their frequency domain characterizations as expressed by Theorem 15.1.1 in graphical form. This comparison rests on the definition of a system as the relationship between two signals.

Note: In the analysis below we will fix the system parameters and vary the signal parameters. It should be remembered, however, that it is the *relationship between* the signal and the system parameters that is the heart of the matter.

15.2 INTERPRETATION OF THE FREQUENCY RESPONSE THEOREM

15.2.1 Introduction

In the last section we discussed the relationship between a sinusoidal input signal, the system, and the output signal in terms of the energy or power spectral density. In this section we will present a graphical interpretation of Theorem 15.1.1. This discussion will highlight the essential tools that are required to analyze the propagation of more complicated signals through a system. These tools will be used extensively in the following sections. Finally, we present a simple example to illustrate how these tools can be used in the analysis of a signal propagating through a system. This example is used to single out the important components of the analysis in the sections that follow.

15.2.2 Fourier Series Interpretation of the Frequency Response Theorem

The effect of the system on a sinusoidal signal propagating through it can be explained graphically using two-sided amplitude and phase spectra[6] of the input and output signals. To construct these graphs, first we write the sinusoidal input signal as an exponential Fourier series

$$x(t) = A_1 \cos(\omega_1 t + \theta_1) = \left(\frac{A_1}{2} e^{j\theta_1}\right) e^{j\omega_1 t} + \left(\frac{A_1}{2} e^{-j\theta_1}\right) e^{-j\omega_1 t} \qquad (15.2.1)$$

$$= X_1 e^{j\omega_1 t} + X_{-1} e^{-j\omega_1 t}.$$

If the frequency response function of the system is $H(\omega)$, by the frequency response theorem the exponential Fourier series of the output signal is

$$y(t) = A_1 |H(\omega_1)| \cos(\omega_1 t + \theta_1) \qquad (15.2.2)$$

$$= \left[\frac{A_1 |H(\omega_1)|}{2} e^{j(\theta_1 + \angle H(\omega_1))}\right] e^{j\omega_1 t} + \left[\frac{A_1 |H(\omega_1)|}{2} e^{-j(\theta_1 + \angle H(\omega_1))}\right] e^{-j\omega_1 t}$$

$$= Y_1 e^{j\omega_1 t} + Y_{-1} e^{-j\omega_1 t}.$$

---

[6] In principle one-sided spectra could also be used, but the two-sided spectra are the type of graph most commonly used in practice. Amplitude and phase spectrums are defined in Section 8.1.

From (15.2.1) and (15.2.2) we see that

$$|Y_1| = |H(\omega_1)||X_1| \quad \text{and} \quad \angle Y_1 = \angle X_1 + \angle H(\omega_1). \tag{15.2.3}$$

The magnitude relationship in (15.2.3) can be seen by plotting on the same graph the amplitude spectrum of the input signal, the magnitude of the frequency response function, and the amplitude spectrum of the output signal. A similar graph is constructed for the phase spectrum. This idea is illustrated by the following example.

Example 15.2.1: This example is a continuation of Example 14.1.3 or 14.2.3. The system is an *RC* network shown in Figure 15.2.1. The frequency response function of this network is

$$H(\omega) = \frac{1}{1+j\omega} = \frac{1}{\sqrt{1+\omega^2}} e^{-j\tan^{-1}(\omega)}. \tag{15.2.4}$$

The input signal is

$$x(t) = \cos(t) = X_{-1}e^{-j\omega_1 t} + X_1 e^{j\omega_1 t} = \frac{1}{2}e^{-jt} + \frac{1}{2}e^{jt}. \tag{15.2.5}$$

The value of the frequency response function at the input signal's frequency $\omega_1 = 1$ rad/sec is

$$|H(\omega_1)| = |H(1)| = \frac{1}{\sqrt{2}} = 0.707. \tag{15.2.6}$$

The magnitude of the output signal is

$$|Y_1| = |H(\omega_1)||X_1| = (0.707)(0.5) = 0.353. \tag{15.2.7}$$

The magnitude of the frequency response function for this network is shown in Figure 15.2.2 along with the amplitude spectrum of the input Fourier series and the output Fourier series. In Figure 15.2.2 the magnitude of the input signal (the black dot) is multiplied by the magnitude of the frequency response at $\omega_1 = 1$ to yield the magnitude of the output signal (the circle).

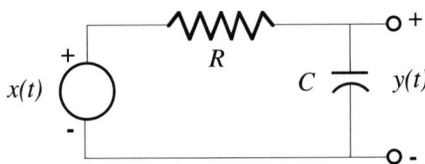

Figure 15.2.1 An *RC* Network With *RC* = 1

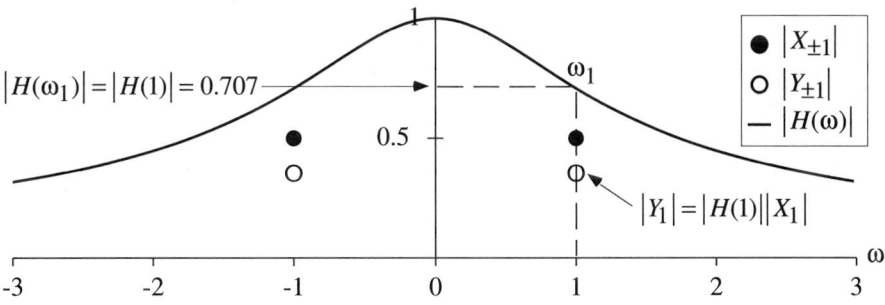

Figure 15.2.2 Magnitude of the Frequency Response Function and Amplitude Spectra of the Input and Output Signals

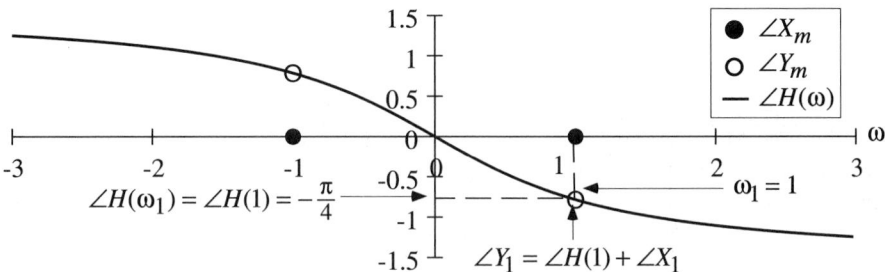

Figure 15.2.3 Phase of the Frequency Response Function of an *RC* Network and Phase Spectra of the Input and Output Signals

The phase of the frequency response function is shown in Figure 15.2.3. The phase of the frequency response function at the frequency of the input signal, $\omega_1 = 1$ rad/sec, is

$$\angle H(\omega_1) = \angle H(1) = -\frac{\pi}{4} \qquad (15.2.8)$$

as shown in Figure 15.2.3. The phase of the output signal is determined from

$$\angle Y_1 = \angle X_1 + \angle H(\omega_1) = \angle X_1 + \angle H(1) = 0 - \frac{\pi}{4} = -\frac{\pi}{4}. \qquad (15.2.9)$$

Figure 15.2.3 clearly shows the phase spectrum of the output signal (the circle) as the sum of the phase of the input signal (the black dot) and the contribution from the system via the frequency response function at $\omega_1 = 1$.

Note: When we consider the magnitude of the input and output signals we *multiply* the quantities in Figure 15.2.2. When we consider the phase of the input and output signals we *add* the quantities in Figure 15.2.3.

▲▲

We also have a power spectrum interpretation of the frequency response theorem using the Fourier series. Recall that the (two-sided) power spectrum[7] of the Fourier series (15.2.1) is

$$S_x(\omega) = \begin{cases} |X_m|^2, & \omega = m\omega_0 \\ 0, & \omega \neq m\omega_0 \end{cases} \tag{15.2.10}$$

Similarly, the power spectrum of the output signal is the graph

$$S_y(\omega) = \begin{cases} |X_m|^2 |H(\omega_m)|^2, & \omega = m\omega_0 \\ 0, & \omega \neq m\omega_0 \end{cases} \tag{15.2.11}$$

$$= \begin{cases} |Y_m|^2, & \omega = m\omega_0 \\ 0, & \omega \neq m\omega_0 \end{cases}$$

Again this information can be displayed graphically as in Example 15.2.1. In this case the magnitude squared of the frequency response function is plotted. The graphical interpretation of this power calculation is described in Section 14.5.

15.2.3 An Example

We conclude this section with an example that illustrates the use of the concepts discussed in this section to analyze a signal propagating through a system. This example emphasizes the relationship between the power spectrums of the input and output signals and the signals themselves. This example also introduces the analysis in the next two sections.

We begin with a relatively simple input signal

$$x(t) = x_1(t) + x_2(t) = 2\cos(0.5t) + (1)\cos(5t). \tag{15.2.12}$$

This signal is shown in Figure 15.2.4. Note that this signal is a Fourier series.

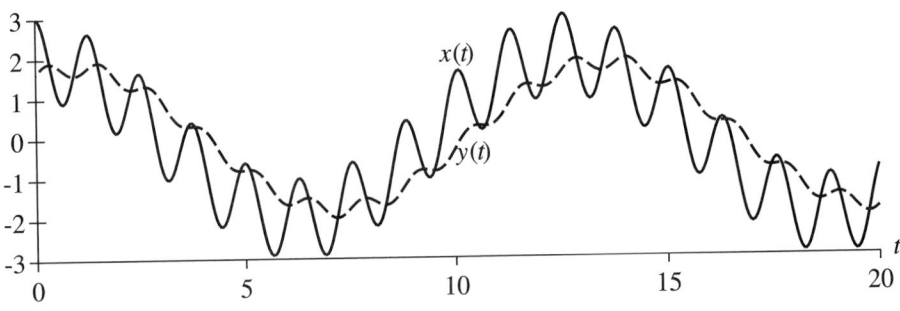

Figure 15.2.4 Input and Output Signals to an *RC* Network

---

7 See Definition 8.5.5.

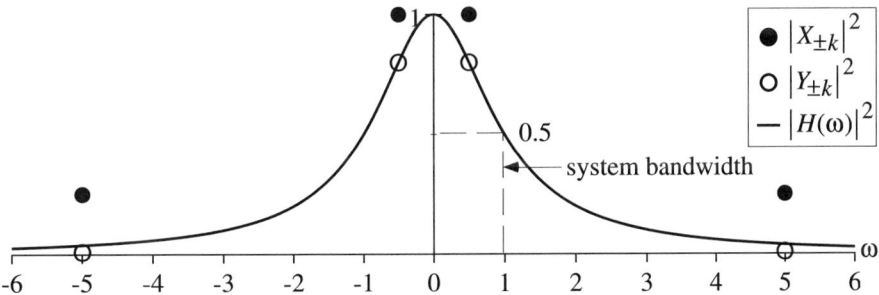

Figure 15.2.5 Magnitude Squared of the Frequency Response Function
of the *RC* Filter

Suppose that this signal is the input signal to the *RC* network shown in Figure 15.2.1.
The magnitude of the frequency response function squared of this system is shown in
Figure 15.2.5 along with the power spectral density of the input and output signals.
This system is a lowpass filter with a half-power bandwidth of $\beta = 1$ rad/sec as shown
in Figure 15.2.5. This frequency divides the passband from the stopband. We have
fixed the bandwidth of this filter at $\beta = 1$ rad/sec. The results below are discussed
with respect to this normalization.

   To study how the signal (15.2.12) propagates through the *RC* network we will use
a Fourier series analysis, the input signal being a simple Fourier series consisting of
two terms. In view of Figure 15.1.4 we first consider each term of the input signal
separately propagating through the system. Then we will reassemble the output
signal from each of these terms. For the input signal $x_1(t)$ with frequency $\omega_1 = 0.5$
rad/sec the corresponding output signal is

$$y_1(t) = (2)|H(0.5)|\cos(0.5t + \angle H(0.5)) \tag{15.2.13}$$

$$= \frac{2}{\sqrt{1+(0.5)^2}}\cos\left(0.5t - \tan^{-1}(0.5)\right) = (1.8)\cos(0.5t - 0.46).$$

For the input signal $x_2(t)$ with frequency $\omega_2 = 5$ rad/sec the corresponding output
signal is

$$y_2(t) = (1)|H(5)|\cos(5t + \angle H(5)) = (0.2)\cos(5t - 1.37). \tag{15.2.14}$$

The output signal is the sum of the two signals in (15.2.13) and (15.2.14)

$$y(t) = y_1(t) + y_2(t) = (1.8)\cos(0.5t - 0.46) + (0.2)\cos(5t - 1.37), \tag{15.2.15}$$

and it is also shown in Figure 15.2.4.

   In Figure 15.2.4 the signal $x_2(t)$ is clearly visible in $x(t)$. The output signal in
Figure 15.2.4 consists mainly of the component due to $x_1(t)$. The component of the
output signal due to $x_2(t)$ has been greatly attenuated. This observation can be

Table 15.2.1 Power in the Input and Output Signals

Input Signal	Power	$|H(\omega_m)|^2$	Output Signal	Power	% Change
$x_1(t)$	2	0.8	$y_1(t)$	1.6	20 %
$x_2(t)$	0.5	0.04	$y_2(t)$	0.02	96 %
$x(t)$	2.5	–	$y(t)$	1.62	35 %

quantified by calculating the power in these signals as shown in Table 15.2.1.

Here the total power in the output signal has been slightly reduced over the power of the input signal as shown in the last line in Table 15.2.1. The power distribution between the two signal components in the output signal is significantly changed relative to the power distribution of these two components in the input signal, however. The power in the output signal $y_1(t)$ has been slightly reduced over the power in the input signal. The output signal $y_2(t)$ is significantly attenuated and the power is substantially reduced. This relative redistribution of the power between the components of the signal accounts for the change in the output signal relative to the input signal in Figure 15.2.4. This change can be traced to the shape of the frequency response function. Note the relative values of the frequency response function at the frequencies of the two components in the input signal in Table 15.2.1. In view of the interpretations of this type of graph presented in the previous subsections of this chapter, it is easy to see that Figure 15.2.5 concisely summarizes the calculations in Table 15.2.1.

The essential points of this example are as follows.

1. The input and output signals are defined by two sinusoids at given frequencies. Each signal is essentially defined by the *relative* amplitudes and phase angles between the components of the signal. The amplitudes determine the relative power distribution in the signal between the two sinusoidal components.

2. The system changes the amplitude and phase of each component of the output signal relative to that particular component of the input signal, but it doesn't change the frequencies of the two components. Hence, we can study the effect of the system on the input signal by comparing each component of the input signal to the corresponding component of the output signal of the same frequency. The change of the amplitude and phase of each component of the output signal relative to that particular component of the input signal is shown in Table 15.2.1. The change in each of the components of the input signal as they pass through the system is determined by the frequency of the input signal and the value of the frequency response function of the system at that frequency.

3. The system changes the amplitude and phase of the components the output signal *relative to each other*. This change resulted in a redistribution of the power between the two components in the output signal when the output signal is compared to the input signal. See Table 15.2.1. This change in the power distribution accounts for the overall change in the output signal compared to the input signal

4. The relative change between the sinusoidal components of the input signal as it passes through the system is determined from the *shape* of the frequency

response function over the range of frequencies of the sinusoidal components in the input signal as shown in Figure 15.2.5. The first sinusoidal component of the input signal, $\omega_1 = 0.5$, is in the passband of the system so the amplitude of that input signal is only slightly attenuated.   The second sinusoidal component of the input signal, $\omega_2 = 5$, is in the stopband of the system so that the amplitude of that component of the input signal is greatly attenuated in the output signal. Combining these two components, we have the output signal shown in Figure 15.2.4.

15.2.4 MATLAB Experiments

The following M-file implements the example in the last subsection. In order to compute the output signal it is necessary to evaluate the frequency response function at the frequency of the sinusoidal components of the input signal. For the simple *RC* network in Figure 15.2.2 this calculation can be done analytically.   For more complicated systems this calculation is far more difficult.   Therefore, this M-file implements this calculation using the **bode** command. This command only requires the transfer function of the system, not the frequency response function. For an explanation of the **bode** command, see Section 16.4.

Example in Subsection 15.2.3

```
clear
% Define system
RC = 1;                % RC time constant
b = [1/RC];            % define numerator
a = [1,1/RC];          % define denominator

% Define time vector
ic = 0.05;             % time increment
tf = 20;               % final time
t = [0:ic:tf];         % define time vector

% Define input Signal
w = [.5,5];                                    % frequencies in input signal
xamp = [2,1];                                  % amplitudes in input signal
x = xamp(1)*cos(w(1)*t) + xamp(2)*cos(w(2)*t);  % input signal

% Calculate output signal
[mag,phs] = bode(b,a,w);                       % calculate amplitude and phase of output signal
phs = (pi/180)*phs;                            % change phase to radians
yamp = xamp.*mag';                             % amplitudes of output signal
y = yamp(1)*cos(w(1)*t + phs(1)) + yamp(2)*cos(w(2)*t + phs(2));  % output signal

% Power calculations
wend = 6;                                      % define frequency interval
ww = linspace(-wend,wend,300);                 % frequency vector for system
[magrc,phsrc] = bode(b,a,ww);                  % frequency response function
pwr = magrc.^2;                                % magnitude square of frequency response function
```

```
% Two-sided power spectrum of input signal
xamp2 = [fliplr(xamp),xamp]./2;          % two-sided amplitude spectrum
xpwr = xamp2.^2;                          % two-sided power spectrum
ws = [-fliplr(w),w];                      % frequency vector for signals


% Two-sided power spectrum of output signal
yamp2 = [fliplr(yamp),yamp]./2;          % two-sided amplitude spectrum
ypwr = yamp2.^2;                          % two-sided power spectrum
```

```
figure(1)                                 figure(2)
plot(t,x,t,y,'--')                        plot(ww,pwr,ws,xpwr,'*',ws,ypwr,'o')
xlabel('Time')                            Title('Power Calculations')
Title('Input and Output Signal of RC Network')   xlabel('Frequency, rad/sec')
```

Exploratory Exercise 15.2.2: In the above M-file vary the frequency of the second component of the input signal between 1 and 10 rad/sec. Observe the change in the output signal and the power spectral density. To correctly plot the power spectra, the variable wend must be modified to select the correct frequency interval for graphing the power spectral density.

▲▲

Exploratory Exercise 15.2.3: Replace the simple *RC* network by the more complicated 3rd-order Butterworth filter. Compare the effect of this system on the input signal with the *RC* network.

▲▲

15.3 PROPAGATION OF A PULSE TRAIN THROUGH A NETWORK

15.3.1 Introduction

In this section we will use the concepts developed in the last section to analyze a pulse train propagating through the *RC* network as shown in Figure 15.3.1. This input signal was selected because it is easy to see the effects of the system on this signal. The results are sufficiently general to extend this analysis to any periodic power signal, however.

We will analyze the signal propagating through this system by the method summarized in Figure 15.1.4. This analysis consists of:

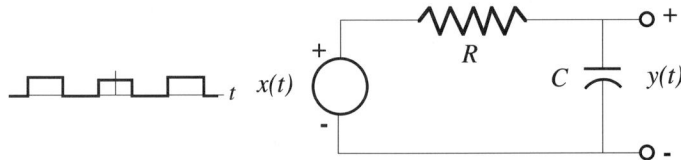

Figure 15.3.1 *RC* Network with a Pulse Train as an Input Signal

1. The periodic pulse train is expressed as a Fourier series.
2. Each term of the Fourier series is passed separately through the system. The frequency response theorem is used to calculate the output signal.
3. The output signal is constructed as a Fourier series from each term of the input signal after it is passed through the system.

   The periodic output signal is compared to the pulse train (input signal) to identify the effect of the system on the pulse train as it passes through the system. This analysis will have two parts. First, the output signal itself will be compared to the input signal. This comparison is useful here, but it doesn't lead to deep insight into the relationship between the signal and the system. In a second comparison the amplitude and phase spectrum as well as the power spectrum of the input and output signals will be compared to each other and to the frequency response function of the system. This analysis in the frequency domain is quite general, and provides deep insight into the propagation of a signal through a system.

   The input signal to the system in Figure 15.3.1 is the pulse train shown in Figure 15.3.2. We will consider three sets of parameters for this pulse train as shown in Table 15.3.1. These signals are shown in Figure 15.3.3. The signal amplitude A is chosen to give convenient scales to the graphs below.

   The Fourier series for the pulse train[8] in Figure 15.3.2 is

$$x(t) = \frac{A\varepsilon}{T_0} + \sum_{\substack{m=1 \\ m\,\text{odd}}}^{\infty} \frac{2A\varepsilon}{T_0} \left| \text{Sa}\left(\frac{m\omega_0\varepsilon}{2} \right) \right| \cos(m\omega_0 t + \theta_m) \tag{15.3.1}$$

$$\theta_m = \begin{cases} 0, & \text{Sa}\left(\dfrac{m\omega_0\varepsilon}{2} \right) \geq 0 \\[2ex] \pi, & \text{Sa}\left(\dfrac{m\omega_0\varepsilon}{2} \right) < 0 \end{cases}$$

When plotting the amplitude and phase spectrum of a Fourier series, we will plot the two-sided amplitude and phase spectrum of the exponential Fourier series

$$x(t) = \sum_{m=-\infty}^{\infty} X_m e^{-jm\omega_0 t} = \sum_{m=-\infty}^{\infty} \left[\frac{A\varepsilon}{T_0} \left| \text{Sa}\left(\frac{m\omega_0\varepsilon}{2} \right) \right| e^{j\theta_m} \right] e^{-jm\omega_0 t}. \tag{15.3.2}$$

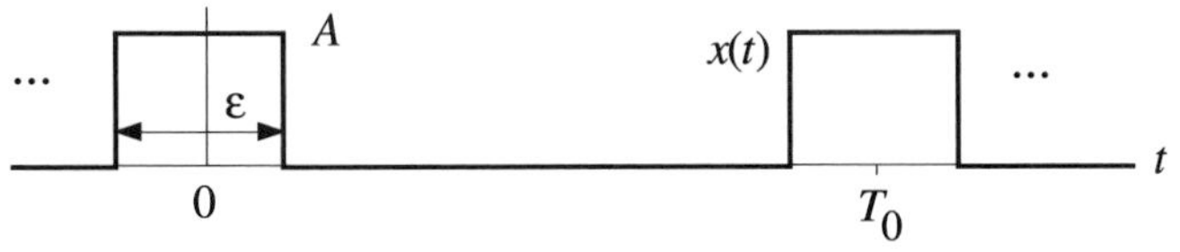

Figure 15.3.2 Periodic Input Signal

---

[8] See Example 7.3.4.

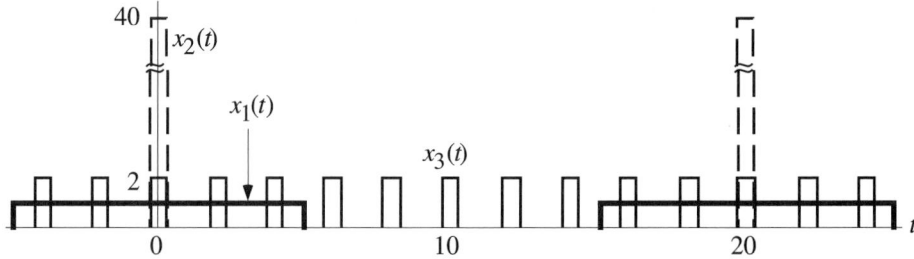

Figure 15.3.3 Input Signals to RC Network

Table 15.3.1 Parameter Values for the Signal in Figure 15.3.2

	Input Signal	Period, T_0	Pulse Width, ε	Amplitude, A
Case 1	$x_1(t)$	20 sec	10	2
Case 2	$x_2(t)$	20 sec	0.5	40
Case 3	$x_3(t)$	2 sec	0.5	4

We will follow the same convention with the output signal. The conversion between these two representations is summarized in Table 7.6.1.

The frequency response of the RC network in Figure 15.3.1 is discussed in Examples 14.1.3 and 14.2.3. The frequency response function for the RC network is

$$H(\omega) = \frac{1}{\sqrt{1+\omega^2}} e^{-j\tan^{-1}(\omega)}. \tag{15.3.3}$$

From the frequency response theorem the steady state response of the network is

$$y(t) = |H(0)|\frac{A\varepsilon}{T_0} \tag{15.3.4}$$

$$+ \sum_{\substack{m=1 \\ m\,\text{odd}}}^{\infty} \left[\frac{1}{\sqrt{1+(m\omega_0)^2}} \right] \left[\frac{2A\varepsilon}{T_0} \text{Sa}\left(\frac{m\omega_0\varepsilon}{2}\right) \right] \cos\left(m\omega_0 t + \theta_m - \tan^{-1}(m\omega_0)\right).$$

15.3.2 Case 1: Input Signal $x_1(t)$

The input signal $x_1(t)$ is shown in Figure 15.3.4 along with a twenty-term approximation of the corresponding output signal $y_{0:20}(t)$.

Notation: In this section we will present several graphs where the output signal is computed by summing several components of a Fourier series as in Figure 15.3.4. We will denote these approximating signals by a "hat" notation as in $\hat{y}_1(t) = y_{0:20}(t)$ without expressly giving the number of terms in the approximation.

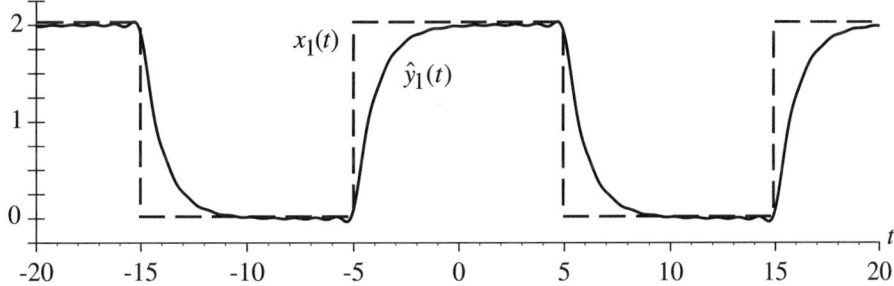

Figure 15.3.4 Case 1: Input and Output Signal

Note: The analysis that follows relies on the approximation of a periodic signal by a finite number of terms of the Fourier series as discussed in Section 8.6. In Figure 15.3.4 the ripples preceding the pulse are due to this method of approximation of the output signal. The system is causal; it doesn't respond before the pulse is switched on.

The amplitude spectrum of the input and output signal along with the magnitude of the frequency response function is shown in Figure 15.3.5. Note that the bandwidth of this system is $\beta = 1$ rad/sec. In Figure 15.3.5 we see that the amplitude spectrum

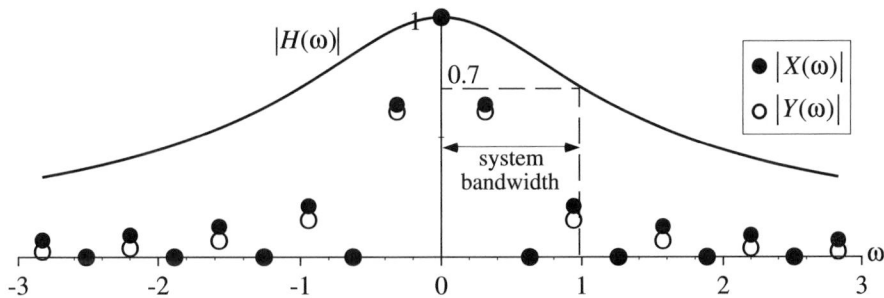

Figure 15.3.5 Case 1: Amplitude Spectrum of the Input and Output Signal

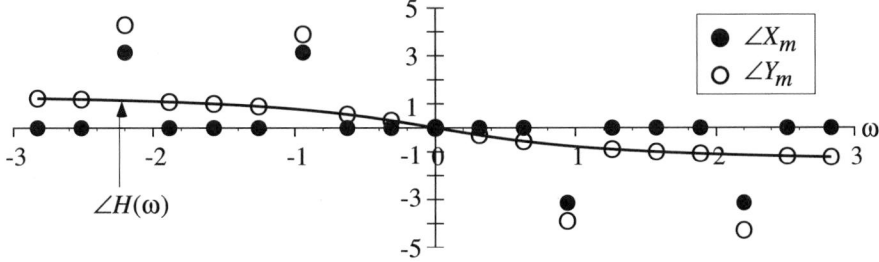

Figure 15.3.6 Case 1: Phase Spectrum of the Input and Output Signal

of the output signal is only a slightly distorted version of the amplitude spectrum input signal. The difference between these two spectrums occurs at high frequencies. These high frequency components contribute mainly to the "corners" of the output signal. Since these components are filtered out by the system, the output signal is rounded as shown in Figure 15.3.4.

The phase spectrum of the input and output signals along with the phase of the frequency response function are shown in Figure 15.3.6. In Figure 15.3.6 we see that the system has introduced phase distortion in the high frequency components. This distortion contributes to the pulse spreading in the output signal shown in Figure 15.3.4.

15.3.3 Case 2: Input Signal $x_2(t)$

When $x_2(t)$ is the input signal, the corresponding output signal is shown in Figure 15.3.7. In comparing $y_2(t)$ in Figure 15.3.7 with $x_2(t)$ we see a different situation. This output signal in Figure 15.3.7 is badly distorted; it is barely recognizable as a copy of the input signal. The amplitude spectrum of the input signal and output signal are shown Figure 15.3.8. In comparing the amplitude spectrum of each term in the Fourier series of the input signal to amplitude spectrum of each term in the Fourier series of the output signal, Figure 15.3.8 shows that some terms with a large amplitude are significantly reduced by the frequency response function. The action of the system is to effectively remove these terms from the output signal. This change in the high frequency amplitude coefficients of the Fourier series of the output signal results in significant signal distortion of the output signal relative to the input signal.

The distortion of the output signal can also be seen by comparing the phase spectrum of the input signal with the phase spectrum of the output signal shown in Figure 15.3.9. The distortion in the phase shown in Figure 15.3.9 is correlated with the significant pulse spreading in the output signal in Figure 15.3.7. The phase distortion also contributes to the delay introduced into the output signal.

15.3.4 Comparison: Case 1 and Case 2

We have considered two periodic signals that are input signals to a *RC* network. The first signal $x_1(t)$ propagated through this system with little distortion. In contrast, the second signal $x_2(t)$ was severely distorted after passing through the system. This difference in the way these signals propagate through this system can be explained

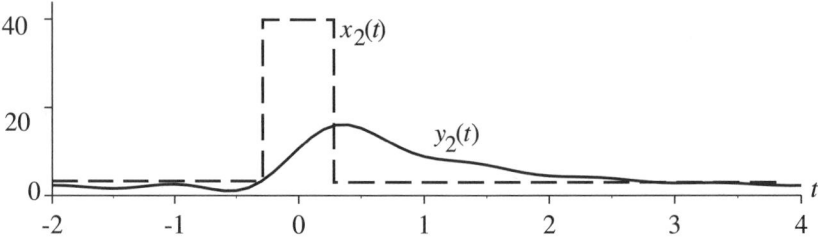

Figure 15.3.7 Output Signal $y_2(t)$ for Input Signal $x_2(t)$

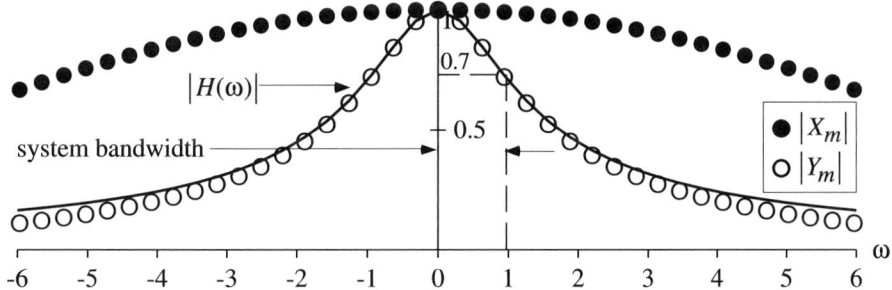

Figure 15.3.8 Case 2: Amplitude Spectrum of the Input and Output Signal

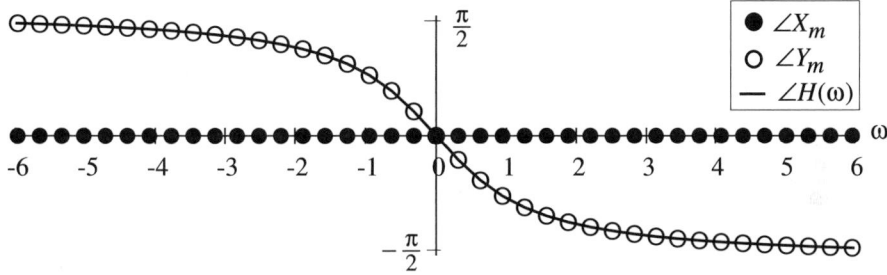

Figure 15.3.9 Case 2: Phase Spectra of the Input and Output Signals

using the power spectra of the input and output signals and the magnitude of the frequency response (squared) of the system. The power spectra of the signals $x_1(t)$ and $y_1(t)$ are shown in Figure 15.3.10. The power spectra of the signals $x_2(t)$ and $y_2(t)$ are shown in Figure 15.3.11. First, note that the period of the two input signals is the same but the width of the pulse has been decreased. As the pulse width is reduced, the power spectrum of the signal in spread over a larger frequency band. This difference is observed by comparing Figure 15.3.10 with Figure 15.3.11.

Note: The frequency scales in Figure 15.3.10 and Figure 15.3.11 are different.

This difference in the output signals for each of these input signals can be explained by examining the relationship between the power distribution of these two input signals with the square of the frequency response function of the system. This system, the *RC* network, is a lowpass filter with a bandwidth of 1 rad/sec.

In Figure 15.3.10 we see that the terms that carry most of the power in the input signal are within the passband of the system. These terms are only slightly attenuated. As a result, the output signal in Figure 15.3.4 is only slightly distorted. The high frequency components that contribute to the "corners" of the input signal are eliminated because the frequency response function goes to zero at high

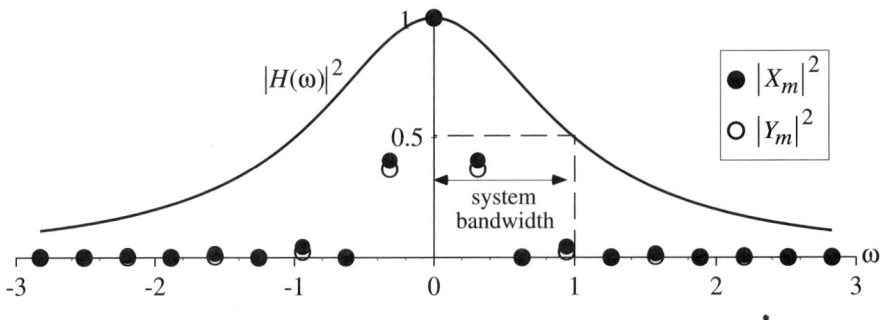

Figure 15.3.10 Power Spectrum of the Signals $x_1(t)$ and $y_1(t)$

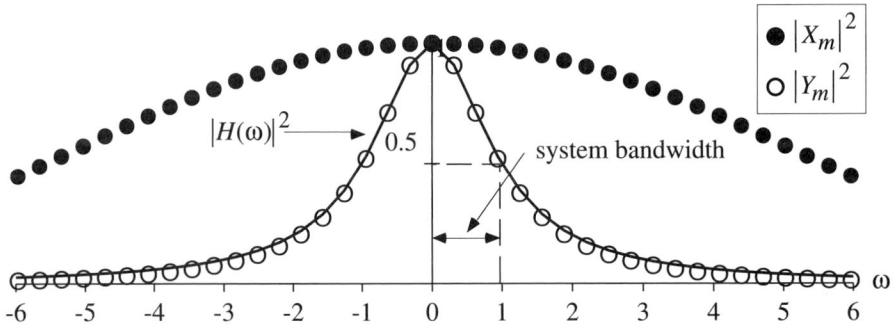

Figure 15.3.11 Power Spectrum of the Signals $x_2(t)$ and $y_2(t)$

frequencies. Hence, the output signal is smoothed out, and a certain amount of signal spreading occurs.

In Figure 15.3.11 we see that the power in the input signal extends significantly beyond the passband of the system. As a result, the power spectrum of the output signal is substantially different from the power spectrum of the input signal. The terms in the input signal that are outside the passband of the system are essentially removed from the output signal. The attenuation of these terms results in the severe signal distortion shown in Figure 15.3.7.

The power spectrum of the output signal $y_2(t)$ conforms to the magnitude of the frequency response function (squared) of the system. In effect, the system will allow only signals with their power in the passband of the system to pass through the system unattenuated. Hence, only signals with their power spectrum in the passband of the system can pass through the system with little attenuation.

This discussion captures the essence of filtering.

15.3.5 A Pulse Train

In the two examples above the pulses were widely separated so that the effect of the system on the pulse train could be studied by investigating the effect of the system on each pulse separately. If we reduce the period of the signal, however, then the pulses can't be considered separately.

Consider the signal $x_3(t)$. This signal is similar to signal $x_2(t)$, but the period of $x_3(t)$ has been reduced. The input and output signals for Case 3 are shown in Figure 15.3.12. Here we see even more distortion than in Figure 15.3.7. This distortion can be explained by a power spectrum analysis. The power spectra of the input and output signals for Case 3 are shown in Figure 15.3.13. By reducing the period of $x_3(t)$, its power spectrum occupies an even larger frequency band than the power spectrum of $x_2(t)$. This difference can be seen by comparing Figure 15.3.13 to Figure 15.3.11 (note the scales in these two figures). Reducing the period of the pulse train forces even more power into the high frequency components of the signal. These components are outside the passband of the system. Hence, the system attenuates more terms in $x_3(t)$ that contain a larger percentage of power of that signal. This filtering leads to larger distortion in the output signal $y_3(t)$.

15.3.6 Summary of the Propagation of the Pulse Train

We have discussed the propagation of a pulse train through a system. When this signal propagates through the system, it experiences distortion. This distortion is reflected in the change in the power of the output signal as compared to the input

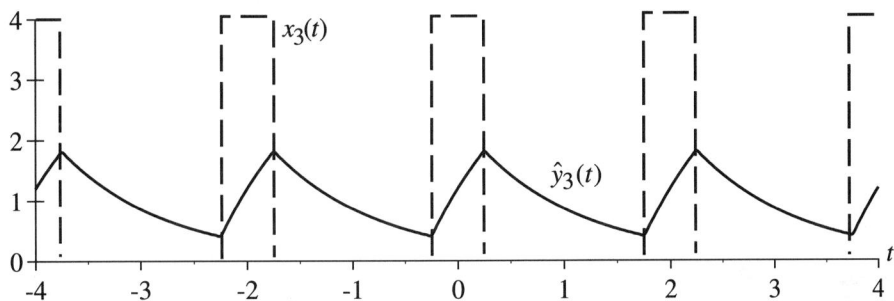

Figure 15.3.12 Input and Output Signals $x_3(t)$ and $y_3(t)$

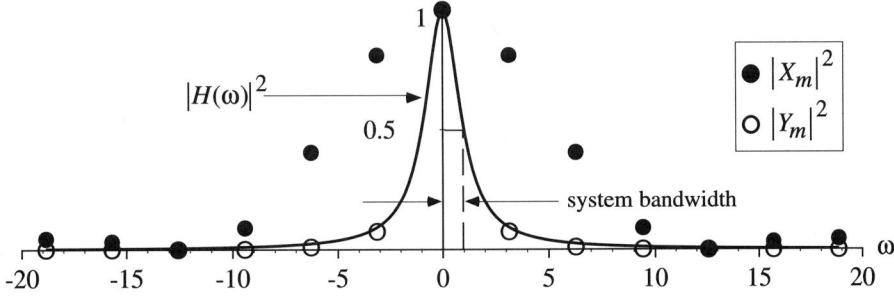

Figure 15.3.13 Case 3: Power Spectra of the Input and Output Signals

Table 15.3.2 Total Power in The Input and Output Signals

Input Signal	Power	Output Signal	Power	% Change
$x_1(t)$	2	$y_1(t)$	1.80	10
$x_2(t)$	40	$y_2(t)$	8.52	78.5
$x_3(t)$	4	$y_2(t)$	1.17	70.1

signal. This change is shown in Table 15.3.2 for the three signals discussed above. To calculate the power in the input signal we use Theorem 8.5.1. The power in the output signal is approximated by summing a large number of the terms in the Fourier series of the output signal (15.3.4).

The signal $x_1(t)$ will pass through the system with little distortion because the power spectrum of the input signal is mostly in the passband of the system. Table 15.3.2 shows that this signal experiences little change in the power. We can modify this signal by decreasing the pulse width, as with $x_2(t)$, or decreasing the period of the signal, as with $x_3(t)$. In either case some of the power in the signal is pushed into the higher frequency terms of the Fourier series of the input signal. As a result, these modified signals are severely distorted when they pass through the system. This change in power is shown in Table 15.3.2. When the power of the signals $y_2(t)$ and $y_3(t)$ are compared to $y_1(t)$, we clearly see the effect of the system. These observations can be summarized as follows.

Signal - System Interaction 15.3.1: Given a system with a periodic power signal propagating through it, the system will change the relative power distribution between the sinusoidal components of the signal when the input signal is compared to the output signal. The distortion induced by the system is determined by the relationship between the power distribution (spectral content) of the input signal and the frequency response function.

▲▲

Signal - System Interaction 15.3.2: If the power spectrum of the input signal is contained primarily within the passband of the system, then the signal will pass through the system with little distortion. If the components of the Fourier series of the input signal with significant power fall outside the passband of the system, then these components will be attenuated and the signal will suffer attenuation and distortion as it passes through the system.

▲▲

15.3.7 MATLAB Experiments

The following M-file implements the example discussed in this section. The overall structure of this M-file follows closely the M-file in Section 15.2. The time domain graph of the input signal follows the example in Section 5.3. The calculation of the Fourier series follows the example given in Section 8.8.

Example in Section 15.3

```
clear
% Define the signal parameters
T0 = 20;                         % period of the signal
ep = 10;                         % pulse width, less than T0
t0 = 0;                          % offset
A = 2;                           % amplitude
w0 = 2*pi/T0;                    % fundamental frequency
nterm = 20;                      % number of terms in the Fourier series computation
npwr = 15;                       % number of terms in power spectrum plots, less than nterm
ws = [-npwr:npwr]*w0;            % frequency axis for power spectrum plots
step = 0.01;                     % increment in time vector
t = [-T0:step:T0];               % time vector for waveforms

% Construct input signal
x = zeros(size(t));
i1 = max(find(t<t0-ep/2));
i2 = min(find(t>t0+ep/2));
x(i1:i2) = A*ones(1,i2-i1+1);

% Calculate coefficients of exponential Fourier series
X0 = A*ep/T0;                    % m = 0  term
for kk = 1:nterm
    X(kk) = X0*(exp(-j*kk*w0*t0))*sinc(kk*ep/T0);   % positive coefficients
end

% Calculate coefficients of cosine Fourier series
Am = 2*abs(X);
qm = angle(X);

% Define system
RC = 1;                          % RC time constant
b = [1/RC];                      % define numerator
a = [1,1/RC];                    % define denominator

% Calculate output signal
% Calculate frequency response of system
wo = [0:nterm]*w0;
[mag,phs] = bode(b,a,wo);        % calculate amplitude and phase of output signal
phs = (pi/180)*phs;              % change phase to radians
Am = [X0,Am];
qm = [0,qm];
Ym = Am.*mag';                   % amplitudes of output signal

% Calculate Fourier series of output signal
y = Ym(1)*ones(size(t));                          % constant term
for jj = 2:nterm
    y = y + Ym(jj)*cos(wo(jj)*t+qm(jj)+phs(jj));      % output signal
end
```

```
% Power calculations
% Calculate magnitude of frequency response squared
ww = linspace(min(ws),max(ws),200);      % frequency vector for system
[magrc,phsrc] = bode(b,a,ww);            % frequency response function
pwrsys = magrc.^2;                       % magnitude square of frequency response function

% Calculate two-sided power spectrum of input signal
Pt = abs(X(1:npwr)).^2;
Pwrx = [fliplr(Pt),X0^2,Pt];             % Power spectrum

% Calculate two-sided power spectrum of output signal
Pty = abs(Ym(2:npwr+1)./2).^2;
Pwry = [fliplr(Pty),Ym(1)^2,Pty];        % Power spectrum

% Plot input and output signals
figure(1)
plot(t,x,t,y,'--')
title('Fourier Series')
xlabel('Time')

% Plot power spectrum
figure(2)
plot(ww,pwrsys,ws,Pwrx,'*',ws,Pwry,'o')
title('Two-Sided Power Spectrum')
xlabel('Frequency, rad/sec')
```

▲▲

Exploratory Exercise 15.3.3: Modify this M-file to plot the amplitude and phase spectra of the input and output signals. Verify the results in this section.

▲▲

Exploratory Exercise 15.3.4: Replace the simple *RC* network by the more complicated 3rd-order Butterworth filter. Compare the effect of this system on the input signal with the *RC* network.

▲▲

Exploratory Exercise 15.3.5: Modify this M-file so that it uses a Fourier series of a triangular wave. Explain your time domain results in terms of the power spectrum plots.

▲▲

Exploratory Exercise 15.3.6: Modify this M-file so that the input signal is the modulated square wave discussed in Section 8.8.3. Explain your time domain results in terms of the power spectrum plots. Next, modify the system by designing a bandpass filer that will allow this signal to propagate through this system with little distortion. Verify these results in the time and frequency domain. Determine the change in the total power of the output signal as a function of the bandwidth of the system.

▲▲

15.4 PROPAGATION OF ENERGY SIGNALS THROUGH A SYSTEM

15.4.1 Introduction

In the last section we characterized the propagation of a periodic power signal through a system. That analysis uses the amplitude and phase spectrum of the Fourier series of the input and output signal and the frequency response function of the system. We found that if many terms of the power spectrum of the signal that carry the power of the signal are contained within the passband of the system, then the signal suffered little distortion when it propagated through the system. If many of the terms of the power spectrum that carry the power of the signal are outside the passband (in the stopband) of the system, then these terms are blocked by the system and the signal suffers severe distortion.

It this section we extend these results to aperiodic signals using the Fourier transform. The method of analysis, which is the same as the last section, is outlined in Section 15.1. We will compare the input and output signal to each other to determine the effect of the system on the signal propagating through it. This comparison is made in both the time and frequency domains. For the frequency domain analysis we will use the concept of energy spectral density of a signal. This concept allows us to use the frequency response theorem to characterize the propagation of a large class of signals through the system.

15.4.2 Propagation of a Pulse Through an *RC* Network

The analysis in this section is based on Theorem 15.1.1. We assume that the system is linear and time invariant. Then if input signal $x(t)$ is an energy signal with a energy spectral density $D_x(\omega)$, then the energy spectral density of the output signal, $D_y(\omega)$, is

$$D_y(\omega) = |H(\omega)|^2 D_x(\omega) \tag{15.4.1}$$

where $H(\omega)$ is the frequency response function of the system. There is a graphical interpretation of (15.4.1) that is closely related to the graphical interpretations given in the last section for periodic power signals. We will illustrate the ideas by way of example. Again we assume the system is an *RC* network, discussed in Examples 14.1.3 and 14.2.3, shown in Figure 15.4.1. The frequency response function of this system is

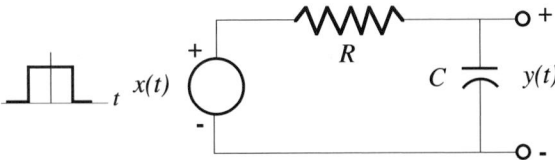

Figure 15.4.1 *RC* Network

$$\frac{Y(\omega)}{X(\omega)} = H(\omega) = \frac{1}{1+j\omega} = \frac{1}{\sqrt{1+\omega^2}} e^{-j\tan^{-1}(\omega)}. \qquad (15.4.2)$$

The square of the magnitude of the frequency response is

$$|H(\omega)|^2 = \frac{1}{1+\omega^2}. \qquad (15.4.3)$$

We will consider an energy input signal of the form

$$x(t) = \frac{1}{\varepsilon}\Pi\left(\frac{t}{\varepsilon}\right). \qquad (15.4.4)$$

(This scaling of the signal is chosen to facilitate the frequency domain graphs below.) The energy spectral density of the input signal is

$$D_x(\omega) = \left|\mathcal{F}\left\{\frac{1}{\varepsilon}\Pi\left(\frac{t}{\varepsilon}\right)\right\}\right|^2 = \left|\left(\frac{1}{\varepsilon}\right)\varepsilon\,\mathrm{Sa}\left(\frac{\omega\varepsilon}{2}\right)\right|^2 = \left|\mathrm{Sa}\left(\frac{\omega\varepsilon}{2}\right)\right|^2. \qquad (15.4.5)$$

The half-energy frequency (Definition 8.3.10) of the input signal, ω_{hp}, is determined from

$$D_x(\omega_{hp}) = \left(\frac{2}{\omega_{hp}\varepsilon}\right)^2 \sin^2\left(\frac{\omega_{hp}\varepsilon}{2}\right) = \frac{1}{2}. \qquad (15.4.6)$$

Solving (15.4.6) numerically we obtain

$$\omega_{hp} \approx \frac{2.8}{\varepsilon}. \qquad (15.4.7)$$

The energy spectral density of the output signal is

$$D_y(\omega) = |H(\omega)|^2 D_x(\omega) = \frac{1}{1+\omega^2} D_x(\omega). \qquad (15.4.8)$$

We will consider two input signals that are pulses of different width. The first input signal is

$$x_1(t) = \left(\frac{1}{10}\right)\Pi\left(\frac{t}{10}\right). \qquad (15.4.9)$$

The output signal for the input signal $x_1(t)$ is shown in Figure 15.4.2. The second signal is

$$x_2(t) = \left(\frac{1}{0.5}\right)\Pi\left(\frac{t}{0.5}\right).$$ (15.4.10)

The output signal for the input signal $x_2(t)$ is shown in Figure 15.4.3. Clearly, these two signals propagate through this system differently.   The first input signal propagates through this system with little distortion.   The corners of the pulse are rounded and there is some pulse spreading.   The second signal is severely distorted as it passes through the system; it is a barely recognizable copy of the input signal with large pulse spreading.   It is the purpose of this section to explain why these two signals experience different distortions as they propagate through the same system. This analysis will be conducted in terms of the energy spectral density of the input signal and the frequency response of the system.
    The energy spectral density of the first input signal $x_1(t)$ along with the energy spectral density of the output signal $y_1(t)$ are shown in Figure 15.4.4.   From (15.4.7) the input signal bandwidth is $B_{i1} = 0.28$.   Its energy spectrum is mostly contained within the bandwidth of the system.   So when the product in (15.4.1) is formed, the energy spectral density of the output signal is very similar to the energy spectral density of the input signal because the frequency response function squared is approximately 1 over the bandwidth of the signal. Hence, there is little distortion in the output signal relative to the input signal.   Also the bandwidth of the output signal, $B_{o1} = 0.28$, is very close to the bandwidth of the input signal, B_{i1}.   There is some flattening of the side lobes in the output signal's energy spectral density

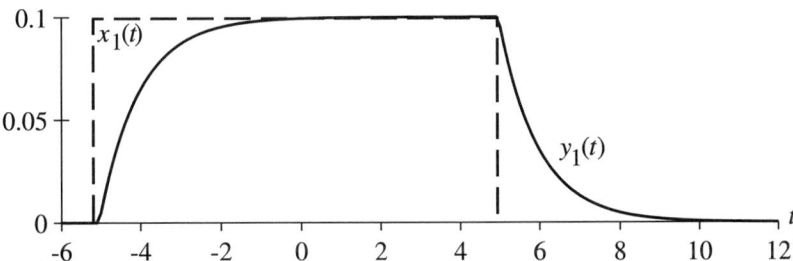

Figure 15.4.2 Input Signal $x_1(t)$ and the Output Signal $y_1(t)$

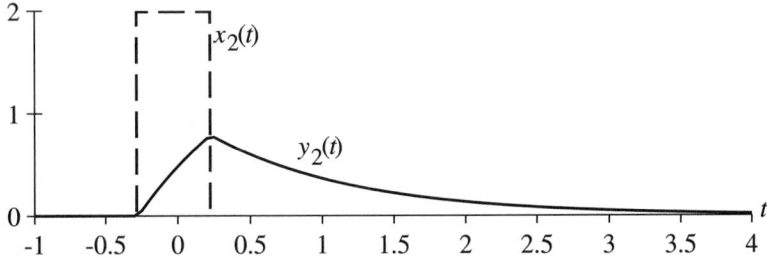

Figure 15.4.3 Input Signal $x_2(t)$ and the Output Signal $y_2(t)$

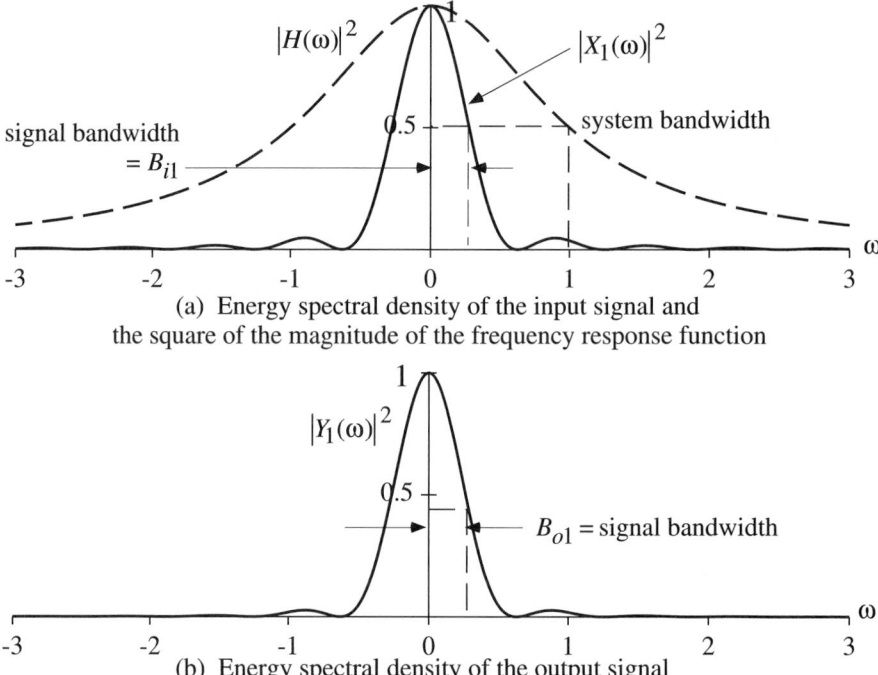

(a) Energy spectral density of the input signal and
the square of the magnitude of the frequency response function

(b) Energy spectral density of the output signal

Figure 15.4.4 Pulse Input Signal of Width 10 Sec

beyond 1 rad/sec. This change in the energy spectral density causes the pulse spreading in Figure 15.4.2.

The energy spectral density of the second input signal $x_2(t)$ along with the energy spectral density of the output signal $y_2(t)$ are shown in Figure 15.4.5. The bandwidth of this input signal is $B_{i2} = 5.6$ rad/sec which is considerably larger than the bandwidth of the system. More specifically, the part of the frequency band where the energy spectral density is large falls outside the passband of the system. As a result, when the product (15.4.1) is formed the energy spectral density of the output signal is zero outside the passband of the system. The energy spectral density of the input signal that is in the stopband of the system has been removed from the output signal. The energy spectral density of the output signal, $|Y(\omega)|^2$, is very different from the energy spectral density of the input signal, $|X(\omega)|^2$. In particular, the bandwidth of the output signal is $B_{o2} = 1$ rad/sec, essentially the same as the bandwidth of the system. So when comparing the output signal to the input signal in Figure 15.4.3, we observe a lot of distortion in the output signal including large pulse spreading.

Recall that the total energy of the input signal is

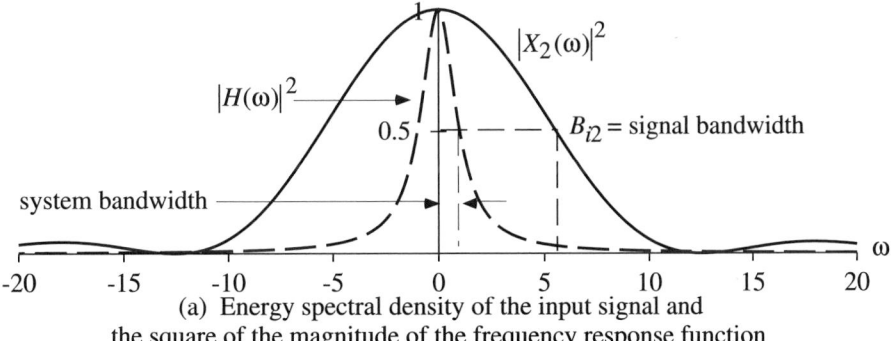

(a) Energy spectral density of the input signal and
the square of the magnitude of the frequency response function

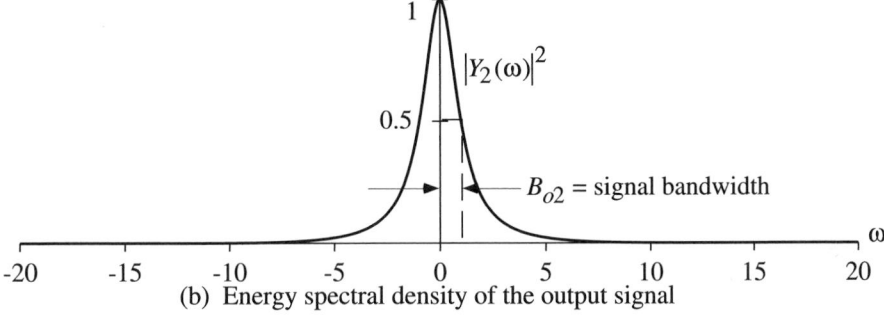

(b) Energy spectral density of the output signal

Figure 15.4.5 Pulse Input Signal of Width 0.5 sec

Table 15.4.1 Total Energy in Input and Output Signals

Input Signal	Energy	Output Signal	Energy	% Change
$x_1(t)$	0.1	$y_1(t)$	0.09	10 %
$x_2(t)$	2	$y_2(t)$	0.43	78.5 %

$$E_x = \int_{-\infty}^{\infty} |x(t)|^2 dt = \frac{1}{2\pi} \int_{-\infty}^{\infty} D_x(\omega)\, d\omega = \frac{1}{2\pi} \int_{-\infty}^{\infty} \left| \mathrm{Sa}\!\left(\frac{\omega\varepsilon}{2}\right) \right|^2 d\omega. \tag{15.4.11}$$

We have a similar expression for the output signal where we use the energy spectral density of the output signal in (15.4.11). The total energy of each of the input and output signals above is shown in Table 15.4.1.

Note: The entries in Table 15.4.1 were obtained by numerical integration.

The total energy in the output signal $y_1(t)$ is approximately the same as the total energy of the input signal $x_1(t)$. The total energy in the output signal $y_2(t)$ is much less than the energy in the input signal $x_2(t)$, which accounts for the reduction in amplitude of the output signal and the pulse distortion.

15.4.3 Summary of the Section

In this section we investigated the propagation of energy signals through a system using the energy spectral density of the signals and the frequency response function of the system. The results of the above two examples can be summarized by the following observations.

Signal - System Interaction 15.4.1:   Given a system with an energy signal propagating through it, the system will change the shape of the energy spectral density of the signal. This change in shape is essentially a redistribution of the energy of the output signal as compared to the input signal. The distortion induced by the system is determined by the relationship between the energy spectral density of the input signal and the frequency response function.

▲▲

Signal - System Interaction 15.4.2:   The components of the input signal with significant energy outside the passband of the system will be removed by the system. If the energy spectrum of the input signal is contained primarily within the passband of the system, then the signal will pass through the system with little distortion. If the energy spectrum of the input signal has significant energy outside the bandwidth of the system, the output signal will suffer distortion when compared to the input signal.

▲▲

It is worthwhile to compare the results of the last section with the results of this section. The similarities in interpretation between the energy spectrum and the power spectrum have already been discussed in Chapter 8. In the last two sections we have described how a system filters a signal in terms of the energy spectrum or power spectrum. In both cases the interpretations are exactly the same in the frequency domain. In fact, this analysis can be extended to aperiodic power signals with no change. This frequency domain interpretation is quite general, very powerful, and widely used.

15.4.4 MATLAB Experiments

The following M-file implements the example given in this section. In order to duplicate the graphs given in this section, the endpoints of both the time and frequency vectors must be adjusted. Unlike the M-files for the Fourier series, the energy calculations in this M-file are specific to the input signals because they use the analytical formulas for the energy. In order to program the energy of a pulse, (15.4.5) was rewritten in terms of a sinc function. This change required the frequency vector to be rewritten in hertz.

Example in Section 15.4

```
clear
% Define the signal parameters
ep = 10;          % pulse width, less than T0
A = 0.1;          % amplitude
ti = -6;          % initial time
tf = 12;          % final time, less than initial time
step = 0.01;      % increment in time vector
```

```
t = [ti:step:tf];      % time vector for waveforms

% Construct input signal
x = zeros(size(t));
i1 = max(find(t<-ep/2));
i2 = min(find(t>ep/2));
x(i1:i2) = A*ones(1,i2-i1+1);

% Define system
RC = 1;              % RC time constant
b = [1/RC];          % define numerator
a = [1,1/RC];        % define denominator

% Calculate output signal
y = lsim(b,a,x,t);

% Calculate energy spectral density of input signal
wend = 3;                          % define frequency interval
w = linspace(-wend,wend,300);      % frequency vector
f = w./(2*pi);
Dx = (sinc(f*ep)).^2;              % energy spectral density of the input signal
mag = bode(b,a,w);                 % calculate frequency response of system
H2 = (mag.^2)';                    % magnitude squared of the frequency response function
Dy = H2.*Dx;                       % energy spectral density of the output signal

% Plot input and output signals
figure(1)
plot(t,x,t,y,'--')
title('Input and Output Signals')
xlabel('Time, sec')
figure(2)
plot(w,Dx,w,H2,'--',w,Dy)
title('Power Spectra')
xlabel('Frequency, rad/sec')
```
▲▲

Exploratory Exercise 15.4.3: Modify this M-file to include a calculation of the total energy in the input and output signal. Use Parseval's theorem for this calculation. Verify the results in Table 15.4.1. The command **trapz** will integrate a function.

▲▲

Exploratory Exercise 15.4.4: Replace the simple *RC* network by the more complicated 3rd-order Butterworth filter. Compare the effect of this system on the input signal with the *RC* network.

▲▲

15.5 TRACKING FOR LINEAR MOTORS

15.5.1 Introduction

In the last two sections we have discussed the propagation of a signal through a system by changing the signal parameters and leaving the system fixed. These results inherently depend on the relationship between the signal parameters and the system parameters. To emphasize this point, in this section we will fix the signal parameters and vary the system parameters.

15.5.2 The Motor

Consider a linear motor introduced in Section 6.5. One use of a linear motor is as a linear positioning device. By applying an appropriate voltage signal to the input of the motor, the proof-mass will move to an appropriate position. Two of these motors could form an x-y plotter, for example.

We also assume that the motor given in Section 6.5 has been modified by the addition of feedback loops as shown in Figure 15.5.1. The two gains, K_{va} and K_{pa}, along with the summing junction in Figure 15.5.1, are implemented electronically. Hence, we are free to set these gains as we like (subject to physical constraints of the electronics). The transfer function of this system is

$$\frac{Y_{pm}(s)}{R_{pm}(s)} = \frac{\dfrac{K_{pa}K_{ef}}{m_{pm}}}{s^2 + \dfrac{K_{pa}K_{ef}K_{va}}{m_{pm}}s + \dfrac{K_{pa}K_{ef}}{m_{pm}}}. \tag{15.5.1}$$

The denominator is a second-order polynomial. Writing this polynomial in standard second-order form[9] we have

$$\frac{Y_{pm}(s)}{R_{pm}(s)} = \frac{\omega_n^2}{s^2 + 2\zeta\omega_n s + \omega_n^2}, \tag{15.5.2}$$

$$\omega_n = \sqrt{\frac{K_{pa}K_{ef}}{m_{pm}}},$$

$$2\zeta\omega_n = \frac{K_{pa}K_{ef}K_{va}}{m_{pm}}, \quad \Rightarrow \quad \zeta = \frac{K_{va}}{2}\sqrt{\frac{K_{pa}K_{ef}}{m_{pm}}}.$$

Given the parameters of the motor, K_{ef} and m_{pm}, we can choose ω_n through the selection of K_{pa}, and we can choose ζ through the selection of K_{va}. The bandwidth of this system is ω_n. By changing the gains K_{pa} and K_{va} we can change the bandwidth of the system. We will consider two sets of gains for the linear motor shown in Table 15.5.1.

---

[9] See Section 3.2.

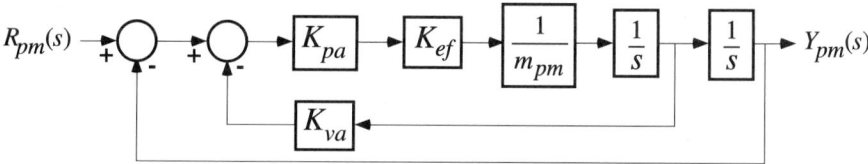

Figure 15.5.1 Block Diagram of a Linear Motor

Table 15.5.1 Parameters of the Linear Motor

		K_{ef}	m_{pm}	K_{pa}	K_{va}	ω_n	ζ
Low Bandwidth Motor	$H_l(s)$	1	10	40	0.7	2	0.7
High Bandwidth Motor	$H_h(s)$	1	10	4000	0.07	20	0.7

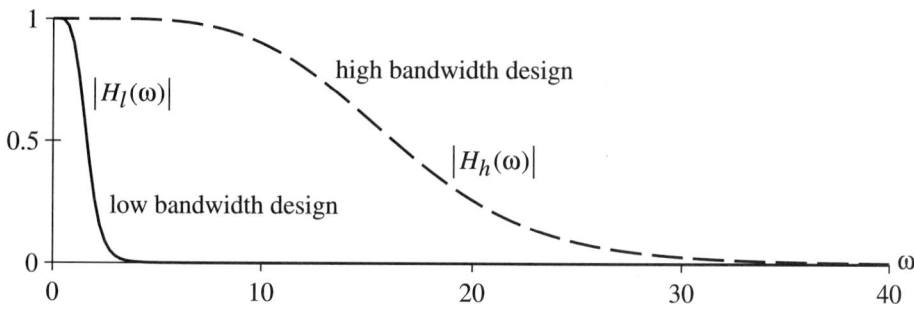

Figure 15.5.2 Bandwidth of the Two Linear Motors

These two motor designs differ because of their bandwidth as shown in Figure 15.5.2.

The stability of the linear motor in Figure 15.5.1 depends on the gains in that block diagram. Some choice of gains will cause the motor to be unstable, while other choices of gains will result in a BIBO stable system. If the closed loop system in Figure 15.5.1 is unstable, most input signals will cause the proof-mass to be driven to its stops, which would render the device inoperable. Hence, we will consider only gains that yield a BIBO stable system, such as those in Table 15.5.1.

This discussion highlights the role of feedback loops in any system. Feedback loops can change the stability of the system, possibly rendering the system useless or destroying it. Feedback loops can also dramatically improve the performance of a system, as shown by the analysis in this section.

The input signal $r_{pm}(t)$ in Figure 15.5.1 is called a <u>reference</u> signal because we want the output signal $y_{pm}(t)$ to exactly follow this input signal. Suppose the reference signal is

$$r_{pm}(t) = \Pi(t). \tag{15.5.3}$$

The reference signal in (15.5.3) doesn't begin at time $t = 0$. The start of the pulse was chosen because the phase of the Fourier transform is convenient for the discussion below.

In the block diagram in Figure 15.5.1 the position of the proof-mass is subtracted from the reference signal to form the <u>error</u> signal

$$e(t) = r_{pm}(t) - y_{pm}(t). \tag{15.5.4}$$

Terminology: The error signal in (15.5.4) comes from the feedback loop with a gain of one (1), and it is subtracted from the reference signal. The feedback configuration in Figure 15.5.1 is called a <u>negative unity feedback loop</u>.[10]

At time $t^+ = -0.5$ the error signal will be positive. This signal is transmitted and amplified as a force on the proof-mass causing it to accelerate in a positive direction. There follows a transient response in the position of the proof-mass. As the difference between the position of the proof-mass and the reference signal decreases, the error signal decreases, and the force applied to the proof-mass decreases. In the long term, the proof-mass must position itself so that the error signal is zero. If the error signal is nonzero, then the proof-mass would continue to accelerate. But this system is BIBO stable, and so all signals are bounded. Hence, the error signal will tend toward zero. The result is that the proof-mass will tend to follow the reference signal. This property of the system in Figure 15.5.1 is called <u>tracking</u>. The two motors defined by the parameters in Table 15.5.1 have different tracking properties. We will explore these properties in this example.

15.5.3 Performance of the Motor

This reference signal (15.5.3) commands the proof-mass to move to a position of 1 unit of displacement for the time $t = -0.5$ from to $t = 0.5$, and then return to the centered position. When this reference signal (15.5.24) is used as the input signal to the low bandwidth motor, the output signal is shown in Figure 15.5.3. When the same reference signal is used as the input signal to the high bandwidth motor, the output signal is shown in Figure 15.5.4.

Clearly, the high bandwidth motor tracks the reference signal better than the low bandwidth motor. The low bandwidth motor is slow to follow the reference signal. By the time the proof-mass has reached the desired position, the reference signal has changed again, and the proof-mass reluctantly follows. For the high bandwidth system, the motor quickly moves to the commanded position, in fact experiencing some overshoot. It returns to the centered position as quickly.

This difference in the motion of the proof-mass can be explained from the frequency response function of the motors and the energy spectral density of the reference signal. Figure 15.5.5 shows the energy spectral densities of the input and output signals along with the square of the frequency response function for the low bandwidth system. These same functions for the high bandwidth motor are shown in Figure 15.5.6.

---

[10] See Definition 10.2.7.

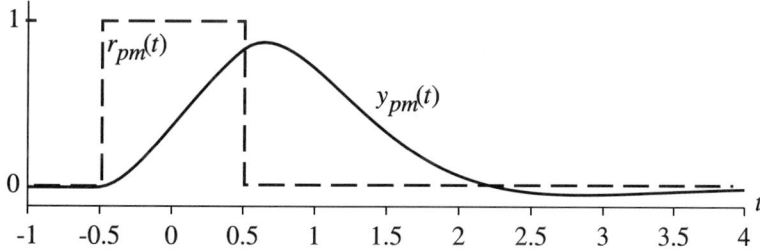

Figure 15.5.3 Input and Output Signals of the Low Bandwidth Motor

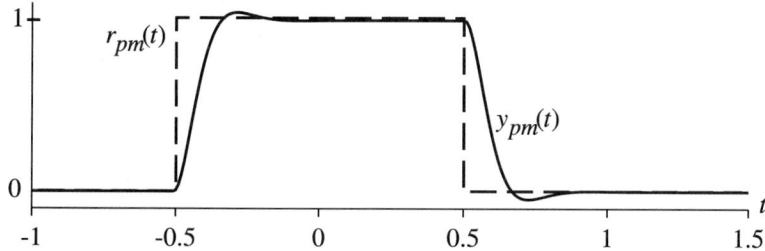

Figure 15.5.4 Input and Output Signals of the High Bandwidth Motor

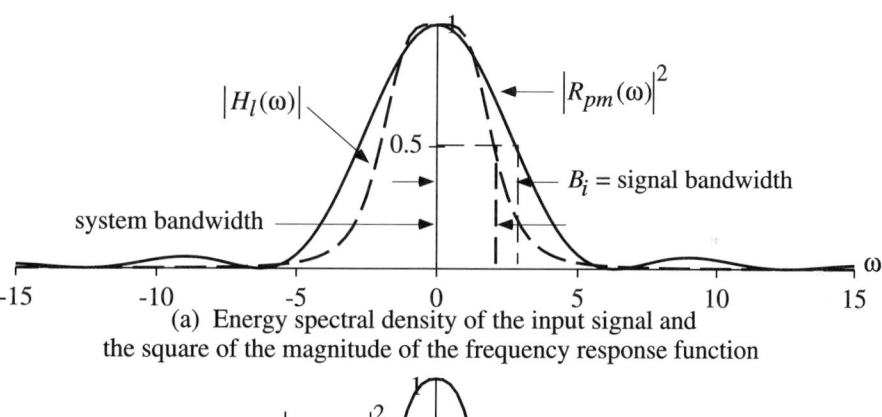

(a) Energy spectral density of the input signal and
the square of the magnitude of the frequency response function

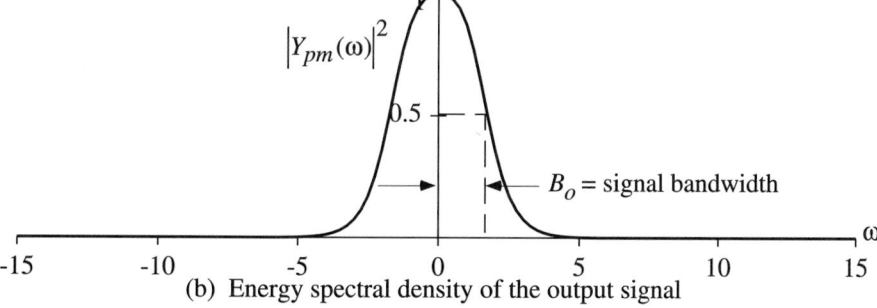

(b) Energy spectral density of the output signal

Figure 15.5.5 Low Bandwidth Motor

Note: The energy spectral density of the reference signal $r_{pm}(t)$ is the same in both figures; the bandwidth of the system has increased causing the scales to change.

In Figure 15.5.5 the square of the frequency response function of the motor is shown as a dashed line. The half-energy bandwidth of the input and output signals are also shown. The passband of the low bandwidth system is 2 rad/sec. The energy spectral density of the reference signal has significant energy content outside the passband of the motor. When the energy spectral density of the output signal is formed by multiplication according to Theorem 15.1.1, the energy spectral density of the output signal is less than the energy spectral density of the reference signal. Since these two energy spectral densities are different, distortion has occurred. Here, the "high frequency components" of the reference signal have been removed resulting in the output signal in Figure 15.5.3.

Quite a different situation occurs with the high bandwidth motor. In Figure 15.5.6 the energy spectral density of the reference signal falls within the passband of the high bandwidth system. As a result when the product of the energy spectral density of the input signal and the magnitude squared of the frequency response function is formed according to Theorem 15.1.1, the energy spectral density of the output signal is virtually identical to the energy spectral density of the input signal. Only minor distortion occurs in the side lobes of the energy spectral density of the

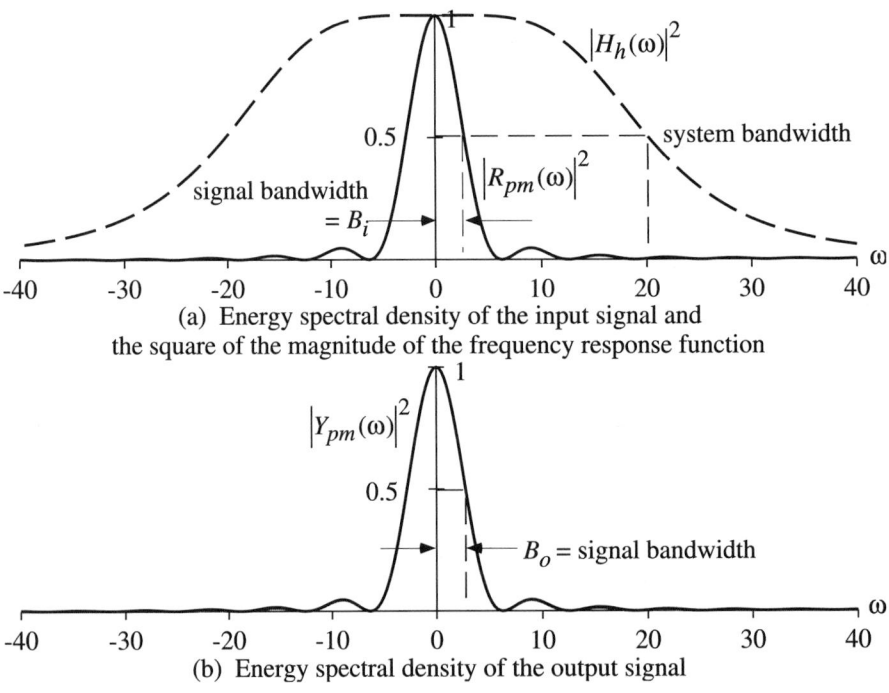

(a) Energy spectral density of the input signal and
the square of the magnitude of the frequency response function

(b) Energy spectral density of the output signal

Figure 15.5.6 High Bandwidth Motor

output signal, $D_y(\omega)$, as compared to the energy spectral density of the reference signal. The distortion of the side lobes leads to the reduced risetime and low overshoot of the output signal $y(t)$ in Figure 15.5.4.

15.5.4 MATLAB Experiments

Exploratory Exercise 15.5.1: Modify the M-file given in Section 15.4 so that the system is the linear motor discussed above instead of the RC network. Verify the results of this section.

▲▲

Exploratory Exercise 15.5.2: Using your modified M-file, vary the gain K_{pa} so that the damping ratio of the motor varies between $0.1 < \zeta < 0.95$. Observe the effect of the changing damping ratio on the output signal.

▲▲

Exploratory Exercise 15.5.3: Change the type of reference signal and observe the effect on the output signal. (Of course, when you change the reference signal, the energy calculations are no longer valid.) Why is the step function used as a test function to benchmark the performance of the system?

▲▲

Exploratory Exercise 15.5.4: Change the reference signal to be a sinusoid. Do several simulations of the input signal and the output signal. At what frequency does the system no longer track this reference signal? Explain your results using the frequency response theorem.

▲▲

15.6 AMPLITUDE MODULATION AND FREQUENCY DIVISION MULTIPLEXING

15.6.1 Introduction

In Section 5.1 we pointed out that many communication systems require the message signal to be impressed onto a carrier signal so that the waveform can be efficiently transmitted over the communication channel. One way the information signal can be embedded into the carrier signal is through modulation of the carrier signal.[11] Suppose the carrier signal is a sinusoid

$$x_c(t) = \cos(\omega_c t). \tag{15.6.1}$$

If the message signal $m(t)$ is impressed into the amplitude of the sinusoid, then the carrier signal is amplitude modulated. The modulated signal is given by

$$s(t) = m(t)\cos(\omega_c t). \tag{15.6.2}$$

---

[11] The terminology and notation of Section 5.1 will be used here.

The modulation process is shown in Figure 15.6.1. In this section we will show how
the frequency domain ideas we have introduced thus far can be used to analyze this
type of signal in the frequency domain. We will show how to extract the message
signal from the amplitude modulated signal. We will also show several message
signals can be multiplexed into a single signal.

15.6.2 Amplitude Modulation

To begin we will develop the amplitude spectrum of an amplitude modulated signal.
We will assume that the message signal $m(t)$ is a baseband signal with a bandwidth
of ω_m. For example, we can represent the amplitude spectrum as shown in Figure
15.6.2a. The Fourier transform of the carrier signal (15.6.1) is given by

$$X_c(\omega) = \pi\big(\delta(\omega + \omega_c) + \delta(\omega - \omega_c)\big). \tag{15.6.3}$$

The amplitude spectrum of the carrier signal is shown in Figure 15.6.2b. The Fourier
transform of the modulated signal can be determined by convolution in the frequency
domain (Property 7.5.9). Taking the Fourier transform of the modulated signal
(15.6.2) we obtain

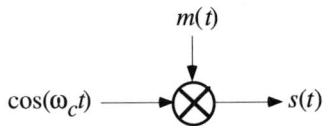

Figure 15.6.1 Amplitude Modulation

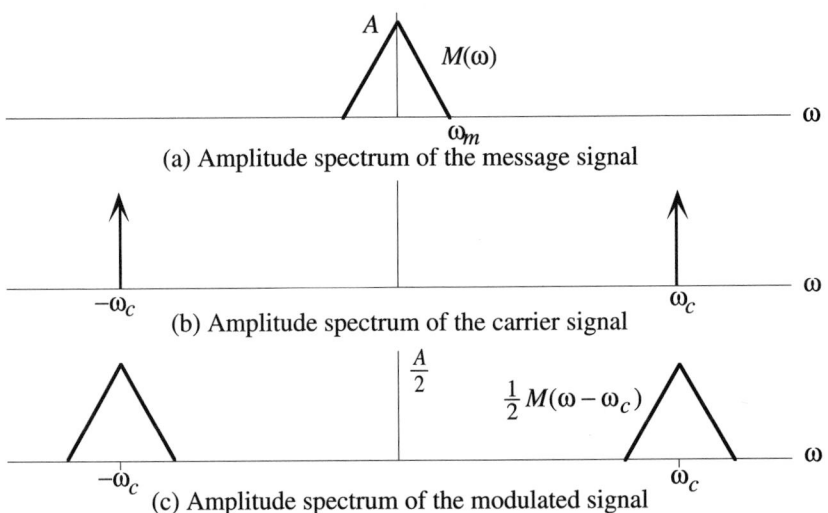

(a) Amplitude spectrum of the message signal

(b) Amplitude spectrum of the carrier signal

(c) Amplitude spectrum of the modulated signal

Figure 15.6.2 Amplitude Spectrum of the Message, Carrier, and Modulated Signal

$$\mathcal{F}\{s(t)\} = \mathcal{F}\{m(t)x_c(t)\} = \int_{-\infty}^{\infty} M(\omega - \lambda)X_c(\lambda)\,d\lambda. \tag{15.6.4}$$

Inserting (15.6.3) into (15.6.4) we get

$$
\begin{aligned}
S(\omega) &= \frac{1}{2\pi}\int_{-\infty}^{\infty} M(\omega - \lambda)X_c(\lambda)\,d\lambda \tag{15.6.5}\\
&= \frac{1}{2}\int_{-\infty}^{\infty} M(\omega - \lambda)\big[\delta(\omega + \omega_c) + \delta(\omega - \omega_c)\big]\,d\lambda\\
&= \frac{1}{2}\big(M(\omega + \omega_c) + M(\omega - \omega_c)\big).
\end{aligned}
$$

The amplitude spectrum of the modulated signal is shown in Figure 15.6.2c.

Figure 15.6.2 contains the key concept behind amplitude modulation. It shows that when we impress the message signal onto the carrier signal, the amplitude spectrum of the message signal is shifted from its original location in the frequency spectrum so that it is centered on the frequency of the carrier signal. This analysis clearly shows that by adjusting the frequency of the carrier signal, ω_c, the amplitude spectrum of the message signal can be relocated in the frequency spectrum at ω_c. (Of course we always require that $\omega_c > 2\omega_m$ so that the amplitude spectrum of the message signal is not distorted in the modulated signal.)

In communication systems, the modulated signal is sent over a communication channel to the receiver where the message signal is extracted. Frequently the communication channel is approximately a bandpass system. Hence, the message signal, which is a baseband signal, could not be transmitted directly over the channel. By amplitude modulating the carrier signal with the message signal, the amplitude spectrum of the transmitted signal is shifted into the passband of the communication channel by the appropriate choice of the carrier frequency. In this way we are able to efficiently transmit a message signal over a bandpass system.

15.6.3 Demodulation

For the message signal to be of use to us we must be able to extract it from the carrier signal at the receiver. The process of extracting the message signal from the modulated signal is called demodulation. There are several ways of demodulating an amplitude modulated signal. Next we will discuss one method of demodulation. Suppose that we multiply the amplitude modulated signal by the carrier wave. The signal we obtain is

$$y(t) = s(t)\cos(\omega_c t) = \big[m(t)\cos(\omega_c t)\big]\cos(\omega_c t). \tag{15.6.6}$$

In order to obtain the amplitude spectrum of this signal, we can simply repeat the analysis given above for the modulated signal in (15.6.2). The amplitude spectrum of $y(t)$ is shown in Figure 15.6.3. Comparing Figure 15.6.3 to Figure 15.6.2a it can be seen that the amplitude spectrum of $y(t)$ contains the amplitude spectrum of the

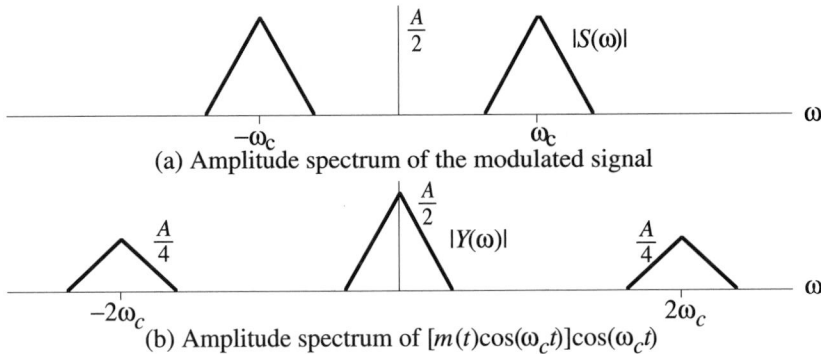

(a) Amplitude spectrum of the modulated signal

(b) Amplitude spectrum of $[m(t)\cos(\omega_c t)]\cos(\omega_c t)$

Figure 15.6.3 Amplitude Spectrum of the Demodulated Signal

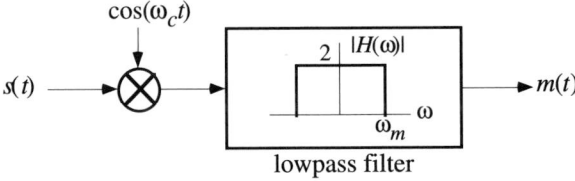

lowpass filter

Figure 15.6.4 Synchronous Demodulator

message signal along with some side lobes. To recover the message signal it is only necessary to lowpass filter $y(t)$. This demodulation scheme is shown in Figure 15.6.4.

Terminology: For this demodulation scheme to work, it is required that the two carrier signals, one used for modulation and one used demodulation, have exactly the same frequency and phase. For this reason, this type of demodulation is called synchronous demodulation.

15.6.4 Frequency Division Multiplexing

It is frequently desirable to impress two information signals onto one carrier signal in such a way that the information signals can be extracted separately at the receiver. This process of encoding more than one message signal into a single waveform is called multiplexing (Definition 5.4.7). Amplitude modulation suggests a way to multiplex two or more signals which we will describe next.

Let $m_1(t)$ and $m_2(t)$ be two bandlimited message signals. The amplitude spectrum of each signal is shown in Figure 15.6.5a. Let $x_{c1}(t)$ and $x_{c2}(t)$ be two sinusoidal carrier signals whose frequencies are separated. Now these two carrier signals are amplitude modulated with the two message signals respectively. These two amplitude modulated signals are summed into a single multiplexed signal

$$s(t) = s_1(t) + s_2(t) = m_1(t)\cos(\omega_{c1}t) + m_2(t)\cos(\omega_{c2}t). \tag{15.6.7}$$

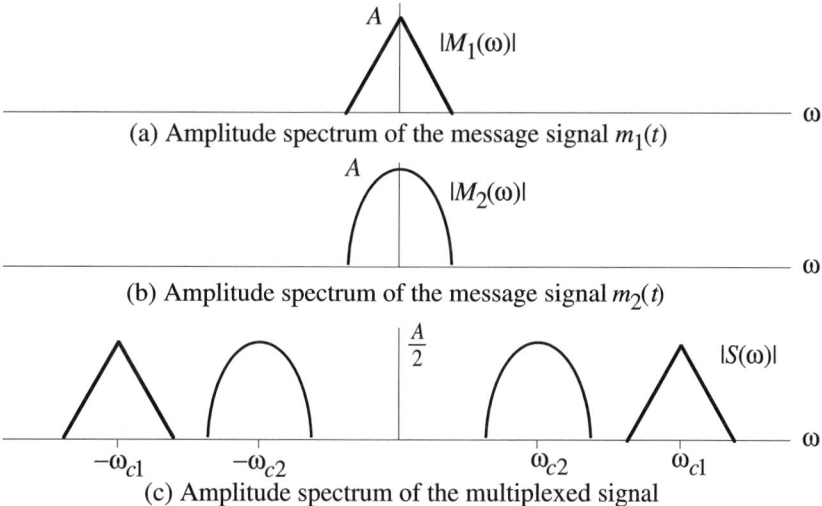

(a) Amplitude spectrum of the message signal $m_1(t)$

(b) Amplitude spectrum of the message signal $m_2(t)$

(c) Amplitude spectrum of the multiplexed signal

Figure 15.6.5   Amplitude Spectra of Multiplexed Signals

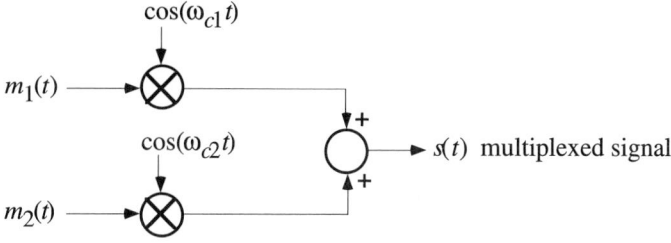

Figure 15.6.6   Frequency Division Multiplexing

This process is shown in Figure 15.6.6. The amplitude spectrum of the multiplexed signal is shown in Figure 15.6.6b. By proper choice of the frequencies of the carrier signals, the spectra of the two message signals appear undistorted in the amplitude spectrum of the multiplexed signal. Hence, these two message signals can be extracted separately from the multiplexed signal by a demodulation technique similar to one described above. In this way two analog message signals can be multiplexed and subsequently demultiplexed with no loss of information.

Definition 15.6.1:   The type of multiplexing shown in Figure 15.6.6 is called <u>frequency division multiplexing</u> (FDM).

▲▲

Suppose that each message signal has a bandwidth of ω_m. Then this message signal, when it is modulated, will occupy $2\omega_m$ rad/sec of bandwidth. Since there is usually a total bandwidth constraint on the multiplexed signal, this bandwidth constraint usually determines how many signals can be multiplexed into one signal.

We have already discussed another method of multiplexing in Section 5.4 called time division multiplexing (TDM).   This approach to multiplexing begins by sampling the message signals. The samples of each of the message signals are then multiplexed by interlacing the symbols that correspond to the samples into a waveform. Thus, in a TDM signal we can separate the message signals by visual inspection of the waveform. If we computed the amplitude spectrum of a TDM signal, however, we would not be able to distinguish the message signals.   The situation is exactly reversed with FDM. In an FDM signal the individual message signals are not distinguishable in the waveform.   However, the message signal amplitude spectra are completely separated in the amplitude spectrum of the FDM signal.

15.7 CHAPTER SUMMARY

In this chapter we have investigated the propagation of a signal through a system using frequency domain concepts. Suppose that a linear, time invariant, BIBO stable system has a frequency response function $H(\omega)$. Suppose that $x(t)$ is an input signal to this system. Then the relationship between the input signal, the output signal, and the system are as follows (Theorem 15.1.1).

1. If the input signal $x(t)$ is an energy signal with a energy spectral density $D_x(\omega)$, then the energy spectral density of the output signal, $D_y(\omega)$, is

$$D_y(\omega) = |H(\omega)|^2 D_x(\omega).$$

2. If the input signal $x(t)$ is a power signal with a power spectral density $S_x(\omega)$, then the power spectral density of the output signal, $S_y(\omega)$, is

$$S_y(\omega) = |H(\omega)|^2 S_x(\omega).$$

This investigation is conducted by an in-depth analysis of several signals propagating through an RC network. These signals included a single sinusoid, a periodic signal, and a pulse. The important concepts used in this analysis are the frequency response function of the system, amplitude spectrum, phase spectrum, energy spectrum, and power spectral density of the signals. The intention of this discussion is to show the interrelationship between all of these concepts and their origin in the frequency response theorem. The following summary of energy signals captures the essence of the analysis in this chapter.

Signal - System Interaction 15.4.1:   Given a system with an energy signal propagating through it, the system will change the shape of the energy spectral density of the signal.   This change in shape is essentially a redistribution of the energy of the output signal as compared to the input signal. The distortion induced by the system is determined by the relationship between the energy spectral density of the input signal and the frequency response function.   ▲▲

Signal - System Interaction 15.4.2: The components of the input signal with significant energy outside the passband of the system will be removed by the system. If the energy spectrum of the input signal is contained primarily within the passband of the system, then the signal will pass through the system with little distortion. If the energy spectrum of the input signal has significant energy outside the bandwidth of the system, the output signal will suffer distortion.

▲▲

In Section 15.6 we developed the spectral characteristics of an amplitude modulated signal. We presented one method of demodulating the signal to recover the message signal. We also showed how amplitude modulation can be used to multiplex several analog signals into one waveform in such a way that they can be individually recovered. This type of multiplexing is called frequency division multiplexing (Definition 15.6.1).

15.8 HOMEWORK FOR CHAPTER 15

Homework Problems for Section 15.2

15.2.1 Let the frequency response of a system be

$$\frac{Y(\omega)}{X(\omega)} = H(\omega) = \frac{\omega_n^2}{(\omega_n^2 - \omega^2) + j2\zeta\omega_n\omega}.$$

Suppose the input signal to this system is $x(t) = \cos(\omega_i t)$. Three sets of parameters for this system and input signal are given below.
(a) Plot the input and output signal for each set of parameters.
(b) Consider the input signal to be a Fourier series.
   (i)   Plot the two-sided amplitude spectrum of the input signal on the same graph as the magnitude of the frequency response function.
   (ii)   Plot the two-sided phase spectrum of the input signal on the same graph as the phase of the frequency response function.
   (iii)   Plot the power spectrum on the same graph as the magnitude of the frequency response squared. Also plot the power spectrum of the output signal.
(c) Use the Fourier transform of $x(t)$ for the following analysis.
   (i)   Plot the amplitude spectrum of the input signal on the same graph as the magnitude of the frequency response function.
   (ii)   Plot the phase spectrum of the input signal on the same graph as the phase of the frequency response function.
   (iii)   Plot the power spectrum on the same graph as the magnitude of the frequency response squared. Also plot the power spectrum of the output signal.
Set 1: $(\zeta, \omega_n) = (0.7, 1)$, $\omega_i = 0.2$
Set 2: $(\zeta, \omega_n) = (0.1, 1)$, $\omega_i = 1$
Set 3: $(\zeta, \omega_n) = (0.7, 1)$, $\omega_i = 4$

15.2.2 Let the frequency response of a system be

$$\frac{Y(\omega)}{X(\omega)} = H(\omega) = \frac{\omega^2}{(1-\omega^2)+j\omega}.$$

(a) Plot the square of the magnitude of the frequency response function. What kind of filter is this system?
(b) Suppose the input signal to this system is $x(t) = \cos(\omega_i t)$. For which frequency ω_i is the power of the output signal maximized? What is the power of that signal?
(c) Suppose the input signal to this system is $x(t) = 1 + \cos(3t)$. Find the output signal.
Note: This type of filter is used to remove the dc component of a signal.

15.2.3 The power spectrum of the input signal to a system is shown in Figure P15.2.3 along with the magnitude of the frequency response function squared.
(a) It is known that the input signal is composed of a sinusoid. What is the frequency of the input signal?
(b) What is the power of the input signal?
(c) What is the power of the output signal?
(d) What frequency of the input signal would maximize the power of the output signal?

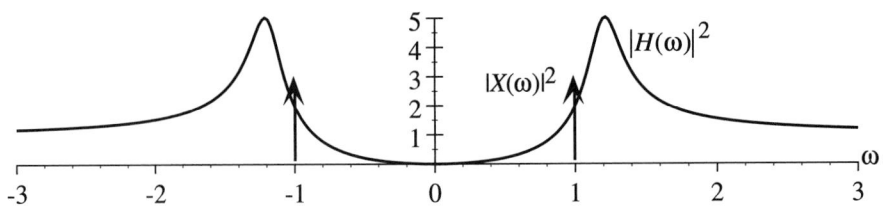

Figure P15.2.3

15.2.4 Consider a system whose magnitude and phase response is shown in Figure P15.2.4a.
(a) Suppose the amplitude and phase spectrum (Fourier transform) of the input signal to this system is shown in Figure P15.2.4b.
(i)   What is the amplitude and phase spectrum of the output signal of the system?
(ii) What is the output signal?
(iii) Plot the input and output signal.
(iv) What is the power in the input and output signals?
(b) Suppose the amplitude and phase spectrum (Fourier series) of the input signal to the system in P15.2.4a is shown in Figure P15.2.4c. Repeat the questions of part (a).

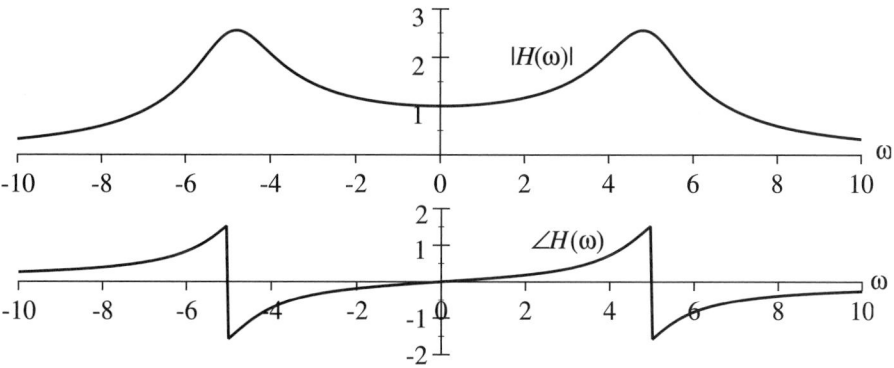

Figure P15.2.4a

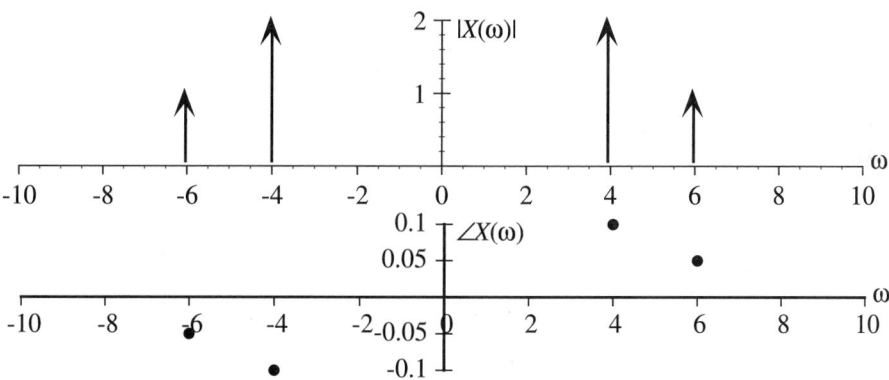

Figure P15.2.4b

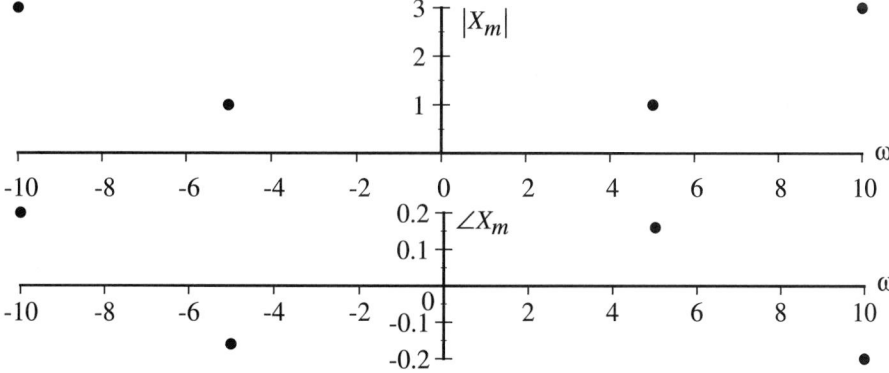

Figure P15.2.4c

Homework Problems for Section 15.3

15.3.1 Consider the Fourier series

$$x(t) = 1 + \sum_{\substack{m=-\infty \\ m \neq 0}}^{\infty} \left[\frac{3}{m\pi} \sin\left(\frac{2m\pi}{3}\right) e^{-j\frac{2m\pi}{3}} \right] e^{-j\frac{2m\pi t}{3}}.$$

(a) Plot the amplitude and phase spectrum of this Fourier series.
(b) Express the series in its trigonometric form.
(c) This signal is the input to an ideal bandpass filter with a bandwidth of 2 rad and centered at 2π rad/sec. In the passband the group delay is $\tau = 2$. Plot the magnitude and phase of the frequency response function of this filter.
(d) Plot the amplitude and phase spectrum of the output signal.
(e) Plot the input and output signal on the same plot.

15.3.2 The power spectrum of a periodic signal $x(t)$ is shown in Figure P15.3.2. Let $x(t)$ be an input signal to an ideal bandpass filter $H(s)$. Suppose that $H(s)$ has zero phase. We want the output signal of this filter to be the two terms of $x(t)$ with the largest power.
(a) Find the passband of this filter.
(b) Sketch the magnitude of the frequency response function of the bandpass filter on the power spectrum of the signal in Figure P15.3.2.
(c) Plot the input and output signal, if possible. If not, why not?

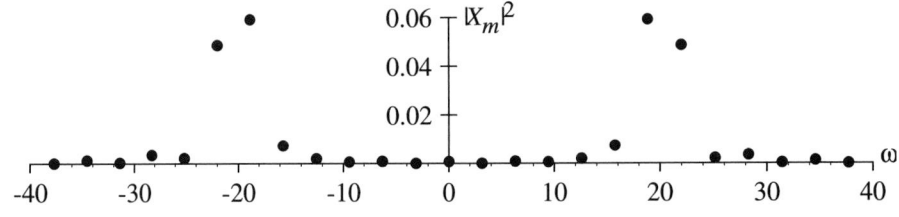

Figure P15.3.2

15.3.3 Consider a system described by the transfer function

$$\frac{Y(s)}{X(s)} = H(s) = \frac{20s}{s^2 + 24s + 100}.$$

Suppose the input signal is

$$x(t) = \begin{cases} 0, & -\dfrac{T_0}{2} \le t < -\dfrac{\tau}{2} \\ 10\cos(\omega_m t), & -\dfrac{\tau}{2} \le t < \dfrac{\tau}{2} \\ 0, & \dfrac{\tau}{2} \le t < \dfrac{T_0}{2} \end{cases}$$

with periodic extension.
(a) Plot the frequency response function.
(b) If $\omega_m = 10$, $\tau = 10$, $T_0 = 20$. If $x(t)$ is the input signal to the system above, plot the input and output two sided amplitude spectra.
(c) Repeat (b) if $\omega_m = 10$, $\tau = 1$, $T_0 = 2$.
(d) Repeat (b) if $\omega_m = 20$, $\tau = 10$, $T_0 = 20$.

15.3.4 Consider the signal shown in Figure P15.3.4. Suppose $x(t)$ is the input signal to a 4th-order Butterworth filter with a bandwidth of 10 rad/sec.
(a) Find the Fourier series of this signal. Assume $\varepsilon = 0.5$.
(b) Plot the two-sided amplitude and phase spectrum for 20 terms.
(c) Find the power in $x(t)$. Plot the two-sided power spectrum.
(d) Find the power in the first 20 nonzero terms.
(e) Plot the first 10 terms in the Fourier series.
(f) Assume $\varepsilon = 0.1$. Repeat (b) - (e)
(g) On the same graph, plot the power spectrum of input signal, the output signal, and the frequency response function of the Butterworth filter.
(h) Plot the output signal for each input signal.

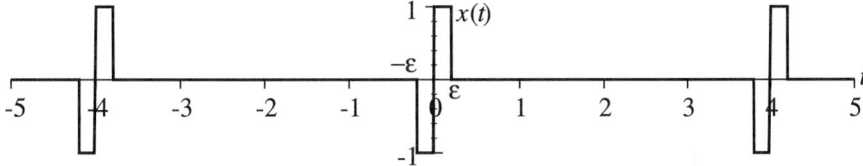

Figure P15.3.4

15.3.5 Consider the signal represented as a Fourier series

$$x(t) = \sum_{m=1}^{\infty} \frac{3}{4m^2} \cos\!\left(2mt - \frac{\pi m}{4}\right).$$

Suppose this signal is an input signal into a system whose frequency response function is shown in Figure P15.3.5.
(a) Find the exponential Fourier series representation of this signal.
(b) Plot the two-sided amplitude and phase spectrum of the output signal.
(c) What is the bandwidth of this system?

(d) What is the output signal of the system?   Plot the input and output
signal of the system on the same graph.

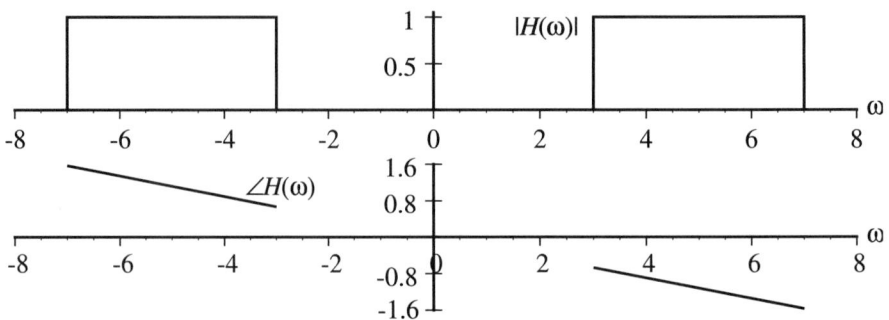

Figure P15.3.5

15.3.6 Consider the signal represented as a the Fourier series

$$x(t) = 1 + \sum_{\substack{m=-\infty \\ m \neq 0}}^{\infty} \left(\frac{-j}{m\pi}\right) e^{jm\pi t}.$$

(a) Plot the amplitude and phase spectrum of this Fourier series.
(b) Express the series in its trigonometric form.
(c) This signal is the input to an ideal lowpass filter with a bandwidth of 10
    rad/sec and a group delay of $\tau = 0.1$. Plot the magnitude and phase of
    the frequency response function of this filter.
(d) Plot the amplitude and phase spectrum of the output signal.
(e) Plot the input and output signal on the same plot.

15.3.7 Suppose $x_1(t)$ is a pulse train with a period of $T_0 = 2$, an amplitude of 1, and
a pulse width of $\tau = 1$. Consider a second signal $x_2(t)$ that is obtained by
periodic extension of $x_1(t)$ where

$$x_2(t) = x_1(t)\cos(20t), \quad -\tfrac{\tau}{2} \leq t < T_0 - \tfrac{\tau}{2}, \quad \text{with periodic extension.}$$

Assume that these two signals are input signals for the system in Figure
P15.3.7.
(a) Plot the frequency response function of this system.
(b) If $x_i(t)$ is an input to the system, plot the power spectrum of the
    corresponding output signal $y_i(t)$ for $i = 1,2$.
(c) Plot an approximation of $y_i(t)$, $i = 1,2$, by choosing four nonzero terms
    of the Fourier series of $y_i(t)$. Justify these calculations.

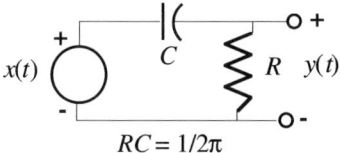

$$RC = 1/2\pi$$

Figure P15.3.7

Homework Problems for Section 15.4

15.4.1 The signal $x(t)$ is the input to a system with the frequency response

$$\frac{Y(\omega)}{X(\omega)} = H(\omega) = \begin{cases} 1e^{-j4\omega}, & 2 \le |\omega| < 6 \\ 0, & \text{elsewhere} \end{cases}$$

Suppose the input signal is given by

$$x(t) = \text{sinc}(t)\cos(4t).$$

(a) Plot the amplitude spectrum of the input signal and the magnitude of the frequency response function on the same graph.
(b) Plot the phase spectrum of the input signal and the phase of the frequency response function on the same graph.
(c) Plot the amplitude and phase spectrum of the output signal.

15.4.2 The amplitude and phase spectrum of a signal $x(t)$ is shown in Figure P15.4.2a. This signal is an input signal to a system whose frequency response function is shown in Figure P15.4.2b.
(a) What is the bandwidth and bandlimit of this signal?
(b) Is this signal a power signal or an energy signal? Find the corresponding power or energy.
(c) What is the bandwidth of the system?
(d) Plot the amplitude and phase spectrum of the output signal. What is the bandwidth of the output signal?
(e) What is the energy or power of the output signal?

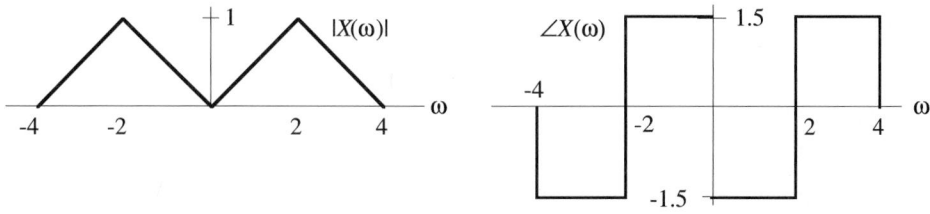

Figure P15.4.2a

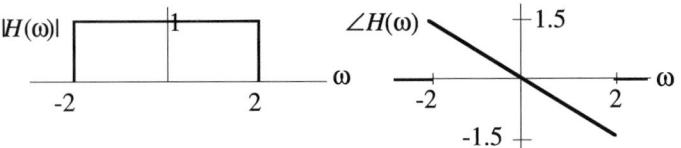

Figure P15.4.2b

15.4.3 The amplitude and phase spectrum of the energy signal $x(t)$ is shown in
Figure P15.4.3a. The magnitude of the frequency response function of this
system is shown in Figure P15.4.3b. Suppose the signal $x(t)$ is an input
signal for this system.

(a) What is the energy of the input signal? What is the bandwidth of the
input signal? What is the bandlimit of the input signal?

(b) What is the bandwidth of this system?

(c) Sketch the amplitude and phase spectrum of the output signal.

(d) What is the bandwidth of the output signal? What is the bandlimit of
the output signal? What is the energy of the output signal?

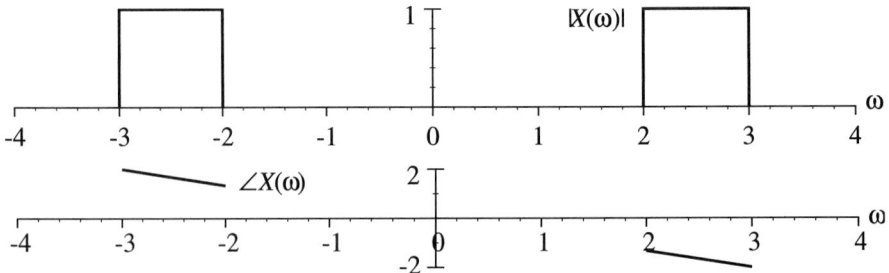

Figure P15.4.3a

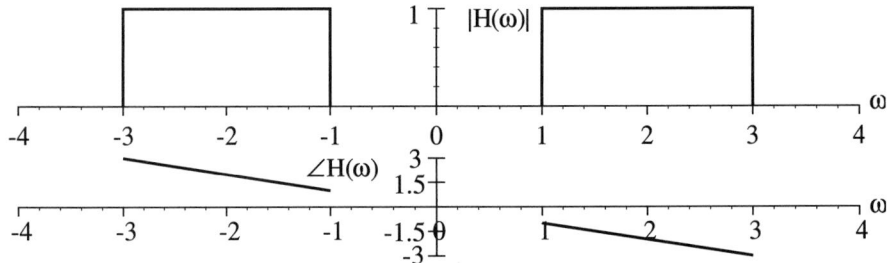

Figure P15.4.3b

15.4.4 Consider the system whose transfer function is

$$\frac{Y(s)}{X(s)} = H(s) = \frac{s^2 + 2(.2)\omega_n s + \omega_n{}^2}{(s + \omega_n)^2}.$$

The power spectrum of the input signal is shown in Figure P15.4.4.
(a) Plot the frequency response function of this system for $\omega_n = 20$ rad/sec.
(b) What is the half-power width of the notch as a function of the damping?
(c) Find ω_n to minimize the power in the output signal above 40 Hz.
(d) Plot the amplitude spectrum of the output signal $y(t)$.
(e) Compute the power of the input and output signals.

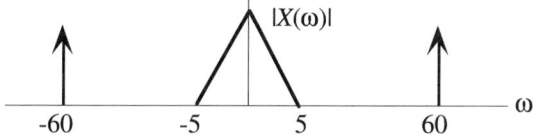

Figure P15.4.4

15.4.5 The signal $x(t)$ in Figure P15.4.5 is the input signal to a system with the frequency response shown in Figure P15.4.5.
(a) Plot the amplitude spectrum of the output signal.
(b) Compute the energy in the input and output signals.
(c) What is the bandwidth of the input and output signals and the system?

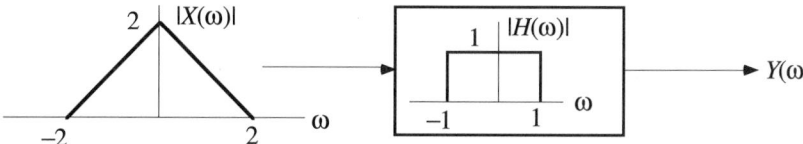

Figure P15.4.5

15.4.6 Consider a signal $x(t)$ whose amplitude spectrum is shown in Figure P15.4.6a.
(a) Consider the signal $x_1(t) = x(t)\cos(100t)$. Find the amplitude spectrum of $x_1(t)$.
(b) Suppose the modulated signal $x_1(t)$ is passed through a filter whose frequency response function is shown in Figure P15.4.6b. Find the amplitude spectrum of the output signal $y(t)$ of this filter.
(c) Find the amplitude spectrum of $x_2(t) = y(t)\cos(100t)$.

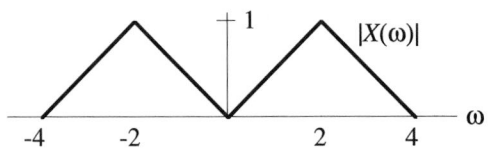

Figure P15.4.6a

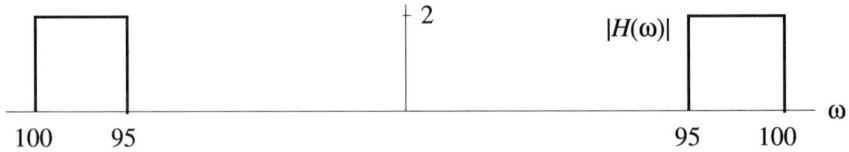

Figure P15.4.6b

15.4.7 Consider a signal whose amplitude spectrum is shown in Figure P15.4.7a.
   (a) What is the bandwidth of this signal? What is the bandlimit of this signal?
   (b) What is the energy of this signal?
   (c) Suppose the signal in Figure P15.4.7b is an input signal for a system whose frequency response function is shown in Figure P15.4.7b. What is the bandwidth of this system?
   (d) Sketch the amplitude spectrum of the output signal from this system. What is the bandwidth and bandlimit of the output signal? What is the energy of the output signal?

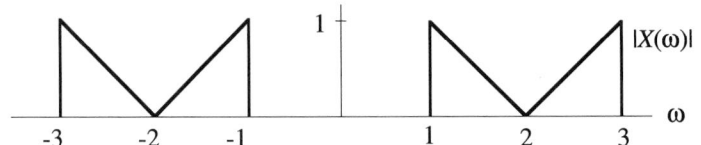

Figure P15.4.7a

Figure P15.4.7b

15.4.8 Consider the *LRC* circuit in Figure P15.4.8a. The parameters are $L = 16\ \mu H$, $C = 30\ \mu F$, and $R = 1\ k\Omega$. The input signal to this system is the supply voltage $v_s(t)$ and the output signal is the current through the capacitor, $i_c(t)$. The input voltage to this network is a square wave of constant frequency,

f_s = 400 kHz, shown in Figure P15.4.8b. The width t_d of each pulse is called the <u>duty cycle</u>. Let $t_d = 0.7$.

(a) Find the cosine Fourier series of this signal.
(b) Show that the dc term of the Fourier series is proportional to the duty cycle t_d.
(c) Determine the frequency response function of this network.
(d) Determine the power in the first three harmonic terms of the current through the capacitor.
(e) Suppose the input voltage to this system is given by

$$v_s(t) = 5\sin((2\pi)1000t) + A_0 + \sum_{m=1}^{\infty} A_m \cos((2\pi m)400,000t + \theta_m)$$

where the coefficients A_m and θ_m are obtained from the Fourier series in (a). Find and plot the Fourier series of the current through the capacitor.
(f) Plot the power spectrum of the input voltage and output current along with the magnitude squared of the frequency response function.

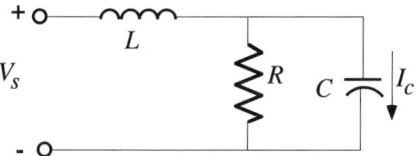

Figure P15.4.8a

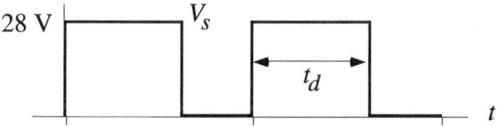

Figure P15.4.8b

15.4.9 Consider the circuit shown in Figure P15.4.9.
(a) Find the transfer function of this system.
(b) If the input signal to this system is $x(t) = u_s(t) - u_s(t-1)$, find $y(t)$ for $RC = 0.2$ and $RC = 2$.
(c) For $RC = 0.2$ plot the magnitude (squared) of the frequency response function and the energy spectral density of the input and output signal. Explain the results in (b) in terms of these graphs.
(d) For $RC = 2$ plot the magnitude (squared) of the frequency response function and the energy spectral density of the input and output signal. Explain the results in (b) in terms of these graphs.

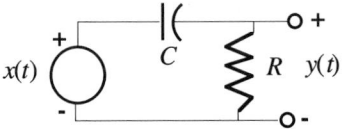

Figure P15.4.9

15.4.10 Consider the parallel resonant circuit in Figure P15.4.10. Assume that $C = 8$ mF, $L = 4$ mH and the Q factor is $Q = 16$. Suppose the input signal is

$$x(t) = \Pi\!\left(\frac{t}{\varepsilon}\right)\cos(\omega_c t)$$

where the carrier frequency is 20 Hz.
(a) For $\varepsilon = 0.01$ plot the input and output signal.
(b) Plot the energy spectrum of the input and output signals and the magnitude of the frequency response function squared.
(c) What is the minimum value of ε such that the output signal is at least 75% of the energy of the input signal?

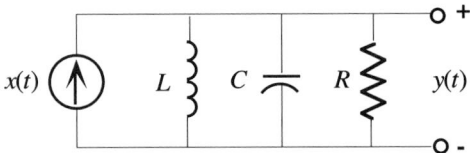

Figure P15.4.10

15.4.11 Suppose the input signal to a system is

$$x(t) = \Pi\!\left(\frac{t}{\varepsilon}\right)\cos(20t).$$

where $\varepsilon = 4$.   Suppose the system is an ideal bandpass filter whose frequency response function is given by

$$|H(\omega)| = \begin{cases} 5, & 18 < |\omega| < 22 \\ 0, & \text{elsewhere} \end{cases}$$

$$\angle H(\omega) = \begin{cases} -2\omega, & 18 < |\omega| < 22 \\ 0, & \text{elsewhere} \end{cases}$$

(a) Is the input signal an energy signal or a power signal? Plot the input signal.
(b) Plot the amplitude and phase spectrum of the input signal. Also plot the energy spectral density of the input signal.

(c) Plot the magnitude and phase of the frequency response function of the system.

(d) Plot the amplitude and phase spectrum of the output signal.

(e) Approximate the ideal bandpass filter with a Butterworth filter. Plot the magnitude and phase of the frequency response function of this filter.

(f) Plot the output signal of the Butterworth filter.

(g) Plot the magnitude and phase spectrum of the output signal of the Butterworth filter. Also plot the energy spectral density.

(h) Repeat this problem for $\varepsilon = 0.1$.

15.4.12 Consider the system

$$\frac{Y(s)}{X(s)} = H(s) = \frac{\omega_n^2}{s^2 + 2(.7)\omega_n s + \omega_n^2}.$$

(a) What is the bandwidth of this system?

(b) Suppose the input signal is $x_1(t) = u_s(t) - u_s(t-1)$. Plot the energy of the output signal, E_1, as a function of ω_n.

(c) Suppose the input signal is $x_2(t) = 0.1\cos(2t)$. Plot the power of the output signal, P_2, as a function of ω_n.

(d) Suppose the input signal into this system is the sum of the two signals above, $x(t) = x_1(t) + x_2(t)$. Suppose we wish to maximize the energy of the output signal due to $x_1(t)$ at the same time minimizing the power of the output signal due to $x_2(t)$. Define the cost function

$$J(\omega_n) = \left(E_1(\omega_n) - 1000 P_2(\omega_n)\right)^2$$

Plot J and determine the ω_n which maximizes this cost function.

(e) Denote the optimal ω_n obtained from part(d) as ω_n^*. Plot the input and output signals for this optimal value of ω_n.

(f) Plot the input and output signal for an ω_n which satisfies $\omega_n < \omega_n^*$.

(g) Plot the input and output signal for an ω_n which satisfies $\omega_n > \omega_n^*$.

15.4.13 Consider the signal whose amplitude spectrum is shown in Figure P15.4.13. This signal is composed of two components: $x_1(t)$ which is an energy signal and $x_2(t)$ which is a sinusoidal interference signal. We wish to remove this interference signal using an ideal notch filter. We also assume that the frequency of this interference signal is not known exactly. Therefore, we want to make the width of the notch wide enough to remove the interference signal, but we don't want to decrease the energy of the output signal due to the input signal $x_1(t)$. Design an ideal notch filter that has a maximum bandwidth to remove the interference signal while only reducing the energy of the output signal by 10%.

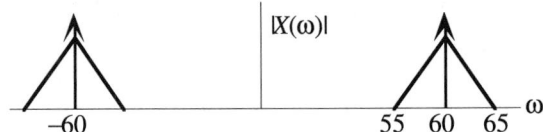

Figure P15.4.13

15.4.14 Consider the system

$$\frac{Y(s)}{X(s)} = H(s) = \frac{40,000}{(s^2 + 5.6s + 2^2)(s^2 + 40s + 100^2)}.$$

(a) Suppose the input signal to this system is $x(t) = \sin(100t)$. Plot the output signal.

(b) Plot the magnitude and phase frequency response function of this system.

(c) Suppose the input signal to this system is $x(t) = u_s(t) - u_s(t - \varepsilon)$. Plot the output signal for $\varepsilon = 1$. Plot the energy spectral density of the input signal and the output signal on the same graph. Also include the plot of the magnitude of the frequency response function squared.

(d) We wish to determine the level of excitation of the two high frequency poles for various values of ε. Plot the magnitude of the energy spectral density of the output signal at the resonant frequency of these poles as a function of ε.

(e) From parts (a) and (b), it can be seen that the output signal will contain a sinusoidal component with a frequency of 100 rad/sec. We wish to filter out this component from the output signal while the components in the output signal due to the other two poles should not be changed. This filter has the transfer function

$$\frac{Y_f(s)}{Y(s)} = H_f(s).$$

(f) Design this filter.

(g) Repeat parts (b) and (c) for the system

$$\frac{Y_f(s)}{X(s)} = H(s)H_f(s).$$

15.4.15 Consider the signal

$$x(t) = \frac{1}{2}\left(1 + \cos\frac{2\pi t}{\tau}\right)\Pi\left(\frac{t}{\tau}\right), \quad \tau = 1.$$

(a) Plot the signal $x(t)$.

(b) Find the Fourier transform of $x(t)$.

(c) Plot the amplitude and phase of $x(t)$.

(d) Suppose that $x(t)$ is an input signal into the system

$$\frac{Y(s)}{X(s)} = \frac{\omega_n^2}{s^2 + 2\omega_n s + \omega_n^2}$$

Let $\omega_n = 1$. Plot the energy spectral density of the input and output signal and the magnitude of the frequency response squared on the same graph. Plot the input and output signals of this system.

(e) Repeat (d) with $\omega_n = 0.1$.

Homework Problems for Section 15.5

15.5.1 An accelerometer is a sensor whose output signal is a voltage that is proportional to the acceleration the device is experiencing. This device is described in Problem 10.1.9. In this problem we will explore further properties of this sensor. For each parameter set in Problem 10.1.9:

(a) Find the frequency response function of this system. Plot the magnitude of the frequency response function.

(b) Suppose the input signal is a sinusoid. What is the highest frequency at which the accelerometer output signal is within 5% of the true value?

(c) Suppose the input signal is the triangular wave shown in Figure P15.5.1 with $T_0 = 1$.

   (i)   Find the Fourier series of the signal in Figure P15.5.1.
   (ii)   Find the Fourier series of the output signal of the accelerometer.
   (iii)   Plot the input and output signals on the same graph using an approximation of the Fourier series.
   (iv)   What is the smallest period of the input signal so that the accelerometer will never be in error more than 10%?

(d) Suppose the input signal is

$$x(t) = \mathrm{Sa}\!\left(\frac{t}{\varepsilon}\right), \quad \varepsilon = 0.1$$

   (i)   Plot the input and output signals on the same graph.
   (ii)   Plot the magnitude of the Fourier transform of the input signal, the magnitude of the frequency response function, and the magnitude of the Fourier transform of the output signal on one graph.
   (iii)   Plot the energy spectral densities of the input and output signals on the same graph as the magnitude of the frequency response function.
   (iv)   What is the smallest value of ε such that the error in the accelerometer's output signal is less that 15%?

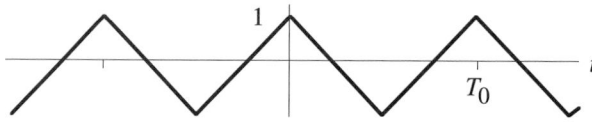

Figure P15.5.1

15.5.2 In this problem we continue the discussion of the accelerometer initiated in Problem 15.5.1. One application of an accelerometer is to provide a feedback signal for the proof-mass actuator discussed in Section 15.5. When the accelerometer is attached to the proof-mass, the output signal of this sensor is a voltage proportional to the acceleration of the proof-mass. The output signal is also corrupted by noise. This noise term is usually modeled as an additive signal at the output signal of the device as shown in Figure P15.5.2a. This noise signal will enter into the power electronics and generate unwanted noise in the actuator signals. Here we investigate the effects of this noise term on the force signal f_{pm} applied to the proof-mass.

The block diagram of the accelerometer attached to the proof-mass actuator is shown in Figure P15.5.2b. This block diagram shows that the velocity feedback loop for the proof-mass actuator is implemented with an accelerometer plus an integrator.

(a) Draw an all-integrator block diagram of the accelerometer. Incorporate this all-integrator block diagram of the accelerometer into the block diagram in Figure P15.5.2b.

(b) In the all-integrator block diagram of the accelerometer plus actuator assign state variables. Develop a state space representation of the all-integrator block diagram.

Answer questions (c) and (d) for each of the parameter sets of the accelerometer shown below:

Set A $m_a = 1$, $c_a = 12$, $k_a = 50$, $K_a = 50$

Set B $m_a = 1$, $c_a = 25$, $k_a = 500$, $K_a = 500$

For questions (c) - (f) assume that the parameter values of the proof-mass actuator are $m_{pm} = 10$ kg, $K_{ef} = 1$, $k_{va} = 0.17$, $k_{pa} = 200$.

(c) Calculate the transfer function $\dfrac{F_{pm}(s)}{N(s)}$.

(d) Plot the magnitude and phase of this frequency response function.

(e) Suppose that the noise is given by $n(t) = \cos(\omega_d t)$. At what frequency ω_d will the steady state response of the force $f_{pm}(t)$ be the largest? What is the corresponding amplitude of the force signal?

(f) Suppose the power spectral density of the noise signal is given by

$$S_n(\omega) = \frac{\omega^4}{\left(\omega_n^2 - \omega^2\right)^2 + (0.1\omega_n\omega)^2}.$$

Plot the power spectral density of the output signal for $\omega_n = 20$.

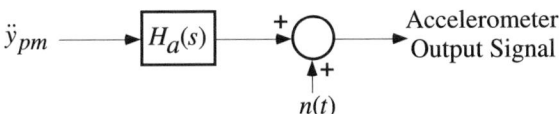

Figure P15.5.2a

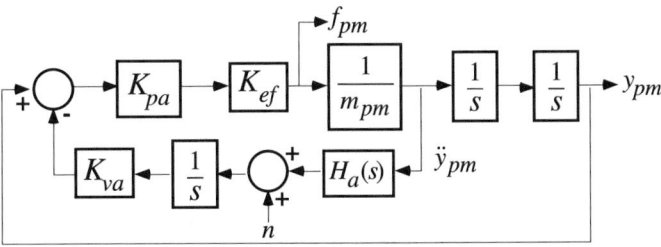

Figure P15.5.2b

15.5.3 In Section 15.5 we discussed the design of a control system for a linear motor. To demonstrate the implications of the difference in the bandwidth of these two feedback systems, consider an x-y plotter. Assume that a linear motor is used to drive each axis. Suppose that the reference signals to each axis are

$$x-axis:\quad r_x(t) = \sin t,\quad t \geq 0$$
$$y-axis:\quad r_y(t) = \cos t,\quad t \geq 0$$

These reference signals describe a circle with radius 1. The reference point rotates at 1 rad/sec and it starts at $(x,y) = (0,1)$. We assume that the x-y plotter starts at $(0,0)$. Each linear motor is at rest and centered at the zero. Hence, when the plotter is turned on, there will be a transient before the signal locks onto the reference signal. Plot the curve in the x-y plane traced out by the linear motors for each of the two control systems for the proof-mass actuator described in Section 15.5.

Homework Problems for Section 15.6

15.6.1 Consider the signal $m(t) = \text{sinc}(t)$. Suppose this message signal is used to amplitude modulate the carrier signal $x_c(t) = \cos(20\pi t)$. Sketch the amplitude spectrum of the modulated signal.

15.6.2 Consider the message signal

$$m(t) = [\text{Sa}(t)]^2$$

Suppose the carrier signal is

$$x_c(t) = e^{j\omega_c t}$$

(a) Sketch the amplitude spectrum of the message signal.
(b) Sketch the amplitude spectrum of the carrier signal.
(c) Sketch the amplitude spectrum of the modulated signal. Compare your result with Figure 15.6.2.

15.6.3 Consider the system shown in Figure P15.6.3a. The amplitude spectrum of the message signal is given in Figure 15.6.2. The magnitude of the frequency response function of the filter is shown in Figure P15.6.3b.
(a) Sketch the amplitude spectrum of the output signal of the filter $y(t)$.
(b) Sketch the amplitude spectrum of the signal $z(t)$. Compare this amplitude spectrum to the amplitude spectrum of $m(t)$.
(c) Compare the bandwidth of the signal $s(t)$ in Figure 15.6.2 and the bandwidth of $y(t)$.

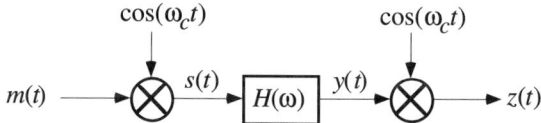

Figure P15.6.3a

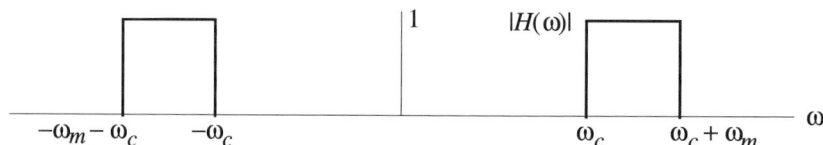

Figure P15.6.3b

15.6.4 Consider the modulated signal $s(t) = m(t)\cos(\omega_c t)$. Suppose that we try to demodulate this signal with the carrier signal $\cos(\omega_c t + \phi)$.
(a) Show that

$$s(t)\cos(\omega_c t + \phi) = \frac{1}{2}m(t)\cos(\phi) + \frac{1}{2}m(t)\cos(2\omega_c t + \phi)$$

(b) What is the output signal of the lowpass filter?
(c) What is the effect of the small phase error ϕ?

15.6.5 Suppose we wish to multiplex N baseband message signals with bandwidths of 5 kHz. If the total bandwidth of the multiplexed signal is 200 kHz, how many signals can be multiplexed?

Chapter 16

Bode Plots

In Chapter 14 we introduced the frequency response theorem and the frequency response function. If the input signal to a system is a sinusoid, the frequency response function relates the steady state response (the output signal) to the system representation. This foundational result relates the structure of the input and output *signals* to the structure of the *system*. This linkage is of immense importance in both system analysis and design. The key quantity in this relationship between signals and systems is the frequency response function, visualized in its graphical form. The shape of the magnitude and phase graphs are key in interpreting this frequency domain information.

Motivated by the frequency response theorem, in this chapter we study the structure of the frequency response function by using the system's (Laplace) transfer function. We relate the general shape of the frequency response function to the poles and zeros of the transfer function. This relationship is established using a traditional method for graphically constructing the Bode plots of the frequency response function. With an understanding of this method, it is easy to visualize the shape of the frequency response function from the poles and zeros of the transfer function.

This information is useful in three ways. First, the computation software that is currently available presumes some understanding of the general shape of the frequency response function and the frequency intervals so that an accurate Bode plot can be obtained. A lack of a clear understanding of the shape of the Bode plots can lead to inaccurate computer graphs for some systems.

Second, the frequency response function plays a central role in the frequency domain synthesis of systems. The general flow of the design methodology is to first construct the frequency response function. Then the transfer function is constructed from a given frequency response function. To effectively construct the transfer function it is essential that the designer understands the relationship between the poles and zeros of the system and the frequency response function. In this section we will develop this relationship between the pole or zero locations of the transfer function and the shape of the corresponding frequency response function. Hence, the designer can go from frequency response function to poles and zeros of transfer functions and vice versa. The results in this chapter lend fundamental insight into the design of systems with specified frequency response function even if the Bode plot is not specifically constructed.

Third, these results are useful for constructing a representation of the system using a measured frequency response function. Here the frequency response function of an existing system is experimentally determined by applying sinusoidal input signals into the system and measuring the steady state response. The gain and phase of the frequency response function is then back-calculated from the input and output signals. By repeating this experiment at various frequencies, a frequency response function is experimentally obtained. Finally, a transfer function is calculated from the experimental frequency response function. The method for constructing Bode plots presented in this chapter is extremely useful for facilitating the construction of the transfer function from frequency response function.

This chapter contains two types of discussion. In the first three sections the Bode plots of elementary factors are derived. These results can be applied directly to simple systems. These results also serve as basic source material for constructing more complex Bode plots. The fourth section gives several detailed examples which illustrate the basic construction procedure for Bode plots of complex transfer functions. The results of this chapter are sufficient to construct the Bode plot of any transfer function.

Summary of Sections

Section 16.1:   We introduce Bode plots.

Section 16.2:   We discuss the construction of straight line approximations to the Bode plots of constant and linear factors.

Section 16.3:   We discuss the construction of straight line approximations to the Bode plots of pairs of complex poles and zeros.

Section 16.4:   We present several examples of the construction of Bode plots.

Section 16.5:   Chapter summary section.

Coverage of the Text

This chapter is motivated by the results of Sections 14.1 - 4.   The construction of the Bode plots only requires the notion of a rational function from Chapter 3.

16.1 INTRODUCTION TO BODE PLOTS

16.1.1 Introduction

Consider the linear, time invariant, BIBO stable system represented using a (Laplace) transfer function

$$\frac{Y(s)}{X(s)} = H(s).$$
(16.1.1)

In Chapter 14 we investigated the output signal of this system when the input signal is the sinusoid

$$x(t) = \left[\cos(\omega_i t)\right] u_S(t).$$
(16.1.2)

The frequency response theorem, Theorem 14.1.1, says that the steady state response of this system is

$$y_{SS}(t) = \left|H(\omega_i)\right| \cos\left(\omega_i t + \angle H(\omega_i)\right).$$
(16.1.3)

The amplitude and phase of the steady state response in (16.1.3) is calculated from the frequency response function $H(\omega)$ which is given by

$$H(s)\big|_{s=j\omega} = H(\omega).$$
(16.1.4)

We have shown that the frequency response function $H(\omega)$ in (16.1.4) completely characterizes the steady state response of the system to these sinusoidal input signals.   The frequency response function plays a central role in the analysis and design of systems as we demonstrated in Chapters 14 and 15.   Because the frequency response function is widely used we would like to understand the structure of the frequency response function in terms of the characteristics of the system.   Unfortunately, the frequency response function is generally a complicated analytical function of ω.   Therefore, it is difficult to study the structure of this function directly.   It turns out, however, that the structure of this function is readily exposed by using the poles and zeros of the transfer function (16.1.1) as we will show in this chapter.   The transfer function is unique among system representations in its ability to expose the structure of the frequency response function.   An understanding of this relationship leads to an intuitive understanding of the relationship between the transfer function and the shape of the frequency response function.

Computer Usage: The frequency response function is frequently plotted using the computer.   Why present a method for hand calculation?   The reason is that the magnitude of the frequency response function can change very rapidly over a very short frequency range followed by relatively smooth behavior over a large frequency band. It is very difficult for the computer to automatically select the frequency band of interest, and then select the number of points required to capture the true behavior of the function. If we understand the general shape of this function, however, we can use the computer very effectively to produce the correct graph by selecting the frequency points over which the graph is plotted. Indeed, most CAD packages assume that the user has this knowledge and is structured to exploit it. The CAD package itself is structured to exploit this knowledge.

16.1.2 Bode Plots

Although the frequency response function is generally a complicated analytical function of ω, it often has a relatively simple graph. So the most common way to present this function is graphically. Therefore, the thrust of this chapter is to develop a graphical method for constructing the magnitude and phase of the frequency response function from the poles and zeros of the transfer function. The method is very easy to apply and it works on very complicated transfer functions. The simplicity of the method resides in the type of graph to be constructed. This chapter will focus on the following graphs.

Definition 16.1.1: The graphs of the magnitude and phase of the frequency response function $H(\omega)$

(1) 20 log $|H(\omega)|$ vs. log ω,
(2) $\angle H(\omega)$ vs. log ω

are called <u>Bode plots</u>.

▲▲

The procedure for constructing the Bode plots is based on the following principle. Consider a transfer function which can be factored into two factors

$$H(s) = H_1(s)H_2(s). \tag{16.1.5}$$

Taking the logarithm[1] of the magnitude of this transfer function we obtain

$$20\log|H(\omega)| = 20\log|H_1(\omega)H_2(\omega)| = 20\log|H_1(\omega)| + 20\log|H_2(\omega)|. \tag{16.1.6}$$

So we see that the magnitude Bode plot of the transfer function is given by a sum of the magnitude Bode plot of each of the factors. Hence, having constructed the Bode plot of each of the factors, we can just sum (graphically) the magnitude Bode plots of each of the factors. Similarly, the phase Bode plot of (16.1.5) is given by

$$\angle H(\omega) = \angle H_1(\omega) + \angle H_2(\omega). \tag{16.1.7}$$

---

[1] This quantity is also directly related to the ratio of the power of the output signal to the power of the input signal. See Section 14.3.

We can construct the phase Bode plot by summing the phase Bode plots for each factor. The reason for plotting this function vs. $\log(\omega)$ will be clear when we derive the method. A logarithmic scale also shows an expanded frequency range on a compressed axis.

As suggested by (16.1.6) and (16.1.7), the Bode plots of complicated functions can be constructed from the Bode plots of <u>elementary factors</u>

$$K, \quad \frac{1}{s}, \quad \frac{\pm a}{s \pm a}, \quad a > 0, \quad \text{and} \quad \frac{\omega_n^2}{s^2 + 2\zeta\omega_n s + \omega_n^2}, \quad -1 < \zeta < 1, \tag{16.1.8}$$

of the transfer function and their inverses. The two numbers a and ω_n in (16.1.8) play a special role in this analysis.

Definition 16.1.2: The numbers a and ω_n in (16.1.8) are called the <u>break frequencies</u> of the factors in (16.1.4), respectively.

▲▲

Notation: We will denote the break frequencies by ω_b.

The break frequencies are always positive. The break frequencies are determined by the poles (and zeros) of the elementary factors. The break frequencies are one of the direct links between the poles and zeros of the transfer function and the Bode plots.

The frequency axis labels for the Bode plots are shown in Figure 16.1.1. The frequency axis in Figure 16.1.1 is a logarithmic scale. Select a frequency, say ω_0, as shown in Figure 16.1.1. The quantity $(\log(\omega_0))$ is located by the linear scale in the second line below the frequency axis. At this frequency (ω_0) we also mark the top frequency axis. This results in the nonuniform tick marks located on the top frequency axis. The corresponding frequencies are labeled by the scale just below this axis. Thus, while the bottom axis is linear in $\log(\omega)$, the custom is to label upper axis in terms of ω.

Definition 16.1.3: The increment in the frequency range from $\omega = 10^x$ to $\omega = 10^{x+1}$ is called a <u>decade</u>.

▲▲

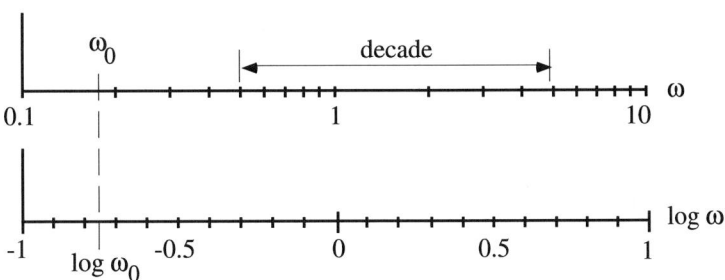

Figure 16.1.1 Axis Scales of a Bode Plot

The decade is a useful unit for explaining the construction of Bode plots.

The method for constructing the Bode plots of $H(s)$ is really a method for evaluating a complex function along the $j\omega$ axis, i.e., for evaluating

$$H(s)\big|_{s=j\omega} = H(\omega), \quad 0 < \omega < \infty. \tag{16.1.9}$$

In this sense, the methodology can be applied whenever the function $H(\omega)$ needs to be constructed. In this text, we will mainly be concerned with the interpretation of the function (16.1.9) according to frequency response theorem. In system theory this function plays other roles, however. Hence, in the discussion below we will refer to the *complex* function $H(\omega)$.

Given a transfer function to construct the Bode plots we first factor the transfer function into elementary factors. Then the Bode plots of each elementary factor are constructed, and the Bode plots of the system are assembled from the Bode plots of each elementary factor. In Section 16.2 we will develop the Bode plots of elementary factors with a real pole or zero. In Section 16.3 we will develop the Bode plots of elementary factors with complex poles and zeros. With this background, we will present the procedure for constructing the Bode plots of the transfer function in Section 16.4.

16.2 BODE PLOTS OF CONSTANTS AND REAL POLES AND ZEROS

16.2.1 Real Poles and Zeros

First, we will develop the Bode plots for a single real pole in the open LHP. This analysis will set the pattern which will be repeated for each of the factors that follow.

A Real Pole in the LHP   Consider the elementary factor

$$H(s) = \frac{a}{s+a} = \frac{1}{\dfrac{s}{a}+1}, \quad a > 0. \tag{16.2.1}$$

Substituting in $j\omega = s$ in (16.2.1), we obtain

$$H(s)\big|_{s=j\omega} = H(\omega) = \frac{a}{j\omega+a} = \frac{1}{j\dfrac{\omega}{a}+1}. \tag{16.2.2}$$

The magnitude and phase of the complex function are

$$|H(\omega)| = \frac{1}{\sqrt{1+\left(\dfrac{\omega}{a}\right)^2}}, \tag{16.2.3a}$$

$$\angle H(\omega) = -\tan^{-1}\frac{\omega}{a}. \hspace{4cm} (16.2.3b)$$

The break frequency for this elementary factor is $\omega = a$. Next we will develop the magnitude and phase Bode plots of the frequency response function in (16.2.3).

Magnitude Plot We will develop the approximation to the magnitude Bode plot of the function in (16.2.3a) by breaking the frequency axis into two intervals divided by the break frequency:

$$I_{p1} = \{\omega \mid 0 < \omega \le a\} \hspace{3.5cm} (16.2.4)$$
$$I_{p2} = \{\omega \mid a < \omega < \infty\}$$

Note that the arrow in Figure 16.2.1 locates these two frequency intervals.

Interval I_{p1}: $\omega < a$ In this frequency interval we can set $\omega = 0$ reducing (16.2.3a) to

$$20\log|H(\omega)| \approx 20\log 1 = 0. \hspace{3cm} (16.2.5)$$

On this interval the magnitude Bode plot is approximately 0. This Bode plot is shown in Figure 16.2.1. The magnitude Bode plot is zero on this frequency interval because we normalized the factor (16.2.1) by the break frequency a. This normalization plays a key role in the construction of the Bode plots as discussed in Section 16.4.

Interval I_{p2}: $\omega > a$ On this frequency interval the squared term is large compared to 1 so we can approximate the radical in (16.2.3a) by

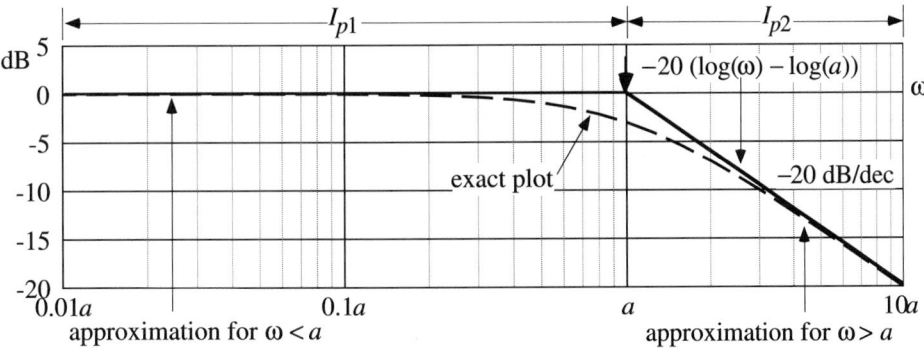

Figure 16.2.1 Approximate and Exact Magnitude Bode Plots for a Real Pole in the LHP

$$\sqrt{1+\left(\frac{\omega}{a}\right)^2} \approx \frac{\omega}{a}. \tag{16.2.6}$$

Inserting (16.2.6) into (16.2.3a) we obtain

$$20\log|H(\omega)| \approx 20\log\left(\frac{1}{\omega/a}\right) = -20\log\left(\frac{\omega}{a}\right) = -20[(\log\omega)-(\log a)]. \tag{16.2.7}$$

The approximation of the magnitude plot comes from (16.2.7). This equation is a straight line in (log ω) with a slope -20 dB/dec and an abscissa intercept of (log a). This straight line approximation for the magnitude Bode plots is shown in Figure 16.2.1.

Remark 16.2.1: Consider the overall structure of the straight line approximation of the Bode plot in Figure 16.2.1 for the elementary factor in (16.2.1). This plot consists of two straight line segments divided by the break frequency $\omega = a$. So we see that the break frequency is one of the three parameters that determines the line in (16.2.7). Another parameter of this Bode plot is the slope of the line. The slope is fixed, however, at -20 dB/dec. Finally, the constant portion of the Bode plot has an ordinate intercept of 0 dB, because of the normalization in (16.2.1). Hence, the break frequency, which completely characterizes the elementary factor, determines the whole line.

This observation suggests the following aid for the construction of the Bode plot. The straight line approximation is characterized by a change in slope from 0 dB/dec to -20 dB/dec. Denote the break frequency a by an arrow pointing downward because the slope of the line is negative. By locating the break frequency, we determine the entire approximation to the Bode plot.

▲▲

Figure 16.2.1 shows the exact and approximate Bode plots for all frequencies. The maximum error between these two graphs occurs at the break frequency. The exact value of the magnitude plot at $\omega = a$ is

$$20\log|H(a)| = 20\log\left(\frac{1}{\sqrt{2}}\right) \approx -3.01 \ dB. \tag{16.2.8}$$

Phase Plot The construction of the phase Bode plot follows along the same lines as the construction of the magnitude Bode plot. We will develop the approximation to the magnitude Bode plot of the function in (16.2.3b) by breaking the frequency axis into three intervals determined by the break frequency:

$$I_{z1} = \left\{\omega \mid 0 < \omega \le 0.1a\right\} \tag{16.2.9}$$
$$I_{z2} = \left\{\omega \mid 0.1a < \omega \le 10a\right\}$$
$$I_{z3} = \left\{\omega \mid 10a < \omega < \infty\right\}$$

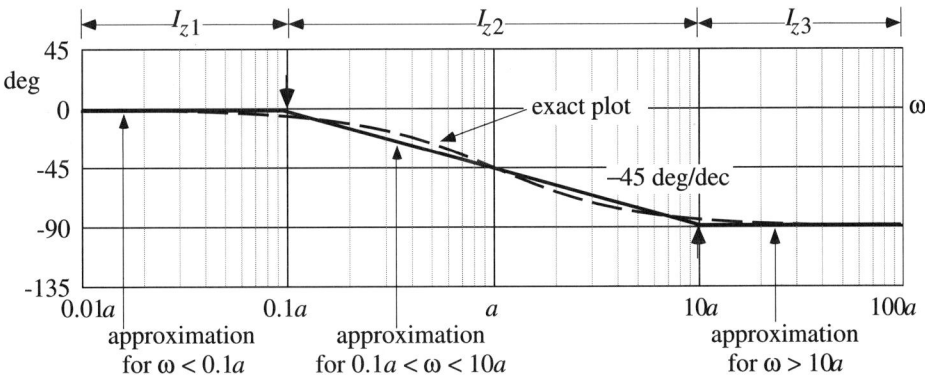

Figure 16.2.2 Approximate and Actual Phase Bode Plots for a Real Pole in the LHP

Note that the arrows in Figure 16.2.2 locate these three frequency intervals.

Interval I_{z1}: $\omega < 0.1a$ In this frequency range we can replace ω by zero reducing (16.2.3b) to

$$\angle H(\omega) \approx -\tan^{-1}(0) = 0. \tag{16.2.10}$$

On this interval the phase Bode plot is approximately 0 as shown in Figure 16.2.2.

Interval I_{z3}: $\omega > 10a$ When the frequency ω is large, the approximation to the phase in (16.2.3b) is just

$$\angle H(\omega) \approx -\tan^{-1}(\infty) = -90°. \tag{16.2.11}$$

This segment of the plot is shown in Figure 16.2.2.

Interval I_{z2}: $0.1a < \omega < 10a$ The straight line approximation to the phase Bode plot is obtained by connecting the approximate curves for $\omega = 0.1a$ and $\omega = 10a$ with a straight line as shown in Figure 16.2.2. The slope of this line is $-45°/\text{dec}$. The exact plot is also shown in Figure 16.2.2. At $\omega = a$, this approximation is exact as can be verified by substituting $\omega = a$ into (16.2.3b).

Remark 16.2.2: Considering the overall structure of the straight line approximation to the phase in Figure 16.2.2, we see that this curve consists of three pieces. The segment on I_{z1} below $\omega = 0.1a$ is constant with a value of $0°$. The segment on I_{z3} above $\omega = 10a$ is constant with a value of $-90°$. Both of these segments are determined by the break frequency. The segment of the curve on the frequency interval I_{z2} is determined by the segments in the frequency intervals above and below it. The slope of this segment of the curse is fixed at $-45°/\text{dec}$. Again we see that the break frequency at $\omega = a$, which completely characterizes the elementary factor in (16.2.1) and determines the entire approximation to the Bode plot.

This observation suggests the following aid for the construction of the Bode plot. At the frequency $\omega = 0.1a$ place an arrow pointing downward. This arrow suggests a change in the slope of the line in the negative direction. At the frequency $\omega = 10a$ place an arrow pointing in the upward direction. This arrow suggests a change in the slope in the positive direction. These arrows, shown in Figure 16.2.2, locate the approximate Bode plot.   ▲▲

A Real Pole in the RHP Next we will consider the construction of the Bode plot of a pole in the RHP. This construction is almost identical to the previous discussion. Consider the elementary factor

$$H(s) = \frac{-a}{s-a} = \frac{-1}{\dfrac{s}{a}-1}, \quad a > 0. \tag{16.2.12}$$

Note: The normalization includes a minus sign.

Substituting in $s = j\omega$ in (16.2.12), we obtain

$$H(s)\big|_{s=j\omega} = H(\omega) = \frac{-a}{j\omega - a} = \frac{1}{1 - j\dfrac{\omega}{a}}. \tag{16.2.13}$$

The break frequency is $\omega = a$.

Magnitude Plot The magnitude of the function in (16.2.13) is identical to the magnitude of a real pole in the LHP shown in (16.2.3a). Hence, the magnitude Bode plot of (16.2.12), shown in Figure 16.2.3a, is the same plot shown in Figure 16.2.1a.

Phase Plot The phase of (16.2.13) is given by

$$\angle H(\omega) = -\tan^{-1}\left(\frac{-\omega}{a}\right) = \tan^{-1}\left(\frac{\omega}{a}\right). \tag{16.2.14}$$

This phase will have the same general shape as the phase of a real pole in the LHP, except it will break up to a $+90°$. The slope of the phase on the frequency interval I_{z2} is $+45°/\text{dec}$. The construction of the straight line approximations of the magnitude and phase proceeds as described for the pole in the LHP. The results are shown in Figure 16.2.3.

The comments in Remark 16.2.2 apply here as well, except that the value of the constant curve on the interval I_{z3} is $+90°$. Hence, the slope of the curve on the frequency interval I_{z2} is positive. Therefore, the arrows in Figure 16.2.3 are reversed. The arrow at $\omega = 0.1a$ points upward, because the slope breaks upward. The arrow at $\omega = 10a$ points downward, because the slope breaks downward.

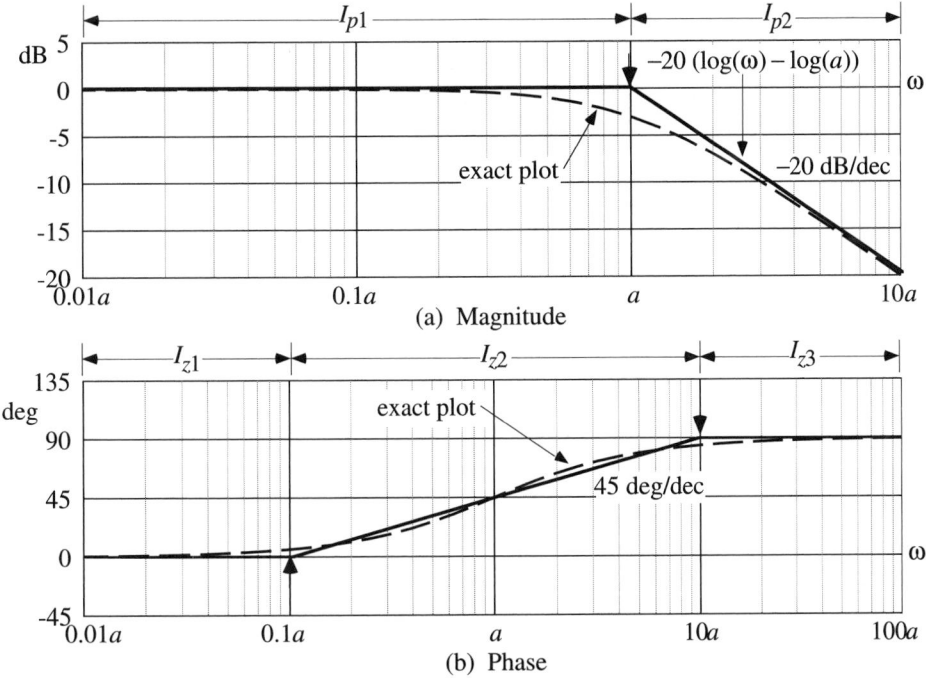

(a) Magnitude

(b) Phase

Figure 16.2.3 Bode Plots for a Pole in the RHP

A Real Zero in the LHP Next we will discuss the Bode plots of real zeros. Consider the elementary factor

$$H(s) = \frac{s+a}{a} = \frac{s}{a} + 1, \quad a > 0. \tag{16.2.15}$$

The complex function is

$$H(\omega) = \frac{j\omega + a}{a} = j\left(\frac{\omega}{a}\right) + 1. \tag{16.2.16}$$

Converting (16.2.16) to polar form we have

$$|H(\omega)| = \sqrt{1 + \left(\frac{\omega}{a}\right)^2}, \qquad \angle H(\omega) = \tan^{-1}\left(\frac{\omega}{a}\right). \tag{16.2.17}$$

We evaluate the functions in (16.2.17) as we did with the real poles.

Magnitude Plot   Again we will divide the frequency axis into two intervals as in (16.2.4). Note that the arrow in Figure 16.2.4a locates these two frequency intervals.

Interval I_{p1}: $\omega < a$ In this frequency range we can replace ω by zero in (16.2.17), reducing (16.2.17) to

$$20\log|H(\omega)| \approx 20\log 1 = 0. \tag{16.2.18}$$

On this frequency interval the magnitude Bode plot is zero as shown in Figure 16.2.4a.

Interval I_{p2}: $a < \omega$ On this frequency band the constant 1 in (16.2.16) is neglected because the frequency is much larger than 1. The magnitude is given by

$$20\log|H(\omega)| \approx 20\log\frac{\omega}{a} = 20\big[(\log\omega) - (\log a)\big]. \tag{16.2.19}$$

The magnitude plot on this frequency band is a straight line in (log ω) with a slope of +20 dB/dec, and an axis intercept of (log a). This line is shown in Figure 16.2.4a. Following the discussion for a real pole, the approximation of the magnitude Bode plot for the real zero is complete. The exact plot is also shown in Figure 16.2.4a.

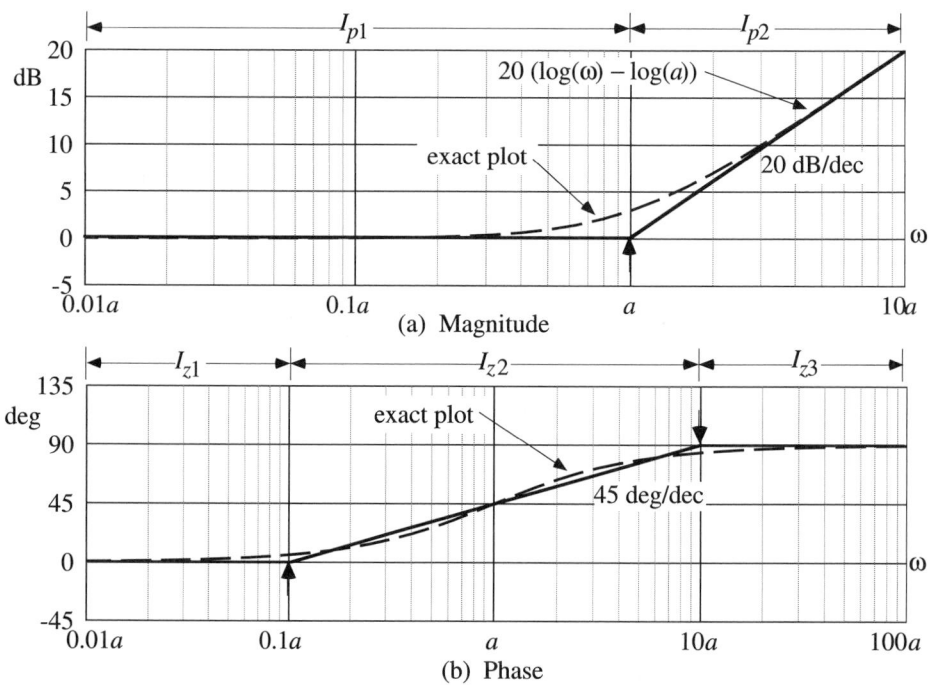

Figure 16.2.4 Approximate and Actual Bode Plots for a Zero Located at $s = -a$

Note: The magnitude plot for a real zero looks like the magnitude plot for a real pole except that the plot for a real zero breaks up, while the plot for a real pole breaks down. In particular, the absolute value of the slope of the magnitude plots is the same.

The comments in Remark 16.2.1 apply here as well, except that the curve breaks upward with a slope of +20 dB/dec on the interval I_{z2}. Therefore, the arrow in Figure 16.2.4 is reversed from the arrow in Figure 16.2.3. The arrow at the break frequency $\omega = a$ points upward, because the slope breaks upward.

Phase Plot Again we evaluate the phase in (16.2.17) on the three frequency intervals determined by the break frequency shown in (16.2.9). Note that the arrows in Figure 16.2.4b locate these three frequency intervals.

Interval I_{z1}: $\omega < 0.1a$ On this interval the argument of the arc tangent is essentially zero. Hence, the phase is approximated by

$$\angle H(\omega) \approx \tan 0 = 0. \tag{16.2.20}$$

This approximation is shown in Figure 16.2.4b.

Interval I_{z3}: $10a < \omega$ For large ω the phase is approximated by

$$\angle H(\omega) = \tan^{-1}\omega \approx 90°. \tag{16.2.21}$$

This approximation is shown in Figure 16.2.4b.

Interval I_{z2}: $0.1a < \omega < 10a$ Following the discussion for a real pole, the phase plot is completed by joining the plot at $\omega = 0.1a$ with the plot at $\omega = 10a$ using a straight line as shown in Figure 16.2.4b. This figure also contains the exact phase Bode plot.

Again we see that the phase approximation is characterized by straight line segments with a change in slope at specific frequencies. These changes in the slope are marked with arrows as shown in Figure 16.2.4b. An up arrow is located at one decade below the break frequency to indicate a change of slope from 0°/dec to 45°/dec. A down arrow is located at one decade above the break frequency to indicate a decrease in the slope from 45°/dec to 0°/dec.

A Real Zero in the RHP Consider the elementary factor with a single zero in the RHP

$$H(s) = \frac{s-a}{-a} = -\frac{s}{a} + 1, \quad a > 0. \tag{16.2.22}$$

Note: The normalization factor contains a minus sign.

Substituting in $s = j\omega$ in (16.2.22), we obtain

$$H(\omega) = \frac{j\omega - a}{-a} = 1 - j\frac{\omega}{a}.$$
(16.2.23)

Magnitude Plot   The magnitude of the complex function in (16.2.23) is identical to the magnitude function in (16.2.17). Hence, the approximations to the magnitude of the Bode plots are the same. The exact and approximate magnitude Bode plots are shown in Figure 16.2.4a.

Phase Plot   The phase of the frequency response function in (16.2.23) is

$$\angle H(\omega) = \tan^{-1}\left(-\frac{\omega}{a}\right) = -\tan^{-1}\frac{\omega}{a}.$$
(16.2.24)

Comparing the phase in (16.2.24) with the phase of a *real pole in the LHP*, we see that they are the same. Hence, the phase plots are the same. The phase Bode plot of this function is shown in Figure 16.2.3.

16.2.2 Constant Factors and Poles and Zeros at the Origin

Introduction   The Bode plots for single real poles and zeros exhibit all of the features typical of Bode plots. The Bode plots of the functions discussed below could be thought of as an extension of the discussion above. When the break frequency goes to infinity, we obtain a constant. When the break frequency goes to zero, we obtain a pole or zero at the origin.

Constant Gain Factors   First, we consider a function which is just a real constant

$$H(s) = K = \begin{cases} |K|e^{j0}, & \text{if } K > 0 \\ |K|e^{j\pi}, & \text{if } K < 0 \end{cases}$$
(16.2.25)

Then the magnitude in decibels is

$$20\log|H(\omega)| = 20\log|K|.$$
(16.2.26)

The phase of $H(s)$ is just

$$\angle H(\omega) = \begin{cases} 0, & \text{if } K > 0 \\ 180°, & \text{if } K < 0 \end{cases}$$
(16.2.27)

One Pole or Zero at the Origin   Next we will consider poles and zeros at the origin.

A Pole at the Origin   Next, consider a pole at the origin

$$H(s) = \frac{1}{s}. \tag{16.2.28}$$

The complex function is

$$H(s)\big|_{s=j\omega} = H(\omega) = \frac{1}{j\omega} = \frac{1}{\omega} e^{-j\frac{\pi}{2}}. \tag{16.2.29}$$

From (16.2.29) we see that the magnitude and phase functions are given by

$$20\log|H(\omega)| = 20\log\left(\frac{1}{\omega}\right) = 20\log 1 - 20(\log\omega) = -20(\log\omega) \tag{16.2.30}$$

$$\angle H(\omega) = -90°.$$

It is clear from (16.2.30) that the magnitude Bode plot is a straight line in (log ω) with a slope of -20 dB/decade which passes through the point (1,0).   This plot is shown in Figure 16.2.5a. The phase Bode plot is shown in Figure 16.2.5b.

A Zero at the Origin   Consider a zero at the origin

$$H(s) = s. \tag{16.2.31}$$

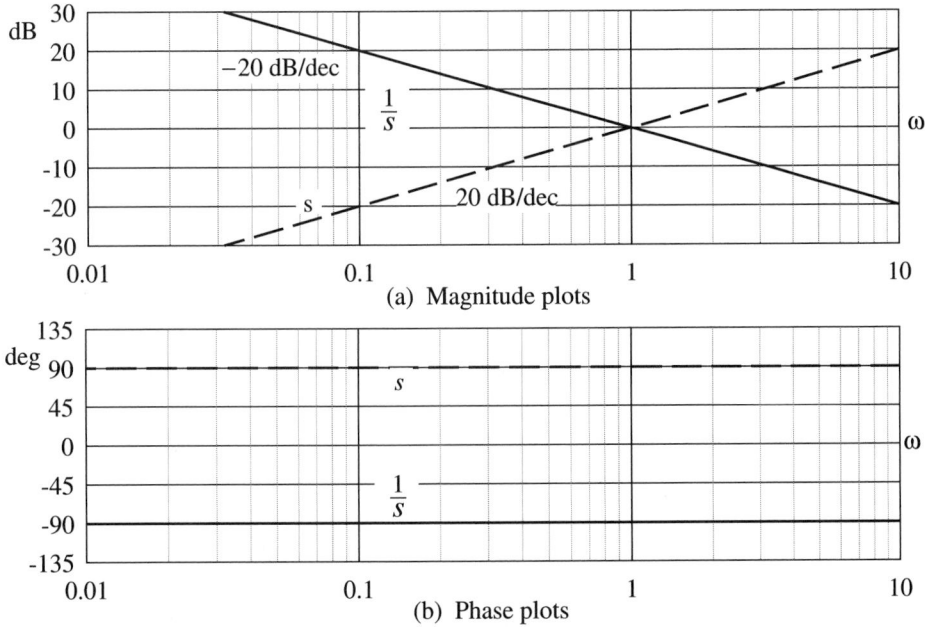

Figure 16.2.5 Bode Plots of a Pole and Zero at the Origin

Then the complex function is

$$H(\omega) = j\omega = \omega e^{j\frac{\pi}{2}}.$$
(16.2.32)

From (16.2.32) we see that the magnitude and phase Bode plots are given by

$$20\log|H(\omega)| = 20(\log\omega),$$
(16.2.33)

$$\angle H(\omega) = 90°.$$

It is clear from (16.2.33) that the magnitude Bode plot is a straight line in (log ω) with a slope of 20 dB/dec which passes through the point (1,0). This plot is shown in Figure 16.2.5a. The phase Bode plot is shown in Figure 16.2.5b.

16.3 BODE PLOTS OF TWO COMPLEX POLES AND ZEROS

16.3.1 Introduction

In the last section the construction of the straight line approximation to the Bode plots for elementary factors with one pole or zero is described. In this section we will give a similar construction for quadratic factors with two complex poles or zeros. The results in these two sections are sufficient to draw the Bode plots of any transfer function.

16.3.2 Two Complex Poles

Consider the elementary factor

$$H(s) = \frac{\omega_n^2}{s^2 + 2\zeta\omega_n s + \omega_n^2} = \frac{1}{\left(\dfrac{s}{\omega_n}\right)^2 + 2\zeta\left(\dfrac{s}{\omega_n}\right) + 1}, \quad -1 < \zeta < 1.$$
(16.3.1)

Note: For the given range for ζ, this factor has a pair of complex poles. If ζ is outside this range this factor has two real poles; it is factored into its linear factors and the results of the last section are used.

The complex function is

$$H(s)\big|_{s=j\omega} = H(\omega) = \frac{1}{\left(\dfrac{j\omega}{\omega_n}\right)^2 + 2\zeta\left(\dfrac{j\omega}{\omega_n}\right) + 1} = \frac{1}{1 - \left(\dfrac{\omega}{\omega_n}\right)^2 + j(2\zeta)\left(\dfrac{\omega}{\omega_n}\right)}.$$
(16.3.2)

This complex function is characterized by the break frequency ω_n and ζ.

Magnitude Plot Again we consider two frequency intervals divided by the break frequency to derive the straight line approximations to the magnitude Bode plot:

$$I_{p1} = \left\{ \omega \mid 0 < \omega \le \omega_n \right\} \tag{16.3.3}$$
$$I_{p2} = \left\{ \omega \mid \omega_n < \omega < \infty \right\}$$

Note that the arrow in Figure 16.3.1 locates these two frequency intervals.

Interval I_{p1}: $\omega < \omega_n$ In this frequency interval ω is small so that we have

$$\frac{\omega}{\omega_n} \approx 0. \tag{16.3.4}$$

Substituting (16.3.4) into (16.3.2) it follows that

$$H(\omega) \approx 1. \tag{16.3.5}$$

On this frequency interval the approximation to the magnitude Bode plot is

$$20 \log |H(\omega)| \approx 0. \tag{16.3.6}$$

The normalization in (16.3.1) is responsible for (16.3.5) and so (16.3.6).

Interval I_{p2}: $\omega > 10\omega_n$ On this frequency interval the square term in (16.3.2) dominates. From (16.3.2) we have

$$H(\omega) \approx \frac{1}{-\dfrac{\omega^2}{\omega_n^2}} = -\frac{\omega_n^2}{\omega^2}. \tag{16.3.7}$$

The straight line approximation to the magnitude Bode plot on this interval is

$$20 \log |H(\omega)| = 20 \log \left(\frac{\omega_n^2}{\omega^2} \right) = 20 \log \omega_n^2 - 20 \log \omega^2 \tag{16.3.8}$$
$$= 40 (\log \omega_n) - 40 (\log \omega) = 40 \left[(\log \omega_n) - (\log \omega) \right].$$

As can be seen from (16.3.8) on the interval I_{p2}, the approximation is a straight line with slope of -40 dB/dec and abscissa intercept $\log \omega_n$. The approximation to the magnitude Bode plot is shown in Figure 16.3.1.

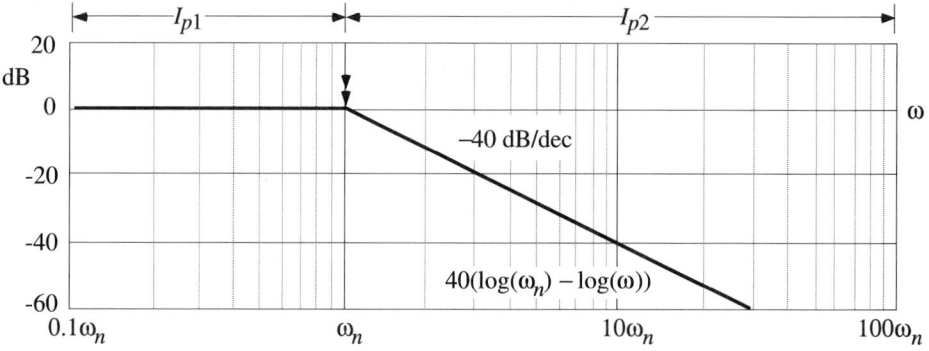

Figure 16.3.1 Straight Line Approximation to the Magnitude Bode Plot
for a Pair of Complex Poles

The slope of the curve in Figure 16.3.1 on the frequency interval I_{p2} is -40 dB/dec.
This slope is twice the slope of a single real pole. We interpret this result to mean
that each pole contributes -20 dB/dec of slope to the straight line approximation in
the frequency interval I_{p2}.

Remark 16.3.1: Consider the overall structure of the straight line approximation of
the Bode plot in Figure 16.3.1. This plot consists of two straight line segments
divided by the break frequency $\omega = \omega_n$ of the elementary factor in (16.3.1). So we
see that the break frequency is one of the three parameters that determines the line
in (16.3.8). Another parameter of this Bode plot is the slope of the line. The slope is
fixed at -40 dB/dec. Finally, the constant portion of the Bode plot has an ordinate
intercept of 0 dB, because of the normalization in (16.3.1). Hence, the break
frequency completely determines the approximation.

This observation suggests the following aid for the construction of the Bode plot.
Denote the break frequency ω_n by a double arrow pointing downward because the
slope of the line is negative and it has a slope of $(2)(20)$ dB/dec $= -40$ dB/dec. ▲▲

When the magnitude function in (16.3.2) is evaluated at the break frequency, the
complex function becomes

$$H(\omega_n) = \frac{1}{-1 + j2\zeta + 1} = \frac{1}{j2\zeta}. \tag{16.3.9}$$

The exact value of the Bode plot is given by

$$20\log|H(\omega_n)| = 20\log\left(\frac{1}{2|\zeta|}\right) = -20\log(2|\zeta|). \tag{16.3.10}$$

For small values of ζ, the magnitude of the Bode plot differs significantly from the straight line approximation in Figure 16.3.1. The exact Bode plots for several values of ζ are shown in Figure 16.3.2. If the damping ratio ζ is small then one correction point is added to the approximate Bode plot at the break frequency $\omega = \omega_n$. One criteria for adding correction points is when $-0.5 < \zeta < 0.5$. Correction points are shown in Figure 16.3.2.

Note: This analysis holds for poles in both the RHP and LHP i.e., for positive and negative values of ζ. The values of ζ shown in Figure 16.3.2 should be interpreted as absolute values.

Phase Plots To derive the phase plot we consider the function (16.3.2) on the three frequency intervals determined by the break frequency

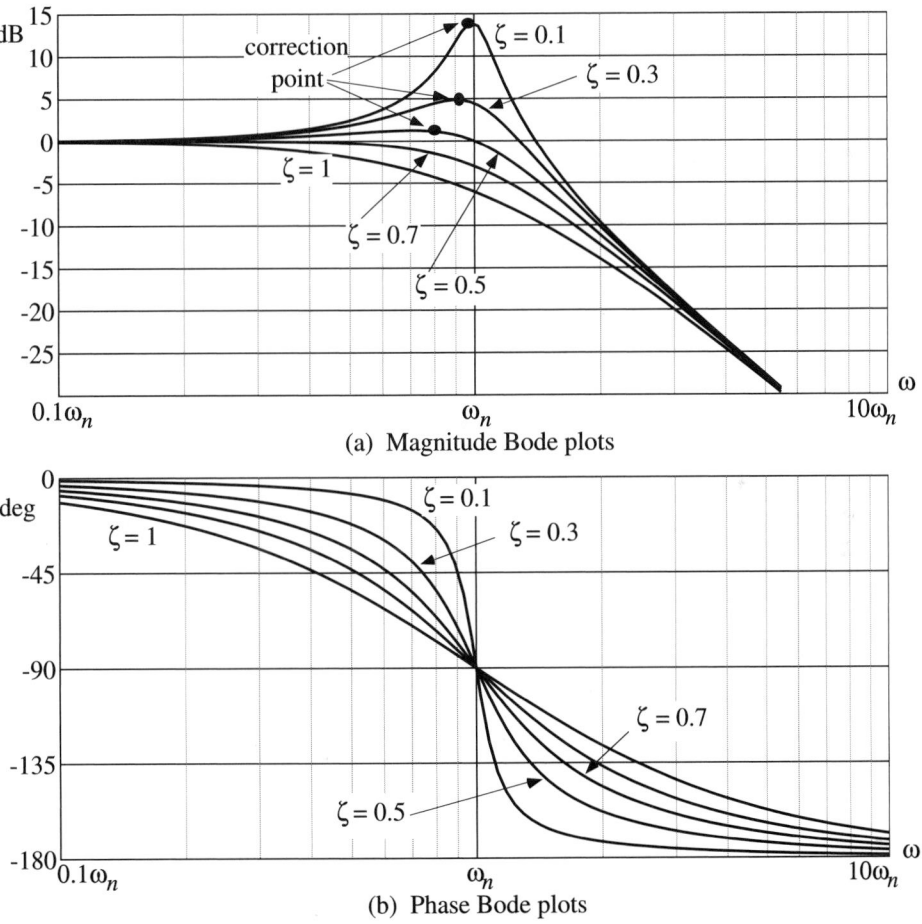

Figure 16.3.2 Exact Bode Plots for the Magnitude for a Pair of Complex Poles

$$I_{z1} = \left\{ \omega \mid 0 < \omega \leq 0.1\omega_n \right\}$$ (16.3.11)
$$I_{z2} = \left\{ \omega \mid 0.1\omega_n < \omega \leq 10\omega_n \right\}$$
$$I_{z3} = \left\{ \omega \mid 10\omega_n < \omega < \infty \right\}$$

Note that the arrows in Figure 16.3.3 locate these three frequency intervals.

Interval I_{z1}: $\omega < 0.1\omega_n$ On this frequency interval the approximation to the phase Bode plot is

$$\angle H(\omega) \approx 0.$$ (16.3.12)

Interval I_{z3}: $\omega > 10\omega_n$ For this frequency band the constant term in (16.3.2) dominates. Here the approximate phase Bode plot is

$$\angle H(\omega) = -180°$$ (16.3.13)

as can be seen in (16.3.7).

Interval I_{z2}: $0.1\omega_n < \omega < 10\omega_n$ Suppose the poles are in the LHP with $0 < \zeta < 1$. The phase of the complex function on the two intervals above is shown in (16.3.12) and (16.3.13). The phase at the break frequency $\omega = \omega_n$ is

$$\angle H(\omega_n) = \begin{cases} -90°, & \text{if } 0 < \zeta < 1 \\ \text{undefined}, & \text{if } \zeta = 0 \end{cases}$$ (16.3.14)

The phase goes from $0°$ to $180°$ through negative angles as shown by (16.3.14). The approximate Bode plot is obtained by connecting the line at $\omega = 0.1\omega_n$ to the line at

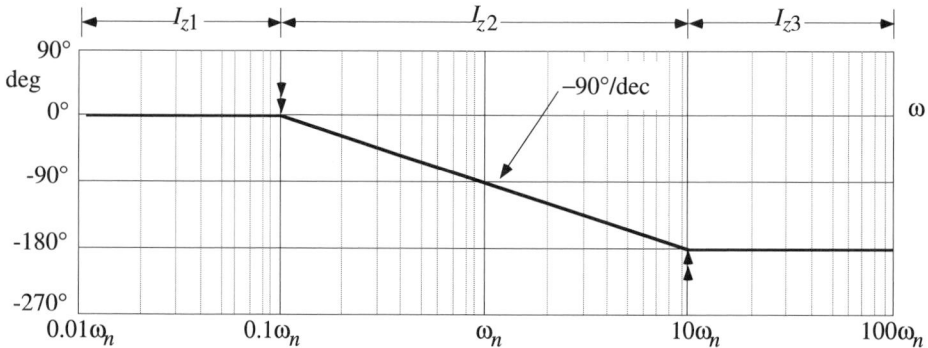

Figure 16.3.3 Approximate Phase Bode Plots for a Pair of Complex Poles in the LHP

$\omega = 10\omega_n$ as shown in Figure 16.3.3. The slope of this line is $-90°$/dec. The exact plots for several values of ζ are shown in Figure 16.3.2. As $\zeta \to 0$ the phase curve approaches a step function as shown in Figure 16.3.2b.

The approximate phase in Figure 16.3.3 on the frequency interval I_{z3} is $-180°$. This value is twice the value of the phase of a real pole, which is $-90°$. The interpretation is that each pole in (16.3.1) contributes $-90°$ of phase on the frequency interval I_{z3}. Also note that the slope of the line on the interval I_{z2}, $-90°$/dec, is twice the slope of a single real pole, $-45°$/dec.

Remark 16.3.2: Considering the overall structure of the straight line approximation to the phase in Figure 16.3.3, we see that this curve consists of three pieces. The segment on I_{z1} below $\omega = 0.1\omega_n$ is constant with a value of $0°$. The segment on I_{z3} above $\omega = 10\omega_n$ is constant with a value of $-180°$. Both of these segments are determined by the break frequency. The segment of the curve on the frequency interval I_{z2} is determined by the segments in the frequency intervals above and below it. Again we see that the break frequency at ω_n determines the whole curve.

This observation suggests the following aid for the construction of the Bode plot. At the frequency $\omega = 0.1\omega_n$ place a double arrow pointing downward. These arrows suggest a change in the slope of the line in the negative direction. At the frequency $\omega = 10\omega_n$ place a double arrow pointing in the upward direction. These arrows suggest a change in the slope in the positive direction. These arrows are shown in Figure 16.3.3.

▲▲

When $\zeta > 0$, the poles are in the RHP. In this case the equation (16.3.12) and (16.3.13) still hold, although we write (16.3.13) as

$$\angle H(\omega) = 180°. \tag{16.3.15}$$

The phase at the break frequency $\omega = \omega_n$ for two poles in the RHP is

$$\angle H(\omega_n) = \begin{cases} 90°, & \text{if } -1 < \zeta < 0, \\ \text{undefined}, & \text{if } \zeta = 0 \end{cases} \tag{16.3.16}$$

Here the phase plot goes from $0°$ to $180°$ as shown in Figure 16.3.4. Keeping in mind Remark 16.3.2, note that the slope of the curve on the frequency interval I_{z2} is $+90°$/dec. The arrows denoting the change in the slope in Figure 16.3.4 have been reversed to accommodate this change.

16.3.3 Two Complex Zeros

Consider the elementary factor

$$H(s) = \frac{s^2 + 2\zeta\omega_n s + \omega_n^2}{\omega_n^2} = \left(\frac{s}{\omega_n}\right)^2 + 2\zeta\left(\frac{s}{\omega_n}\right) + 1, \quad -1 < \zeta < 1. \tag{16.3.17}$$

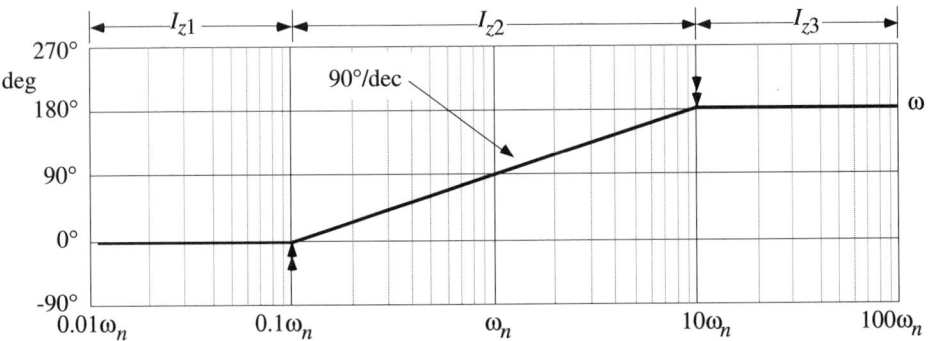

Figure 16.3.4 Approximate Bode Phase Plot for a Pair of Complex Poles
in the RHP

This factor has two complex zeros for these values of ζ. If ζ is outside this range, this factor has two real zeros, it is factored into linear factors, and the results of the last section are applied. Substituting in $s = j\omega$ in (16.3.17), we obtain

$$H(\omega) = \left(\frac{j\omega}{\omega_n}\right)^2 + 2\zeta\left(\frac{j\omega}{\omega_n}\right) + 1 = 1 - \left(\frac{\omega}{\omega_n}\right)^2 + 2\zeta\left(\frac{j\omega}{\omega_n}\right). \tag{16.3.18}$$

Comparing the function in (16.3.18) to the function in (16.3.2), we see that they are inverse of each other. The situation is analogous to the discussion of elementary factors with real poles and zeros.

The construction of the straight line approximations to the Bode plot for the complex function in (16.3.18) repeats the construction of the other Bode plots above.

Magnitude Plot Again we split the frequency axis into two intervals divided by the break frequency as in (16.3.3).

Interval I_{p1}: $\omega < \omega_n$ In this frequency range ω is small so that we have

$$\frac{\omega}{\omega_n} \approx 0. \tag{16.3.19}$$

It follows from (16.3.19) that

$$H(\omega) \approx 1, \quad \text{and} \quad 20\log|H(\omega)| = 0. \tag{16.3.20}$$

Interval I_{p2}: $\omega > a$ On this frequency interval the square term in (16.3.18) dominates. From (16.3.18) we have

$$H(\omega) \approx -\frac{\omega^2}{\omega_n^2}. \tag{16.3.21}$$

The straight line approximation to the magnitude Bode plot on this interval is

$$20\log|H(\omega)| = 20\log\left(\frac{\omega^2}{\omega_n^2}\right) = 20\log\omega^2 - 20\log\omega_n^2 = 40\left[(\log\omega) - (\log\omega_n)\right]. \tag{16.3.22}$$

The equation in (16.3.22) is a straight line with a slope of 40 dB/dec and an abscissa intercept of $(\log\omega_n)$ on the interval I_{p2}. The straight line approximation to the magnitude Bode plot is shown in Figure 16.3.5. At the break frequency $\omega = \omega_n$, the complex function (16.3.18) is

$$H(\omega_n) = -1 + j2\zeta + 1 = j2\zeta. \tag{16.3.23}$$

The value of the Bode plot at this break frequency is

$$20\log|H(\omega_n)| = 20\log(2|\zeta|). \tag{16.3.24}$$

Since $\zeta < 1$, this value is negative. The exact Bode plots for several values ζ are shown in Figure 16.3.6.

Remark 16.3.3: If the function has a pair of complex zeros with a small ζ, usually one correction point is added to the approximation to the Bode plot at $\omega = \omega_n$. The criteria for adding this correction point are essentially the same as explained in Remark 16.3.1. Several correction points are indicated in Figure 16.3.6.

We also have similar observations above the magnitude plot of the complex zeros in Figure 16.3.5 as we did about the magnitude plot of the complex poles

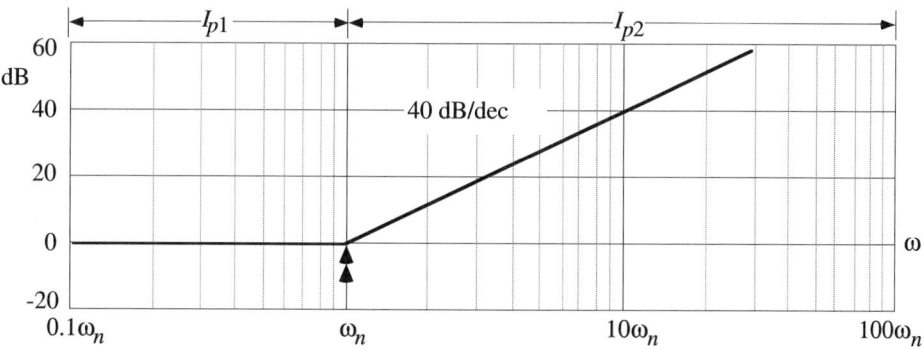

Figure 16.3.5 Straight Line Approximations to the Magnitude Bode Plot
for a Pair of Complex Zeros

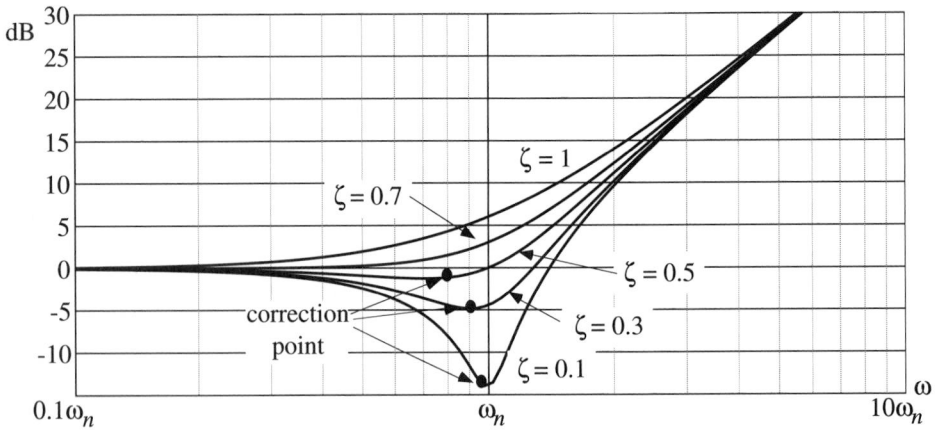

Figure 16.3.6 Exact Magnitude Bode Plot for a Pair of Complex Zeros for Several Values of the Damping Ratio

explained in Remark 16.3.2 (with obvious changes).   The appropriate arrows are shown in Figure 16.3.5.

▲▲

Phase Plot The construction of the phase plot proceeds in a similar manner to the phase plots for the complex poles described above.   Again we construct the phase on three intervals as in (16.3.11).

Interval I_{z1}: $\omega < 0.1\omega_n$ It follows from (16.3.19) that the phase Bode plot in this frequency range is

$$\angle H(\omega) \approx 0.$$
(16.3.25)

Interval I_{z3}: $\omega > 10\omega_n$ On this frequency interval the square term in (16.3.19) dominates.   We have

$$H(\omega) \approx -\frac{\omega^2}{\omega_n^2}.$$
(16.3.26)

The straight line approximation to the phase Bode plot on this interval is

$$\angle H(\omega) \approx 180°.$$
(16.3.27)

Interval I_{z2}: $0.1\omega_n < \omega < 10\omega_n$ At this frequency, the complex function is

$$H(\omega_n) = -1 + j2\zeta j + 1 = j2\zeta.$$
(16.3.28)

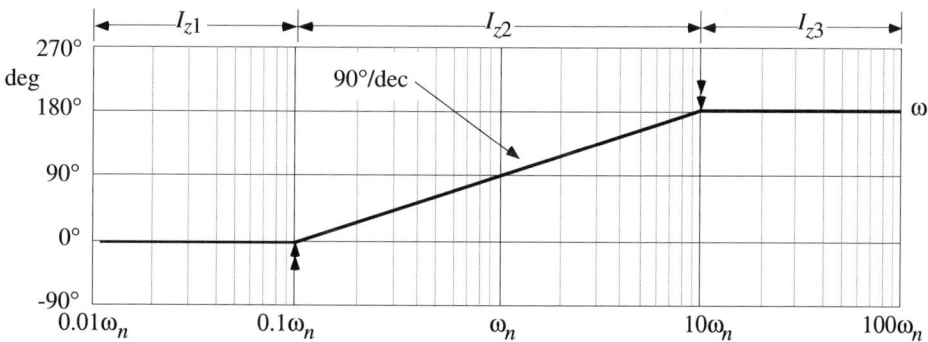

Figure 16.3.7 Straight Line Approximations to the Phase Bode Plot for
a Pair of Complex Zeros in the LHP

Suppose that $0 < \zeta < 1$ so that the zeros are in the LHP. Then the phase satisfies

$$\angle H(\omega_n) = \begin{cases} 90°, & \text{if} \quad 0 < \zeta < 1, \\ \text{undefined}, & \text{if} \quad \zeta = 0 \end{cases} \qquad (16.3.29)$$

The phase Bode plot goes from $0°$ to $180°$ through positive values as indicated by
(16.3.29). In this case the approximation to the phase Bode plot is shown in Figure
16.3.7. The slope of the phase plot on the interval I_{z2} is $+90°$/dec. The exact phase
plots for this case are similar in shape to those shown in Figure 16.3.2b.

Remark 16.3.4: Considering the overall structure of the straight line approximation
to the phase in Figure 16.3.7, we see that this curve consists of three pieces. The
segment on I_{z1} below $\omega = 0.1\omega_n$ is constant with a value of $0°$. The segment on I_{z3}
above $\omega = 10\omega_n$ is constant with a value of $180°$. Both of these segments are
determined by the break frequency. The segment of the curve on the frequency
interval I_{z2} is determined by the segments in the frequency intervals above and
below it. Again we see that the break frequency at ω_n determines the whole curve.

This observation suggests the following aid for the construction of the Bode plot.
At the frequency $\omega = 0.1\omega_n$ place a double arrow pointing upward. These arrows
suggests a change in the slope of the line in the positive direction. At the frequency
$\omega = 10\omega_n$ place a double arrow pointing in the downward direction. These arrows
suggests a change in the slope in the negative direction. These arrows are shown in
Figure 16.3.3.

▲▲

If the complex zeros are in the RHP then $-1 < \zeta < 0$,

$$\angle H(\omega_n) = \begin{cases} -90°, & \text{if} \quad 0 > \zeta > -1 \\ \text{undefined}, & \text{if} \quad \zeta = 0 \end{cases} \qquad (16.3.30)$$

The phase Bode plot goes from $0°$ to $-180°$ through negative values as indicated by (16.3.30). In this case the straight line approximations are the same as a pair of complex poles in LHP as shown in Figure 16.3.3. All comments that apply to that figure also apply here.

16.4 GRAPHICAL CONSTRUCTION OF BODE PLOTS

16.4.1 Construction Procedure

Let a system be represented by the transfer function

$$\frac{Y(s)}{X(s)} = H(s). \tag{16.4.1}$$

The Bode plots of the frequency response function of this system are

(1) $20 \log |H(\omega)|$ vs. $\log \omega$, and
(2) $\angle H(\omega)$ vs. $\log \omega$.

In this section we will give several examples that illustrate the graphical construction of Bode plots. Recalling the discussion in Section 16.1, suppose the transfer function in (16.4.1) can be factored into two factors as $H(s) = H_1(s)H_2(s)$. The magnitude Bode plot is given by

$$20 \log|H(\omega)| = 20 \log|H_1(\omega)H_2(\omega)| = 20 \log|H_1(\omega)| + 20 \log|H_2(\omega)|. \tag{16.4.2}$$

Similarly, the phase Bode plot is given by

$$\angle H(\omega) = \angle H_1(\omega) + \angle H_2(\omega). \tag{16.4.3}$$

From (16.4.2) and (16.4.3) we see that the Bode plots of the system can be constructed by constructing the Bode plots of each factor and then graphically adding these plots. From the previous two sections we know how to construct the Bode plots of elementary factors. The idea, then, is to factor the transfer function in (16.4.1) into elementary factors, draw the Bode plot of each factor based on our prior knowledge, and then assemble the Bode plot of the system from Bode plots of each of the factors. This idea is explained in detail in the following procedure.

Procedure 16.4.1: Given the transfer function $H(s)$:
Step 1 Factor the transfer function into constant, linear, and quadratic factors. The poles or zeros of the linear factors should be real and the poles or zeros of the quadratic factors should be complex. Also, the factors should be appropriately normalized.

Magnitude Bode Plot
Step 2 Identify all break frequencies on the frequency axis (if any). Mark the break frequency of a pole by an arrow pointing down. Mark the break frequency of a zero

by an arrow pointing up. If the break frequency corresponds to a complex pair of zeros or poles, add a second arrow.

Step 3   Draw the straight line approximations of the magnitude for each factor. Include correction points as necessary for quadratic poles, and sketch the corrected approximation to the Bode plot.

Step 4   Construct the magnitude Bode plot of $H(s)$ by adding together the straight line approximations of each factor. When adding together the plots of each factor, the slope of the composite plot at each frequency is the sum of the slopes of the individual factors at that frequency.

Phase Bode Plot

Step 5   For each break frequency of each factor labeled in (2), identify the corresponding frequencies at one decade below the break frequency $0.1\omega_b$ and one decade above the break frequency $10\omega_b$. If the phase plot breaks down through negative frequencies, mark the frequency $0.1\omega_b$ with an arrow that points down. Also label the frequency $10\omega_b$ with an arrow that points up. If the phase plot breaks up through positive frequencies, reverse the arrows at the two frequencies. If the phase plot corresponds to a quadratic factor, add a second set of arrows.

Step 6   Draw the straight line approximations of the phase for each factor.

Step 7   Construct the phase Bode plot of $H(s)$ by adding together the straight line approximations of each factor. When adding together the plots of each factor, the slope of the composite plot at each frequency is the sum of the slopes of the individual factors at that frequency.

▲▲

Based on the developments in Sections 16.1 we observe several characteristics of the Bode plots of the elementary factors. These characteristics are summarized in Table 16.5.1 in the summary section of this chapter. Frequent reference to this table in the discussion below may prove advantageous.

16.4.2 Examples

In this subsection we will present two examples that show how to use Procedure 16.4.1 to construct the Bode plots of a transfer function.

Example 16.4.2: Consider the transfer function

$$H(s) = \frac{1}{s(s+2)}. \tag{16.4.4}$$

Using Procedure 16.4.1, we construct the Bode plot as follows.

Step 1   First we factor the transfer function into factors of the type discussed in the last two sections. For the transfer function (16.4.4) we have

$$H(s) = \frac{1}{s(s+2)} = \left(\frac{1}{s}\right)\left(\frac{1}{s+2}\right) = \left(\frac{1}{s}\right)\left[\frac{1}{2}\left(\frac{2}{s+2}\right)\right] = \left(\frac{1}{2}\right)\left(\frac{1}{s}\right)\left(\frac{2}{s+2}\right). \tag{16.4.5}$$

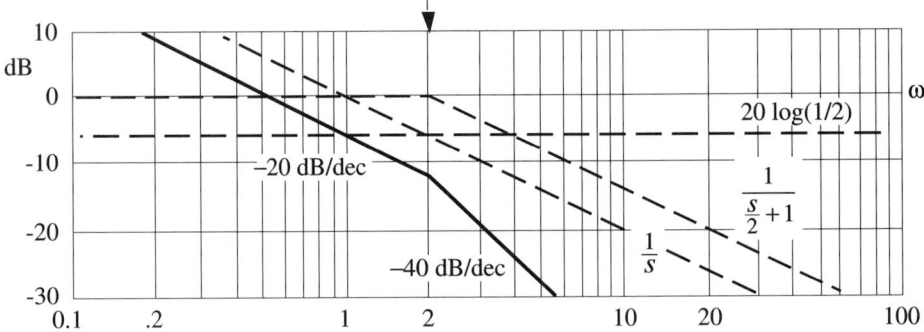

Figure 16.4.1 Approximate Magnitude Bode Plots of Example 16.4.2

Note: A constant 2 was required to normalize the last factor in (16.4.5). This normalization resulted in a constant factor of 1/2.

Step 2 The break frequency at $\omega_b = 2$ of the second elementary factor is marked with a downward arrow as shown in Figure 16.4.1. There is no break frequency associated with the pole at the origin.

Step 3 Next we draw the straight line approximations for each factor. Consider first the pole at $s = -2$. The magnitude plot is zero until the break frequency $\omega_b = 2$. Then it breaks down at -20 dB/dec as shown in Figure 16.4.1 as indicated by the arrow.

The magnitude Bode plot for the pole at the origin is constructed by recalling that this plot has a slope of -20 dB/dec. At $\omega = 1$ it passes through 0 dB. This line is shown in Figure 16.4.1.

The magnitude of the constant factor 1/2 is

$$20\log|K| = 20\log(0.5) = -6. \tag{16.4.6}$$

This Bode plot is a straight line with zero slope as shown in Figure 16.4.1.

Step 4 To construct the magnitude Bode plot, the straight line approximations of each factor are added together. First consider the frequencies less than the break frequency $\omega < 2$. In this frequency range the pole at $s = -2$ does not contribute to the composite plot. The plot is constructed from the constant and the Bode plot of the pole at the origin. At the frequency $\omega = 1$, the straight line of the pole intersects the Bode plot for the constant gain 1/2. This composite plot is shown as a solid line in Figure 16.4.1.

For frequencies $\omega > 2$, there is a contribution from the approximate Bode plot of the pole at $s = -2$. Starting at the break frequency $\omega_b = 2$ the straight line approximation to the pole at $s = -2$ adds an additional slope of -20 dB/dec to the Bode plot of $H(s)$ as shown in Figure 16.4.1. So at the break frequency, the slope of the Bode plot of the function (16.4.5) changes from a -20 dB/dec to a slope of -40 dB/dec. This change in slope is indicated by the arrow at the break frequency.

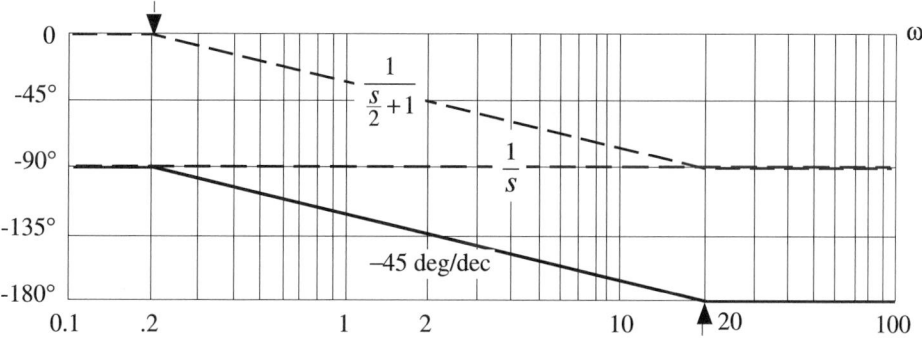

Figure 16.4.2 Approximate Phase Bode Plots of Example 16.4.2

With no other break frequencies, the magnitude Bode plot is completed.

Step 5 The phase Bode plot is shown in Figure 16.4.2. For the pole at $s = -2$, we locate the frequency one decade below the break frequency at $0.1\omega_b = 0.2$ with a downward arrow. We also locate the frequency one decade above the break frequency at $\omega = 10\omega_b = 20$ with a upward arrow. These arrows are located in Figure 16.4.2.

Step 6 Next the phase plot is constructed for each factor. The constant factor contributes $0°$ phase because it is positive. The pole at the origin contributes a constant $-90°$ of phase.

The phase plot for the pole at $s = -2$ is zero for the frequency interval below $\omega < 0.2$. For the frequency interval above $\omega > 20$ the phase is $-90°$. Having located these two lines, the phase plot is completed by connecting the line at $\omega = 0.2$ with the line at $\omega = 20$. This phase plot is shown in Figure 16.4.2 as a dashed line.

Step 7 Next the two phase plots are added together. When the phase plot of the pole at $s = -2$ is added to the constant $-90°$, the whole plot is shifted down by $-90°$. This shift gives the phase Bode plot in Figure 16.4.2.

Check At $\omega = 0.5$ rad/sec, the contribution of each factor is -6 dB and 6 dB respectively. These two contributions sum to 0 dB on the composite plot. At $\omega = 4$, the contributions are -6 dB, -6 dB, and -12 dB, which total to -24 dB on the composite plot.

This transfer function has two poles and no zeros. Therefore, at frequencies above the break frequency, the slope of the magnitude plot should be -40 dB/dec. The phase plot in this frequency range should be a $-180°$, where each pole contributes $-90°$ and the constant contributes $0°$.

The transfer function has one pole at the origin. Therefore, the magnitude plot at frequencies below the lowest break frequency, below $\omega_b = 2$, should be -20 dB/dec. The phase should be $-90°$ in this frequency range.

▲▲

The following example is rather complicated, showing how to handle a RHP pole as well as a pair of complex zeros with a small damping ratio.

Example 16.4.3: Suppose we want to construct the Bode plots for the system

$$\frac{Y(s)}{X(s)} = \left(\frac{2}{250}\right)\frac{s^2 + 30s + 2500}{(s-2)(s+10)}. \tag{16.4.7}$$

Step 1 First, we factor the transfer function in (16.4.7) into the required form. We obtain

$$\frac{Y(s)}{X(s)} = \left(\frac{2}{250}\right)\left[\left(-\frac{1}{2}\right)\left(\frac{-2}{s-2}\right)\right]\left[\left(\frac{1}{10}\right)\left(\frac{10}{s+10}\right)\right] \tag{16.4.8}$$

$$\times \left(\left(\frac{2500}{1}\right)\frac{s^2 + 2(0.3)(50)s + 50^2}{50^2}\right) .$$

$$= (-1)\left(\frac{-2}{s-2}\right)\left(\frac{10}{s+10}\right)\left(\frac{s^2 + 2(0.3)(50)s + 50^2}{50^2}\right).$$

Note: To construct the Bode plots for the quadratic factor in (16.4.7), we were required to express the poles in terms of $\zeta = 0.3$ and $\omega_n = 50$.

Note: The RHP pole required the introduction of a negative sign in the normalization constant.

Step 2 The three break frequencies are located at $\omega_{b1} = 2$, $\omega_{b2} = 10$, and $\omega_{b3} = 50$. These frequencies are marked in Figure 16.4.3a. Since the third break frequency corresponds to a pair of complex zeros, they are marked by a double arrow and they point up.

Step 3 The straight line approximations of the magnitude Bode plots of the three factors in (16.4.8) are shown in Figure 16.4.3a as dashed lines. The magnitude plot of the pole in the RHP with a break frequency of $\omega_{b1} = 2$ is zero below the break frequency. Above the break frequency, the magnitude plot is a straight line with a slope of -20 dB/dec like a pole in the LHP. The magnitude plot of the pole with the break frequency of $\omega_{b2} = 10$ has the same construction. The break frequency for the quadratic factor is $\omega_{b3} = 50$. Below the break frequency the magnitude plot is zero. Above the break frequency the plot is a straight line with a slope of $+40$ dB/dec and a frequency axis intercept at the break frequency. This line breaks up as indicated by the double arrows.

Note: The slope is 40 dB/dec because there are two zeros associated with this factor.

Finally, we check the value of the damping ratio $\zeta = 0.3$ of the quadratic factor to see if a correction point is needed. Referring to Figure 16.3.6, we see that there is a -5 dB decrease in the Bode plot near the break frequency for this quadratic factor. Therefore, we add a -5 dB correction point to the magnitude Bode plot as shown in

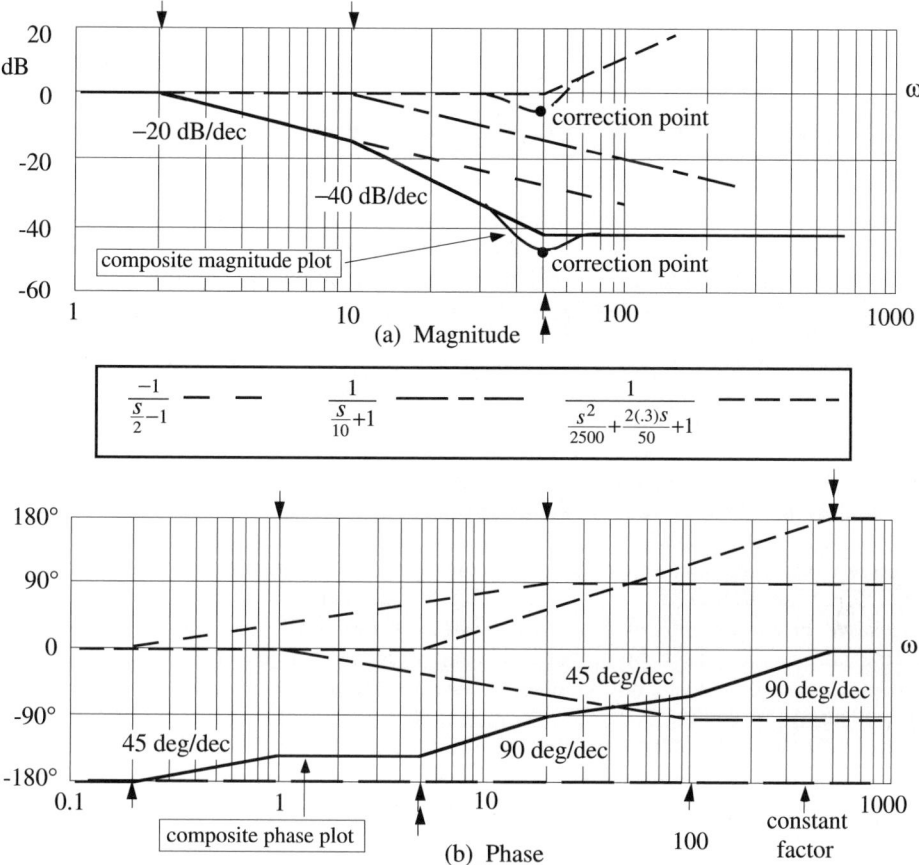

Figure 16.4.3 Magnitude Bode Plot for Example 16.4.3

Figure 16.4.3a. Then the magnitude plot is adjusted as shown. The constant factor has a magnitude of 0 dB so there is no need to include it in the magnitude Bode plot.
Step 4 Finally, the Bode plot is obtained by adding together the three curves in Figure 16.4.3a.

Below the first break frequency at $\omega_{b1} = 2$ both plots are zero, so the composite plot has the value of the constant. The composite plot is shown by the solid line in Figure 16.4.3a.

At the first break frequency at $\omega_{b1} = 2$, the slope of the composite Bode plot changes from 0 dB/dec to −20 dB/dec as indicated by the arrow pointing downward. Starting at the second break frequency, $\omega_{b2} = 10$, the slope changes to −40 dB/dec as indicated by the arrow at that break frequency.

Starting at the third break frequency, $\omega_{b3} = 50$, the quadratic factor contributes an additional 40 dB/dec (20 dB/dec for each pole) to the slope of the composite magnitude plot, as shown in Figure 16.4.3a. The double arrows at this break frequency indicate that the slope of the composite plot breaks upward by 40 dB/dec. This leaves the slope of the composite plot flat at 0 dB/dec.

Since we have accounted for all the factors of the function (16.4.8), the straight line magnitude plot is complete. Finally, we must account for the correction point from the quadratic factor. At the break frequency of the quadratic factor, the value of the straight line approximation is −5 dB. Here we add a correction of −5 dB to the composite plot. This point is shown in Figure 16.4.3a. The Bode plot is completed by sketching in a correction in the neighborhood of the correction point.

Check: Since this function has two zeros and two poles, the slope of the Bode plot at high frequencies should be 0 dB/dec. Since this function has no poles or zeros at the origin, the slope of the composite plot should be constant at low frequencies.

Step 5 The construction of the phase plot for this transfer function starts by marking the frequencies one decade above and below the break frequencies as shown in Figure 16.4.3b. The frequencies where the arrows are inserted into the Bode plot are shown in Table 16.4.1.

Note: We have added double arrows for the quadratic factor. The phase of the RHP pole breaks up to 90° so the arrows are reversed from the arrows of the LHP pole.

Step 6 Next, the approximations for each factor are drawn as shown in Figure 16.4.3b by the dashed lines. The phase plot of the single pole in the LHP is drawn as discussed in the previous example. The phase of the RHP pole has the same shape as the phase of the LHP pole, but it breaks up to 90°. The constant factor of (-1) contributes a constant −180° of phase. The approximation for the quadratic factor breaks up at 90°/dec.

Step 7 To form the approximation of the filter, we add the plots of each factor shown in Figure 16.4.3b.

For the frequency interval where $0 < \omega \leq 0.1\omega_{b1} = 0.2$, the phase is −180° from the constant factor. The phase plot is −180° on this interval as shown by the solid line in Figure 16.4.3b.

On the frequency interval $0.2 < \omega \leq 0.1\omega_{b2} = 1$, the phase from the single pole in the RHP contributes the composite phase plot. Here the phase plot has a slope 45°/dec as indicated by the single upward arrow.

On the frequency interval $1 < \omega \leq 0.1\omega_{b3} = 5$, the phase from both single poles contribute to the composite phase plot. On this frequency band the plot is flat. The slope of the composite plot is decremented by 45°/dec as indicated by the downward arrow at the second break frequency.

On the frequency interval $5 < \omega \leq 10\omega_{b1} = 20$, the contribution to the phase by the complex zeros is 90°/dec. Here the slope of the composite plot changes from 0°/dec to 90°/dec as indicated by the double upward pointing arrows.

Table 16.4.1 Break Frequencies of Example 16.4.3

$0.1\omega_{b1}$	$10\omega_{b1}$	$0.1\omega_{b2}$	$10\omega_{b2}$	$0.1\omega_{b3}$	$10\omega_{b3}$
0.2	20	1	100	5	500

On the frequency interval $20 < \omega \le 10\omega_{b2} = 100$, the contribution to the phase by the pole in the RHP is a constant $90°$. The contribution of the complex zeros to the slope is $90°/\text{dec}$, and the contribution from the single pole is $-45°/\text{dec}$. Here the slope of the composite plot changes from $90°/\text{dec}$ to $45°/\text{dec}$ as indicated by the downward pointing arrow.

On the frequency interval $100 < \omega \le 10\omega_{b3} = 500$, the contribution to the slope of the phase by the complex zeros is $90°/\text{dec}$. The $90°$ contributions from the two single poles cancel out each other. The slope of the composite plot increases by $45°/\text{dec}$ as indicated by the upward pointing arrow.

On the frequency interval $\omega > 500$, the contribution by the pair of complex zeros is a constant $180°$, and the contribution from the constant (-1) is $-180°$. On this frequency interval, the phase plot is a constant $0°$. The slope of the composite plot decreases by $90°/\text{dec}$ as indicated by the double downward pointing arrow.

Check: The function in (16.4.8) has no poles or zeros at the origin, and the constant factor is negative. Hence, for those frequencies less than a decade below the lowest break frequency, the phase plot is $-180°$. This function has one LHP pole that contributes $-90°$ to the phase plot at frequencies above the third break frequency. On this frequency band, the pole in the RHP contributes $+90°$. The pair of complex zeros contributes $+180°$. Hence, for those frequencies a decade above the largest break frequency, the phase Bode plot is a constant $(-180°) + (1)(-90°) + (1)(90°) +(180°) = 0°$.

▲▲

16.4.3 MATLAB Experiments

The MATLAB command **bode** generates a frequency response function from a transfer function or state space representation. The **bode** command used without left hand will generate two Bode plots. This feature is useful for quick viewing of the Bode plots, but it has two drawbacks. First, the plots may not be accurate if the system has poles close to the imaginary axis or if it is very complicated. Second, the frequency axis and/or scaling may not be suitable for the analysis task at hand. In either case it is desirable to generate the Bode plots according to a user selected scaling.

The frequency points at which the frequency response function is computed can be specified by a frequency vector which is included in the right-hand arguments of the **bode** command. The command **logspace** will generate a vector of logarithmically spaced frequency points. When the **bode** command is invoked with left hand arguments, the magnitude and phase of the frequency response function are returned. The magnitude is returned in linear units that must be converted to dB using the **log10** command. The Bode plots are then graphed using the **semilogx** command.

The **bode** command requires a system for the command arguments. The system may be specified either by a transfer function or a state space representation. The syntax of the bode command is slightly different for these two different input arguments as explained in the MATLAB manual. If an LTI system name is used in the **bode** command, the output arguments will be arrays. These arrays must be converted to vectors before being plotted.

The **freqs** command will also calculate the frequency response function of a system. The input arguments are the system and the frequency vector. The output argument of this command, however, is not the magnitude and phase, but a vector containing the complex frequency response function at the specified frequency points in the frequency vector.

The following M-file generates the Bode plots for the Example 16.4.3.

Example 16.4.3: (*continued*) Compare the exact Bode plots to the approximations shown in Figure 16.4.3.

```
clear
% Generate transfer function
num = (2/250)*[1,30,2500];         % numerator polynomial
dem = conv([1,-2],[1,10]);         % denominator polynomial
% Generate frequency response function
w = logspace(-1,3,400);            % frequency vector
[mag,phs] = bode(num,dem,w);       % frequency response function
dbmag = 20*log10(mag);             % magnitude in dB

figure(1)                          figure(2)
semilogx(w,dbmag)                  semilogx(w,phs)
grid                               grid
title('Magnitude Bode Plot')       title('Phase Bode Plot')
xlabel('frequency, rad/sec')       xlabel('frequency, rad/sec')
ylabel('dB')                       ylabel('deg')
```

▲▲

Exploratory Exercise 16.4.4: Verify all of the Bode plots in Example 16.4.2. Compare the exact Bode plots to the approximations shown in Figures 16.4.1 and 16.4.2.

▲▲

16.5 CHAPTER SUMMARY

16.5.1 Constructing Bode Plots

Given a system

$$\frac{Y(s)}{X(s)} = H(s) \tag{16.5.1}$$

the frequency response function is

$$H(\omega) = H(s)\big|_{s=j\omega} \tag{16.5.2}$$

The frequency response theorem describes the output signal relative to the input signal in terms of the frequency response function evaluated at the frequency of the input signal. The relevant quantities are easily read from the magnitude and phase plots of the frequency response function. Therefore, the frequency response function is usually given in graphical form. The <u>Bode plots</u> (Definition 16.1.1) of the system are

(1) 20 log |$H(\omega)$| vs. log ω
(2) $\angle H(\omega)$ vs. log ω

where the frequency, ω, $0 < \omega < \infty$ is in rad/sec and the logarithm is in base 10. The first Bode plot, (1), is called the <u>magnitude</u> Bode plot. The second Bode plot, (2), is called the <u>phase</u> Bode plot.

In this chapter we have developed a procedure for construction the Bode plots of any transfer function. This procedure begins by factoring the transfer function into elementary factors. Sections 16.2 - 3 provide a detailed description for the hand construction of Bode plots for constant, linear, and quadratic factors. This results are summarized in Table 16.5.1. Procedure 16.4.1 describes the methodology for constructing the Bode plots of a transfer function. This methodology is illustrated by several examples in Section 16.4.

Table 16.5.1. Summary of Bode Plot Construction

			magnitude**, slope		phase**
linear		Figure		Figure	
	LHP pole	16.2.1	−20 dB/dec	16.2.2	−90°
	RHP pole	16.2.3	−20 dB/dec	16.2.3	90°
	LHP zero	16.2.4	20 dB/dec	16.2.4	90°
	RHP zero	16.2.5	20 dB/dec	16.2.5	−90°
	pole at the origin*	16.2.6	−20 dB/dec	16.2.6	−90°
	zero at the origin*	16.2.6	20 dB/dec	16.2.6	90°
constants					
	positive constant		0 dB/dec		0°
	negative constant		0 dB/dec		180°
quadratic					
	LHP poles	16.3.2	−40 dB/dec	16.3.2	−180°
	RHP poles	16.3.2	−40 dB/dec	16.3.4	180°
	LHP zeros	16.3.6	40 dB/dec	16.3.6	180°
	RHP zeros	16.3.6	40 dB/dec	16.3.3	−180°

** One decade above the break frequency.
* The magnitude Bode plot is a line with the given slope.

16.5.2 Summary

The frequency response function plays a fundamental role in the analysis and design of systems as evidenced by the discussion in Chapters 14 and 15. Therefore, it is important to understand the relationship between the system representation and the shape of the frequency response function. Bode plots provide this insight. So a detailed knowledge of the construction procedure for Bode plots is also indirectly useful in a great variety of situations where the frequency response of the system plays an important role.

A review of the construction procedure for Bode plots identifies how Bode plots relate the shape of the frequency response function to the system representation. The key link is provided by the relationship between the poles and zeros of the (Laplace) transfer function and the elementary factors used to construct the Bode plots. The poles and zeros of the transfer function define the break frequencies of the elementary factors that are used to construct the Bode plot. The break frequencies determine the frequencies where the frequency response function of the system changes its shape. Furthermore, the frequency response function of the elementary factors have a specific shape summarized in Table 16.5.1. The frequency response functions of the elementary factors of the transfer function are added together to form the frequency response function of the system. In many cases, the contribution of each elementary factor is obvious in the frequency response function of the system. Herein lies the usefulness of the construction procedure for Bode plots.

16.6 HOMEWORK FOR CHAPTER 16

Homework Problems for Section 16.4

16.4.1 (a) Draw the straight line approximations of the magnitude and phase Bode plots for each of the following transfer functions. Use graph paper. Add a correction point where necessary.

(b) Verify the approximations with MATLAB.

(c) For each transfer function determine the slope of the magnitude Bode plot below the lowest break frequency and the slope above the highest break frequency. Also determine the value of the phase Bode plot below the lowest break frequency and above the highest break frequency.

(i) $H(s) = \dfrac{10(s+20)}{(s+50)}$

(ii) $H(s) = \dfrac{-4(s+10)}{s^2 + 0.4s + 1}$

(iii) $H(s) = \dfrac{(12)(s+0.5)}{s(s^2 + 0.4s + 4)}$

(iv) $H(s) = \dfrac{(s+3)(s-8)}{s(s+16)}$

(v) $H(s) = \dfrac{(s+0.8)}{(s+0.5)(s^2 + 2.4s + 4)}$

(vi) $H(s) = \dfrac{(s+3)(s+8)}{s(s+16)}$

(vii) $H(s) = \dfrac{s+4}{s(s+1)}$

(viii) $H(s) = \dfrac{s}{s^2 + 0.4s + 4}$

(ix) $H(s) = \dfrac{s^2 - 1}{(s^2 + 0.15s + 0.25)(s^2 + 0.4s + 4)}$

16.4.2 Draw the straight line approximations to the Bode plots for the proof-mass actuator described in Section 6.5. Assume that $K_{ef} = 1$ and $m_{pm} = 10$.

16.4.3 Consider the mass-spring-damper system discussed in Section 6.4. Suppose the input signal is the force on the mass and the output signal is the displacement of the mass. Assume that the nominal values are $m_{st} = 400$ kg, $c_{st} = 80$ N/m/sec, and $k_{st} = 600$ N/m.
   (a) Using MATLAB plot several Bode plots on the same graph as the stiffness varies between $10 < k_{st} < 1500$.
   (b) Repeat (a) if the damping varies as $10 < c_{st} < 1000$.
   (c) Repeat (a) if the mass varies as $100 < m_{st} < 2000$.
   (d) Compare the results of this exercise with Exploratory Exercise 10.1.11.

16.4.4 Consider the RC network discussed in Section 6.3 where the output signal is the voltage across the capacitor. Suppose that $1/RC = 2$.
   (a) Draw the straight line Bode plots of this system.
   (b) Consider the integrator with a gain

$$\frac{Y_1(s)}{X_1(s)} = \frac{K}{s}.$$

   Find a value of the gain K such that the magnitude Bode plot of the integrator matches the magnitude Bode plot of the RC network as closely as possible.
   (c) Over what frequency range does the RC network act as an integrator?
   (d) What are the advantages of using the RC filter in (a) rather than the integrator in (b)?

16.4.5 (a) Consider the system

$$\frac{Y(s)}{X(s)} = H(s) = \frac{s}{s^2 + 1.414s + 1}$$

   Plot the approximate Bode plots of this system.

   (b) The system

$$\frac{Y(s)}{X(s)} = Ks$$

   is a pure differentiator. Find a value of K such that the magnitude Bode plot of the differentiator matches the magnitude Bode plot of the filter in (a) as closely as possible.
   (c) Over what frequency band are these two systems approximately the same?

(d) What are the advantages of using the filter in (a) rather than the differentiator in (b)?

(e) Suppose that the input signal is $x(t) = u_s(t)$ to the system in (a). Identify the amplitude of the output signal using the frequency response theorem. (This input signal is a zero frequency sinusoid.)

16.4.6 Consider the frequency response function discussed in Problem 14.4.1 and shown in Figure P14.4.1.

(a) What is the pole-zero excess of the transfer function?

(b) Are there any poles or zeros at the origin?

(c) Find a transfer function for this system.

16.4.7 Consider the frequency response function discussed in Problem 14.4.2 and shown in Figure P14.4.2.

(a) What is the pole-zero excess of the transfer function?

(b) Are there any poles or zeros at the origin?

(c) Find a transfer function for this system.

16.4.8 Suppose we wish to design a lowpass filter with a 10 kHz bandwidth.

(a) If the transfer function of this filter has one pole, find the transfer function. Plot the Bode plots.

(b) If the transfer function of this filter has two poles, find the transfer function. Plot the Bode plots.

(c) Suppose we define the stopband to be 20 dB down. Compare the transition band of the two filters in (a) and (b).

16.4.9 Suppose we want to find a filter whose phase is $0°$ at $\omega = 0$, $+85°$ at $\omega = 0.25$ rad/sec, and $-158°$ at $\omega = 4$ rad/sec. Find a 4th-order transfer function that meets these specifications as closely as possible.

Chapter 17

Introduction to Discrete-Time Signals and Systems

Chapter Outline

Introduction to Chapters 17 to 23 In the previous chapters we have analyzed physical processes that evolve in real time. Our basic approach is to model the observed physical process with a mathematical function. This approach is systematically presented through the development of the following concepts:

1. The definition of a signal and a system.
2. Signal representations.

653

3.   System representations, their interrelationships, and their properties.
4.   The propagation of a signal through a system in the frequency domain.

The signal representations we used to model these physical processes are functions that depend on a real variable. These signal models lead naturally to differential equations as system models. We also used the Fourier series and Fourier transform, and the Laplace transform, analysis tools that can be applied to real functions, for alternative signal and system representations.

Beginning with this chapter we turn our attention to physical processes that can't be observed at every instant of time. We can effectively observe them only at discrete instants of time. Some of these processes are naturally discrete, such as the operation of a computer, or economic processes. Other discrete-time processes are generated from continuous-time processes by sampling.

The definition of a signal and a system still applies to these discrete-time processes, but the representation of a discrete-time signal must be a discrete-time function. This starting point launches us into a separate but parallel development of discrete-time signals and systems. In fact, the parallels are so strong that from an abstract mathematical viewpoint, the results can be said to be the same (with a few exceptions). Nonetheless, different analysis tools are required for discrete functions, and the interpretations of the results in physical terms are different. Therefore, a discussion of discrete-time theory is given in the coming chapters.

Discrete functions require a different set of transform tools. The mathematical details of these transforms are given in Chapter 18. The z-transform is introduced as the discrete-time equivalent of the Laplace transform. The discrete-time Fourier transform is introduced as the discrete-time equivalent of the Fourier transform. The basic properties of these transforms are also given in this chapter. In keeping with the organizational philosophy of the text, this chapter is conceived as a mathematical background chapter. The discussion of system theory begins in the following chapters.

Discrete-time process often arises by sampling a continuous-time process. At the signal level, a discrete-time signal is derived from a continuous-time signal by sampling. The question then arises as to the relationship between the continuous-time signal and the discrete-time signal. In particular, can the continuous-time signal be recovered from the discrete-time signal? This question is discussed fully in Chapter 19. These results in this chapter, including the Nyquist sampling theorem, form the foundation of sampling theory. The practical reconstruction of a signal from its samples is also considered.

The development of continuous-time signal representations included the development of Fourier series, amplitude and phase spectra, and energy and power spectral densities. These results are extended to discrete-time signals by using the discrete-time Fourier transform in Chapter 20. The interpretation of these quantities is nearly identical to the continuous-time concepts. We are able to extend these concepts by one additional step, however. Discrete-time signals lend themselves quite naturally to computational algorithms. By introducing some minor modifications to the discrete-time Fourier transform, we are able to compute the (discrete-time) Fourier transform directly from the signal samples. This algorithm (widely known as the FFT) is of great practical importance because it allows us to compute the Fourier transform of realistic signals.

Discrete-time signals give rise to discrete-time systems. As with continuous-time systems, it is necessary to develop system representations to make this concept practically useful. Four discrete-time system representations are introduced in Chapter 21. These system representations are completely analogous to the continuous-time system representations: the convolution summation (convolution integrals), difference equations and state space representations (differential equations), z-transform transfer function (Laplace transfer function), and discrete-time Fourier transform transfer function (Fourier transfer function). The properties of these system representations are developed in Chapter 22. Again, these properties are the exact parallels of the properties of continuous-time time systems. Using these properties, we develop the relationship between all of these system representations. Chapter 22 also explains how a discrete system is constructed by sampling the input and output signals of a continuous-time system.

A key result in system theory is the frequency response theorem. If the input signal to a system is a sinusoid, this theorem expresses the output signal in terms of the amplitude and phase spectra of the input signal and the frequency response of the system. This result lays the ground work for characterizing the propagation of a signal through a system in terms of the energy or power spectral density of the input signal and the frequency response of the system. These results, which are explained in depth in Chapters 14 and 15 for continuous-time systems, form a cornerstone of system theory. They are extended to discrete-time signals and systems in Chapter 23, with little change. The relationship of these results to digital filter design is briefly mentioned.

Introduction to Chapter 17 The purpose of this chapter is to introduce the basic definitions associated with discrete-time signals and systems. We also introduce, briefly, the four system representations we will study in the coming chapters.

Discrete-time signals very commonly arise by sampling continuous-time signal electronically. In this chapter we give an introduction to the sampling process. We briefly describe the electronic hardware that is used to effect the sampling including some of the complexities that arise in the processes. We then describe some of the applications where these devices are used. These descriptions are used as a background for the theory that is developed in the succeeding chapters.

Summary of Sections

Section 17.1:   We introduce discrete-time signals.

Section 17.2:   We discuss the basic terminology of sampling.

Section 17.3:   We discuss quantization and coding.

Section 17.4:   We discuss digital-to-analog converters.

Section 17.5:   We introduce discrete-time systems.

Section 17.6:   We introduce digital filters.

Coverage of the Text

This chapter requires only the first sections of Chapters 5 and 6. Background material on discrete-time functions can be found in Section 2.3.

17.1 INTRODUCTION TO DISCRETE-TIME SIGNALS

17.1.1 Introduction

A continuous-time process can be observed at every instant of real time and a real number can be assigned to that observation. In this way we construct a continuous-time signal, a model of the continuous-time process. There is a large class of physical processes for which we can't use this modeling process to construct a representation of a signal, however. The problem is that we can't observe the process at every instant of real time; only at certain instants of time can we make a measurement. Or the process is undefined in between the instants of time when it can be measured. One important example of this type of process is the contents of a computer register. The contents of this computer register change on every clock cycle of the computer. At certain instants of time we can definitely say what the contents of this register are. In between these times, however, the contents are being updated so the actual contents are meaningless. Hence, if we were to construct a signal which models the contents of the computer register, we could only observe the contents of the register at discrete instants of time, and our signal would only be defined at discrete instants of time.

Discrete processes commonly occur in another way. Suppose we wish to build a signal model of a physical process by experimental measurements. There are many ways of recording a signal depending on the process that is observed and the type of sensor used. We can divide all of these measurement methods into two classes, however. The first class of recording media produces a signal that is directly proportional to the observed process *at all times*. Hence, the recorded signal is a continuous-time signal. Two examples of this type of measurement technique would be an analog voltmeter and a strip recorder used to record earthquake data or heartbeats. This signal can be analyzed using the techniques introduced in the continuous-time part of this text.

The second class of measurement methods records the signal only at discrete instants of time. This class of measurement methods is called <u>sampling</u>. Each measurement is called a <u>sample</u>. Instead of having a record of the process at every instant of time, we have a record of the process only at discrete points in time. This record of discrete observations constitutes a discrete-time signal. It is with these samples that we develop our understanding of the physical processes. We can also think of the discrete-time signal as being extracted from the continuous-time signal which models the underlying process rather than extracting the discrete-time signal directly from the physical process.

In this section we introduce the definitions of a discrete-time signals and several other types of signals associated with sampling. We also give several examples of discrete-time signals.

17.1.2 Signal Definitions

Our starting place in the development of the theory of discrete-time signals and systems is with a signal, defined at the beginning of this text, whose definition we recall here.

Definition 5.1.1: A <u>signal</u> is a function which represents the time variation of a physical variable.

▲▲

It is worth emphasizing that the signal is the function that represents the physical variable, not the physical variable itself.

Definition 5.1.2: The function which is chosen to describe a signal is called the <u>representation</u> of the signal. The process of building a representation of the signal is called signal <u>modeling</u>.

▲▲

Terminology: The signal representation is also called the signal <u>model</u>.

In this part of the text we are interested in physical processes that are observed at discrete instants of time. In order to model these processes, it is necessary to introduce a new representation of time. Since we can only observe discrete processes at discrete times, it is convenient to assign integers to these observation times. Functions that depend on the integers are called <u>discrete functions</u>.[1] This class of functions is used to define discrete-time signals.

Definition 5.1.4: A <u>discrete (-time)</u> signal $x(n)$ is a function which depends on a discrete-time variable n and which represents the discrete-time variation of a physical variable, i.e., $x: I \to \mathcal{R}$.

▲▲

NOTATION CHANGE: In the discussion below n is assumed to be an integer unless specifically stated otherwise. We reserve n to stand for discrete-time. This notation represents a *notation change* from continuous-time systems where n was used to denote the order of the system.[2]

Notation: Continuous-time and discrete-time signals are distinguished from each other by their arguments t and n, respectively.

Terminology: When discussing discrete-time signals and systems, it is common to shorten "discrete-time" to "discrete." This terminology is consistent with the terminology for continuous (-time) signals and systems.

---

[1] The basic properties of discrete functions are described in Section 2.3. The definitions in this section will be used extensively in the coming chapters.

[2] This notation change is driven by the widespread usage of n in these two conflicting ways.

In more advanced analysis of signals, which is more abstract, it is sometimes advantageous to define complex-valued signals. That is, a signal is defined as a mapping from the integers into the complex numbers. This extension of the definition of a signal is rather straightforward, and few modifications of the results given in this text are required to accommodate these signals.

In Definition 5.1.4 a discrete-time signal is a mapping that takes the integers into the real numbers. We can also define two other classes of signals that are closely related to discrete-time signals.

Definition 17.1.1: A <u>quantized</u> signal $x_q(t)$ is a signal that maps the real numbers into the integers, i.e., $x_q: \mathcal{R} \to I$.

▲▲

Definition 17.1.2: A <u>digital</u> signal $x_d(n)$ is a signal that maps the integers into the integers, i.e., $x_d: I \to I$.

▲▲

Quantized signals occur in the conversion of a continuous signal to a discrete signal. The signals which most accurately model the operation of a computer are digital signals. Unfortunately, it is difficult, mathematically, to work with quantized and digital signals. Hence, it is more common to model these signals with discrete-time signals and then study the quantization effects separately. The systematic study of quantized signals is beyond the scope of this text although we will introduce some of the basic concepts in Section 17.4. Our attention will focus on discrete-time signals for which a large number of analysis techniques exist.

The theory which we will develop is based on discrete functions as models for discrete-time signals. Basic definitions of discrete functions can be found in Section 2.3. The functions found in this section are adapted for modeling discrete-time signals in exactly the same way that real functions are adapted to model continuous-time signals as described in Chapter 5. Almost all of the techniques developed in Chapter 5 can be translated in a straightforward way to discrete-time signal modeling.

17.1.3 Examples of Discrete-Time Signals

Discrete-time signals occur in many contexts. In this subsection we will give a few of the more common appearances of discrete signals.

Electronic Sampling One of the most common occurrences of discrete signals is the sampling of an electronic waveform, usually a voltage. An analog-to-digital converter (described in the following sections) will sample a voltage signal and translate that signal into a digital signal that can be stored in a computer. The digital waveforms that form the discrete signal are described in Section 5.4. This electronic conversion of a continuous-time signal to a discrete-time signal is the heart of many devices. For example, it is the essence of a digital voltmeter. It is also used wherever a sensor is used to experimentally measure a physical process. (A sensor produces at its output a voltage that is proportional to a measured physical variable.) This process is embedded in the examples to follow.

Speech Processing The recording and playback of human speech has been a part of modern life for most of this century. Recently, digital electronics have made possible the storage of the spoken word in computer memory. The speech is measured by a microphone, the output of the microphone is electronically sampled, and the discrete signal is stored in memory, for example. There are now many possibilities for processing the speech signal with computer algorithms. A voice recognition algorithm can be applied to determine the identity of the speaker. The discrete signal can be transmitted over a communication system to a listener in a different location. In addition the transmitted signal can be scrambled to protect the content of the message. By analyzing the discrete speech signal, we are able to gain insight into the structure of these signals. This analysis leads to the possibility of speech synthesis, or the synthesis of other sounds such as in an electronic synthesizer. All of these possibilities for the manipulation and synthesis of speech originate from our ability to measure and analyze a speech signal. The analysis is based on the discrete-time signal used to model the speech signal.

Image Processing In much the same way that speech can be digitally recorded and processed, so can images be digitally recorded and stored. These images are stored by associating with each discrete point in the image a number. Thus we obtain a two-dimensional discrete signal. This image can be processed using many of the same concepts that are used to process one-dimensional signals. Photo enhancements are quite common these days. Obviously, these photos can be transmitted from one place to another as discrete signals, even from places as distant as Jupiter.

Geological Sciences Sudden movements of underground rock formations can give rise to elastic waves in the earth's crust. These waves, called seismic waves, propagate in all directions from the source of the waves. If these waves are large enough they can cause widespread destruction (earthquakes). These waves can be recorded as discrete-time signals at the earth's surface by sampling the output signal of an accelerometer. Post-processing of these signals can yield the magnitude and location of the source of the disturbance. This information can be very useful for analyzing the characteristics of an underground nuclear test, for example. Alternatively, seismic waves can be deliberately induced by controlled explosions to map the geological structure below the earth's surface. The composition and geometry of the rock formations can be deduced by analyzing the reflected waves. This analysis is typically performed by computer analysis of the sampled signals.

Electromagnetics Another very important source of discrete-time signals originates from the sensing of electromagnetic waves. A device, often an antenna, is used to detect the presence of electromagnetic waves. The output signal is usually a voltage proportional to the amplitude of the electromagnetic wave and at the same frequency. This output signal is sampled to produce a discrete-time signal which a subsequently analyzed to extract the information from the electromagnetic wave. This process is fundamental to wireless communication, for example. It also plays an integral part in modern radar, where the reflected electromagnetic signal is sampled and digitally processed to determine the location of the target. Another important application of the processing of electromagnetic wave occurs in biomedical applications. Brainwaves are an important indicator of the functioning of the brain. Brainwaves are typically analyzed by sampling the output signal of the detector and digitally processing the sampled signal.

There are, of course, hundreds of other types of discrete signals. The above overview is designed give an indication of the diverse nature of the physical processes that give rise to discrete signals.

17.2 INTRODUCTION TO SAMPLING

17.2.1 Introduction

Sampling is the time honored way of observing nature scientifically. The scientist observes an event and records it. This record was a set of discrete numbers, a discrete-time signal. Since the advent of solid state electronics, this measurement process has been automated with digital electronics and computer processing. For example, a digital voltmeter measures at evenly spaced time intervals the voltage in a circuit; it doesn't measure the voltage in continuous-time. The same is true when sounds are recorded in digital storage media. A microphone changes the sound pressure into a continuous-time electrical signal. This electrical signal is then periodically measured by the digital electronics and stored in computer memory. The continuous-time electrical signal is not recorded, however.

This electronic data collection offers a number of advantages over traditional analog recording of signals. The primary advantage is that electronic samples can be submitted to computer processing to uncover hidden characteristics. Advanced algorithms can be applied to the sampled signal which could not be applied to data recorded in analog media. This computer processing of electronically captured measurements provides a major motivation for the development of the theory of discrete-time signals and systems. It is important to understand the measurement process and the relationship between the physical process and the measurements if the theory is to have relevance.

In this section we will introduce some of the basic ideas connected with sampling. We will begin with the mathematical definition of sampling which will be used extensively in the following chapters. Then we will briefly explain the operation of the electronic device that is widely used for sampling.

17.2.2 Definitions

Sampling is the process of extracting a discrete-time signal from a continuous-time signal. The mathematical definition of sampling is given next.

Definition 17.2.1: Let $x(t)$ be a continuous-time signal and let $T_s > 0$ be a fixed number. From $x(t)$ we derive the discrete signal

$$\hat{x}(n) = x(nT_s) \quad \text{for} \quad n = \dots, -2, -1, 0, 1, 2, \dots$$

The number $T_s > 0$ is called the <u>sample period</u>. The <u>sampling frequency</u> f_s, in hertz, is defined as $f_s = \dfrac{1}{T_s}$. In radians the sampling frequency is $\omega_s = 2\pi f_s$. The time

$t = nT_S$ is called a <u>sample instant</u>. The interval $nT_S \leq t < (n+1)T_S$ is called the <u>sample interval</u>. The numbers $x(nT_S)$ are called the <u>sample values</u> of $x(t)$. The discrete signal $\hat{x}(n)$ is called the <u>discrete-time sampled version</u> of $x(t)$. The process of extracting the sample values from $x(t)$ is called <u>sampling</u>.

▲▲

Notation: The "hat" symbol of the sampled version of $x(t)$, $\hat{x}(n)$, is to indicate that the discrete signal is derived from $x(t)$. The sample period is suppressed in $\hat{x}(n)$ in this text.

Assumption: Whenever we analyze sampled signals in this text, first we will select the sample period and fix it. Furthermore, we assume that one of the samples falls at time $t = 0$. The sample period will then remain fixed throughout the analysis. This assumption also implies that the sample values are evenly spaced (sometimes called <u>harmonic sampling</u>). This assumption is essential to the derivation of most of the results presented below.

The sampling process is shown pictorially in Figure 17.2.1. Figure 17.2.1a shows a continuous-time signal $x(t)$. This signal is sampled with a sample period of T_S. The corresponding sample values $x(nT_S)$ are also shown in Figure 17.2.1a. From these sample values the sampled version of $x(t)$, $\hat{x}(n)$, is constructed as in Definition 17.2.1.

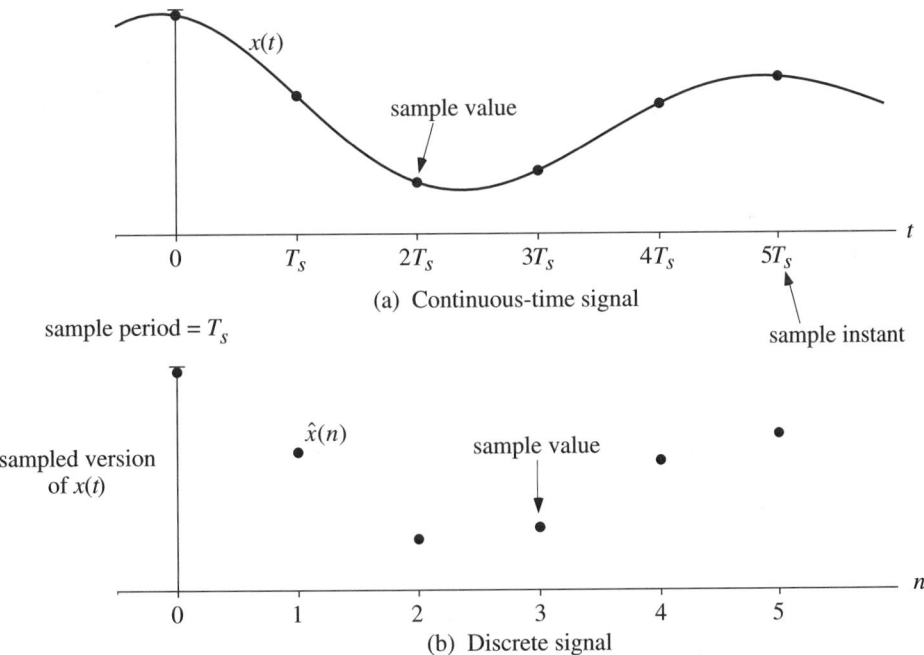

Figure 17.2.1 Graphical Representation of Sampling

$$x(t) \longrightarrow \qquad \qquad \longrightarrow \hat{x}(n)$$

switch closes instantaneously every T_S seconds

Figure 17.2.2 System Representation of Sampling

This discrete-time signal is shown in Figure 17.2.1b. The process of constructing the discrete-time signal in Figure 17.2.1b from the continuous-time signal in Figure 17.2.1a is called sampling.   Note that the time scale in Figure 17.2.1a is in continuous-time, and the time scale in Figure 17.2.1b is in discrete-time. A discrete-time axis shows only integers.

The sampling process can be thought of as a system. The input signal to this system is the continuous-time signal $x(t)$. The output signal is the discrete-time sampled version of $x(t)$, $\hat{x}(n)$. This system is shown in Figure 17.2.2. Sampling as a system is not a continuous system because the output signal is a discrete signal. Similarly this system is not a discrete system because the input signal is not a discrete signal.

In Figure 17.2.2 the input signal is a continuous-time signal and the output signal is a discrete-time signal. We can build a model of the sampling process either in the continuous domain by using continuous-time signals, or in the discrete domain by using discrete-time signals. The continuous-time model of the sampled signal allows the signal to be analyzed using the tools of continuous-time system theory.   Using the discrete-time model of the sampled signal, we can analyze the way a computer processes a signal. Both approaches will be explored.

Example 17.2.2: Consider the continuous-time signal

$$x(t) = e^{\alpha t}. \tag{17.2.1}$$

If we sample this signal with a sample period of T_S, we obtain the discrete-time signal

$$\hat{x}(n) = e^{\alpha T_s n} = (e^{\alpha T_s})^n = a^n \quad \text{where} \quad a = e^{\alpha T_s}. \tag{17.2.2}$$

The sampled signal in (17.2.2) motivates our definition of a discrete-time exponential in Definition 2.3.6.

▲▲

Example 17.2.3: Consider the continuous-time signal

$$x(t) = \sin(\omega_c t). \tag{17.2.3}$$

If we sample this signal with a sample period of T_S, we obtain the signal

$$\hat{x}(n) = \sin(\omega_c T_s n) = \sin(\Omega_c n), \qquad \omega_c T_s = \Omega_c. \tag{17.2.4}$$

The discrete-time signal in (17.2.4) motivates our definition of a discrete-time sinusoid in Definition 2.3.9.

▲▲

A sampled sinusoid is given in (17.2.4). The frequency of the sinusoid in continuous-time is ω_c and the frequency in discrete-time is $\Omega_c = \omega_c T_s$. It turns out that this relationship is of fundamental importance in sampling theory as will become clear in the Chapter 19.

Definition 17.2.4: The frequency variable ω is called the <u>continuous</u> frequency variable. Let T_s be the sample period. Define the <u>discrete</u> frequency variable Ω by

$$\Omega = \omega T_s.$$

▲▲

Notation: In some of the signal processing literature ω is used for the discrete frequency and Ω is used for the continuous frequency. (Our notation for discrete frequency also conflicts with the notation for an ohm as a measure of electrical resistance.)

Remark 17.2.5: The units on the continuous frequency variable, ω, are radians/second. From Definition 17.2.4 we see that the units on the discrete frequency variable Ω are radians or radians/sample. The interpretation of the discrete frequency is the number of radians between sample instants.

▲▲

17.2.3 Analog-to-Digital Converters

There are many ways of sampling a signal depending on the physical process that is being observed. One of the most frequently used methods employs electronic sampling. Here a sensor is used to observe the physical process that produces an electronic output signal. This output signal is then passed through a device that samples the output signal of the sensor and converts it into a digital signal. The device which accomplishes this task is called an <u>analog-to-digital converter</u> (ADC). In this subsection we will briefly describe the operation of this device and relate it to the definition of sampling above. There are many different types of ADC converters available today. And there are many issues which must be addressed to discuss an ADC in depth. For the purposes of this text, however, it is only necessary to convey the basic concept behind an ADC.

An ADC is shown in Figure 17.2.3. An ADC is an integrated circuit which accepts an analog voltage (a continuous-time signal) shown as $x(t)$ in Figure 17.2.3. The ADC also accepts a timing signal that provides an interface between the un-clocked outside world with the synchronized world inside a data processing unit (harmonic sampling). The output of the ADC is a set of pins which the ADC sets to a high voltage or a low voltage over one sample period. (This sample period usually consists of several or many clock cycles.) The high and low voltages on the pins represent, respectively, a logical 1 or a logical 0. That is, each pin represents a bit. Collectively, these pins represent a binary word that corresponds to the quantized

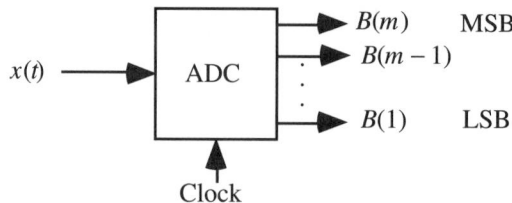

Figure 17.2.3 Analog-to-Digital Converter

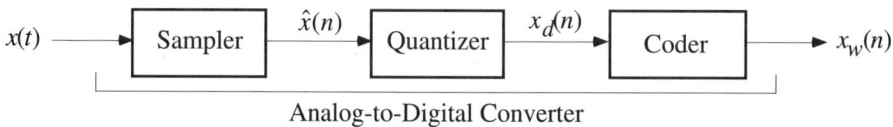

Analog-to-Digital Converter

Figure 17.2.4 Components of an ADC

sample value at that sample instant. The data bus of the processing unit then acquires the data from the ADC.

Conceptually, an ADC consists of the three components shown in Figure 17.2.4. The components in Figure 17.2.4 don't necessarily correspond to physical components in the solid state device. Rather they correspond to the signal level operations that take place within the device.

Sampler   The sampler picks off the signal value at a given instant in time. This operation is repeated at each sample time. The signal $\hat{x}(n)$ is a discrete-time signal.

Quantization   Quantization converts a real-valued sample into a discrete-valued sample. The signal $x_d(n)$ is a quantized signal.

Coder   The coder converts the discrete-valued number obtained from the quantizer into a binary form that can be directly stored by the computer. The signal $x_w(n)$ might be a 8-bit binary word, for example. The signal $x_w(n)$ is a digital signal.

In this section we will describe the operation of the sampler. In the next section we will describe the operation of the quantizer and coder.

Sample and Hold The sampler is actually a sample and hold (S/H) device that accepts a (rapidly) changing analog input, samples the input signal at the sample instant nT_s, and holds that analog value over the sample period $nT_s \leq t < (n + 1)T_s$. A typical configuration of a S/H and its operation is shown in Figure 17.2.5. After receiving a clock signal, the switch in Figure 17.2.5a closes for a fraction of the sample period charging the capacitor. When the switch is opened, the capacitor holds its charge for the remaining sample period. The output signal of the S/H is a

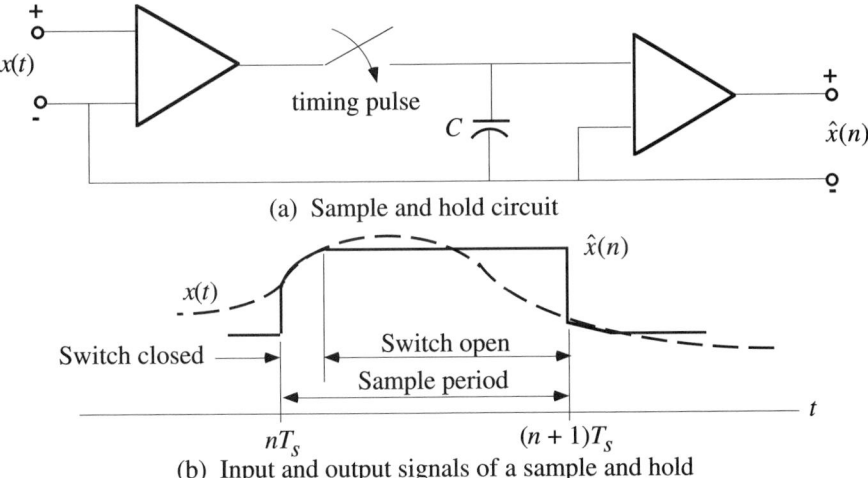

(a) Sample and hold circuit

(b) Input and output signals of a sample and hold

Figure 17.2.5 Sample and Hold Device

constant value equal to the last value of the input signal just before the switch opened.

We can associate the sample and hold with the sampler in the conceptual block diagram of ADC in Figure 17.2.4. The input signal to the sample and hold is a continuous-time signal. If we look at the output signal of the sample and hold over the sample period in Figure 17.2.5b, we see that after the switch opens, the output signal is constant. If we consider the switch to be closed only for a very short time or if the input signal does not vary while the switch is closed, the constant value of the output signal will be the value of the input signal at the sample instant nT_s.

Again, ignoring the period of time during which the switch is closed, we can assign a single discrete value to this sample period and index it to the sample instant at time n. That is, we can model the output signal of the sample and hold as a discrete signal $x(nT_s) = \hat{x}(n)$. From this viewpoint, we see that the sample and hold device is basically the sampler in the conceptual block diagram of ADC in Figure 17.2.4.

The discrete signal $\hat{x}(n)$ must be translated into binary form for processing by a computer. This translation is accomplished by the two remaining blocks in Figure 17.2.4. Their operation is described in the next section. For most of the analysis in this text, however, we will work directly with the discrete signal $\hat{x}(n)$, leaving the more difficult analysis of the digital signals to more advanced texts.

17.3 CODING AND QUANTIZATION

17.3.1 Introduction

In Section 17.2 we pointed out that a ADC conceptually consists of three operations shown in Figure 17.2.4. In the last section we described the basic operation of the sampler. In this section we will describe the operation of the last two blocks:

quantizer and coder. It turns out that the most convenient description begins with coding and progresses in the reverse direction of the arrows in Figure 17.2.4 back to the sampler.

17.3.2 Coding

The purpose of an ADC is to sample a continuous-time electronic waveform and convert that sample into binary number that can be stored and/or processed in digital format in a computer. The sample value of the input electronic waveform $x(t)$ is a real number. The coding process is the part of that operation that converts a real number into binary format. So we begin with a description of binary representation of a real number.

Definition 17.3.1: The basic units of information in a computer are logical 0's and 1's called <u>bits</u> (binary digits). A <u>binary</u> sequence is a sequence of bits; a sequence of 0's and 1's. The number of bits is the <u>length</u> of the sequence.

▲▲

Physically, bits are commonly represented as a high and a low level of voltage. In computer memory, bits are stored in registers of fixed length which imposes a further structure on the binary sequences.

Definition 17.3.2: A <u>(binary) word w of length m</u> is a sequence of m bits.

▲▲

Notation: We denote each distinct binary word by w_i.

Next we consider each word separately. To each word we will assign a real number. In order to make that assignment, we must rank order the bits as follows.

Definition 17.3.3: The first bit in a binary word $B(m)$ we call the <u>most significant bit</u> (MSB). Beginning with the next bit we assign the positions $B(m-1)$ to $B(1)$ as shown here.

$$
\begin{array}{cccc}
& \overbrace{\qquad word \qquad} & & \\
\hline
1 & 0 & \cdots & 1 \\
B(m) & B(m-1) & \cdots & B(1) \\
\Uparrow & & & \Uparrow \\
MSB & & & LSB
\end{array}
$$

The bit $B(1)$ is called the <u>least significant bit</u> (LSB).

▲▲

The definition of a binary word (in terms of the number of bits) and the assignment of the bits is application specific. This discussion is intended only to outline the framework of transcribing information into an electronic signals. A detailed treatment of this subject is beyond the scope of this text.

Table 17.3.1 Coding of a Binary Sequence of Length 3

w_i	011	010	001	000/100	101	110	111
$\bar{c}_i$	3/4	1/2	1/4	0	−1/4	−1/2	−3/4

Using the ordering of the bits in the word as suggested by Definition 17.3.3, for each binary sequence of length m there are 2^m distinct words. To each word we can assign a real number.

Definition 17.3.4: The process of assigning a real number c_i to each word is called <u>coding</u>.

▲▲

Terminology: There are many other definitions of coding.

Conceptually, the simplest way to code real numbers into binary words is to first assign ± 1 to the extreme limits of the words and fractions to the intermediate words using a uniform distribution. Table 17.3.1 shows such a distribution for a binary sequence of length three.

Terminology: The assignment of binary words to fractions shown in Table 17.3.1 is called <u>sign-magnitude format</u> because the MSB keeps track of the sign of the real number.

A more careful analysis of binary coding treats each binary word as a decimal fraction in base 2. This level of detail is necessary for the analysis of truncation and roundoff error in arithmetic operations. For our purposes, it is enough to recognize that each real number is assigned a binary word. Table 17.3.1 gives one way of assigning a binary word to a real number. There are several other ways of making this assignment. The particular sequence of bits assigned to the real numbers has implications for the way the digital electronics is configured to carry out arithmetic operations. There are advantages and disadvantages to the various coding schemes. This discussion is beyond the scope of this text.

The code that is displayed in Table 17.3.1 allows only numbers less than one to be assigned to the binary words. If we wish to assign numbers larger than one to a binary word, the fractions can be multiplied by a positive number FS so that we obtain

$$c_i = FS\bar{c}_i. \tag{17.3.1}$$

It is clear from Table 17.3.1 that there are only a finite number of real numbers c_i that can be assigned exactly to the binary words w_i. Exactly which real numbers are chosen depends on: (a) the positive number FS and (b) the length of the word. We give these critical parameters names.

Definition 17.3.5: The numbers c_i are called <u>quantization levels</u>. The number FS is called the <u>full-scale</u> factor.

▲▲

Consider again the conceptual diagram of an ADC in Figure 17.2.4. The digital output signal of the coder is

$$x_w(n) = w_i \qquad (17.3.2)$$

where each sample value is a binary word from Table 17.3.1, say. The coder produces this coded signal by translating the quantization level c_i into a binary word. Hence, the input signal into the coder is the digital signal

$$x_q(n) = c_i \qquad (17.3.3)$$

where each sample value is a quantization level. Next we will discuss the generation of the quantized sampled signal $x_q(n)$.

17.3.3 Quantization

Binary words are used to represent real numbers in computer memory. For our purposes, the real numbers which we wish to store are the sample values of a signal. The sample values can take on any real values while we have available only a finite number of binary words or, equivalently, a finite number of quantization levels. Therefore, we must have a rule for assigning a range of real values to a single quantization level.

Definition 17.3.6: The process of assigning a real number to a quantization level is called <u>quantization</u>. A device that performs the quantization process is called a <u>quantizer</u>.

▲▲

The quantizer is the middle block shown in Figure 17.3.1. Next we describe how the quantized sampled signal $x_q(n)$ is generated by the quantizer from the sampled signal $\hat{x}(n)$. There are two important variables for this description.

Definition 17.3.7: The <u>dynamic range</u> of the quantizer is plus and minus the full scale factor $\pm FS$. The <u>resolution</u> r of the quantizer is the difference between the quantization levels $r = |c_i - c_{i-1}|$.

▲▲

If $|\hat{x}(n)| < FS$, then this signal will satisfy

$$c_i - \frac{r}{2} \le \hat{x}(n) < c_i + \frac{r}{2} \qquad (17.3.4)$$

for some i. In this case we make the assignment

$$\hat{x}(n) \leftrightarrow c_i = x_q(n). \qquad (17.3.5)$$

If $\hat{x}(n) \geq \pm FS$, then the quantizer would assign $\pm FS$ to this sample value no matter how large $\hat{x}(n)$ is. (Practically, we try to keep the magnitude of the sample values less than FS.)

Notation: We represent the quantization process with the mathematical expression

$$Q[\hat{x}(n)] = c_i = x_q(n). \tag{17.3.6}$$

The quantization process can be represented as a static nonlinearity.[3] For the process described in (17.3.2) - (17.3.3), the corresponding static nonlinearity is shown in Figure 17.3.1.

It is obvious that the quantization process is nonlinear. An accurate representation of the analog signal is only obtained when the analog signal occupies most of the dynamic range without saturating the quantizer. There are several variations on the quantization process to correct for the signal distortions introduced by the quantizer. Again, this discussion is beyond the scope of this text.

17.3.4 Summary

From the discussion in the last two sections it can be seen that an ADC accepts a continuous-time signal and through sampling extracts a discrete-time signal. The quantizer then changes this discrete-time signal into a digital signal. Finally, the coder represents the digital signal in binary form. The analysis of digital signals is not straightforward; in fact it is quite difficult. Specialized techniques must be used.

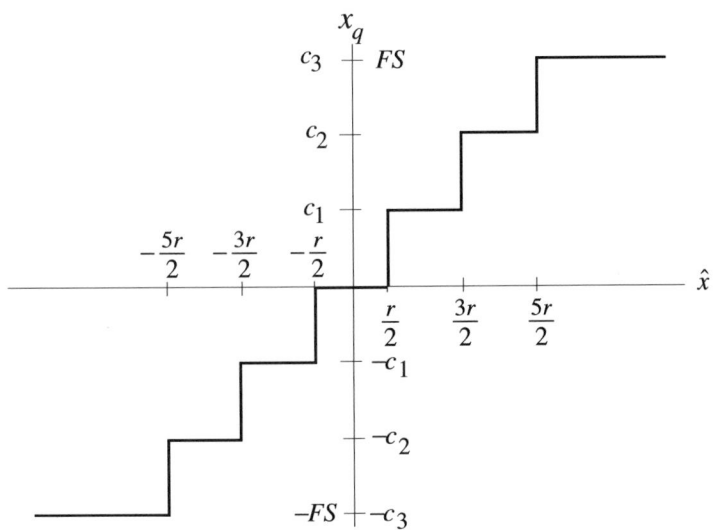

Figure 17.3.1 Quantizer Nonlinearity

---

[3] Static nonlinearities are discussed in Section 13.6.

Discrete signals, the output signal of the sampler, can be analyzed using a number of powerful techniques, however. We will undertake this analysis in the following chapters. The general assumption, which is widely used, is that the quantization levels are fine enough so that we can treat the signal as a discrete signal, ignoring the quantization. We will make this assumption in this text.

17.4 DIGITAL-TO-ANALOG CONVERTERS

17.4.1 Introduction

In the previous two sections we have discussed the process of sampling. This process converts a continuous-time signal into a discrete-time signal. In this section we will consider the reverse process of converting a discrete-time signal into a continuous-time signal. First, we will given a mathematical description of this process. Then we will briefly describe an electronic device that implements the basic concept.

17.4.2 Zero-Order Hold

Suppose we are given a discrete-time signal and a sampling period T_s. We want to construct a continuous-time signal from the discrete-time signal such that if we re-sample the continuous-time signal we will recover the discrete-time signal. A simple and practical method is to hold the sample value over the sample interval. This concept is illustrated in Figure 17.4.1. The given discrete-time signal is shown in Figure 17.4.1a. (This signal is taken from Figure 17.2.1.) To construct an approximation to the original continuous-time signal, we simply define a continuous-time signal that is constant over each sample interval. The value of this signal is the sample value at the beginning of the sample interval. The reconstructed signal is shown in Figure 17.4.1b along with the original continuous-time signal from Figure 17.2.1. The reconstructed signal in Figure 17.4.1b can be compared to the original signal also shown in Figure 17.4.1b. The approximation gets better as the sample period is reduced compared to the variations in the continuous-time signal. Note that the reconstructed signal $x_o(t)$ is a continuous-time signal as is the original signal $x(t)$. The reconstructed signal, however, can take on only certain real values. It is a quantized signal, Definition 17.1.1.

We view the reconstruction process as a system. The input signal is $\hat{x}(n)$ and the output signal is $x_o(t)$. The system is shown in Figure 17.4.2.

Definition 17.4.1: The system defined by $\hat{x}(n)$ as the input signal and $x_o(t)$ as the output signal is called a <u>zero-order hold</u> (ZOH).

▲▲

Notation: The subscript o means that the signal $x_o(t)$ has been reconstructed from its sample values by using a zero-order hold.

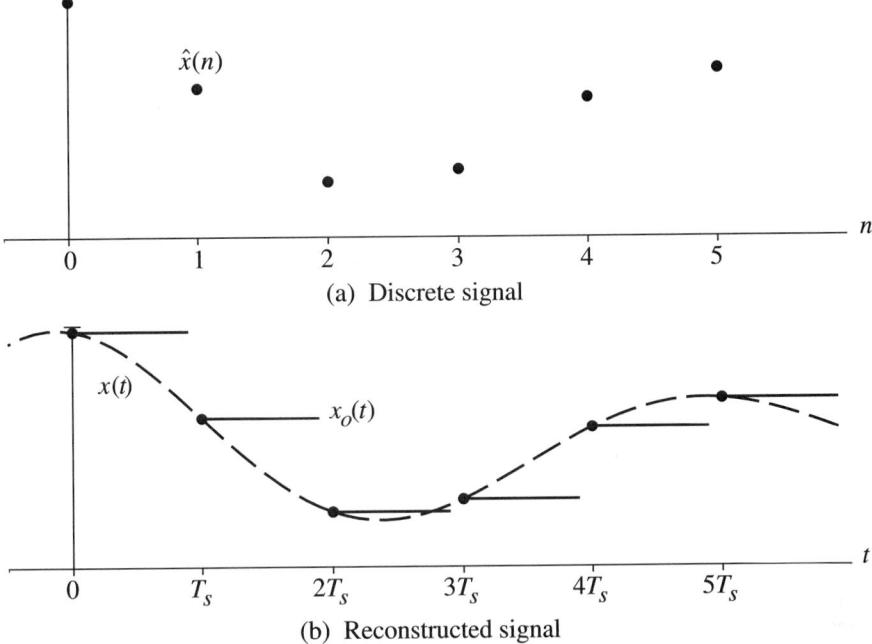

$\hat{x}(n)$

(a) Discrete signal

$x(t)$

$x_o(t)$

(b) Reconstructed signal

Figure 17.4.1 Signal Reconstruction Using a Zero-Order Hold

$$\hat{x}(n) \longrightarrow \boxed{\textbf{ZOH}} \longrightarrow x_o(t)$$

Figure 17.4.2 "Block Diagram" of a Zero-Order Hold

This discussion suggests several interesting questions. Suppose we sample a continuous-time signal. Can we exactly reconstruct the signal from its samples? The answer to this question is contained in Nyquist's sampling theorem which is developed in Chapter 19. This theorem lends deep insight into the structure of a sampled signal. Suppose that we use the ZOH in an attempt to reconstruct the continuous-time signal from its samples. What is the relationship between the reconstructed signal and the original signal? This question is also answered in Chapter 19 with the help of a frequency domain analysis.

17.4.3 Digital-to-Analog Conversion

A <u>digital-to-analog converter</u> (DAC) is an integrated circuit which accepts m voltage signals at high and low levels (corresponding to a binary word) and produces an analog output which is proportional to the real number represented by the binary word at the input of the DAC. Hence, the DAC is an electronic implementation of the ZOH which is widely used.

The inputs and outputs of a DAC are shown in Figure 17.4.3.

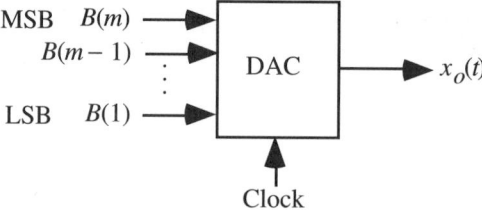

Figure 17.4.3 Digital-to-Analog Converter

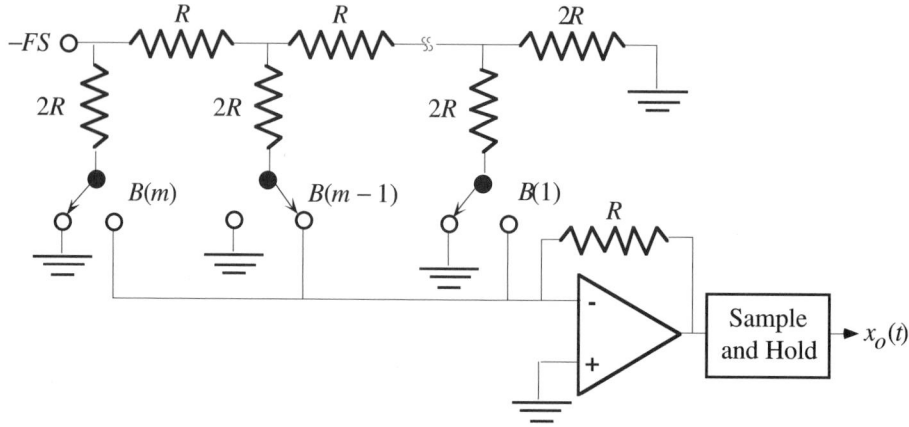

Figure 17.4.4 Inverted R-$2R$ Ladder Digital-to-Analog Converter

The input to the device is a binary word which is made available (by the clock signal) at the sample instant nT_S. The output signal of the DAC is a constant voltage over the time interval $nT_S \le t < (n+1)T_S$. A popular circuit configuration of a DAC, called an inverted R-$2R$ ladder, is shown in Figure 17.4.4. The output voltage $x_o(t)$ of this DAC is given by

$$x_o(t) = FS\left[B(n)2^{-1} + B(n-1)2^{-2} + \cdots + B(1)2^{-m}\right] \tag{17.4.1}$$

where $B(m)$ is the most significant bit and $B(1)$ is the least significant bit. (The correct realization of the output voltage $x(t)$ that corresponds to the binary word at the input of the DAC depends on the coding scheme used to generate the binary word and the voltage levels of the digital message signal used to deliver the binary word to the input of the DAC. These issues are straightforward but beyond the scope of this text.)

The switching in a DAC often causes transients that corrupt the output of the summing op amp in Figure 17.4.4. Many DAC's also contain a zero-order hold to isolate the output signal from the transients as shown in Figure 17.4.4.

A DAC is a system which accepts a digital signal and produces a quantized signal. Furthermore, the output signal consists of a series of steps, the width of the steps being the sample period T_s. We often need a model of this device.   To develop the model, it is necessary to ignore the quantization inherent in the input and output of the DAC. Therefore, we assume that the input signal is a discrete signal and the output signal is a continuous-time signal consisting of steps. Now we are describing the zero-order hold (ZOH) introduced above. A detailed development of the transfer function of the ZOH is given in Section 19.4.

17.5 INTRODUCTION TO DISCRETE-TIME SYSTEMS

17.5.1 Introduction

To this point in this chapter we have been discussing discrete signals. From discrete signals we progress easily to discrete-time systems using the basic definition of a system. Discrete-time systems are just relationships between discrete-time signals. One important example of discrete-time systems is associated with the computer processing of data. We are given a data vector which we process according to a certain algorithm. The algorithm will yield another data vector. There are many ways to look at this process. In this text we will adopt the following point of view: the data vector which we give the computer constitutes a discrete signal. The data vector which the computer returns to us is a discrete signal; the computer produces an output signal given an input signal. This description fits the definition of a system. Hence, we can identify the computer processing of data with system theory. This viewpoint is quite useful, and it motivates much of the theory that is developed in the following chapters.

As we have emphasized in the previous chapters, the system representation plays a key role in the analysis of the system. The system we have just described, the computer processing of data, is quite different from the systems we have studied in the previous chapters. In fact, because the input and output signals are discrete, none of the systems representations we have introduced thus far can be used to describe this system. It is a primary purpose of the following chapters to introduce system representations that can be used for this example. Naturally, these system representations are called discrete-time systems.

Although discrete-time system representations have a different mathematical form than continuous-time system representations, we will show that these system representations have the *same* system properties. Hence, most of the concepts developed for continuous-time systems, such as frequency response, extend directly to discrete-time systems. In some cases, the calculations for discrete-time system representations are slightly different than continuous-time systems, but these differences will be highlighted. In a few instances the interpretations of the concepts for discrete-time systems must be modified but they remain essentially the same as the concepts introduced in the previous chapters of the text.

In this section we formally define discrete systems and then we give a preview of the system representations we will investigate in the coming chapters.   The similarities to continuous-time systems are evident.

17.5.2 Definitions

All systems are relationships between signals. The basic definition of a system we introduced in Chapter 6 is still the starting place.

Definition 6.1.1: A <u>system</u> generates a response, or output signal, for a given input signal.

▲▲

In the coming chapters we want to specialize this definition to systems whose input and output signals are discrete. (In fact, we have already introduced this definition.)

Definition 6.1.5: If the input and output signals of a system are discrete-time signals, then we call the system a <u>discrete(-time)</u> system.

▲▲

The definition of a system is easily remembered by the abstract expression

$$y(n) = \mathcal{H}[x(n)].$$
(17.5.1)

This expression simply says that the output signal $y(n)$ is determined from the system by the input signal $x(n)$. Since the input and output signals are discrete signals, this must be a discrete system. As with continuous systems, we can also represent the concept of a system with the cartoon shown in Figure 17.5.1.

We can also define systems where the input signal is a continuous-time signal and the output signal is a discrete-time signal as

$$y(n) = \mathcal{H}[x(t)].$$
(17.5.2)

An analog-to-digital converter is a system of this type. Similarly there are systems where the input signal is a discrete-time signal and the output signal is a continuous-time signal, such as digital-to-analog converters. These systems are neither continuous-time nor discrete-time systems. We don't give a name to these systems as we will not study them in depth.

The implication of the cartoon in Figure 17.5.1 is that there exists a mathematical expression of a system. This mathematical expression gives the relationship between the input and output signals.

Definition 6.1.8: We call an explicit mathematical expression for (17.5.1) a system <u>representation</u>.

▲▲

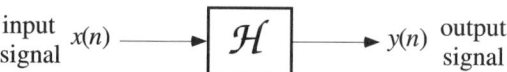

Figure 17.5.1 Cartoon of a Discrete System

Terminology: A system representation is also called a <u>model</u>.

Definition 6.1.9:   The process of deriving a system representation is called <u>modeling</u>.

▲▲

The premise of system theory is that insight into a physical process can be obtained by studying a mathematical model for that system. The insight gained is explicitly or implicitly contained in the model we choose for the system. Typically, one system can have several representations.   Each representation is useful in its own way and gives a different insight into the physical process.

17.5.3 Discrete-Time System Representations

We will investigate four system representations: difference equations, transfer functions using the z-transform, transfer functions using the discrete-time Fourier transform, and convolution sums. As suggested by the names, these discrete system representations share many properties with the continuous system representations we introduced in previous chapters. In Chapters 21 and 22 we will study each of these representations in turn.

Difference Equations   Difference equations are the differential equations of discrete-time systems. An example of a difference equation is

$$y(n+N) + a_1 y(n+(N-1)) + \cdots + a_N y(n) \tag{17.5.3}$$
$$= b_0 x(n+N) + b_1 x(n+(N-1)) + \cdots + b_N x(n)$$
$$y(0) = y_0, \quad y(1) = y_1, \quad y(2) = y_2, \ldots, y(N-1) = y_{N-1}.$$

By a change of variables (17.5.3) can be written as

$$y(n) + a_1 y(n-1) + \cdots + a_N y(n-N) = b_0 x(n) + b_1 x(n-1)) + \cdots + b_N x(n-N). \tag{17.5.4}$$

The difference equation in (17.5.3) is said to be in <u>advance form</u>. The difference equation in (17.5.4) is said to be in <u>delay form</u>. The delay form is very convenient for the implementation of the difference equations with digital electronics.   The advance form is useful for solving initial value problems. We will also represent (17.5.3) using the discrete-time state space equations.

Transfer Functions The z-transform is used to construct a transfer function for the difference equations above. Consider (17.5.3) or (17.5.4) and suppose that the initial conditions are zero. If we take the z-transform of (17.5.3) and solve for the ratio of the output over the input, we get the transfer function

$$\frac{Y(z)}{X(z)} = \frac{b_0 z^N + b_1 z^{N-1} + \cdots + b_N}{z^N + a_1 z^{N-1} + \cdots + a_N}. \tag{17.5.5}$$

This transfer function can also be written as

$$\frac{Y(z)}{X(z)} = \frac{b_0 + b_1 z^{-1} + \cdots + b_M z^{-N}}{1 + a_1 z^{-1} + \cdots + a_N z^{-N}} = H(z). \tag{17.5.6}$$

We will most often work with the form of the transfer function in (17.5.6).

NOTATION CHANGE: We have changed the notation of indexing the coefficients of the polynomials in (17.5.5) and (17.5.6) from continuous-time transfer functions. Here the index of the coefficient matches the negative power of z. This notation represents a *notation change* from continuous-time signals. When indexing the coefficients of polynomials of positive powers of s, the index of the coefficient matches the positive power of s.

Under certain conditions, the transfer function can also be expressed using the discrete-time Fourier transform. Using this transform the transfer function becomes

$$\frac{Y(\Omega)}{X(\Omega)} = \frac{b_0 + b_1 e^{-j\Omega} + \cdots + b_M e^{-j\Omega N}}{1 + a_1 e^{-j\Omega} + \cdots + a_N e^{-j\Omega N}} = H(\Omega). \tag{17.5.7}$$

Generally, this transfer function is displayed in graphical form in terms of its magnitude and phase vs. frequency (the frequency response of the system).

Convolution Summations   The fourth system representation is the analog of the convolution integral for continuous-time systems. The convolution representation is closely related to the transfer function in (17.5.5). We rewrite (17.5.5) as

$$Y(z) = H(z)X(z). \tag{17.5.8}$$

If we invert (17.5.8) using the convolution property of the (two-sided) z-transform we obtain

$$y(n) = \sum_{k=-\infty}^{\infty} h(n-k)x(k). \tag{17.5.9}$$

This system representation is the convolution representation of (17.5.1). The function $h(n)$, which is obtained by taking the inverse z-transform of the transfer function, is called the <u>impulse response function</u>. The convolution representation plays an important role in discrete system theory.

17.5.4 System Response to Standard Inputs

In our study of continuous-time systems we frequently found it convenient to characterize a system by its response to a standard input signal. Hence, we defined the impulse response and the step response. We will also find these characterizations useful for discrete systems.

Definition 17.5.1: The <u>impulse response</u> of a system is the output signal of that system when the system is at rest and the input signal is an impulse function. ▲▲

Suppose that the system is modeled by the convolution representation in (17.5.9). Let the input signal be an impulse function

$$x(n) = \delta(n). \tag{17.5.10}$$

Then the output signal is given by

$$y_i(n) = \sum_{k=-\infty}^{\infty} h(n-k)x(k) = \sum_{k=-\infty}^{\infty} h(n-k)\delta(k) = h(n). \tag{17.5.11}$$

When the input signal of the system is an impulse function, the output signal of the system is the impulse response function.

Terminology: The term "impulse response function" is often shortened to "impulse response" based on Definition 17.5.1.

Again consider a system at rest and suppose the input signal is a unit step function.

Definition 17.5.2: The <u>step response</u> of a system is the output signal of that system when the system is at rest and the input signal is a unit step function. ▲▲

17.6 INTRODUCTION TO DIGITAL FILTERS

17.6.1 Digital Filters

In this previous chapters we have studied continuous-time systems shown in Figure 17.6.1a. We have characterized how a continuous-time system produces an output signal $y(t)$ for a given input signal $x(t)$ in the frequency domain. Loosely speaking, we can say that the system acts as a filter, with the frequency response function of the system shaping the Fourier transform of the input signal to produce the output signal. Indeed, in classical filter design the system (filter) is constructed to have a specified frequency response function (frequency selective filters). The frequency response function of the filter ensures that the output signal will have certain spectral properties, or the output signal will retain the desirable spectral properties of the input signal.

In this text we have focused on the analysis of the system models. In practice, however, we want to translate these transfer functions into hardware for implementation. Historically, a filter has been constructed with electronic components, op amps, capacitors, and resistors, to achieve the desired system function.

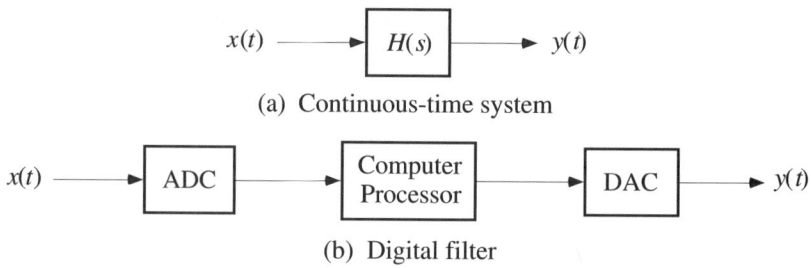

Figure 17.6.1 Analog and Digital Filters

Terminology: This type of filter is called an <u>analog</u> filter because it is a continuous-time system which accepts continuous-time input signals and produces continuous-time output signals.

   With the advent of modern digital electronics and computer processing, an alternative system configuration has emerged. This configuration is shown in Figure 17.6.1b. The basic idea is to sample the continuous-time signal to translate it into the domain of the computer using an ADC. Then the computer processes the sampled signal to achieve the desired system function. In many applications, this processing is accomplished by using a specially designed digital signal processing chip. Finally, the processed signal is converted back into a continuous-time output signal using a DAC. This entire digital system is to be constructed so that between the input and output signal it has exactly the same characteristics as the continuous-time system over a specified frequency range.[4]

Terminology: We call the system in Figure 17.6.1b a <u>digital</u> filter.

   While at first glance it may not be obvious why a digital system would be desirable over an analog system, there are, in fact, several advantages. It turns out that the digital system can give better performance than an analog system, and the digital filter can be programmed to have characteristics that are not permitted by an analog system. Furthermore, the digital processor has all of the desirable characteristics associated with computer processing such as the ability to easily reprogram the processing algorithm, to run multiple algorithms at once, to store the signals for future analysis (off-line processing), and the possibility for VLSI implementations with the advantages of size, power, and cost. The disadvantages of these digital filters lie in their speed and cost. These disadvantages are disappearing as cheaper, faster hardware becomes available.

   The following chapters on discrete-time signals and systems provide the basic tools to analyze a digital filter. In Chapter 19 we discuss sampling in some detail. This discussion highlights the limitations placed on the digital processing by the sampling process. We also present a model of the DAC. This chapter presents the basics of how a continuous-time signal is translated into the digital domain, and how

---

[4] Most digital filters will also include prefilters for sampling and postfilters to remove distortions introduced by the DAC.

it is brought back out to the continuous-time domain. The frequency content of a signal plays a fundamental role in analyzing the propagation of a discrete signal through the digital filter. The Fourier transform, used for the analysis of continuous-time signal, is extended to discrete-time signals in Chapter 20 and its basic properties are developed. To analyze the action of the digital system on the signals we require a representation of the system. In Chapters 21 and 22 we develop several representations of discrete systems. We also show how these discrete-time system models are related to continuous-time system models. Finally, we develop the frequency response of a discrete-time system in Chapter 23. This frequency domain analysis provides deep insight into the propagation of a discrete-time signal through a discrete system and motivates the design of digital filters.

The results in the following chapters form the core of discrete-time signal and system analysis. As such they have applications to all facets of the analysis and synthesis of discrete-time signals and systems. We don't mean to imply that the results here are narrowly focused on digital filters. It's just that the digital filter paradigm forms a convenient framework for organizing these results.

17.6.2 Examples of Discrete-Time Systems

The advent of digital electronics that makes possible a digital filter increased significantly the number of discrete systems. In this subsection we discuss several types of common discrete systems. Some of these systems rely on digital electronics and some don't. These examples give the reader a feeling for the range of applications of this theory.

Digital Processing of Sound   One of the most visible applications of digital filtering is the electronic processing of sound and speech. An everyday example is the CD player. A CD player, of course, replicates the music stored on a compact disk (CD). In fact, the music is stored on the CD as sampled data in the form of bits. When the music is recorded, the voltage from the microphone is sampled and translated into a digital signal. Typically, this digital signal is filtered by a discrete system to improve its quality and then it is stored optically on the CD. The CD player uses an optical system to read the digital signal that is stored on the CD. After additional filtering of this digital signal in the CD player, the digital signal is reconstructed and amplified before it is transmitted to the speakers.

There are many other examples of the digital processing of speech. For example, a voice recognition system functions by sampling the voltage from the microphone, and processing the sampled signal. In a similar way, the voice signal can be scrambled for security. Or it can be modified for special effects.

Signal Processing of Sampled Data   As we have seen, a signal can carry much information about a physical process. To analyze the physical process, we must extract the information from the process. Unfortunately, the signal that we measure may be a corrupted version of the signal that represents the process due to sensor inaccuracies and nonlinearities. Therefore, prior to the signal processing the measured signal is passed through a digital filter to "clean it up." Frequently, this filter is used to remove the noise. The digital filter may also be used to remove specific corrupting signals that entered through the measurement device. For example, 60 Hz line noise is often removed from electrocardiogram (ECG) signals and other signals that arise in medical applications.

Alternatively, a sampled signal is digitally filtered to extract the information. For example, the presence of a sinusoid of a certain frequency in a measured signal may carry certain information. If the sampled signal is passed through a bandpass filter, this sinusoid can be detected. This idea can used to detect the presence of a certain dial tone in a touch-tone phone, for example.

Applications in Communications   A communication system can be defined as a system for the transmission of data from one place to another. This data can be derived from a voice signal, but the data could also be computer data. The purpose of the communication system is to embed the information into a signal that can be transmitted, effect the transmission of that signal, and extract the information at the receiver. Most modern communication systems are digital. In a digital communication system, the data is represented in digital form. Hence, many of the signals internal to the communication system can be viewed as discrete-time signals. The operations performed by the communication system on the discrete-time signals can be described in terms of discrete-time systems. These operations include the shaping of the quantization process to reduce noise, various filtering operations to improve the quality of voice transmission, and the compression of data to increase throughput in channels with limited capacity.

Applications in Economics   Discrete-time systems are part of the fundamental underlying mathematical theory used for the study of economics. Financial variables are typically measured on a daily, monthly, or yearly basis. Hence, these variables naturally form discrete-time signals. If we choose to study a particular economy, the signals are all interrelated and we can develop a system to express this interrelationship. For example, consider a bank account. There are deposits and withdrawals which we can define to be input signals. Define the output signal to be the average daily balance. When we include the interest paid on the balance in the account, we obtain a system. It is an active area of research to extend this idea to complex systems such as the national economy and the stock exchange.

Applications in Biology   Discrete-time systems occur naturally in the study of insect and animal populations. For example, consider the population of rabbits and foxes in an ecosystem. We can model the population of both the rabbits and the foxes by counting the number of individuals in each population each week, say. In this way we obtain a discrete-time signal which models the populations. Clearly, these two populations are dynamically related. The foxes eat the rabbits, causing the rabbit population to decline while the fox population increases due to a healthy diet. If the rabbit population is low, the fox population will decline due to a shortage in the food supply. The rabbit population will subsequently increase due to an absence of predators. With a little effort a discrete-time system can be developed that will capture the fluctuations in these two populations. In a similar way, the theory of discrete-time systems can be applied to many other areas in biology.

17.7 HOMEWORK FOR CHAPTER 17

Homework Problems for Section 17.1

17.1.1 (a) Give an example of a discrete signal.
(b) Give an example of a quantized signal.
(c) Give an example of a digital signal.

17.1.2 Using MATLAB plot the following signals.

(i) $x(n) = \sin(0.2\pi n)$

(ii) $x(n) = \sin(0.02\pi n)$

(iii) $x(n) = \Pi_3(2n) + \sin(0.2\pi n)$

(iv) $x(n) = \sin\!\left((0.2\pi)^n\right)$

(v) $x(n) = \sin(0.5\pi \Pi_3(n))$

(vi) $x(n) = \Pi_3(5\sin(0.2\pi n))$

(vii) $x(n) = (1.3)^n \cos(0.2\pi(n))\Pi_{20}(n)$

(viii) $x(n) = \Pi_2(n)$
$\qquad\qquad + \Pi_2(n-5)\cos(0.2\pi(n-5))$

Homework Problems for Section 17.2

17.2.1 Suppose each of the following signals is sampled using the sample period shown. Plot the continuous-time and sampled signal in MATLAB.

(i) $x(t) = \text{Sa}(t), \quad T_s = 0.5$

(ii) $x(t) = \sin(2\pi t), \quad T_s = 0.1$

(iii) $x(t) = \sin(2\pi t), \quad T_s = 0.5$

(iv) $x(t) = \sin(2\pi t), \quad T_s = 1.2$

(v) $x(t) = \Pi(t) - \Pi(t-1) + \Pi(t-2), \quad T_s = 0.3$

(vi) $x(t) = u_s(t) - u_s(t-3), \quad T_s = 1.2$

17.2.2 The following discrete sinusoids are obtained by sampling a continuous-time sinusoid with a sampling frequency of $f_s = 2$ Hz. In each case determine the frequency of the continuous-time sinusoid.

(i) $x(t) = \sin(0.2n)$

(ii) $x(t) = \cos(2n)$

(iii) $x(t) = \cos(0.7n + 0.03)$

(iv) $x(t) = \sin(0.5\pi n)$

17.2.3 Suppose the continuous-time signal

$$x(t) = \cos(20\pi t)$$

is sampled with the sampling frequency given in the following table. Plot the sampled signal on the given interval for the sampling frequency shown.

(i) $0 \le t \le 2 \qquad f_s = 100$ Hz

(ii) $0 \le t \le 2 \qquad f_s = 27$ Hz

(iii) $0 \le t \le 10 \qquad f_s = 9.8$ Hz

Homework Problems for Section 17.3

17.3.1 Suppose we are allowed three bits to code the quantization levels between ± 1 using the sign-magnitude format. Develop a table showing the quantization levels and the binary words.

17.3.2 Suppose that the signal

$$x(t) = \sin(2\pi t + 0.02)$$

is sampled with a sampling frequency of $f_s = 10$ Hz over the time interval $0 \le t \le 1$. Suppose that these samples are then coded according to Problem 17.3.1.
(a) What are the quantized values of each sample?
(b) What are the binary codes for each sample?

17.3.3 Consider the function shown in Figure P17.3.3a. This function, defined on the interval $-T_0/2 < t < T_0/2$, is odd and symmetric about $T_0/4$ on the interval $0 < t < T_0/2$. Define the signal $y(t)$ as the periodic extension of the function, $y_f(t)$, in Figure P17.3.3a.
(a) Show that the Fourier series of $y(t)$ is given by

$$y(t) = \sum_{\substack{k=-\infty \\ k \ne 0}}^{\infty} Y_k e^{jk\omega_0 t}$$

where

$$Y_k = \frac{j(A-B)}{k\pi}\left[\cos(k\omega_0\tau_2) - \cos\left(k\omega_0\left(\tau_2 - \frac{T_0}{2}\right)\right)\right]$$
$$+ \frac{jA}{k\pi}\left[\cos\left(k\omega_0\left(\tau_1 - \frac{T_0}{2}\right)\right) - \cos(k\omega_0\tau_1)\right]$$

(b) Suppose that the input signal to the quantizer nonlinearity in Figure P17.3.3b is

$$x(t) = \sin(2\pi t)$$

Find and plot the output signal $y(t)$ of the quantizer.
(c) Plot the two-sided amplitude spectrum of $y(t)$.
(d) Compute the power in $y(t)$, P_y.
(e) Compute the power in fundamental of $y(t)$, P_f.
(f) Compute the power in the harmonics of $y(t)$, $P_{har} = P_y - P_f$.

17.3.4 Consider an ADC with input $x(t)$. Suppose the input signal is restricted to 0 to 1 V and this range is quantized by three bits. If the input signal is a constant voltage $x(t) = 0.3$ V, what is the output binary number?

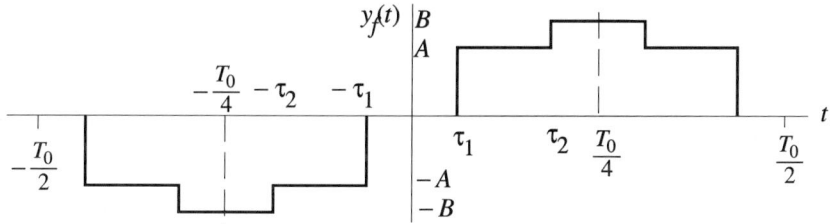

Figure P17.3.3a

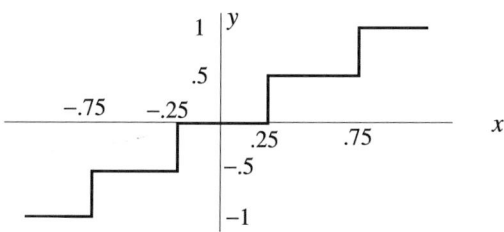

Figure P17.3.3b

17.3.5 Suppose the quantizer has 7 levels: c_{-3}, c_{-2}, c_{-1}, c_0, c_1, c_2, c_3. Let the resolution be $r = 0.25$.
  (a) Sketch the quantization nonlinearity that relates the input signal to the (quantized) output signal.
  (b) Assign a three digit binary number to each quantization level.
  (c) Sketch the output signal from the quantizer if the input signal is

$$x(t) = \Lambda(t).$$

Homework Problems for Section 17.4

17.4.1 Suppose the following discrete signals are the input signal to a ZOH. Using MATLAB plot the output signal if:
  (a) $T_s = 1$
  (b) $T_s = 0.001$

  (i) $x(n) = (0.7)^n u_s(n)$ (iii) $x(n) = \sin(0.5\pi n)$
  (ii) $x(n) = \sin(0.02\pi n)$ (iv) $x(n) = \sin(\pi n)$

Homework Problems for Section 17.5

17.5.1   Consider the system

$$y(n+2) + 2y(n+1) + 3y(n) = u_s(n)$$
$$y(0) = 2, \quad y(1) = 1$$

(a) Find $y(n)$ by direct calculation.
(b) Write a MATLAB M-file to calculate and plot $y(n)$.

Homework Problems for Section 17.6

17.6.1   Give an example of a discrete system.

Chapter 18

The *z*-Transform and the Discrete-Time Fourier Transform

Chapter Outline

The analysis and design of continuous signals and systems depends heavily on the Fourier and Laplace transforms. We would like to have similar tools available for the analysis and design of discrete signals and systems. It is the purpose of this chapter to develop two transforms that play the role of the Laplace and Fourier transforms for discrete-time signals and systems.

The first important transform is the two-sided *z*-transform. A number of basic properties of the *z*-transform are discussed. The one-sided *z*-transform is also presented. It is shown how to solve difference equations using the one-sided *z*-transform. There is a striking resemblance between the *z*-transform and the Laplace transform. The concepts that underlie these two mathematical objects are the same, and there are parallels between many of their properties. The presentation of the *z*-transforms is made with this observation in mind; it parallels the presentation of the Laplace transform in Chapter 9.

The second important transform is the discrete-time Fourier transform (DTFT). This transform is defined in this chapter and a number of its important properties are introduced. The DTFT plays the role of the Fourier transform for discrete-time signals and systems. This transform allows us to define the amplitude and phase spectrum and the energy or power spectrum of a discrete-time signal (Chapter 20). This transform also ultimately leads to a method for computing the Fourier transform from the samples of the signal (Chapter 20).

The motivation of this chapter is to present compactly some mathematical tools we will need for discrete-time signal and system analysis. The discussion of discrete-time signals and systems begins in earnest in the next chapter.

Summary of Sections

Section 18.1:   We introduce the two-sided *z*-transform.

Section 18.2:   We develop the basic properties of the *z*-transform.

Section 18.3:   We discuss the one-sided *z*-transform and its application to the solution of difference equations.

Section 18.4:   We define and discuss the discrete-time Fourier transform (DTFT).

Section 18.5:   Chapter summary section.

Coverage of the Text

The chapter requires only the basic math background in Chapters 2 and 3, and partial fraction expansion from Chapter 9. The introduction to discrete-time signals and systems in Chapter 17 is helpful. Most of Section 18.4 is independent of the rest of the chapter.

18.1 THE TWO-SIDED *z*-TRANSFORM

18.1.1 Definition

In this section we introduce the two-sided *z*-transform. We begin with its definition.

Definition 18.1.1: Let $x(n)$ be a discrete function that satisfies

(a) $x(n)$ is defined for $-\infty < n < \infty$

(b)
$$\sum_{n=-\infty}^{\infty} |x(n)|\rho^{-n} \leq M < \infty \quad \text{for} \quad 0 \leq \rho_{min} < \rho < \rho_{max} < \infty$$

For these functions the (two-sided) z-transform is defined as

$$\mathcal{Z}\{x(n)\} = \sum_{n=-\infty}^{\infty} x(n)z^{-n} = X(z).$$

The inverse z-transform is defined by

$$x(n) = \mathcal{Z}^{-1}\{X(z)\} = \frac{1}{2\pi j} \oint X(z)z^{n-1} \, dz, \quad n = \ldots, -2, -1, 0, 1, 2, \ldots$$

▲▲

The contour integral is performed by integrating in the complex plane counterclockwise on a circular contour centered at the origin and with a radius ρ, $\rho_{max} > \rho > \rho_{min}$. Like the previous transforms, there is a one-to-one correspondence between a discrete function and its z-transform.

Definition 18.1.2: The correspondence

$$x(n) \leftrightarrow X(z) = \mathcal{Z}\{x(n)\}$$

is called a z-transform pair.

▲▲

Table 18.5.2 contains several z-transform pairs.
    The next two examples illustrate the calculation of the z-transform for a given discrete function.

Example 18.1.3: Consider the unit impulse function $x(n) = \delta(n)$. To find the z-transform of this function we can use Definition 18.1.1 directly. We have

$$\mathcal{Z}\{\delta(n)\} = \sum_{n=-\infty}^{\infty} \delta(n)z^{-n} = 1 \tag{18.1.1}$$

since $\delta(n) = 0$ for $n \neq 0$.

▲▲

Example 18.1.4: Consider the discrete exponential function

$$x(n) = a^n u_s(n). \tag{18.1.2}$$

From the definition of the z-transform we have

$$\mathcal{Z}\{a^n u_s(n)\} = \sum_{n=-\infty}^{\infty} a^n u_s(n) z^{-n} = \sum_{n=0}^{\infty} a^n z^{-n} = \sum_{n=0}^{\infty} (az^{-1})^n. \tag{18.1.3}$$

In order for the geometric series in (18.1.3) to converge we must have

$$|az^{-1}| < 1, \quad \text{or} \quad |a| < |z|. \tag{18.1.4}$$

When (18.1.4) is satisfied, the series in (18.1.3) converges to

$$\sum_{n=0}^{\infty} (az^{-1})^n = \frac{1}{1 - (az^{-1})} = \frac{z}{z - a}. \tag{18.1.5}$$

For the special case, $a = 1$, we have

$$\mathcal{Z}\{u_s(n)\} = \frac{z}{z - 1}. \tag{18.1.6}$$

▲▲

Mathematically, the z-transform is a complex function mapping the complex plane into the complex plane. Many of the properties of z-transforms can be explained by graphically displaying the domain of a z-transform.

Definition 18.1.5: The domain of a z-transform, which is a copy of the complex plane, is called the z-plane.

▲▲

The z-plane plays a role in z-transforms similar to the role the s-plane plays for Laplace transforms.

18.1.2 Region of Convergence

From Definition 18.1.1 we see that the z-transform is written formally as an infinite sum. For this mathematical expression to have meaning we must show that this infinite sum converges (has a finite value). Since the z-transform is a function convergence of the infinite series means that there must exist at least one complex number, say z_0, such that when substituted into the infinite sum, it converges. In Example 18.1.4 the z-transform of the function in (18.1.2) is the infinite sum in (18.1.3). This sum converges for any z with modulus greater than $|a|$ as shown in (18.1.4). For this example the infinite sum converges not only for a single z, but for a whole set of complex numbers given in (18.1.4). The set of those elements of the complex plane where the infinite sum converges plays an important role in the use of the z-transform.

Definition 18.1.6: Consider the z-transform $X(z)$. Let ROC be the set of all complex numbers for which the z-transform $X(z)$ in Definition 18.1.1 converges. Then the set ROC is called the <u>region of convergence</u>.

▲▲

Notation: When we want to specify the region of convergence for a particular function $x(n)$ we write ROC_x.

The next two examples calculate the ROC for the previous two examples of a z-transform.

Example 18.1.4: (*continued*) The ROC of the discrete exponential in (18.1.2) is given in (18.1.4) and is shown in Figure 18.1.1 in the z-plane. The ROC is determined by drawing a circle centered on the origin and with a radius the same as the modulus of the pole of the z-transform (18.1.5). The ROC is then all of those complex numbers outside the disk determined by this circle. The ROC doesn't contain the circle because of the strict inequalities in (18.1.4).

To derive the ROC in Figure 18.1.1, write the complex number z in polar form as

$$z = \rho e^{j\omega}. \tag{18.1.7}$$

Substituting (18.1.7) into (18.1.4) we get

$$|a| < |z| = \left|\rho e^{j\omega}\right| = \rho. \tag{18.1.8}$$

Hence, we have

$$ROC = \left\{z \in C \,\middle|\, z = \rho e^{j\omega}, \rho > |a| \quad \text{for all } \omega\right\} \tag{18.1.9}$$

which is the set shown in Figure 18.1.1. In the notation of Definition 18.1.1

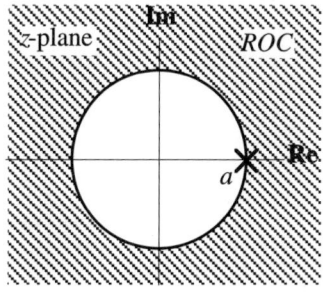

Figure 18.1.1 Region of Convergence of the z-Transform of Example 18.1.4

$\rho_{min} = |a|$ and $\rho_{max} = \infty$. The *ROC* in Figure 18.1.1 is typical of the *ROC* for any right-hand sequence.[1]

▲▲

Example 18.1.7:   The *z*-transform for the impulse function (18.1.1) converges everywhere in the *z*-plane. A little thought shows that any finite length sequence will have a *ROC* that is the whole *z*-plane except possibly $z = 0$ or $z = \infty$.

▲▲

For the *z*-transform of a discrete signal to exist, the infinite series in Definition 18.1.1 must converge. As the previous examples show, the convergence of the infinite series is equivalent to the existence of a nonempty *ROC*. Hence, the existence of a *z*-transform is equated to the existence of a nonempty *ROC* for that transform.

Two discrete functions which are completely different can have a *z*-transform with the same functional form. These two *z*-transforms have different *ROC*'s, however. The *ROC* differentiates the *z*-transforms of these two sequences. The next example illustrates this point.

Example 18.1.8:   Next consider the left-hand sequence

$$x(n) = \begin{cases} 0, & n \geq 0 \\ -b^n, & n < 0 \end{cases} \tag{18.1.10}$$

From the definition of the *z*-transform we have

$$X(z) = \sum_{n=-\infty}^{\infty} -b^n u_s(-n-1)z^{-n} = -\sum_{n=-\infty}^{-1} \left[bz^{-1} \right]^n. \tag{18.1.11}$$

Now let $n = -k$ be a change of variables in (18.1.11). Again using the formula for a geometric series we have

$$X(z) = -\sum_{k=1}^{\infty} \left[bz^{-1} \right]^{-k} = 1 - \sum_{k=0}^{\infty} \left[b^{-1}z \right]^k = 1 - \frac{1}{1 - zb^{-1}} = \frac{z}{z - b}. \tag{18.1.12}$$

The *ROC* for this series is

$$\left| zb^{-1} \right| < 1 \quad \text{or} \quad |z| < |b|. \tag{18.1.13}$$

This *ROC* is shown in Figure 18.1.2.   In the notation of Definition 18.1.1 $\rho_{min} = 0$ and $\rho_{max} = |b|$. The *ROC* in Figure 18.1.2 is typical of the *ROC* for any left-hand sequence.

---

[1] This terminology for discrete functions is discussed in Section 2.3.

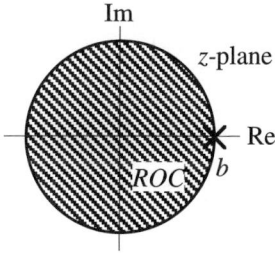

Figure 18.1.2 Region of Convergence of the z-Transform of (18.1.12)

The form of the z-transform of each of the functions in (18.1.2) and (18.1.10) is the same as can be seen by comparing (18.1.5) and (18.1.12). The *ROC* for (18.1.5), however, is outside the circle defined by the pole of (18.1.5) and the *ROC* of (18.1.12) is inside the circle defined by the pole of (18.1.12). We distinguish between the z-transforms of these two discrete signals by attaching the *ROC* to each z-transform.

▲▲

If a discrete signal is a right-hand or left-hand sequence, then the existence of the z-transform is ascertained from the convergence of the infinite series that defines the z-transform. If the discrete function is a two-sided sequence, then a two-step procedure is usually employed to determine if the two-sided z-transform exists for this signal. The following example illustrates this point.

Example 18.1.9: Consider the discrete signal

$$x(n) = a^n u_s(n) - b^n u_s(-n-1). \tag{18.1.14}$$

which is a two-sided sequence. This sequence has already been divided into a right-hand sequence and a left-hand sequence. This division is the first step in the construction of a z-transform for this signal.

This signal is a sum of the signals in the previous two examples. We can calculate the z-transform of this sequence by calculating the z-transform of each of the two sequences separately and then adding the results. Using the previous examples, we obtain (formally)

$$X(z) = \left[\sum_{n=0}^{\infty} a^n z^{-n} \right] - \left[\sum_{n=-\infty}^{-1} b^n z^{-n} \right] = \frac{z}{z-a} + \frac{z}{z-b}, \tag{18.1.15}$$

$$ROC = \left\{ z \,\middle|\, |z| < a \ \text{and} \ |z| > b \right\}.$$

The *ROC* for (18.1.14) is that set of complex numbers for which each of the infinite sums in (18.1.15) converges. That is, for each $z_0 \in ROC$ both of the summations in (18.1.15) must converge. Hence, the *ROC* is the intersection of the *ROC* for each of the terms in (18.1.15). Therefore, we determine the *ROC* of each

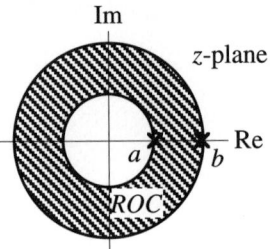

Figure 18.1.3 *ROC* of the Signal in (18.1.14)

of the right-hand and left-hand sequences separately, and determine if they have a nonzero intersection. The *ROC* of the right-hand sequence is given in Figure 18.1.1 and the *ROC* of the left-hand sequence is given in Figure 18.1.2. If $|b| \leq |a|$, the intersection of the two *ROC*s in Figure 18.1.1 and 18.1.2 is empty. Therefore, the *ROC* for the two-sided sequence (18.1.14) is empty and the *z*-transform does not exist since the infinite series will not converge for any *z*. For $|b| > |a|$, the intersection of the two *ROC*s in Figure 18.1.1 and 18.1.2 is shown in Figure 18.1.3. The *ROC* of the two-sided sequence shown in Figure 18.1.3, and the *z*-transform for (18.1.14) is given by (18.1.15). In the notation of Definition 18.1.1. $\rho_{min} = |a|$ and $\rho_{max} = |b|$. The *ROC* in Figure 18.1.3 is typical of the *ROC* for two-sided sequence.

▲▲

18.1.3 MATLAB Experiments

It is clear from the discussion above that the *ROC* of a *z*-transform can be obtained from the pole locations, if the *z*-transform is a proper rational function. MATLAB can be used to determine the pole locations using the **roots** command to factor the denominator polynomial, or the poles and zeros can be plotted using the **zplane** command.

Consider the finite length signal

$$\{x(0) \quad x(1) \quad x(2) \quad x(3)\}. \tag{18.1.16}$$

The *z*-transform of this signal is

$$x(0) + x(1)z^{-1} + x(2)z^{-2} + x(3)z^{-3} = x(0)z^3 + x(1)z^2 + x(2)z^1 + x(3). \tag{18.1.17}$$

Furthermore, the signal in (18.1.16) can be arranged in a vector

$$\vec{q} = \begin{bmatrix} x(0) & x(1) & x(2) & x(3) \end{bmatrix}^T. \tag{18.1.18}$$

Hence, finite length signals can be represented as either polynomials or vectors. It is useful to keep this fact in mind. Of course, MATLAB makes use of this observation to do polynomial calculations.

18.2 PROPERTIES OF THE TWO-SIDED z-TRANSFORM

18.2.1 Introduction

Most often the properties of the z-transform, rather than the definition of the z-transform, are used to carry out the analysis of discrete signals and systems. Therefore, in this section we will give several of the basic properties of the z-transform. Many of these properties bear great resemblance to the properties of the Laplace and Fourier transform. For example, the usefulness of the Laplace transform lies, in part, in our ability to compute the inverse transform by partial fraction expansion and table lookup. The z-transform also has these properties, making it a useful computational tool. In this section we also present a method for computing the inverse z-transform that is very similar to the procedure used for Laplace transforms.

18.2.2 Basic Properties

We present some of the properties of the z-transform next. In most cases these properties parallel the properties of the continuous-time transforms. Note that unlike the continuous-time transforms we have discussed in previous chapters, we must check for the existence of the *ROC* in each property. In the statement of each property we assume that the z-transform exists for each z-transform pair.

The first property is of fundamental importance.

Property 18.2.1:   (Linearity)   Given the z-transform pairs $x_1(n) \leftrightarrow X_1(z)$ and $x_2(n) \leftrightarrow X_2(z)$, and the real numbers a_1 and a_2, then

$$\mathcal{Z}\{a_1 x_1(n) + a_2 x_2(n)\} = a_1 X_1(z) + a_2 X_2(z), \qquad ROC_{x_1} \cap ROC_{x_2}.$$

▲▲

One of the implications of Property 18.2.1 is that z-transforms can be inverted by partial fraction expansion. See the discussion of inverse z-transforms below. Another consequence of Property 18.2.1 is that the z-transform can be used to solve difference equations. This application of the z-transform is discussed in the next section.

The next property of the z-transform plays a fundamental role in the representation of discrete systems.

Property 18.2.2: (Discrete Convolution) Given the z-transforms $x(n) \leftrightarrow X(z)$ and $h(n) \leftrightarrow H(z)$, then

$$\mathcal{Z}\left\{ \sum_{k=-\infty}^{\infty} h(n-k)x(k) \right\} = H(z)X(z), \qquad ROC_h \cap ROC_x.$$

▲▲

The convolution property shows that the product of two z-transforms is a convolution sum in the time domain. The following property is a converse statement of Property 18.2.2 which gives the z-transform of the product of two signals in the time domain.

Property 18.2.3: Given the z-transforms $x_1(n) \leftrightarrow X_1(z)$ with $ROC = \{z | \underline{r}_1 < |z| < \bar{r}_1\}$ and $x_2(n) \leftrightarrow X_2(z)$ with $ROC = \{z | \underline{r}_2 < |z| < \bar{r}_2\}$, then

$$Z\{x_1(n)x_2(n)\} = \frac{1}{2\pi j} \oint_\Gamma X_2\left(\frac{z}{v}\right) X_1(v) v^{-1} \, dv, \qquad ROC \supset \left\{z \, | \, \underline{r}_1\underline{r}_2 < |z| < \bar{r}_1\bar{r}_2\right\}$$

where Γ is a contour that lies in the *ROC*.

▲▲

Given two signals as in Property 18.2.3, if the *ROC* contains the unit circle, then we can introduce the change of variables

$$v = e^{j\phi}, \quad \text{and} \quad z = e^{j\theta}. \tag{18.2.1}$$

Substituting (18.2.30) into the integral in Property 18.2.3 we obtain

$$W\left(e^{j\theta}\right) = \frac{1}{2\pi} \int_{-\pi}^{\pi} Y\left(e^{j(\theta-\phi)}\right) X\left(e^{j\phi}\right) d\phi. \tag{18.2.2}$$

Terminology: Since the integration in (18.2.2) is carried out around the unit circle, it is called <u>circular convolution</u>.

From Property 18.2.3 we can easily obtain the extension of Parseval's theorem to discrete signals. Let $x(n)$ be a discrete signal whose *ROC* includes the unit circle. Then we have

$$Z\{x(n)\bar{x}(n)\}_{z=1} = \sum_{n=-\infty}^{\infty} x(n)\bar{x}(n)z^{-n}\Bigg|_{z=1} = \sum_{n=-\infty}^{\infty} |x(n)|^2. \tag{18.2.3}$$

Using (18.2.3) in Property 18.2.3 we obtain the following property.

Property 18.2.4: **Parseval's Theorem**   Given the z-transform $x(n) \leftrightarrow X(z)$ where the *ROC* of $x(n)$ contains the unit circle, then

$$\sum_{n=-\infty}^{\infty} |x(n)|^2 = \frac{1}{2\pi j} \oint_\Gamma X\left(\frac{1}{v}\right) X(v) v^{-1} dv$$

where $\bar{\Gamma} = e^{j\phi}, 0 \le \phi < 2\pi$.

▲▲

A property that is quite similar to convolution is correlation.

Property 18.2.5:   (Correlation) Given the z-transforms $x_1(n) \leftrightarrow X_1(z)$ and $x_2(n) \leftrightarrow X_2(z)$, then

$$Z\left\{\sum_{k=-\infty}^{\infty} x_1(k)x_2(k-m)\right\} = X_1(z)X_2(z^{-1}), \qquad ROC_{x_1} \cap ROC_{x_2}.$$

▲▲

The following property of the z-transform will be used below to solve difference equations.

Property 18.2.6: (Right and Left Shift) Given the z-transform $x(n) \leftrightarrow X(z)$ and an integer κ,

$$Z\{x(n-\kappa)\} = z^{-\kappa} X(z), \quad ROC_x \text{ except possibly } 0 \text{ and } / \text{ or } \infty.$$

▲▲

The following three properties of the z-transform are generally useful in deriving the z-transform of various signals.

Property 18.2.7: (Frequency Scaling) Given the z-transform $x(n) \leftrightarrow X(z)$ with $ROC = \{z|\, r_1 < |z| < r_2\}$ and a constant a (real or complex), then

$$Z\{a^n x(n)\} = X\left(\frac{z}{a}\right), \qquad ROC = \left\{z \mid |a|r_1 < |z| < |a|r_2\right\}$$

▲▲

Property 18.2.8:   (Time Reversal)   Given the z-transform $x(n) \leftrightarrow X(z)$ with $ROC = \{z|\, r_1 < |z| < r_2\}$, then

$$Z\{x(-n)\} = X\left(\frac{1}{z}\right), \qquad ROC = \left\{z \mid \frac{1}{r_2} < |z| < \frac{1}{r_1}\right\}$$

▲▲

Property 18.2.9:   (Differentiation in the z-Domain)   Given the z-transform $x(n) \leftrightarrow X(z)$ then

$$Z\{nx(n)\} = -z\frac{dX(z)}{dz}.$$

▲▲

18.2.3 Inverse z-Transforms

Partial Fraction Expansion The usefulness of z-transforms (in part) is our ability to calculate the signal that corresponds to a z-transform. This calculation can be accomplished in several ways, including using Definition 18.1.1. The procedure we give next is very similar to the inversion of the Laplace transform using partial fraction expansion and table lookup. z-Transforms can be inverted in a similar manner. In fact, the following procedure uses partial fraction expansion that is discussed in Section 9.3.

Procedure 18.2.10 : Let $X(z)$ be a rational z-transform.
Step 1 Divide the z-transform by z.
Step 2 Expand $X(z)/z$ using partial fractions.
Step 3 Multiply by z.
Step 4 Invert each term separately using the tables taking into account the *ROC*. ▲▲

Procedure 18.2.10 is based on the same concept that the inversion of Laplace transforms is founded. If we examine Table 18.5.2 of z-transform pairs, however, we observe that many of the z-transforms contain a z in the numerator. As a practical matter, we would like in the end to have a z-transform we can read from the table. A straightforward partial fraction expansion will not give us z-transform with a z in the numerator. Hence, the partial fraction expansion is modified as in Procedure 18.2.10.
The next three examples illustrate Procedure 18.2.10. These examples will show how the *ROC* enters into the calculations.

Example 18.2.11: Consider the z-transform

$$X(z) = \frac{z}{4(z-1)\left(z-\frac{1}{4}\right)}, \qquad ROC = \{|z| > 1\}. \tag{18.2.4}$$

Following Procedure 18.2.10, first we divide both sides of (18.2.4) by z. The result is

$$\frac{X(z)}{z} = \frac{1}{4(z-1)\left(z-\frac{1}{4}\right)}. \tag{18.2.5}$$

Second, the right-hand side of (18.2.5) is expanded into a partial fraction expansion

$$\frac{1}{4(z-1)\left(z-\frac{1}{4}\right)} = \frac{\frac{1}{3}}{z-1} + \frac{-\frac{1}{3}}{z-\frac{1}{4}}. \tag{18.2.6}$$

Third, combining (18.2.6) with (18.2.5) and multiplying through by z we get

$$X(z) = \frac{z}{4(z-1)\left(z-\frac{1}{4}\right)} = \left(\frac{1}{3}\right)\frac{z}{z-1} - \left(\frac{1}{3}\right)\frac{z}{z-\frac{1}{4}}. \tag{18.2.7}$$

Fourth, we invert each term of the partial fraction expansion using Table 18.5.2. From (18.2.4) we see that the *ROC* has the form required for right-hand signals. Hence, both terms in (18.2.7) invert to right-hand signals as in Example 18.1.4. We have

$$x(n) = \left(\frac{1}{3}\right)u_S(n) + \left(\frac{-1}{3}\right)\left(\frac{1}{4}\right)^n u_S(n). \tag{18.2.8}$$

▲▲

Example 18.2.12: Consider the z-transform

$$X(z) = \frac{1}{4(z-1)\left(z-\frac{1}{4}\right)}, \qquad ROC = \{|z| > 1\}. \tag{18.2.9}$$

First, we divide both sides of (18.2.9) by z. The result is

$$\frac{X(z)}{z} = \frac{1}{4z(z-1)\left(z-\frac{1}{4}\right)}. \tag{18.2.10}$$

Second, when we expand (18.2.10) into a partial fraction expansion we obtain

$$\frac{1}{4z(z-1)\left(z-\frac{1}{4}\right)} = \frac{1}{z} + \frac{\frac{1}{3}}{z-1} + \frac{-\frac{4}{3}}{z-\frac{1}{4}}. \tag{18.2.11}$$

Third, combining (18.2.11) with (18.2.10) and multiplying through by z we get

$$X(z) = 1 + \left(\frac{1}{3}\right)\frac{z}{z-1} - \left(\frac{4}{3}\right)\frac{z}{z-\frac{1}{4}}. \tag{18.2.12}$$

Fourth, we invert each term of the partial fraction expansion, as in Example 18.2.11, which gives

$$x(n) = \delta(n) + \left(\frac{1}{3}\right)u_S(n) - \left(\frac{4}{3}\right)\left(\frac{1}{4}\right)^n u_S(n). \tag{18.2.13}$$

▲▲

Example 18.2.13: In this example we will show how the *ROC* affects the inversion of the z-transform. Consider again Example 18.2.11. Suppose, however, that the *ROC* has changed to

$$X_2(z) = \frac{z}{4(z-1)\left(z-\frac{1}{4}\right)} = \left(\frac{1}{3}\right)\frac{z}{z-1} - \left(\frac{1}{3}\right)\frac{z}{z-\frac{1}{4}}, \qquad ROC = \left\{1 > |z| > \frac{1}{4}\right\}. \tag{18.2.14}$$

The partial fraction expansion in (18.2.7) still holds and it is shown in (18.2.14).

   To use Table 18.5.2 to invert each term in (18.2.14), we consider each term separately. The inverse of the second term is still

$$\left(-\frac{1}{3}\right)\frac{z}{z-\frac{1}{4}} \leftrightarrow \left(-\frac{1}{3}\right)\left(\frac{1}{4}\right)^n u_s(n) \tag{18.2.15}$$

because the *ROC* is outside the circle through $z = 1/4$. The other term in (18.2.14) must be inverted differently from the examples above because the *ROC* in (18.2.14) is inside the circle through $z = 1$. Using Example 18.1.8, we invert this term as

$$\left(\frac{1}{3}\right)\frac{z}{z-1} \leftrightarrow \left(\frac{1}{3}\right)(-u_s(-n-1)). \tag{18.2.16}$$

Combining (18.2.15) and (18.2.16) the inverse *z*-transform of (18.2.14) is

$$X_2(z) \leftrightarrow -\left(\frac{1}{3}\right)u_s(-n-1) - \left(\frac{1}{3}\right)\left(\frac{1}{4}\right)^n u_s(n). \tag{18.2.17}$$

▲▲

    This partial fraction expansion procedure has an alternative interpretation when the *z*-transform is written in terms of powers of z^{-1}. In this representation, the partial fraction expansion of the proper, rational function $X(z)$ has the form

$$X(z) = \frac{b_0 + b_1 z^{-1} + \cdots + b_M z^{-M}}{1 + a_1 z^{-1} + \cdots + a_N z^{-N}} = \sum_{k=1}^{N} \frac{c_k}{1 - p_k z^{-1}}. \tag{18.2.18}$$

The residues are evaluated from the formula

$$c_k = (1 - p_k z^{-1})X(z)\Big|_{z=p_k}. \tag{18.2.19}$$

Note that each term in the partial fraction expansion has the form

$$\frac{c_k}{1 - p_k z^{-1}} = \frac{z c_k}{z - p_k} \tag{18.2.20}$$

so that this expansion is equivalent to the expansion procedure given above.

Long Division Last we mention that the inverse *z*-transform can be obtained by long division if the *z*-transform is a rational function. Indeed, carrying out the long division we obtain

$$X(z) = \frac{b_0 + b_1 z^{-1} + \cdots + b_M z^{-M}}{1 + a_1 z^{-1} + \cdots + a_N z^{-N}} = 1 + x(1)z^{-1} + x(2)z^{-2} + \cdots. \tag{18.2.21}$$

18.2.4 MATLAB Experiments

MATLAB will compute both the partial fraction expansion (**residuez** command) and perform the long division (**impz** command). The basic input data are the numerator and denominator polynomials. The output data of the **residuez** command are ordered vectors of the residues with the corresponding poles and the constants from the improper part of the rational function. The output data from the **impz** command are two ordered vectors containing the expansion of the rational function and the corresponding time vector.

The **residuez** command output data is given in terms of the notation in (18.2.18). The **impz** command output data is given in terms of the notation in (18.2.21). Both of these commands assume that the polynomials are written and entered as functions of z^{-1}. It is assumed that the right-most entry in the vector defining the polynomial corresponds to the constant term in (18.2.18) and (18.2.19) of both the numerator and denominator polynomials. Consequently, the defining vectors of the polynomials must be of the same length with zero placeholders in the numerator if some of the b_k coefficients are zero. See the examples below.

MATLAB generally assumes that the *z*-transforms correspond to right-hand sequences. Appropriate adjustments must be made when dealing with left-hand sequences.

The following M-files illustrate the use of these commands.

Example 18.2.11: (*continued*)
```
clear
% Enter rational function
b = [0,0.25,0];
a = [1,-1.25,0.25];

% Calculate residues
[r,p,k] = residuez(b,a);

% Display answers
disp('residue calculation')
disp('residues'); disp(r');
disp('poles'); disp(p');
disp('constants'); disp(k)

% Long division
[x,n] = impz(b,a);

% Display answers
disp('function, x'); disp(x');
disp('time'); disp(n');
```
▲▲

Exploratory Exercise 18.2.14: Duplicate the results of Example 18.2.12.

▲▲

Exploratory Exercise 18.2.15: Plot the output data of the long division calculation using the stem command.

▲▲

18.3 THE ONE-SIDED z-TRANSFORM

18.3.1 Introduction

In this section we introduce the one-sided z-transform. This z-transform is closely related to the two-sided z-transform, and this relationship is discussed in detail along with properties of the one-sided z-transform. The one-sided z-transform is similar to the Laplace transform. The two-sided z-transform, which is widely used in signal processing, is similar to the two-sided Laplace transform (which is not discussed in this text). The one-sided z-transform can be used to solve difference equations with initial conditions which are not solvable with two-sided z-transforms. The solution of such problems is described in this section.

18.3.2 Definition of the One-Sided z-Transform

We begin with the definition of the one-sided z-transform.

Definition 18.3.1: Let $x(n)$ be a discrete signal that satisfies

(a) $x(n) = 0$ for $n < 0$

(b)
$$\sum_{n=0}^{\infty} |x(n)| \rho^{-n} \le M < \infty \quad \text{for} \quad 0 \le \rho_{min} < \rho < \infty$$

For these signals the (one-sided) z-transform is defined as

$$\mathcal{Z}\{x(n)\} = \sum_{n=0}^{\infty} x(n) z^{-n} = X(z).$$

The inverse z-transform is defined by

$$x(n) = \mathcal{Z}^{-1}\{X(z)\} = \frac{1}{2\pi j} \oint X(z) z^{n-1} \, dz, \quad n = 0, 1, 2, \ldots$$

▲▲

Terminology: The *one-sided* z-transform is so named because the restrictions on the signal in (a) in Definition 18.3.1 lead to a one-sided sum in the definition of $X(z)$. This observation also explains the term *two-sided* z-transform.

In comparing the one-sided z-transform in Definition 18.3.1 with the two-sided z-transform in Definition 18.1.1, we see two differences between these two transforms. First, the one-sided z-transform only applies to right-hand sequences as shown in (a) of Definition 18.3.1. Thus the defining summation begins at zero. This change in summation limit only affects certain properties of the two-sided z-transform (to be explained below). Many of the properties of the two-sided z-transform are the same as the properties of the one-sided z-transform.

The second difference, which is much less obvious, is found in the region of convergence. Consider the signal

$$x(n) = a^n u_s(n) + b^n u_s(n). \tag{18.3.1}$$

The region of convergence of the z-transform of the first term is shown in Figure 18.1.1. The *ROC* shown in this figure is the only type of *ROC* that occurs for the one-sided z-transform, because only right-hand sequences are allowed. That is, the types of *ROC* that are shown in Figures 18.1.2-3 never occur because left-hand sequences are not allowed. The *ROC* of the first sequence in (18.3.1) is outside the circle that passes through $z = a$, and the *ROC* of the second sequence in (18.3.1) is outside the circle that passes through $z = b$. Hence, the intersection of these two *ROC*'s is *always* nonempty. So we see that the *ROC* of the signal in (18.3.1) *always* exists, because all of the *ROC*'s have the form found in Figure 18.1.1. Hence, when using the one-sided z-transform, it is not necessary to keep track of the *ROC*.

In general, the one-sided z-transform is used like the Laplace transform. Conditions (a) and (b) in Definition 18.3.1 correspond to conditions (a) and (b) in the Definition 9.1.1. Similarly, conditions (a) and (b) of the definition of the two-sided z-transform, Definition 18.1.1, are very similar to conditions (a) and (b) of the Fourier transform, Definition 7.4.1. The two-sided z-transform is used like the Fourier transform.

18.3.3 Properties of the One-Sided z-Transform

The one-sided z-transform inherits almost all of the properties of the two-sided z-transform. In particular, the inversion of the one-sided z-transform is similar to the methods used for inverting the two-sided z-transform as discussed in the last section. With one-sided z-transforms only right-hand sequences are allowed, however. Hence, the inversion of each term in the partial fraction expansion can be accomplished without reference to the *ROC*.

The shift properties of one-sided z-transforms, which are given next, are different from the shift property for two-sided z-transforms. The shift property of the one-sided z-transform depends on whether the shift is to the right or the left.

Property 18.3.2: (Right Shift)   Given the z-transform pair $x(n) \leftrightarrow X(z)$ and a positive integer, $\kappa > 0$,

$$\mathcal{Z}\{x(n - \kappa)\} = z^{-\kappa} X(z).$$

▲▲

When a discrete function is right shifted by κ, the elements corresponding to $n = 0, 1, \dots, \kappa - 1$ are filled with zeros. When these terms are dropped out of the summation, the factor $z^{-\kappa}$ can be factored out. The left shift property given next works out differently.

Property 18.3.3: (Left Shift) Given the z-transform pair $x(n) \leftrightarrow X(z)$ and a positive integer, $\kappa > 0$,

$$\mathcal{Z}\{x(n+\kappa)\} = z^{\kappa}X(z) - x(0)z^{\kappa} - x(1)z^{\kappa-1} - \cdots - x(\kappa-1)z.$$

▲▲

When the discrete function is shifted to the left, the initial values of the function must be explicitly included in the summation. This attribute of the left shift makes it useful for solving initial value problems of difference equations.

The following two properties of the one-sided z-transform are useful for checking the answer to a z-transform calculation.

Property 18.3.4: (Initial Value Theorem) Given the z-transform pair $x(n) \leftrightarrow X(z)$, then

$$x(0) = \lim_{z \to \infty} X(z).$$

▲▲

Property 18.3.5: (Final Value Theorem) Given the z-transform pair $x(n) \leftrightarrow X(z)$, then

$$\lim_{n \to \infty} x(n) = \lim_{z \to 1} (z-1)X(z)$$

provided that the ROC of $(z-1)X(z)$ contains the unit circle.

▲▲

18.3.4 Solution of Difference Equations in Advance Form

Consider the difference equation in advance form

$$y(n+N) + a_1 y(n+(N-1)) + \cdots + a_N y(n) \qquad (18.3.2)$$
$$= b_0 x(n+N) + b_1 x(n+(N-1)) + \cdots + b_N x(n)$$
$$y(0) = y_0, \quad y(1) = y_1, \quad y(2) = y_2, \ldots, y(N-1) = y_{N-1}.$$

Difference equations are the differential equations of discrete-time systems. One-sided z-transforms can be used to solve difference equations with initial conditions as in (18.3.2) just as Laplace transforms can be used to solve differential equations. The application of z-transforms to difference equations requires the use of the left shift Property. The technique is demonstrated by example.

Example 18.3.6: Consider the difference equation

$$y(n+1) - \frac{1}{4}y(n) = \frac{1}{4}x(n), \qquad (18.3.3)$$
$$y(0) = 0, \quad x(n) = u_s(n).$$

Taking the z-transform of (18.3.3) and using the left shift, Property 18.3.3, we get

$$\left[zY(z) - zy(0) \right] - \frac{1}{4} Y(z) = \frac{1}{4} \frac{z}{z-1}. \tag{18.3.4}$$

Solving (18.3.4) for $Y(z)$ we have

$$Y(z) = \frac{z}{4(z-1)\left(z - \frac{1}{4}\right)}. \tag{18.3.5}$$

This particular z-transform appeared in Example 18.2.11 where the inverse z-transform was calculated. The corresponding discrete-time function is

$$y(n) = \left(\frac{1}{3}\right) u_S(n) + \left(\frac{-1}{3}\right)\left(\frac{1}{4}\right)^n u_S(n). \tag{18.3.6}$$

▲▲

When applying z-transforms to solve difference equations, some care must be exercised so that the z-transform is applied correctly. Since the one-sided z-transform has been defined only on the nonnegative time interval, the one-sided z-transform cannot be applied if the initial conditions are given in negative time. Also the difference equation must be transformed into the advance form (18.3.2) before the one-sided z-transform can be applied.

If the initial conditions of the difference equation are zero, then either the one-sided or two-sided z-transform may be employed to find the solution. In this event, the difference equation is usually expressed in delay form. Also note that if the input signal is a left-hand sequence, then the two-sided z-transform must be used.

18.4 DISCRETE-TIME FOURIER TRANSFORM

18.4.1 Introduction

The Fourier transform is particularly useful for signal and system analysis as we have shown in previous chapters. We would like to have such a tool for analyzing discrete-time signals and systems. Such a transform would have to apply to discrete-time functions. The discrete-time Fourier transform fills this role. In this section we will define this transform and give several of its properties. We will also relate the discrete-time Fourier transform to the z-transform.

18.4.2 Definition

We begin with the definition of the discrete-time Fourier transform.

Definition 18.4.1: Let $x(n)$ be a discrete-time function that satisfies

(a) $x(n)$,   is defined for $-\infty < n < \infty$

(b)
$$\sum_{n=-\infty}^{\infty} |x(n)| < M < \infty.$$

Then the <u>discrete-time Fourier transform</u> (DTFT) is defined as

$$X(\Omega) = \mathcal{DTFT}\{x(n)\} = \sum_{n=-\infty}^{\infty} x(n)e^{-j\Omega n}$$

The <u>inverse discrete-time Fourier transform</u> (IDTFT) is defined as

$$x(n) = \mathcal{IDTFT}\{X(\Omega)\} = \frac{1}{2\pi} \int_{2\pi} X(\Omega)e^{j\Omega n} \, d\Omega, \quad -\infty < n < \infty.$$

▲▲

Notation: The integration in the IDTFT is over any interval of length 2π.

Terminology: In Definition 18.4.1, the DTFT is called the <u>analysis equation</u> and the IDTFT is called the <u>synthesis equation</u>.

As with other transforms, each function is uniquely associated with its DTFT.

Definition 18.4.2:   Let the DTFT of the function $x(n)$ be $X(\Omega)$.   Then the relationship

$$x(n) \leftrightarrow X(\Omega)$$

is called a <u>DTFT pair</u>.

▲▲

A table of DTFT transform pairs is given in the summary section of this chapter.

Notation: The DTFT is sometimes written as $X(e^{j\Omega})$.   We will not use this notation.

Notation: The DTFT depends on a real variable Ω which is used to denote frequency in discrete-time. This frequency variable will play the role in discrete-time that ω plays in continuous-time. This distinction between these two frequency variables and their relationship via the sampling period will be discussed in detail in Chapter 20.

The units on the discrete frequency variable Ω are radians or radians/sample. By way of comparison, the units on the continuous-time frequency variable ω are radians/sec. See Remark 17.2.5.

The next example illustrates the use of the definition of the DTFT.

Example 18.4.3: Consider the discrete-time function

$$x(n) = a^n u_s(n). \tag{18.4.1}$$

Using Definition 18.4.1 the DTFT of this function is

$$\mathcal{DTFT}\{x(n)\} = \sum_{n=-\infty}^{\infty} x(n)e^{-j\Omega n} = \sum_{n=0}^{\infty} a^n e^{-j\Omega n} \tag{18.4.2}$$

$$= \frac{1}{1 - ae^{-j\Omega}} = \frac{e^{j\Omega}}{e^{j\Omega} - a}, \quad |a| < 1$$

because this series is a geometric series. The DTFT in (18.4.2) depends on the function $e^{j\Omega}$. This function is periodic with period 2π. Hence, the DTFT is also periodic with period 2π.

▲▲

The last example shows quite clearly that the DTFT is periodic with period 2π. In fact, this property is true of the DTFT in general. From Definition 18.4.1 $X(\Omega)$ is an infinite sum of the functions $e^{j\Omega n}$. These functions are all periodic with periods that are multiples of 2π. Hence, the DTFT is periodic with a period of 2π. This property is of fundamental importance for interpreting and using the DTFT.

18.4.3 Properties of the DTFT

The DTFT has a number of properties that make it useful for the analysis of discrete-time signals and systems. These properties are mostly analogous to the properties of the other transforms presented in this text. In the properties discussed below, we assume that the DTFT pair exists.

As with the other transforms we have studied, the DTFT is linear. This property allows it to be applied to a wide range of problems in signals and systems.

Property 18.4.4:   (Linearity)   Given the DTFT pairs $x_1(n) \leftrightarrow X_1(\Omega)$ and $x_2(n) \leftrightarrow X_2(\Omega)$ along with the real numbers a_1 and a_2, then

$$\mathcal{DTFT}\{a_1 x_1(n) + a_2 x_2(n)\} = a_1 X_1(\Omega) + a_2 X_2(\Omega).$$

▲▲

Like the Fourier transform, we would like to use the DTFT to study systems. The next property is used in this respect.

Property 18.4.5:   (Convolution)   Given the DTFT pairs $x_1(n) \leftrightarrow X_1(\Omega)$ and $x_2(n) \leftrightarrow X_2(\Omega)$, then

$$\mathcal{DTFT}\left\{ \sum_{k=-\infty}^{\infty} x_1(k)x_2(n-k) \right\} = X_1(\Omega)X_2(\Omega).$$

▲▲

Closely related to the convolution property is the correlation property. (Note the subtle difference between these two properties in the appearance of the minus sign in the argument of the functions and transforms.)

Property 18.4.6:   (Correlation)   Given the DTFT pairs $x_1(n) \leftrightarrow X_1(\Omega)$ and $x_2(n) \leftrightarrow X_2(\Omega)$, then

$$\mathcal{DTFT}\left\{ \sum_{k=-\infty}^{\infty} x_1(k)x_2(n+k) \right\} = X_1(\Omega)X_2(-\Omega).$$

▲▲

The next property can be seen as a kind of converse to convolution. This property is important in determining the spectral properties of functions and in the design of digital filters.

Property 18.4.7:   (Windowing)   Given the DTFT pairs $x_1(n) \leftrightarrow X_1(\Omega)$ and $x_2(n) \leftrightarrow X_2(\Omega)$, then

$$\mathcal{DTFT}\{x_1(n)x_2(n)\} = \frac{1}{2\pi} \int_{2\pi} X_1(\lambda)X_2(\Omega - \lambda)\, d\lambda$$

▲▲

Once again we encounter Parseval's theorem. As we have seen in previous chapters, Parseval's theorem allows us to quantify the notion of the frequency content of a signal. Those interpretations for continuous signals extend to discrete signals as will be shown in the coming chapters.

Property 18.4.8: Parseval's Theorem Given the DTFT pairs $x_1(n) \leftrightarrow X_1(\Omega)$ and $x_2(n) \leftrightarrow X_2(\Omega)$, then

$$\sum_{n=-\infty}^{\infty} x_1(n)x_2(n) = \frac{1}{2\pi} \int_{2\pi} X_1(\Omega)\overline{X_2(\Omega)}\, d\Omega$$

▲▲

The DTFT can also be used for system analysis. The following property allows it to be applied to difference equations, exactly as two-sided z-transforms are applied to difference equations.

Property 18.4.9: (Time Shift) Given the DTFT pair $x(n) \leftrightarrow X(\Omega)$, then

$$\mathcal{DTFT}\{x(n-\kappa)\} = e^{-j\Omega\kappa} X(\Omega).$$

▲▲

The following properties are used for signal analysis.

Property 18.4.10: (Time Reversal) Given the DTFT pair $x(n) \leftrightarrow X(\Omega)$, then

$$\mathcal{DTFT}\{x(-n)\} = X(-\Omega).$$

▲▲

Property 18.4.11: (Frequency Shift) Given the DTFT pair $x(n) \leftrightarrow X(\Omega)$, then

$$\mathcal{DTFT}\left\{e^{j\Omega_0 n} x(n)\right\} = X(\Omega - \Omega_0).$$

▲▲

Property 18.4.12: (Modulation) Given the DTFT pair $x(n) \leftrightarrow X(\Omega)$, then

$$\mathcal{DTFT}\{x(n)\cos(\Omega_0 n)\} = \frac{1}{2}\left[X(\Omega + \Omega_0) + X(\Omega - \Omega_0)\right].$$

▲▲

In applying the modulation property it is useful to remember that the frequency of a discrete sinusoid is usually restricted to be between $0 \leq \Omega_0 \leq \pi$. See Remark 2.3.10.

18.4.4 The Generalized DTFT

The definition of the DTFT requires that the function $x(n)$ to be absolutely summable, condition (b) of Definition 18.4.1. The DTFT, however, can be extended to other functions in two ways. First, we note that the DTFT is a sum of periodic functions, i.e., $e^{-j\Omega n}$. Hence, $X(\Omega)$ is also periodic with period 2π. As a result, it has a Fourier series. In fact, the Fourier series of $X(\Omega)$ is the DTFT. That is, we can think of the definition of the DTFT of the function as a Fourier series which converges for functions which are square summable. (See Chapter 7 for a review of the exponential Fourier series and its convergence conditions.) For the Fourier series to converge to the function, the function must satisfy

$$\sum_{n=-\infty}^{\infty} |x(n)|^2 < M < \infty.$$ (18.4.3)

Since we have that

$$\left[\sum_{n=-\infty}^{\infty}|x(n)|\right]^2 \geq \sum_{n=-\infty}^{\infty}|x(n)|^2 \tag{18.4.4}$$

the DTFT, as a Fourier series, could exist for functions which are not absolutely summable. (The summation on the left in (18.4.4) could be infinite while the summation on the right could be finite.) The following example is one such function.

Example 18.4.13: Consider the function $x(n)$ which has the DTFT

$$X(\Omega) = \begin{cases} 1, & 0 \leq |\Omega| < \Omega_c \\ 0, & \Omega_c \leq |\Omega| < \pi \end{cases} \tag{18.4.5}$$

From Definition 18.4.1, the inverse DTFT is given by

$$x(n) = \frac{1}{2\pi}\int_{-\pi}^{\pi} X(\Omega)e^{j\Omega n}\,d\Omega = \frac{1}{2\pi}\int_{-\Omega_c}^{\Omega_c} e^{j\Omega n}\,d\Omega = \frac{\sin\Omega_c n}{\pi n}, \quad n \neq 0. \tag{18.4.6}$$

For $n = 0$, we have

$$x(n) = \frac{1}{2\pi}\int_{-\Omega_c}^{\Omega_c} X(\Omega)e^{j\Omega n}\,d\Omega\bigg|_{n=0} = \frac{1}{2\pi}\int_{-\pi}^{\pi} d\Omega = \frac{\Omega_c}{\pi}. \tag{18.4.7}$$

Therefore,

$$x(n) = \begin{cases} \dfrac{\Omega_c}{\pi}, & n = 0 \\ \dfrac{\sin(\Omega_c n)}{\pi n}, & n \neq 0 \end{cases} \tag{18.4.8}$$

The graph of the function $x(n)$ and its DTFT are shown in Figure 18.4.1. It is left to the reader to verify that the sequence (18.4.8) is not absolutely summable, but it is square summable. Hence, this sequence doesn't satisfy condition (b) of Definition 18.4.1. Nonetheless, we use this DTFT pair.

▲▲

The second way in which this transform can be extended is by defining the transform for certain functions and using this DTFT pair to transform other functions, as was done for the Fourier transform in Section 7.5. The following example illustrates a most useful DTFT pair.

Example 18.4.14: Consider the DTFT

$$X(\Omega) = 2\pi\delta(\Omega), \quad -\pi < \Omega \leq \pi. \tag{18.4.9}$$

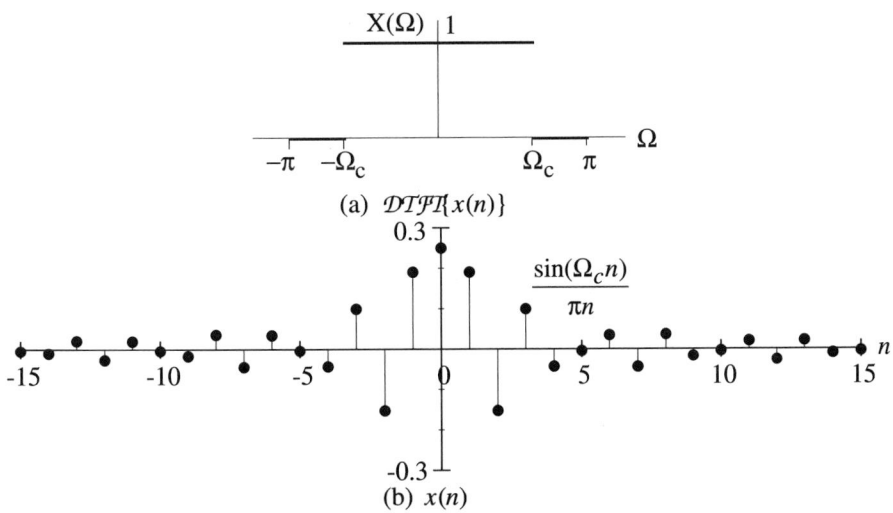

Figure 18.4.1 The Function in (18.4.8) and Its DTFT

Computing the inverse DTFT of (18.4.9) we have

$$x(n) = \frac{1}{2\pi} \int_{-\pi}^{\pi} X(\Omega) e^{j\Omega n} d\Omega = \frac{1}{2\pi} \int_{-\pi}^{\pi} 2\pi\delta(\Omega) e^{j\Omega n} d\Omega = 1, \quad -\infty < n < \infty. \tag{18.4.10}$$

Based on (18.4.10) we define the DTFT pair as

$$x(n) = 1 \leftrightarrow X(\Omega) = 2\pi\delta(\Omega). \tag{18.4.11}$$

Note that

$$\lim_{m \to \infty} \sum_{n=-m}^{m} |x(n)| = \lim_{m \to \infty} \sum_{n=-m}^{m} |1| = \infty \tag{18.4.12}$$

so the function in (18.4.11) doesn't satisfy condition (b) of Definition 18.4.1. We use the pair in (18.4.11) nonetheless.

▲▲

Because the DTFT is periodic, we have a choice of how we represent these functions. We can graph these functions over one period (usually $[-\pi, \pi]$), or we can graph the function over the entire time interval. Hence, the DTFT in (18.4.9) is equivalent to the representation

$$X(\Omega) = \sum_{n=-\infty}^{\infty} 2\pi\delta(\Omega - 2\pi n). \tag{18.4.13}$$

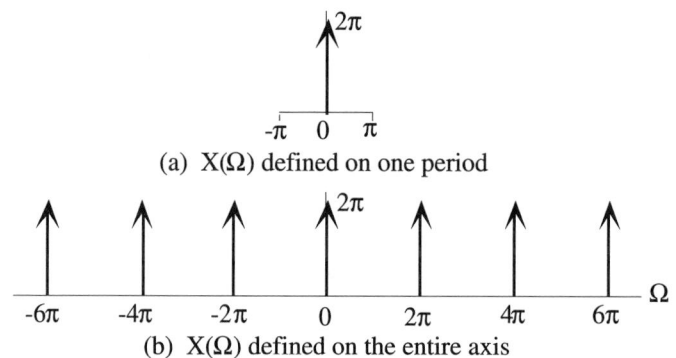

(a) $X(\Omega)$ defined on one period

(b) $X(\Omega)$ defined on the entire axis

Figure 18.4.2 Two Representations of a DTFT

These two representations are shown in Figure 18.4.2. The periodicity of the DTFT appears again and again when we use it for signal and system analysis.

18.4.5 Relationship to the z-Transform

There is a close relationship between the DTFT and the two-sided z-transform. Recall that the z-transform is defined as

$$\mathcal{Z}\{x(n)\} = \sum_{n=-\infty}^{\infty} x(n)z^{-n}. \tag{18.4.14}$$

Comparing (18.4.14) with Definition 18.4.1, we would like to set $z = e^{j\Omega}$. Then it follows from (18.4.14) that

$$X(z)\big|_{z=e^{j\Omega}} = \sum_{n=-\infty}^{\infty} x(n)e^{-j\Omega n}. \tag{18.4.15}$$

This equivalence is in fact valid if the two-sided z-transform converges for points on the unit circle. We summarize this observation in the following theorem.

Theorem 18.4.15: Let $x(n)$ be a discrete-time function with a z-transform $X(z)$ whose *ROC* includes the unit circle. Then the DTFT exists for this function, and it is given by

$$X(z)\big|_{z=e^{j\Omega}} = X(\Omega).$$

▲▲

If the z-transform is a rational function, then the poles can be used to determine if the DTFT can be computed from $X(z)$. These results rely on the discussion of

BIBO stability to be presented in Chapter 22. So these results will be presented without proof.

Theorem 18.4.16: (a) Let $x_r(n)$ be a right-handed sequence with z-transform $X_r(z)$. Then the DTFT of this function exists if the poles of $X_r(z)$ are inside the open unit circle.
(b) Let $x_l(n)$ be a left-handed sequence with z-transform $X_l(z)$. Then the DTFT of this function exists if the poles of $X_l(z)$ are outside the open unit disk.

▲▲

Theorem 18.4.16 can be extended to cover two-sided sequences in an obvious way. Also note that the two-sided z-transform applied to right-hand sequences is the one-sided z-transform.

Example 18.4.17: Consider the discrete-time function

$$x(n) = a^n u_s(n). \tag{18.4.16}$$

The z-transform of this function is

$$X(z) = \frac{z}{z - a}. \tag{18.4.17}$$

If $|a| < 1$ then its pole is inside the unit circle and the DTFT of this function can be computed from $X(z)$. We have

$$\mathcal{DTFT}\{x(n)\} = X(z)|_{z=e^{j\Omega}} = \frac{e^{j\Omega}}{e^{j\Omega} - a}. \tag{18.4.18}$$

This result agrees with Example 18.4.3.

▲▲

18.5 CHAPTER SUMMARY

In Section 18.1 the two-sided z-transform is introduced and discussed.

Definition 18.1.1: Let $x(n)$ be a discrete function that satisfies

(a) $x(n)$ is defined for $-\infty < n < \infty$
(b)
$$\sum_{n=-\infty}^{\infty} |x(n)| \rho^{-n} \leq M < \infty \quad \text{for} \quad 0 \leq \rho_{min} < \rho < \rho_{max} < \infty$$

For these functions the <u>(two-sided) z-transform</u> is defined as

$$\mathcal{Z}\{x(n)\} = \sum_{n=-\infty}^{\infty} x(n)z^{-n} = X(z).$$

The <u>inverse *z*-transform</u> is defined by

$$x(n) = \mathcal{Z}^{-1}\{X(z)\} = \frac{1}{2\pi j} \oint X(z)z^{n-1}\, dz, \quad n = \dots, -2, -1, 0, 1, 2, \dots$$

▲▲

The *z*-transform has associated with it a region of convergence (*ROC*) (Definition 18.1.2) which is used to uniquely associate the *z*-transform with the function. The function and its *z*-transform form a <u>*z*-transform pair</u>. Several *z*-transform pairs are given in Table 18.5.2.

In Section 18.3 the one-sided *z*-transform is introduced and discussed. It is basically the same as the two-sided *z*-transform except that the functions are restricted to right-hand sequences.

A number of properties of the one-sided and two-sided *z*-transforms are discussed in this chapter. These properties are summarized in Table 18.5.1 at the end of this section. The left shift property of the one-sided *z*-transform can be used to solve initial value problems of difference equations as discussed in Section 18.3.

In Section 18.4 the discrete-time Fourier transform is introduced.

Definition 18.4.1: Let $x(n)$ be a discrete-time function that satisfies

(a) $x(n)$,   is defined for $-\infty < n < \infty$

(b) $\sum_{n=-\infty}^{\infty} |x(n)| < M < \infty.$

Then the <u>discrete-time Fourier transform</u> (DTFT) is defined as

$$X(\Omega) = \mathcal{DTFT}\{x(n)\} = \sum_{n=-\infty}^{\infty} x(n)e^{-j\Omega n}$$

The <u>inverse discrete-time Fourier transform</u> (IDTFT) is defined as

$$x(n) = \mathcal{IDTFT}\{X(\Omega)\} = \frac{1}{2\pi} \int_{2\pi} X(\Omega)e^{j\Omega n}\, d\Omega \quad -\infty < n < \infty.$$

▲▲

The function and its DTFT form a <u>DTFT pair</u>. This transform behaves very much like the Fourier transform, and it is used to analyze signals. The properties of this transform are summarized in Table 18.5.3 at the end of this section. A table of DTFT pairs is given in Table 18.5.4.

The DTFT is closely related to the two-sided z-transform. According to Theorem 18.4.15 if the z-transform $X(z)$ of a signal $x(n)$ has a ROC that includes the unit circle, then the DTFT exists for this signal, and it is given by

$$X(z)\big|_{z=e^{j\Omega}} = X(\Omega).$$

Conditions for the ROC of a z-transform to contain the unit circle are given in Theorem 18.4.16 in terms of the poles of the z-transform.

Table 18.5.1 z-Transform Properties

Two-Sided z-Transform

Location	Property	Expression
18.2.1	Linearity	$\mathcal{Z}\{a_1 x_1(n) + a_2 x_2(n)\} = a_1 X_1(z) + a_2 X_2(z),\ ROC_{x_1} \cap ROC_{x_2}$
18.2.2	Discrete Convolution	$\mathcal{Z}\left\{\displaystyle\sum_{k=-\infty}^{\infty} h(n-k)x(k)\right\} = H(z)X(z),\quad ROC_h \cap ROC_x$
18.2.3	Circular Convolution	$\mathcal{Z}\{x_1(n)x_2(n)\}$ $= \dfrac{1}{2\pi j}\displaystyle\oint_\Gamma X_2\!\left(\dfrac{z}{v}\right)X_1(v)v^{-1}dv,\quad ROC \supset \left\{z \mid \underline{r}_1\underline{r}_2 < \|z\| < \bar{r}_1\bar{r}_2\right\}$
18.2.4	Parseval's Theorem	$\displaystyle\sum_{n=-\infty}^{\infty}|x(n)|^2 = \dfrac{1}{2\pi j}\oint_\Gamma X\!\left(\dfrac{1}{v}\right)X(v)v^{-1}dv$
18.2.5	Correlation	$\mathcal{Z}\left\{\displaystyle\sum_{k=-\infty}^{\infty} x_1(k)x_2(k-m)\right\} = X_1(z)X_2(z^{-1}),\quad ROC_{x_1} \cap ROC_{x_2}$
18.2.6	Shift	$\mathcal{Z}\{x(n-\kappa)\} = z^{-\kappa}X(z),\quad ROC_x - \{0,\infty\}$
18.2.7	Frequency Scaling	$\mathcal{Z}\{a^n x(n)\} = X\!\left(\dfrac{z}{a}\right),\quad ROC = \left\{z \mid |a|r_1 < \|z\| < |a|r_2\right\}$
18.2.8	Time Reversal	$\mathcal{Z}\{x(-n)\} = X\!\left(\dfrac{1}{z}\right),\quad ROC = \left\{z \mid \dfrac{1}{r_2} < \|z\| < \dfrac{1}{r_1}\right\}$
18.2.9	Differen-tiation	$\mathcal{Z}\{nx(n)\} = -z\dfrac{dX(z)}{dz},\quad ROC_x$ except possibly 0 or ∞

One-Sided z-Transform

Location	Property	Expression
18.3.2	Right Shift	$\mathcal{Z}\{x(n-\kappa)\} = z^{-\kappa}X(z)$
18.3.3	Left Shift	$\mathcal{Z}\{x(n+\kappa)\} = z^{\kappa}X(z) - x(0)z^{\kappa} - x(1)z^{\kappa-1} - \cdots - x(\kappa-1)z$

Table 18.5.2 z-Transform Pairs

Function	z-Transform	ROC
$\delta(n) \leftrightarrow$	1	z-plane
$\delta(n-\kappa), \quad \kappa>0 \leftrightarrow$	$\dfrac{1}{z^\kappa}, \quad \kappa>0$	$\lvert z\rvert>0$
$u_s(n) \leftrightarrow$	$\dfrac{z}{z-1}$	$\lvert z\rvert>1$
$u_s(n)-u_s(n-\kappa), \quad \kappa>0 \leftrightarrow$	$\dfrac{z^\kappa-1}{z^{\kappa-1}(z-1)}, \quad \kappa>0$	$\lvert z\rvert>1$
$nu_s(n) \leftrightarrow$	$\dfrac{z}{(z-1)^2}$	$\lvert z\rvert>1$
$a^n u_s(n) \leftrightarrow$	$\dfrac{z}{z-a}$	$\lvert z\rvert>\lvert a\rvert$
$na^n u_s(n) \leftrightarrow$	$\dfrac{az}{(z-a)^2}$	$\lvert z\rvert>a$
$-b^n u_s(-n-1) \leftrightarrow$	$\dfrac{z}{z-b}$	$\lvert z\rvert<\lvert b\rvert$
$-nb^n u_s(-n-1) \leftrightarrow$	$\dfrac{bz}{(z-b)^2}$	$\lvert z\rvert<b$
$\sin(\Omega_0 n)u_s(n) \leftrightarrow$	$\dfrac{z\sin\Omega_0}{z^2-(2\cos\Omega_0)z+1}$	$\lvert z\rvert>1$
$\cos(\Omega_0 n)u_s(n) \leftrightarrow$	$\dfrac{z(z-\cos\Omega_0)}{z^2-(2\cos\Omega_0)z+1}$	$\lvert z\rvert>1$
$a^n\sin(\Omega_0 n)u_s(n) \leftrightarrow$	$\dfrac{z(a\sin\Omega_0)}{z^2-(2a\cos\Omega_0)z+a^2}$	$\lvert z\rvert>a$
$a^n\cos(\Omega_0 n)u_s(n) \leftrightarrow$	$\dfrac{z(z-a\cos\Omega_0)}{z^2-(2a\cos\Omega_0)z+a^2}$	$\lvert z\rvert>a$
$a^n\cos(\Omega_0 n+\theta)u_s(n) \leftrightarrow$	$\dfrac{z(z\cos\theta-a\cos(\theta-\Omega_0))}{z^2-(2a\cos\Omega_0)z+a^2}$	$\lvert z\rvert>a$
	$=\dfrac{0.5ze^{j\theta}}{z-ae^{j\Omega_0}}+\dfrac{0.5ze^{-j\theta}}{z-ae^{-j\Omega_0}}$	

Table 18.5.3 Properties of the Discrete-Time Fourier Transform

Location	Property	Expression
Property 18.4.4	Linearity	$\mathcal{DTFT}\{a_1 x_1(n) + a_2 x_2(n)\} = a_1 X_1(\Omega) + a_2 X_2(\Omega)$
Property 18.4.5	Convolution	$\mathcal{DTFT}\left\{\displaystyle\sum_{k=-\infty}^{\infty} x_1(k) x_2(n-k)\right\} = X_1(\Omega) X_2(\Omega)$
Property 18.4.6	Correlation	$\mathcal{DTFT}\left\{\displaystyle\sum_{k=-\infty}^{\infty} x_1(k) x_2(n+k)\right\} = X_1(\Omega) X_2(-\Omega)$
Property 18.4.7	Windowing	$\mathcal{DTFT}\{x_1(n) x_2(n)\} = \dfrac{1}{2\pi}\displaystyle\int_{2\pi} X_1(\lambda) X_2(\Omega - \lambda)\, d\lambda$
Property 18.4.8	Parseval's Theorem	$\displaystyle\sum_{n=-\infty}^{\infty} x_1(n) x_2(n) = \dfrac{1}{2\pi}\int_{2\pi} X_1(\Omega)\overline{X_2(\Omega)}\, d\Omega$
Property 18.4.9	Time Shift	$\mathcal{DTFT}\{x(n-\kappa)\} = e^{-j\Omega\kappa} X(\Omega)$
Property 18.4.10	Time Reversal	$\mathcal{DTFT}\{x(-n)\} = X(-\Omega) = X^*(\Omega)$
Property 18.4.11	Frequency Shift	$\mathcal{DTFT}\{e^{j\Omega_0 n} x(n)\} = X(\Omega - \Omega_0)$
Property 18.4.12	Modulation	$\mathcal{DTFT}\{x(n)\cos(\Omega_0 n)\} = \dfrac{1}{2}[X(\Omega + \Omega_0) + X(\Omega - \Omega_0)]$

Table 18.5.4 Discrete-Time Fourier Transform Pairs

Time Signal		DTFT
$x(n) = 1$	$\leftrightarrow$	$\delta(\Omega)$
$\text{sgn}(n)$	$\leftrightarrow$	$\dfrac{2}{1 - e^{-j\Omega}}$
$u_s(n)$	$\leftrightarrow$	$\dfrac{1}{1 - e^{-j\Omega}} + \pi\delta(\Omega)$
$\delta(n)$	$\leftrightarrow$	1
$\delta(n - \kappa) \quad \kappa \in I$	$\leftrightarrow$	$e^{-j\Omega\kappa} \quad \kappa \in I$
$a^n u_s(n)$	$\leftrightarrow$	$\dfrac{e^{j\Omega}}{e^{j\Omega} - a}, \quad |a| < 1$
$e^{j\Omega_0 n}$	$\leftrightarrow$	$2\pi\delta(\Omega - \Omega_0)$
$\Pi_\kappa(n)$	$\leftrightarrow$	$\dfrac{\sin\left(\left(\kappa + \frac{1}{2}\right)\Omega\right)}{\sin\left(\dfrac{\Omega}{2}\right)}$
$\dfrac{BL}{\pi}\text{Sa}(BLn)$	$\leftrightarrow$	$\Pi\left(\dfrac{\Omega}{2BL}\right) \quad 0 \leq BL < \pi$

Table 18.5.4 Discrete-Time Fourier Transform Pairs *(continued)*

$$\cos(\Omega_0 n) \quad \leftrightarrow \quad \pi\left[\delta(\Omega+\Omega_0)+\delta(\Omega-\Omega_0)\right]$$

$$\sin(\Omega_0 n) \quad \leftrightarrow \quad j\pi\left[\delta(\Omega+\Omega_0)-\delta(\Omega-\Omega_0)\right]$$

$$\cos(\Omega_0 n+\theta) \quad \leftrightarrow \quad \pi\left[e^{-j\theta}\delta(\Omega+\Omega_0)+e^{j\theta}\delta(\Omega-\Omega_0)\right]$$

18.6 HOMEWORK FOR CHAPTER 18

Homework Problems for Section 18.1

18.1.1 Find the z-transform of the following signals and sketch the *ROC*.

(i) $(0.1)^n u_s(n) - 2^n u_s(-n-1)$ (v) $\{-1 \quad 2 \quad 1\}$
                                                                       $\uparrow$

(ii) $(0.8)^n u_s(n) - (2)^n u_s(-n-1)$ (vi) $\{-1 \quad 2 \quad 1\}$
                                                                          $\uparrow$

(iii) $(-0.2)^n u_s(n) + 5(0.5)^n u_s(-n-1)$ (vii) $\left[(0.3)^n \sin(1.2n)\right]u_s(n)$

(iv) $\{-1 \quad 2 \quad 1\}$ (viii) $2\delta(n-3) - 2\delta(n+3)$
             $\uparrow$

18.1.2 Suppose that the *ROC* of the signal $x(n)$ is an annulus bounded by the circles a and $1/a$. Is this signal a right-hand sequence? Left-hand sequence? Neither? Give an example of such a sequence.

18.1.3 For each of the following signals:
(a) Determine the parameter values for which the z-transform will exist.
(b) Find the z-transform.
(c) Plot the *ROC*.

(i) $x(n) = u_s(-n-1) + a^n u_s(n)$

(ii) $x(n) = -b^n u_s(-n-1) + 0.5^n u_s(n)$

(iii) $x(n) = (0.3)^n u_s(n) - (0.4)^n u_s(-n-1) + (-.7)^n u_s(n)$

(iv) $x(n) = \left[(ja)^n \cos\left(\frac{\pi n}{2}\right) - j(jb)^n \sin\left(\frac{\pi n}{2}\right)\right]u_s(n)$

18.1.4 Find the z-transform of each of the following functions. Why are these z-transforms all the same?

(i) $x(n) = \left[\sin(4n)\right]u_s(n)$

(ii) $x(n) = \left[\sin((4+2\pi)n)\right]u_s(n)$

(iii) $x(n) = \left[\sin((2\pi-4)n)\right]u_s(n)$

18.1.5 Let the sequence $x(n)$ have a rational z-transform $X(z)$. For each of the following ROC's determine whether the z-transform must possess, may possess, or cannot possess one of the poles $\{0.2, \ 0.4, \ 0.5, \ 0.8\}$.

Determine whether the sequence $x(n)$ is a right-hand, left-hand, or two-sided sequence, if possible.

(i) $ROC = \{0.4 < \lvert z \rvert < 0.5\}$	(v) $ROC = \{\lvert z \rvert < 0.8\}$
(ii) $ROC = \{0.2 < \lvert z \rvert < 0.8\}$	(vi) $ROC = \{\lvert z \rvert < 0.2\}$
(iii) $ROC = \{0.4 < \lvert z \rvert < 0.5\}$	(vii) $ROC = \{\lvert z \rvert > 0.5\}$
(iv) $ROC = \{\lvert z \rvert > 0.8\}$	(viii) $ROC = \{\lvert z \rvert < 0.4\}$

Homework Problems for Section 18.2

18.2.1 Prove each of the z-transform properties in Table 18.5.1.

18.2.2 Using the linearity and time shift properties find the z-transform of the signal $x(n) = u_S(n) - u_S(n - M)$.

18.2.3 Prove the following properties of the z-transform.

(a) $\mathcal{Z}\{\bar{x}(n)\} = \bar{X}(\bar{z})$

(c) $\mathcal{Z}\{\operatorname{Im}(x(n))\} = \dfrac{1}{2j}\left[X(z) - \bar{X}(\bar{z})\right]$

(b) $\mathcal{Z}\{\operatorname{Re}(x(n))\} = \dfrac{1}{2}\left[X(z) + \bar{X}(\bar{z})\right]$

18.2.4 Show that the integral in (18.2.2) follows from Property 18.2.3.

18.2.5 Find the inverse z-transform of each of the following functions.

(i) $X(z) = \dfrac{z}{(z - 0.5)(z + 0.7)}, \quad ROC = \{0.5 < \lvert z \rvert < 0.7\}$

(ii) $X(z) = \dfrac{z}{(z - 0.5)(z + 0.7)}, \quad ROC = \{\lvert z \rvert < 0.5\}$

(iii) $X(z) = \dfrac{z}{(z - 0.5)(z + 0.7)}, \quad ROC = \{\lvert z \rvert > 0.7\}$

18.2.6 Find the inverse z-transform of

$$X(z) = \dfrac{z}{(z + 2)(z - 3)}$$

when the ROC is

(i) $ROC = \{\lvert z \rvert > 3\}$

(ii) $ROC = \{\lvert z \rvert < 2\}$

(iii) $ROC = \{2 < \lvert z \rvert < 3\}$

18.2.7   Find the inverse z-transform of each of the following functions.

(i) $X(z) = \dfrac{z^2}{z^2 - 0.36}$, $ROC = \{|z| > 0.6\}$

(ii) $X(z) = \dfrac{z^2}{(1 - az)(z - a)}$, $ROC = \left\{ a < |z| < \dfrac{1}{a} \right\}$

(iii) $X(z) = 2 + 3z^{-1} - z^{-2} - 4z^{-3}$, $ROC = \{\text{entire } z \text{ plane except } z = 0\}$

(iv) $X(z) = 2 + 3z^{1} - z^{2} - 4z^{3}$, $ROC = \{\text{entire } z \text{ plane except } z = \infty\}$

(v) $X(z) = \dfrac{z}{(z - 0.5)(z - 0.25)}$, $ROC = \{|z| > 0.5\}$

(vi) $X(z) = \dfrac{z}{(z - 0.5)(z - 0.25)}$, $ROC = \{0.25 < |z| < 0.5\}$

(vii) $X(z) = \dfrac{z^2}{(z + 0.5)(z - 0.5)(z - 1)}$, $ROC = \{0.25 < |z| < 1\}$

(viii) $X(z) = \dfrac{z^2}{(z^2 + 1)(z - 1)}$, $ROC = \{|z| > 1\}$

18.2.8   Consider the signal

$$x_1(n) = \begin{cases} a^n, & n \ge 0 \\ a^{-n}, & n < 0 \end{cases}$$

Define the signal

$$y(n) = x_1(n)x_2(n)$$

(a) What assumptions must be made about the ROC of $x_2(n)$ so that the z-transform of $y(n)$ exists?   Is $x_2(n)$ a right-hand sequence?   Left-hand sequence?   Two-sided sequence?

(b) Suppose

$$x_2(n) = nu_s(n).$$

Find $Y(z)$ if possible.   If not, why not?

18.2.9   Consider the signals

$$x_1(n) = \{1, 2, 3\}, \quad x_2(n) = \{-1, 0, 1, 2\}$$

(a) Find the convolution sum of these two signals using the time domain summation.

(b) Find the z-transform of these two signals.   Find the convolution of these two signals by multiplying the two z-transforms and taking the inverse z-transform.

Note: This calculation shows why the MATLAB command **conv** is used both for convolution and polynomial multiplication.

Homework Problems for Section 18.3

18.3.1 Suppose that $x(n)$ is a right-hand sequence with exactly κ nonzero values in negative time.
(a) Modify the right shift property of the one-sided z-transform to accommodate signals of this type.
(b) Using the results of (a) show that:
$$\kappa = 1: \quad \mathcal{Z}\{x(n-1)\} = z^{-1}X(z) + x(-1)$$
$$\kappa = 2: \quad \mathcal{Z}\{x(n-2)\} = z^{-2}X(z) + z^{-1}x(-1) + x(-2)$$
(c) Use this modified property to solve the following difference equation.

$$y(k) = -3y(k-1) - 2y(k-2) + x(k-2),$$
$$y(-1) = 1, \ y(-2) = 2, \text{ and } x(k) = \delta(k)$$

18.3.2 For each of the following systems find $y(n)$.
(i) $\quad y(n+2) + 2y(n+1) + 3y(n) = u_s(n)$

$\quad\quad y(0) = 2, \quad y(1) = 1$
(ii) $\quad y(n) = y(n-1) - y(n-2) + x(n) - x(n-1),$

$\quad\quad y(0) = y(-1) = 0, \quad x(n) = \delta(n)$
(iii) $\quad y(n) - 2.5y(n-1) + 2y(n-2) - 0.5y(n-3) = \delta(n) - \delta(n-1) + 0.5\delta(n-2)$

$\quad\quad y(0) = y(-1) = y(-2) = y(-3) = 0$
(iv) $\quad 4y(n+2) - 4y(n+1) + 4y(n) = x(n)$

$\quad\quad y(1) = -1, \ y(0) = 0, \quad x(n) = \delta(n)$

18.3.3 Prove the right and left shift properties of the one-sided z-transform.

18.3.4 Consider the two discrete functions $x_1(n) = 3^n u_s(n)$ and $x_2(n) = 6^n u_s(n)$.
(a) Find the z-transform of each function.
(b) Draw the *ROC* of each function in the same z-plane.
(c) What is the *ROC* of $x(n) = x_1(n) + x_2(n)$?

18.3.5 Suppose the right-hand sequence $x(n)$ has a rational z-transform with two poles. This sequence has the properties that

$$x(0) = 0.3 \text{ and } x(\infty) = 0.5$$

(a) Find $X(z)$.
(b) Find $x(n)$. Confirm the properties above.

18.3.6   Consider the signal

$$x(n) = \begin{cases} a^n u_s(n), & n \text{ even} \\ b^n u_s(n), & n \text{ odd} \end{cases}$$

Find the z-transform of this signal using the definition.

18.3.7   Consider the difference equation

$$y(n+2) + 0.4y(n+1) - 0.21y(n) = x(n+1) + 0.5x(n)$$

(a) If $x(n) = 0$ and the initial conditions are $y(0) = -2$ and $y(1) = 1$, find $y(n)$.

(b) Find $y(n)$ if the initial conditions are zero and

$$x(n) = 3(-0.9)^n u_s(n).$$

Note: This problem shows that the solution of a difference equation can be decomposed into a component due to the initial conditions and a component due to the forcing function.

Homework Problems for Section 18.4

18.4.1   Prove each of the DTFT properties in Table 18.5.3.

18.4.2   Determine the DTFT of the following functions.
(i) $u_s(-n-1)$ (iii) $\sin(\Omega_0 n + \theta)$
(ii) $\text{sgn}(n)$
                                       (iv) $\left[\cos^2\left(\frac{\pi n}{10}\right)\right]^k, \quad k = 0,1,2$

18.4.3   Can the z-transform be used to find the DTFT of the following functions? Why? If so, find the DTFT.
(i) $\cos(\Omega_0 n) u_s(n)$ (iii) $na^n u_s(n)$

(ii) $a^n \cos(\Omega_0 n) u_s(n)$ (iv) $-b^n u_s(-n-1)$

18.4.4   Let $x(n)$ be a real signal with a DTFT of $X(\Omega)$. Since the DTFT is a complex function, we can write it as

$$X(\Omega) = X_r(\Omega) + jX_i(\Omega).$$

where $X_r(\Omega)$ and $X_i(\Omega)$ are real functions.
(a) Show that

$$X_r(\Omega) = \sum_{n=-\infty}^{\infty} x(n)\cos(\Omega n), \quad \text{and} \quad X_i(\Omega) = -\sum_{n=-\infty}^{\infty} x(n)\sin(\Omega n)$$

(b) Show that $X_r(\Omega)$ is an even function and $X_i(\Omega)$ is an odd function.

(c) Show that

$$x(n) = \frac{1}{\pi} \int_0^\pi \left(X_r(\Omega) \cos(\Omega n) - X_i(\Omega) \sin(\Omega n) \right) d\Omega$$

(d) Suppose that $x(n)$ is an even function. Show that

$$X_r(\Omega) = x(0) + 2 \sum_{n=1}^\infty x(n) \cos(\Omega n)$$

$$X_i(\Omega) = 0$$

$$x(n) = \frac{1}{\pi} \int_0^\pi X_r(\Omega) \cos(\Omega n) \, d\Omega$$

(e) Suppose that $x(n)$ is an odd function. Show that

$$X_r(\Omega) = 0$$

$$X_i(\Omega) = -2 \sum_{n=1}^\infty x(n) \sin(\Omega n)$$

$$x(n) = -\frac{1}{\pi} \int_0^\pi X_i(\Omega) \sin(\Omega n) \, d\Omega$$

18.4.5 Consider the signal

$$x(n) = \begin{cases} \dfrac{\Omega_c}{\pi}, & n = 0 \\ \dfrac{\sin(\Omega_c n)}{\pi n}, & n \neq 0 \end{cases}$$

Define the partial sums

$$X_{-K:K}(\Omega) = \sum_{n=-K}^K x(n) e^{j\Omega n}$$

Using MATLAB plot several of these partial sums. Note that there is a constant overshoot near the discontinuity of $X(\Omega)$.

18.4.6 Consider the signals

$$x_1(n) = u_s(n) - u_s(n-5), \quad x_2(n) = \cos(\Omega_0 n)$$

Using these two signals define the signal

$$y(n) = \sum_{k=-\infty}^{\infty} x_1(n-k)x_2(k)$$

(a) Find the DTFT of $x_1(n)$ and $x_2(n)$.
(b) Find the DTFT of $y(n)$.
(c) For what frequencies Ω_0 will $y(n) = 0$ for all n?

18.4.7 Find the DTFT from the following one-sided z-transforms, if possible. If not possible, give a reason.

(i) $X(z) = \dfrac{z}{(z-0.5)(z+0.7)}$

(ii) $X(z) = \dfrac{z}{z-1}$

(iii) $X(z) = \dfrac{0.42z}{z^2 - 0.85z + 0.36}$

(iv) $X(z) = \dfrac{0.3z}{(z-0.3)^2}$

(v) $X(z) = \dfrac{z}{(z+1.2)(z+0.7)}$

(vi) $X(z) = \dfrac{z^4 - 1}{z^3(z-1)}$

Chapter 19

Sampling

In Chapter 17 we introduced the concept of sampling. We also briefly described analog-to-digital converters and digital-to-analog converters, electronic devices which sample an electronic signal and reconstruct a signal from its samples. In this chapter we will explore this concept, sampling, in more depth. We will explore the relationship between a signal and its samples by developing the conditions under which it is possible to exactly reconstruct a signal from its samples. It is straightforward to extract the sampled signal from the original continuous-time signal. The exact relationship between the sampled version of the signal and the signal itself is more complicated, however. In this chapter we will give a complete discussion of the relationship between a signal and its samples.

In this chapter we address the following question: "Can we recover a continuous-time signal from its samples?" The answer to this question is contained in the

Nyquist sampling theorem. We will fully develop the concepts behind this theorem. This theorem will tell us what information is lost in the sampling process, and what information about the original signal can be extracted from the sample version of the signal. Hence, we can characterize those signals which can be exactly reconstructed from their samples. We can also characterize how the sampled signal is corrupted by those signal components which can't be reconstructed (aliasing). This characterization explains how we can filter a signal before sampling so that the signal components of interest can be recovered from the (filtered) signal samples. These results are fundamental to the understanding of the computer processing of sampled signals.

Nyquist's sampling theorem suggests how we can exactly reconstruct a signal from its samples (when such reconstruction is possible). This reconstruction formula is generally not practical for the real-time reconstruction of a signal from its samples. Usually a DAC is used, which in signals and systems terminology is a zero-order hold (ZOH). In this chapter we develop a transfer function for a ZOH. This transfer function allows us to evaluate the distortion that is introduced by a ZOH when it is used to reconstruct a signals from its samples. This analysis also shows us how to design a prefilter or postfilter to remove the distortion from the reconstructed signal.

Summary of Sections

Section 19.1:   We introduce the basic continuous-time model of a sampled signal and derive the Fourier transform of this signal.

Section 19.2:   We develop the concept that allows us to exactly reconstruct a signal from its samples.

Section 19.3:   We discuss aliasing and the Nyquist sampling theorem.

Section 19.4:   We discuss the zero-order hold as a practical method for reconstructing a signal from its samples.

Section 19.5:   We present an example illustrating the concepts developed in this chapter.

Section 19.6:   Chapter summary section.

Coverage of the Text

This chapter requires the definition of a discrete-time signal and system from Chapter 17. The development of the results in this chapter require the Fourier transform from Chapter 7 and the concept of the amplitude and phase spectrum (spectral content) of a signal developed in Section 8.1. Section 19.4 requires the concept of transfer function from Section 10.1. Some reference is made to the simple concepts of filtering from Chapter 14. The discussion of anti-aliasing and smoothing filters does require an understanding of filters from Chapter 15, but these concepts are not essential to the main results of the chapter.

19.1 FOURIER TRANSFORM OF A SAMPLED SIGNAL

19.1.1 Introduction

In this chapter we will develop the conditions under which it is possible to reconstruct a signal from its samples. These conditions are given in the Nyquist sampling theorem. The fundamental concept behind this theorem is contained in the frequency domain description of the sampled signal. To develop this idea we first require a continuous-time model of a sampled signal. Then we need the Fourier transform of the sampled signal. Both of these results are developed in this section.

We begin by reviewing the definition of sampling and the associated terminology from Section 17.2.

Definition 17.2.1: Let $x(t)$ be a continuous-time signal and let $T_S > 0$ be a fixed number. From $x(t)$ we derive the discrete signal

$$\hat{x}(n) = x(nT_S) \quad \text{for} \quad n = \ldots - 2, -1, 0, 1, 2, \ldots$$

The number $T_S > 0$ in (19.1.31) is called the <u>sample period</u>. The <u>sampling frequency</u> f_S in hertz is defined as

$$f_S = \frac{1}{T_S}.$$

In radians the sampling frequency is $\omega_S = 2\pi f_S$. The time $t = nT_S$ is called a <u>sample instant</u>. The interval $nT_S \le t < (n+1)T_S$ is called the <u>sample interval</u>. The numbers $x(nT_S)$ are called the <u>sample values</u> of $x(t)$. The discrete signal $\hat{x}(n)$ is called the <u>discrete-time sampled version</u> of $x(t)$. The process of extracting the sample values from $x(t)$ is called <u>sampling</u>.

▲▲

Assumption: Whenever we analyze sampled signals and systems in this text, first we will select the sample period and fix it. The sample period will then remain fixed throughout the analysis. This assumption is essential to the derivation of most of the results presented below.

19.1.2 Ideal Sampling

In order to analyze a sampled signal we must develop a representation of this signal that is amenable to our analysis tools, namely the Fourier transform. Next we introduce a representation of a sampled signal that uses continuous-time functions to which Fourier analysis can be applied. An alternative viewpoint is to represent the sample signal as a discrete signal and apply the DTFT. This approach will be developed in Chapter 20.

The continuous-time model of the sampled signal requires the following function. Given the sample period T_S define the function

$$\sigma(t) = \sum_{n=-\infty}^{\infty} \delta(t - nT_s). \tag{19.1.1}$$

Multiplying the signal $x(t)$ by the function (19.1.1) we obtain

$$x*(t) = x(t)\sigma(t) = \sum_{n=-\infty}^{\infty} x(t)\delta(t - nT_s) = \sum_{n=-\infty}^{\infty} x(nT_s)\delta(t - nT_s). \tag{19.1.2}$$

In (19.1.2) we have used the sifting property of the impulse function, Remark 2.2.2. The model of the sampled signal in which we are interested is given in (19.1.2).

Definition 19.1.1: The signal $x*(t)$ in (19.1.2) is called the <u>continuous-time sampled version</u> of the continuous-time signal $x(t)$.

▲▲

The key to the model of the sampled signal is the function in (19.1.1).

Definition 19.1.2:   The function $\sigma(t)$ in (19.1.1) is called the <u>ideal sampling function</u>.

▲▲

Notation: The * notation attached to the signal means that $x*(t)$ is the continuous-time sampled version of $x(t)$.

The sampling process in (19.1.2) can be represented as a multiplication of the sampling function $\sigma(t)$ with the signal $x(t)$ as shown in Figure 19.1.1.   The construction of the continuous-time sampled version of $x(t)$ is shown in Figure 19.1.2. The continuous-time signal and its sampled version are shown Figure 19.1.2a. The ideal sampling function is shown in Figure 19.1.2b. When this function is multiplied by $x(t)$ in Figure 19.1.2a, we obtain the continuous-time sampled version of $x(t)$ in Figure 19.1.2c.

19.1.3 Fourier Transform of a Continuous-Time Sampled Signal

The analysis of sampling proceeds by deriving the Fourier transform of the continuous-time sampled version of $x(t)$ in (19.1.2).   Since the sampled signal in (19.1.2) is a continuous-time signal, not a discrete-time signal, we can take its Fourier transform.   To derive this Fourier transform, we simply take the Fourier transform of the continuous-time sampled signal.   We have

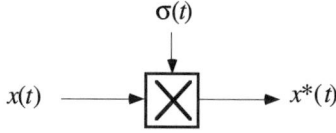

Figure 19.1.1 A Model of Sampling Process

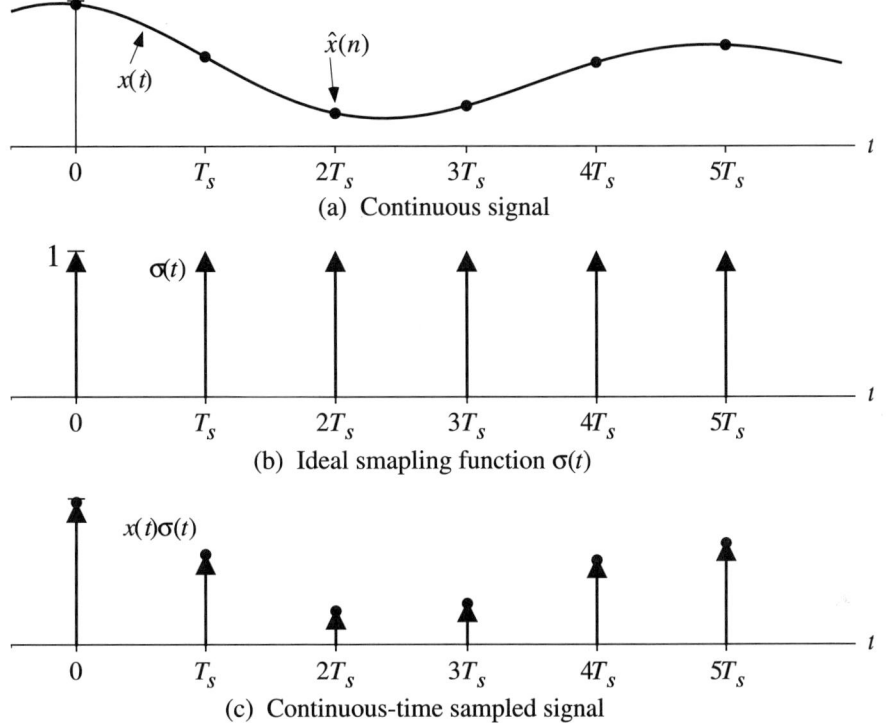

(a) Continuous signal

(b) Ideal smapling function $\sigma(t)$

(c) Continuous-time sampled signal

Figure 19.1.2 Continuous-Time Model of a Sampled Signal

$$\mathcal{F}\{x*(t)\} = X^*(\omega) = \mathcal{F}\left\{\sum_{n=-\infty}^{\infty} x(nT_s)\delta(t-nT_s)\right\} = \sum_{n=-\infty}^{\infty} x(nT_s)e^{-j\omega T_s n} \qquad (19.1.3)$$

where we have used Property 7.5.12 of the Fourier transform

$$\mathcal{F}\{x(nT_s)\delta(t-nT_s)\} = x(nT_s)e^{-j\omega T_s n}. \qquad (19.1.4)$$

The expression in (19.1.3) is sometimes used, but there is another expression for this Fourier transform which is more useful. This second expression is derived from the ideal sampling function. The ideal sampling function (19.1.1) is a periodic function with a Fourier series

$$\sigma(t) = \sum_{n=-\infty}^{\infty} S_n e^{jn\omega_s t}, \quad \omega_s = \frac{2\pi}{T_s} \qquad (19.1.5)$$

where the coefficients S_n are given by

$$S_n = \frac{1}{T_s} \int_{-\frac{T_s}{2}}^{\frac{T_s}{2}} \delta(t) e^{-jn\omega_s t} \, dt = \frac{1}{T_s}. \qquad (19.1.6)$$

The period of the Fourier series (19.1.5) is exactly the sampling period T_s. Using (19.1.5) in (19.1.2) leads to

$$x*(t) = x(t)\sigma(t) = \sum_{n=-\infty}^{\infty} x(t) S_n e^{jn\omega_s t} = \frac{1}{T_s} \sum_{n=-\infty}^{\infty} x(t) e^{jn\omega_s t}. \qquad (19.1.7)$$

The Fourier transform of (19.1.7) is

$$\mathcal{F}\{x*(t)\} = \int_{-\infty}^{\infty} \left[\sum_{n=-\infty}^{\infty} x(t) S_n e^{jn\omega_s t} \right] e^{-j\omega t} \, dt \qquad (19.1.8)$$

$$= \sum_{n=-\infty}^{\infty} S_n \int_{-\infty}^{\infty} x(t) e^{-j(\omega - n\omega_s)t} \, dt.$$

In the last line we recognize the definition of the Fourier transform of $x(t)$ (Definition 7.4.1) for a shifted frequency scale ($\omega - n\omega_s$), i.e.,

$$\int_{-\infty}^{\infty} x(t) e^{-j(\omega - n\omega_s)t} \, dt = X(\omega - n\omega_s). \qquad (19.1.9)$$

Substituting (19.1.9) into (19.1.8) and using (19.1.6) we have established the following theorem.

Theorem 19.1.3: Let $x(t)$ be a continuous-time signal with Fourier transform $X(\omega)$. If the sample period is T_s, then the Fourier transform of the sampled signal $x*(t)$ is

$$X*(\omega) = \mathcal{F}\{x*(t)\} = \sum_{n=-\infty}^{\infty} \frac{1}{T_s} X(\omega - n\omega_s) = \sum_{n=-\infty}^{\infty} x(nT_s) e^{-j\omega T_s n}.$$

▲▲

Theorem 19.1.3 gives the Fourier transform of the *continuous-time* sampled version of the signal $x(t)$. This theorem relies on a representation of the sampled signal in continuous-time. If the sampled signal is represented in discrete-time, then we can derive the *discrete-time* Fourier transform of the sampled signal. The DTFT of the sampled signal is closely related to the result of Theorem 19.1.3. This relationship is established in Section 20.3.

Theorem 19.1.3 plays a central role in the analysis in the rest of this chapter. It is important that this result be fully understood. The following example begins the interpretation of Theorem 19.1.3.

Example 19.1.4: Let $x(t)$ be a bandlimited signal. The amplitude spectrum of $x(t)$, $|X(\omega)|$, shown in Figure 19.1.3a.

Note: The signal $x(t)$ in Figure 19.1.3a is a bandlimited signal (Definition 8.3.4) because its amplitude spectrum is zero for frequencies $|\omega| > BL$. Many of the results below including the Nyquist sampling theorem are derived for bandlimited signals, one example of which is shown in Figure 19.1.3a. The shape of the amplitude spectrum is not important; it is important that the amplitude spectrum is zero for frequencies $|\omega| > BL$.

The sampling frequency is also identified in Figure 19.1.3a. In this example we have assumed a specific relationship between the bandlimit BL and the sampling frequency $\omega_s/2 > BL$. This relationship was selected to expose the fundamental character of the Fourier transform of the sampled signal. We will study the opposite relationship $\omega_s/2 < BL$ when we discuss aliasing in Section 19.3.

The Fourier transform of $x^*(t)$, $X^*(\omega)$, is shown in Figure 19.1.3b where the $n = -1, 0, 1$ terms in the sum are identified. The $n = 0$ term $X(\omega)$ in the summation is a copy of the amplitude spectrum scaled by T_s and centered at the origin as shown in Figure 19.1.3b. The $n = 1$ term, $X(\omega - \omega_s)$, is a copy of the amplitude spectrum scaled by T_s and centered at the sampling frequency ω_s. The $n = -1$ term, $X(\omega + \omega_s)$, is a copy of the amplitude spectrum scaled by T_s and centered at the sampling frequency $-\omega_s$. And so on.

▲▲

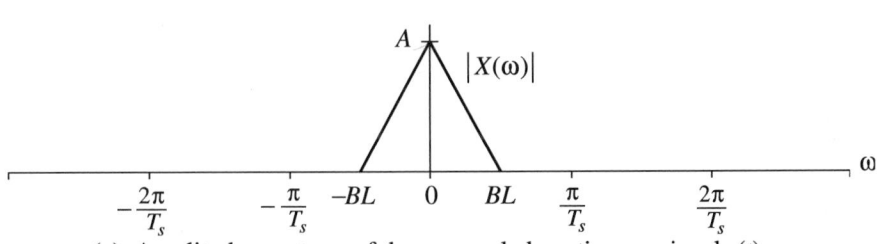

(a) Amplitude spectrum of the unsampled continuous signal $x(t)$

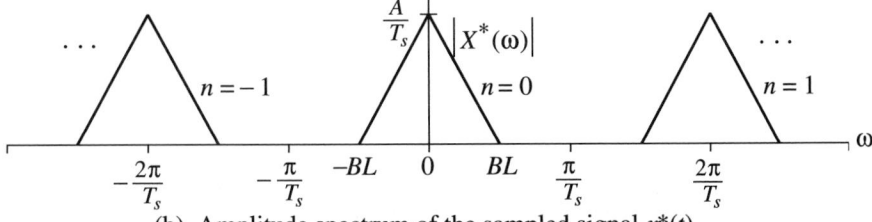

(b) Amplitude spectrum of the sampled signal $x^*(t)$

Figure 19.1.3 Amplitude Spectrum of a Signal and Its Continuous-Time Sampled Version

19.2 RECONSTRUCTION OF SIGNALS FROM THEIR SAMPLES

Suppose we are given the continuous-time version of a sampled signal. The question with which we are concerned in this chapter is "Can we uniquely recover the original signal from its samples?" If the answer is yes, as we will assume in this section, then we face the next question "How can we *exactly* reconstruct the original signal from its samples?" The answer to this question is not obvious. In this section we will answer this question using the Fourier transform of the continuous-time sampled version of the signal developed in the last section.

Let $x(t)$ be a bandlimited signal with Fourier transform $X(\omega)$ and let T_s be the sample period. Then from Theorem 19.1.3 the continuous-time sampled version of this signal is given by

$$x^*(t) = \sum_{n=-\infty}^{\infty} x(nT_s)\delta(t - nT_s) \tag{19.2.1}$$

and its Fourier transform is

$$X^*(\omega) = \sum_{n=-\infty}^{\infty} f_s X(\omega - n\omega_s). \tag{19.2.2}$$

We propose to develop a formula for reconstructing $x(t)$ from $x^*(t)$ by working with the Fourier transform of $x^*(t)$ in (19.2.2) rather than the sampled signal directly. Let $H_{con}(\omega)$ be a Fourier transform that will be defined below. Multiplying $X^*(\omega)$ by $H_{con}(\omega)$ we obtain

$$X_{con}(\omega) = H_{con}(\omega)X^*(\omega). \tag{19.2.3}$$

The inverse Fourier transform of $X_{con}(\omega)$, $x_{con}(t)$, is a new signal which we will show is equal to $x(t)$ under certain assumptions. By taking the inverse Fourier transform of (19.2.3) we will obtain a formula for reconstructing $x(t)$ from its samples.

The analysis now proceeds by identifying an appropriate Fourier transform $H_{con}(\omega)$. Before defining this Fourier transform, however, we give a system interpretation to (19.2.3). If we define $X^*(\omega)$ as the input signal and $X_{con}(\omega)$ as the output signal, then $H_{con}(\omega)$ is the Fourier transfer function of a system which relates these two signals. This relationship is visualized in Figure 19.2.1. The system interpretation of (19.2.3) motivates the terminology of the following definition of the Fourier transform $H_{con}(\omega)$.

$$x^*(t) = \sum_{n=-\infty}^{\infty} x(nT_s)\delta(t - nT_s) \longrightarrow \boxed{H_{con}(\omega)} \longrightarrow x_{con}(t)$$

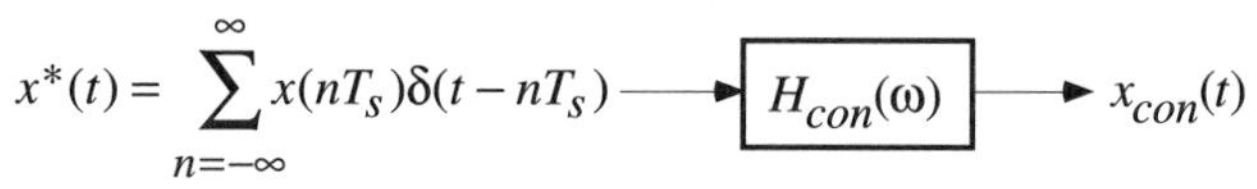

Figure 19.2.1 Reconstruction of a Signal from Its Samples

Definition 19.2.1:   Let $T_s > 0$ be   the   sampling   period.   An <u>ideal   lowpass</u> <u>reconstruction filter</u> $H_{con}(\omega)$ is defined as

$$|H_{con}(\omega)| = \begin{cases} T_s, & |\omega| \leq \dfrac{\pi}{T_s} = \dfrac{\omega_s}{2} \\ \\ 0, & |\omega| > \dfrac{\pi}{T_s} = \dfrac{\omega_s}{2} \end{cases}$$

$$\angle H(\omega) = 0, \quad \text{for all } \omega.$$

▲▲

The amplitude spectrum of $H_{con}(\omega)$ is shown in Figure 19.2.2.   On the interval $-\dfrac{\omega_s}{2} < \omega < \dfrac{\omega_s}{2}$ the gain is T_s.   Outside this frequency interval the amplitude spectrum is zero.   The Fourier transform $H_{con}(\omega)$ is completely defined by the sampling period T_s.

Terminology: If we interpret $H_{con}(\omega)$ as a transfer function of a system, then the graph in Figure 19.2.2 is the magnitude of the frequency response function.   This frequency response function has the same shape as a lowpass filter (Definition 14.6.2).   Furthermore, the stopband is identically zero.   Hence, the adjective *ideal*.

This system terminology is for the convenience of the description of the mathematical development that follows.   We could just speak about multiplying Fourier transforms.   At no time do we propose to build this filter.   This filter is for analysis only.

   Having defined the Fourier transform $H_{con}(\omega)$ in Definition 19.2.1, we can proceed to the reconstruction of the signal from its samples.   Our analysis is carried out in the frequency domain, and it is illustrated in Figure 19.2.3.   The Fourier transform of a signal $x(t)$ is shown in Figure 19.2.3a.   In this section we will confine ourselves to signals which have this type of Fourier transform, i.e., signals that are bandlimited and their bandlimit is related to the sampling frequency as shown in Figure 19.2.3a.   It turns out that these signals can be exactly reconstructed from their samples.   The general case is discussed in the next section.

   The Fourier transform of the continuous-time sampled signal, (19.2.2), is shown in Figure 19.2.3b.   The construction of this amplitude spectrum is discussed in the last section in Figure 19.1.3.

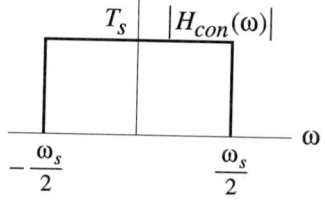

Figure 19.2.2 Amplitude Spectrum of $H_{con}(\omega)$

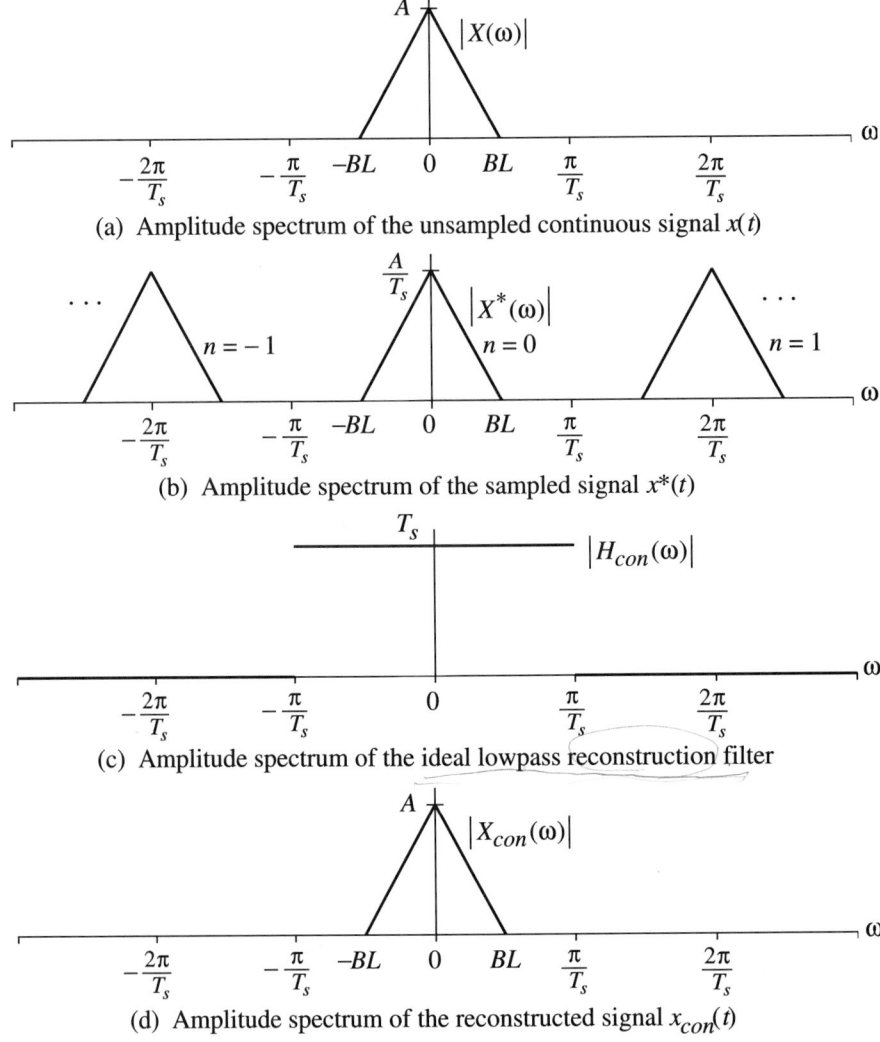

Figure 19.2.3 Reconstruction of a Signal from Its Sampled Version

The signal $X_{con}(\omega)$ is obtained from (19.2.3). In particular, the amplitude spectrum of this signal is given by

$$|X_{con}(\omega)| = |H_{con}(\omega)||X*(\omega)|. \tag{19.2.4}$$

This amplitude spectrum can be obtained graphically from Figure 19.2.3. The amplitude spectrum of $H_{con}(\omega)$ is shown in Figure 19.2.3c. Multiplying this graph by the graph of $|X*(\omega)|$ in Figure 19.2.3b we obtain the graph of $|X_{con}(\omega)|$ in Figure

19.2.3d. Since the amplitude spectrum of $H_{con}(\omega)$ is zero for frequencies greater than $|\pi/T_s|$, only the $n = 0$ term of $|X^*(\omega)|$ is retained in the graph of $|X_{con}(\omega)|$. Also note that the amplitude of $|X^*(\omega)|$ is multiplied by T_s when it is passed through the ideal lowpass reconstruction filter so that the peak amplitude of $|X_{con}(\omega)|$ is the same as the peak amplitude of $|X(\omega)|$.

Figure 19.2.3 explains the amplitude relationship in (19.2.4). The phase relationships in (19.2.3) are trivial. Since the phase of $H_{con}(\omega)$ is zero, the phase of $X_{con}(\omega)$ is the same as the phase of $X^*(\omega)$.

We can summarize the analysis of (19.2.3) and Figure 19.2.3 as follows. Comparing Figure 19.2.3a and 19.2.3d, we see that the amplitude spectrum of the signal $x(t)$ and $x_{con}(t)$ are the same. The phase spectra of $x(t)$ and $x_{con}(t)$ are also the same. We conclude that the Fourier transforms of the signals $x(t)$ and $x_{con}(t)$ are the same. Hence, $x(t)$ and $x_{con}(t)$ must be the same signal. Hence, Figure 19.2.3 is a frequency domain description of the ideal reconstruction of a signals from its samples.

We can obtain a formula for reconstructing the signal $x_{con}(t)$ from the sampled version of $x(t)$ from the Fourier transforms in (19.2.3). Using the convolution property of Fourier transforms, Property 7.5.7, the inverse Fourier transform of (19.2.3) is given by

$$x_{con}(t) = \int_{-\infty}^{\infty} h_{con}(t-\lambda)x^*(\lambda)\, d\lambda. \tag{19.2.5}$$

The impulse response function of the ideal lowpass reconstruction filter is

$$\mathcal{F}^{-1}\left\{H_{con}(\omega)\right\} = h_{con}(t) = \text{sinc}\left(\frac{t}{T_s}\right). \tag{19.2.6}$$

Using (19.2.6) and (19.2.1) in (19.2.5) we obtain the formula

$$x_{con}(t) = \sum_{n=-\infty}^{\infty} x(nT_s)\left[\text{sinc}\left(\frac{t-nT_s}{T_s}\right)\right]. \tag{19.2.7}$$

So knowing the sample values of $x(t)$, $x(nT_s)$, the original signal can be reconstructed via the infinite sum in (19.2.7). An example of the use of this formula is given in the last section.

Terminology: Interpolation is the fitting of a continuous function to a given set of sample values. The formula in (19.2.7) is called <u>bandlimited</u> interpolation because the interpolation function in (19.2.6) is derived from a bandlimited filter.

19.3 ALIASING AND THE NYQUIST SAMPLING THEOREM

19.3.1 Introduction

In this section we continue our discussion of the reconstruction of a signal $x(t)$ from its samples. We will use the continuous-time sampled version of $x(t)$

$$x^*(t) = \sum_{n=-\infty}^{\infty} x(nT_s)\delta(t - nT_s).$$

(19.3.1)

The Fourier transform of the sampled signal, as given in Theorem 19.1.3, is

$$X*(\omega) = \frac{1}{T_s} \sum_{n=-\infty}^{\infty} X(\omega - n\omega_s).$$

(19.3.2)

In the last section we derived conditions under which the signal could be exactly reconstructed from its samples. In this section we will show when exact reconstruction of the signal from its samples is not possible. When this analysis is combined with the analysis in the last section, we have Nyquist's sampling theorem, which we state at the end of this section.

19.3.2 A Sinusoid

To begin this discussion we will consider a sinusoidal signal

$$x(t) = \cos(\omega_0 t).$$

(19.3.3)

Let the sample period be T_s. The continuous-time sampled version of (19.3.3) is

$$x*(t) = \sum_{n=-\infty}^{\infty} \cos(\omega_0 t)\delta(t - nT_s) = \sum_{n=-\infty}^{\infty} \cos(\omega_0 T_s n)\delta(t - nT_s).$$

(19.3.4)

Suppose that the (continuous) frequency of (19.3.3) and the sample period satisfy

$$2\pi < \omega_0 T_s < 3\pi.$$

(19.3.5)

Define

$$0 < \tilde{\omega}_0 T_s < \pi, \quad \text{where} \quad \tilde{\omega}_0 T_s = \omega_0 T_s - 2\pi.$$

(19.3.6)

It is shown in Remark 2.3.10 that if the frequency of a discrete-time sinusoid satisfies (19.3.6), then

$$\cos(\omega_0 T_s n) = \cos(\tilde{\omega}_0 T_s n).$$

(19.3.7)

Using (19.3.7), (19.3.4) becomes

$$x*(t) = \sum_{n=-\infty}^{\infty} \cos(\omega_0 T_s n)\delta(t - nT_s) = \sum_{n=-\infty}^{\infty} \cos(\tilde{\omega}_0 T_s n)\delta(t - nT_s). \tag{19.3.8}$$

To reconstruct the signal $x(t)$ from its samples, the process of going from (19.3.3) to (19.3.4) is reversed. Comparing (19.3.8) to (19.3.4), we would reconstruct the continuous signal from (19.3.8) as

$$x_a(t) = \cos(\tilde{\omega}_0 t). \tag{19.3.9}$$

Comparing (19.3.9) to (19.3.3) we see that these two continuous sinusoidal signals have different frequencies yet their sampled versions are identical as shown in (19.3.8). The signal (19.3.9) is called an <u>alias</u> of the signal in (19.3.3). (See Definition 19.3.2 below.) The alias appears because of the property of discrete-time sinusoids demonstrated in (19.3.7).

Example 19.3.1: Consider the signal

$$x(t) = \cos(\omega_0 t) = \cos(12\pi t). \tag{19.3.10}$$

Suppose that the sample period is $T_s = 0.2$ sec so that the sampling frequency is $\omega_s = (2\pi/T_s) = 10\pi$ rad/sec. The sampled version of (19.3.10) is

$$x*(t) = \sum_{n=-\infty}^{\infty} x(nT_s)\delta(t - nT_s) = \sum_{n=-\infty}^{\infty} \left[\cos((12\pi)(0.2)n)\right]\delta(t - 0.2n). \tag{19.3.11}$$

The discrete frequency in (19.3.11) satisfies (19.3.5). That is,

$$\tilde{\omega}_0 = \omega_0 - 2\pi = 2.4\pi - 2\pi = 0.4\pi. \tag{19.3.12}$$

Hence, (19.3.7) holds. We have

$$\cos(\omega_0 T_s n) = \cos(12\pi(0.2)n) = \cos(2.4\pi n) = \cos(0.4\pi n) = \cos(\tilde{\omega}_0 T_s n). \tag{19.3.13}$$

Using (19.3.13) in (19.3.11) and we have

$$x*(t) = \sum_{n=-\infty}^{\infty} (\cos(2.4\pi n))\delta(t - 0.2n) = \sum_{n=-\infty}^{\infty} \left[\cos(0.4\pi n)\right]\delta(t - 0.2n). \tag{19.3.14}$$

Using the discrete frequency in the second summation in (19.3.14) to reconstruct a continuous sinusoid from the sampled signal in (19.3.14) we have

$$\tilde{\omega}_0 = \frac{0.4\pi}{T_S} = \frac{0.4\pi}{0.2} = 2\pi. \tag{19.3.15}$$

Finally, the reconstructed signal is

$$x_{con}(t) = \cos(2\pi t). \tag{19.3.16}$$

The signal in (19.3.16) is known as an alias. Figure 19.3.1 shows the original signal (19.3.10) along with the sample values and the alias (19.3.16). We can interpret the discussion above using the frequency domain concepts developed in the last section. Employing the ideal lowpass reconstruction filter we can restate the development above using the amplitude spectra of the various signals. The entire development is summarized in Figure 19.3.2. Note that Figure 19.3.2 is similar to Figure 19.2.3. The amplitude spectrum of $x(t)$ in (19.3.10) is shown in Figure 19.3.2a. The amplitude spectrum of sampled version of $x(t)$ is

$$X*(\omega) = \frac{1}{T_S} \sum_{n=-\infty}^{\infty} X(\omega - n\omega_s) \tag{19.3.17}$$

$$= \frac{1}{T_S} \sum_{n=-\infty}^{\infty} \left[\pi\big(\delta(\omega - 12\pi - 20\pi n) + \delta(\omega + 12\pi - 20\pi n)\big)\right].$$

The complete amplitude spectrum of this signal is shown in Figure 19.3.2b.[1]

In an attempt to reconstruct the signal $x(t)$ from its samples, we pass the sampled version of $x(t)$, $x*(t)$, through an ideal lowpass reconstruction filter as discussed in Section 19.2. This filter is shown in Figure 19.3.2b overlaid on $X*(\omega)$. The amplitude spectrum of the reconstructed signal $X_{con}(\omega)$ is shown in Figure 19.3.2c. Clearly, amplitude spectrum of the signal $X_{con}(\omega)$ in Figure 19.3.2c is not identical to the amplitude spectrum of the signal $x(t)$ in Figure 19.3.2a. Hence, the reconstructed signal in (19.3.16) is not the same as the original signal in

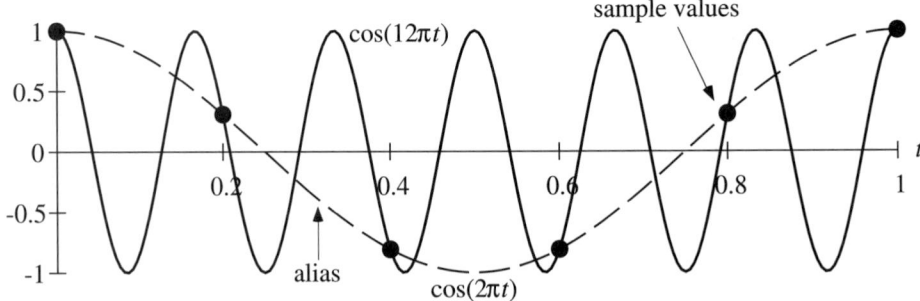

Figure 19.3.1 Aliasing in the Time Domain

---

[1] The reader is encouraged to verify this construction step by step.

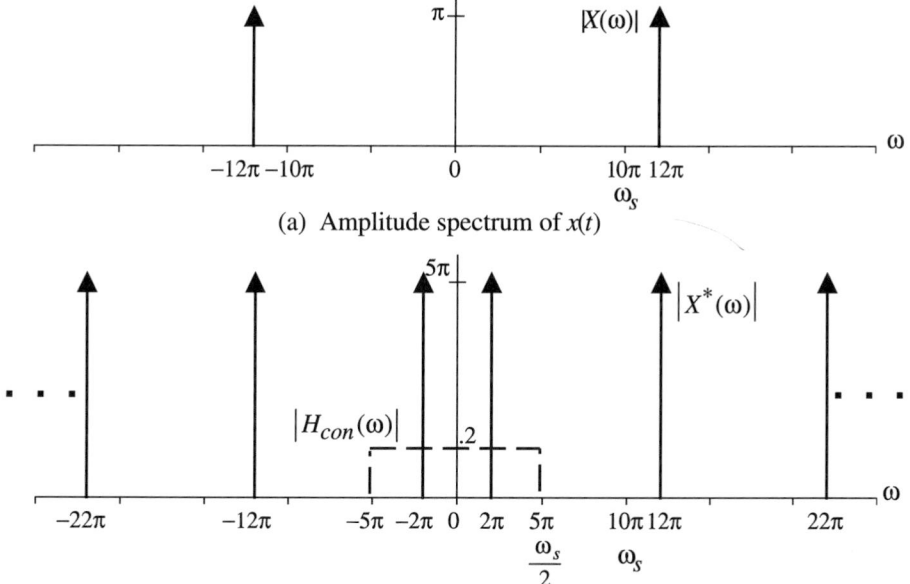

(a) Amplitude spectrum of $x(t)$

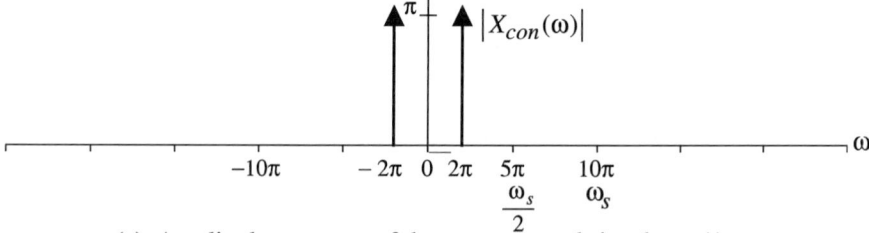

(b) Amplitude spectrum of $x^*(t)$ and the ideal lowpass reconstruction filter

(c) Amplitude spectrum of the reconstructed signal $x_{con}(t)$

Figure 19.3.2 Frequency Domain Interpretation of Aliasing

(19.3.10). In fact this amplitude spectrum corresponds to the alias signal in (19.3.16) and shown in Figure 19.3.1.

▲▲

19.3.3 General Case of Aliasing

Aliasing can occur also with signals that have continuous spectra. Again we appeal to frequency domain arguments using the ideal lowpass reconstruction filter developed in Section 19.2. Consider a signal whose amplitude spectrum is shown in Figure 19.3.3a.

Note: Figure 19.3.3 is similar to Figure 19.2.3. We assume that the Fourier transform is real.

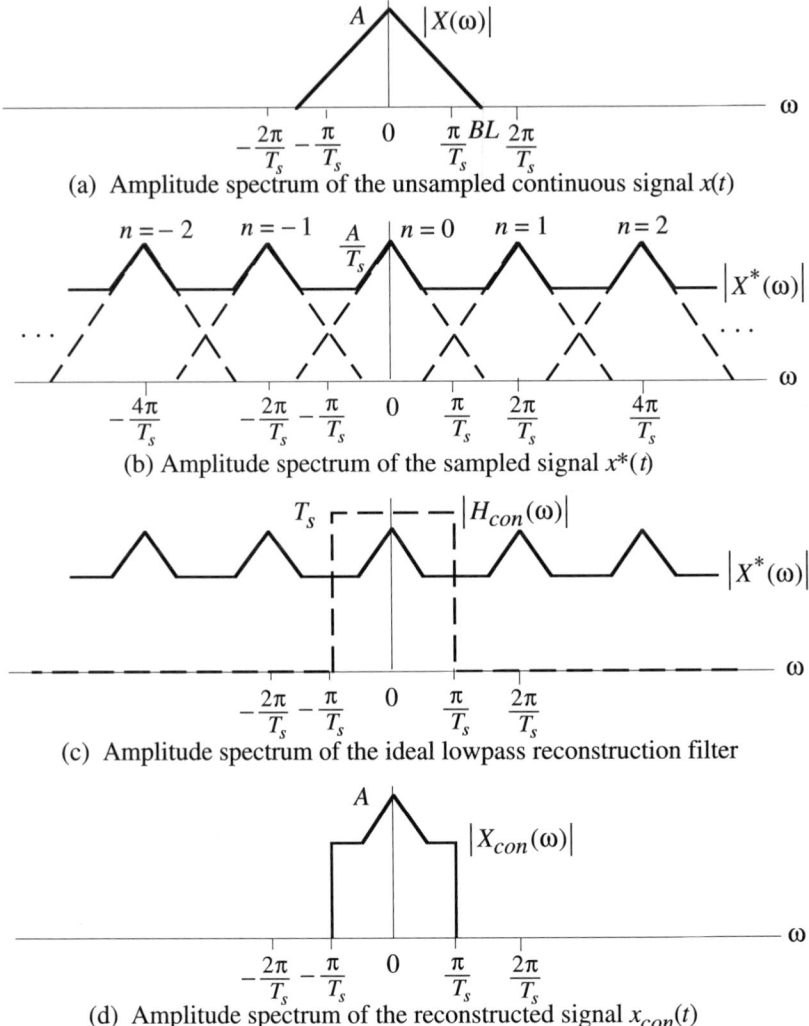

(a) Amplitude spectrum of the unsampled continuous signal $x(t)$

(b) Amplitude spectrum of the sampled signal $x*(t)$

(c) Amplitude spectrum of the ideal lowpass reconstruction filter

(d) Amplitude spectrum of the reconstructed signal $x_{con}(t)$

Figure 19.3.3 Analysis of a Signal Sampled Below the Nyquist Rate

The Fourier transform of the sampled signal is given in (19.3.2). Figure 19.3.3b shows the amplitude spectrum of the terms $n = -2,-1,0,1,2$ in the sum (19.3.2). Because of the relationship between the bandlimit and sampling frequency, the amplitude spectrum of each term overlaps, unlike Figure 19.2.3. For example, the shifted amplitude spectrum corresponding to $n = 1$ in (19.3.2) overlaps with the $n = 0$ term. This overlap is extracted from Figure 19.3.3b and shown in Figure 19.3.4.

Definition 19.3.2: The overlap of the Fourier transform of each of the terms of the sampled signal in (19.3.2) is called <u>aliasing</u>. The corresponding signal in the time domain caused by this overlap is also called an <u>alias</u>.

▲▲

On the frequency band $0 < \omega < 2\pi/T_s$ it appears as if the amplitude spectrum has been folded back and added to itself. Hence, the frequency content of the reconstructed signal will be changed on the region of overlap as shown in Figure 19.3.4. This idea can be extended to signals that are not bandlimited.

   The amplitude spectrum of the sampled version of $x(t)$, $X^*(\omega)$, is constructed by adding together all of the terms in Figure 19.3.3b. This amplitude spectrum is also shown in Figure 19.3.3b and in Figure 19.3.3c.

   Next we consider the reconstruction of a signal in the presence of aliasing. The amplitude spectrum of the ideal lowpass reconstruction filter is shown in Figure 19.3.3c. The amplitude spectrum of the reconstructed signal $x_{con}(t)$ shown in Figure 19.3.3d, is obtained by multiplying the two amplitude spectrums in Figure 19.3.3c. The amplitude spectrum of the reconstructed signal $x_{con}(t)$ does not match the amplitude spectrum of the original signal $x(t)$. In fact, the reconstructed signal is a distorted approximation of the signal $x(t)$. The amount of distortion is determined by the overlap between the copies of the amplitude spectrum of $x(t)$.

   If we sample any signal that is not bandlimited, then some aliasing will occur. Furthermore, real signals are not bandlimited due to the presence of noise. Often a signal is lowpass filtered before it is sampled to prevent aliasing.

19.3.4 Nyquist Sampling Theorem

The question of uniquely reconstructing a signal from its samples is essentially resolved by Figures 19.3.3 and 19.2.3. Two concepts play a central role in this analysis. First, the signal must be bandlimited. Let $x(t)$ be a signal with a bandlimit of BL. The second concept is aliasing. Aliasing is defined as the overlap between individual amplitude spectrums in the Fourier transform of $x^*(t)$. This overlap is determined by the relationship between the bandlimit and the sampling frequency. Consider again Figure 19.3.4. It is clear in this figure that the two amplitude spectra will not overlap if

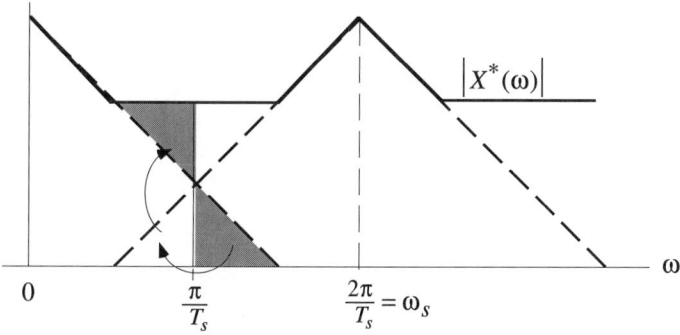

Figure 19.3.4 Aliasing

$$2BL < \omega_s = 2\pi f_s.\tag{19.3.18}$$

The bandlimit determines the lower limit on the sampling frequency so that aliasing doesn't occur. We formalize this relationship with a definition.

Notation: To state this definition, we translate the bandlimit in radians into hertz by defining $BL = 2\pi f_{bl}$.

Definition 19.3.3: Let $x(t)$ be a signal with a bandlimit of f_{bl} Hz. Then the frequency f_{ny}, defined as $2 f_{bl} = f_{ny}$ Hz, is called the <u>Nyquist sampling frequency</u> <u>for the signal $x(t)$</u>.

▲▲

The discussion above shows that if there is aliasing, the signal can't be uniquely reconstructed from its samples. On the other hand, if there is no aliasing, as in Figure 19.2.3, then the signal can be reconstructed uniquely using the formula (19.2.7). Now we can state Nyquist's theorem.

Theorem 19.3.4: **Nyquist Sampling Theorem** Let $x(t)$ be a signal with a Nyquist frequency of f_{ny} hertz. Then this signal can be uniquely reconstructed from its sample values if the sampling frequency is greater than the Nyquist frequency of the signal. That is, the sampling frequency f_s satisfies $f_{ny} < f_s$. In this case the signal $x(t)$ is uniquely determined by

$$x_{con}(t) = \sum_{n=-\infty}^{\infty} x(nT_s) \left[\operatorname{sinc}\left(\frac{t - nT_s}{T_s} \right) \right].$$

▲▲

The Nyquist frequency *for a signal* $x(t)$ is twice the bandlimit of the signal. In words, Theorem 19.3.4 says that in order to uniquely recover a signal from its samples, the signal must be sampled at a rate greater than twice its bandlimit.

The term "Nyquist sampling frequency" is also commonly used in reference to a given sampling frequency. Given the sampling frequency f_s, the Nyquist frequency is said to be $f_{ny} = f_s/2$. In this definition, the Nyquist frequency characterizes *all* signals that can be reconstructed if a particular signal is sampled at the frequency f_s.

Since both definitions of the Nyquist sampling frequency are routinely used, we introduce a second definition of this frequency. Definition 19.3.3 is referenced to a signal. Definition 19.3.5 is referenced to a sampling frequency.

Definition 19.3.5: Let f_s be a given sampling frequency. Then the frequency f_{ny}, defined as $\frac{1}{2} f_s = f_{ny}$ Hz, is called the <u>Nyquist sampling frequency for the sampling</u> <u>frequency f_s</u>.

▲▲

Terminology: In many practical systems, the signals are sampled above their Nyquist rate, to minimize the aliasing. The selection of the sampling frequency to be much greater than the Nyquist rate is called <u>oversampling</u>. If the signal is sampled at less than the Nyquist rate, the signal is said to be <u>undersampled</u>.

Example 19.3.6: Suppose that the amplitude spectrum of the signal $x_1(t)$ is as shown in Figure 19.3.5. The bandlimit of this signal is $f_{bl} = 10$ Hz. Therefore the Nyquist frequency is

$$f_{ny} = 2f_{bl} = 2(10) = 20 \text{ Hz.} \tag{19.3.19}$$

   Suppose that the magnitude of the Fourier transform of the signal $x_2(t)$ is as shown in Figure 19.3.6. The bandlimit of this signal is $f_{bl} = 100$ Hz. Therefore the Nyquist frequency is

$$f_{ny} = 2f_{bl} = 2(100) = 200 \text{ Hz.} \tag{19.3.20}$$

Terminology: In Figure 19.3.5 the bandlimit is equal to the bandwidth. Hence, the Nyquist frequency can be computed from the bandwidth. In Figure 19.3.6 the bandlimit is 100 Hz but the bandwidth of the signal is 50 Hz. This example shows that Theorem 19.3.4 must be expressed in terms of the "bandlimit." Nonetheless, this theorem is often stated with the word "bandwidth."

   The signal in Figure 19.3.6 is a bandpass signal. There exist alternative sampling techniques for reducing the sampling rate for a bandpass signal. The sampling rate using these schemes is approximately twice the bandwidth of the signal.

▲▲

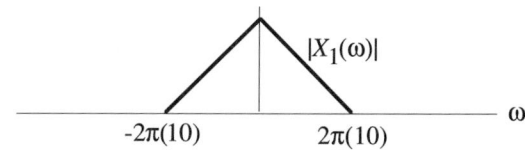

Figure 19.3.5 Amplitude Spectrum of $x_1(t)$

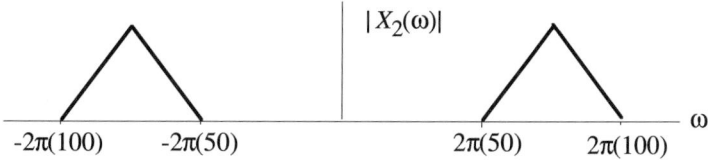

Figure 19.3.6 Amplitude Spectrum of $x_2(t)$

19.4 ZERO-ORDER HOLD

19.4.1 The Definition

Suppose we have the discrete-time sampled version $\hat{x}(n)$ of a signal $x(t)$ and we wish to reconstruct the original signal from its samples $\hat{x}(n) = x(nT_s)$. If the signal is sampled above its Nyquist rate, then the signal can be exactly reconstructed from its samples using the formula

$$x_{con}(t) = \sum_{n=-\infty}^{\infty} x(nT_s)\left[\text{sinc}\left(\frac{t - nT_s}{T_s}\right)\right]. \tag{19.4.1}$$

This formula, however, is rather cumbersome and not well suited to digital implementations. Observe that the sinc functions are defined over all time. Hence, to exactly reconstruct the signal at time t_0 we must know *all* of the sample values including those for $nT_s > t_0$. Clearly, this type of reconstruction is unacceptable for real-time applications because we don't know the sample values in the future. The formula in (19.4.1) is used in applications where all of the sample values are stored in a computer before the processing begins, such as image processing.

A simple and practical method is to hold the sample value over the sample interval. This process is illustrated in Figure 19.4.1. The original continuous-time signal and the sampled signal are shown in Figure 19.4.1. To construct an approximation to the original continuous-time signal, we simply define a continuous-time signal that is constant over each sample interval. The value of this signal is the sample value at the beginning of the sample interval. The reconstructed signal is shown in Figure 19.4.1b along with the original continuous-time signal. An analytical expression for the reconstructed signal is

$$x_o(t) = \sum_{n=-\infty}^{\infty} x(nT_s)\left[u_s(t - nT_s) - u_s\left(t - (n+1)T_s\right)\right]. \tag{19.4.2}$$

The term in brackets is just a pulse function which is one over the sample interval $nT_s \leq t < (n+1)T_s$ and zero elsewhere. Hence, over this sample interval the reconstructed signal, $x_o(t)$, has a value of $x(nT_s)$.

In digital electronics, the device that converts a digital signal to a continuous-time signal is called a digital-to-analog converter (DAC). Many different types of DAC's implement the reconstruction process in Figure 19.4.1.

19.4.2 Transfer Function of the ZOH

It is often necessary to model this reconstruction process. The easiest way to proceed is to use the Laplace transform. Therefore, we truncate the summation in (19.4.2) so we can apply the Laplace transform. Then we obtain

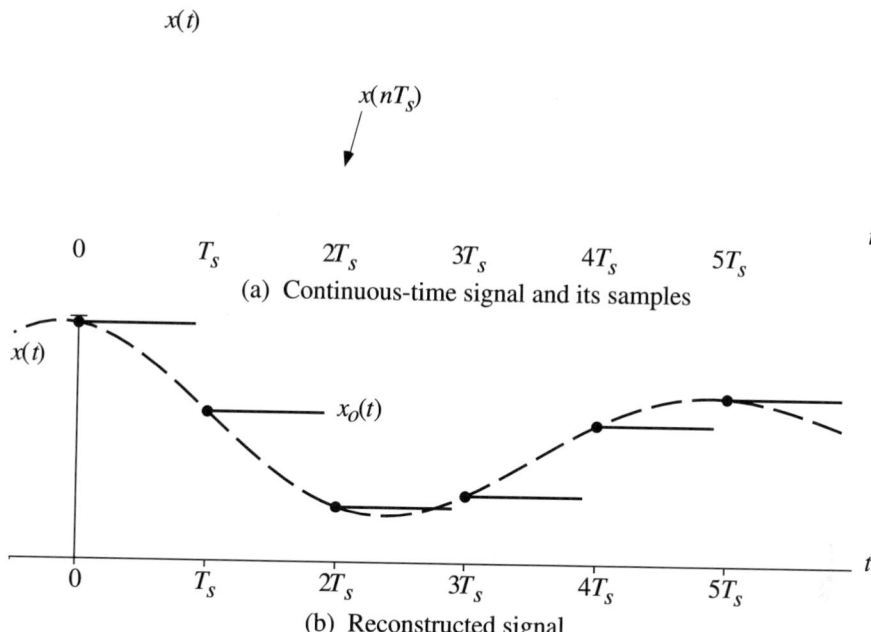

Figure 19.4.1 Signal Reconstruction Using a Zero-Order Hold

$$X_o(s) = \mathcal{L}\{x_0(t)\} = \sum_{n=0}^{\infty} x(nT_s)\mathcal{L}\{[u_s(t-nT_s)-u_s(t-(n+1)T_s)]\} \qquad (19.4.3)$$

$$= \sum_{n=0}^{\infty} x(nT_s)\left[\frac{e^{-snT_s}}{s}-\frac{e^{-s(n+1)T_s}}{s}\right] = \left[\frac{1-e^{-sT_s}}{s}\right]\sum_{n=0}^{\infty} x(nT_s)e^{-snT_s}.$$

Using Property 9.2.9 of Laplace transforms, it is easily shown that the Laplace transform of the continuous-time sampled version of $x(t)$ is

$$\mathcal{L}\{x^*(t)\} = X^*(s) = \sum_{n=0}^{\infty} x(nT_s)\mathcal{L}\{\delta(t-nT_s)\} = \sum_{n=0}^{\infty} x(nT_s)e^{-snT_s}. \qquad (19.4.4)$$

Comparing (19.4.4) into (19.4.3) we obtain

$$\frac{X_o(s)}{X^*(s)} = \frac{1-e^{-sT_s}}{s} = H_o(s). \qquad (19.4.5)$$

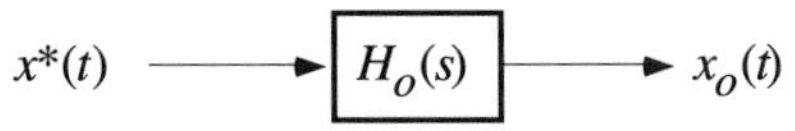

Figure 19.4.2 "Block Diagram" of a Zero-Order Hold

This analysis has shown that we can view the process in Figure 19.4.1 as a system whose input signal is the continuous-time sampled version of $x(t)$, $x^*(t)$, and whose output signal is the signal $x_o(t)$ in (19.4.2). The system is shown in Figure 19.4.2. We give this system a name.

Definition 19.4.1: The system defined by $x^*(t)$ as the input signal and $x_o(t)$ in (19.4.2) as the output signal is called a <u>zero-order hold</u> (ZOH).

▲▲

Notation: The subscript o on $x_o(t)$ means that the signal $x(t)$ has been reconstructed from its sample values by using a zero-order hold.

19.4.3 Frequency Response of a ZOH

In the discussion of the reconstruction of a signal from its samples, we introduced the concept of an ideal lowpass reconstruction filter. This concept leads to the reconstruction formula in (19.4.1). In this section we have introduced the ZOH. Now the question arises as to the relationship between these two methods for reconstructing a signal. It is clear that the ZOH is an approximation method, and that it will reproduce the signal exactly only for very special signals while (19.4.2) gives exact reconstruction when the Nyquist criteria is satisfied. Nonetheless for many practical situations this approximation produced by the ZOH is acceptable. So we would like to have deeper insight into the ZOH. Again we will turn to a frequency domain analysis.

We will compare the frequency response function of both the ideal lowpass reconstruction filter and the ZOH. To derive the frequency response function for the ZOH insert $s = j\omega$ into (19.4.5), we obtain

$$H_o(\omega) = \frac{1 - e^{-j\omega T_s}}{j\omega}\left(\frac{e^{j\frac{\omega T_s}{2}}}{e}\right)\left(e^{-j\frac{\omega T_s}{2}}\right) = \frac{e^{j\frac{\omega T_s}{2}} - e^{-j\frac{\omega T_s}{2}}}{2j}\left(\frac{2e^{-j\frac{\omega T_s}{2}}}{\omega}\right) \qquad (19.4.6)$$

$$= \left(\frac{2T_s}{\omega T_s}\right)\left(e^{-j\frac{\omega T_s}{2}}\right)\sin\left(\frac{\omega T_s}{2}\right).$$

By using

$$\frac{\omega T_s}{2} = \frac{\omega}{2}\left(\frac{2\pi}{\omega_s}\right) = \frac{\pi\omega}{\omega_s}, \qquad (19.4.7)$$

in (19.4.6), the frequency response function becomes

$$H_0(\omega) = T_s \left[\frac{\sin\left(\dfrac{\pi\omega}{\omega_s}\right)}{\dfrac{\pi\omega}{\omega_s}} \right] e^{-j\frac{\pi\omega}{\omega_s}} = T_s e^{-j\frac{\pi\omega}{\omega_s}} \operatorname{sinc}\left(\frac{\omega}{\omega_s}\right).$$

(19.4.8)

The magnitude and phase of (19.4.8) is shown in Figure 19.4.3. Also shown is the magnitude of the frequency response function of the ideal lowpass reconstruction filter labeled as $|H_{con}(\omega)|$. The phase of this filter is zero.

In comparing the two frequency responses in Figure 19.4.3 we see that the ZOH distorts the reconstructed signal in two ways. First, compare the two frequency response functions on the interval $0 < \omega < 0.5\omega_s$. On this interval the ZOH introduces some distortion because the frequency response function deviates from the straight line of the ideal reconstruction filter. If this distortion is unacceptable, then the ZOH can be followed by an equalization filter. Second, the ZOH introduces higher frequency components into the output signal. These components can be explained as the result of the step approximation to the signal. We can also give a frequency domain explanation of these components.

Suppose the input signal to the ZOH is a discrete sinusoid

$$x(n) = A\cos(\Omega_0 n).$$

(19.4.9)

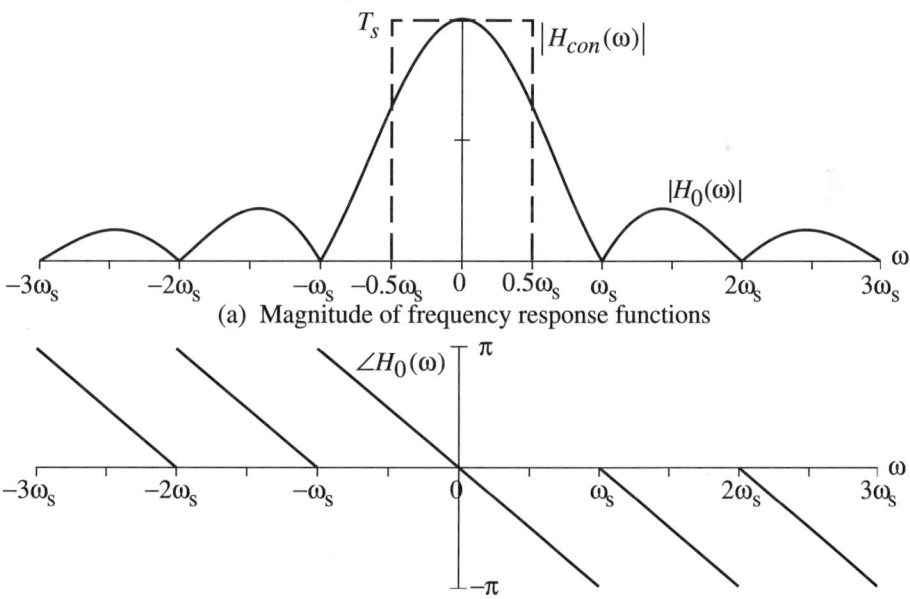

(a) Magnitude of frequency response functions

(b) Phase of the frequency response function of a zero-order hold

Figure 19.4.3 Frequency Response of a Zero-Order Hold

To study the effect of the ZOH on this signal, we translate this signal into the continuous-time domain and take its Fourier transform. We obtain

$$X^*(\omega) = \sum_{n=-\infty}^{\infty} \frac{A\pi}{T_s} \left[\delta(\omega - \omega_0 - n\omega_s) + \delta(\omega + \omega_0 - n\omega_s) \right].$$

(19.4.10)

The amplitude spectrum of this signal along with the frequency response function of the ZOH is shown in Figure 19.4.5a. The amplitude spectrum of the output signal is obtained by multiplying the amplitude spectrum with the frequency response function, and it is shown in Figure 19.4.5b. Instead of a pure sinusoid at the output of the ZOH we see a signal with higher order harmonics. These higher order harmonics are due, of course, to the step approximation to the sinusoid as shown in Figure 19.4.1.

19.4.4 Smoothing Filter

Quite often the higher order harmonics are removed by filtering the output of the ZOH as shown in Figure 19.4.5.

Terminology: This filter in Figure 19.4.5 is often called a <u>smoothing</u> filter because it tends to smooth out the corners on the step approximations generated by the ZOH.

The cutoff frequency of the smoothing filter should be chosen as no greater than half of the sampling frequency. For this choice of cutoff frequency the smoothing filter would remove all of the harmonics for $x_o(t)$ in Figure 19.4.4 except the fundamental harmonic. In addition, the smoothing filter can compensate for the droop in the magnitude of the frequency response function shown in Figure 19.4.4. The magnitude of the frequency response function in the passband of the smoothing filter should be the "inverse" of the magnitude of the ZOH shown in Figure 19.4.4.

19.5 AN EXAMPLE

19.5.1 Introduction

In this section we will present an example that illustrates some of the concepts developed in this chapter. Consider the signal

$$x(t) = e^{-\alpha t} u_s(t), \quad \alpha > 0.$$

(19.5.1)

The continuous-time sampled version of (19.5.1) is

$$x*(t) = \sum_{n=0}^{\infty} e^{-\alpha n T_s} \delta(t - nT_s) = \sum_{n=0}^{\infty} \left(e^{-\alpha T_s} \right)^n \delta(t - nT_s) = \sum_{n=0}^{\infty} a^n \delta(t - nT_s).$$

(19.5.2)

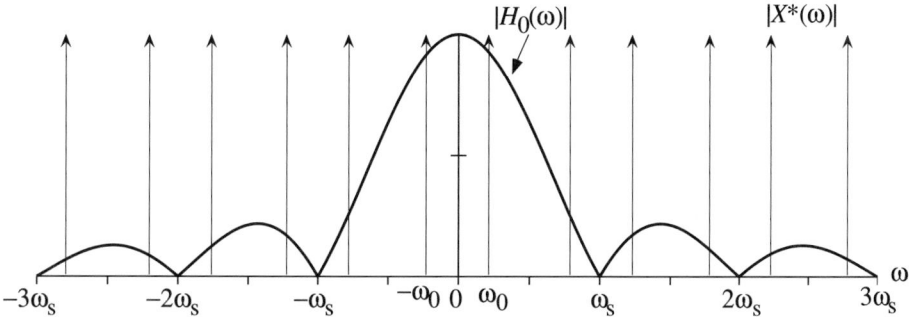

Figure 19.4.4 Distortion of a ZOH

(a) Amplitude spectrum of the input signal and
frequency response function of the ZOH

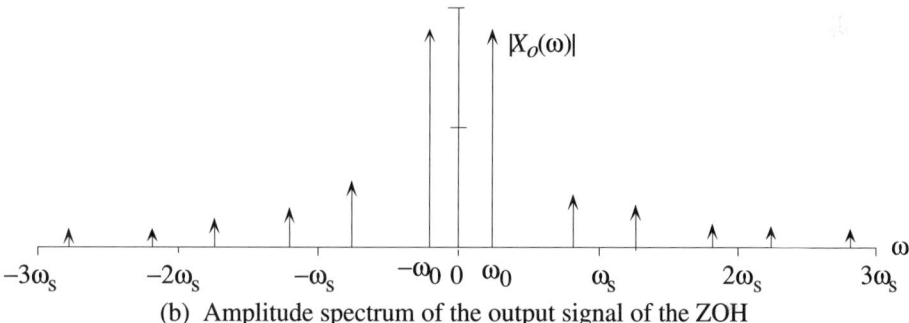

(b) Amplitude spectrum of the output signal of the ZOH

Figure 19.4.5 Smoothing Filter

We assume that $\alpha < 0$ so that the Fourier transform of (19.5.1) exists.   This assumption implies $|a| < 1$ so the Fourier transforms of the sampled signals below also exist.   In this example we will sample this signal, construct the continuous-time Fourier transform of the sampled signal, and apply the interpolation formula to recover the signal from its samples.   We will also show how to use a filter to reduce the aliasing in the sampled signal.   The M-files for generating this example are contained at the end of the section.   This example illustrates many aspects of Nyquist's sampling theorem, Theorem 19.3.5.

19.5.2 Fourier Transform of the Sampled Signal

First we will construct the continuous-time Fourier transform of the sampled signal. For this particular example, this Fourier transform can be computed exactly using a geometric series as

$$X^*(\omega) = \sum_{n=-\infty}^{\infty} a^n u_s(n) e^{-j\omega T_s n} = \sum_{n=0}^{\infty} a^n e^{-j\omega T_s n} = \frac{1}{1 - a e^{-j\omega T_s}}. \tag{19.5.3}$$

By Theorem 19.1.4 the Fourier transform of the sampled signal is also given by

$$X^*(\omega) = \sum_{n=-\infty}^{\infty} \frac{1}{T_s} X(\omega - n\omega_s) \tag{19.5.4}$$

where the Fourier transform of the signal (19.5.1) is

$$X(\omega) = \mathcal{F}\{e^{-\alpha t} u_s(t)\} = \frac{1}{\alpha + j\omega}. \tag{19.5.5}$$

Assume that $\alpha = 1$ and $f_s = 2$ Hz. Then the Fourier transform of the continuous-time sampled signal (19.5.3) is shown in Figure 19.5.1. The exact Fourier transform $|X^*(\omega)|$ shown in Figure 19.5.1 is calculated from (19.5.3). Also shown in Figure 19.5.1 are the amplitude spectra of the three terms series in (19.5.4) $|X(\omega - n\omega_s)|$ $n = -1, 0, 1$. Each of these terms is centered on a multiple of the sampling frequency $n\omega_s = n4\pi$. This figure indicates how each of these terms contributes to the Fourier transform in (19.5.4). Note, however, that the amplitude spectra of the three individual terms don't add up to the amplitude spectrum of the exact Fourier transform. The infinite sum (19.5.4) takes into account the phase relationship between the terms of the sum as well as the magnitude of each term. The amplitude spectrum of each of the individual terms in Figure 19.5.1 doesn't contain any phase information, of course. Hence, the individual amplitude spectra don't add to the amplitude spectrum of the exact Fourier transform because the phase information is not in the amplitude spectrum.

Aliasing is defined as the overlap of the magnitude between the terms in the infinite sum in (19.5.4). The aliasing in this sampled signal is evident in Figure 19.5.1. This signal is typical of practical signals in that the amplitude spectrum

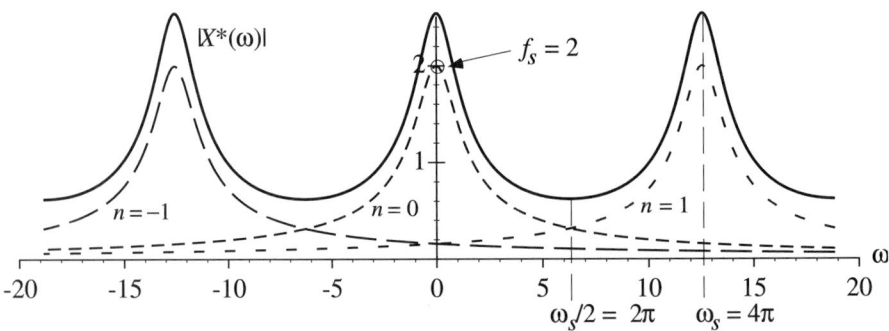

Figure 19.5.1 Fourier Transform of the Sampled Signal in (19.5.3)

normally extends to infinity. It is clear from this figure that aliasing always occurs for practical signals. See also the discussion of folding in Subsection 19.3.3.

19.5.3 Reconstruction of the Signal

Next we consider the reconstruction of the signal from its samples using the formula derived in Section 19.2

$$x_{con}(t) = \sum_{n=-\infty}^{\infty} x(nT_s)\left[\text{sinc}\left(\frac{t - nT_s}{T_s} \right) \right]. \tag{19.5.6}$$

In Figure 19.5.2 the interpolation formula in (19.5.6) is applied to the sample values of the signal in (19.5.1) assuming $\alpha = 1$ and $f_s = 2$ Hz. This figure shows the original signal, its sample values, each of the interpolation functions in (19.5.6), and the sum of the interpolation functions. Note that at each sample instant only one of the interpolation functions is nonzero. Thus at each sample instant the reconstructed signal is equal to the original signal.

The original signal and the reconstructed signal are extracted from Figure 19.5.2 and shown in Figure 19.5.3. We see that the original signal is not exactly reconstructed from its samples. We don't achieve exact reconstruction because of the aliasing shown in Figure 19.5.1.

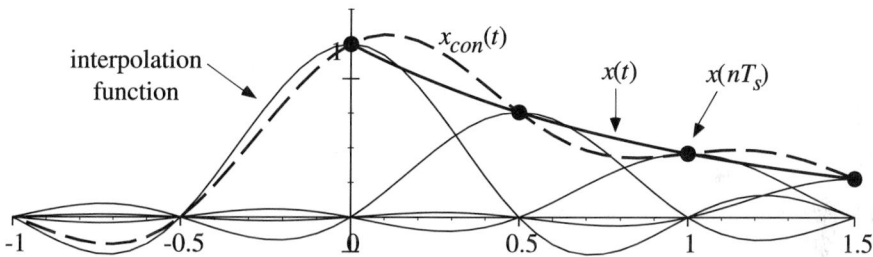

Figure 19.5.2 Reconstruction of the Signal from Its Sample Values

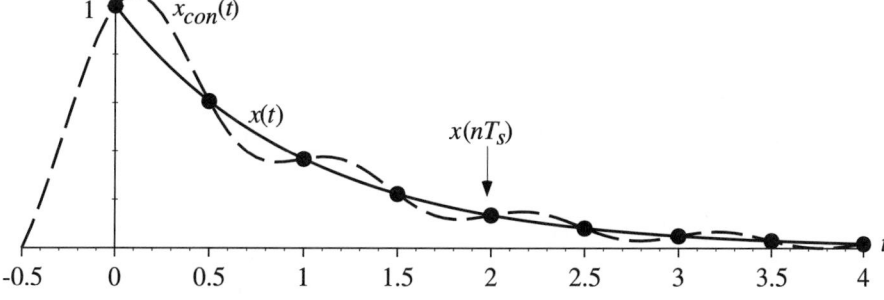

Figure 19.5.3 Original and Reconstructed Signal

19.5.4 Anti-Aliasing Filter

Nyquist's sampling theorem says that a signal can be perfectly reconstructed if it is bandlimited. In practice, no signal is bandlimited, if for no other reason than the presence of noise. Therefore, when a signal is sampled, aliasing of the form shown in Figure 19.5.1 is typically encountered. The amplitude spectrum of the shifted copies of the Fourier transform of the continuous-time signal tend to trail off and overlap with the amplitude spectrum of the term centered at the origin creating aliasing. An obvious way to mitigate this problem is to lowpass filter the signal before it is sampled to reduce the overlap between the terms after sampling. Such a filter is shown in Figure 19.5.4. As shown in Figure 19.5.4 the output signal of the filter $x_f(t)$ is sampled rather than the original signal $x(t)$.

Terminology: A filter that is used to lowpass filter a signal before it is sampled to reduce the effects of aliasing is called an <u>anti-aliasing</u> filter.

Next we will design an anti-aliasing filter

$$\frac{X_f(\omega)}{X(\omega)} = H_{as}(\omega) \tag{19.5.7}$$

for the signal in (19.5.1). In order to avoid aliasing, the amplitude spectrum of the signal should be zero for frequencies above half of the sampling frequency. Therefore, we should chose the bandwidth of the anti-aliasing filter so as to drive the amplitude spectrum of the output signal toward zero for frequencies above half of the sampling frequency as shown in Figure 19.5.5. We also don't want to distort the amplitude spectrum of $x(t)$ for those frequencies that are not aliased. Therefore, the ripple in the passband of the filter should be small. (These specifications also suggest a small transition band, although the width of the transition band depends on the sampling frequency and the bandwidth of the signals to be sampled. See the discussion below.) Therefore, we propose a 4th order Chebyshev 2 filter so that we achieve a significant reduction in the amplitude spectrum in the frequency band above half of the sampling frequency. The Fourier transform of the $x(t)$, the output signal $x_f(t)$, and the magnitude of the frequency response function of the anti-aliasing filter are shown in Figure 19.5.5. The frequency domain interpretation of a signal propagating through a system, as shown in Figure 19.5.5, is discussed extensively in Chapters 14 and 15. Figure 19.5.5 clearly shows that the amplitude spectrum of the filtered signal has been significantly reduced above one-half of the sampling frequency. The filter has also caused some undesirable distortion in the amplitude spectrum below one-half of the sampling frequency as well.

The amplitude spectrum of the sampled output signal of the anti-aliasing filter is shown in Figure 19.5.6. The amplitude spectrum of the unfiltered sampled signal

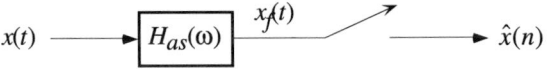

Figure 19.5.4 Anti-Aliasing Filter

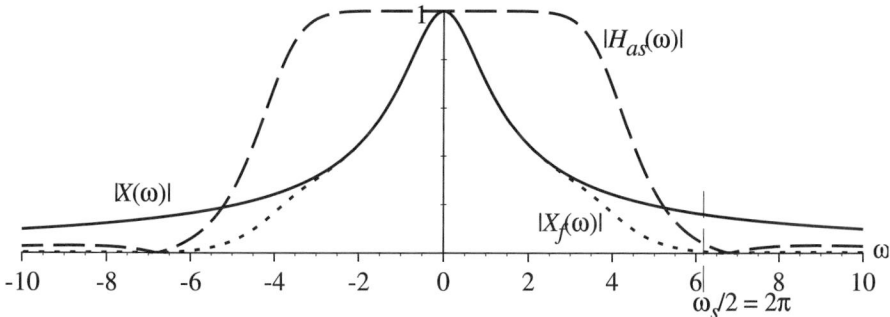

Figure 19.5.5 Action of the Anti-Aliasing Filter

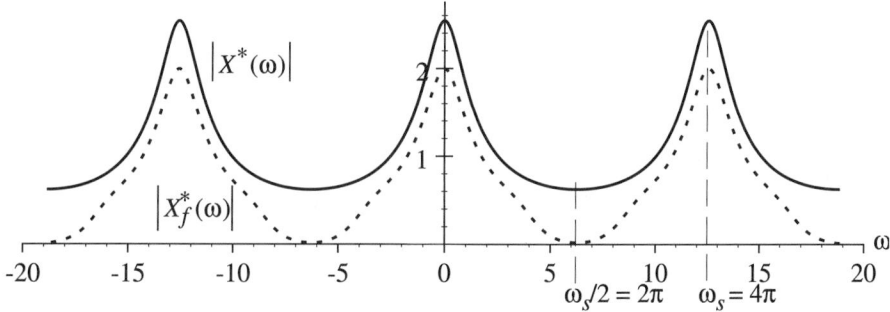

Figure 19.5.6 Amplitude Spectra of the Filtered and Unfiltered Sampled Signals

$|X^*(\omega)|$ from Figure 19.5.1 is also shown in Figure 19.5.6. It is clear that the aliasing has been significantly reduced in the filtered signal when Figure 19.5.6 is compared to Figure 19.5.1.

Quite often the sampling frequency is chosen based on the aliasing requirements. For noncritical applications, the sampling frequency is chosen to be twice the passband edge frequency of the anti-aliasing filter. When the aliasing must be minimized, the sampling frequency is chosen to be 3 to 4 times the passband edge frequency.

19.5.5 MATLAB Experiments

The following M-files generate the figures discussed with the example above.

Example 19.5.1: The MATLAB code that generated Figure 19.5.1 is shown below.

```
clear
% Define parameters
fs = 2;                         % sample frequency, Hz
Ts = 1/fs;                      % sampling period
ws = 2*pi*fs;                   % sampling frequency, rad
alpha = 1;                      % continuous signal constant
```

```
a = exp(-alpha*Ts);              % discrete signal constant
w0 = ws+ws/2;                    % define frequency interval
w = linspace(-w0,w0,400);        % define frequency axis
X0 = 1./(alpha+j*w);             % n = 0 term
Xp = 1./(alpha+j*(w-ws));        % n = 1 term
Xn = 1./(alpha+j*(w+ws));        % n = -1 term
X = fs*[Xn;X0;Xp];               % components of amplitude spectrum
Xmag = abs(X);                   % amplitude spectrum of components
Xstar = 1./(1-a*exp(-j*w*Ts));   % exact Fourier transform
Xsmag = abs(Xstar);              % amplitude spectrum of signal
% Insert code for anti-aliasing filter from Example 19.5.7 here
figure(1)
plot(w,Xmag,w,Xsmag)
title('Fourier Transform of the Sampled Signal')
xlabel('frequency, rad')
ylabel('magnitude')
```

▲▲

Exploratory Exercise 19.5.2: In the M-file in Example 19.5.1 vary the sampling frequency between $0.5 < f_s < 10$. For the sampling frequency fixed at $f_s = 2$, vary the signal parameter between $0.1 < \alpha < 5$. What is the bandwidth of this signal? What can you conclude about the relationship between the bandwidth of the signal and the sampling frequency with regard to aliasing?

▲▲

Exploratory Exercise 19.5.3: Modify the M-file in Example 19.5.1 to compute the Fourier transform of the continuous time sampled version of

$$x(t) = e^{\alpha|t|}. \tag{19.5.8}$$

▲▲

Example 19.5.4: The M-file that was used to generate Figures 19.5.2 and 19.5.3 is given below.

```
clear
fs = 2;                          % sampling frequency, Hz
Ts = 1/fs;                       % sampling period
alpha = 1;                       % signal parameter
Ns = 8;                          % number of samples
npt = 10;                        % number of time steps in between samples
t1 = [-2*Ts:(Ts/npt):0-(Ts/npt)];  % negative time interval
t2 = [0:(Ts/npt):Ns*Ts];         % positive time interval
t = [t1,t2];                     % time vector
x1 = zeros(size(t1));            % continuous-time signal
x2 = exp(-alpha*t2);
x = [x1,x2];
xhat = x2(1:npt:length(x2));     % sampled signal
that = t2(1:npt:length(t2));     % time vector for sampled signal
% Calculate interpolation function for each sample value
Nsample = [0:Ns];                % vector of indices of sample points
```

```
for n = 1:length(xhat)
    xcn(n,:) = xhat(n)*sinc((t-Nsample(n)*Ts)/Ts);
end
xcon = ones(size(xhat))*xcn;        % sum interpolation functions
figure(1)
plot(t,xcn,t,xcon,t,x,that,xhat,'*')
title('Interpolation  Functions')
xlabel('Time')
figure(2)
plot(that,xhat,'*',t,xcon,t,x)
title('Original and Reconstructed Signal')
xlabel('Time')
```

▲▲

Exploratory Exercise 19.5.5: In the M-file in Example 19.5.4 vary the sampling frequency between $0.5 < f_s < 10$. How does the sampling frequency affect the interpolation?

▲▲

Exploratory Exercise 19.5.6: Modify the M-file in Example 19.5.4 for the signal in (19.5.7).

▲▲

Example 19.5.7: The M-file code that was used to generate Figures 19.5.5 and 19.5.6 is given below. This code should be inserted into the M-file that is in Example 19.5.1 at the appropriate place. In this code the Fourier transform of the filtered, sampled signal is computed by summing three terms of the infinite sum (19.5.4) rather than by exact calculation as in Example 19.5.1.

```
clear
% generate anti-aliasing filter
[bf,af] = cheby2(4,30,ws/2,'s');      % anti-aliasing filter
h = freqs(bf,af,w);                   % frequency response
hp = freqs(bf,af,w-ws);               % shifted (n = 1) frequency response
hn = freqs(bf,af,w+ws);               % shifted (n = -1) frequency response
% create filtered signal
X0f = X0.*h;                          % filtered Fourier transform (n = 0)
Xpf = Xp.*hp;                         % filtered Fourier transform (n = 1)
Xnf = Xn.*hn;                         % filtered Fourier transform (n = -1)
Xf = fs*[Xnf;X0f;Xpf];
Xstarf = sum(Xf);                     % Fourier transform of filtered sampled signal
X0fmag = abs(X0f);                    % amplitude spectrum of the filtered signal
Xsfmag = abs(Xstarf);                 % amplitude spectrum of the sampled filtered signal
figure(2)
plot(w,abs(X0),'-',w,abs(h),':',w,X0fmag,'--')
title('Fourier Transform of the Signal and Filtered Signal')
xlabel('frequency, rad')
ylabel('magnitude')
figure(3)
plot(w,Xsmag,'-',w,Xsfmag,'--')
```

title('Fourier Transform of the Sampled Signal and Filtered Sampled Signal')
xlabel('frequency, rad')
ylabel('magnitude')

▲▲

Exploratory Exercise 19.5.8: Vary the bandwidth of the anti-aliasing filter above and below half of the sampling frequency and observe the effect on the amplitude spectrum of the filtered, sampled signal. Also change the type of filter and observe the effect on the amplitude spectrum of the filtered, sampled signal.   Try Butterworth, Chebyshev, and elliptic filters. Finally, vary the order of the filter and observe the effect on the amplitude spectrum of the filtered, sampled signal.

▲▲

Exploratory Exercise 19.5.9: Modify the M-file in Example 19.5.4 to reconstruct the output signal of the anti-aliasing filter. Compare your results with the results in Figure 19.5.3.

▲▲

19.6 CHAPTER SUMMARY

In this chapter we have investigated sampling of continuous-time signals.   This investigation started with the basic definition of sampling (Definition 17.2.1). Given a continuous-time signal $x(t)$ and the sample period $T_S > 0$, we derive the discrete-time sampled version of $x(t)$

$$\hat{x}(n) = x(nT_S) \quad \text{for} \quad n = \ldots - 2, -1, 0, 1, 2, \ldots \tag{19.6.1}$$

The sampling frequency f_s in hertz is defined as

$$f_s = \frac{1}{T_S}. \tag{19.6.2}$$

In radians the sampling frequency is $\omega_s = 2\pi f_s$. The process of extracting the sample values from $x(t)$ is called sampling.

The main result of this chapter is the Nyquist sampling theorem (Theorem 19.3.4). Given a signal $x(t)$ with a bandlimit of f_{bl} Hz, then the frequency f_{ny}, $2f_{bl} = f_{ny}$ Hz, is called the Nyquist sampling frequency for the signal $x(t)$ (Definition 19.3.3). (See also Definition 19.3.5.) According to Theorem 19.3.4 this signal can be uniquely reconstructed from its sample values if the sampling frequency f_s satisfies $f_{ny} < f_s$. The proof of this theorem relies on the Fourier transform of the sampled signal which is of interest in its own right. The signal

$$x * (t) = x(t)\sigma(t) = \sum_{n=-\infty}^{\infty} x(t)\delta(t - nT_s) = \sum_{n=-\infty}^{\infty} x(nT_s)\delta(t - nT_s) \tag{19.6.3}$$

is called the <u>continuous-time sampled version</u> of the continuous-time signal $x(t)$. The Fourier transform of this signal is given by (Theorem 19.1.3)

$$X*(\omega) = \mathcal{F}\{x*(t)\} = \sum_{n=-\infty}^{\infty} \frac{1}{T_s} X(\omega - n\omega_s) = \sum_{n=-\infty}^{\infty} x(nT_s)e^{-j\omega T_s n}. \tag{19.6.4}$$

Theorem 19.1.3 shows that the Fourier transform consists of a sum of shifted copies of the Fourier transform of $x(t)$. If the signal $x(t)$ is bandlimited and the sampling frequency is high enough, then the contribution of each of the terms in the sum can be isolated on the frequency axis. In this case, the original signal can be exactly recovered. If the terms in this sum overlap, then we say aliasing has occurred. Aliasing is discussed at length in Sections 19.3 and 19.5.

Theorem 19.3.4 gives the interpolation formula

$$x_{con}(t) = \sum_{n=-\infty}^{\infty} x(nT_s)\left[\text{sinc}\left(\frac{t - nT_s}{T_s}\right)\right] \tag{19.6.5}$$

for the reconstruction of a signal from its samples. This formula requires that all of the samples be known before the formula can be applied. A more practical way to reconstruct the signal is to simply hold the sample value over the sample period. This reconstruction method is known as a zero-order hold, and it is discussed in Section 19.4. The transfer function and frequency response function of a ZOH is also derived in this section.

19.7 HOMEWORK FOR CHAPTER 19

Homework Problems for Section 19.1

19.1.1   The amplitude spectrum of a continuous time signal is shown in Figure P19.1.1. Suppose that the signal in the figure above is sampled with a sampling frequency of $\omega_s = 300$ rad/sec. Sketch the amplitude spectrum of the Fourier transform of the sampled signal.

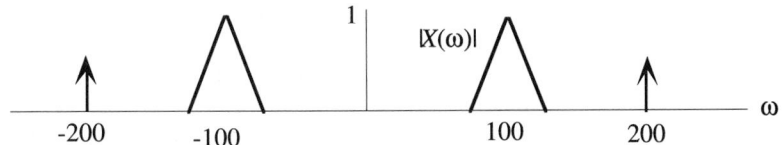

Figure P19.1.1

19.1.2   The amplitude spectrum of a sampled signal is shown in Figure P19.1.2.
(a) What is the sampling frequency?
(b) Sketch the amplitude spectrum of the original signal.

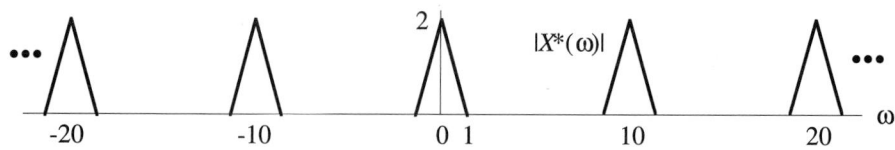

$$|X^*(\omega)|$$

Figure P19.1.2

19.1.3   The amplitude spectrum of $x(t)$ is shown in Figure P19.1.3. This signal is amplitude modulated by $\cos(100\pi t)$. If the modulated signal is sampled with sample period $T_s = 1$ msec, sketch the amplitude spectrum of the sampled signal.

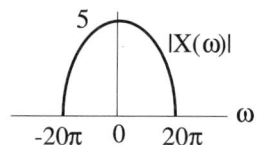

Figure P19.1.3

Homework Problems for Section 19.2

19.2.1   Verify the expression in (19.2.7).

19.2.2   Consider the signal

$$x(t) = \left[\mathrm{Sa}\!\left(\frac{t}{2}\right) \right]^2$$

(a) Find the Fourier transform of this signal. What is its bandwidth?
(b) Suppose this signal is sampled with a sampling frequency of 10 rad/sec. Find the Fourier transform of the sampled version of $x(t)$.
(c) Sketch the amplitude spectrum of the sampled signal.
(d) Find the ideal lowpass reconstruction filter for this sample rate. Can this signal be exactly reconstructed?

19.2.3   The amplitude spectrum of a sampled signal is shown in Figure P19.2.3.
(a) What is the sampling frequency?
(b) Sketch the ideal lowpass reconstruction filter on this amplitude spectrum.
(c) Sketch the amplitude spectrum of the reconstructed signal if exact reconstruction is used.

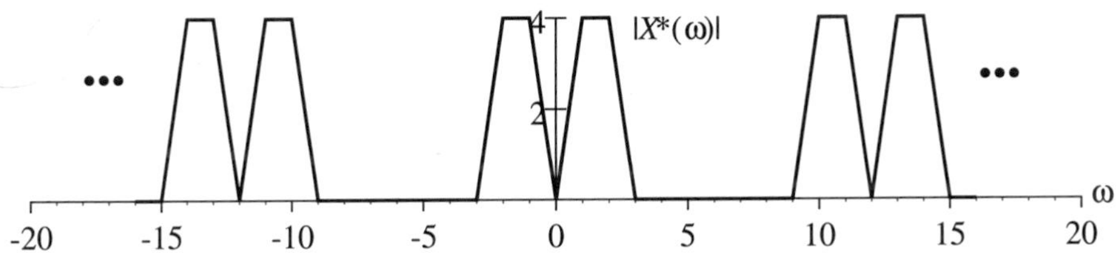

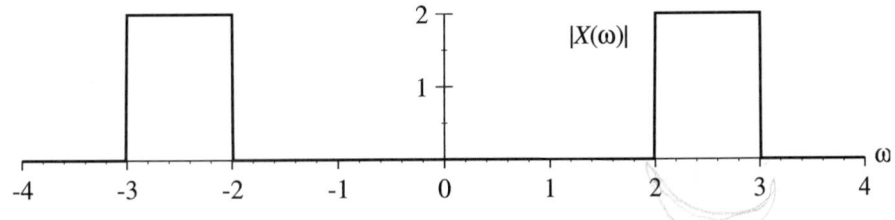

Figure P19.2.3

Homework Problems for Section 19.3

19.3.1 The amplitude and phase spectrum of the energy signal $x(t)$ is shown in Figure P19.3.1.

(a) What is the bandwidth of this signal? What is the bandlimit of this signal?

(b) What is the Nyquist sampling frequency of this signal?

(c) If this signal is sampled with a sample frequency of $\omega_s = 10$, sketch the amplitude spectrum of the sampled signal.

(d) What is the energy of this signal?

(e) What is the energy of the continuous-time sampled version of this signal? Is this an energy or power signal?

Figure P19.3.1

19.3.2 Consider a signal $x(t)$ whose amplitude spectrum is shown in Figure P19.3.2. Also consider the signal

$$x_1(t) = x(t)\cos(100t).$$

(a) Sketch the amplitude spectrum $x_1(t)$.

(b) Suppose that the signals $x(t)$ and $x_1(t)$ are sampled with a sampling frequency of $\omega_s = 40$ rad/sec. Sketch the amplitude spectrum of each of these signals.

(c) Suppose these signals are to be reconstructed using an ideal lowpass construction filter. Show this filter in the sketch of the amplitude spectrum of the sampled signal. Also show the Fourier transform of the reconstructed signal (the output of the ideal lowpass reconstruction filter).

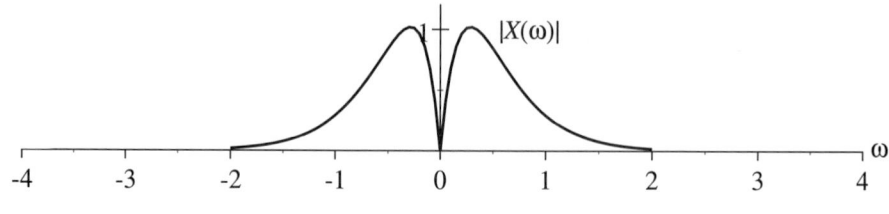

Figure P19.3.2

19.3.3   The amplitude spectrum of the signal $x(t)$ is shown in Figure P19.3.3.
(a) What is the Nyquist frequency of this signal?
(b) If this signal is sampled at a rate of $\omega_s = 3$ rad/sec, sketch the amplitude spectrum of the sampled signal. Sketch the ideal reconstruction filter on the amplitude spectrum of the sampled signal.
(c) If this signal is sampled at a rate of $\omega_s = 0.5$ rad/sec, sketch the amplitude spectrum of the sampled signal. Sketch the ideal reconstruction filter on the amplitude spectrum of the sampled signal.

Figure P19.3.3

19.3.4   Consider the signal $x(t)$ whose amplitude spectrum is shown in Figure P19.3.4. Also define the signals

$$x_1(t) = x(t)\cos(10t) \quad \text{and} \quad x_2(t) = x(t)\cos(15t).$$

(a) What is the Nyquist frequency of each signal?
(b) If this signal is sampled at a rate $\omega_s = 10$ rad/sec, sketch the amplitude spectrum of each of the sampled signals.
(c) Can we recover $x(t)$ from each of these sampled signals?

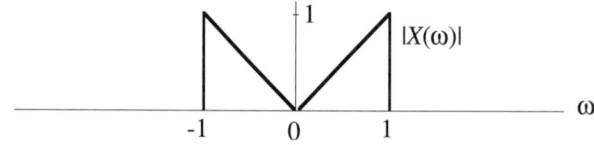

Figure P19.3.4

19.3.5   Let the Fourier transform of the signal $x(t)$ be

$$X(\omega) = \Pi\!\left(\frac{\omega}{\frac{\pi}{2}}\right)\!(\cos(\omega))$$

(a) Find $x(t)$.
(b) Plot the amplitude spectrum of $x(t)$ and $x^*(t)$.
(c) What is the Nyquist frequency of $x(t)$?
(d) Let $\omega_0 > \pi$. Then define $Y(\omega) = X(\omega - \omega_0) + X(\omega + \omega_0)$. Find $y(t)$.
(e) What is the Nyquist frequency of $y(t)$?

19.3.6 Consider the signal

$$x(t) = \left[\mathrm{Sa}(3t)\right]\cos(7t)$$

(a) Find the Fourier transform of this signal.
(b) Sketch the amplitude spectrum of this signal.
(c) What is the Nyquist rate of this signal?
(d) Suppose this signal is sampled at a sampling frequency of $\omega_s = 50$ rad/sec. Plot the amplitude spectrum of the sampled signal.

19.3.7 Consider the signal

$$x(t) = \left[\mathrm{Sa}\!\left(\tfrac{t}{2}\right)\right]^2$$

(a) What is the Nyquist rate of this signal?
(b) Suppose this signal is sampled with a sampling frequency of 10 rad/sec. Sketch the amplitude spectrum of the sampled signal.

19.3.8 Consider the signal

$$x(t) = \cos(20t).$$

(a) Suppose this signal is sampled with a sampling frequency 50 rad/sec. Plot the sampled signal on top of the continuous-time signal.
(b) Sketch the amplitude spectrum of the sampled signal.
(c) Sketch the ideal lowpass reconstruction filter on the amplitude spectrum of the sampled signal.
(d) If we reconstruct the sampled signal what signal do we obtain?
(e) Suppose this signal is sampled with a sampling frequency 15 rad/sec. Repeat parts (a) - (d). Also plot the reconstructed signal on the sampled signal and original continuous-time signal.

Homework Problems for Section 19.4

19.4.1 The amplitude spectrum of the signal $x(t)$ is shown in Figure P19.4.1. The phase spectrum is zero.
(a) Find the signal $x(t)$.

(b) Suppose this signal is sampled with a sampling frequency of $\omega_s = 8$ rad/sec. Sketch the amplitude spectrum of the sampled signal.

(c) Suppose this signal is reconstructed from its samples using an ideal reconstruction filter. What is the amplitude spectrum of the reconstructed signal? What is the reconstructed signal?

(d) Suppose this signal is reconstructed using a zero-order hold (ZOH). What is the amplitude spectrum of the reconstructed signal? What is the reconstructed signal?

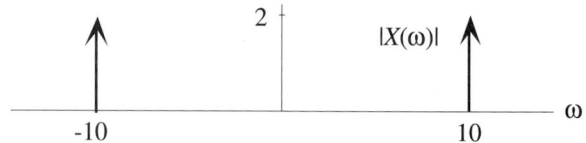

Figure P19.4.1

19.4.2 The amplitude spectrum of a sampled signal is shown in Figure P19.4.2.

(a) The sampled signal is to be reconstructed with a zero-order hold. Sketch the frequency response of the zero-order hold on the amplitude spectrum in Figure P19.4.2.

(b) Sketch the amplitude spectrum of the reconstructed signal below.

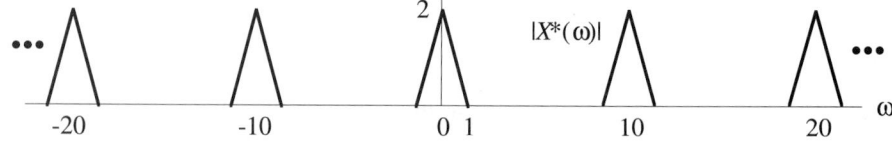

Figure P19.4.2

19.4.3 Suppose that the signal

$$x(t) = \sin(2\pi t + 0.02)$$

is sampled with a sampling frequency of $f_s = 10$ Hz. The sampled signal is then reconstructed using a ZOH.

(a) Sketch the output signal of the ZOH on the interval $0 \le t \le 1$.

(b) Repeat (a) if the sampling frequency is $f_s = 20$ Hz.

19.4.4 A signal $x(t)$ has a bandlimit of 3 kHz. This signal is sampled with a sampling frequency of ω_s and converted by to continuous-time using a ZOH.

(a) At what sampling frequency will the ZOH attenuate the magnitude of $\left| X^*(\omega) \right|$ by no more than 1 dB?

(b) With this sampling frequency, what is the maximum attenuation in the frequency band $[0, \omega_s]$?

19.4.5 Suppose the amplitude spectrum of the signal $x(t)$ is

$$|X(\omega)| = \Pi\left(\frac{\omega}{2\beta}\right).$$

The signal $x(t)$ is sampled with a sampling frequency of $\omega_s = 4\beta$, and then converted to continuous-time with a ZOH. To filter out the unwanted harmonics the output signal of the ZOH is filtered by the filter

$$H(\omega) = \frac{1}{1 + a_1(j\omega) + a_2(j\omega)^2}.$$

Find the filter coefficients a_1 and a_2 (as functions of β) which satisfy:
(i) $|X_o(\beta)||H(\beta)| = |X_o(0)||H(0)|$.
(ii) At least 10 dB attenuation above the sampling frequency.
(iii) Reasonably flat passband.

19.4.6 The amplitude spectrum of the signal $x(t)$ is shown in Figure P19.4.6. This signal is sampled at the Nyquist rate and converted back into a continuous-time signal using a ZOH. The output signal is filtered with a Butterworth filter to remove the harmonics. Design this Butterworth filter such that
(i)   The attenuation at half of the sampling frequency doesn't exceed 1.5 dB.
(ii) The attenuation at the sampling frequency is greater than 10 dB.
(iii) The order is minimized.
Plot the amplitude spectrum of the output signal of the ZOH and the amplitude spectrum of the output signal of the filter on the same axis.

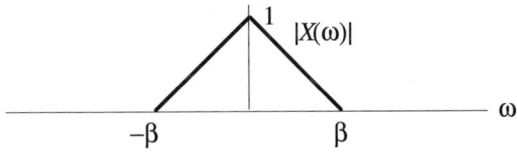

Figure P19.4.6

Homework Problems for Section 19.5

19.5.1 Suppose the following signal is sampled with a sampling rate of f_s Hz.

$$x(t) = e^{-a|t|}$$

(a) What is the Fourier transform of the sampled signal?
(b) For $a = 1$ plot the amplitude spectrum of the sampled signal for $f_s = 1$ Hz.
(c) Plot the amplitude spectrum of the sampled signal for $f_s = 10$ Hz.

(d) Fix the sampling frequency at $f_s = 1$ Hz. Plot the magnitude of the Fourier transform of the sampled signal for various values of a.

19.5.2 The amplitude spectrum of a continuous-time signal is shown Figure P19.5.2. This signal is to be lowpass filtered to remove the sinusoidal component before sampling.

(a) What should be the bandwidth of the analog anti-aliasing filter?

(b) Sketch the magnitude of an ideal lowpass filter.

(c) Suppose that we must use a second-order lowpass Butterworth filter. The magnitude of the signal must not be attenuated more than 1 dB at 100 rad/sec. Select the bandwidth of the Butterworth filter to achieve maximum attenuation of the sinusoidal signal. How much is the sinusoidal signal attenuated?

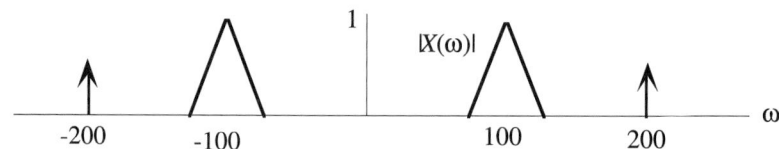

Figure P19.5.2

19.5.3 Consider the signal

$$x(t) = \omega_c e^{-\alpha t} \sin(\omega_c t), \quad \omega_c = 8, \quad \alpha = 0.2.$$

Suppose that this signal is sampled with a sampling frequency of $f_s = 50$ Hz. Before sampling this signal is passed through an anti-aliasing filter. Design this filter so that the amplitude spectrum of the signal is reduced by 30 dB at the Nyquist frequency of the sampling frequency.

(a) State your filter type, bandwidth, order, etc.

(b) Plot the amplitude spectrum of the input and output signals on the same plot as the magnitude Bode plot of the filter.

19.5.4 The signal

$$x(t) = Ae^{-\alpha t} u_s(t)$$

is sampled at a frequency of $\omega_s = 10\pi$ rad/sec. The magnitude of the frequency response of the sampled signal has the values

$$\left| X^*(0) \right| = 6 \quad \text{and} \quad \left| X^*\!\left(\frac{\omega_2}{2}\right) \right| = 2.$$

Determine the exponential factor α and the scalar multiplier A.

Chapter 20

Spectral Content of Discrete Signals

Chapter Outline

One of the central themes of this text is the representation of signals and the characterization of the properties of the signal based on the representation. In Chapter 8 we showed that one of the most important representations of a continuous-time signal is the Fourier transform of that signal. The Fourier transform provides unique insight into the characteristics of the signal through the amplitude and phase spectra and the energy spectral density. These characterizations provided by the Fourier transform are extended to periodic signals through the Fourier series representation. Similar concepts also exist for aperiodic power signals. The essential role of these frequency domain signal concepts in the analysis of signals propagating through systems is explained in Chapter 15.

One purpose of this chapter is to extend these concepts to discrete-time signals. The discrete-time Fourier transform (DTFT) provides a frequency domain representation of a discrete-time signal that is analogous to the Fourier transform of a continuous-time signal. Similarly, the exponential Fourier series is defined for periodic discrete-time signals. Using these two representations we can easily extend all of the frequency domain characterizations from continuous-time signals to discrete-time signals. These extensions are accomplished in the first two sections of this chapter. The similarities and differences between the continuous-time and discrete-time concepts are highlighted.

We have seen that the Fourier transform (or the discrete-time Fourier transform) of a signal lends great insight into the properties of the signal and how that signal propagates through a system. The amplitude and phase spectra along with the energy spectral density provide information about the signal that isn't obvious from the time history. So it is obvious that we would like to determine the Fourier transform of signals that we observe in practice. In the examples presented thus far, we have based our Fourier analysis on the analytical representation of the signal. Using a functional representation of the signal we have calculated the Fourier transform using the mathematical properties of the transform. Unfortunately, most experimentally observed signals don't come with an analytical representation nor is it reasonable to expect such a representation to exist. Fortunately, many signals can be sampled and the samples stored digitally. It is natural, then, to look for a method for computing the Fourier transform of a signal from its samples. In this chapter we will present such a method for computing an approximation to the Fourier transform of the signal from its sample values. This algorithm is called the discrete Fourier transform (DFT).

The DFT as an algorithm for computing the Fourier transform of a continuous-time signal is developed in two stages. In the first stage, the continuous-time signal is sampled. The relationship between the Fourier transform of the continuous-time signal and the Fourier transform of the continuous-time sample version of the signal is developed in Chapter 19. The results of Chapter 19 are essential to the interpretation of the DFT. In the second stage, the development of the DFT shifts to the discrete domain. First, the relationship between the Fourier transform of the continuous-time sampled version of the signal and the DTFT of the sampled version of the signal is developed. Then the DFT is developed as a numerical approximation of the DTFT. The nature of this approximation is discussed through an extended example.

While the presentation of the DFT in this chapter focuses on the computation of the DTFT of the signal, the DFT has many other applications in signal processing and digital filtering. These applications are explored in advanced treatments of this subject.

Summary of Sections

Section 20.1:   We introduce the basic concepts of energy and power of a signal. Energy spectral density and signal bandwidth are also defined.

Section 20.2:   We introduce discrete periodic signals and the power spectral density of a signal.

Section 20.3:   We define the discrete Fourier transform (DFT).

Section 20.4:   We give two examples of the DFT and discuss zero padding.

Section 20.5:   Chapter summary section.

Coverage of the Text

This chapter requires the Fourier transform from Chapter 7 and the DTFT in Section 18.4. A deep understanding of the DFT requires an understanding of the results in Chapter 19. The signal concepts in Sections 20.1 and 20.2 are similar to the signal concepts in Chapter 8.

20.1 DISCRETE-TIME ENERGY SIGNALS

20.1.1 Introduction

One of the central themes of this text is the representation of signals and the characterization of the properties of the signal based on the representation. In Chapter 8 we showed that one of the most important representations of a continuous-time signal is the Fourier transform of that signal. The Fourier transform provides unique insight into the characteristics of the signal through the amplitude and phase spectra and the energy or power spectral density. The role of signal energy in the analysis of signals propagating through systems is explained in Chapter 15.

  In this section these important frequency domain characterizations of a continuous-time signal are extended to discrete-time signals through the use of the discrete-time Fourier transform (DTFT). We define amplitude and phase spectra of a signal using the DTFT. Then the energy of a discrete signal is introduced. This concept is extended into the frequency domain through the use of Parseval's theorem. This theorem allows us to define the notion of signal bandwidth. The bandwidth concept is then used to classify signals.

20.1.2 Amplitude and Phase Spectra

Up to this point we have discussed discrete-time signals $x(n)$. An alternative representation of these signals is provided by the DTFT. Recall that this transform is defined[1] by

$$x(n) \;\leftrightarrow\; X(\Omega) = \sum_{n=-\infty}^{\infty} x(n) e^{-j\Omega n} \tag{20.1.1}$$

when it exists. The complex function $X(\Omega)$ completely describes the signals and so we can use this function as a representation of the signal.

---

[1] See Definition 18.4.1.

Terminology: Since this function depends on the frequency variable Ω, we say that the DTFT is a <u>frequency domain representation</u> of a discrete signal.

Since $X(\Omega)$ is a complex function, it is completely defined by its magnitude and phase.

Definition 20.1.1: Let $x(n)$ be a discrete signal with the DTFT $X(\Omega)$. The function

$$|X(\Omega)| \quad \text{vs.} \quad \Omega$$

is called the <u>amplitude spectrum</u> of this signal and the function

$$\angle X(\Omega) \quad \text{vs.} \quad \Omega$$

is called the <u>phase spectrum</u> of this signal.

▲▲

The amplitude and phase spectra are most often displayed as graphs. These graphs exhibit many of the same properties as the amplitude and phase spectra of continuous signals. The amplitude spectrum has even symmetry and the phase spectrum has odd symmetry.[2] These facts follow from the observation

$$X(-\Omega) = \sum_{n=-\infty}^{\infty} x(n)e^{-j(-\Omega)n} = \overline{X}(\Omega). \tag{20.1.2}$$

These graphs also have some characteristic properties that are not seen in the amplitude and phase spectra of continuous signals. From (20.1.1) we see that the DTFT is a sum of periodic functions. In fact, it is a Fourier series in Ω. Therefore, the DTFT is itself a periodic function with period 2π. (The period of the $n = 1$ term in (20.1.1) usually determines the period of the function.) It follows that the amplitude and phase are both periodic functions. These observations are illustrated by the following example.

Example 20.1.2: Consider the signal

$$x(n) = a^n u_s(n), \quad -1 < a < 1. \tag{20.1.3}$$

The DTFT of this signal, derived in Example 18.4.3, is

$$X(\Omega) = \frac{1}{1 - ae^{-j\Omega}}. \tag{20.1.4}$$

The amplitude spectrum of this signal is

---

[2] This observation assumes that the corresponding signal is real.

$$|X(\Omega)| = \frac{1}{\sqrt{(1 - a\cos(\Omega))^2 + (a\sin(\Omega))^2}}. \tag{20.1.5}$$

The phase spectrum is

$$\angle X(\Omega) = -\tan^{-1}\left(\frac{a\sin(\Omega)}{1 - a\cos(\Omega)}\right). \tag{20.1.6}$$

The amplitude and phase spectra of this signal is shown in Figure 20.1.1 for $a = 0.8$. We have extended the scales in Figure 20.1.1 to clearly show the periodic nature of the amplitude and phase spectra. The amplitude and phase spectra of the same signal is shown in Figure 20.1.2 for $a = -0.8$. The graphs in Figure 20.1.2 show that the amplitude spectrum is an even function about $\Omega = \pi$, and the phase is an odd function about $\Omega = \pi$. These properties follow from the periodicity of the DTFT. Usually, we will graph these functions on the interval $[0, \pi]$.

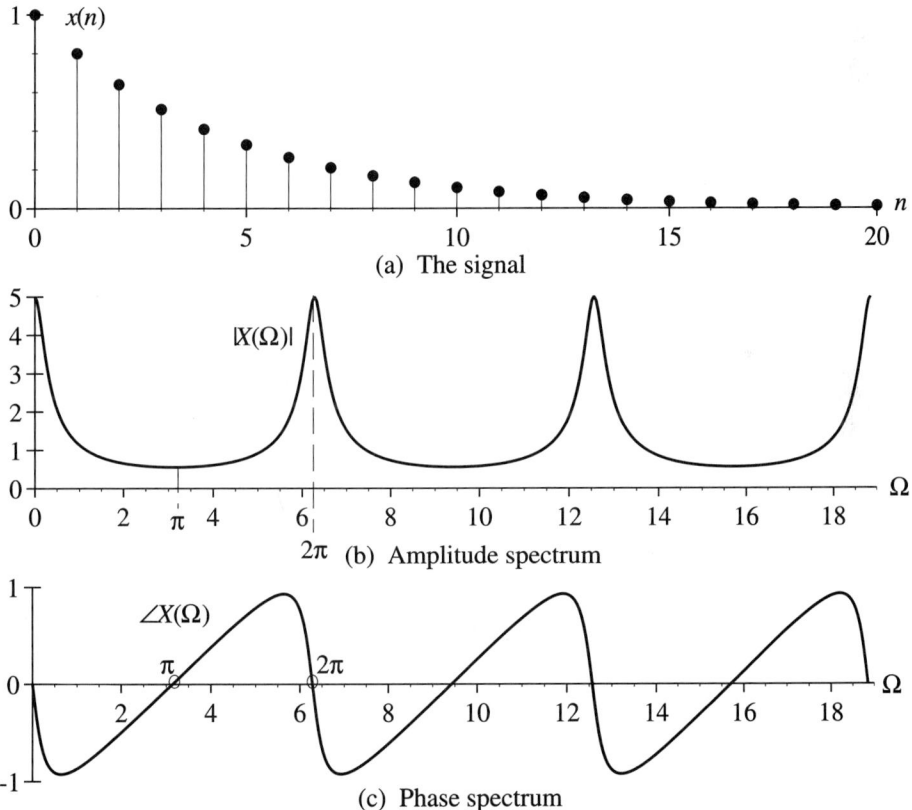

Figure 20.1.1 Amplitude and Phase Spectra of (20.1.3) for $a = 0.8$

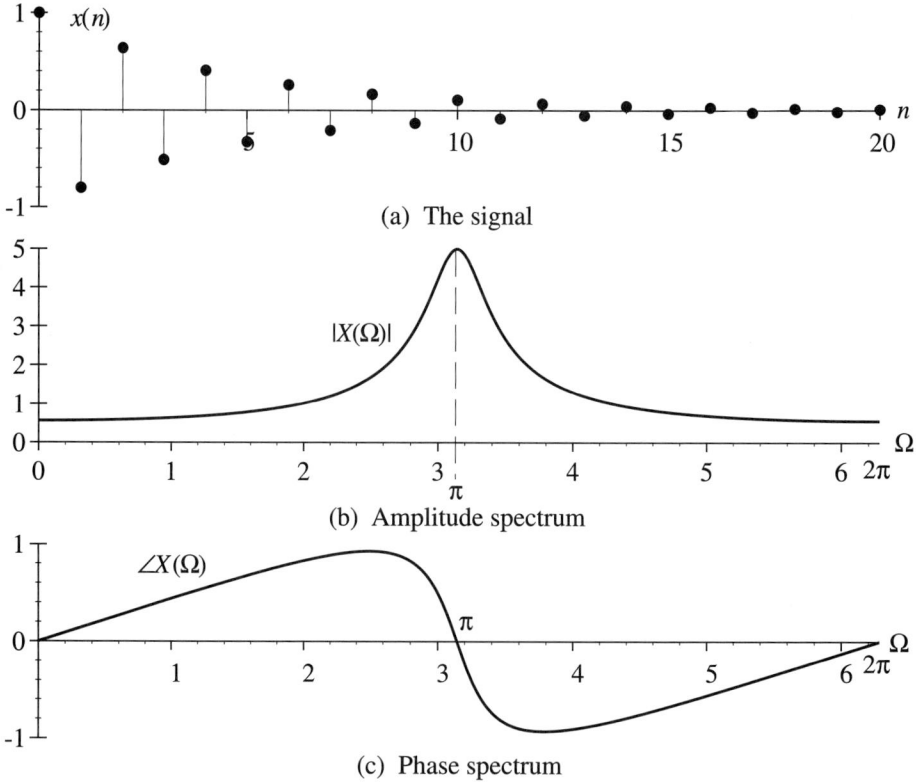

Figure 20.1.2 Amplitude and Phase Spectra of (20.1.3) for $a = -0.8$

It is interesting to compare the time response of these two signals and the shape of their amplitude and phase spectra on the interval $[0, \pi]$. Note that the shapes are quite different.

▲▲

20.1.3 Signal Energy

Next we turn our attention to the energy in the signal. We begin with a definition.

Definition 20.1.3: Given the signal $x(n)$ we define the <u>energy</u> of this signal as

$$E_x = \sum_{n=-\infty}^{\infty} |x(n)|^2$$

provided this sum exists and $E_x > 0$. If the signal has finite, nonzero energy, the signal is said to be an <u>energy</u> signal.

▲▲

Example 20.1.4: Consider the signal

$$x(n) = \begin{cases} A, & 0 \le n \le M-1 \\ 0, & \text{otherwise} \end{cases} \tag{20.1.7}$$

Since $x(n)$ is bounded and has finite duration, it is an energy signal. The energy is

$$E_X = \sum_{n=-\infty}^{\infty} |x(n)|^2 = \sum_{n=0}^{M-1} |A|^2 = MA^2. \tag{20.1.8}$$

▲▲

For the DTFT to exist for a discrete signal, the signal has to be absolutely summable. If $x(n)$ is an energy signal, then this condition is satisfied. Hence, we can use the DTFT and its properties to analyze the concept of energy. The energy of the signal can be related to its DTFT using a discrete-time version of Parseval's theorem for the DTFT, Property 18.4.8.

Theorem 20.1.5: Parseval's Theorem   Suppose that $x(n)$ is an energy signal. Then the energy in this signal is given by

$$E_X = \sum_{n=-\infty}^{\infty} |x(n)|^2 = \frac{1}{2\pi} \int_{2\pi} |X(\Omega)|^2 \, d\Omega.$$

▲▲

Based on this theorem we introduce the following definition.

Definition 20.1.6: The <u>energy spectral density</u> of an energy signal $x(n)$ is defined as

$$D_X(\Omega) = |X(\Omega)|^2.$$

▲▲

Terminology: The energy spectral density is sometimes called the <u>energy spectrum</u> or <u>energy density spectrum</u>.

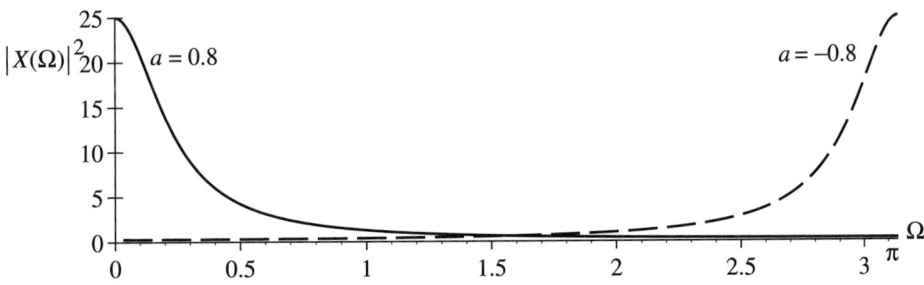

Figure 20.1.3 Energy Spectrum of the Signal (20.1.3)

The energy spectral density is always nonnegative and even.

The next two examples illustrate the calculation of the amplitude spectrum and energy spectral density.

Example 20.1.2: (*continued*) Consider the signal in (20.1.3). This signal is an energy signal for $-1 < a < 1$. Using the DTFT in (20.1.4), the energy spectral density of this signal is

$$D_x(\Omega) = |X(\Omega)|^2 = \overline{X}(\Omega)X(\Omega) = \frac{1}{\left(1 - ae^{-j\Omega}\right)\left(1 - ae^{j\Omega}\right)} = \frac{1}{1 - 2a\cos(\Omega) + a^2}. \qquad (20.1.9)$$

The energy spectra of this signal for $a = 0.8$ and $a = -0.8$ are shown in Figure 20.1.3.
▲▲

20.1.4 Classification of Signals

Parseval's theorem, Theorem 20.1.5, relates the energy in a signal to the integral of the energy spectral density. This theorem gives us an alternative to calculating the total energy in the signal, but it also provides additional insight into the characteristics of the signal. Using Parseval's theorem we can calculate the energy of the signal contained in a specific frequency band. Consider the energy spectral density shown in Figure 20.1.4. Suppose we chose a frequency interval $[\Omega_1, \Omega_2]$ as shown in Figure 20.1.4. When we calculate the energy over the frequency interval $[\Omega_1, \Omega_2]$ we must also define a frequency interval that is symmetric with $\Omega = \pi$, calculate the energy over this interval, and include this energy with the energy from the defining interval. Hence, the energy in this frequency band is the area under the energy spectral density

$$E = (2)\frac{1}{2\pi}\int_{\Omega_1}^{\Omega_2} |X(\Omega)|^2 \, d\Omega. \qquad (20.1.10)$$

The reason for this calculation can be explained from the properties of a discrete sinusoid discussed in Remark 2.3.10. If a discrete frequency satisfies $0 < \Omega_0 < \pi$,

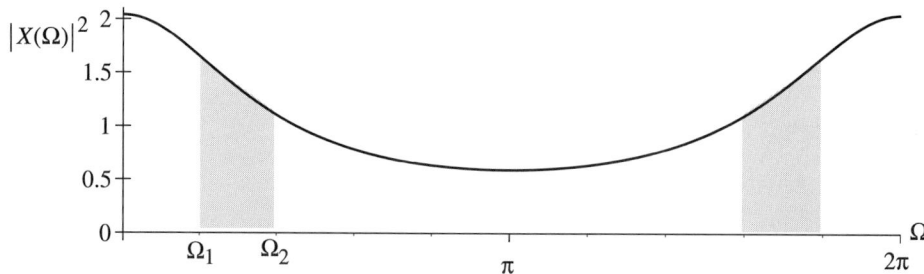

Figure 20.1.4 Energy in a Frequency Band

then

$$x(n) = \cos\big((2\pi - \Omega_0)n\big) = \cos(\Omega_0 n). \tag{20.1.11}$$

Since these two sinusoids are identical, it makes no sense to define discrete frequencies outside the interval $0 < \Omega < \pi$. Hence, we always work over this frequency interval in the discrete domain. The DTFT, however, is defined[3] over the frequency interval $0 < \Omega < 2\pi$. To make the calculations work out in the frequency domain, we must include the frequency interval that is symmetric with $\Omega = \pi$ in the energy calculation.

Because of (20.1.10) we can talk about those frequency bands that contain most of the energy of the signal. This idea leads to the notion of bandwidth of a signal, and also to the classification of signals based on the shape of their energy spectral density. These concepts, that are defined next, are almost identical to the same concepts defined for continuous-time signals in Section 8.3. The main difference is that in the discrete domain discrete frequency is restricted to the frequency interval $0 < \Omega < \pi$ whereas continuous frequency is defined over the frequency interval $0 < \Omega < \infty$.

Definition 20.1.7: Let $x(n)$ be a signal with an amplitude spectrum $|X(\Omega)|$. If the amplitude spectrum satisfies

$$|X(\Omega)| = 0, \quad BL < |\Omega| < \pi$$

the signal is said to be <u>bandlimited</u> with a <u>bandlimit</u> of BL (rad).

▲▲

Next we define the interval where the amplitude spectrum of a bandlimited signal is nonzero. Let

$$I_s = \big\{\Omega \mid |X(\Omega)| \neq 0, 0 \leq \Omega \leq \pi\big\}. \tag{20.1.12}$$

Assume that I_s is a single connected interval. Now we define the extremes of this interval. Let

$$\min_{0 \leq \Omega < \pi} I_s = \Omega_{min}, \quad \text{and} \quad \max_{0 < \Omega \leq \pi} I_s = \Omega_{max}. \tag{20.1.13}$$

Definition 20.1.8: Let $x(n)$ be a signal with an amplitude spectrum $|X(\Omega)|$. Define the frequency interval in (20.1.12). The <u>bandwidth</u> B of the signal is defined as

$$B = \Omega_{max} - \Omega_{min}.$$

▲▲

---

[3] If we had defined the energy spectral density over the frequency interval $-\pi < \Omega < \pi$, then the interval would be symmetric about the origin.

Of course, no signal is strictly bandlimited. Therefore, in practice the definition of bandlimit and bandwidth is relaxed to mean the frequency interval where "most" of the signal energy resides. There are several definitions for the bandwidth that depend on the application at hand.

We can carry this idea a bit further and classify the signal based on where on the frequency axis the energy resides.

Definition 20.1.9: Let $x(n)$ be a signal with an amplitude spectrum $|X(\Omega)|$. Define the frequency interval in (20.1.12) and the extremes of this interval in (20.1.13).

If Ω_{min} and Ω_{max} satisfy	we say the signal is a
$\Omega_{min} = 0$ and $\Omega_{max} < \pi$	lowpass signal
$0 < \Omega_{min}$ and $\Omega_{max} < \pi$	bandpass signal
$0 < \Omega_{min}$ and $\Omega_{max} = \pi$	highpass signal

If the minimum and maximum frequencies satisfy

$$(10)(\Omega_{max} - \Omega_{min}) < \frac{\Omega_{max} + \Omega_{min}}{2}$$

we say the signal is <u>narrowband</u>. Otherwise, the signal is <u>wideband</u>.

▲▲

Example 20.1.2: (*continued*) The signal in Figure 20.1.1 is a lowpass signal. The signal in Figure 20.1.2 is a highpass signal. Note that only the signal parameter has changed. This behavior is not typically observed in continuous signals.

▲▲

It is interesting to note that highpass energy signals don't exist in continuous-time, because they would have infinite energy. Continuous-time energy signals can only be lowpass or bandpass signals.

20.1.5 MATLAB Experiments

The following M-file is used for generating the figures in Example 20.1.2. The signal parameter is restricted to the interval $-1 \le a \le 1$. The DTFT can also be computed with the **freqz** command.

Example 20.1.2 (*continued*)
```
clear
a = -0.8;                    % define signal parameter
N = 20;                      % length of time vector
n = [0:N];                   % time vector
x = a.^n;                    % generate signal
W = linspace(0,2*pi,400);    % frequency vector
X = 1./(1-a*exp(-j*W));      % DTFT
Xmag = abs(X);               % magnitude spectrum
Xphs = angle(X);             % phase spectrum
Dx = Xmag.^2;                % energy spectral density
```

figure(1)
plot(W,Xmag)
title('Magnitude Spectrum')
xlabel('frequency, rad')

figure(2)
plot(W,Xphs)
title('Phase Spectrum')
xlabel('frequency, rad')

figure(3)
plot(W,Dx)
title('Energy Spectrum')
xlabel('frequency, rad')

figure(4)
plot(n,x,'o')
title('Signal')
xlabel('time')

Exploratory Exercise 20.1.10: Change the signal parameter between the values $-1 \le a \le 1$. Observe the change in the amplitude and phase spectra as well as the energy spectral density. For each value of a classify the signal according to Definition 20.1.9.

▲▲

20.2 DISCRETE-TIME POWER SIGNALS

20.2.1 Introduction

In this section we introduce periodic discrete signals and we define the exponential Fourier series for these signals along with the amplitude and phase spectra. It is easily shown that periodic signals have infinite energy. We introduce the definition of a power signal to describe these signals. It is shown that a version of Parseval's theorem holds for power signals and so we are able to define the power spectral density of a discrete power signal. The power spectral density allows us to extend to power signals the notions of bandwidth etc. that we developed for energy signals. Finally, we introduce the concept of correlation functions that is used to extend the power concepts to aperiodic power signals. Most of the basic interpretations of Fourier series and power remain the same as those discussed in Chapters 7 and 8.

20.2.2 Discrete-Time Periodic Signals

The definition of a periodic discrete-time signal is the same as the definition of a continuous-time periodic signal.[4]

Definition 20.2.1: The discrete-time signal $x(n)$ is <u>periodic</u> if there exists a constant integer $N_0 > 0$ such that

$$x(n) = x(n + N_0), \quad \text{for all } n.$$

The smallest number N_0 for which this relationship holds is called the (<u>fundamental</u>) <u>period</u>. All other signals are <u>aperiodic</u>.

▲▲

---

[4] See Definition 7.1.1.

An easy way to define a periodic signal is by using periodic extension[5]. The following example illustrates the idea.

Example 20.2.2: Consider the discrete-time signal

$$x_p(n) = \begin{cases} A, & 0 \le n \le M-1 \\ 0, & M \le n \le N_0 - 1 \end{cases} \tag{20.2.1}$$

Define the periodic signal $x(n)$ by periodic extension of (20.2.1). This signal is shown in Figure 20.2.1.

▲▲

The last example may suggest that a discrete-time periodic signal can be obtained by sampling a continuous-time periodic signal. In general this is not true, unless there is a specific relationship between the sampling period and the fundamental period of the continuous-time periodic signal.

We can also define a Fourier series for a periodic discrete-time signal as we did for continuous-time, periodic signals. The following definition is based on the exponential representation[6] of a continuous-time Fourier series.

Definition 20.2.3: The <u>Fourier series</u> for a periodic discrete-time signal is defined as

$$x(n) = \sum_{k=0}^{N_0-1} c_k e^{j\left(\frac{2\pi k}{N_0}\right)n}$$

where

$$c_k = \frac{1}{N_0} \sum_{n=0}^{N_0-1} x(n) e^{-j\left(\frac{2\pi k}{N_0}\right)n}.$$

▲▲

The Fourier series is another representation for a periodic discrete-time signal. This

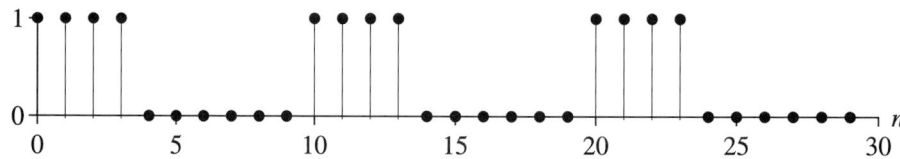

Figure 20.2.1 Discrete Periodic Square Wave with $M = 4$ and $N_0 = 10$

---

[5] See Definition 7.1.3.

[6] See Definition 7.2.4.

representation lends additional insight into the characteristics of periodic signals.

Example 20.2.4: This example illustrates the Fourier series of a discrete periodic signal. Consider the periodic signal defined by the periodic extension of the signal in (20.2.1). This signal has a Fourier series whose coefficients are computed using Definition 20.2.3. We have

$$c_k = \frac{1}{N_0} \sum_{n=0}^{N_0-1} x(n) e^{-j\frac{2\pi kn}{N_0}} = \frac{1}{N_0} \sum_{n=0}^{M-1} A e^{-j\frac{2\pi kn}{N_0}} . \tag{20.2.2}$$

The coefficients in (20.2.2) are computed for each value of k. The series in (20.2.2) is a geometric series, so we have

$$c_k = \begin{cases} \dfrac{AM}{N_0}, & k = 0 \\[2ex] \dfrac{A}{N_0} \dfrac{1 - e^{-j\frac{2\pi Mk}{N_0}}}{1 - e^{-j\frac{2\pi k}{N_0}}}, & k = 1, 2, \ldots, N_0 - 1 \end{cases} \tag{20.2.3}$$

This expression can be simplified to

$$c_k = \begin{cases} \dfrac{AM}{N_0}, & k = 0 \\[2ex] \dfrac{A}{N_0} \left(e^{-j\frac{2\pi k(M-1)}{N_0}} \right) \dfrac{\sin\left(\dfrac{\pi kM}{N_0}\right)}{\sin\left(\dfrac{\pi k}{N_0}\right)}, & k = 1, 2, \ldots, N_0 - 1 \end{cases} \tag{20.2.4}$$

▲▲

Example 20.2.4 illustrates a basic difference between a Fourier series of a continuous signal and the Fourier series of a discrete signal. The Fourier series has only a finite number of terms $k = 0, 1, \ldots, N_0 - 1$. To see why, let $\tilde{k} = k + N_0$. Then we have

$$e^{-j\frac{2\pi n\tilde{k}}{N_0}} = e^{-j\frac{2\pi n(k+N_0)}{N_0}} = e^{-j\frac{2\pi nk}{N_0}} e^{-j\frac{2\pi nN_0}{N_0}} = e^{-j\frac{2\pi nk}{N_0}} . \tag{20.2.5}$$

Since the coefficients of each term in the Fourier series is unique, there must be only N_0 unique terms in the Fourier series. If we would plot the coefficients of each of the terms for $-\infty < k < \infty$, these coefficients c_k would be periodic with period N_0.

This observation is directly related to the fact that the frequencies[7] of a discrete sinusoid must be between 0 and 2π.

In the Fourier series representation of a periodic signal, each coefficient c_k is associated with a complex exponential defined by a specific discrete frequency. Hence, each Fourier coefficient is associated with a specific discrete frequency

$$c_k \leftrightarrow \frac{2\pi k}{N_0} = \Omega_k. \tag{20.2.6}$$

Therefore, the Fourier series is completely specified if we know the coefficients and the frequencies associated with them. As with continuous-time Fourier series we can define a function of frequency that captures this information. Since the coefficients are complex, we must specify both magnitude and phase to completely define the coefficient.

Definition 20.2.5: Let $x(n)$ be a periodic discrete signal with a Fourier series. The function

$$|c_k| \quad \text{vs.} \quad \Omega_k = \frac{2\pi k}{N_0}$$

is called the <u>amplitude spectrum</u> of the Fourier series. The function

$$\angle c_k \quad \text{vs.} \quad \Omega_k = \frac{2\pi k}{N_0}$$

is called the <u>phase spectrum</u> of the Fourier series.   ▲▲

The amplitude and phase spectra are frequently displayed graphically. The next example illustrates this idea.

Example 20.2.6: The amplitude and phase spectra of the periodic signal in Example 20.2.4 is shown in Figure 20.2.2. As we discussed above, the frequency range is confined to be between 0 and 2π. We also see that the amplitude spectrum has even symmetry about π and the phase spectrum has odd symmetry about π. We made the same observations about the amplitude and phase spectra of an energy signal in the last section.   ▲▲

20.2.3 Power Spectral Density

In the last section we introduced the notion of the energy in a signal. For some signals, periodic signals for instance, the energy is infinite. Hence, all of the concepts based on energy can't be applied to these signals. For these signals it is often useful to define the average energy over time.

---

[7] See Remark 2.3.10.

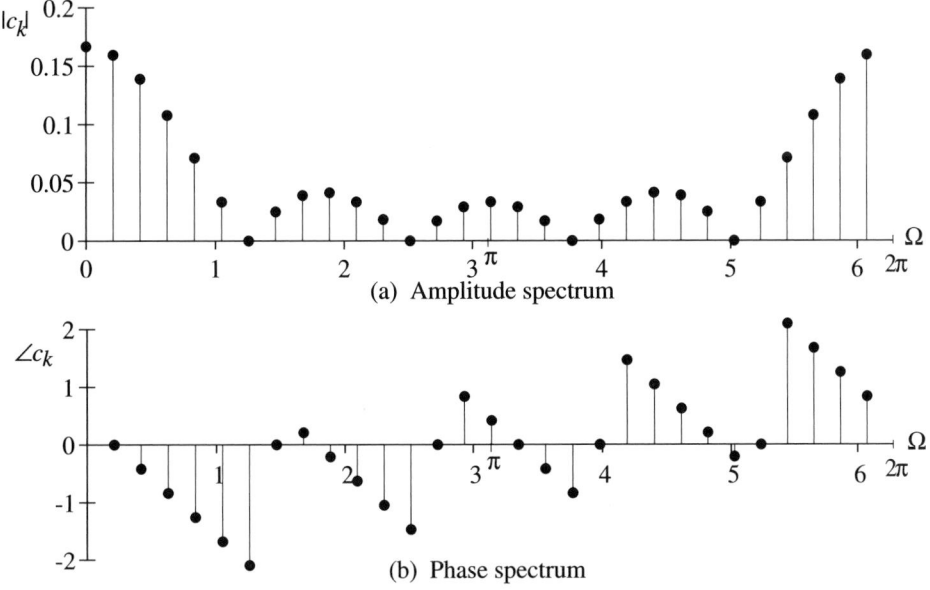

Figure 20.2.2 Amplitude and Phase Spectra for $M = 5$ and $N_0 = 30$

Definition 20.2.7: Given the signal $x(n)$ we define the <u>power</u> of this signal as

$$P_x = \lim_{\kappa \to \infty} \frac{1}{2\kappa + 1} \sum_{n=-\kappa}^{\kappa} |x(n)|^2$$

provided this limit exists and $P_x > 0$. If the signal has finite, nonzero power, the signal is said to be a <u>power</u> signal.

▲▲

The power of a signal is just the average energy of the signal over time. Hence, if the signal is an energy signal, then the power in the signal is zero. On the other hand, the energy in a power signal is infinite. Hence, power signals and energy signals are two mutually exclusive sets. There are signals that are neither energy or power signals.

The following theorem, which is a parallel to Theorem 8.5.1, gives an easy formula for computing the power in a period signal.

Theorem 20.2.8: Let $x(n)$ be a periodic signal. Then $x(n)$ is a power signal. Furthermore, the power of this signal is given by

$$P_x = \lim_{\kappa \to \infty} \frac{1}{2\kappa + 1} \sum_{n=-\kappa}^{\kappa} |x(n)|^2 = \frac{1}{N_0} \sum_{n=0}^{N_0-1} |x(n)|^2.$$

▲▲

This theorem says that the power in a periodic signal can be calculated from one period of the signal. The following example illustrates the application of Theorem 20.2.8.

Example 20.2.9: Consider the periodic signal defined in Example 20.2.2. According to Theorem 20.2.8 the total power in this signal is

$$P = \frac{1}{N_0} \sum_{n=0}^{N_0-1} |x(n)|^2 = \frac{1}{N_0} \sum_{n=0}^{M-1} A^2 = \frac{MA^2}{N_0}. \qquad (20.2.7)$$

▲▲

The power in a periodic signal can also be computed from its Fourier series. The following theorem parallels Theorem 8.5.1.

Theorem 20.2.10: Parseval's Theorem   Let $x(n)$ be a periodic signal with a Fourier series. Then the power in this signal is given by

$$P_x = \frac{1}{N_0} \sum_{n=0}^{N_0-1} |x(n)|^2 = \frac{1}{N_0} \sum_{n=0}^{N_0-1} |c_k|^2.$$

▲▲

Theorem 20.2.10 shows that the total power in the signal is the sum of the power of each periodic component. Since this sum is finite, the power can be computed exactly, unlike the calculation for a continuous-time Fourier series.

Each Fourier coefficient determines the power associated with the periodic component of the signal. Since each coefficient can be associated with a discrete frequency as in (20.2.6), the Fourier coefficients determine the distribution of power as a function of the frequency of each periodic component. Based on this interpretation of Theorem 20.2.10, we introduce the following definition.

Definition 20.2.11: Let $x(n)$ be a periodic signal with a Fourier series. Then the power spectral density of this signal is defined as

$$S_x(\Omega) = \begin{cases} |c_k|^2, & \Omega = \frac{2\pi k}{N_0}, \quad k = 0,1,\ldots,N_0 - 1 \\ 0, & \text{otherwise} \end{cases}$$

▲▲

Using the power spectral density, all of the concepts defined for energy signals in the last section (bandwidth, etc.) can be extended to power signals. The following example illustrates the calculation of the power spectral density of a periodic signal.

Example 20.2.12: Consider the periodic signal defined in Example 20.2.2. The coefficients of the Fourier series are given in (20.2.4). According to Definition 20.2.11 the power spectrum of this signal is

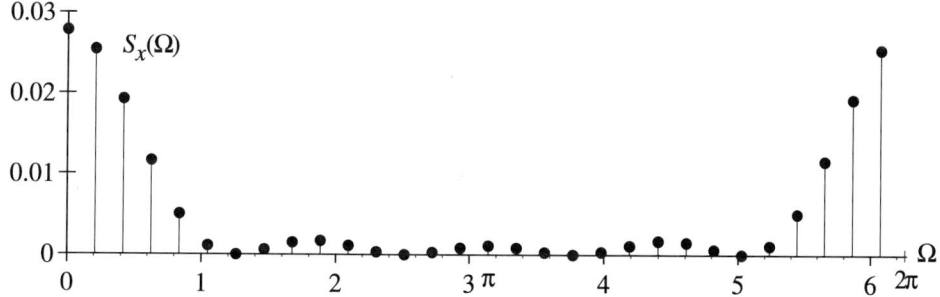

Figure 20.2.3 Power Spectrum in (20.2.7) for $M = 5$ and $N_0 = 30$

$$S_x(\Omega) = \begin{cases} \left(\dfrac{AM}{N_0}\right)^2, & \Omega = 0 \\[4mm] \left(\dfrac{A}{N_0}\right)^2 \left[\dfrac{\sin\left(\dfrac{\pi M k}{N_0}\right)}{\sin\left(\dfrac{\pi k}{N_0}\right)}\right]^2, & \Omega = \dfrac{2\pi k}{N_0}, \quad k = 1,2,\ldots,N_0 - 1 \\[4mm] 0, & \text{otherwise} \end{cases} \tag{20.2.8}$$

This power spectrum is shown in Figure 20.2.3. The power spectrum exhibits the same symmetry properties that the amplitude possesses.

▲▲

20.2.4 Correlation Functions

We can extend the notion of the power spectral density to aperiodic power signals. To accomplish this extension, we introduce the idea of correlation of two signals. Correlation is also used in other contexts in signal analysis. We begin with the definition.

Definition 20.2.13:   (a)   Let $x(n)$ and $y(n)$ be two energy signals.   The <u>crosscorrelation</u> of these two energy signals is defined as

$$r_{xy}(k) = \sum_{n=-\infty}^{\infty} x(n+k)y(n).$$

The <u>autocorrelation</u> of the energy signal $x(n)$ is defined as

$$r_{xx}(k) = \sum_{n=-\infty}^{\infty} x(n+k)x(n).$$

(b) Let $x(n)$ and $y(n)$ be two power signals. The <u>crosscorrelation</u> of these two power signals is defined as

$$r_{xy}(k) = \lim_{K \to \infty} \frac{1}{2K+1} \sum_{n=-K}^{K} x(n+k)y(n)$$

provided the limit exists. The <u>autocorrelation</u> of the power signal $x(n)$ is defined as

$$r_{xx}(k) = \lim_{K \to \infty} \frac{1}{2K+1} \sum_{n=-K}^{K} x(n+k)x(n)$$

provided the limit exists.   ▲▲

Terminology: The index k is called the <u>lag parameter</u> or <u>time shift parameter</u>.

The definition of the correlation functions for energy and power signals are closely related. The only difference is the presence of the limit for the power signal. Hence, if we state a result using correlation functions for energy signals, this result can be translated to power signals by inserting the limit. Of course, when we use limits we must be sure they exist.

The correlation functions are directly related to the energy or power in a signal. We state the following theorem without proof.

Theorem 20.2.14: (a) Let $x(n)$ be an energy signal. Then the energy spectral density of this signal is given by

$$D_x(\Omega) = \mathcal{DTFT}\{r_{xx}(k)\}.$$

Furthermore, the energy in the signal is given by

$$E_x = \frac{1}{2\pi} \int_{-\pi}^{\pi} D_x(\Omega)\, d\Omega = r_{xx}(0).$$

(b) Let $x(n)$ be a power signal. Then the power spectral density of this signal is given by

$$S_x(\Omega) = \mathcal{DTFT}\{r_{xx}(k)\}.$$

Furthermore, the power in the signal is given by

$$P_x = \frac{1}{2\pi} \int_{-\pi}^{\pi} S_x(\Omega)\, d\Omega = r_{xx}(0).$$

                                                                          ▲▲

20.2.5 MATLAB Experiments

The following MATLAB M-file generated the power spectrum of the signal in Example 20.2.2 shown in Figure 20.2.3.

Example 20.2.12: (*continued*)

```
clear
N0 = 30;                                          % period of square wave
M = 5;                                            % pulse width
k = [0:N0-1];                                     % time vector
c = (1/N0)^2*((sin(pi*k*M/N0))./(sin(pi*k/N0))).^2;   % calculation of coefficients
c(1) = (M/N0)^2;                                  % k = 0  coefficient
W = (2*pi/N0)*k;                                  % frequency vector
plot(W,c,'o')
title('Power Spectral Density')
xlabel('Frequency, rad')
```

Exploratory Exercise 20.2.15: Change the parameters of the signal and observe how the power spectral density changes. First, vary M between 2 and 20. Then vary N between 10 and 50. Compare the results to Exploratory Exercise 8.8.1.

▲▲

Exploratory Exercise 20.2.16: Rewrite the MATLAB code to calculate the coefficients from the complex representation in (20.2.3). Then calculate and plot the amplitude and phase spectra as well as the power spectrum. Again vary the signal parameters and observe the change in the plots. Also plot all of these graphs on the frequency interval $[-\pi,\pi]$.

▲▲

20.3 COMPUTING THE FOURIER TRANSFORM: THE DFT

We have seen that the Fourier transform (or the discrete-time Fourier transform) of a signal lends great insight into the properties of the signal. The amplitude and phase spectra along with the energy spectral density provide information about the signal that isn't obvious from the time history. In practice we can often experimentally measure the signal through sampling and record the samples digitally. It is natural, then, to look for a method for computing the Fourier transform of a signal from its samples. In this section we will present a method for computing an approximation to the Fourier transform of the signal $x(t)$ using the sample values of $x(t)$. This algorithm is called the discrete Fourier transform (DFT).

The DFT, the algorithm, is based on several concepts. First, it uses sampling theory, developed in Chapter 19, because we are processing the samples of the signal. The algorithm itself is developed in the discrete domain using the DTFT. Therefore, we must understand the relationship between the continuous- and discrete-time Fourier transform. The DTFT is based on the samples of the signal, but it requires an infinite number of samples. It is also a continuous function of Ω. A realistic computation, of course, must use only a finite number of samples and it must discretize the frequency axis. The DFT is derived from the DTFT by extracting a finite number of samples from the signal and sampling the DTFT in frequency.

This development is carefully explained in this section. In the next section we will present several examples that illustrate the calculation of the DFT.

To reverse the algorithm and compute a discrete signal from samples of the DTFT is also of interest. The inverse DFT requires knowledge of the Fourier series of a discrete signal as discussed in the last section.

Terminology: The discrete-time Fourier transform (DTFT) is the discrete-time equivalent of the Fourier transform. In this section we develop the discrete Fourier transform (DFT) which is related to, but different from, the discrete-time Fourier transform (DTFT). Care should be taken to distinguish between these transforms whose acronyms are so similar. In practice, the DTFT is used for analysis while the DFT is used for computation.

Relationship Between Continuous-Time Transforms   Let $x(t)$ be given with a Fourier transform $X(\omega)$. The first step in computing the Fourier transform of this signal is to sample the signal. Let the sampling period be T_S for the sampling frequency, $f_S = \dfrac{1}{T_S}$. The continuous-time sampled version of the signal $x(t)$ is

$$x^*(t) = \sum_{n=-\infty}^{\infty} x(nT_S)\delta(t - nT_S).$$

(20.3.1)

The (continuous-time) Fourier transform of (20.3.1) is

$$X^*(\omega) = \frac{1}{T_S} \sum_{n=-\infty}^{\infty} X(\omega - n\omega_S).$$

(20.3.2)

Since we are processing the samples of $x(t)$, the best that we can do is compute $X^*(\omega)$ in (20.3.2). That is, we can't compute $X(\omega)$ directly. The transform $X(\omega)$ is embedded in $X^*(\omega)$ as can be seen from (20.3.2). This transform is inevitably distorted by the aliasing. The amount of distortion, however, depends on the choice of the sampling frequency f_S. The relationship between $X(\omega)$ and $X^*(\omega)$ is explored in depth in Sections 19.2 and 19.3.

Relationship of the Fourier Transform to the DTFT   We wish to work in the domain of discrete frequency. Therefore, we express the continuous-time Fourier transform of the sampled signal, (20.3.2), in terms of discrete-time Fourier transform (DTFT) of the sampled version of $x(t)$. Recall from Theorem 19.1.3 that the continuous-time Fourier transform can be expressed as

$$\mathcal{F}\{x^*(t)\} = X^*(\omega) = \mathcal{F}\left\{ \sum_{n=-\infty}^{\infty} x(nT_S)\delta(t - nT_S) \right\} = \sum_{n=-\infty}^{\infty} x(nT_S)e^{-j\omega T_S n}.$$

(20.3.3)

Direct comparison of the expression in (20.3.3) with the definition of the DTFT, Definition 18.4.1, shows that substitution of the discrete frequency,[8] $\Omega = T_s\omega$, into (20.3.3) gives the DTFT. That is,

$$\sum_{n=-\infty}^{\infty} x(nT_s)e^{-j\omega T_s n}\bigg|_{\omega T_s = \Omega} = \sum_{n=-\infty}^{\infty} x(nT_s)e^{-j\Omega n}. \tag{20.3.4}$$

This substitution leads to the following theorem.

Theorem 20.3.1: Let $x(t)$ be a continuous-time signal, and suppose that it is sampled with a sampling period of T_s. Then the Fourier transform of the continuous-time sampled signal in (20.3.3) is related to the DTFT of the discrete-time sampled signal, $\hat{x}(n) = x(nT_s)$, by

$$X^*(\omega)\bigg|_{\Omega = T_s\omega} = \hat{X}(\Omega).$$

▲▲

Notation: The hat notation $\hat{X}(\Omega)$ indicates that this DTFT is obtained from sample values of $x(t)$.

From Theorem 20.3.1 it is clear that the Fourier transform of the continuous-time sampled signal is related to the DTFT of the discrete-time sampled signal by simply rescaling the frequency axis. The functional form of these two transforms is the same. This result establishes the connection between frequency analysis of sampled signals in the continuous-time domain and sampled signals in the discrete-time domain. This theorem is illustrated by the following example.

Example 20.3.2: In Section 19.5 we considered the signal

$$x(t) = e^{-\alpha t}u_s(t), \quad \alpha > 0. \tag{20.3.5}$$

The continuous-time sampled version of (20.3.5) is

$$x^*(t) = \sum_{n=0}^{\infty} e^{-\alpha n T_s}\delta(t - nT_s) = \sum_{n=0}^{\infty} a^n\delta(t - nT_s). \tag{20.3.6}$$

We assume that $\alpha < 0$ so that the Fourier transform of (20.3.5) exists. This assumption implies $|a| < 1$ so the Fourier transform of the sampled signal also exists and is given by

---

[8] See Definition 17.2.4.

$$X^*(\omega) = \sum_{n=-\infty}^{\infty} a^n u_s(n) e^{-j\omega T_s n} = \sum_{n=0}^{\infty} a^n e^{-j\omega T_s n} = \frac{1}{1 - ae^{-j\omega T_s}}. \tag{20.3.7}$$

According to Theorem 20.3.1 the DTFT of the discrete-time sampled signal is given by

$$\hat{X}(\Omega) = X^*(\omega)\Big|_{\Omega = T_s\omega} = \frac{1}{1 - ae^{-j\omega T_s}}\Big|_{\Omega = T_s\omega} = \frac{1}{1 - ae^{-j\Omega}}. \tag{20.3.8}$$

The amplitude spectrum of both the Fourier transform of the continuous-time sampled signal (also given in Figure 19.5.1) and the DTFT of the discrete-time sampled signal is shown in Figure 20.3.1.

▲▲

Because we are working with the samples of the signal, the best we can hope to compute is the DTFT of the signal in Theorem 20.3.1. There are two computational problems, however. First, the summation is infinite because there are an infinite number of samples of the signal. Second, the discrete frequency variable Ω is continuous which again means that the function would have to be evaluated at an infinite number of points. To obtain a computationally tractable formula, we must limit the number of sample points and only evaluate the transform at a finite number of frequencies.

Truncation of the Sampled Signal To reduce the infinite summation in the DTFT to a finite sum, we introduce an approximation of the sampled signal into the DTFT by simply extracting the first N_s samples from the sampled signal. Formally, we define a new signal

$$\tilde{x}(n) = \begin{cases} \hat{x}(n), & n = 0, 1, \ldots, N_s - 1 \\ 0, & \text{otherwise} \end{cases} \tag{20.3.9}$$

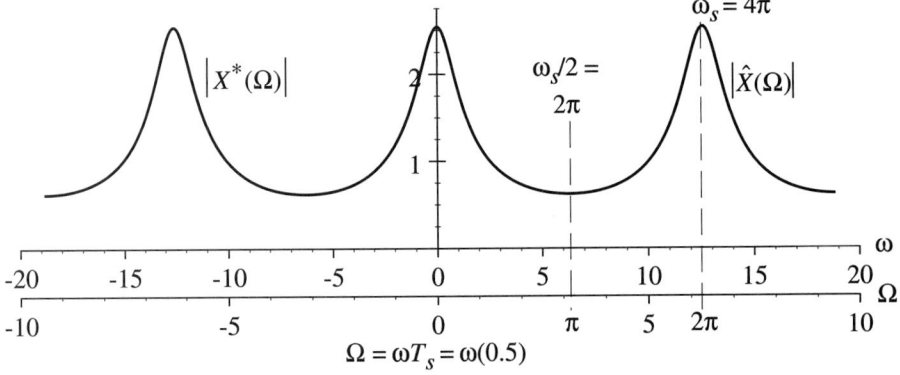

Figure 20.3.1 Amplitude Spectrum of the Sampled Signal

Notation: The tilde on $\tilde{x}(n)$ indicates that a finite number of samples are extracted from the sample values $\hat{x}(n)$ of the continuous-time signal $x(t)$.

The DTFT of the new signal (20.3.9) is

$$\tilde{X}(\Omega) = \sum_{n=-\infty}^{\infty} \tilde{x}(n)e^{-j\Omega n} = \sum_{n=0}^{N_s-1} \tilde{x}(n)e^{-j\Omega n} \qquad (20.3.10)$$

because the signal (20.3.9) is essentially of finite length. The DFT is the numerical evaluation of this DTFT. This summation can't be evaluated numerically yet because it is a continuous function of the frequency Ω.

Sampling the DTFT   The next step in the development of the DFT is to numerically compute the function $\tilde{X}(\Omega)$ in (20.3.10) by evaluating the summation at discrete frequency points. In particular, we will compute the DTFT for the discrete frequency sample points

$$\Omega_k = \frac{2\pi k}{N_s}, \quad k = 0, 1, 2, \ldots, N_s - 1. \qquad (20.3.11)$$

Note that the number of frequency points is the same as the number of points in the truncated signal (20.3.9). The frequency sampled DTFT is

$$\tilde{X}(k) = \sum_{n=0}^{N_s-1} \tilde{x}(n)e^{-j\Omega_k n} = \sum_{n=0}^{N_s-1} \tilde{x}(n)e^{-j\frac{2\pi n k}{N_s}} \qquad (20.3.12)$$

$$= \sum_{n=0}^{N_s-1} \tilde{x}(n)W^{kn}, \quad k = 0, 1, 2, \ldots, N_s - 1.$$

where

$$W = e^{-j\frac{2\pi}{N_s}}. \qquad (20.3.13)$$

The formula in (20.3.12) is the DFT.

Notation: The summation in (20.3.12) is simplified by the notation in (20.3.13) as shown in the last equality of (20.3.12). This notation is widely used.

Terminology: The factor W is sometimes called a <u>twiddle</u> factor.

Now we see that for a given frequency Ω_k we can compute the value of the transform $\tilde{X}(k)$ in (20.3.12) because there is only a finite number of terms in the sum because the sampled signal was truncated. Also we will compute the DFT at a finite number of frequency points (20.3.11) that are evenly spaced around the unit circle. This observation is illustrated in Example 20.4.1.

Inverse Discrete Fourier Transform   The formula in (20.3.12) allows us to compute an approximation to the DTFT from the signal sample values. We would also like to reverse this process, and compute the signal $\tilde{x}(n)$ which corresponds to the samples of the DTFT $\tilde{X}(k)$.   To accomplish this inversion, let n_0, $0 \le n_0 \le N_s - 1$ be fixed. Now multiply (20.3.12) by W^{-kn_0} and sum on k from 0 to $N_s - 1$. We obtain

$$\sum_{k=0}^{N_s-1} \tilde{X}(k) W^{-kn_0} = \sum_{k=0}^{N_s-1} \left[\sum_{n=0}^{N_s-1} \tilde{x}(n) W^{kn} \right] W^{-kn_0}. \tag{20.3.14}$$

Reversing the order of summation in (20.3.14) we obtain

$$\sum_{k=0}^{N_s-1} \left[\sum_{n=0}^{N_s-1} \tilde{x}(n) W^{kn} \right] W^{-kn_0} = \sum_{n=0}^{N_s-1} \tilde{x}(n) \left[\sum_{k=0}^{N_s-1} W^{k(n-n_0)} \right]. \tag{20.3.15}$$

Now it can be shown that

$$\sum_{k=0}^{N_s-1} W^{k(n-n_0)} = \begin{cases} N_s, & n = n_0 \\ 0, & n \ne n_0 \end{cases} \tag{20.3.16}$$

Combining (20.3.14) - (20.3.16) we obtain

$$\sum_{k=0}^{N_s-1} \tilde{X}(k) W^{-kn_0} = \sum_{n=0}^{N_s-1} \tilde{x}(n) \left[\sum_{k=0}^{N_s-1} W^{k(n-n_0)} \right] = N_s \tilde{x}(n_0). \tag{20.3.17}$$

From this analysis we conclude that

$$\tilde{x}(n) = \frac{1}{N_s} \sum_{k=0}^{N_s-1} \tilde{X}(k) W^{-kn}, \quad n = 0, 1, \dots, N_s - 1. \tag{20.3.18}$$

The equations (20.3.12) and (20.3.14) represent a reciprocal relationship between two sequences of numbers $\tilde{x}(n)$, $n = 0, 1, \dots, N_s - 1$ and $\tilde{X}(k)$, $k = 0, 1, \dots, N_s - 1$.   This relationship is unique because the Fourier series is unique.   This relationship between (20.3.12) and (20.3.18) can be identified as a transform and we do so next.

Definition 20.3.3: Let $\tilde{x}(n)$, $n = 0, 1, \ldots, N_s - 1$ be a sequence of real numbers. Then the <u>discrete Fourier transform</u> (DFT) of this signal is defined as

$$\tilde{X}(k) = \sum_{n=0}^{N_s-1} \tilde{x}(n) W^{kn}, \quad k = 0, 1, \ldots, N_s - 1, \qquad W = e^{-j\frac{2\pi}{N_s}}$$

The <u>inverse discrete Fourier transform</u> (IDFT) is defined as

$$\tilde{x}(n) = \frac{1}{N_s} \sum_{k=0}^{N_s-1} \tilde{X}(k) W^{-kn}, \quad n = 0, 1, \ldots, N_s - 1.$$

▲▲

The DFT in Definition 20.3.3 is an algorithm that can be used to compute an approximation of the Fourier transform of a continuous signal from the samples of the signal. This algorithm is the object of our search. It is widely used.

Terminology: When the number of points in the DFT is important, we say the DFT is an <u>N_s point</u> DFT.

Notation: We have chosen to represent the argument of the DFT by the index k, i.e., $\tilde{X}(k)$. This notation helps us to distinguish the DFT from the DTFT.

20.3.1 Summary of the DFT Development

The development above proceeded from the Fourier transform of the continuous-time signal to the DFT. To trace these steps in reverse, we have:

(1)   The DFT $\tilde{X}(k)$ in Definition 20.3.3 is derived by sampling in the frequency domain the DTFT of the truncated signal $\tilde{X}(\Omega)$ in (20.3.10).

(2)   The truncated signal $\tilde{x}(n)$ is extracted from the sampled signal $\hat{x}(n)$.

(3)   The DTFT of the sampled signal $\hat{X}(\Omega)$ is related to the continuous-time Fourier transform of the sampled signal $X^*(\omega)$ in Theorem 20.3.1.

(4)   The continuous-time Fourier transform of the sampled signal $X^*(\omega)$ is related to the Fourier transform of the continuous-time signal $X(\omega)$.

The hope is that the DFT approximates the continuous-time Fourier transform up to the Nyquist sampling frequency of f_s.

20.3.2 Properties of the DFT

The DFT has several properties that make it useful for signal and system analysis. In particular, the DFT plays a central role digital signal processing and digital filters. The DFT is derived from the DTFT and so some of its properties are inherited from the DTFT. For example, we can define an amplitude and phase spectra for the DFT exactly like the DTFT (we won't give a separate definition here). There is a major difference between the DFT and the DTFT, however. The DFT requires that the signal be of finite length. The finite length of the signal imposes added complexity into many of the properties of the DFT. These properties are beyond the scope of this text. We will be content to give two of the most important properties for the use of the DFT for signal analysis.

Because the DTFT is linear, the DFT is also linear.

Property 20.3.4:   (Linearity)   Given the DFT pairs $\tilde{x}_1(n) \leftrightarrow \tilde{X}_1(k)$ and $\tilde{x}_2(n) \leftrightarrow \tilde{X}_2(k)$ along with the real numbers a_1 and a_2 then

$$\mathcal{DFT}\{a_1\tilde{x}_1(n) + a_2\tilde{x}_2(n)\} = a_1\tilde{X}_1(k) + a_2\tilde{X}_2(k), \quad k = 0, 1, \ldots, N_s$$

▲▲

The definition of the energy of a discrete signal, Definition 20.1.1, extends directly to the truncated signals used to define the DFT. These truncated, bounded signals are always energy signals, because they have finite length and so the defining sum is always finite. As expected, Parseval's theorem also extends to the DFT.

Property 20.3.5: Parseval's Theorem Let $\tilde{x}(n)$ be a sequence of length N_s. Then the energy in this signal is given by

$$E_{\tilde{x}} = \sum_{n=0}^{N_s-1} |\tilde{x}(n)|^2 = \frac{1}{N_s} \sum_{k=0}^{N_s-1} |\tilde{X}(k)|^2.$$

▲▲

The definition of the energy spectral density, Definition 20.1.6, can easily be extended to the DFT via Property 20.3.5.

Parseval's theorem for the DTFT states that the energy in the signal can be calculated from the DTFT of the signal. This calculation, however, involves an infinite sum which can never be computed exactly. When calculating the energy using the DFT, the sum is finite and, hence, computable. The DFT is widely used to compute the energy in a signal. It must be remembered that the DFT requires the signal to be truncated. Hence, the energy calculated from the DFT is only an approximation to the original signal energy.

20.4 EXAMPLES OF THE DFT

20.4.1 Introduction

The other transforms we have introduced in this text were stated in terms of transform pairs. In particular, for previous transforms there was an emphasis on analytical representations of the function and its transform. The DFT, however, is used as a numerical tool. The purpose of this section is to show how to do the numerical calculations involved in the DFT. We will present two examples illustrating the use of the DFT. The first example is purely numerical. It illustrates the basic structure of the DFT algorithm. The second example computes the DFT of a simple discrete exponential function. Since we know the DTFT of this function, it is easy to show how the DFT is related to the DTFT for this particular function.

Frequently, the samples from which we compute the DFT are obtained by sampling a continuous signal. Then the DFT can be related to the Fourier transform of the continuous signal as discussed in the last section. In this section we give an explicit discussion of the relationship between the frequency scales of the DFT and the frequency scale of the Fourier transform of the continuous signal.

Finally, we know that the DFT computes an approximation to the Fourier transform at given points. To get a better representation out of the DFT, one way is to compute the DFT at more points. The standard method for increasing the number of points is to zero pad the DFT calculation. Zero padding is discussed at the end of this section.

20.4.2 Numerical Example

The following example illustrates the numerical calculation of the DFT of a specific signal.

Example 20.4.1: Consider the signal

$$\tilde{x}(n) = \{2 \quad 1 \quad -1 \quad 0 \quad 1 \quad -1\}. \tag{20.4.1}$$
$$\uparrow$$

Notation: The notation

$$\tilde{x}(n) = \{2 \quad 1\} \tag{20.4.2}$$
$$\uparrow$$

means that

$$\tilde{x}(0) = 2, \quad \tilde{x}(1) = 1, \quad \text{etc.} \tag{20.4.3}$$

The arrow indicates the $n = 0$ term.

To compute the DFT for this signal we will use Definition 20.3.3. For this signal, $N_s = 6$. So the twiddle factor is

$$W = e^{-j\frac{2\pi}{N_s}} = e^{-j\frac{2\pi}{6}} = e^{-j\frac{\pi}{3}} = 0.5 - j\frac{\sqrt{3}}{2} = a - jb. \tag{20.4.4}$$

Suppose that $\hat{k} = N_s + k, \quad 0 < k < N_s$. It then follows that

$$W^{\hat{k}} = W^{(k+N_s)} = W^k W^{N_s} = W^k \quad \text{because} \quad W^{N_s} = e^{-j\frac{2\pi N_s}{N_s}} = 1. \tag{20.4.5}$$

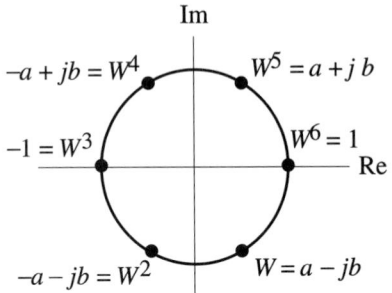

Figure 20.4.1 Locations of the Twiddle Factors for $N_s = 6$

Hence, we need only consider twiddle factors for $0 \le k < N_s$. The twiddle factors are all evenly spaced around the unit circle as shown in Figure 20.4.1. The twiddle factors are the complex frequencies at which we evaluate the DTFT to form the DFT. Figure 20.4.1 clearly shows that the DFT is computed at points evenly spaced around the unit circle. Now it is simply a matter of computing the sums in Definition 20.3.3. These calculations are shown next.

<u>k = 0</u>

$$\tilde{X}(0) = \sum_{n=0}^{N_s-1} \tilde{x}(n)W^{(0)n} = \sum_{n=0}^{5} \tilde{x}(n) = 2 \tag{20.4.6}$$

<u>k = 1</u>

$$\tilde{X}(1) = \sum_{n=0}^{5} \tilde{x}(n)W^{(1)n} \tag{20.4.7}$$

$$= \tilde{x}(0) + \tilde{x}(1)W + \tilde{x}(2)W^2 + \tilde{x}(3)W^3 + \tilde{x}(4)W^4 + \tilde{x}(5)W^5$$

$$= 2 + (1)(a - jb) + (-1)(-a - jb) + (0)(-1) + 1(-a + jb) + (-1)(a + jb) = 2$$

<u>k= 2</u>

$$\tilde{X}(2) = \sum_{n=0}^{5} \tilde{x}(n)W^{2n} \tag{20.4.8}$$

$$= \tilde{x}(0) + \tilde{x}(1)W^2 + \tilde{x}(2)W^4 + \tilde{x}(3)W^6 + \tilde{x}(4)W^8 + \tilde{x}(5)W^{10}$$

$$= \tilde{x}(0) + \tilde{x}(1)W^2 + \tilde{x}(2)W^4 + \tilde{x}(3)W^0 + \tilde{x}(4)W^2 + \tilde{x}(5)W^4$$

$$= 2 + (1)(-a - jb) + (-1)(-a + jb) + (0)(1) + (1)(-a - jb) + (-1)(-a + jb)$$

$$= 2 - j4b$$

The calculations for $k = 3, 4, 5$ are similar.

$\underline{k = 3}$

$$\tilde{X}(3) = \sum_{n=0}^{5} \tilde{x}(n)W^{3n} = 2 \qquad\qquad (20.4.9)$$

$\underline{k = 4}$

$$\tilde{X}(4) = \sum_{n=0}^{5} \tilde{x}(n)W^{4n} = 2 + j4b \qquad\qquad (20.4.10)$$

$\underline{k = 5}$

$$\tilde{X}(5) = \sum_{n=0}^{5} \tilde{x}(n)W^{5n} = 2 \qquad\qquad (20.4.11)$$

Now the DFT of (20.4.1) is

$$\tilde{X}(k) = \left\{ \underset{\uparrow}{2}, \quad 2, \quad 2 - j2\sqrt{3}, \quad 2, \quad 2 + j2\sqrt{3}, \quad 2 \right\}. \qquad (20.4.12)$$

The magnitude and phase spectra of the DFT in (20.4.12) are shown in Figure 20.4.2.
▲▲

20.4.3 Frequency Axis Scaling of the DFT

Quite often the discrete signal is obtained by sampling a continuous signal with a sample period of T_s. It is of interest to relate the DFT to the Fourier transform of the continuous signal. The most important issue is to relate the independent variable of the DFT, the index, to continuous frequency. This relationship can be traced through the development of the DFT given in the last section. The DFT is derived from the DTFT by sampling discrete frequency. The relationship between the index k and the discrete frequency sample points is

$$\Omega_k = \frac{2\pi k}{N_s}. \qquad\qquad (20.4.13)$$

Next we must relate discrete frequency to continuous frequency.   But this relationship is, by Definition 17.2.4,

$$\omega = \frac{\Omega}{T_s}. \qquad\qquad (20.4.14)$$

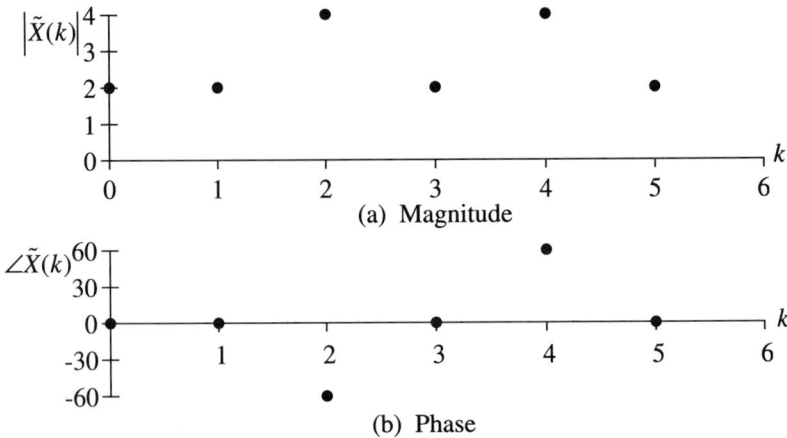

Figure 20.4.2 Magnitude and Phase Spectra of the Signal in Example 20.4.2

Using these two relationships the argument of the DFT can be alternatively expressed as

$$\underset{\substack{\uparrow \\ \text{index}}}{\tilde{X}(k)} = \tilde{X}\left(\underset{\substack{\uparrow \\ \Omega_k \\ \text{discrete} \\ \text{frequency}}}{\frac{2\pi k}{N_s}}\right) = \tilde{X}\left(\underset{\substack{\uparrow \\ \omega_k \\ \text{continuous} \\ \text{frequency}}}{\frac{2\pi k}{N_s T_s}}\right) \qquad (20.4.15)$$

The particular scale that is used to graph the DFT depends on the application at hand. The following example shows how to rescale the frequency axis.

Example 20.4.2: Consider again the DFT computed in Example 20.4.1. The graph of this DFT is shown in Figure 20.4.2 where the frequency axis is shown in terms of the index k. We can also graph the DFT in terms of discrete frequency. The frequency scaling is explained in (20.4.15). In this example $N_s = 6$, so that the discrete frequencies are

$$\Omega_k = \frac{2\pi k}{N_s} = \frac{2\pi k}{6} = \frac{\pi k}{3}. \qquad (20.4.16)$$

The magnitude of the DFT is shown in Figure 20.4.3 where the axis is ruled in discrete frequency.

Suppose that the signal $\tilde{x}(n)$ was obtained by sampling a continuous-time signal with a sampling frequency of $\omega_s = 6000$ rad/sec. This sampling frequency corresponds to a sampling period of

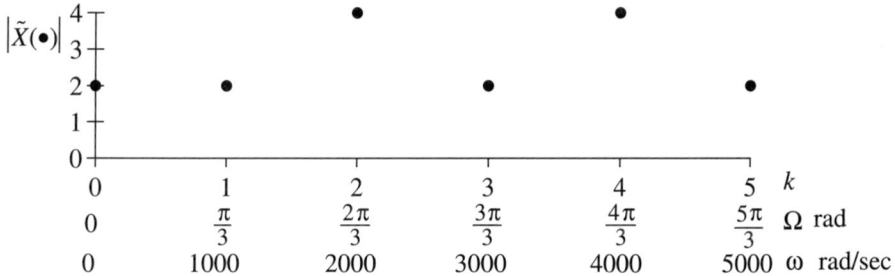

Figure 20.4.3 Frequency Axes of the DFT

$$T_s = \frac{2\pi}{\omega_s} = \frac{2\pi}{6000}. \tag{20.4.17}$$

Then the scale on the frequency axis in terms of continuous frequency is

$$\frac{2\pi k}{T_s N_s} = \frac{2\pi k}{\left(\dfrac{2\pi}{6000}\right)(6)} = 1000k. \tag{20.4.18}$$

This scale is also shown in Figure 20.4.3.

▲▲

20.4.4 An Analytical Example

The following example computes the DFT from the samples of a discrete exponential. Because of the simplicity of the signal, we are able to relate the DFT directly to the DTFT of the discrete signal. This example gives us some feel for the approximations involved with the DFT.

Example 20.4.3: This example is also discussed in Section 19.5 and in Example 20.1.2. Consider the signal

$$x(t) = e^{-\alpha t} u_s(t), \quad \alpha > 0. \tag{20.4.19}$$

The discrete-time sampled version of the exponential signal (20.4.19) is

$$\hat{x}(n) = a^n u_s(n), \quad a = e^{-\alpha T_s}. \tag{20.4.20}$$

The DTFT of this sampled signal is

$$\hat{X}(\Omega) = \sum_{n=-\infty}^{\infty} a^n u_s(n) e^{-j\Omega n} = \sum_{n=0}^{\infty} a^n e^{-j\Omega n} = \frac{1}{1 - ae^{-j\Omega}}. \tag{20.4.21}$$

We consider the use of the DFT to compute an approximation to the DTFT of this signal. To that end we extract a finite length sequence from the discrete-time sampled version of $x(t)$

$$\tilde{x}(n) = \hat{x}(n)\big[u_s(n - N_s) - u_s(n)\big] = a^n\big[u_s(n - N_s) - u_s(n)\big]. \tag{20.4.22}$$

Using these samples the DFT of the finite length sequence is

$$\tilde{X}(k) = \sum_{n=0}^{N_s-1} \tilde{x}(n) W^{kn} = \sum_{n=0}^{N_s-1} a^n e^{-j\frac{2\pi kn}{N_s}} = \frac{1 - a^{N_s} e^{-j\frac{2\pi k N_s}{N_s}}}{1 - ae^{-j\frac{2\pi k}{N_s}}} = \frac{1 - a^{N_s}}{1 - ae^{-j\frac{2\pi k}{N_s}}}. \tag{20.4.23}$$

Comparing (20.4.23) to the DTFT of this signal in (20.4.21) we see that they are different. This difference between (20.4.23) and (20.4.21) is the factor $1 - a^{N_s}$ which accounts for the fact that we extracted a finite length sequence $\tilde{x}(n)$ from the sampled signal (20.4.20). Note that as the number of sample points increases, the approximation to the DTFT gets better, because $a^{N_s} \to 0$ as $N_s \to \infty$.

Figure 20.4.4 shows the magnitude of the DTFT of the sampled signal along with the magnitude of the DFT for two different numbers of sample points. Figure 20.4.4

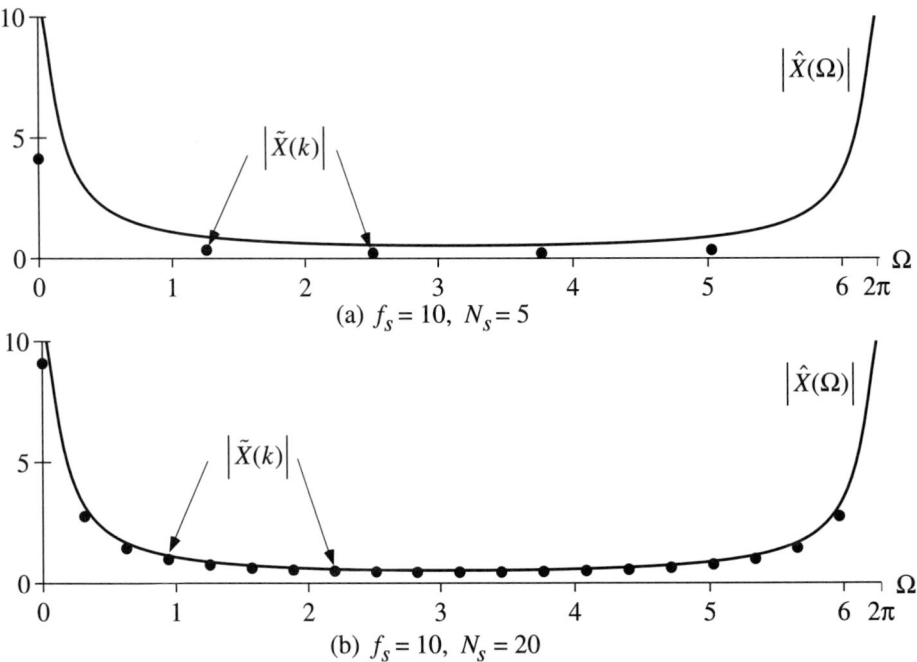

Figure 20.4.4 Magnitude of the DTFT of the Discrete Sampled Signal for $a = 1$ and the Magnitude of the DFT of This Signal

shows that as the numbers of points in the DFT increases, the approximation to the DTFT gets better, and the distance between the points decreases (interpolation gets better). Of course, the computational burden increases with more points.

▲▲

20.4.5 Zero Padding

Suppose we wish to calculate the DFT of the signal

$$\tilde{x}(n) = \left\{\tilde{x}(0), \quad \ldots, \quad \tilde{x}(N_s - 1)\right\} \tag{20.4.24}$$

Recall that the DFT is computed at N_s evenly spaced frequency points between 0 and 2π. Therefore, if we increase the number of points in the DFT we will obtain a better representation of the DTFT of the sampled signal. One way to increase the number of points is to add zeros to the end of the signal samples as

$$\tilde{x}_{pd}(n) = \left\{\tilde{x}(0), \quad \ldots, \quad \tilde{x}(N_s - 1), \quad 0, \quad \ldots, \quad 0\right\}. \tag{20.4.25}$$

If we have added L zeros to the signal, then the padded signal will be of length $L + N_s = N_p$. The DFT is now computed using the new total number of points. In this case the frequency spacing is

$$\Omega_k = \frac{2\pi k}{N_p} = \frac{2\pi k}{N_s + L} < \frac{2\pi k}{N_s}. \tag{20.4.26}$$

Clearly, the DFT is unchanged because we are just adding zero to the sum in the DFT. By using a large number of points, however, we are interpolating between the points of an N_s point DFT. The increase in the resolution in frequency gives a better graph. The next example illustrates zero padding.

Example 20.4.3: (*continued*) Consider again the signal in Example 20.4.3. The magnitude of the DTFT of the sampled signal along with the DFT of the truncated signal are shown in Figure 20.4.4a for $f_s = 10$ and $N_s = 5$. Suppose we zero pad the truncated signal in (20.4.22) to a length of $N_p = 20$. Then the magnitude of the DFT of the zero padded signal is shown in Figure 20.4.5 along with the two magnitudes shown in Figure 20.4.4. Note that the DFT of the zero padded signal passes through the DFT of the truncated signal in (20.4.22), and it interpolates in between the points of an $N_s = 5$ point DFT. In practice a signal is zero padded with enough points so that the interpolation can be shown as a continuous line.

▲▲

20.4.6 The FFT

To compute the DFT we must multiply the sample values by a complex number W^k and sum the results. Because these twiddle factors are points on the unit circle, these points have several axis of symmetry if the number of frequency samples is a power of 2; $N_p = 2^N$. In this case many of the computations are redundant and/or

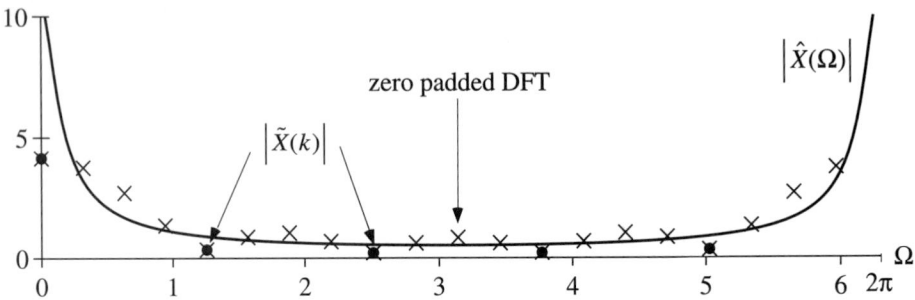

Figure 20.4.5 Zero Padding

can be obtained from the previous calculations by simple conjugation. This observation is the basis of the FFT, a fast *algorithm* to compute the DFT using the symmetry of the unit circle. The FFT yields the DFT of the sampled function. Most computer packages, such as MATLAB, provide an implementation of the FFT and denote it as such.

20.4.7 MATLAB Experiments

The command in MATLAB that calculates the DFT of a signal is **fft**. The input data to this command is the signal vector. The output data is a complex vector containing the DFT of the signal. The magnitude and phase of this vector must be computed independently. Also the frequency vector must be generated independently. In addition, the number of points in the FFT, say zp, can be specified. If zp is longer than the signal vector, then the signal vector is automatically zero padded. In practice, zp is chosen to be a power of 2.

Example 20.4.3 (*continued*) This M-file generates the graphs in Figure 20.4.4.

```
clear
fs = 10;                    % sampling frequency
Ts = 1/fs;                  % sampling period
Ns = 20;                    % number of points in DFT
n = 0:Ns-1;                 % time vector
alpha = 1;                  % continuous signal parameter
a = exp(-alpha*Ts);         % discrete signal parameter
x = a.^n;                   % discrete signal
wd = 2*pi/Ns;               % frequency spacing of DFT
ft = 40;                    % interpolation points for frequency axis
W = 0:wd/ft:2*pi;           % discrete frequency vector
X = 1./(1-a*exp(-j*W));     % DTFT
Xdft = fft(x);             % DFT
wdft = n*wd;                % frequency vector for DFT
plot(W,abs(X),wdft,abs(Xdft),'o')
title('Comparison of DTFT and DFT')
xlabel('frequency, rad')
ylabel('amplitude spectrum')
```

▲▲

Exploratory Exercise 20.4.4: For a fixed sampling frequency, vary the number of points in the DFT between $N_S = 5$ and 30. How does the amplitude spectrum of the DFT compare to the amplitude spectrum of the DTFT? Repeat this experiment for sampling frequencies between $f_s = 5$ and 30 Hz.

▲▲

Exploratory Exercise 20.4.5: Modify the MATLAB M-file above to include zero padding. (See MATLAB help on the command **fft**.) Modify the plot statement to include the points calculated from the zero padded DFT. To plot the magnitude of the DFT a frequency axis for the interval $[0, \pi]$ must be generated of points spaced by $(2\pi/zp)$.

Suppose the discrete signal is zero padded to a length of 32. Plot the amplitude spectrum for $N_S = 5, 10, 20$. How does the zero padded amplitude spectrum differ from the original amplitude spectrum? Specifically, how do the two amplitude spectra differ at the frequencies $\Omega_k = 2\pi k/N_s$?

▲▲

Exploratory Exercise 20.4.6: Modify the MATLAB M-file to calculate the Fourier transform of the continuous-time signal $x(t)$. Plot the amplitude spectrum of this Fourier transform on the plot with the amplitude spectrum of the DTFT and the DFT. Then repeat Exploratory Exercise 20.4.4.

▲▲

20.5 CHAPTER SUMMARY

In this chapter we have introduced several definitions pertaining to discrete-time signals and established their interrelationship. These results are summarized in Table 20.5.1.

Over the course of the text we have introduced the Fourier transform for continuous-time signals, the discrete-time Fourier transform for discrete signals, and the discrete Fourier transform. The continuous-time signal with its Fourier transform is

Continuous-Time Signal	Fourier Transform
$x(t)$	$X(\omega) = \dfrac{1}{2\pi} \displaystyle\int_{-\infty}^{\infty} x(t)e^{-j\omega t}\, d\omega$
	Definition 7.4.1

When we sample the continuous-time signal we obtain

Continuous-Time Signal	Fourier Transform
$x^*(t) = \displaystyle\sum_{n=-\infty}^{\infty} x(nT_s)\delta(t - nT_s)$	$X^*(\omega) = \dfrac{1}{T_s} \displaystyle\sum_{n=-\infty}^{\infty} X(\omega - n\omega_s)$
Definition 19.1.1	Theorem 19.1.3

Table 20.5.1 Signal Properties

Definition 20.1.1	amplitude spectrum	$\|X(\Omega)\|$
	phase spectrum	$\angle X(\Omega)$
Definition 20.1.3	energy	$E_x = \displaystyle\sum_{n=-\infty}^{\infty} \|x(n)\|^2$
Theorem 20.1.5	Parseval's theorem	$E_x = \displaystyle\sum_{n=-\infty}^{\infty} \|x(n)\|^2 = \frac{1}{2\pi}\int_{2\pi} \|X(\Omega)\|^2 \, d\Omega$
Definition 20.1.6	energy spectral density	$D_x(\Omega) = \|X(\Omega)\|^2$
Definition 20.1.7	bandlimit	$\|X(\Omega)\| = 0, \quad \pi \ge \|\Omega\| > BL$
Definition 20.2.1	periodic	$x(n) = x(n + N_0) \quad \text{for } N_0 > 0$
Definition 20.2.3	Fourier series	$x(n) = \displaystyle\sum_{k=0}^{N_0-1} c_k e^{j\frac{2\pi nk}{N_0}}$
		$c_k = \dfrac{1}{N_0} \displaystyle\sum_{n=0}^{N_0-1} x(n) e^{-j\frac{2\pi kn}{N_0}}$
Definition 20.2.5	amplitude spectrum	$\|c_k\| \quad \text{vs.} \quad \Omega_k = \dfrac{2\pi k}{N_0}$
	phase spectrum	$\angle c_k \quad \text{vs.} \quad \Omega_k = \dfrac{2\pi k}{N_0}$
Definition 20.2.7	power	$P_x = \displaystyle\lim_{K\to\infty} \frac{1}{2K} \sum_{n=-K}^{K} \|x(n)\|^2$
Theorem 20.2.10	Parseval's theorem	$P_x = \dfrac{1}{N_0} \displaystyle\sum_{n=0}^{N_0-1} \|c_k\|^2$
Definition 20.2.11	power spectral density	$\|c_k\|^2, \quad k = 0, 1, \ldots, N_0 - 1$
Definition 20.2.13	autocorrelation function	$r_{xx}(k) = \displaystyle\lim_{\kappa\to\infty} \frac{1}{2\kappa+1} \sum_{n=-\kappa}^{\kappa} x(n+k)x(n)$
Theorem 20.2.14	power spectral density	$S_x(\Omega) = \mathcal{DFT}\{r_{xx}(k)\}$

We can also think of the sampled signal as a discrete-time signal.

Discrete-Time Signal

$$\hat{x}(n) = x(nT_s)$$

Definition 17.2.1

Discrete-Time Fourier Transform

$$\hat{X}(\Omega) = \sum_{n=-\infty}^{\infty} \hat{x}(n)e^{-j\Omega n}$$

Definition 18.4.1

The relationship between the Fourier transform of the sampled signal and the DTFT of that signal is (Theorem 20.3.1)

$$X^*(\omega)\Big|_{\Omega=T_s\omega} = \hat{X}(\Omega)$$

where T_s is the sampling period. Given a discrete signal $\hat{x}(n)$ we extract a finite length sequence to define the DFT.

Discrete-Time Signal

$$\tilde{x}(n) = \begin{cases} \hat{x}(n), & n = 0,1,\ldots,N_s-1 \\ 0, & \text{otherwise} \end{cases}$$

Equation (20.3.4)

Discrete Fourier Transform

$$\tilde{X}(k) = \sum_{n=0}^{N_s-1} \tilde{x}(n)e^{-j\frac{2\pi nk}{N_s}}, k = 0,\ldots,N_s-1$$

Definition 20.3.3

In practice, we sample the continuous-time signal, extract a finite number of samples, and calculate the DFT. We try to select the sampling frequency and the number of data points such that the DFT will faithfully reproduce the Fourier transform of the continuous-time signal. The DFT can be expressed in terms of the index, discrete frequency, or continuous frequency. This relationship can be summarized by

$$\underset{\substack{\uparrow \\ \text{index}}}{\tilde{X}(k)} = \underset{\substack{\uparrow \\ \Omega_k \\ \text{discrete} \\ \text{frequency}}}{\tilde{X}\left(\frac{2\pi k}{N_s}\right)} = \underset{\substack{\uparrow \\ \omega_k \\ \text{continuous} \\ \text{frequency}}}{\tilde{X}\left(\frac{2\pi k}{N_s T_s}\right)} \qquad (20.4.15)$$

20.6 HOMEWORK FOR CHAPTER 20

Homework Problems for Section 20.1

20.1.1 The energy spectral density (calculated from the DFT) of a signal is shown
in Figure P20.1.1.
(a) What is the bandwidth of this signal is discrete frequency?
(b) Suppose that the discrete signal was obtained by sampling a continuous
signal with a sample rate of $f_s = 10$ Hz. What is the bandwidth of the
continuous-time signal?
(c) Estimate the energy in this signal.

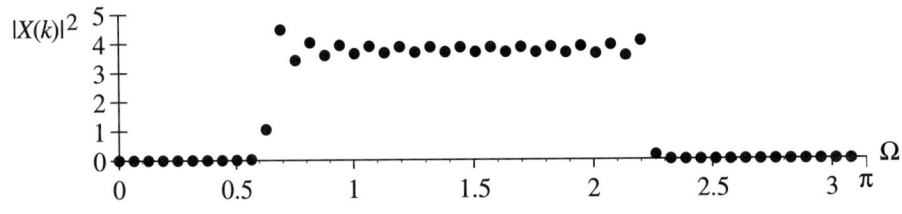

Figure P20.1.1

20.1.2 The energy spectral density (calculated from the DFT) of a signal is shown
in Figure P20.1.2.
(a) What is the bandwidth of this signal in discrete frequency?
(b) Suppose that the discrete signal was obtained by sampling a continuous
signal with a sample rate of $f_s = 30$ kHz. What is the bandwidth of the
continuous-time signal?
(c) Estimate the energy in this signal.

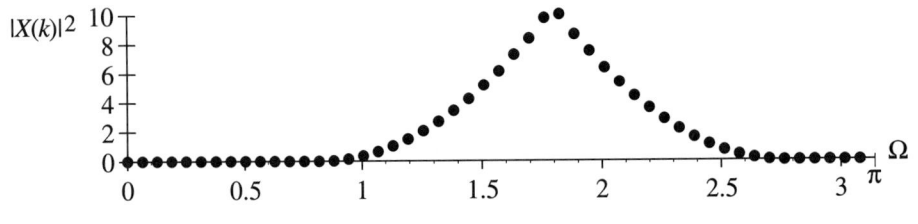

Figure P20.1.2

20.1.3 For each of the following signals:
(a) Using MATLAB calculate the FFT of this signal using the time interval
and the number of points shown.
(b) Plot the energy spectral density of the signal.
(c) Estimate the energy in the signal using MATLAB.
  (i) $x(t) = [Sa(3t)]^2 \cos(50t), \quad -8 \leq t \leq 8, \quad N_s = 100$
  (ii) $x(t) = [Sa(5t)]\cos(30t), \quad -8 \leq t \leq 8, \quad N_s = 100$

20.1.4 Consider the signal

$$x(n) = \begin{cases} A, & 0 \le n \le M-1 \\ 0, & \text{otherwise} \end{cases}$$

(a) Show that the DTFT of this signal is given by

$$X(\Omega) = A\frac{1-e^{-j\Omega M}}{1-e^{-j\Omega}} = Ae^{-j\Omega\frac{M-1}{2}}\left[\frac{\sin\left(\dfrac{\Omega M}{2}\right)}{\sin\left(\dfrac{\Omega}{2}\right)}\right]$$

(b) Plot the amplitude spectrum of this signal for $M = 10$.
(c) Plot the phase spectrum of this signal for $M = 10$.
(d) Plot the energy spectrum of this signal for $M = 10$.

20.1.5 Determine whether the following signals are energy signals. If they are, find their energy.

(i) $x(n) = 0.5^n u_s(n)$

(v) $x(n) = \sin\left(\dfrac{\pi n}{4}\right)$

(ii) $x(n) = 2^n u_s(n)$

(vi) $x(n) = \dfrac{4}{\pi n}\sin\left(\dfrac{\pi n}{4}\right)$

(iii) $x(n) = \left(\dfrac{1}{n}\right)u_s(n-1)$

(vii) $x(n) = \left(\dfrac{4}{\pi n}\right)^{\frac{1}{2}}\sin\left(\dfrac{\pi n}{4}\right)$

(iv) $x(n) = \left(\dfrac{1}{n}\right)^{\frac{1}{2}}u_s(n-1)$

(viii) $x(n) = \cos\left(\dfrac{\pi n}{4}\right)(u_s(n) - u_s(n-9))$

20.1.6 Suppose that the signal

$$x(t) = Ae^{-\alpha t}u_s(t), \quad \alpha > 0$$

is sampled with a sampling frequency of f_s. Suppose we know that the energy in the continuous-time signal is $E_{cont} = 30$ and the energy in the discrete-time signal is $E_{dis} = 10$. Find a bound on the time constant of the signal and the sampling frequency.

20.1.7 Consider the signal

$$x(n) = \text{sinc}\left(\dfrac{\pi n}{30}\right)\cos(\Omega_0 n).$$

(a) Find the DTFT of this signal.
(b) Plot the amplitude spectrum for $\Omega_0 = 0.5\pi$.

(c) Plot the bandwidth as a function of Ω_0.

(d) For what frequencies is this signal narrowband?

Homework Problems for Section 20.2

20.2.1 Calculate the power in the following signals.

(i) $x(n) = u_s(n)$

(ii) $x(n) = Ae^{j\Omega_0 n}$

(iii) $x(n) = (-1)^n$

(iv) $x_p(n) = r_p(n), \quad 0 \le n < 5$

$x(n) = $ periodic extension of $x_p(n)$

20.2.2 (a) Find the exponential Fourier series of the following signals.

(b) Plot the amplitude and phase spectra.

(c) Plot the power spectrum.

(d) Calculate the power in the signal using Parseval's theorem.

(i) $x(n) = \cos\left(\frac{\pi n}{3}\right)$

(ii) $x(n) = \sum_{k=1}^{5} \frac{1}{k} \cos\left(\frac{\pi k n}{6}\right)$

(iii) $x(n) = \sum_{k=-\infty}^{\infty} \delta(n - 10k)$

(iv) $x_p(n) = r_p(n), \quad 0 \le n < 5$

$x(n) = $ periodic extension of $x_p(n)$

20.2.3 Let $x(t)$ be a periodic signal with period T_0. Suppose that this signal is sampled with a sampling period of T_s. Find the relationship between the period of the periodic signal and the sampling period such that the sampled signal is periodic.

20.2.4 Let $x(n)$ be a periodic signal with a period of N_0. Suppose that this periodic signal has a Fourier series with coefficients c_k. Verify the following properties of this Fourier series.

signal	Fourier coefficients
$x(n - n_0)$	$c_k e^{-j\frac{2\pi n_0 k}{N_0}}$
$x(-n)$	c_{-k}
$x^*(n)$	c_{-k}^*
$e^{j\frac{2\pi n M}{N_0}} x(t)$	c_{k-M}
$x(n) - x(n-1)$	$(1 - e^{-j\frac{2\pi k}{N_0}})c_k$
$\sum_{k=-\infty}^{n} x(k), \quad c_0 = 0$	$(c_k)\left[1 - e^{-j\frac{2\pi k}{N_0}}\right]^{-1}$

20.2.5   Prove Theorem 20.2.10.

20.2.6   Given two energy signals, show that their autocorrelation function can be expressed in terms of a certain convolution; i.e.,

$$r_{xy}(k) = \sum_{n=-\infty}^{\infty} x(n)y(n-k) = x(n) * y(-n)$$

20.2.7   Find the energy spectral density of the following signals using the autocorrelation function.

(i) $x(n) = Aa^n u_s(n)$ (ii) $x(n) = \Pi_2(n)$

20.2.8   Suppose that $x(n)$ is a periodic signal. Show that the power spectral density is the same whether it is calculated using Theorem 20.2.14 or Definition 20.2.11. Hint: $c_k = \bar{c}_{N_0 - k}$

20.2.9   Consider the signal

$$x(n) = (0.8)^n u_s(n).$$

This signal is truncated to $N_s = 5$ samples to create the signal $\tilde{x}(n)$.
(a) Calculate the autocorrelation function $\tilde{r}_{xx}(n)$ of the truncated signal.
(b) Calculate the energy spectral density of the truncated signal from the autocorrelation function. Compare this energy spectral density to the energy spectral density of the original signal.
(c) Repeat these calculations for truncated signals of lengths $N_s = 10, 20$.

Homework Problems for Section 20.3

20.3.1   Consider the DFT in Definition 20.3.3. Define two vectors

$$\vec{x} = \left[\tilde{x}(0), \ldots, \tilde{x}(N_s - 1) \right]^T$$
$$\vec{X} = \left[\tilde{X}(0), \ldots, \tilde{X}(N_s - 1) \right]^T$$

Show that the definition of the DFT implies that these two vectors are related by the matrix equation

$$\vec{X} = \Phi \vec{x}$$

where Φ is a $N \times N$ matrix whose elements are powers of the twiddle factors.

20.3.2 Show that the DFT of the signal

$$x(n) = \begin{cases} 1, & k = 0, 1, 2 \\ 0, & k = 3, \ldots, 7 \end{cases}$$

is given by

$$\tilde{X}(k) = \left[e^{-j\frac{k\pi}{4}} \right] \frac{\sin\left(\frac{k\pi}{3}\right)}{\sin\left(\frac{k\pi}{8}\right)}$$

Compare this answer with Example 20.2.4. Why are they the same?

20.3.3 Show that the twiddle factors satisfy

$$\sum_{k=0}^{N_s-1} W^{k(n-n_0)} = \begin{cases} N_s, & n = n_0 \\ 0, & n \neq n_0 \end{cases}$$

20.3.4 Suppose the continuous-time signal

$$x(t) = Ae^{-\alpha t} u_s(t)$$

is sampled with a sampling frequency of f_s to yield the discrete-time signal

$$\hat{x}(n) = A(a)^n u_s(n).$$

(a) Find the constant that relates the energy in the continuous-time signal to the energy in the discrete-time signal.
(b) Suppose we truncate the discrete-time signal to N_s samples. How many samples should we retain so that the energy in the truncated signal is 90% of the energy in $\hat{x}(n)$?

Homework Problems for Section 20.4

20.4.1 For the following discrete signals:
(a) Plot the twiddle factors W^k in the z-plane.
(b) Find the DFT of each signal.
(c) Plot the magnitude and phase of the DFT.
   (i) $\tilde{x}(n) = \{1, \ 2, \ -1, \ -3, \ 1\}$ (iii) $\tilde{x}(n) = \{1, \ -2, \ -3\}$
   (ii) $\tilde{x}(n) = \{-2, \ 3, \ 4\}$ (iv) $\tilde{x}(n) = \{-1, \ 0, \ 3, \ -2, \ 1, \ 0\}$

20.4.2 Consider the following discrete signals.
(a) Plot the twiddle factors W^k in the z-plane.
(b) Find the DFT of each signal.

(c) Plot the magnitude and phase of the DFT.

Note: These signals are zero padded signals of Problem 20.4.1.

(i) $\tilde{x}(n) = \{1, \quad 2, \quad -1, \quad -3, \quad 1, \quad 0, \quad 0, \quad 0\}$

(ii) $\tilde{x}(n) = \{-2, \quad 3, \quad 4, \quad 0, \quad 0, \quad 0\}$.

(iii) $\tilde{x}(n) = \{1, \quad -2, \quad -3, \quad 0, \quad 0, \quad 0, \quad 0, \quad 0\}$.

20.4.3 Consider the signal

$$x(t) = \left[e^{-2t} \sin(8t) \right] u_s(t)$$

Suppose that this signal sampled, and we want to calculate its DFT from the samples.

(a) Find the Fourier transform of this signal. (Use the Laplace transform.)

(b) Find the DTFT of this signal.

(c) Write a MATLAB M-file to do the following calculations.

(i)   Plot the time history of this sampled signal in a separate figure window.

(ii)   Plot the amplitude spectrum of the continuous signal and the amplitude spectrum of the DTFT of the signal in continuous frequency.

(iii)   Using a finite number of samples of this signal, calculate the DFT. (Start with $f_s = 5$ and $N_s = 5$.)   Plot the amplitude spectrum of the DTFT of the sampled signal, and the amplitude spectrum of the DFT on the same plot using discrete frequency. Also plot this graph using discrete frequency. Investigate the effect of changing the number of samples in the DFT.

(iv)   Zero pad the sampled signal. Plot the amplitude spectrum of the DTFT and the zero padded DFT on the same graph in discrete frequency. Determine the effect of the number of zeros in the zero padding.

(d) Through numerical experimentation, investigate the relationship between:

(i)   The sampling frequency and the frequency of the sinusoid

(ii)   The sampling frequency and the damping coefficient

(iii)   The number of points in the DFT and the damping coefficient

(iv)   The number of points in the DFT and the frequency of the sinusoid

20.4.4 Consider the following discrete signal.

$$\tilde{x}(n) = \{-1, \quad 0, \quad 2, \quad 1, \quad 0, \quad -1, \quad 0, \quad 0\}$$

(a) Find the twiddle factors W^k for this signal.

(b) Calculate $\tilde{X}(5)$ of the DFT. Clearly show your work.

20.4.5 Modify the M-file in Section 20.4 to calculate the DFT of the following signals.

(i) $x(t) = e^{-\alpha|t|}$, $\alpha > 0$ (iii) $x(t) = e^{-\alpha(t-t_0)}u_s(t-t_0)$, $\alpha > 0$

(ii) $x(t) = e^{-\alpha|t-t_0|}$, $\alpha > 0$ (iv) $x(t) = e^{\alpha t}$, $\alpha > 0$

20.4.6 Consider the signal $\hat{x}(n) = a^n u_s(n)$, which is discussed in Example 20.4.3. Let $a = 0.9$. Find the number of samples N_s such that the $\left\| \hat{X}(\Omega_k) \right| - \left| \tilde{X}(k) \right\| < 10^{-6}$.

20.4.7 Let $\tilde{x}(n)$ be a length 4 sequence. Denote the DFT of this sequence by $\tilde{X}_4(k)$. Suppose we wish to zero pad this sequence with one zero. Denote the DFT of the zero padded sequence by $\tilde{X}_5(k)$. If these DFTs are arranged into vectors, then they are related as

$$\vec{\tilde{X}}_5 = \Psi \vec{\tilde{X}}_4$$

(a) Find the matrix Ψ.
(b) Write a MATLAB M-file to construct the matrix Ψ.
(c) Consider the signal

$$\tilde{x}(n) = \{0,\quad 0.8,\quad 0.64,\quad 0.512\}$$

Calculate the DFT of this signal. From this DFT calculate the DFT of a zero padded signal using the method above.

20.4.8 Consider the finite length sequence

$$x(n) = \{1,\quad -0.2,\quad 0.4,\quad -0.1\}.$$

(a) How will the DFT differ from the DTFT?
(b) How will zero padding affect the relationship between the DFT and the DTFT?

20.4.9 For each of the following DFTs:
(a) Plot the twiddle factors.
(b) Find $x(n)$ using the IDFT.
   (i) $X(k) = \{1.875,\quad 0.75 - j0.375,\quad 0.625,\quad 0.75 + j0.375\}$
   (ii) $X(k) = \{5,\quad -j2.4142,\quad 1,\quad -j0.4142,\quad 1,\quad j0.4142,\quad 1,\quad j2.4142\}$
   (iii) $X(k) = \{9,\quad -2.12 - j1.54,\quad 0.12 + j0.36,\quad 0.12 - j0.36,\quad -2.12 + j1.54\}$
   (iv) $X(k) = \{-4,\quad 4.5 + j2.6,\quad 0.5 + j0.866,\quad 0,\quad 0.5 - j0.866,\quad 4.5 - j2.6\}$

20.4.10 (a) Find the DTFT of the signal

$$\hat{x}(n) = \cos(\Omega_0 n).$$

(b) Find the DTFT of the signal

$$\hat{x}_w(n) = \cos(\Omega_0 n)\big[u_s(n) - u_s(n - N_s)\big]$$

Plot the amplitude spectrum of this DTFT for $\Omega_0 = 0.5\pi$ and label Ω_0.

(c) Suppose that the DFT is used to compute the DTFT in (b). Find the relationship between Ω_0 and N_s such that

$$\tilde{X}(k) = \begin{cases} A, & \dfrac{2\pi k}{N_s} = \Omega_0 \\ 0, & \text{otherwise} \end{cases}$$

(d) When $N_s = 90, 94, 98, 104$, calculate the DFT of

$$x(n) = \cos\left(\frac{2\pi n}{15}\right).$$

Note: This calculation shows how to use the DFT to identify the frequency of a sinusoid if it is not known.

20.4.11 Consider the signal

$$\hat{x}(n) = \cos(\Omega_1 n) + \cos(\Omega_2 n) = \cos(0.5\pi n) + \cos(0.55\pi n).$$

(a) If $N_s = 20$, calculate the DFT and plot the amplitude spectrum.
(b) If $N_s = 100$, calculate the DFT and plot the amplitude spectrum.
(c) In (a) and (b) attempt to estimate the frequencies in $\hat{x}(n)$ from the amplitude spectrum. Is it possible in both cases?
(d) Find the DTFT of this signal using the result in Problem 20.4.10(b).
(e) Show that to resolve the two frequencies in $\hat{x}(n)$ the number of sample points must satisfy

$$|\Omega_1 - \Omega_2| > \frac{2\pi}{N_s}.$$

(f) Suppose that the continuous-time signal

$$x(t) = \cos(\omega_1 t) + \cos(\omega_2 t)$$

is sampled with a sampling period of T_s to give the discrete signal shown above. Extend the inequality in (e) to include this sampling period.

20.4.12 Consider the signal

$$x(n) = \sin\left(\frac{\pi n}{20}\right)\sin\left(\frac{\pi n}{20} + \phi\right).$$

(a) If $N_S = 100$ and $\phi = 0$, calculate the autocorrelation function of this signal.

(b) Calculate and plot the energy spectral density from the autocorrelation function.

(c) Repeat (a) and (b) for $\phi = \frac{\pi}{4}, \frac{\pi}{2}, \frac{\pi}{3}, \pi$. Comment on the changes in the energy spectral density.

20.4.13 When a keypad is pushed on a standard telephone keypad two tones (sinusoids at specified frequencies) are generated. The frequencies for each keypad are shown in the following table. Note that each dual tone consists of one low frequency and one high frequency. The problem is to determine which frequencies are in the received signal. Assume that the sampling frequency is 8 kHz.

(a) Based on the resolution limit of the DFT, determine the minimum number of samples necessary to resolve the two tones in all keypads using an N_S point DFT where N_S is a power of 2.

(b) Calculate the energy in the two DFT samples at each frequency as a percentage of the total energy in the signal.

(c) Write an M-file to automatically detect each dual tone using a $4N_S$ point DFT. (Zero pad the DFT to $4N_S$.)

Table P20.4.13 Dual Tone Frequencies

frequencies, Hz	1209	1336	1477	1633
697	1	2	3	A
770	4	5	6	B
852	7	8	9	C
941	*	0	#	D

Discrete-Time System Representations

In Chapter 17 we introduced five system representations: convolution summations, difference equations, transfer functions, state space representations, and the DTFT transfer function. In this chapter we will discuss each of these system representations in more detail. Each system representation is defined and several examples of these systems are given.

The whole development of this chapter is entirely analogous to the discussion of continuous-time systems in Chapters 10 to 12. Our presentation accents the similarities between continuous and discrete system. There are several important differences between these two classes of systems, and these differences are pointed out as well.

We also introduce block diagrams as another representation for transfer functions. Using block diagrams we show how difference equations correspond to all-delay block diagrams. These results are analogous to the canonical all-integrator block diagrams discussed in Chapter 11. The presentation of this material is slightly different, however. We adopt the viewpoint of digital filters, and develop the system structures within that framework.

As with continuous systems, these discrete system representations are all closely related to each other. These relationships are developed in the next chapter.

Summary of Sections

Section 21.1:   We discuss discrete convolution.

Section 21.2:   We discuss difference equations and transfer functions.

Section 21.3:   We introduce block diagrams for discrete systems.

Section 21.4:   We introduce the DTFT transfer function.

Section 21.5:   We develop state space representations from all-delay block diagrams.

Section 21.6:   We show how to decompose a system into a cascade and parallel interconnection structure.

Section 21.7:   Chapter summary section.

Coverage of the Text

The results in this chapter require only the definition of discrete-time signals and systems in Chapter 17 and z-transforms in Chapter 18. The development of the concepts for discrete-time systems parallels the development of continuous-time systems, and frequent references are made to Chapters 10 - 12; however, this material is not strictly required except for Section 21.6. Section 21.6 requires the solution to state space equations developed in Section 12.3 and the notion of a ZOH.

21.1 DISCRETE CONVOLUTION

21.1.1 Introduction

In this chapter we consider the discrete-time system

$$y(n) = \mathcal{H}\big[x(n)\big].$$

(21.1.1)

This abstract expression for a system is used primarily to remind us of the definition of a system. Recall that the definition of a system, Definition 6.1.1, is in terms of the input signal and output signal. Given an input signal the system produces an output signal. In order to apply this concept we must have a concrete mathematical expression for this system, called a representation, Definition 6.1.8. It is the purpose of this chapter to introduce several representations of systems, and develop some of their properties.   In this section we will introduce the first representation, the convolution representation.

21.1.2 Definitions

The first representation of the system (21.1.1) which we will discuss in this chapter is

$$y(n) = \sum_{k=-\infty}^{\infty} h(n-k)x(k). \tag{21.1.2}$$

Definition 21.1.1: The system representation (21.1.2) is called the <u>convolution representation</u> of the system (21.1.1).

▲▲

Terminology: The convolution representation is also called a <u>convolution summation</u>.

Notation: The system representation in (21.1.2) expresses a relationship between an input signal $x(n)$ and an output signal $y(n)$. Comparing the convolution summation in (21.1.2) to the expression for a system in (21.1.1) we see that this relationship is expressed by: 1) the function $h(n)$, and 2) the summation sign. It is important to remember that $x(n)$ and $y(n)$ are signals and $h(n)$ is related to the system.

This discrete-time convolution representation of a system is very similar to the convolution integral used as a representation of a continuous-time system. In general the results for the continuous-time convolution representation in Sections 12.1 and 12.2 have equivalent concepts in the discrete-time.
If the input signal to the convolution integral is the unit impulse signal $x(n) = \delta(n)$ then the system output is

$$y_i(n) = \sum_{k=-\infty}^{\infty} h(n-k)x(k) = \sum_{k=-\infty}^{\infty} h(n-k)\delta(k) = h(n). \tag{21.1.3}$$

Based on this calculation we have the following definition.

Definition 21.1.2: The function $h(n)$ in the convolution representation is called the <u>impulse response function</u> of the system.

▲▲

The following example illustrates the direct evaluation of the convolution sum.

Example 21.1.3: Consider the system

$$y(n) = \sum_{k=-\infty}^{\infty} h(k)x(n-k) = \sum_{k=-\infty}^{\infty} \left[a^k u_s(k) \right] x(n-k) = \sum_{k=0}^{\infty} \left[a^k \right] x(n-k). \tag{21.1.4}$$

If the input signal is a unit step function, $x(n) = u_s(n)$, then the output signal is

$$y(n) = \sum_{k=0}^{\infty} \left[a^k \right] u_s(n-k). \tag{21.1.5}$$

In the summation in (21.1.5) the unit step function is zero for $k > n$. Hence, (21.1.5) can be written as

$$y(n) = \sum_{k=0}^{\infty} \left[a^k \right] u_s(n-k) = \sum_{k=0}^{n} a^k. \tag{21.1.6}$$

Because (21.1.6) is a geometric series we have

$$y(n) = \sum_{k=0}^{n} a^k = \frac{1 - a^{n+1}}{1 - a}. \tag{21.1.7}$$

▲▲

21.1.3 FIR Systems

We can further classify systems based on the impulse response function.

Definition 21.1.4: Let $h(n)$ be an impulse response for a system. Then this system is a <u>finite impulse response (FIR) system</u> if $h(n) = 0$ for all but a finite number of n's. Otherwise, this system is an <u>infinite impulse response (IIR) system</u>.

▲▲

The similarity between the mathematical theory of continuous and discrete signals and systems is striking. In many cases the results from continuous systems translates directly to discrete systems. There are a few essential differences between continuous and discrete systems, however. Discrete FIR systems have no important equivalent in continuous-time systems. FIR systems also play an important role in digital filtering.

Example 21.1.5: (Averaging) One of the simplest discrete systems is averaging over the past N input samples. The difference equation is

$$y(n) = \frac{1}{N} \sum_{k=0}^{N-1} x(n-k). \tag{21.1.8}$$

The convolution representation can be found directly from (21.1.8). Equating (21.1.8) to (21.1.2) we see the impulse response function should satisfy

$$h(n) = \begin{cases} \frac{1}{N}, & 0 \le n \le N-1 \\ 0, & \text{elsewhere} \end{cases} \tag{21.1.9}$$

Hence, the difference equation (21.1.8) is already a convolution representation. The impulse response function in (21.1.8) has only N nonzero values. Hence, this system is an FIR system.

▲▲

The convolution summation lends itself to direct computation, particularly for FIR systems. The next example illustrates this computation.

Example 21.1.6: Suppose that the impulse response function of an FIR system is given by

$$h(n) = (2-n)\left[u_s(n) - u_s(n-2)\right]. \tag{21.1.10}$$

Suppose also that the input signal is given by

$$x(n) = (n+1)\left[u_s(n) - u_s(n-2)\right]. \tag{21.1.11}$$

Then the output signal can be determined by direct calculation as follows:

$$y(0) = \cdots + h(1)x(-1) + h(0)x(0) + h(-1)x(1) + h(-2)x(2) + \cdots \tag{21.1.12}$$
$$= \cdots + (1)(0) + (2)(1) + (0)(2) + (0)(0) + \cdots = 2$$
$$y(1) = \cdots + h(2)x(-1) + h(1)x(0) + h(0)x(1) + h(-1)x(2) + \cdots$$
$$= \cdots + (0)(0) + (1)(1) + (2)(2) + (0)(0) + \cdots = 5$$
$$y(2) = \cdots + h(3)x(-1) + h(2)x(0) + h(1)x(1) + h(0)x(2) + \cdots$$
$$= \cdots + (0)(0) + (0)(1) + (1)(2) + (2)(0) + \cdots = 2$$
$$y(3) = \cdots + h(4)x(-1) + h(3)x(0) + h(2)x(1) + h(1)x(2) + \cdots$$
$$= \cdots + (0)(0) + (0)(1) + (0)(2) + (1)(0) + \cdots = 0$$

▲▲

21.1.4 Other Forms of the Convolution Summation

The form of the convolution summation in (21.1.2) can be simplified under various assumptions. If the input signal $x(n)$ begins at a finite time, $n = 0$ say, then the convolution summation is

$$y(n) = \sum_{k=0}^{\infty} h(n-k)x(k). \tag{21.1.13}$$

If we assume further that the impulse response function satisfies $h(n-k)=0,\ \ k>n$, then the summation in (21.1.13) becomes

$$y(n) = \sum_{k=0}^{n} h(n-k)x(k). \tag{21.1.14}$$

The summation in (21.1.14) is finite for finite values of n and it always exists.

Consider again the convolution summation in (21.1.2). It is easily shown that by a change of variables of summation

$$y(n) = \sum_{k=-\infty}^{\infty} h(n-k)x(k) = \sum_{i=-\infty}^{\infty} h(i)x(n-i). \tag{21.1.15}$$

There is a simplified mathematical expression for the convolution representation based on the equivalence of the expressions in (21.1.15).

Notation: Since, the functions $x(n)$ and $h(n)$ enter into the summation in (21.1.15) symmetrically, we can write

$$y(n) = \sum_{k=-\infty}^{\infty} h(n-k)x(n) = \sum_{i=-\infty}^{\infty} h(i)x(n-i) = h(n) * x(n). \tag{21.1.16}$$

The usefulness of this notation is illustrated with a cascade connection of two systems shown in Figure 21.1.1. The cascade system in Figure 21.1.1 has a convolution representation obtained by composing the convolution representation of the two system. This expression is rather messy, however. Using the notation in (21.1.16) the composite system in Figure 21.1.1 can be represented as

$$y(n) = h_2(n) * h_1(n) * x_1(n). \tag{21.1.17}$$

The order of the filters in Figure 21.1.1 or in (21.1.17) can be reversed, but the relationship between the input and output signals will remain the same.

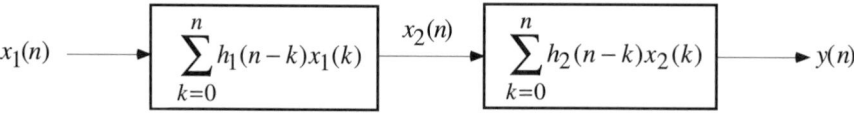

Figure 21.1.1 Cascade System Interconnection

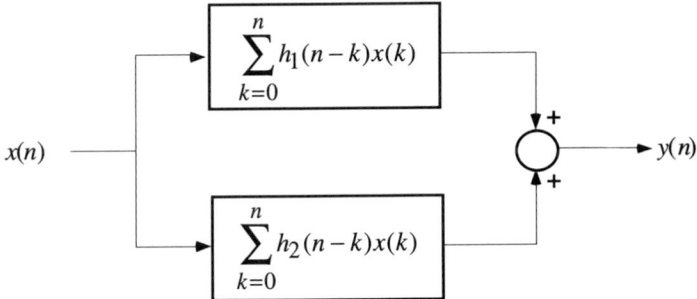

Figure 21.1.2 Parallel System Interconnection

The convolution representation can also be used to represent the parallel interconnection of two systems as shown in Figure 21.1.2. It is easily shown that the convolution representation for the system in Figure 21.1.2 is given by

$$y(n) = [h_2(n) + h_1(n)] * x(n). \tag{21.1.18}$$

21.1.5 MATLAB Experiments

The MATLAB command for performing discrete convolution is **conv**. The input data is the two signals to be convolved. The output data is the convolved signal. The following M-file illustrates the use of this command through the implementation of the averaging system in Example 21.1.5 above.

Example 21.1.5: (*continued*) Averaging

```
clear
n = [0:20];           % time vector for input signal
x = n;                % input signal

N = 3;                % averaging parameter
h = (1/N)*[0:N-1];    % impulse response function
y = conv(h,x);        % convolution command

ny =[0:length(y)-1];  % time vector for output signal
plot(n,x,'o',ny,y,'*')
xlabel('time')
title('Averaging')
```

Exploratory Exercise 21.1.7: Calculate by hand a few values of the output signal to verify the computations of the computer. Explain the behavior of the output signal near the beginning and end of that signal. ▲▲

Exploratory Exercise 21.1.8: Change the value of the averaging parameter between 2 and 10. What is the effect on the output signal? Also, try other input signals such as sinusoids. ▲▲

21.2 DIFFERENCE EQUATIONS AND TRANSFER FUNCTIONS

21.2.1 Introduction

In this section we will continue the discussion of representations for the discrete system

$$y(n) = \mathcal{H}\big[x(n)\big] \tag{21.2.1}$$

with the introduction of difference equations and the (z-transform) transfer function. The difference equation is an important representation of discrete systems which is the counterpart of the differential equation as a representation for continuous systems. Difference equations are frequently used to implement a discrete system in a computer.

Another system representation is obtained by taking the z-transform of the difference equation, and forming a transfer function. These transfer functions are the discrete time counterpart of Laplace transfer functions for continuous time systems. The transfer function exposes many important properties of the difference equation that are not immediately obvious from the difference equation. There are obvious parallels in the material in this section with the continuous-time systems in Chapter 10.

In this section we will introduce the basic terminology of difference equations which is somewhat more complex than differential equations. We also introduce the transfer function. Then we discuss the properties of both of these representations.

21.2.2 Difference Equations

Difference equations are the differential equations of discrete-time systems. Let N be a fixed positive integer. Consider the equation

$$y(n) + a_1 y(n-1) + \cdots + a_N y(n-N) = b_0 x(n) + b_1 x(n-1)) + \cdots + b_M x(n-M). \tag{21.2.2}$$

Definition 21.2.1: The equation in (21.2.2) is called a <u>difference equation</u>. The <u>order</u> of this equation is N.

▲▲

The order of the difference equation is the maximum difference between the arguments of the output signal $y(n)$. By a change of variables, (21.2.2) can be written as

$$y(n+N) + \cdots + a_{N-1} y(n+1) + a_N y(n) \tag{21.2.3}$$
$$= b_0 x(n+N) + \cdots + b_{N-1} x(n+1) + b_N x(n),$$
$$y(0) = y_0, \ y(1) = y_1 \ y(2) = y_2, \ldots, y(N-1) = y_{N-1}$$

where we have selected $N = M$ for convenience.

NOTATION CHANGE: The indices on the coefficients of the difference equation in (21.2.2) are keyed to the delay of the signal. This notation is different from the notation used for differential equations, but it is standard in the signal processing literature.

Definition 21.2.2: In (21.2.2) and (21.2.3) the <u>present value of the signal</u> y is $y(n)$. <u>Future values of the signal</u> y are $y(n + n_0)$ and <u>past values of the signal</u> y are $y(n - n_0)$ for a fixed integer $n_0 > 0$.

▲▲

In the difference equation in (21.2.3) the present value of the output signal y, $y(n)$ depends on future values of the signal y, $y(n + n_0)$, where as in (21.2.2) the present value of the signal y, $y(n)$ depends on past values of the signal y, $y(n - n_0)$. This observation leads to the following definition.

Definition 21.2.3: The difference equation in (21.2.2) is said to be in <u>delay form</u>. The difference equation in (21.2.3) is said to be in <u>advance form</u>.

▲▲

The delay form is very convenient for the implementation of the difference equations with digital electronics. The advance form is useful for solving initial value problems.

Example 21.2.4: The difference equation[1]

$$y(n+2) - 2r\xi y(n+1) + r^2 y(n) = x(n+2) - r\xi x(n+1), \quad \xi = \cos(\Omega_o) \tag{21.2.4}$$

is a system if we select $x(n)$ as the input signal and $y(n)$ as the output signal. This difference equation is in advance form. The order of this system is $N = 2$. In delay form, this system is

$$y(n) - 2r\xi y(n-1) + r^2 y(n-2) = x(n) - r\xi x(n-1). \tag{21.2.5}$$

▲▲

Two special forms of the difference equation in (21.2.3) frequently occur in signal processing. The first special form of (21.2.3) is

$$y(n) + a_1 y(n-1) + \cdots + a_N y(n-N) = b_0 x(n). \tag{21.2.6}$$

Another special form is

$$y(n) = b_0 x(n) + b_1 x(n-1)) + \cdots + b_{M-1} x(n-(M-1)) + b_M x(n-M). \tag{21.2.7}$$

---

[1] This system is the discrete-time equivalent of the standard second-order system in continuous-time systems.

Definition 21.2.5:   The model of a discrete system in (21.2.6) is called an underline{autoregressive} (AR) model. The model of the system shown in (21.2.7) is called a underline{moving average} (MA) model. The model of the system shown in (21.2.2) is called a underline{autoregressive moving average} (ARMA) model.
▲▲

Terminology: Since the present value of y in (21.2.6) depends on past values of y, this system is said to be recursive. In (21.2.7) the present value of y depends not on past values of y, but only on past values of x. This system is said to be nonrecursive. See Definitions 21.3.6 and 21.3.7.

The MA model is closely related to an FIR convolution representation. Indeed, if we identify $h(n) = b_n$, then (21.2.7) can be written as

$$y(n) = \sum_{k=0}^{M} b_k x(n-k) = \sum_{k=0}^{M} h(k)x(n-k). \qquad (21.2.8)$$

This system is clearly an FIR system. It turns out that AR and ARMA models are IIR systems.

21.2.3 Transfer Functions

As we have seen, the Laplace transfer function is useful for the analysis of continuous-time systems. We would like to have a similar tool for discrete-time systems, in particular, systems described by difference equations. The z-transform can be used in this role.

Definition 21.2.6: Consider the system (21.2.1). The (z-transform) transfer function $H(z)$ is defined as the ratio of the z-transform of the output signal over the z-transform of the input signal:

$$H(z) = \frac{Z\{y(n)\}}{Z\{x(n)\}}.$$

▲▲

Terminology: The transfer function is sometimes called the *system function*.

We have introduced the one-sided and two-sided z-transform in Chapter 17. We will employ both of these transforms here. In fact, it is often not necessary to specify which transform we are using; the appropriate transform is apparent from context. So we will just refer to the z-transforms. When one of these transforms is required, we will be explicit.

Remark 21.2.7: Definition 21.2.6 requires that the z-transform of the input and output signals exist. More importantly, it requires that the ratio of *every* input/output signal pair be the same.
▲▲

This definition implies that the transfer function can be calculated directly from the z-transform of the input and output signals. In fact, this approach to calculating the transfer function is often used. This topic falls under the heading of system identification. More often, however, the transfer function is obtained from a difference equation.

Theorem 21.2.8: Suppose that a system is represented by the difference equation in (21.2.2). *Assume that all of the initial conditions are zero.* Then the transfer function is given by

$$\frac{Y(z)}{X(z)} = \frac{b_0 + b_1 z^{-1} + \cdots + b_M z^{-M}}{1 + a_1 z^{-1} + \cdots + a_N z^{-N}}$$

▲▲

NOTATION CHANGE: The index of the coefficient of the polynomials matches the negative of the power of z. This notation is different from the notation of polynomials in s associated with continuous-time systems.

Theorem 21.2.8 requires the initial conditions to be zero. The transfer function describes only the input/output behavior of the system whereas initial conditions model internal structure of the system. The same requirement is imposed on the differential equations when calculating transfer functions for continuous-time systems.

The transfer function in Theorem 21.2.8 is derived from the delay form of the difference equations in (21.2.2). If we compute the transfer function from the advance form of the difference equations in (21.2.3) (with the initial conditions zero) we get

$$\frac{Y(z)}{X(z)} = \frac{b_0 z^N + b_1 z^{N-1} + \cdots + b_N}{z^N + a_1 z^{N-1} + \cdots + a_N}. \tag{21.2.9}$$

These two transfer functions are related to each other by multiplying the top and bottom of (21.2.9) by z^{-N}. We obtain

$$\frac{Y(z)}{X(z)} = \left[\frac{z^{-N}}{z^{-N}}\right] \frac{b_0 z^N + b_1 z^{N-1} + \cdots + b_N}{z^N + a_1 z^{N-1} + \cdots + a_N} = \frac{b_0 + b_1 z^{-1} + \cdots + b_N z^{-N}}{1 + a_1 z^{-1} + \cdots + a_N z^{-N}}. \tag{21.2.10}$$

Example 21.2.9: The transfer function of the system in Example 21.2.4 is

$$Z\left\{y(n+2) - 2r\xi y(n+1) + r^2 y(n) = x(n+2) - r\xi x(n+1)\right\} \tag{21.2.11}$$

$$\leftrightarrow z^2 Y(z) - 2r\xi z Y(z) + r^2 Y(z) = z^2 X(z) - r\xi z X(z).$$

From (21.2.11) we obtain the transfer function

$$\frac{Y(z)}{X(z)} = H(z) = \frac{z^2 - r\xi z}{z^2 - 2r\xi z + r^2}. \tag{21.2.12}$$

Since the difference equation in (21.2.11) is in advance form, we obtain a transfer function as a function of z. If we take the z-transform of the delay form in (21.2.5), we obtain

$$Y(z) - 2r\xi z^{-1}Y(z) + r^2 z^{-2}Y(z) = X(z) - r\xi z^{-1}X(z). \tag{21.2.13}$$

Now solving (21.2.13) for the transfer function we find

$$\frac{Y(z)}{X(z)} = H(z) = \frac{1 - r\xi z^{-1}}{1 - 2r\xi z^{-1} + r^2 z^{-2}}. \tag{21.2.14}$$

▲▲

Almost all of the terminology used for transfer functions for continuous-time systems applies to transfer function for discrete-time system. For example, we have the following definition.

Definition 21.2.10:   The <u>poles</u> of the transfer function are the roots of the denominator polynomial. The <u>zeros</u> of the transfer function are the roots of the numerator polynomial.

▲▲

Terminology: As with continuous systems, we will plot the poles and zeros of the transfer function in the complex plane. Because the transfer function of a discrete system depends on the complex variable z, we refer to the complex plane as the <u>z-plane</u>. See the discussion in Section 18.1.

Example 21.2.11: Consider the transfer function in (21.2.14), which is in delay form. To compute the poles and zeros of this transfer function, we must first convert it to advance form in (21.2.12). Now the zeros are 0 and $r\xi$. The poles are

$$\text{poles} = \frac{-(-2r\xi) \pm \sqrt{(-2r\xi)^2 - 4r^2}}{2} = r\zeta \pm jr\sqrt{\zeta^2 - 1}. \tag{21.2.15}$$

If $-1 < \xi < 1$, then we can write $\xi = \cos(\Omega_0)$. This observation leads to a polar representation of the poles as shown in Figure 21.2.1.

Notation: It is customary to draw the unit circle in the z-plane.

In this representation $\pm\Omega_0$ is the angle from the positive real axis. The radial distance to the poles is given by r. The zeros of the transfer function (21.2.15) are also shown. Compare Figure 21.2.1 with the discussion of a standard second-order polynomial in Subsection 3.2.3.

▲▲

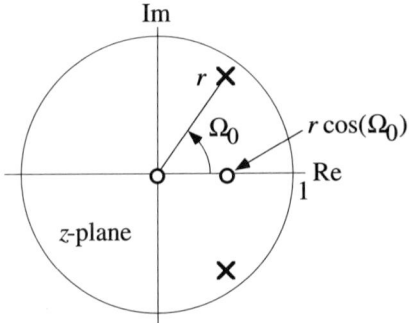

Figure 21.2.1 Poles and Zeros of the Transfer Function (21.2.12)

Example 21.2.12: (Averaging) One of the simplest discrete systems is averaging over the past N input samples. The difference equation is

$$y(n) = \frac{1}{N} \sum_{k=0}^{N-1} x(n-k). \tag{21.2.16}$$

The z-transform of this system is

$$\mathcal{Z}\{y(n)\} = \frac{1}{N} \sum_{k=0}^{N-1} X(z)z^{-k}. \tag{21.2.17}$$

Solving (21.2.17) for the transfer function we obtain

$$\frac{Y(z)}{X(z)} = \frac{1}{N} \sum_{k=0}^{N-1} z^{-k} = \frac{1-z^{-N}}{N\left(1-z^{-1}\right)}. \tag{21.2.18}$$

The transfer function of this discrete system in (21.2.18) is not proper. Not proper

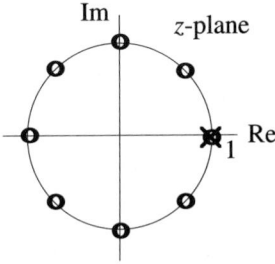

Figure 21.2.2 Poles and Zeros of the Transfer Function in (21.2.18)

transfer functions occur in discrete system while they are rarely encountered in continuous systems.

The zeros of this system are shown in Figure 21.2.2 for $N = 8$.   The zeros are evenly spaced around the unit circle.   They are the so-called roots of unity.   The factor in the denominator in (21.2.18) cancels with the zero at $z = 1$, and so this transfer function has no poles.   The system in (21.2.16) is a FIR system.   It is also a moving average (MA) system.

▲▲

21.2.4 MATLAB Experiments

The entry of system representations (transfer functions as ratios of polynomials **tf** and transfer functions as zeros and poles **zpk**) is discussed in Subsection 10.1.5. Discrete-time systems are entered in exactly the same way, except that the sampling period is also included in the definition of the system.   If the sampling period is not defined, then this input argument is set to -1.   These LTI objects, however, use the notation convention for polynomials in s.   This notation convention is used in Chapters 1-16 in this text, but it is not used for z-transform transfer functions. Therefore, MATLAB 5 includes an additional LTI object called **filt**.   This system representation does conform to the notation convention of writing the transfer functions in powers of z^{-1}.   When using this LTI object, the sample period is specified only if it is defined.   Example 21.2.9 illustrates the use of this command, however, the MATLAB manual should also be consulted.

To simulate a discrete system, the **lsim** command is used.   This command is described in Section 10.1.5.   The command recognizes the input argument is a discrete system.   For discrete systems the time vector is not used in the **lsim** command.

The zeros and poles of a discrete system can be plotted is the z-plane using the **zplane** command.   The input arguments are the numerator and denominator of the transfer function.

MATLAB 4 The versions of MATLAB before MATLAB 5 don't represent systems as objects.   In these versions the system is represented by numerator and denominator polynomials and the user must keep track of the system.   The command which accomplishes discrete system simulation is **filter**.   The input data is the system and the input signal.   The output data is the output signal of the system.   This command is easiest to use if the system is represented as a difference equation in delay form or as the corresponding transfer function in powers of z^{-k} as in Theorem 21.2.8.   Care must be taken when entering the vectors of filter coefficients b and a. The vectors defining the polynomials in the numerator and denominator are assumed to begin with the constant term and they must be of the same length.   This format is slightly different than just entering polynomials.   See in particular the second M-file and the MATLAB help file on the command **filter**.   At this writing, these commands still work in MATLAB 5.   They are illustrated in the second M-file below.

The first M-file calculates the response of the standard second-order discrete system in Example 21.2.9 to a sinusoidal input signal.   This M-file also plots the poles and zeros of the transfer function so that the relationship between the poles and zeros and the output of the system can be observed.

Example 21.2.9: (*continued*) Standard Second-Order Discrete System

```
clear
% define pole locations
r = 0.95;                        % radius
W0 = 45;                         % angle
zeta = cos((pi/180)*W0);
b = [1,-r*zeta,0];               % numerator polynomial
a = [1,-2*r*zeta,r^2];           % denominator polynomial
h = filt(b,a);
figure(1)
zplane(b,a)
title('Zero/Pole Plot')
% Calculate the time responses
n = [0:40];                      % define time axis
wi = pi/5;                       % frequency of the input sinusoid
x = cos(wi*n);                   % input signal
y = lsim(h,x);                   % calculate output signal
figure(2)
plot(n,x,'o',n,y,'+')            % plot the input and output signals
xlabel('time')
title('Time Histories')
legend('input signal','output signal')
```

▲▲

Exploratory Exercise 21.2.13: In the M-file above change the frequency of the input signal between $\omega_i = \pi/3, \pi/4, \pi/5$. How does the output signal change relative to the input signal? Comment on the relationship between the magnitude and phase of the steady state output signal and the input signal. This behavior will be explained in Section 23.1.

▲▲

Exploratory Exercise 21.2.14: In the M-file above change the location of the poles by varying the radial distance r between 0.8 and 1.05. Also change the angle of the poles between 0 and 180 deg. How does the output signal change? How do the poles change? When is the output signal bounded?

▲▲

The next M-file calculates the output signal of a discrete system with two real poles when the input signal is a unit step function.

System with Two Real Poles

```
clear
% define pole locations
p1 = 0.25;                       % first pole
p2 = 0.85;                       % second pole
b = [0,1];                       % numerator polynomial
a = conv([1,-p1],[1,-p2]);       % denominator polynomial
```

```
figure(1)
zplane(b,a)
title('Zero/Pole  Plot')
% Calculate the time responses
n = [0:40];                      % define time axis
x = ones(size(n));               % input signal
y = filter(b,a,x);               % calculate output signal
figure(2)
stem(y)                          % plot the input and output signals
xlabel('time')
title('Time  Histories')
```

▲▲

Exploratory Exercise 21.2.15: In the M-file above vary the second pole location between p2 = 0.65 and 1.05. Characterize the rise time of the output signal as a function of pole location. When is the system BIBO stable?

▲▲

Exploratory Exercise 21.2.16: In the M-file above place the first pole on the negative real axis between $-1 < p1 < 0$. What happens to the output signal? Compare this discrete system behavior to continuous system behavior. Is it possible for a continuous system with two real poles to have oscillatory behavior when the input signal is a unit step function? Try placing the other pole on the negative real axis.

▲▲

Exploratory Exercise 21.2.17: In the M-file above place both poles at the origin p1 = p2 = 0. Characterize the output signal. Compare this discrete system behavior to continuous system behavior. Is it possible for a continuous system with two real poles to reach 1 exactly in a finite amount of time when the input signal is a unit step function?

▲▲

Exploratory Exercise 21.2.18: In the first M-file above replace the sinusoidal input signal with a unit step function. Observe the output signal for various pole locations.

▲▲

21.3 BLOCK DIAGRAMS AND NETWORK STRUCTURES

21.3.1 Introduction

In this section we continue our discussion from the last section of representations of discrete systems

$$y(n) = \mathcal{H}\big[x(n)\big] \tag{21.3.1}$$

that can be represented by the transfer function

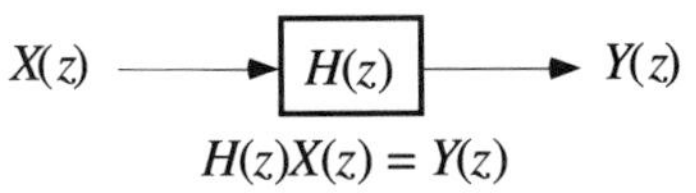

$$X(z) \longrightarrow \boxed{H(z)} \longrightarrow Y(z)$$
$$H(z)X(z) = Y(z)$$

Figure 21.3.1 Block Diagram of a Discrete System

$$\frac{Y(z)}{X(z)} = H(z). \tag{21.3.2}$$

The transfer function is an algebraic expression useful for computationally analyzing the system. For complex systems, however, the algebraic expression can mask the structure of the system. In this section we introduce block diagrams as an alternative way to represent a block diagram. Block diagrams, of course, are a pictorial representation of a transfer function as shown in Figure 21.3.1.

Definition 21.3.1: The pictorial representation in Figure 21.3.1 of the transfer function in (21.3.2) is called a <u>block diagram</u>.

▲▲

Block diagram representations of discrete systems are identical to block diagram representations of continuous systems in that the interpretation and manipulation of the pictures are identical. Block diagrams of continuous systems are discussed in detail in Sections 10.2 - 10.4. All of the definitions and manipulations of block diagrams for continuous systems are also allowed for discrete systems described by block diagrams. In particular the following concepts are fundamental to the understanding of block diagrams.

1. The interconnection of the blocks through the use of branch points (Definition 10.2.2) and summing point (Definition 10.2.3).
2. Cascade interconnection of two subsystems (System Structure 10.2.4).
3. Parallel interconnection of two subsystems (System Structure 10.2.5).
4. Feedback interconnection of two subsystems (System Structure 10.2.6).
5. Block diagram manipulation as summarized in Figure 10.6.1.

We assume familiarity with these block diagram concepts and block diagram reduction in this section. In this section we will focus on the discrete systems interpretation of the block diagram. We will introduce the unit delay and use this block to develop two block diagrams (network structures) that play an important role in the implementation of discrete systems in microprocessors.

Terminology: In this section we will use the terminology from signal processing and digital filtering. This terminology includes several terms for block diagrams including *network structures* and *filters*.

21.3.2 Basic Building Blocks

To illustrate the use of block diagrams in discrete systems we will present several examples of systems which play a central role in signal processing. These examples illustrate the relationship between the block diagram and the underlying difference equations. They also show simple block diagram manipulations.

We begin with two elementary blocks which play a central role in the development of more complicated block diagrams. Consider the transfer function

$$\frac{Y(z)}{X(z)} = H(z) = z. \tag{21.3.3}$$

Cross-multiplying in (21.3.3) and taking the inverse z-transform using Property 18.2.6 (left shift), we obtain the difference equation

$$y(n) = x(n+1). \tag{21.3.4}$$

Thus, the output signal of this system is the input signal advanced by one unit of time. The corresponding block diagram is shown in Figure 21.3.2 where the time domain signals are also shown.

Definition 21.3.2: The transfer function in (21.3.3) and its block diagram in Figure 21.3.2 are called a <u>unit advance</u>.

▲▲

Consider the transfer function

$$\frac{Y(z)}{X(z)} = H(z) = z^{-1}. \tag{21.3.5}$$

Cross multiplying in (21.3.5) and taking the inverse z-transform using Property 18.2.6 (right shift), we obtain the difference equation

$$y(n) = x(n-1). \tag{21.3.6}$$

Thus, the output signal of this system is the input signal delayed by one unit of time. The corresponding block diagram is shown in Figure 21.3.3.

Definition 21.3.3: The transfer function in (21.3.6) and its block diagram in Figure 21.3.3 are called a <u>unit delay</u>.

▲▲

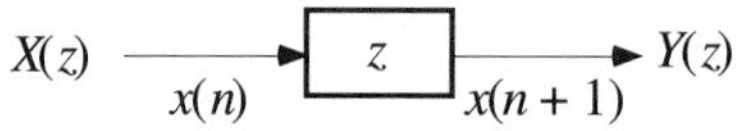

Figure 21.3.2 Block Diagram of a Unit Advance

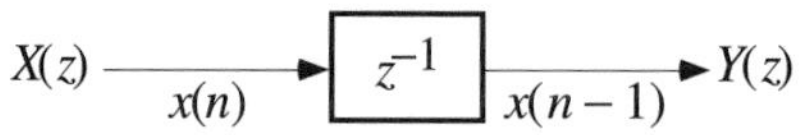

Figure 21.3.3 Block Diagram of a Unit Delay

The unit delay in discrete system theory plays a role analogous to the role the integrator plays in continuous system theory. The next two examples illustrate the use of unit delays. These particular systems are stepping stones to the network structures that play a central role in the implementation of digital filters.

Example 21.3.4: Consider the system

$$y_1(n) = x_1(n) - a_1 y_1(n-1) \tag{21.3.7}$$

where we have indexed the input signal, output signal, and the coefficient for purposes that will become clear below. The transfer function for this system is

$$\frac{Y_1(z)}{X_1(z)} = \frac{1}{1 + a_1 z^{-1}} = H_1(z). \tag{21.3.8}$$

From the difference equation in (21.3.7) we obtain the block diagram of this system as shown in Figure 21.3.4.

▲▲

Example 21.3.5: Consider the system

$$y_2(n) = b_0 x_2(n) + b_1 x_2(n-1). \tag{21.3.9}$$

The transfer function of this system is

$$\frac{Y_2(z)}{X_2(z)} = b_0 + b_1 z^{-1} = H_2(z). \tag{21.3.10}$$

From the difference equation in (21.3.9) we obtain the block diagram of this system as shown in Figure 21.3.5.

▲▲

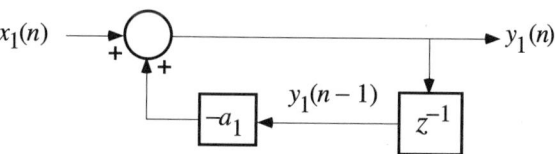

Figure 21.3.4 Block Diagram of Example 21.3.4

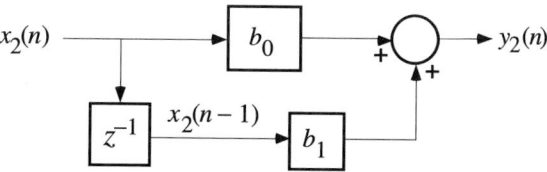

Figure 21.3.5 Block Diagram of Example 21.3.5

In (21.3.7) the present value of the output signal $y_1(n)$ depends on past values of the output signal, i.e., $y_1(n-1)$. This dependence leads to the feedback loop in the block diagram in Figure 21.3.4.

Definition 21.3.6: Discrete systems represented by difference equations in delay form in which the present value of the output signal depends on the past values of the output signal are called <u>recursive</u>.   ▲▲

In the system (21.3.9) the present value of the output signal does not depend on past values of the output signal. As a result, the block diagram in Figure 21.3.5 does not contain any feedback loops.

Definition 21.3.7: In a discrete system represented by difference equations in delay form if the present value of the output signal does not depend on the past values of the output signal, then the system is said to be <u>nonrecursive</u>.   ▲▲

The impulse response of the system in Example 21.3.4 is

$$(21.3.11)$$

$$z^{-1}\left\{\frac{1}{1+a_1z^{-1}}\right\} = \left(a_1^n\right)u_s(n).$$

The impulse response function in (21.3.11) has infinite length and it is an IIR system. The block diagram of this system in Figure 21.3.4 has a feedback loop and it is a recursive system.

The impulse response of the system in Example 21.3.6 is

$$(21.3.12)$$

$$z^{-1}\left\{b_0 + b_1 z^{-1}\right\} = b_0\delta(n) + b_1\delta(n-1).$$

The impulse response function in (21.3.12) has finite length ($= 2$) and it is an FIR system. The block diagram of the FIR system in Figure 21.3.5 doesn't contain any feedback loops and it is a nonrecursive system.

The block diagram of an IIR system can only be drawn with feedback loops. A block diagram of the FIR system is naturally drawn without feedback loops; however, it can be drawn with feedback loops. The terms recursive and nonrecursive generally refer to the presence of feedback loops in the implementation of the discrete system.

21.3.3 Network Structures

The two simple examples above show how a transfer function can be represented as an all-delay block diagram. These block diagrams capture the structure of the difference equation which is lost in the transfer function. This idea of using an all-delay block diagram to expose the internal structure of a system can be extended to more complex transfer functions. As can be expected, there is more than one way to decompose a complex transfer function. We will discuss two decompositions of an IIR system and one decomposition of an FIR system next. These decompositions are built up from the two examples above. This idea is expanded in the last section of this chapter.

Direct Form I Consider the systems (21.3.7) and (21.3.9). Suppose we set the input signal of the system (21.3.7) equal to the output signal of the system (21.3.9); i.e.,

$$x_1(n) = y_2(n). \tag{21.3.13}$$

To construct a block diagram for this system, we cascade the two systems in Figures 21.3.4 and 21.3.5. The result is shown in Figure 21.3.6a. In Figure 21.3.6a we have reversed the order of the unit delay and the scalar gain from Figures 21.3.4 and 21.3.5. Next notice that the two signals on the lower paths are first delayed and then summed. These two operations can be reversed; that is, the signals can first be summed and then delayed as shown in Figure 21.3.6b. The advantage of the block diagram in Figure 21.3.6b is that one delay has been eliminated.

   The difference equation obtained by connecting the system (21.3.7) to the system (21.3.9) using (21.3.13) is

$$y_1(n) = -a_1 y_1(n-1) + b_0 x_2(n) + b_1 x_2(n-1). \tag{21.3.14}$$

The transfer function for this system is calculated directly from the difference equation in (21.3.14) or from the block diagram in Figure 21.3.6. We have

$$\frac{Y_1(z)}{X_2(z)} = H(z) = \frac{b_0 + b_1 z^{-1}}{1 + a_1 z^{-1}}. \tag{21.3.15}$$

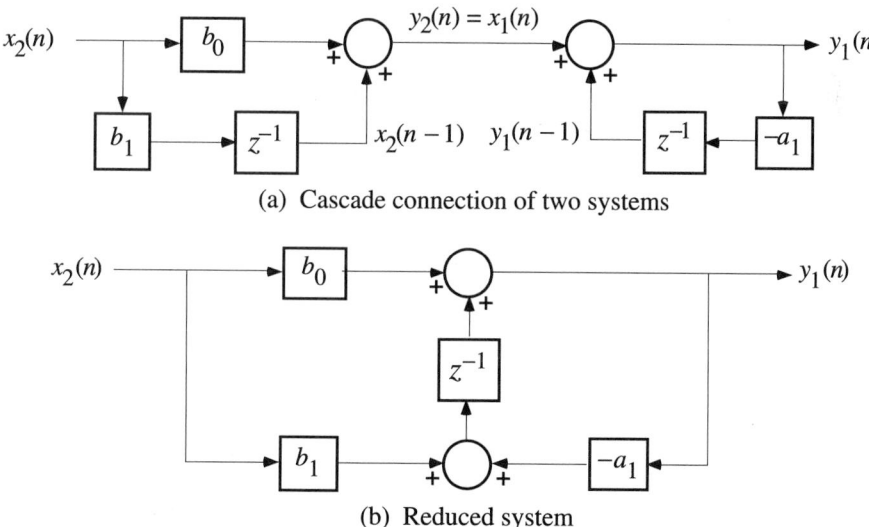

(a) Cascade connection of two systems

(b) Reduced system

Figure 21.3.6 Block Diagram of a System in Direct Form I

If we compare the transfer function in (21.3.15) to the block diagram in Figure 21.3.6b, we see that to reconstruct the block diagram, we only need to insert the coefficients from the transfer function. Hence, we see how to construct block diagrams from transfer functions with orders larger than one. Consider the transfer function

$$\frac{Y(z)}{X(z)} = H(z) = \frac{b_0 + b_1 z^{-1} + b_2 z^{-2}}{1 + a_1 z^{-1} + a_2 z^{-2}}. \tag{21.3.16}$$

The direct form I block diagram for this transfer function is shown in Figure 21.3.7.

Definition 21.3.8: Any system represented by a block diagram with the structure shown in Figure 21.3.7 is said to be in <u>direct form I</u>.

▲▲

We can write a set of difference equations which describe the system in the direct form I block diagram in Figure 21.3.7 using the characterization of the unit delay in Figure 21.3.3. To the input signals of the delays in Figure 21.3.7 we assign the signals $w_i(n)$. Then according to Figure 21.3.3 the output signals are $w_i(n - 1)$. These signals are shown in Figure 21.3.7. Now we can write the difference equations from the block diagram. We have

$$y(n) = b_0 x(n) + w_1(n-1), \tag{21.3.17}$$
$$w_1(n) = b_1 x(n) - a_1 y(n) + w_2(n-1),$$
$$w_2(n) = b_2 x(n) - a_2 y(n).$$

These equations are used to implement the direct form I system structure in a microprocessor.

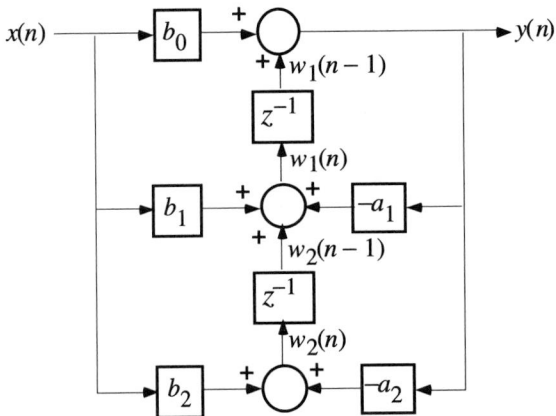

Figure 21.3.7 Second-Order Direct Form I Network

Direct Form II We can create a second type of system if we set the output signal from the system in (21.3.7) equal to the input signal in (21.3.9),

$$y_1(n) = w(n) = x_2(n). \tag{21.3.18}$$

The block diagram of this system is obtained by interconnecting the block diagrams in Figures 21.3.4 and 21.3.5. The result is shown in Figure 21.3.8a. In Figure 21.3.8a both systems require $w(n)$ and $w(n-1)$; i.e., the delay element is redundant. Therefore, these two systems can be combined as in Figure 21.3.8b. Figure 21.3.8a shows that the transfer function of the system in Figure 21.3.8b is

$$\frac{Y(z)}{X(z)} = H(z) = H_1(z)H_2(z) = \frac{b_0 + b_1 z^{-1}}{1 + a_1 z^{-1}}. \tag{21.3.19}$$

The difference equation which is constructed from the transfer function (21.3.19) is

$$y(n) = \left[b_0 x(n) + b_1 x(n-1) \right] - a_1 y(n-1). \tag{21.3.20}$$

By comparing the transfer function in (21.3.19) to the block diagram in Figure 21.3.8b, we see that only the coefficients of the transfer function are needed to construct the block diagram. Hence, this structure can be extended to higher-order transfer functions. Consider the transfer function

$$\frac{Y(z)}{X(z)} = H(z) = \frac{b_0 + b_1 z^{-1} + b_2 z^{-2}}{1 + a_1 z^{-1} + a_2 z^{-2}}. \tag{21.3.21}$$

A block diagram of this system is shown in Figure 21.3.9.

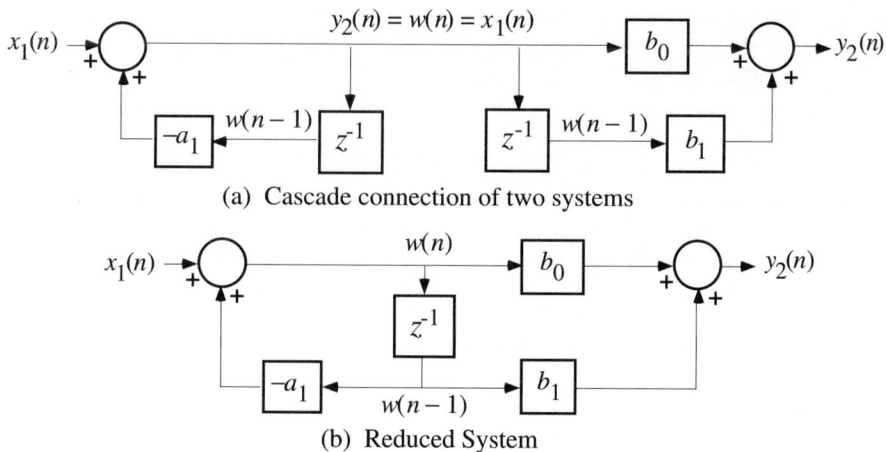

(a) Cascade connection of two systems

(b) Reduced System

Figure 21.3.8 A Block Diagram in First-Order Direct Form II Network

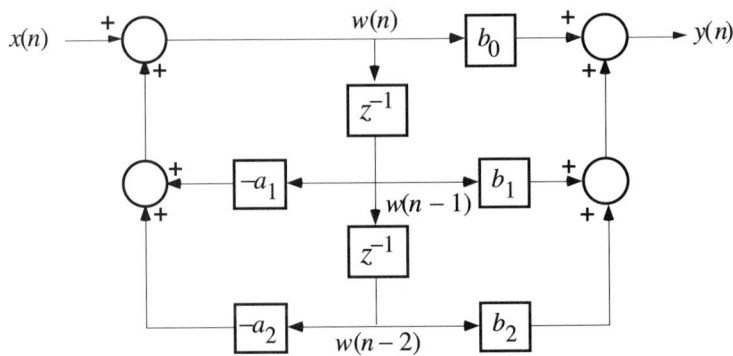

Figure 21.3.9 A Block Diagram of a System in Direct Form II

Definition 21.3.9: Any system represented by a block diagram with the structure shown in Figure 21.3.9 is said to be in <u>direct form II</u>.

▲▲

Terminology: This terminology, direct form I and direct form II, comes from the signal processing literature.

We can also write a set of difference equations for the direct form II using the procedure discussed for direct form I. To the input signal of the uppermost delay in Figure 21.3.9 we assign $w(n)$. The output signal from this delay is $w(n-1)$. This signal is also the input signal into the second delay. Hence, the output signal from the second delay is $w(n-2)$. These signals are labeled in Figure 21.3.9. From the block diagram in Figure 21.3.9 we obtain the difference equations

$$w(n) = -a_1 w(n-1) - a_2 w(n-2) + x(n), \qquad (21.3.22)$$
$$y(n) = b_0 w(n) + b_1 w(n-1) + b_2 w(n-2).$$

When we compare the difference equations in (21.3.22), which describe the internal structure of the direct form II, to the difference equations in (21.3.17), which describe the internal structure of the direct form I, we see that they are different. So when the same transfer function is implemented in two different microprocessors using these two different structures, the performance of the algorithms within the microprocessors will be different. Advanced signal processing texts investigate these ideas.

Consider the block diagram in Figure 21.3.9. To construct this block diagram, we only require the coefficients of the transfer function in (21.3.21). The same observation is also true of Figure 21.3.7.

Definition 11.2.3: The block diagrams in Figure 21.3.7 and 21.3.8 are called <u>canonical</u> block diagrams.

▲▲

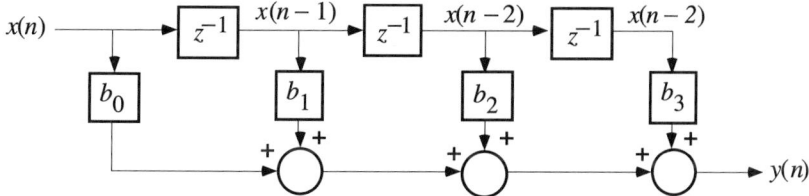

Figure 21.3.10 Network Structure for an FIR System

Terminology: We identified several canonical structures in our discussion of state space realizations in Section 11.2. The usage of canonical here is the same as the usage in Section 11.2.

The direct form I and direct form II canonical structures are slightly different from the canonical structures that were introduced for continuous time systems. These differences are discussed in Section 21.5.

FIR or Transversal Filter The last basic network structure we will discuss is the FIR filter, say,

$$y(n) = b_0 x(n) + b_1 x(n-1) + b_2 x(n-2) + b_3 x(n-3). \tag{21.3.23}$$

The transfer function for this system is

$$\frac{Y(z)}{X(z)} = b_0 + b_1 z^{-1} + b_2 z^{-2} + b_3 z^{-3}. \tag{21.3.24}$$

A block diagram of this system is shown in Figure 21.3.10.

Terminology: FIR filters are also called <u>transversal filters</u> and <u>tapped delay lines</u>.

21.4 DTFT TRANSFER FUNCTION

In this section we will continue the discussion of representations for the discrete system

$$y(n) = \mathcal{H}[x(n)]. \tag{21.4.1}$$

Suppose that this system is represented by a convolution representation

$$y(n) = \sum_{k=-\infty}^{\infty} h(n-k)x(k). \tag{21.4.2}$$

Suppose we take the DTFT of the convolution representation in (21.4.2). Using the convolution property, Property 18.4.5, we get

$$\mathcal{DTFT}\left\{ y(n) = \sum_{k=-\infty}^{\infty} h(n-k)x(k) \right\} \Rightarrow Y(\Omega) = H(\Omega)X(\Omega) \qquad (21.4.3)$$

or

$$\frac{Y(\Omega)}{X(\Omega)} = H(\Omega).$$

Definition 21.4.1: The system representation in (21.4.3) is called the <u>DTFT transfer function</u>.

▲▲

Of course, it is implicitly assumed that the all of the transforms in (21.4.3) exist. The definition of the DTFT, Definition 18.4.1, requires that each of the signals and the impulse response function be absolutely summable. This requirement is a real restriction on the system. We will return to this requirement in Section 22.3.

Example 21.4.2: Consider the system

$$y(n) = \sum_{k=\infty}^{\infty} h(k)x(n-k) = \sum_{k=\infty}^{\infty} \left(a^k u_s(k) \right) x(n-k) = \sum_{k=0}^{\infty} \left(a^k \right) x(n-k). \qquad (21.4.4)$$

The impulse response function is

$$h(n) = a^n u_s(n). \qquad (21.4.5)$$

The DTFT transfer function is

$$\frac{Y(\Omega)}{X(\Omega)} = \mathcal{DTFT}\{h(n)\} = \sum_{n=-\infty}^{\infty} a^n u_s(n) e^{-j\Omega n} = \frac{e^{j\Omega}}{e^{j\Omega} - a}, \quad |a| < 1. \qquad (21.4.6)$$

where we have used Example 18.4.3. Note the restriction on the value of a which defines the impulse response function.

▲▲

The DTFT transfer function can also be derived directly from the difference equation representation of a system using the time shift property of the DTFT. The following example illustrates this idea.

Example 21.4.3: Consider the discrete system

$$y(n) - 2r\zeta y(n-1) + r^2 y(n-2) = x(n) - 2r\zeta x(n-1) + x(n-2). \qquad (21.4.7)$$

Taking the DTFT of this system using the time shift property, Property 18.4.9, we obtain

$$Y(\Omega) - 2r\zeta e^{-j\Omega} Y(\Omega) + r^2 e^{-j2\Omega} Y(\Omega) \qquad (21.4.8)$$
$$= X(\Omega) - 2r\zeta e^{-j\Omega} X(\Omega) + e^{-j2\Omega} X(\Omega).$$

Solving for the transfer function yields

$$\frac{Y(\Omega)}{X(\Omega)} = \frac{1 - 2r\zeta e^{-j\Omega} + e^{-j2\Omega}}{1 - 2r\zeta e^{-j\Omega} + r^2 e^{-j2\Omega}}. \qquad (21.4.9)$$

▲▲

Of course, the DTFT transfer function can also be calculated for FIR filters as illustrated by the next example.

Example 21.4.4: Consider the averaging system

$$y(n) = \frac{1}{N} \sum_{k=0}^{N-1} x(n-k). \qquad (21.4.10)$$

Using the time shift property of the DTFT the transfer function of (21.4.10) is given by

$$\frac{Y(\Omega)}{X(\Omega)} = \frac{1}{N} \sum_{k=0}^{N-1} e^{-jk\Omega} = \frac{1}{N} \frac{1 - e^{-j\Omega N}}{1 - e^{-j\Omega}}. \qquad (21.4.11)$$

▲▲

The DTFT transfer function is mainly of theoretical value; it is seldom used for computations. When direct computations are required, the DFT is used. The DTFT transfer function also plays an important role in the discussion of the frequency response of a system.

21.5 DISCRETE STATE SPACE REPRESENTATIONS

21.5.1 Introduction

In this section we will continue the discussion of representations for the discrete system

$$y(n) = \mathcal{H}[x(n)]. \qquad (21.5.1)$$

In Section 21.3 we introduced difference equations as a representation of the discrete system in (21.5.1). In this section we will develop an alternative form of this difference equation which can be expressed in compact matrix notation: the state space representation. State space representations for discrete systems share many properties with state space representations for continuous systems discussed in Chapters 10 and 11. Therefore, one goal in this section is to connect the ideas of discrete-time to the concepts introduced for continuous-time state space representations, and point out some differences.

21.5.2 All-Delay Block Diagrams and State Space Equations

We will develop discrete state space representations from block diagrams, in particular, unit delays. The development in this section parallels the development in Section 10.5. In that section integrators and all-integrator block diagrams play a central role in the development of state space representations. The discrete-time equivalent of an integrator is a unit delay, introduced in Figure 21.4.3 and shown in Figure 21.5.1. In Figure 21.5.1 we have labeled the output signal of this block as $q(n)$. Since this block is a pure delay, the input signal must be $q(n + 1)$. The signal $q(n)$ is, of course, a state variable. State space representations are naturally associated with block diagrams which only contain unit delays.[2]

Definition 21.5.1: If in a block diagram the only blocks in which z appears are a unit delay of the form in Figure 21.5.1, the block diagram is called an <u>all-delay</u> block diagram.

▲▲

The direct form I and direct form II systems in Figures 21.4.7 and 21.4.9 in Section 21.4 are all-delay block diagrams.

State space representations can be derived from all-delay block diagrams. The procedure for writing the state equations from the all-delay block diagram follows Procedure 10.5.9 for writing state equations from all-integrator block diagrams.

Procedure 21.5.2: Given an all-delay block diagram we can construct a state space representation as follows.
Step 1 Denote the output signal of each delay with the state variable $q(n)$. Then the input signal to each delay is labeled $q(n + 1)$.
Step 2 From the block diagram write the equations which express the input signal to each delay in terms of the other delay output signals and the input signal to the system. Also express the output in terms of these variables.
Step 3 Define the vector whose elements are the state variables.
Step 4 Using the vector defined in Step 3, write the equations from Step 2 in matrix form.

▲▲

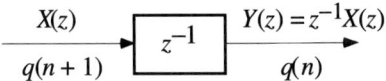

Figure 21.5.1 Delay Block Diagram

---

[2] All-delay block diagrams are the discrete-time equivalent of all-integrator block diagrams. See the discussion in Section 10.5.

The next example illustrates this procedure.

Example 21.5.3: Consider again the direct form I all-delay block diagram in Figure 21.5.2. To translate this block diagram into state equations we use Procedure 21.5.2.
Step 1 First, the output signals of the delays are labeled with state variables $q_k(n)$ as shown in Figure 21.5.3. The input signals are labeled as $q_k(n + 1)$.
Step 2 Next, from the block diagram we can write the following equations:

$$q_1(n+1) = q_2(n) - a_1 y(n) + b_1 x(n), \tag{21.5.2}$$
$$q_2(n+1) = q_3(n) - a_2 y(n) + b_2 x(n),$$
$$q_3(n+1) = -a_3 q_3(n) + b_3 x(n),$$
$$y(n) = q_1(n) + b_0 x(n).$$

To develop the state space equations, each of the equations in (21.5.2) must be in terms of only $q_k(n)$ and the input signal. Substituting for $y(n)$ in the state equations we obtain

$$q_1(n+1) = -a_1 q_1(n) + q_2(n) + (b_1 - a_1 b_0) x(n), \tag{21.5.3}$$
$$q_2(n+1) = -a_2 q_1(n) + q_3(n) + (b_2 - a_2 b_0) x(n),$$
$$q_3(n+1) = -a_3 q_1(n) + (b_3 - a_3 b_0) x(n),$$
$$y(n) = q_1(n) + b_0 x(n).$$

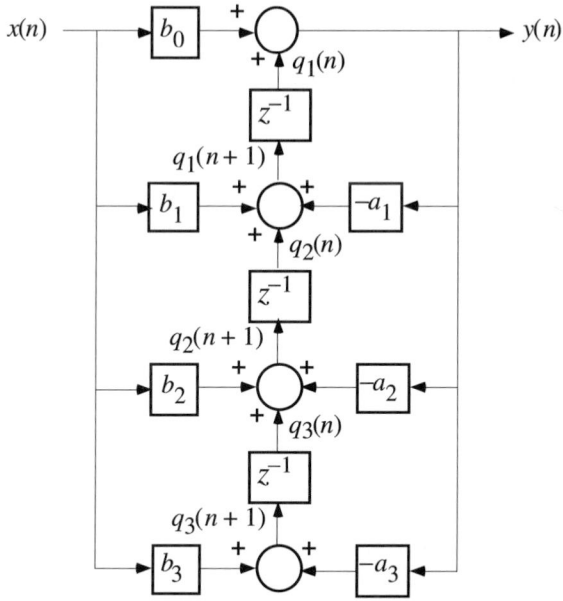

Figure 21.5.2 All-Delay Block Diagram

To simplify the notation define

$$\beta_1 = b_1 - a_1 b_0, \quad \beta_2 = b_2 - a_2 b_0, \quad \beta_3 = b_3 - a_3 b_0. \tag{21.5.4}$$

Step 3 From the state variables identified in Figure 21.5.3, we construct the state vector

$$\vec{q}(n) = \begin{bmatrix} q_1(n) \\ q_2(n) \\ q_3(n) \end{bmatrix}. \tag{21.5.5}$$

Step 4 Rewriting the equations in (21.5.2) using (21.5.5) we obtain

$$\vec{q}(n+1) = \begin{bmatrix} -a_1 & 1 & 0 \\ -a_2 & 0 & 1 \\ -a_3 & 0 & 0 \end{bmatrix} \vec{q}(n) + \begin{bmatrix} \beta_1 \\ \beta_2 \\ \beta_3 \end{bmatrix} x(n), \tag{21.5.6}$$

$$y(n) = \begin{bmatrix} 1 & 0 & 0 \end{bmatrix} \vec{q}(n) + b_0 x(n).$$

The state space model for the system in Figure 21.5.3 is given by (21.5.6). (Compare this state space model to the continuous-time state space representation of a system with a direct feedthrough term in Subsection 11.2.4. Recall that we use different indexing for the coefficients for continuous-time and discrete-time transfer functions.)
▲▲

The system representation in (21.5.6) is of the form

$$\vec{q}(n+1) = A\vec{q}(n) + Bx(n), \quad \vec{q}(0) = q_0 \tag{21.5.7}$$

$$y(n) = C\vec{q}(n) + Dx(n).$$

Definition 10.5.8: The vector $\vec{q}(n)$ is called the <u>state vector</u>, or the <u>state</u> of the system (21.5.7). The components of the state vector $q_k(n)$ are called the <u>state variables</u>. The set of equations (21.5.7) are called the <u>state space equations</u>. The number of state variables, the dimension of the state vector, is called the <u>order</u> of the system.
▲▲

The state space representation in (21.5.7) not only describes the relationship between the input and output signal, but it also describes the "internal" behavior of the system through the state variables.

Procedure 21.5.2 is completely general. It will translate any all-delay diagram into the state space difference equations (21.5.6). Of course, we can work backwards to translate a state space representation into an all-delay block diagram. This procedure is essentially identical to Procedure 10.5.11.

Example 21.5.4: Consider the system in Example 21.5.3. To draw the all-delay block diagram from the state equations, first the state equations in (21.5.6) are multiplied out into the scalar equations in (21.5.3). Next, one delay block is drawn for each state variable in (21.5.3). Then these blocks are connected according to the equations in (21.5.3). The result is shown in Figure 21.5.3.

▲▲

If we compare the block diagram in Figure 21.5.3 with the block diagram in Figure 21.5.2, we see that they are different. The difference is in the way the direct feedthrough term is incorporated in the block diagram. To establish the relationship between the canonical state space representations and direct form I and II, first note that the transfer function of the system in Figure 21.5.2 is

$$\frac{Y(z)}{X(z)} = \frac{b_0 + b_1 z^{-1} + b_2 z^{-2} + b_3 z^{-3}}{1 + a_1 z^{-1} + a_2 z^{-2} + a_3 z^{-3}}. \tag{21.5.8}$$

This transfer function is derived from Figure 21.5.2. If we perform long division on (21.5.8) we obtain

$$\frac{Y(z)}{X(z)} = b_0 + \frac{\beta_1 z^{-1} + \beta_2 z^{-2} + \beta_3 z^{-3}}{1 + a_1 z^{-1} + a_2 z^{-2} + a_3 z^{-3}} \tag{21.5.9}$$

where the β_i's are defined in (21.5.4). The transfer function in (21.5.9) corresponds to

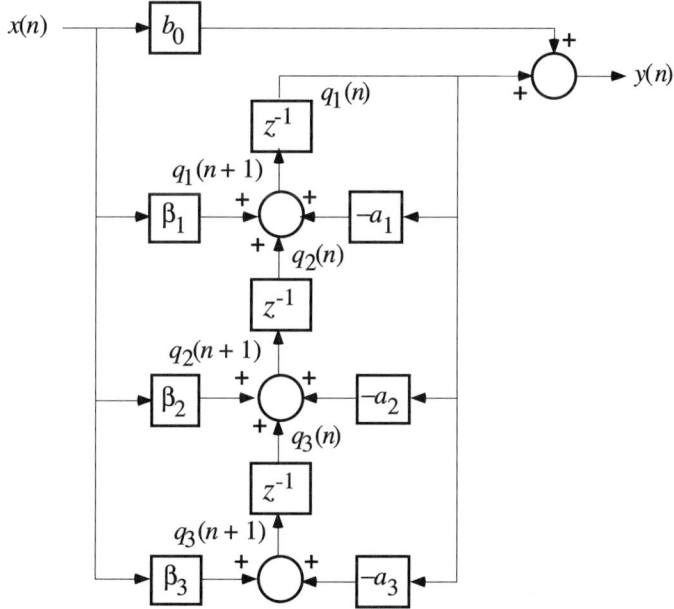

Figure 21.5.3 All-Delay Block Diagram for the State Equations in (21.5.6)

the all-delay block diagram in Figure 21.5.3 and to the state space equations in (21.5.6).

Astute readers will recognize the all-delay block diagram in Figure 21.5.3 as having the same form as the all-integrator block diagram corresponding to the first canonical realization in Figure 11.2.3. Several methods for translating a transfer function of a continuous system into an all-integrator block diagram were discussed in Section 11.2. Those methods apply equally well to transfer functions for discrete systems. For discrete systems, the integrators are replaced by delay blocks.

It is important to recognize the difference between the block diagram in Figure 21.5.3, which corresponds to the state space equations in (21.5.6), and the block diagram in Figure 21.5.2, which is in direct form I. The state space equations in (21.5.6) have the same canonical form as canonical realization 1 for continuous state space equations. Hence, the underlying all-delay block diagram and all-integrator block diagram have the same structure. All-delay block diagrams in direct form I or II, however, don't have a direct counterpart in the continuous-time canonical all-integrator block diagrams discussed in Chapters 10 and 11.

Example 21.5.5: Consider the system represented by the transfer function

$$\frac{Y(z)}{X(z)} = \frac{1 - r\zeta z^{-1}}{1 - 2r\zeta z^{-1} + r^2 z^{-2}}. \tag{21.5.10}$$

Suppose that we wish to translate this transfer function into a state space representation. This translation is accomplished as illustrated by Example 21.5.4. First, we perform long division to reduce the transfer function in (21.5.10) into the form in (21.5.9). We obtain

$$\frac{Y(z)}{X(z)} = \frac{1 - r\zeta z^{-1}}{1 - 2r\zeta z^{-1} + r^2 z^{-2}} = 1 + \frac{r\zeta z^{-1} - r^2 z^{-2}}{1 - 2r\zeta z^{-1} + r^2 z^{-2}}. \tag{21.5.11}$$

Now comparing (21.5.11), (21.5.9), and the state space representation (21.5.6), we get

$$\bar{q}(n+1) = \begin{bmatrix} 2r\zeta & 1 \\ -r^2 & 0 \end{bmatrix} \bar{q}(n) + \begin{bmatrix} r\zeta \\ -r^2 \end{bmatrix} x(n), \tag{21.5.12}$$

$$y(n) = \begin{bmatrix} 1 & 0 \end{bmatrix} \bar{q}(n) + x(n).$$

▲▲

21.5.3 Solution to State Space Equations

One of the advantages of the state space representation over transfer functions is that it can easily accommodate initial conditions, as shown in (21.5.7). Next we will develop an expression for the solution of state space equations which includes the initial conditions as well as the input signal. To begin, suppose the input signal is zero, while the initial condition is nonzero. By direct substitution it can be verified that the states of the system evolve as

$$\vec{q}(0) = q_0, \tag{21.5.13}$$

$$\vec{q}(1) = A\vec{q}(0) = A\vec{q}_0,$$

$$\vec{q}(2) = A\vec{q}(1) = A^2\vec{q}_0,$$

$$\vdots$$

$$\vec{q}(n) = A^n q_0.$$

From (21.5.13) the output signal is

$$y(n) = CA^n\vec{q}_0. \tag{21.5.14}$$

Using the terminology of Definition 10.1.1, the system output signal in (21.5.14) is called the <u>zero input response</u>. In a similar manner we can compute the output signal for a nonzero input signal when the initial condition is zero. We have

$$\vec{q}(0) = 0, \tag{21.5.15}$$

$$\vec{q}(1) = A\vec{q}(0) + Bx(0),$$

$$\vec{q}(2) = A\vec{q}(1) + Bx(1) = A\big(A\vec{q}(0) + Bx(0)\big) + Bx(1)$$

$$= A^2\vec{q}(0) + ABx(0) + Bx(1),$$

$$\vec{q}(3) = A\vec{q}(2) + Bx(2) = A\big(A^2\vec{q}(0) + ABx(0) + Bx(1)\big) + Bx(2)$$

$$= A^3\vec{q}(0) + A^2Bx(0) + ABx(1) + Bx(2),$$

$$\vdots$$

$$\vec{q}(n) = \sum_{k=0}^{n-1} A^{n-1-k} Bx(k).$$

The output signal is obtained by substituting (21.5.15) into the output equation in (21.5.7). We get

$$y(n) = C\vec{q}(n) + Dx(n) = \sum_{k=0}^{n-1} CA^{n-1-k} Bx(k) + Dx(n). \tag{21.5.16}$$

Using Definition 10.1.1, the system output signal in (21.5.14) is called the <u>zero state response</u> because the initial condition on the state vector is zero.

In comparing the output signals in (21.5.14) and (21.5.16), we see that the (matrix) function A^{n-1} plays a key role.

Definition 21.5.6: The function A^{n-1}, $n \geq 0$ is called the <u>state transition matrix</u> of the discrete-time state space equations (21.5.7).

▲▲

Note: The state transition matrix for continuous-time systems is defined in Definition 12.3.2.

These results are summarized in the following theorem.

Theorem 21.5.7: Given the state space representation

$$\vec{q}(n+1) = A\vec{q}(n) + Bx(n), \quad \vec{q}(0) = \vec{q}_0$$
$$y(n) = C\vec{q}(n) + Dx(n)$$

and the input signal $x(n)$, $n \geq 0$, the output signal of this system is

$$y(n) = CA^n \vec{q}_0 + \sum_{k=0}^{n-1} CA^{n-1-k} Bx(k) + Dx(n).$$

▲▲

The following example illustrates Theorem 21.5.7.

Example 21.5.8: Consider the (trivial) one-dimensional state space representation

$$q(n+1) = \lambda q(n) + bx(n), \quad q(0) = q_0, \tag{21.5.17}$$
$$y(n) = q(n).$$

Suppose that the input signal is zero. Then according to Theorem 21.5.7, the output signal is

$$y(n) = \lambda^n q_0. \tag{21.5.18}$$

Note that

$$y(n) = \lambda^n q_0 \to 0 \quad \text{if} \quad |\lambda| < 1, \tag{21.5.19}$$
$$y(n) = \lambda^n q_0 \to \infty \quad \text{if} \quad |\lambda| > 1.$$

We can compare this behavior to a similar continuous system

$$\dot{q}(t) = aq(t) + bx(t), \quad q(0) = q_0, \tag{21.5.20}$$
$$y(t) = q(t).$$

If the input signal is zero, the output signal for the continuous system is

$$y(t) = e^{at} u_s(t). \tag{21.5.21}$$

Here

$$y(t) = e^{at}u_s(t) \rightarrow 0 \quad \text{if} \quad a < 0, \tag{21.5.22}$$

$$y(t) = e^{at}u_s(t) \rightarrow \infty \quad \text{if} \quad a > 0.$$

Comparing the behavior of the output signal of discrete system in (21.5.19) with the output signal of the continuous system in (21.5.22), we see that the parameter values for which the output signal goes to zero are different. The output signal of the continuous system goes to zero if the parameter a is negative while the output signal of the discrete system goes to zero if the parameter λ has an absolute value less than one. This distinction represents one of the fundamental differences between continuous and discrete systems.

▲▲

21.5.4 MATLAB Experiments

The representation of a state space system in MATLAB 5 is described in Subsection 10.5.5. Discrete-time state space systems are also entered a LTI objects using the command **ss** exactly like continuous-time systems except that the sample period is attached to the input arguments to identify the system as discrete. These systems are simulated using the **lsim** command without a time vector as an input argument. These commands are illustrated by the M-file below for Example 21.5.5.

MATLAB 4   In previous versions of MATLAB discrete state space systems are defined by its four defining matrices A, B, C, and D. These four matrices are then used as input data for commands that operate on a system. To simulate a system represented by state equations, the **dlsim** command is used. The **dlsim** command works just like the **lsim** command except that no time vector is needed.

Example 21.5.5: (*continued*) The following M-file constructs the state space model of the system in this example and simulates this system.

```
clear
% Define pole locations
r = 0.95;                          % radius
W0 = 45;                           % angle
zeta = cos((pi/180)*W0);
% Define state space matrices
A = [2*r*zeta,1;-r^2,0];           % state matrix
B = [r*zeta;-r^2];                 % input matrix
C = [1,0];                         % output matrix
D = 1;                             % direct feedthrough matrix
h = ss(A,B,C,D,-1);                % define system, unspecified sample period
% Calculate the time responses
n = [0:80];                        % define time axis
wi = pi/5;                         % frequency of the input sinusoid
x = cos(wi*n);                     % input signal
y = lsim(h,x);                     % calculate output signal
% Plot figures
plot(n,x,'o',n,y,'+')              % plot the input and output signals
xlabel('time')
```

title('Time Histories')
legend('input signal','output signal')   ▲▲

Exploratory Exercise 21.5.9: Modify the above M-file to include initial conditions on the system. Find initial conditions on the system to remove the transient response.   ▲▲

21.6 NETWORK INTERCONNECTION STRUCTURES

21.6.1 Introduction

Suppose we are given a transfer function

$$\frac{Y(z)}{X(z)} = H(z). \tag{21.6.1}$$

One of the fundamental problems in digital filter design is how to translate this transfer function into difference equations that can be implemented in a digital computer. One way to accomplish this implementation is discussed in Section 21.3. Given the transfer function (21.6.1), we can realize this transfer function in direct form I, an all-delay block diagram. Then from the block diagram we can write the difference equations. These equations, then, would be implemented in the digital computer. Note that the construction of the all-delay block diagram is equivalent to writing the difference equations.

Following this procedure, we could derive many different all-delay block diagrams for one transfer function. Based on our results thus far, we consider all of these block diagrams to be equivalent because they all yield the same transfer function. To implement these difference equations in the microprocessor, however, the coefficients must be entered into storage registers which have a finite bit length. That is, the coefficients of the transfer function are truncated when they are entered into the digital processor. This truncation alters the transfer function in a nontrival way. In some cases the properties of the transfer function can be significantly altered by small truncation errors in the coefficients of the transfer function. While a complete analysis of this phenomena is beyond the scope of this text, we note that it has been found that when the block diagram of the transfer function (21.6.1) is decomposed into smaller blocks, its implementation in the digital processor is less sensitive than when the transfer function is implemented as one difference equation. This decomposition is routinely used in practice.

21.6.2 Cascade Form and Parallel Form

In Section 21.3 we introduced the idea of decomposing a transfer function using an all-delay block diagram. We introduced the direct form I, direct form II, and the transversal FIR filter for this purpose. In this section we will further develop this idea using the notion of system interconnection structures introduced in Section 10.2. The basic idea is to not only decompose the systems in terms of all-delay block diagrams, but to also decompose the system into a cascade or parallel interconnection structure. Consider the two subsystems

$$\frac{Y_i(z)}{X_i(z)} = H_i(z), \quad i = 1, 2. \tag{21.6.2}$$

The Cascade Interconnection Structure 10.2.4 of these two systems would lead to the block diagram in Figure 21.6.1. The transfer function of the cascade interconnection structure in Figure 21.6.1 is given by

$$\frac{Y(z)}{X(z)} = H_1(z)H_2(z). \tag{21.6.3}$$

Consider again the two systems in (21.6.2). A Parallel Interconnection Structure 10.2.5 of these two systems would lead to the block diagram in Figure 21.6.2. The transfer function of the system in Figure 21.6.2 is given by

$$\frac{Y(z)}{X(z)} = H_1(z) + H_2(z). \tag{21.6.4}$$

The basic approach is to decompose the transfer function (21.6.1) into either a cascade or parallel interconnection structure. First, note that the cascade interconnection structure in Figure 21.6.1 corresponds to a product of the subsystem transfer functions in (21.6.3). Hence, to decompose (21.6.1) into a cascade interconnection structure we simply have to factor the transfer function as

$$\frac{Y(z)}{X(z)} = H(z) = \frac{b(s)}{a(s)} = \frac{b_1(s)b_2(s)}{a_1(s)a_2(s)} = \left(\frac{b_1(s)}{a_1(s)}\right)\left(\frac{b_2(s)}{a_2(s)}\right) = H_1(s)H_2(s). \tag{21.6.5}$$

Of course, this factorization is easily accomplished by factoring the numerator and denominator into linear factors and then grouping these factors into the desired subsystems. The groupings of the poles and zeros are nonunique.[3] However, it is generally assumed that the polynomials forming the numerator and denominator polynomials should have real coefficients. Therefore, complex poles and zeros should be grouped together into separate subsystems. It has also been shown that lower-order subsystems are preferable. Hence, the transfer function is generally decomposed into first- or second-order blocks adhering to the requirement for real coefficients.

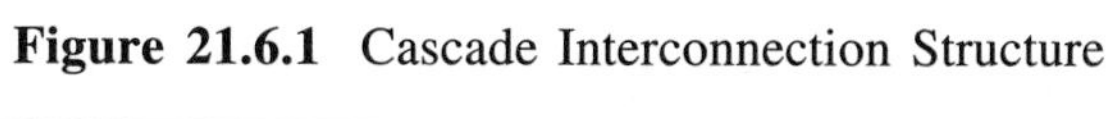

$X(z) \longrightarrow \boxed{H_1(z)} \longrightarrow \boxed{H_2(z)} \longrightarrow Y(z)$

Figure 21.6.1 Cascade Interconnection Structure

---

[3] Some guidance is provided by advanced analysis of scaling effects.

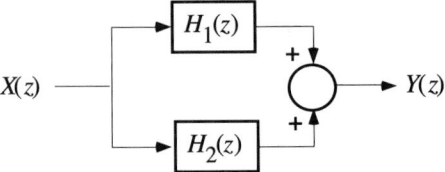

Figure 21.6.2 Parallel Interconnection Structure

To illustrate this decomposition, suppose that the two subsystems in (21.6.5) are given by

$$\frac{Y_i(z)}{X_i(z)} = H_i(z) = \frac{b_{0i} + b_{1i}z^{-1} + b_{2i}z^{-2}}{1 + a_{1i}z^{-1} + a_{2i}z^{-2}}, \quad i = 1,2. \tag{21.6.6}$$

Notation: The second index on the coefficients refers to the subsystem.

An all-delay block diagram corresponding to the decomposition in (21.6.6) (assuming a 4th-order transfer function) which uses (21.6.5) is given in Figure 21.6.3.

Terminology: In Figure 21.6.3 each individual system is shown in direct form I. Hence, this network structure is called <u>cascade direct form I</u>. If direct form II is used to realize the second-order blocks, the overall structure is called <u>cascade direct form II</u>. This terminology is repeated for other network structures as well.

A second decomposition is based on the parallel interconnection structure. This decomposition is obtained by expanding the given transfer function (21.6.1) in a partial fraction expansion

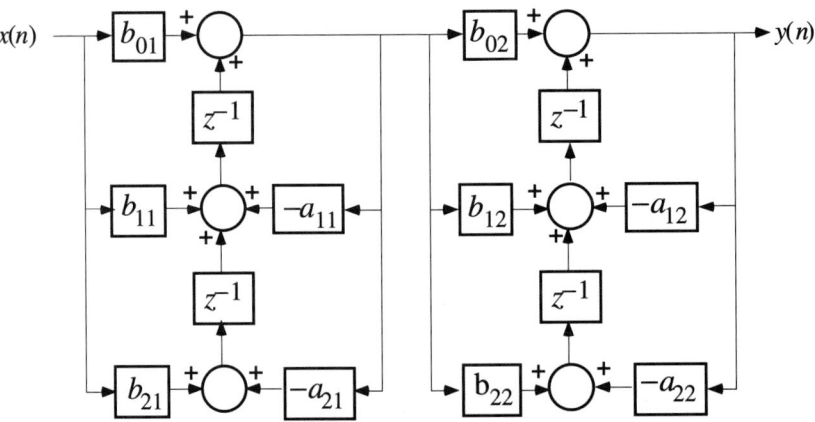

Figure 21.6.3 Cascade Decomposition of (21.6.5)

$$H(z) = d + \frac{c_1}{1 - p_1 z^{-1}} + \frac{c_2}{1 - p_2 z^{-1}} + \cdots + \frac{c_N}{1 - p_N z^{-1}} \qquad (21.6.7)$$

$$= \sum_{i=1}^{\overline{N}} \frac{b_{0i} + b_{1i} z^{-1} + b_{2i} z^{-2}}{a_{0i} + a_{1i} z^{-1} + a_{2i} z^{-2}}.$$

Since we wish to avoid complex coefficients again we group complex poles into second-order factors as shown in (21.6.7). We have also grouped the pairs of real poles together for convenience. If there are an odd number of poles, then the appropriate coefficients of one the terms in (21.6.7) are zero. Also the constant has been absorbed into one of the second-order factors.

Example 21.6.1: Consider the transfer function

$$\frac{Y(z)}{X(z)} = \frac{0.2489 + 0.7466z^{-1} + 0.7466z^{-2} + 0.2489z^{-3}}{1 + 0.5043z^{-1} + 0.5289z^{-2} - 0.0423z^{-3}} \qquad (21.6.8)$$

$$= \frac{(z-1)^3}{(z-0.074)\left((z+0.2892)^2 + 0.6986^2\right)}.$$

A cascade decomposition of this transfer function is given by

$$\frac{Y(z)}{X(z)} = \left[\frac{(1 - z^{-1})}{(1 - 0.074z^{-1})}\right]\left[\frac{1 - 2z^{-1} + z^{-2}}{1 + 0.5783z^{-1} + 0.5717z^{-2}}\right]. \qquad (21.6.9)$$

A parallel decomposition of this system is found by first doing a partial fraction expansion of (21.6.8). We find that

$$H(z) = \frac{-0.2943 + j0.0332}{1 - (-0.2892 + j0.6986)z^{-1}} \qquad (21.6.10)$$

$$+ \frac{-0.2943 - j0.0332}{1 - (-0.2892 - j0.6986)z^{-1}} + \frac{6.7222}{1 + 0.0740z^{-1}} - 5.8846.$$

Combining the terms with complex poles and the constant term we obtain

$$H(z) = \frac{-6.4733 - 3.6197z^{-1} - 3.3643z^{-2}}{1 + 0.5783z^{-1} + 0.5717z^{-2}} + \frac{6.7222}{1 + 0.0740z^{-1}}. \qquad (21.6.11)$$

A direct form II parallel decomposition is shown in Figure 21.6.4.

▲▲

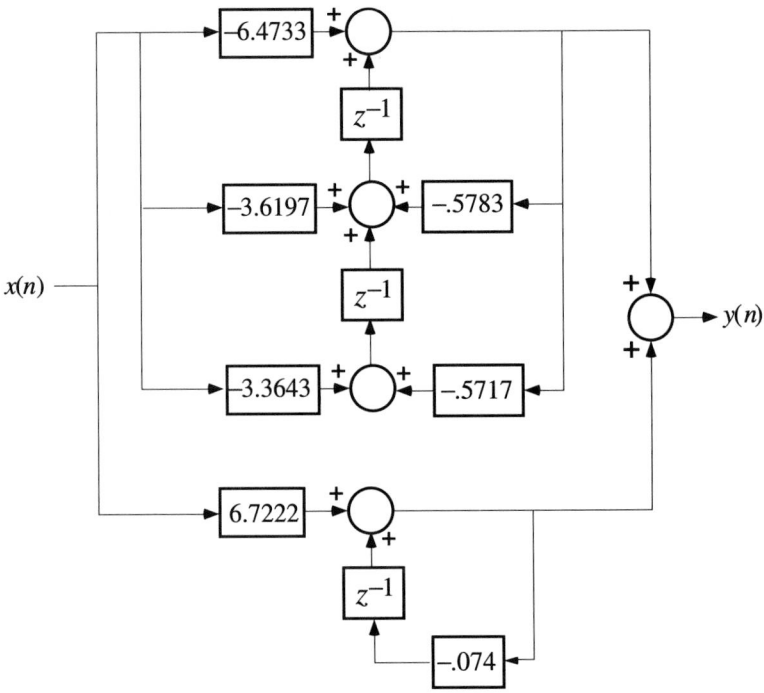

Figure 21.6.4 Direct Form II Parallel Decomposition of (21.6.8)

21.6.3 MATLAB Experiments

The following M-file calculates the coefficients of the parallel decomposition of the transfer function in Example 21.6.1.

Example 21.6.1: (*continued*)

```
clear
% generate transfer function
[b,a] = cheby1(3,.5,.6);
[zeros,poles,const] = tf2zp(b,a); % zeros and poles of transfer function
% calculate parallel decomposition
[r,p,k]=residuez(b,a);
p1 = [1,-p(1)];              % complex pole
p2 = [1,-p(2)];              % complex conjugate pole
b1 = r(1)*p2 + r(2)*p1;      % numerator of second-order term
a1 = real(conv(p1,p2));      % denominator of second-order term
b2 = [b1,0] + k*a1;          % numerator of linear term with constant coefficient
```

▲▲

The complex poles obtained from the residue command p1 and p2 must be correlated manually with their locations in the vector of poles p.

Exploratory Exercise 21.6.2: Generate a fourth-order transfer function of a standard filter and determine its cascade and parallel decompositions.

▲▲

21.7 CHAPTER SUMMARY

In this chapter we introduced five system representations. These representations include: (1) the convolution representation, Definition 21.1.1, (2) difference equations, Definition 21.2.1, (3) the z-transform transfer function, Definition 21.2.6, (4) the DTFT transfer function, Definition 21.4.1, and (5) state space equations (21.5.7), Definition 10.5.8.

The convolution representation is characterized by the impulse response function $h(n)$ (Definition 21.1.2). If the impulse response function is zero for all but a finite number of values of $h(n)$, it is called a finite impulse response (FIR) system. Otherwise it is an infinite impulse response (IIR) system (Definition 21.1.4).

The poles of the transfer function are the roots of the denominator polynomial (Definition 21.2.10). The zeros of the transfer function are the roots of the numerator polynomial.

Block diagrams are introduced in Section 21.3 based on the transfer function (Definition 21.3.1). The most important elementary building block is a unit delay (Definition 21.3.3). Block diagrams are characterized as recursive and nonrecursive (Definition 21.3.7) depending on the presence or absence of feedback loops, respectively. Two special block diagrams called direct form I (Figure 21.3.7) and direct form II (Figure 21.3.9) are introduced. Also an FIR network structure (Figure 21.3.10) is discussed.

These block diagrams are examples of a class of block diagrams called all-delay block diagrams (Definition 21.5.1). State space representations are developed from all-delay block diagrams using Procedure 21.5.2. An explicit solution to the state space equations (Theorem 21.5.7) is given by

$$y(n) = CA^n \vec{q}_0 + \sum_{k=0}^{n-1} CA^{n-1-k} Bx(k) + Dx(n)$$

where the function A^{n-1}, $n \geq 0$ is called the <u>state transition matrix</u> (Definition 21.5.6).

In Section 21.6 the decomposition of a system into a cascade or parallel interconnection structure is discussed. This decomposition is usually combined with the direct form I and direct form II network structures of Section 21.3.

21.8 HOMEWORK FOR CHAPTER 21

Homework Problems for Section 21.1

21.1.1 Convolve the following signals by hand. Show all of your work.

(i) $h(n) = (2 - n)(u_s(n) - u_s(n - 2))$, $x(n) = n(u_s(n) - u_s(n - 2))$.

(ii) $h(n) = u_S(n) - u_S(n-2)$, $x(n) = n\big(u_S(n) - u_S(n-2)\big)$

(iii) $h(n) = 2\big(u_S(n+1) - u_S(n-2)\big)$, $x(n) = u_S(n) - u_S(n-2)$

(iv) $h(n) = u_S(n+1) - u_S(n-2)$, $x(n) = (-1)^n\big(u_S(n) - u_S(n-2)\big)$

21.1.2 Suppose the impulse response function for a system is

$$h(n) = \big[u_S(n) - u_S(n-5)\big] + (-1)\big[u_S(n-5) - u_S(n-10)\big].$$

We consider two input signals given by

$$x_1(n) = \big[u_S(n) - u_S(n-5)\big] + (-1)\big[u_S(n-5) - u_S(n-10)\big],$$
$$x_2(n) = (-1)\big[u_S(n) - u_S(n-5)\big] + \big[u_S(n-5) - u_S(n-10)\big].$$

(a) Using MATLAB find the output signal for each input signal using **conv**. Plot both output signals. (Pad the input signals and the impulse response function with 5 zeros, before and after.)

(b) Develop a criteria based on the output signal for determining which input signal is present.

(c) Define a noise signal *ns* by

$$ns = A * \textbf{randn}(\textbf{size}(x_1))$$

Find the output signal of the system when the input signal is $x_1 + ns$. Determine the largest value of A for which x_1 can be detected in the output signal. Document your answer.

21.1.3 Suppose we are given two systems with the following impulse response functions.

$$h_1(n) = (0.25)^n u_S(n), \qquad h_2(n) = \delta(n) + 0.2\delta(n-1)$$

(a) Find the impulse response function of a cascade interconnection of these systems.

(b) Find the impulse response function of a parallel interconnection of these systems.

21.1.4 Suppose the impulse response function for a system is

$$h(n) = \begin{cases} \sin(0.5n), & 0 \le n \le 20 \\ 0, & \text{elsewhere} \end{cases}$$

We consider two input signals to this system. The first input signal is

$$d(n) = A * (\text{'}\textbf{randn'}).$$

The MATLAB command **randn** will generate a random number with Gaussian distribution, 0 mean, and unit variance. The second input signal is

$$x(n) = h(n) + d(n)$$

(a) Compute the output signal of the system for each of these two input signals (with $A = 0.1$) using the convolution command in MATLAB. Plot both output signals on one graph. (Pad the input signals and the impulse response function with 5 zeros, before and after.)
(b) Determine the largest value of A for which $x(n)$ can be detected in the output signal. Document your answer.

21.1.5   Suppose the impulse response function of a system is given by

$$h(n) = 2[u_S(n+N-1) - u_S(n-N)]$$

Let the input signal be given by

$$x(n) = u_S(n) - u_S(n-M)$$

Plot the output signal using the MATLAB **conv** command. Let $M = 10$ and $N = 5$. Be sure to include the proper time scaling on the plot.

21.1.6   Calculate $y(6)$ if the impulse response function of a discrete system is

$$h(n) = \left((0.5)^n \sin(0.003n)\right)u_S(n),$$

and the input signal is

$$x(n) = u_S(n+1) - u_S(n-2).$$

21.1.7   Let $h(n)$ be the impulse response function of a system. Suppose that $h(n)$ has only a finite number of nonzero values. Show that the convolution sum is of fixed finite length.

21.1.8   Using MATLAB evaluate all of the convolution integrals in the Section 12.2 Problems.

21.1.9   (a) Is the impulse response of an FIR filter an energy or power signal?
(b) Is the impulse response of an IIR filter an energy or power signal?

21.1.10 (a) Suppose we form a cascade connection of an IIR filter and an FIR filter. Is the composite system an FIR or an IIR system?
(b) Suppose we form a parallel connection of an IIR filter and an FIR filter. Is the composite system an FIR or an IIR system?
(c) Answer (a) and (b) if both systems are IIR systems.
(d) Answer (a) and (b) if both systems are FIR systems.

Homework Problems for Section 21.2

21.2.1 For each of the following transfer functions
  (a) Find the difference equation in delay form that corresponds to each system.
  (b) Find the difference equation in advance form that corresponds to each system.
  (c) Draw the pole-zero diagram.

(i) $\dfrac{Y(z)}{X(z)} = \dfrac{0.6 - 0.5z^{-1} + 0.4z^{-2}}{1 + 0.3z^{-1} + 0.2z^{-2}}$ (iv) $\dfrac{Y(z)}{X(z)} = \dfrac{(1 + z^{-1})^3}{1 - 0.98z^{-1} + 0.66z^{-2}}$

(ii) $\dfrac{Y(z)}{X(z)} = \dfrac{0.6z - 0.5 + 0.4z^{-1}}{1 + 0.3z^{-1} + 0.2z^{-2}}$ (v) $\dfrac{Y(z)}{X(z)} = \dfrac{1 - z^{-4}}{1 - 0.25z^{-4}}$

(iii) $\dfrac{Y(z)}{X(z)} = \dfrac{2 - 0.3z^{-1}}{1 - 0.7z^{-1} - 0.12z^{-2}}$ (vi) $\dfrac{Y(z)}{X(z)} = \dfrac{z^2 - 3z^{-1}}{1 - 0.5z^{-2}}$

21.2.2 For each of the following transfer functions find and plot the impulse response function.

(i) $\dfrac{Y(z)}{X(z)} = H(z) = \dfrac{z^2}{(z - 0.36)(z + 0.5)}$, $ROC = \{|z| > 0.5\}$

(ii) $\dfrac{Y(z)}{X(z)} = H(z) = \dfrac{z^2}{(z - 0.36)(z + 0.5)}$, $ROC = \{0.36 < |z| < 0.5\}$

(iii) $\dfrac{Y(z)}{X(z)} = H(z) = \dfrac{z^2}{(z - 0.36)(z + 0.5)}$, $ROC = \{|z| < 0.36\}$

21.2.3 For each of the following discrete systems
  (a) Find the transfer function of this system.
  (b) Plot the pole-zero diagram of the transfer function.
  (i) $y(n) + 0.5043y(n-1) + 0.5289y(n-2) - 0.0423y(n-3)$
       $= 0.2489x(n) + 0.7466x(n-1) + 0.7466x(n-2) + 0.2489x(n-3)$
  (ii) $y(n) - y(n-1) - 0.24y(n-2) = 0.24x(n)$

21.2.4 For each of the following FIR systems
  (a) Find the difference equation that corresponds to this transfer function.
  (b) Draw the pole-zero diagram.

(i) $\dfrac{Y(z)}{X(z)} = H(z) = 1 - 1.91z^{-1} + z^{-2}$

(ii) $\dfrac{Y(z)}{X(z)} = H(z) = \dfrac{1}{3}\left(z + 1 + z^{-1}\right)$

21.2.5 Show that the roots of $a(s) = z^2 - 2\cos(\Omega_0)z + 1$ are $p_{1,2} = e^{\pm j\Omega_0}$.

21.2.6 Find the transfer function of a system whose impulse response function is

$$h(n) = \begin{cases} \sin(0.5n), & 0 \le n \le 20 \\ 0, & \text{elsewhere} \end{cases}$$

21.2.7 For each of the following discrete systems
(a) Write this difference equation in advance form.
(b) Find the transfer function in terms of z.
  (i) $y(n) + 0.5043y(n-1) + 0.5289y(n-2) - 0.0423y(n-3)$
        $= 0.2489x(n) + 0.7466x(n-1) + 0.7466x(n-2) + 0.2489x(n-3)$
  (ii) $y(n) - y(n-1) - 0.24y(n-2) = 0.24x(n)$

21.2.8 Determine the transfer function of the system whose step response is

$$y(n) = 2 - (0.5)^n, \quad n \ge 0$$

Homework Problems for Section 21.3

21.3.1 For each of the following transfer functions
(a) Draw the direct form I block diagram of this system.
(b) Draw the direct form II block diagram of this system.
(c) Write the difference equations that correspond to each block diagram.
  (i) $\dfrac{Y(z)}{X(z)} = \dfrac{0.6 - 0.5z^{-1} + 0.4z^{-2}}{1 + 0.3z^{-1} + 0.2z^{-2}}$
  (ii) $\dfrac{Y(z)}{X(z)} = \dfrac{(0.885)(1 - 1.53z^{-1} + z^{-2})}{1 - 1.35z^{-1} + 0.77z^{-2}}$
  (iii) $\dfrac{Y(z)}{X(z)} = \dfrac{(0.9147)(1 - 0.868z^{-1} + z^{-2})}{1 - 0.761z^{-1} + 0.8347z^{-2}}$
  (iv) $\dfrac{Y(z)}{X(z)} = \dfrac{0.057 + 0.1709z^{-1} + 0.1709z^{-2} + 0.057z^{-3}}{1 - 1.1905z^{-1} + 1.0452z^{-2} - 0.3990z^{-3}}$

21.3.2 For each of the following discrete systems
(a) Draw the direct form I block diagram of this system.
(b) Draw the direct form II block diagram of this system.
(c) For each block diagram, write the difference equations that correspond to this block diagram.
  (i) $y(n) + 0.5043y(n-1) + 0.5289y(n-2) - 0.0423y(n-3)$
        $= 0.2489x(n) + 0.7466x(n-1) + 0.7466x(n-2) + 0.2489x(n-3)$
  (ii) $y(n) - 2.11y(n-1) + 2.36y(n-2) - 1.32y(n-3) + 0.34y(n-4)$
        $= 0.07x(n) + 0.01x(n-1) + 0.1x(n-2) + 0.01x(n-3) + 0.07x(n-4)$

21.3.3 Find the transfer function of the block diagram in Figure P21.3.3 by block diagram reduction.

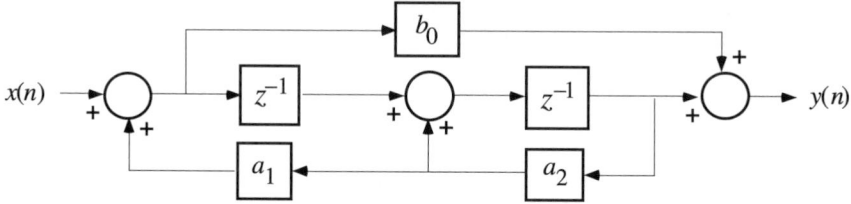

Figure P21.3.3

21.3.4   Find the transfer function of the system shown in Figure P21.3.4.

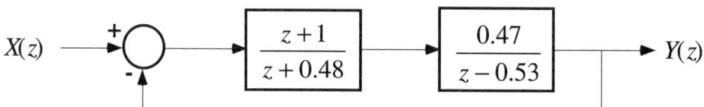

Figure P21.3.4

21.3.5   Find the transfer function of the system shown in Figure P21.3.5.

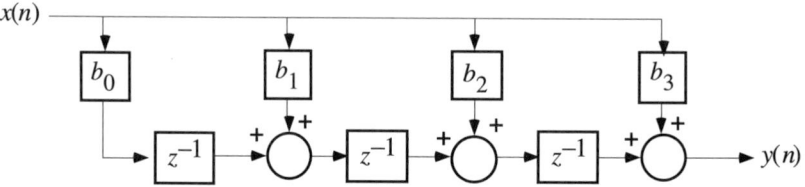

Figure 21.3.5

21.3.6   Find the transfer function of the system in Figure P21.3.6.

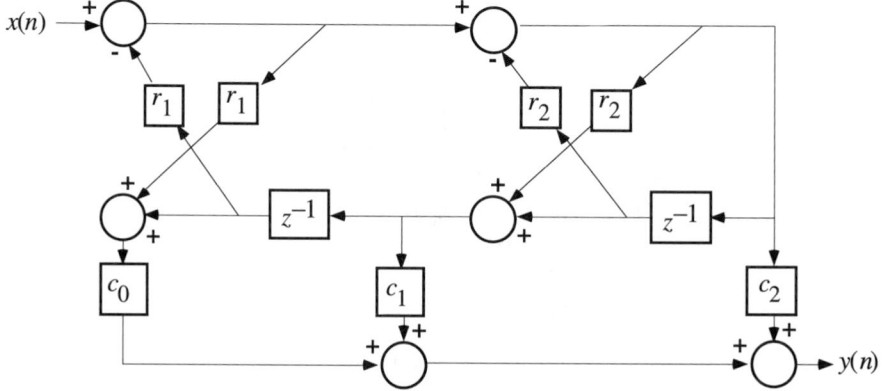

Figure P21.3.6

21.3.7   Consider the FIR system

$$\frac{Y(z)}{X(z)} = 1 + b_1 z^{-1} + b_2 z^{-2} = \frac{1 + (1 - b_1)z^{-1} + (b_2 - b_1)z^{-2} - b_2 z^{-3}}{1 - z^{-1}}$$

(a) Draw the direct form II realization of the first transfer function. (Compare to Figure 21.3.10.)

(b) Draw the direct form II realization of the second transfer function.

Note: The first block diagram is a nonrecursive system while the second block diagram is a recursive system.

Homework Problems for Section 21.4

21.4.1   Find the DTFT transfer function of the following systems.
   (i) $y(n) = x(n) - 1.91x(n-1) + x(n-2)$
   (ii) $y(n) - y(n-1) - 0.24y(n-2) = 0.24x(n)$
   (iii) $y(n) = \frac{1}{3}(x(n+1) + x(n) + x(n-1))$

21.4.2   For each of the following DTFT transfer functions and input signals, find the corresponding output signals.
   (i) $\dfrac{Y(\Omega)}{X(\Omega)} = H(\Omega) = \tan^{-1}(\sin(\Omega)), \quad x(n) = \cos\left(\dfrac{\pi n}{4}\right)$
   (ii) $\dfrac{Y(\Omega)}{X(\Omega)} = H(\Omega) = \sin(\Omega) - \dfrac{1}{8}\sin(2\Omega), \quad x(n) = \cos^2\left(\dfrac{\pi n}{4}\right)$

Homework Problems for Section 21.5

21.5.1   Write the state space equations for the block diagram in Figure P21.3.3.

21.5.2   Find a state space representation of the system shown in Figure P21.3.6.

21.5.3   Draw the all-delay block diagram of the following systems.

   (i) $\vec{q}(n+1) = \begin{bmatrix} 0 & 1 & 0 \\ 0 & 0 & 1 \\ -a_3 & -a_2 & -a_1 \end{bmatrix} \vec{q}(n) + \begin{bmatrix} 0 \\ 0 \\ 1 \end{bmatrix} x(n)$

   $y(n) = \begin{bmatrix} b_0 & b_1 & b_2 \end{bmatrix} \vec{q}(n)$

   (ii) $\vec{q}(n+1) = \begin{bmatrix} \lambda_1 & 0 \\ 0 & \lambda_2 \end{bmatrix} \vec{q}(n) + \begin{bmatrix} b_1 \\ b_2 \end{bmatrix} x(n)$

   $y(n) = \begin{bmatrix} c_1 & c_2 \end{bmatrix} \vec{q}(n)$

   (iii) $\vec{q}(n+1) = \begin{bmatrix} \sigma & \omega \\ -\omega & \sigma \end{bmatrix} \vec{q}(n) + \begin{bmatrix} 0 \\ b_1 \end{bmatrix} x(n)$

   $y(n) = \begin{bmatrix} c_1 & 0 \end{bmatrix} \vec{q}(n)$

21.5.4 Find the state transition matrix of the following systems.

(i) $\vec{q}(n+1) = \begin{bmatrix} \lambda_1 & 0 \\ 0 & \lambda_2 \end{bmatrix}\vec{q}(n) + \begin{bmatrix} b_1 \\ b_2 \end{bmatrix}x(n)$

$y(n) = \begin{bmatrix} c_1 & c_2 \end{bmatrix}\vec{q}(n)$

(ii) $\vec{q}(n+1) = \begin{bmatrix} 0 & -0.5 \\ 0.5 & 0 \end{bmatrix}\vec{q}(n) + \begin{bmatrix} 0 \\ 1 \end{bmatrix}x(n)$

$y(n) = \begin{bmatrix} 2 & 0 \end{bmatrix}\vec{q}(n)$

21.5.5 For each of the following systems
(a) Solve for the states and the output signal of the following systems.
(b) What are the eigenvalues of the state matrices?

(i) $\vec{q}(n+1) = \begin{bmatrix} \lambda_1 & 0 \\ 0 & \lambda_2 \end{bmatrix}\vec{q}(n), \quad \vec{q}(0) = \begin{bmatrix} q_{10} \\ q_{20} \end{bmatrix}$

$y(n) = \begin{bmatrix} c_1 & c_2 \end{bmatrix}\vec{q}(n)$

(ii) $\vec{q}(n+1) = \begin{bmatrix} 0 & 0.5 \\ 0.5 & 0 \end{bmatrix}\vec{q}(n), \quad \vec{q}(0) = \begin{bmatrix} 1 \\ 2 \end{bmatrix}$

$y(n) = \begin{bmatrix} 2 & 0 \end{bmatrix}\vec{q}(n)$

(iii) $\vec{q}(n+1) = \begin{bmatrix} 0 & 1 & 0 \\ 0 & 0 & 1 \\ 0 & 0 & 0 \end{bmatrix}\vec{q}(n), \quad \vec{q}(0) = \begin{bmatrix} 0 \\ 0 \\ 1 \end{bmatrix}$

$y(n) = \begin{bmatrix} 1 & 0 & 0 \end{bmatrix}\vec{q}(n)$

21.5.6 Consider the transfer function

$$H(z) = \frac{b_0 + b_1 z^{-1} + b_2 z^{-2}}{1 + a_1 z^{-1} + a_2 z^{-2}}.$$

(a) Draw the direct form I and II block diagrams of this system using unit advances instead of unit delays.
(b) Derive the state space equations for these two block diagrams.

21.5.7 Consider the state space representation

$$\vec{q}(n+1) = \begin{bmatrix} -0.7 & 1 & 0 \\ 0.25 & 0 & 1 \\ 0.175 & 0 & 0 \end{bmatrix}\vec{q}(n) + \begin{bmatrix} -0.3 \\ -0.71 \\ 0.175 \end{bmatrix}x(n),$$

$$y(n) = \begin{bmatrix} 1 & 0 & 0 \end{bmatrix}\vec{q}(n) + x(n).$$

(a) Find the impulse response of this system.
(b) Draw the all-delay block diagram of this system.

Homework Problems for Section 21.6

21.6.1 For each of the following discrete systems
  (a) Draw a block diagram of this system in parallel direct form I using first- and second-order sections.
  (b) Write the difference equations that correspond to this block diagram.
  (i) $y(n) + 0.5043y(n-1) + 0.5289y(n-2) - 0.0423y(n-3)$

   $$= 0.2489x(n) + 0.7466x(n-1) + 0.7466x(n-2) + 0.2489x(n-3)$$
  (ii) $y(n) - 2.11y(n-1) + 2.36y(n-2) - 1.32y(n-3) + 0.34y(n-4)$

   $$= 0.07x(n) + 0.01x(n-1) + 0.1x(n-2) + 0.01x(n-3) + 0.07x(n-4)$$

21.6.2 For each of the following transfer functions
  (a) Draw a cascade direct form I realization of this system using first- and second-order sections.
  (b) Draw a parallel direct form I realization of this system using first- and second-order sections.
  (c) Write the difference equations that correspond to this block diagram.
  (i) $$\frac{Y(z)}{X(z)} = \frac{(1+z^{-1})^3}{(1-0.6z^{-1})(1-0.98z^{-1}+0.66z^{-2})}$$
  (ii) $$\frac{Y(z)}{X(z)} = \frac{1-z^{-4}}{1-0.25z^{-4}}$$

21.6.3 (a) Find the transfer function of the system shown in Figure P21.6.3.
  (b) Write the difference equations for this system.

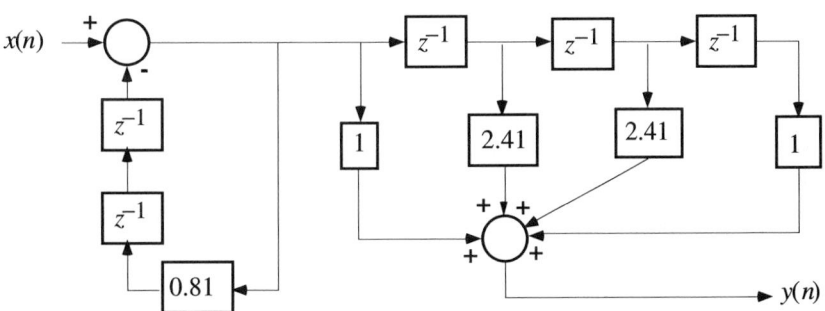

Figure P21.6.3

21.6.4   For each of the following transfer functions
(a) Draw a cascade direct form II realization of this filter using first- and second-order sections.
(b) Draw a parallel direct form II realization of this filter using first- and second-order sections.
(c) Write the difference equations that correspond to this block diagram.

(i) $\dfrac{Y(z)}{X(z)} = \dfrac{z^{-2}}{1 - 1.4z^{-1} + 0.48z^{-2}}$

(ii) $\dfrac{Y(z)}{X(z)} = \dfrac{1 + z^{-1} + 0.25z^{-2}}{1 - 1.4z^{-1} + 0.48z^{-2}}$

(iii) $\dfrac{Y(z)}{X(z)} = \dfrac{0.8 - 5.6z^{-1} + 14.112z^{-2} - 15.136z^{-3} + 5.824z^{-4}}{1 - 1.75z^{-1} + 1.0625z^{-2} - 0.2656z^{-3} + 0.0234z^{-4}}$

(iv) $\dfrac{Y(z)}{X(z)} = 1 + 1.2z^{-1} - 0.15z^{-2} - 0.586z^{-3} - 0.168z^{-4}$

Chapter 22

Properties of Discrete-Time Systems

Chapter Outline

The primary purpose of this chapter is twofold. First, we will introduce several basic properties of systems. These properties serve to characterize a system in terms of easily checked criteria. This classification of systems determines which tools can be used for the analysis of the system, for example. Hence, these properties are of fundamental importance in understanding systems. After we have defined these system properties, we will determine the system properties of each of the system representations we introduced in the last chapter.

Once again we find that the theory we develop for discrete-time systems is an exact parallel to the theory developed for continuous-time systems. In fact, the definitions of properties of systems we introduce in this chapter are exactly the same as the properties defined in Chapter 13. Most of the results in this chapter are parallels to the results in Chapter 13, with the notable exception of the stability results. Comparisons with the continuous-time results are made throughout the chapter.

Second, we will develop the interrelationships between the system representations we have introduced in the previous chapter. These interrelationships are used in the routine analysis of systems. System analysis typically proceeds in several stages. At each stage, different tools may be applied to further enhance the understanding of the system. The particular analysis tool depends on the system representation. To switch between analysis tools requires that we switch between system representations. Hence, the interrelationships between the representations play a key role in system analysis. For example, all computer analysis tools, such as MATLAB, have built-in commands to effect these system conversions. The properties of systems developed in the first part of the chapter are used in the development of the interrelationships between the systems.

There are many systems which consist of a combination of discrete systems and continuous systems. The input and output signals of these systems are interconnected by sampling and analog-to-digital converters. In order to analyze these systems it is necessary to translate the entire system into the discrete-time domain. In the last section of this chapter we will give a method for translating a continuous system into a discrete system.

Summary of Sections

Section 22.1:   We define the basic system properties.

Section 22.2:   We discuss the properties of the system representations introduced in Chapter 21.

Section 22.3:   We define BIBO stability. We also develop several stability tests.

Section 22.4:   We develop the relationship between the various system representations.

Section 22.5:   We show how to construct a discrete system from a continuous system when the input signal to the system is the output signal from a zero-order hold.

Section 22.6:   Chapter summary section.

Coverage of the Text

The results in this chapter continue the development of discrete-time systems in Chapter 21. The definitions of the properties of discrete-time systems are the same as the properties of continuous-time systems presented in Chapter 13. Comparisons with this material is drawn, but this material is not strictly required. The relationship between system representations is similar to the material in Chapters 10 - 12, but, again, this material is not strictly required except for the realizations from Section 11.2. Section 22.5, however, requires knowledge of state space equations including Section 12.3 and the ZOH in Section 19.4.

22.1 PROPERTIES OF SYSTEMS

22.1.1 Introduction

In the last chapter we developed several representations for discrete-time systems of the form

$$y(n) = \mathcal{H}[x(n)]. \tag{22.1.1}$$

In this section we will define several properties for this abstract system. We will then show that the specific system representations introduced in the last chapter possess these properties. This information is very useful in two ways. First, the fact that all of the system representations share common properties implies that these system representations are in many ways equivalent. These equivalencies will be developed later in this chapter. The properties that are not shared also serve to define how these system representations differ. Second, these properties determine the physical processes that these representations can be used to model. This information can be very useful when developing a model of a discrete system.

The results in this chapter parallel the results for continuous systems presented Chapters 6 and 10 - 13 with a few exceptions. In particular the definitions of the system properties in this section are exactly the same as the definitions given for continuous systems. We will highlight the similarities and differences between continuous and discrete systems throughout the chapter.

22.1.2 Definitions

We begin with a definition that has renewed importance for discrete systems.

Definition 22.1.1: A system is <u>causal</u> if for all input signals $x_1(n)$ and $x_2(n)$ that satisfy

$$x_1(n) = x_2(n) \quad \text{for} \quad n \le n_0$$

the corresponding output signals satisfy

$$y_1(n) = y_2(n) \quad \text{for} \quad n \le n_0$$

for all n_0.

▲▲

Very roughly, a system is causal if it doesn't respond until it receives an input signal. The same definition was given for continuous-time systems. Continuous-time systems are almost always causal; this definition plays a very small role in the analysis of continuous systems. Noncausal discrete systems do occur in practice, however. Noncausal systems are associated with off-line computer processing of stored data, such as in image processing.

Example 22.1.2: Consider the difference equation

$$y(n) + a_1 y(n-1) = x(n+2). \tag{22.1.2}$$

The present value of the output signal $y(n)$ depends on a future value of the input signal $x(n+2)$. This system isn't causal.

▲▲

The next two definitions are of fundamental importance in describing the system representations in the last chapter.

Definition 22.1.3: A system is <u>linear</u> if for two input signals $x_1(n)$ and $x_2(n)$ with corresponding output signals $y_1(n)$ and $y_2(n)$ and two real numbers a_1 and a_2,

$$\mathcal{H}\left[a_1 x_1(n) + a_2 x_2(n)\right] = a_1 y_1(n) + a_2 y_2(n).$$

Otherwise, we say the system is <u>nonlinear</u>.

SS

Example 22.1.4: The following systems are linear:

$$y(n) - ny(n-1) = x(n), \tag{22.1.3}$$
$$y(n+2) + 2y(n) = x(n+1).$$

The following systems are nonlinear:

$$y(n) - x(n)y(n-1) = x(n), \tag{22.1.4}$$
$$y(n+2) + 2y(n) = x(n+1) + 2.$$

▲▲

In general, we can't apply any of the transforms (z-transforms or the DTFT) to nonlinear systems. Since these transforms are the basis of many of our analysis tools, we find the analysis of nonlinear systems difficult. In particular, the frequency analysis of a system (introduced in the next chapter) requires the system to be linear.

The next property is just as important as linearity.

Definition 22.1.5: A system is <u>time invariant</u> if for all n_0

$$\mathcal{H}\left[x(n-n_0)\right] = y(n-n_0).$$

Otherwise, the system is <u>time-varying</u>.

▲▲

Example 22.1.6: The following systems are time invariant:

$$y(n) - x(n)y(n-1) = x(n), \tag{22.1.5}$$

$$y(n+2) + 2y(n) = x(n+1) + 2,$$

$$y(n-1) - x(n-1)y(n-2) = x(n).$$

The following system is time-varying:

$$y(n+2) + ny(n) = x(n+1). \tag{22.1.6}$$

▲▲

In general, we can't apply any of the transforms (z-transforms or the DTFT) to time-varying systems just as we can't apply these transforms to nonlinear systems. Nonlinear and time-varying systems can exhibit extremely bizarre behavior including chaos, behavior which is not observed in linear, time invariant systems.

The next property is used to distinguish between system representations which capture only the input/output behavior, and system representations which can incorporate some internal characteristics of the system.

Definition 22.1.7: A system is <u>relaxed</u> at time n_0 if the output signal $y(n)$ for $n_0 \le n$ is solely and uniquely excited by the input signal $x(n)$ for $n_0 \le n$.

▲▲

Basically, a system is not relaxed in two ways. First, the system has stored internal "energy." This energy is usually represented in the form of initial conditions on the difference equation. Second, the system may have an input signal which has been present since $t = -\infty$. Since the system is continually stimulated, it can't be relaxed.

Definition 22.1.8: If a system is relaxed, then the output signal for a given input signal is called the <u>zero state response</u>. If the input signal is zero, but the system is not relaxed, then the output signal is called the <u>zero input response</u>.

▲▲

22.2 PROPERTIES OF SYSTEM REPRESENTATIONS

22.2.1 Introduction

In the last section we introduced four fundamental properties of systems: causal, linear, time invariant, and relaxed. In this section we are going to discuss the properties of the system representations that were introduced in Chapter 21: the convolution summation, transfer functions, difference equations, and state space representations. It turns out that all of these system representations are linear and time invariant. Because all of these representations possess these common properties, they are in some sense equivalent. The relationships between these representations are established in Section 22.4. These system representations also differ in some respects. These differences are highlighted by the properties of causality and relaxedness. Understanding the similarities and differences between these system representations is required for the correct usage of these representations.

22.2.2 Derivation of the Convolution Representation

To begin, we will show that we can derive the convolution representation of a system using only the properties of linearity and time invariance. This derivation is given for continuous systems in Section 13.1. The derivation that follows is somewhat more straightforward in discrete time.

Consider the input signal

$$x(n), \quad -M \le n \le N. \tag{22.2.1}$$

Using the impulse function this input signal can be written as

$$x(n) = \sum_{k=-M}^{N} x(k)\delta(n-k). \tag{22.2.2}$$

Note that this representation of the signal is a linear combination of functions $\delta(n-k)$ weighted by the real numbers $x(k)$. The output signal of the system is given by

$$y(n) = \mathcal{H}[x(n)] = \mathcal{H}\left[\sum_{k=-M}^{N} x(k)\delta(n-k)\right] \tag{22.2.3}$$

where we have used the input signal representation in (22.2.2).

To continue, we need the impulse response function of the system. We can imagine that this function is obtained by using an impulse function as the input signal $x(n) = \delta(n)$ and "measuring" the output signal $y_i(n) = h(n)$. We assume that the impulse response function of the system is $h(n)$. This assumption requires that the system be relaxed when we measure the impulse response function.

We observe that the input signal to the system in (22.2.3) is a linear combination of time-shifted impulse functions. Assuming this system is *linear*, we can write (22.2.3) as

$$y(n) = \mathcal{H}\left[\sum_{k=-M}^{N} x(k)\delta(n-k)\right] = \sum_{k=-M}^{N} x(k)\mathcal{H}[\delta(n-k)]. \tag{22.2.4}$$

Next we observe that the output signal in (22.2.4) is a linear combination of the output signals corresponding to the input signals $x(n) = \delta(n-k)$. Assuming that the system is *time invariant*, we can write (22.2.4) as

$$y(n) = \sum_{k=-M}^{N} x(k) \mathcal{H}[\delta(n-k)] = \sum_{k=-M}^{N} x(k)h(n-k). \tag{22.2.5}$$

This system representation is, of course, the convolution representation.

This result is highly significant. It says that to know that a convolution representation exists, we need only verify that the system is linear and time invariant. Often we can easily verify these properties independently of writing equations for a model. Knowing that the convolution representation exists, we immediately know that the other representations exist (as will be shown below). Hence, if we need to find a model, we can find the model in the most convenient form. Furthermore, we can immediately apply any of the analysis techniques we will develop below, such as frequency response methods. On the other hand, if we verify that the system is nonlinear or time-varying, then, in general, these analysis techniques can't be applied to this system.[1]

The analysis above didn't require the system to be causal. Also, the system was required to be relaxed in only an indirect way, to find the impulse response function. The following theorem summarizes the properties of the convolution representation.

Theorem 22.2.1: Consider the system

$$y(n) = \sum_{k=-\infty}^{\infty} h(n-k)x(k).$$

This system is causal if and only if $h(n) = 0$ for $n < 0$. This system is linear and time invariant. This system is relaxed if $x(n) = 0$ for $n < k_0$.

▲▲

It is easy to see that if $h(n) \neq 0$ for $n < 0$, then the system will respond before the input signal $x(n) = \delta(n)$ is applied. In this case the system can't be causal. The fact that this representation is linear and time invariant comes from the derivation of the convolution summation given above. This model is an input-output model which makes no provision for modeling internal energy storage elements. Hence, this system is relaxed if the input signal is zero at negative infinity.

22.2.3 Properties of Difference Equations

We can also derive the properties of difference equations. In order to state the results clearly, we write the difference equation in delay form as

$$y(n) + a_1 y(n-1) + \cdots + a_{N-1} y(n-N+1) + a_N y(n-N) \tag{22.2.6}$$
$$= b_{-M_+} x(n+M_+) + \cdots + b_{-1} x(n+1) + b_0 x(n) + \cdots + b_M x(n-M_-).$$

The properties of the difference equation (22.2.6) are summarized in the following theorem.

---

[1] For some nonlinear systems with specialized forms, a variant of frequency analysis can be used.

Theorem 22.2.2: Consider the difference equation in (22.2.6) where the coefficients are all real numbers. Then this system is linear and time invariant. If the initial conditions are zero and the input signal satisfies $x(n) = 0$ for $n < n_0$, then this system is relaxed. If $M_+ = 0$, then this system is causal. Otherwise, the system is not causal.

▲▲

The properties of linearity and time invariance are verified by direct calculation. It is easy to see that if $M_+ \neq 0$, then the present value of the output signal $y(n)$ depends on a future value of the input signal $x(n + M_+)$. Hence, the system is noncausal. The initial conditions are used to model internal stored energy. If these initial conditions are zero and the input signal is turned on at some finite time, the system is relaxed.

22.2.4 Properties of Transfer Functions

Next we turn our attention to the properties of the transfer function which are summarized in the following theorem.

Theorem 22.2.3: Consider the system represented by the two-sided z-transform transfer function which is a rational function

$$\frac{Y(z)}{X(z)} = H(z) = \frac{b(z)}{a(z)}.$$

This system is linear and time invariant. This system is causal if and only if the ROC of the transfer function is exterior to a circle of finite radius and includes the point at infinity. This system is relaxed if the input signal is zero at time equals negative infinity.

▲▲

We can obtain the transfer function by taking the z-transform of the convolution summation. If the transfer function comes from the convolution representation, then this representation must be linear and time invariant to apply the convolution property, Property 18.2.2 of the z-transform. When we discussed the *ROC* of z-transforms in Section 18.1 we showed that a left-hand sequence has a *ROC* which is interior to a circle; i.e., the *ROC* of such a sequence doesn't include the point of infinity. By Theorem 22.2.1 we know that if the impulse response function is a left-hand sequence, then it is not causal. Hence, we are led to the characterization of causality in Theorem 22.2.3. Since the convolution representation is relaxed in that internal stored energy must be zero, the transfer function must also be relaxed. If the two-sided z-transform is used, then the input signal can be a two-sided sequence. In this case the system is relaxed only if the input signal is zero at negative infinity.

Quite frequently, the transfer function is obtained by taking the z-transform of the difference equation in (22.2.6). In this case we can think of the properties of the transfer function as being inherited directly from the underlying difference equation. (Just cross-multiply and take the inverse z-transform to obtain the difference equation.) Note that in order to apply the linear and shift properties, Properties

18.2.1 and 18.2.3 of the z-transform, in the first place, the system must be linear and time invariant. Furthermore, to calculate the transfer function from the difference equation, the initial conditions must be zero. Hence, the system must be relaxed if the input signal is zero at $t = -\infty$.

 If the transfer function is calculated from the difference equation (22.2.6), we obtain a simplification in the characterization of the causality of the system stated in the following corollary.

Corollary 22.2.4: Suppose the transfer function

$$\frac{Y(z)}{X(z)} = H(z) = \frac{b(z)}{a(z)}$$

is calculated from the difference equation in (22.2.6). This system is causal if the transfer function is proper. If the transfer function is nonproper,[2] the system is not causal.

▲▲

The causality properties of the transfer function follow from the fact that the order of the polynomials in the transfer function are directly related to the order of the advance of the signals in the difference equation.

 The transfer function can also be calculated using the DTFT. The following theorem summarizes the properties of these transfer functions.

Theorem 22.2.5: Consider the DTFT transfer function

$$\frac{Y(\Omega)}{X(\Omega)} = H(\Omega).$$

This system is linear and time invariant. This system is relaxed if the input signal is zero at $t = -\infty$.

▲▲

 If the DTFT transfer function comes from the convolution representation, then this representation must be linear and time invariant to apply the convolution property of the DTFT, Property 18.4.5. The DTFT transfer function doesn't allow the representation of internal stored energy, so the system must be relaxed in this sense. The DTFT can be applied to signals that are two-sided sequences. Then the system may not be relaxed in that the input signal is nonzero at $t = -\infty$. The DTFT transfer function can be applied to systems that are noncausal. There is no easy way to identify such systems from the DTFT transfer function, however. Hence, the theorem is silent on this point.

22.2.5 Properties of State Space Representations

Next we consider these properties of a state space representation. A state space representation is basically an alternative representation of a difference equation.

---

[2] The order of the numerator is greater than the order of the denominator.

Hence, the properties of state space representations are similar to difference equations with one difference. State space representations are usually confined to causal systems. These observations are summarized in the following theorem.

Theorem 22.2.6: Consider the system

$$\vec{q}(n+1) = A\vec{q}(n) + Bx(n), \quad \vec{q}(0) = \vec{q}_0$$

$$y(n) = C\vec{q}(n) + Dx(n)$$

This system is causal, linear, and time invariant. This system is relaxed if the initial conditions are zero $\vec{q}(0) = 0$.

▲▲

22.2.6 Summary

A review of the previous five theorems will show that all the system representations are linear and time invariant.

Terminology: The system representations discussed in this chapter are sometimes called <u>LTI</u> systems because they are all linear and time invariant. MATLAB, for example, uses this terminology.

All these system representations can be used for a noncausal system, except for state space representations. If the system is noncausal, the two-sided z-transform must be used.

Of these system representations, only difference equations and state space equations can be used to model internal stored energy; these systems are not relaxed.

22.3 BIBO STABILITY

22.3.1 Definitions

In the first section of this chapter we introduced several properties of the discrete-time system

$$y(n) = \mathcal{H}\big[x(n)\big]. \tag{22.3.1}$$

In this section we will discuss one further property of this system: stability. Again, the definition of stability is exactly the same for both continuous and discrete systems. The motivations for studying stability of discrete systems are the same as motivations for studying the stability of continuous systems as discussed in Section 13.3.

The basic idea behind stability is that a "small" input signal should yield a "small" output signal. Suppose the discrete system is a digital filter implemented in digital hardware. If the output signal grows without bound for small input signals, the registers in the processor will overflow, and the filter will cease to perform correctly.

To introduce a mathematical description of a stable system, we need a measure of a "small" signal. This measure is provided by the following definition.

Definition 22.3.1: A signal $x(n)$ is <u>bounded</u> if there exists a constant M, $0 < M < \infty$, such that

$$|x(n)| \leq M < \infty \quad \text{for all } n.$$

▲▲

In particular, a bounded signal does not go to infinity as $n \to \infty$.

Based on this measure of small, we can characterize a system for which small input signals lead to small output signals.

Definition 22.3.2: A system is <u>bounded-input-bounded-output (BIBO) stable</u> if every bounded input signal results in a bounded output signal. Otherwise, the system is (BIBO) <u>unstable</u>.

▲▲

This definition of stability captures the property that we want a system to possess. The next question is how to determine if a given system is BIBO stable. The definition of BIBO stability is not useful for determining the stability of a system, because every single input signal must be checked, usually an impossibility. Suppose, however, that we are given a representation of the system. We would like to develop a simple test that we can apply to the system representation to check the stability of the system. Various tests for stability for the system representations introduced in the last chapter will be developed next.

We will not develop explicit stability criteria for either difference equations or DTFT transfer functions. In fact, there exist no simple stability criteria for these representations. They are usually converted into another representation, the z-transform transfer function or a state space representation, to test the stability of the system.

22.3.2 The Convolution Representation

There is a simple test for the BIBO stability of a system represented by a convolution summation

$$y(n) = \sum_{k=-\infty}^{\infty} h(k)x(n-k). \tag{22.3.2}$$

Theorem 22.3.3: Suppose a system is represented by a convolution representation (22.3.2) with an impulse response function $h(n)$. Then this system is BIBO stable if and only if

$$\sum_{n=-\infty}^{\infty} |h(n)| < \infty.$$

▲▲

Theorem 22.3.3 is the discrete counterpart to Theorem 13.3.7. The proof of this theorem parallels the proof of Theorem 13.3.7 and will not be repeated here. This test for stability is better than the definition, but it is usually difficult to apply to IIR systems. We will develop more convenient tests below. Theorem 22.3.3 is an important theoretical analysis tool, however.

Example 22.3.4: The impulse response function of an FIR system has a finite number of nonzero values. That is, an FIR system is given by

$$y(n) = \sum_{k=0}^{N} h(k)x(n-k). \tag{22.3.3}$$

Applying Theorem 22.3.3 to test the stability of this system we have

$$\sum_{k=0}^{N} |h(k)|. \tag{22.3.4}$$

Hence, the summation in (22.3.4) is always finite. Thus, FIR systems are always BIBO stable. This property of FIR filters makes them attractive for implementation of digital filters.

▲▲

Example 22.3.5: Consider the system with an impulse response function

$$h(n) = a^n u_S(n). \tag{22.3.5}$$

Applying Theorem 22.3.3, we have

$$\sum_{n=-\infty}^{\infty} |a^n u_S(n)| = \sum_{n=0}^{\infty} |a|^n = \frac{1}{1-|a|} \quad \text{if} \quad |a| < 1. \tag{22.3.6}$$

For $|a| < 1$ the series is finite and the system is BIBO stable. If $|a| \geq 1$, the series in (22.3.6) is infinite and the system is unstable.

Next consider a system with an impulse response function

$$h(n) = -b^{-n} u_S(-n-1). \tag{22.3.7}$$

Applying Theorem 22.3.3 we have

$$\sum_{n=-\infty}^{\infty} |-b^{-n} u_S(-n-1)| = 1 - \sum_{n=0}^{\infty} \left|\frac{1}{b}\right|^n = \frac{1}{1-b} \quad \text{if} \quad |b| > 1. \tag{22.3.8}$$

The infinite summation is bounded if $|b| > 1$. In this case the system is BIBO stable. Otherwise the system is unstable.

▲▲

In writing the summation in (22.3.2) we implicitly assumed that the summation is finite; otherwise the sum has no meaning. If the input signal $x(n)$ begins at a finite time, $n = 0$, say, then the convolution summation is

$$y(n) = \sum_{k=0}^{\infty} h(n-k)x(k). \tag{22.3.9}$$

If we assume further that the impulse response function is causal, then $h(n-k) = 0$, $k > n$. In this case the summation in (22.3.9) becomes

$$y(n) = \sum_{k=0}^{n} h(n-k)x(k). \tag{22.3.10}$$

The summation in (22.3.10) is finite for finite values of n, and it always exists.

If we don't assume that the system is causal or that the input signal begins at a finite time, then we must put a further restriction on the system, namely, that it must be BIBO stable. Assume that the system (22.3.2) is BIBO stable, and that the input signal is bounded. Then the output signal exists.

22.3.3 Transfer Functions

Next we turn our attention to the stability properties of the transfer function representation

$$\frac{Y(z)}{X(z)} = H(z) = \frac{b(z)}{a(z)}. \tag{22.3.11}$$

Because of mathematical technicalities, it is easier to discuss the stability of causal and noncausal systems separately. The first theorem is the mostly widely used theorem.

Theorem 22.3.6: Suppose a causal system is represented by a rational transfer function in (22.3.11). Then this system is BIBO stable if and only if all poles of the transfer function are in the open unit disk.

▲▲

Theorem 22.3.6 says that the poles of the transfer function must have magnitude less than one; $|p| < 1$. That is, the poles must be inside the unit circle.[3] The proof of this theorem parallels the proof of Theorem 13.4.1, and is omitted.

---

[3] See Definition 3.1.15.

Example 22.3.7: Consider the causal system

$$\frac{Y(z)}{X(z)} = \frac{z}{z-a}, \quad ROC = \{z \,|\, |z| > a\}. \tag{22.3.12}$$

The pole p of this system is $p = a$. According to Theorem 22.3.6, this system is stable when $|a| < 1$. If $a = 0.5$, this system is BIBO stable even though the pole is in the RHP.

▲▲

Example 22.3.8: Consider the system

$$y(n) - 2r\zeta y(n-1) + r^2 y(n-2) = x(n) - r\zeta x(n-1). \tag{22.3.13}$$

The transfer function for this system is

$$\frac{Y(z)}{X(z)} = H(z) = \frac{z^2 - r\zeta z}{z^2 - 2r\zeta z + r^2}. \tag{22.3.14}$$

If we take $r = 0.9$ and $\Omega_0 = (\pi/4)$ rad, then the poles are shown in Figure 22.3.1a. Since these poles are within the unit circle, this system is BIBO stable. If we take $r = 1.3$ and $\Omega_0 = (\pi/4)$ rad, then the poles are also shown in Figure 22.3.1b. For these parameters the system is unstable.

▲▲

Theorem 22.3.6 is the discrete analog of Theorem 13.4.1 for continuous systems. A continuous system is BIBO stable if the poles are in the open LHP. The stability criteria for continuous and discrete systems is different. Figure 22.3.2 shows the stability regions in the s-plane and the z-plane. The difference between the stability criteria is one of the essential differences between continuous and discrete systems.

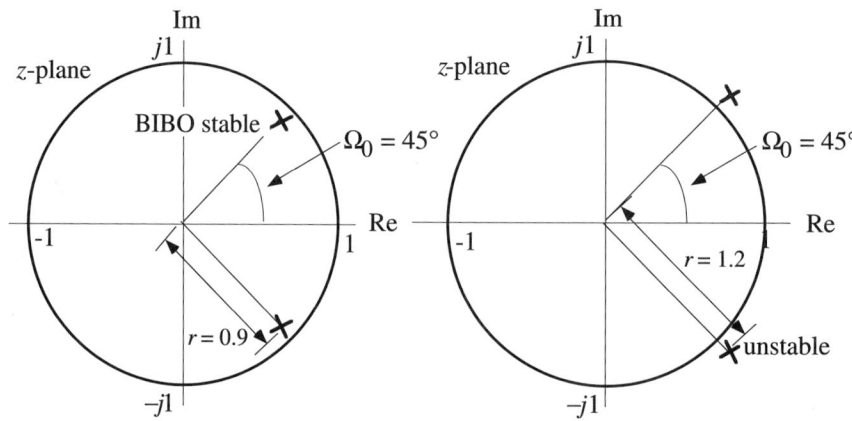

Figure 22.3.1 Poles for the Systems in Example 22.3.8

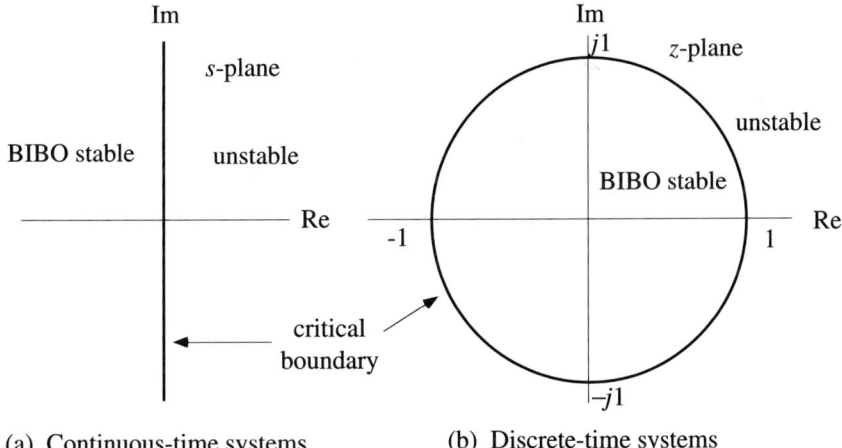

(a) Continuous-time systems   (b) Discrete-time systems

Figure 22.3.2 Stability Regions for Continuous and Discrete Systems

In Figure 22.3.2 the critical boundary is the $j\omega$-axis, which divides the two stability regions. Turning to the z-plane, it is the unit circle that divides the stability regions. These stability boundaries play other important roles in the analysis of systems, particularly in the frequency response. The difference between these boundaries in s-plane and z-plane is one of the essential differences between continuous and discrete systems.

Notation: It is customary to draw the unit circle in the z-plane. The reason is that the unit circle denotes the stability boundary which is critical when working with discrete systems.

Noncausal systems require a separate stability analysis. The following example illustrates the essential idea.

Example 22.3.9:   Next, consider a noncausal system defined by the impulse response function

$$h(n) = \begin{cases} 0, & n \geq 0 \\ -b^n, & n < 0 \end{cases}$$

(22.3.15)

In Example 22.3.5 we have shown that this system is stable for $|b| > 1$. The z-transform of this impulse response function is (see Example 18.1.8)

$$\frac{Y(z)}{X(z)} = \frac{z}{z-b}, \quad ROC = \{z \mid |z| < b\}.$$

(22.3.16)

This system, which is noncausal, is BIBO stable even though it has a pole outside the unit circle. Clearly, Theorem 22.3.6 must be modified for noncausal systems. ▲▲

The *ROC* of the transfer function in the last example includes the unit circle. Note also that the *ROC* of the transfer function of the causal system in (22.3.12) includes the unit circle when the system is BIBO stable. These observations lead to the following theorem.

Theorem 22.3.10: Let a system be represented by its transfer function $H(z)$. If the *ROC* of the function $H(z)$ contains the unit circle, then the system is BIBO stable. ▲▲

Theorem 22.3.10 applies to both causal and noncausal systems. Because causal systems are the type of system most commonly encountered, however, Theorem 22.3.6 is more frequently used.

22.3.4 State Space Representations

A state space representation can be used to check the BIBO stability of a system. A complete discussion of BIBO stability for state space representations is beyond the scope of this text, however. (This theorem is closely related to the BIBO stability criteria given for continuous-time transfer functions in Theorem 13.4.8. See the discussion there.) The following stability criteria is often useful, however, as it easily checked using MATLAB.

Theorem 22.3.11:   Consider the system

$$\vec{q}(n+1) = A\vec{q}(n) + Bx(n), \quad \vec{q}(0) = 0$$
$$y(n) = Cx(n) + Dx(n)$$

This system is BIBO stable if all the eigenvalues of A are inside the unit circle. ▲▲

Note that the implication in this theorem goes only one direction. There exist systems which have eigenvalues outside the unit circle, but that are BIBO stable.

This theorem requires the eigenvalues of the state matrix A to be strictly inside the unit circle; the modulus of the eigenvalues must be less than one. If the eigenvalues are located on the unit circle, the system is potentially BIBO unstable.

Example 22.3.12: Consider the trivial system

$$q(n+1) = \lambda q(n) + bx(n), \tag{22.3.17}$$
$$y(n) = cq(n).$$

According to Theorem 22.3.11, this system is BIBO stable if $|\lambda| < 1$. (Note that λ is an eigenvalue of the state matrix of this system.) This discrete system can be compared to the continuous-time system

$$\dot{q}(t) = aq(t) + bx(t), \tag{22.3.18}$$
$$y(t) = cq(t).$$

This continuous system is BIBO stable if $a < 0$. (Note that a is an eigenvalue of the state matrix of this system.)

▲▲

Example 22.3.13: Consider the system

$$\vec{q}(n+1) = \begin{bmatrix} 0 & -r^2 \\ 1 & 2r\zeta \end{bmatrix} \vec{q}(n) + \begin{bmatrix} -r^2 \\ r\zeta \end{bmatrix} x(n), \tag{22.3.19}$$

$$y(n) = \begin{bmatrix} 0 & 1 \end{bmatrix} \vec{q}(n) + x(n),$$

which was developed in Example 21.5.5. Assume that $|\zeta| < 1$. The eigenvalues of the state matrix are calculated from

$$\det(\lambda I - A) = \det\left(\begin{bmatrix} \lambda & r^2 \\ -1 & \lambda - 2r\zeta \end{bmatrix}\right) = \lambda^2 - 2r\zeta\lambda + r^2. \tag{22.3.20}$$

This polynomial is in standard second-order form if we identify $\omega_n = r$. Hence, the radial distance to the poles will be less than one if $|r| < 1$. In this case, the system is BIBO stable.

▲▲

22.4 RELATIONSHIPS BETWEEN SYSTEM REPRESENTATIONS

22.4.1 Introduction

We have been studying the discrete-time system

$$y(n) = \mathcal{H}[x(n)]. \tag{22.4.1}$$

In Chapter 21 we introduced four system representations for this system: the convolution summation, the z-transform transfer function, the DTFT transfer function, and state space equations. In the last section we showed that all these system representations are linear and time invariant. This result suggests that there is a close relationship between these representations. We shall develop these relationships in this section.

The relationships between the system representations allow us to easily translate between the system representations. Hence, we can select the representation that is most suited for our application, or for the type of analysis we wish to conduct. For example, to design a digital filter we often begin by selecting the magnitude of the DTFT transfer function to have a certain shape. Then with the help of computer computation, MATLAB say, this function is translated into z-transform transfer function. Finally, this transfer function is implemented digital hardware by using the convolution summation or difference equations. Inherent in this design process is the fact that all these system representations are essentially equivalent. We will develop these basic results in this section.

We will consider the following four system representations of the system (22.4.1). Our starting place is the convolution representation of a system

$$y(n) = \sum_{k=-\infty}^{\infty} h(n-k)x(k). \tag{22.4.2}$$

This system also has a z-transform transfer function

$$\frac{Y(z)}{X(z)} = H(z), \tag{22.4.3}$$

and a DTFT transfer function

$$\frac{Y(\Omega)}{X(\Omega)} = H(\Omega). \tag{22.4.4}$$

We also assume that this system has a state space representation

$$\vec{q}(n+1) = A\vec{q}(n) + Bx(n), \quad \vec{q}(0) = \vec{q}_0 \tag{22.4.5}$$
$$y(n) = C\vec{q}(n) + Dx(n).$$

22.4.2 Relationships Between System Representations

We begin with the relationship between the convolution representation and the transfer function. This relationship follows from the convolution property of z-transforms, Property 18.2.2.

Theorem 22.4.1: Suppose that the impulse response function of the system in (22.4.2) has a z-transform. Then the transfer function of this system is given by

$$\frac{Y(z)}{X(z)} = H(z)$$

where

$$H(z) = \mathcal{Z}\{h(n)\} \quad \text{and} \quad h(n) = \mathcal{Z}^{-1}\{H(z)\}. \qquad \blacktriangle\blacktriangle$$

Theorem 22.4.2 applies to either the one-sided or two-sided z-transform. If the system is noncausal, the two-sided z-transform must be used.

A similar relationship exists between the impulse response function and the DTFT transfer function. Of course, the DTFT must exist. According to Definition 18.4.1, the impulse response function in (22.4.2) must satisfy

$$\sum_{n=-\infty}^{\infty} |h(n)| < M < \infty. \tag{22.4.6}$$

This condition also implies that the system is BIBO stable according to Theorem 22.3.3. This observation is summarized in the next theorem. This theorem applies to both causal and noncausal systems.

Theorem 22.4.2: Suppose that a system modeled by the convolution representation (22.4.2) is BIBO stable. Then the DTFT transfer function is related to the impulse response function by

$$\mathcal{DTFT}\{h(n)\} = H(\Omega) \quad \text{and} \quad \mathcal{IDTFT}\{H(\Omega)\} = h(n).$$

▲▲

Next we consider the relationship between the DTFT transfer function and the z-transform transfer function. This relationship is based on Theorem 22.3.10. The key idea is that the *ROC* of the transfer function should contain the unit circle; i.e., the system is BIBO stable. Then the DTFT transfer function can be calculated from the transfer function $H(z)$ by applying Theorem 18.4.15.

Theorem 22.4.3: Let the system (22.4.1) have a transfer function (22.4.3). Then if this system is BIBO stable, the transfer function (22.4.3) is related to the DTFT transfer function (22.4.4) by

$$H(z)\big|_{z=e^{j\Omega}} = H(\Omega).$$

▲▲

Next we consider the calculation of the transfer function from the state space equations. This calculation, using z-transforms, parallels exactly the calculation of the transfer function from continuous-time state space equations given in Section 11.1. To compute the transfer function (22.4.3) from the state equations (22.4.5) we apply the z-transform to (22.4.5). Assuming the initial conditions are zero, the transformed equations are

$$z\vec{Q}(z) = A\vec{Q}(z) + BX(z), \tag{22.4.7}$$

$$Y(z) = C\vec{Q}(z) + DX(z).$$

Solving the first equation for $\vec{Q}(z)$ we obtain

$$\vec{Q}(z) = (zI - A)^{-1} BX(z). \tag{22.4.8}$$

Then using (22.4.8) in the second equation in (22.4.7) we have the following result.

Theorem 22.4.4: Given the state space representation with zero initial conditions

$$\vec{q}(n+1) = A\vec{q}(n) + Bx(n), \qquad \vec{q}(0) = 0$$
$$y(n) = C\vec{q}(n) + Dx(n),$$

the transfer function of this system is given by

$$\frac{Y(z)}{X(z)} = C(zI - A)^{-1}B + D.$$

▲▲

Once again, the development parallels the continuous-time development almost exactly. The transfer function in Theorem 22.4.4 is the same formula derived for continuous systems using Laplace transforms in Theorem 11.1.2. In general, all the results related to state space representations of the continuous-time systems also hold for discrete-time systems.

The relationship between state space equations and transfer functions is called realization theory. An introduction to this theory is given in Chapter 11. The centerpiece of this theory is several canonical forms that allow for the direct translation of transfer functions into state space equations. All those results translate to discrete time. The following realization is a translation into discrete-time of Realization 11.2.2.

Realization 22.4.5: Given the difference equation

$$y(n) + a_1 y(n-1) + a_2 y(n-2) + a_3 y(n-3) = b_1 x(n-1) + b_2 x(n-2) + b_3 x(n-3)$$

with the transfer function

$$\frac{Y(z)}{X(z)} = \frac{b_1 z^{-1} + b_2 z^{-2} + b_3 z^{-3}}{1 + a_1 z^{-1} + a_2 z^{-2} + a_3 z^{-3}} = \frac{b(z)}{a(z)} = H(z),$$

a realization of this transfer function is

$$\vec{q}(n+1) = \begin{bmatrix} q_1(n+1) \\ q_2(n+1) \\ q_3(n+1) \end{bmatrix} = \begin{bmatrix} 0 & 0 & -a_3 \\ 1 & 0 & -a_2 \\ 0 & 1 & -a_1 \end{bmatrix} \vec{q}(n) + \begin{bmatrix} b_3 \\ b_2 \\ b_1 \end{bmatrix} x(n),$$

$$y(t) = \begin{bmatrix} 0 & 0 & 1 \end{bmatrix} \vec{q}(t) + \begin{bmatrix} 0 \end{bmatrix} x(t).$$

The block diagram corresponding to this state space representation is shown in Figure 22.4.3.

▲▲

Several comments are in order. First, we have written the difference equation in delay form, because it clearly identifies the indexing on the coefficients. The state space equations are written in advance form as is customary.

Second, recall that we have changed the indexing notation on the coefficients of the polynomials to conform with standard usage in discrete system theory. This change has led to a change in the indexing of the coefficients in the state space equations in the canonical form. Care must be taken to use a consistent indexing method among the difference equations, the transfer function, and the state space equations.

Third, the transfer function is required to be strictly proper. If the transfer function is proper, but not strictly proper, then long division must be carried out to reduce it to a strictly proper part with a direct feedthrough term. Then Realization 22.4.5 can be used. Subsection 11.2.4 illustrates this procedure.

Fourth, we can obtain another realization by reordering the state variables. By simply re-ordering the state variables we don't change the transfer function, as shown in Section 11.3.

The state space representation (22.4.5) can also be directly related to the convolution representation (22.4.2). Since the convolution representation doesn't account for internal initial conditions, the initial conditions of the state space equations must be set to zero. Then from Theorem 21.5.7 the output signal for a given input signal is

$$y(n) = \sum_{k=0}^{n-1} CA^{n-1-k} Bx(k) + Dx(n). \tag{22.4.9}$$

Comparing (22.4.9) to (22.4.2) we can identify the impulse response function. This observation is summarized in the following theorem.

Theorem 22.4.6: Given the state space representation with zero initial conditions

$$\vec{q}(n+1) = A\vec{q}(n) + Bx(n), \qquad \vec{q}(0) = 0$$
$$y(n) = C\vec{q}(n) + Dx(n),$$

the impulse response of this system is given by

$$h(n) = CA^{n-1}B + D\delta(n).$$

▲▲

22.4.3 Summary

The following example illustrates the relationships between the system representations.

Example 22.4.7: Consider the system

$$\frac{Y(z)}{X(z)} = H(z) = \frac{1 - r\xi z^{-1}}{1 - 2r\xi z^{-1} + r^2 z^{-2}}. \tag{22.4.10}$$

The relationship of this system to the convolution representation is given in Theorem 22.4.1. From the table of z-transforms, Table 18.5.2, the impulse response function is

$$\mathcal{Z}^{-1}\{H(z)\} = h(n) = \mathcal{Z}^{-1}\left\{\frac{1 - r\xi z^{-1}}{1 - 2r\xi z^{-1} + r^2 z^{-2}}\right\} = \left(r^n \cos(n\Omega_0)\right)u_s(n). \tag{22.4.11}$$

We can also compute the DTFT transfer function from the z-transfer transfer function in (22.4.10) using Theorem 22.4.3 if we assume that the system is BIBO stable. In that case we have

$$\frac{Y(\Omega)}{X(\Omega)} = H(z)\big|_{z=e^{j\Omega}} = \frac{1 - r\xi e^{-j\Omega}}{1 - 2r\xi e^{-j\Omega} + r^2 e^{-j2\Omega}}. \tag{22.4.12}$$

To develop a state space representation of the transfer function in (22.4.10) we will use one of the canonical forms given above, say Realization 11.2.6. The transfer function in (22.4.10) is proper. To write a state space representation, we must reduce it to a strictly proper transfer function by long division. We have

$$\frac{Y(z)}{X(z)} = H(z) = \frac{1 - r\xi z^{-1}}{1 - 2r\xi z^{-1} + r^2 z^{-2}} = 1 + \frac{r\xi z^{-1} - r^2 z^{-2}}{1 - 2r\xi z^{-1} + r^2 z^{-2}}. \tag{22.4.13}$$

Now using Realization 11.2.6 we obtain

$$\vec{q}(n+1) = \begin{bmatrix} 0 & 1 \\ -r^2 & 2r\zeta \end{bmatrix}\vec{q}(n) + \begin{bmatrix} 0 \\ 1 \end{bmatrix}x(n), \tag{22.4.14}$$

$$y(n) = \begin{bmatrix} -r^2 & r\zeta \end{bmatrix}\vec{q}(n) + 1x(n).$$

▲▲

22.4.4 MATLAB Experiments

In MATLAB 5 the various system representations can be converted from one form to another by the commands **tf**, **zpk**, and **ss** as explained in Subsection 11.1.4. The system name is used as the input argument to one of these commands and the conversion is implemented internally. This conversion is illustrated by the M-file below.

MATLAB 4 In previous versions of MATLAB the model conversions were implemented by a specific set of commands. For example, the command **tf2ss** converts a transfer function into a state space representation. The input data are the polynomials that define the transfer function. The output data are the matrices that define the state space representation. The other commands have a similar syntax.

The following M-file illustrates the conversion between the various system representations of the system in Example 22.4.7. Note that MATLAB 5 stores the transfer functions in positive powers of z rather than negative powers of z as in the text.

Example 22.4.7: (*continued*)

```
clear
% define pole locations
r = 0.95;                       % radius
W0 = 45;                        % angle
zeta = cos((pi/180)*W0);
b = [1,-r*zeta,0];              % numerator polynomial
a = [1,-2*r*zeta,r^2];          % denominator polynomial
htf = tf(b,a,-1);               % define system
htf
hss = ss(htf);                  % convert to a state space representation
hss
hzp = zpk(hss);                 % convert to zero pole gain form
hzp
```

▲▲

Exploratory Exercise 22.4.8: Verify that the state space representation produced by MATLAB is the same as the representation given in Example 22.4.7, but with the states reordered.

▲▲

22.5 CONTINUOUS-TO-DISCRETE SYSTEM TRANSFORMATIONS

22.5.1 Introduction

In this section we will study the system shown in Figure 22.5.1. Here we have a continuous-time system whose input signal is generated by a zero-order hold (ZOH). Typically the input signal to this ZOH is generated by a computer. The output signal of the continuous-time system is sampled to produce a discrete signal. If we define a system by the input signal to the ZOH and the sampled output signal, then the system is a discrete-time system. In this section we will derive the discrete-time representation of this system in terms of the continuous-time system and the sampling frequency.

This type of system occurs when a computer is used to control a continuous-time system. The sampled output signal from the continuous-time system is sent to the computer. Based on the measured signal and possibly additional information, the computer generates a discrete input signal to the ZOH. In order for the computer to make an intelligent decision, the computer must know how the continuous-time

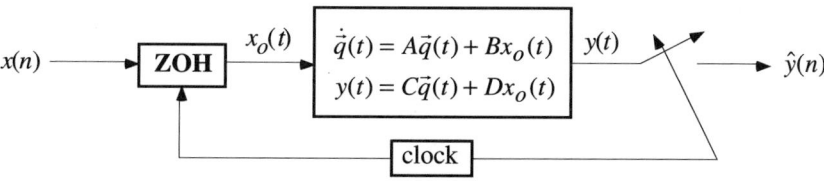

Figure 22.5.1 Continuous System with a Zero-Order Hold Input Signal

system will respond to a given input signal.   For the computer to predict this response it must have a discrete-time model of how the continuous-time system behaves. It is this model we propose to develop in this section.

22.5.2 Discretization

It is assumed that the input signal into the system in Figure 22.5.1 is a discrete signal $x(n)$.   During each sample period the discrete signal sample value $x(nT_s)$ is converted into a continuous-time signal $x_o(t)$ by a ZOH.[4] The output signal of the ZOH is used to drive the continuous system. The output signal of the continuous system $y(t)$ is also sampled to yield the discrete signal

$$\hat{y}(n) = y(nT_s). \tag{22.5.1}$$

It is assumed that the ZOH and the sampler on the output signal of the continuous system are synchronized. Our goal is to derive a system representation for the system with $x(n)$ as an input signal and $\hat{y}(n)$ as an output signal.

We will assume that the continuous system is represented by a state space representation

$$\dot{\vec{q}}(t) = A\vec{q}(t) + Bx_o(t), \tag{22.5.2}$$
$$y(t) = C\vec{q}(t) + Dx_o(t).$$

We assume that we know the state of the system at time $t = nT_s$, and we want to the know the state of the system at time $t = (n + 1)T_s$.   The computation is accomplished by using the solution of the state equations, Theorem 12.3.4,

$$\vec{q}(t) = e^{A(t-t_0)}\vec{q}(0) + \int_{t_0}^{t} e^{A(t-\lambda)}Bx_o(\lambda)\,d\lambda. \tag{22.5.3}$$

We take the state $\vec{q}(nT_s)$ as an initial condition at time $t = nT_s$. Now using (22.5.3) we can calculate the state of the system at time $t = (n + 1)T_s$ as

$$\vec{q}\big((n+1)T_s\big) = e^{A((n+1)-n)T_s}\vec{q}(nT_s) + \int_{nT_s}^{(n+1)T_s} e^{A((n+1)T_s-\lambda)}Bx_o(\lambda)\,d\lambda. \tag{22.5.4}$$

Because the input signal to this continuous system is generated by the ZOH we can model the input signal as

$$x_o(t) = x(n), \quad nT_s \le t < (n+1)T_s. \tag{22.5.5}$$

Notice that the input signal is constant over the integration period. Using this fact and introducing the change of variables $\gamma = (n+1)T_s - \lambda$, (22.5.4) becomes

---

[4] For a discussion of a zero-order hold see Section 19.4.

$$\hat{\vec{q}}(n+1) = e^{AT_s}\hat{\vec{q}}(n) + \left[\int_0^{T_s} e^{A\gamma}\, d\gamma\right] Bx(n). \qquad (22.5.6)$$

The complete state space model can be constructed by including the output equation from (22.5.2). We summarize these results in the following theorem.

Theorem 22.5.1: Suppose we are given a system (22.5.2) whose input signal is generated by a ZOH and whose output signal is sampled as in Figure 22.5.1. Then the discrete state space representation for this system is given by

$$\hat{\vec{q}}(n+1) = F\hat{\vec{q}}(n) + Gx(n),$$

$$\hat{y}(n) = C\hat{\vec{q}}(n) + Dx(n),$$

where

$$F = e^{AT_s}, \qquad G = \left[\int_0^{T_s} e^{A\gamma}\, d\gamma\right] B.$$

▲▲

Terminology: When we compute a discrete system in Theorem 22.5.1 from the continuous system in (22.5.2), we say that continuous system has been <u>discretized</u>.

Note that the discrete system describes the relationship between the discrete input signal and the sampled output signal of the continuous system at the sample points $t = nT_s$. We don't know what the continuous system output signal $y(t)$ is doing in between the sample points.

The following example illustrates the use of Theorem 22.5.1.

Example 22.5.2: In Example 10.5.6 the state space model of a proof-mass actuator is given as

$$\dot{\vec{q}}(t) = \begin{bmatrix} 0 & 1 \\ 0 & 0 \end{bmatrix}\vec{q}(t) + \begin{bmatrix} 0 \\ \dfrac{K_{ef}}{m_{pm}} \end{bmatrix} v_{pm}(t), \qquad (22.5.7)$$

$$y_{pm}(t) = \begin{bmatrix} 1 & 0 \end{bmatrix}\vec{q}(t) + 0v_{pm}(t).$$

Suppose that we wish to control this actuator with a computer. The input signal $v_{pm}(t)$ is to be generated with a ZOH. We assume that the input signal to the ZOH is given by $\hat{v}_{pm}(n)$. We want to discretize this system using Theorem 22.5.1.

First, we must compute the state transition matrix e^{At}. The state transition matrix for (22.5.7) is developed in Example 12.3.3. It is given by

$$e^{At} = \mathcal{L}^{-1}\left\{(sI - A)^{-1}\right\} = \mathcal{L}^{-1}\left\{\begin{bmatrix} \dfrac{1}{s} & \dfrac{1}{s^2} \\ 0 & \dfrac{1}{s} \end{bmatrix}\right\} = \begin{bmatrix} u_s(t) & tu_s(t) \\ 0 & u_s(t) \end{bmatrix}. \tag{22.5.8}$$

Now the state matrix of the discrete system is given by

$$e^{AT_s} = \begin{bmatrix} u_s(T_s) & T_s u_s(T_s) \\ 0 & u_s(T_s) \end{bmatrix} = \begin{bmatrix} 1 & T_s \\ 0 & 1 \end{bmatrix}. \tag{22.5.9}$$

To compute the input matrix for the discrete system we must evaluate the integral in (22.5.6). For the state transition matrix in (22.5.8) we have

$$\int_0^{T_s} e^{A\gamma}\, d\gamma = \begin{bmatrix} \displaystyle\int_0^{T_s} u_s(\gamma)\, d\gamma & \displaystyle\int_0^{T_s} \gamma u_s(\gamma)\, d\gamma \\ 0 & \displaystyle\int_0^{T_s} u_s(\gamma)\, d\gamma \end{bmatrix} = \begin{bmatrix} T_s & \dfrac{T_s^2}{2} \\ 0 & T_s \end{bmatrix}. \tag{22.5.10}$$

Now the input matrix is given by

$$G = \left[\int_0^{T_s} e^{A\gamma}\, d\gamma\right] B = \begin{bmatrix} T_s & \dfrac{T_s^2}{2} \\ 0 & T_s \end{bmatrix} \begin{bmatrix} 0 \\ \dfrac{K_{ef}}{m_{pm}} \end{bmatrix} = \begin{bmatrix} \dfrac{T_s^2 K_{ef}}{2m_{pm}} \\ \dfrac{T_s K_{ef}}{m_{pm}} \end{bmatrix}. \tag{22.5.11}$$

Now the discrete system is given by

$$\hat{\bar{q}}(n+1) = \begin{bmatrix} 1 & T_s \\ 0 & 1 \end{bmatrix} \hat{\bar{q}}(n) + \begin{bmatrix} \dfrac{T_s^2 K_{ef}}{2m_{pm}} \\ \dfrac{T_s K_{ef}}{m_{pm}} \end{bmatrix} \hat{v}_{pm}(n), \tag{22.5.12}$$

$$\hat{y}(n) = \begin{bmatrix} 1 & 0 \end{bmatrix} \hat{\bar{q}}(n).$$

▲▲

22.5.3 MATLAB Experiments

The MATLAB command which implements Theorem 22.5.1 is **c2d**. The input arguments for this command are the LTI system and the sample period. The command returns the discrete-time system. The following M-file illustrates the use of this command.

Example 22.5.2: (*continued*)   This M-file calculates the discrete state space equations for the proof-mass in Example 22.5.2.

```
clear
mpm = 10;                     % proof-mass mass
Kef = 1;                      % gain
% define state space system
Apm = [0,1;0,0];              % state matrix
Bpm = [0;Kef/mpm];            % input matrix
Cpm = [1,0];                  % output matrix
Dpm = 0;                      % direct feedthrough matrix
hpm = ss(Apm,Bpm,Cpm,Dpm);    % define system
% convert to discrete system
Ts = 1;  % sample period
hpmd = c2d(hpm,Ts);
hpmd
% simulate system
n = [0:100];                  % time vector
w0 = 1;                       % frequency of input signal
vpm = sin(w0*Ts*n);           % sinusoidal input signal
ypm = lsim(hpmd,vpm);         % simulate discrete system
% plot output signals
plot(n*Ts,ypm,'*')
xlabel('time, sec')
ylabel('displacement')
```

▲▲

Exploratory Exercise 22.5.3:   Verify that the MATLAB calculation agrees with (22.5.12).

▲▲

Exploratory Exercise 22.5.4:   Examine the effect of the sample period on the eigenvalues of the discrete state space system. How is the sample period related to the frequency of the input signal?

▲▲

22.6 CHAPTER SUMMARY

In this chapter we have considered a discrete-time system

$$y(n) = \mathcal{H}[x(n)].\tag{22.6.1}$$

We defined four properties of this system listed in Table 22.6.1.

In this chapter we have developed the properties of each representation introduced in the last chapter. These results are summarized in Table 22.6.2.

Table 22.6.1 Properties of Systems

Causality	Definition 22.1.1
Linearity	Definition 22.1.3
Time Invariance	Definition 22.1.5
Relaxed	Definition 22.1.7

Table 22.6.2   Discrete-Time System Representations

Theorem	Model	Properties
Theorem 22.2.1	$y(n) = \displaystyle\sum_{k=-\infty}^{\infty} h(n-k)x(k)$	linear, time invariant
Theorem 22.2.2	$y(n) + a_1 y(n-1) + \cdots + a_N y(n-N)$ $= b_0 x(n) + b_1 x(n-1)) + \cdots + b_M x(n-M)$	linear, time invariant
Theorem 22.2.3	$\dfrac{Y(z)}{X(z)} = H(z) = \dfrac{b_0 + \cdots + b_M z^{-M}}{1 + \cdots + a_N z^{-N}}$	linear, time invariant, relaxed
Theorem 22.2.5	$\dfrac{Y(\Omega)}{X(\Omega)} = H(\Omega)$	linear, time invariant, BIBO stable
Theorem 22.2.6	$\vec{q}(n+1) = A\vec{q}(n) + Bx(n), \quad \vec{q}(0) = q_0$ $y(n) = C\vec{q}(n) + Dx(n)$	linear, time invariant, causal

In Section 22.5 a discrete-time equivalent system of a continuous system whose input signal is the output signal of a ZOH is developed (Theorem 22.5.1). Given the system

$$\dot{\vec{q}}(t) = A\vec{q}(t) + Bx_o(t),$$
$$y(t) = C\vec{q}(t) + Dx_o(t),$$

then the discrete state space representation for this system is given by

$$\hat{\vec{q}}(n+1) = F\hat{\vec{q}}(n) + Gx(n),$$
$$\hat{y}(n) = C\hat{\vec{q}}(n) + Dx(n),$$

where

$$F = e^{AT_s}, \qquad G = \left[\int_0^{T_s} e^{A\gamma}\, d\gamma\right]B.$$

We also defined the concept of stability for a system.

Definition 22.3.2: A system is <u>bounded-input-bounded-output stable (BIBO) stable</u> if every bounded input signal results in a bounded output signal. Otherwise, the system is (BIBO) <u>unstable</u>.

▲▲

A stability criteria was developed for each system representation. These results are summarized in Table 22.6.3.

The relationship between the system representations is developed in Section 22.4. These results are summarized in Figure 22.6.1.

Table 22.6.3 BIBO Stability Criteria

Theorem	Representation	Criteria
Theorem 22.3.3	$$y(n) = \sum_{k=-\infty}^{\infty} h(n-k)x(k).$$	$$\sum_{n=-\infty}^{\infty} |h(n)| < \infty.$$
Theorem 22.3.6	$$\frac{Y(z)}{X(z)} = \frac{b_0 + \cdots + b_M z^{-M}}{1 + \cdots + a_N z^{-N}}, M \leq N$$	all poles are inside the unit circle
Theorem 22.3.10	$$\frac{Y(z)}{X(z)} = \frac{b_0 + \cdots + b_M z^{-M}}{1 + \cdots + a_N z^{-N}}$$	ROC contains the unit circle
Theorem 22.3.11	$$\vec{q}(n+1) = A\vec{q}(n) + Bx(n), \quad \vec{q}(0) = q_0$$ $$y(n) = C\vec{q}(n) + Dx(n).$$	eigenvalues of A are inside the unit circle

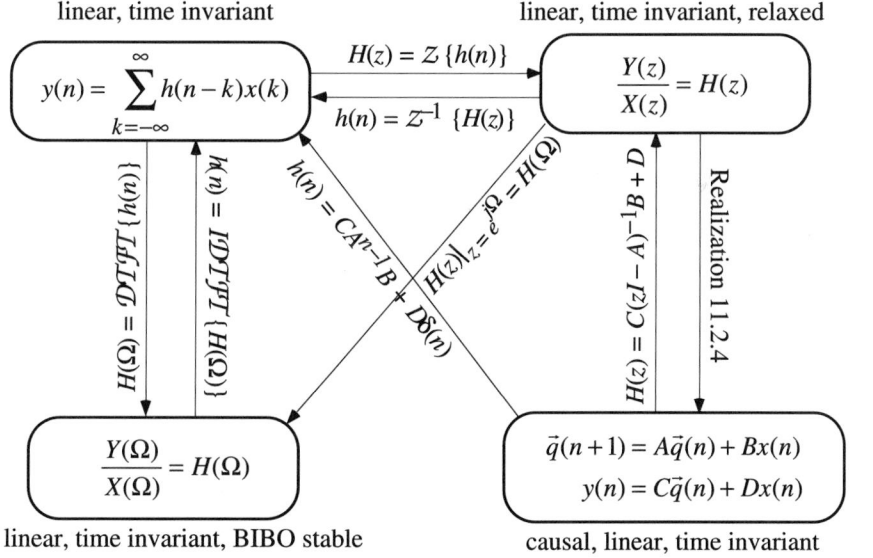

Figure 22.6.1 Relationship Between System Representations

22.7 HOMEWORK FOR CHAPTER 22

Homework Problems for Section 22.1

22.1.1   Recall from Chapter 17 that $Q[x]$ is the quantizer function.   Consider the system

$$y(n) = a_1 Q[y(n-1)] + b_0 x(n).$$

Is this system causal? Linear? Time invariant?

22.1.2   Consider sampling as a system as shown in Figure 17.2.2.   Show that this system is causal but it is not linear or time invariant.

22.1.3   Determine whether the following systems are causal, linear, or time invariant.
   (i) $y(n) = x(5n)$
   (ii) $y(n) = 3x(n) + 4$
   (iii) $y(n) = 4x(n^3)$
   (iv) $y(n) = \displaystyle\sum_{k=-\infty}^{n} x(k)$

   (v) $y(n) = [x(n)]^3$
   (vi) $y(n) = (\sin(0.2\pi n)) x(n)$
   (vii) $y(n) = |x(n)|$

22.1.4   The system

$$y(n) = x(n) - 0.8x(n-1) + 4$$

is not linear.   By a change of variables, make this system linear.

Homework Problems for Section 22.2

22.2.1   Suppose that the following functions are impulse response functions of a system.   Determine if each system is causal.
   (i) $h(n) = (0.6)^n u_s(n)$
   (ii) $h(n) = (0.6)^n u_s(-n-1)$
   (iii) $h(n) = 0.8^{|n|}$

   (iv) $h(n) = 2[u_s(n+N-1) - u_s(n-N)]$
   (v) $h(n) = \dfrac{\sin(\beta n)}{\pi n}$

22.2.2   Suppose that a transfer function derived from a difference equation is not proper.   Show that the ROC doesn't contain the point at infinity (and so it is not causal).

22.2.3 Suppose two systems are represented by their impulse response functions $h_1(n)$ and $h_2(n)$. Prove or disprove the following statements.

(a) The system formed by the cascade interconnection of these two systems is causal iff each of the two given systems is causal.

(b) The system formed by the parallel interconnection of these two systems is causal iff each of the two given systems is causal.

22.2.4 Determine if each of the following systems are causal.

(i) $\dfrac{Y(z)}{X(z)} = H(z) = \dfrac{z^2}{(z-0.36)(z+0.5)}$, $\qquad ROC = \{|z| > 0.5\}$

(ii) $\dfrac{Y(z)}{X(z)} = H(z) = \dfrac{z^2}{(z-0.36)(z+0.5)}$, $\qquad ROC = \{0.5 < |z| < 0.36\}$

(iii) $\dfrac{Y(z)}{X(z)} = H(z) = \dfrac{z^2}{(z-0.36)(z+0.5)}$, $\qquad ROC = \{|z| < 0.36\}$

22.2.5 For each of the following systems
(a) Is the system causal? Why?
(b) Is the system linear? Why?
(c) Is the system time invariant? Why?

(i) $y(n) - y(n-1) - 0.24y(n-2) = 0.24x(n)$
(ii) $y(n) = 0.5x(n+1) + 0.5x(n-1)$
(iii) $y(n)x(n) + x(n) = x(n)x(n-1)$
(iv) $y(n) - y(n-1) - 0.24y(n-2) = 0.24x(n+2)$
(v) $y(n) - 0.5y(n-1) = nx(n)$
(vi) $y(n) - 0.5y(n-1) = x(n) + n$

Homework Problems for Section 22.3

22.3.1 Determine if each of the systems is BIBO stable.

(i) $\dfrac{Y(z)}{X(z)} = \dfrac{0.6 - 0.5z^{-1} + 0.4z^{-2}}{1 + 0.3z^{-1} + 0.2z^{-2}}$ $\qquad$ (iii) $\dfrac{Y(z)}{X(z)} = 1 + 0.5z^{-1} + 2z^{-2}$

(ii) $\dfrac{Y(z)}{X(z)} = \dfrac{0.6z - 0.5 + 0.4z^{-1}}{1 + 0.3z^{-1} + 0.2z^{-2}}$ $\qquad$ (iv) $\dfrac{Y(z)}{X(z)} = \dfrac{2 - 0.3z^{-1}}{1 - 0.7z^{-1} - 0.12z^{-2}}$

22.3.2 Suppose that the following functions are impulse response functions of a system. Determine if each system is BIBO stable and give a reason for your answer.

(i) $h(n) = (0.6)^n u_s(n)$ $\qquad$ (v) $h(n) = 0.8^{|n|}$

(ii) $h(n) = (0.6)^n u_s(-n-1)$ $\qquad$ (vi) $h(n) = (0.8)^n u_s(n) - (2)^n u_s(-n-1)$

(iii) $h(n) = (1.01)^n u_s(-n-1)$ $\qquad$ (vii) $h(n) = 2[u_s(n+N-1) - u_s(n-N)]$

(iv) $h(n) = (1.01)^n u_s(n)$ $\qquad$ (viii) $h(n) = \dfrac{\sin(\beta n)}{\pi n}$

22.3.3 Suppose two causal systems are represented by their transfer functions $H_1(z)$ and $H_2(z)$. Prove or disprove the following statements.

  (a) The system formed by the cascade interconnection of these two systems is BIBO stable iff each of the two given systems is BIBO stable.

  (b) The system formed by the parallel interconnection of these two systems is BIBO stable iff each of the two given systems is BIBO stable.

  (c) The system formed by the feedback interconnection of these two systems is BIBO stable iff each of the two given systems is BIBO stable.

22.3.4 Determine if each of the following systems is BIBO stable.

  (i) $\dfrac{Y(z)}{X(z)} = H(z) = \dfrac{z^2}{(z-0.36)(z+0.5)}, \qquad ROC = \{|z| > 0.5\}$

  (ii) $\dfrac{Y(z)}{X(z)} = H(z) = \dfrac{z^2}{(z-0.36)(z+0.5)}, \qquad ROC = \{0.5 < |z| < 0.36\}$

  (iii) $\dfrac{Y(z)}{X(z)} = H(z) = \dfrac{z^2}{(z-0.36)(z+0.5)}, \qquad ROC = \{|z| < 0.36\}$

22.3.5 Determine if each of the following systems is BIBO stable.

  (i) $y(n) - y(n-1) - 0.24y(n-2) = 0.24x(n)$

  (ii) $y(n+1) + 0.3y(n) = 0.5x(n)$

  (iii) $y(n) = y(n-1) - y(n-2) + x(n) + x(n-1)$

  (iv) $y(n) = -0.2y(n-1) + x(n+1) - 0.5x(n-1)$

22.3.6 Prove that the ROC of a causal, BIBO stable system includes the unit circle.

22.3.7 Consider a system whose impulse response function is

$$h(n) = a^n \big(u_s(n) - u_s(n-N) \big), \quad |a| < 1$$

Prove that the system defined by the transfer function $H^{-1}(z)$ is BIBO stable.

22.3.8 Determine if each of the following systems is BIBO stable.

  (i) $\vec{q}(n+1) = \begin{bmatrix} -1.2 & 0 \\ 0 & 0.4 \end{bmatrix} \vec{q}(n) + \begin{bmatrix} 2 \\ 5 \end{bmatrix} x(n)$

       $y(n) = \begin{bmatrix} -2 & 3 \end{bmatrix} \vec{q}(n)$

  (ii) $\vec{q}(n+1) = \begin{bmatrix} 0 & 0.5 \\ -0.5 & 0 \end{bmatrix} \vec{q}(n) + \begin{bmatrix} 1 \\ 2 \end{bmatrix} x(n),$

       $y(n) = \begin{bmatrix} 2 & 0 \end{bmatrix} \vec{q}(n)$

(iii) $\vec{q}(n+1) = \begin{bmatrix} 0 & 1 & 0 \\ 0 & 0 & 1 \\ 0 & 0 & 0 \end{bmatrix} \vec{q}(n) + \begin{bmatrix} 0 \\ 0 \\ 1 \end{bmatrix} x(n),$

$y(n) = \begin{bmatrix} 1 & 0 & 0 \end{bmatrix} \vec{q}(n)$

22.3.9 Consider the system

$$\vec{q}(n+1) = \begin{bmatrix} 0 & 1 & 0 \\ 0.4\lambda & 0 & 0 \\ 0 & 0 & 0.6\lambda \end{bmatrix} \vec{q}(n) + \begin{bmatrix} 0 \\ 1 \\ 1 \end{bmatrix} x(n),$$

$y(n) = \begin{bmatrix} 1 & 0 & 1 \end{bmatrix} \vec{q}(n)$

Find the range of λ for which this system is BIBO stable.

Homework Problems for Section 22.4

22.4.1 Suppose that a discrete system is represented by the convolution representation

$$y(n) = \sum_{k=-\infty}^{\infty} \left((0.5)^n \sin(0.003n) \right) u_s(n-k) x(k).$$

(a) Find the z-transform transfer function of this system.
(b) Find a state space representation of this system.
(c) Is this system causal? Why?
(d) Is this system BIBO stable? Why?

22.4.2 Consider the system

$$\frac{Y(z)}{X(z)} = H(z) = \frac{0.2z}{(z+0.4)(z-0.2)}, \qquad ROC = \{|z| > 0.4\}$$

(a) Find the impulse response function of this system.
(b) Why does the DTFT transfer function exist for this system?
(c) Find the DTFT transfer function of this system. (Don't simplify the expression.)

22.4.3 For each of the following systems
(a) Find the transfer function of each system.
(b) Find the impulse response function of each system.
(c) For what parameter values does the DTFT transfer function exist for each system?

(i) $\vec{q}(n+1) = \begin{bmatrix} \lambda_1 & 0 \\ 0 & \lambda_2 \end{bmatrix} \vec{q}(n) + \begin{bmatrix} b_1 \\ b_2 \end{bmatrix} x(n)$

$y(n) = \begin{bmatrix} c_1 & c_2 \end{bmatrix} \vec{q}(n)$

(ii) $\vec{q}(n+1) = \begin{bmatrix} \sigma & \omega \\ -\omega & \sigma \end{bmatrix} \vec{q}(n) + \begin{bmatrix} 0 \\ b_1 \end{bmatrix} x(n)$

$y(n) = \begin{bmatrix} c_1 & 0 \end{bmatrix} \vec{q}(n)$

(iii) $\vec{q}(n+1) = \begin{bmatrix} 0 & 1 & 0 \\ 0 & 0 & 1 \\ 0 & 0 & 0 \end{bmatrix} \vec{q}(n) + \begin{bmatrix} 0 \\ 0 \\ 1 \end{bmatrix} x(n)$

$y(n) = \begin{bmatrix} 1 & 0 & 0 \end{bmatrix} \vec{q}(n)$

22.4.4 Find the transfer function of the following systems.

(i) $\vec{q}(n+1) = \begin{bmatrix} 0 & 1 & 0 \\ 0 & 0 & 1 \\ -a_3 & -a_2 & -a_1 \end{bmatrix} \vec{q}(n) + \begin{bmatrix} 0 \\ 0 \\ 1 \end{bmatrix} x(n)$

$y(n) = \begin{bmatrix} b_0 & b_1 & b_2 \end{bmatrix} \vec{q}(n)$

(ii) $\vec{q}(n+1) = \begin{bmatrix} \sigma & \omega & 0 \\ -\omega & \sigma & 0 \\ 0 & 0 & \lambda \end{bmatrix} \vec{q}(n) + \begin{bmatrix} 0 \\ b_2 \\ b_3 \end{bmatrix} x(n)$

$y(n) = \begin{bmatrix} c_1 & 0 & c_3 \end{bmatrix} \vec{q}(n)$

(iii) $\vec{q}(n+1) = \begin{bmatrix} 0 & 1 & 0 \\ 0.4\lambda & 0 & 0 \\ 0 & 0 & 0.6\lambda \end{bmatrix} \vec{q}(n) + \begin{bmatrix} 0 \\ 1 \\ 1 \end{bmatrix} x(n),$

$y(n) = \begin{bmatrix} 1 & 0 & 1 \end{bmatrix} \vec{q}(n)$

22.4.5 Find a state space representation of the following transfer functions.

(i) $\dfrac{Y(z)}{X(z)} = \dfrac{0.6 - 0.5z^{-1}}{1 + 0.3z^{-1} + 0.2z^{-2}}$

(ii) $\dfrac{Y(z)}{X(z)} = \dfrac{1 - z^{-4}}{1 - 0.25z^{-4}}$

(iii) $\dfrac{Y(z)}{X(z)} = \dfrac{0.2489 + 0.7466z^{-1} + 0.7466z^{-2} + 0.2489z^{-3}}{1 + 0.5043z^{-1} + 0.5289z^{-2} - 0.0423z^{-3}}$

(iv) $\dfrac{Y(z)}{X(z)} = \dfrac{1}{z^{-4}}$

22.4.6 Find a BIBO stable state space representation with an eigenvalue outside of the unit circle.

22.4.7   Find a state space representation of the following system using Realization 11.2.6.

$$\frac{Y(z)}{X(z)} = \frac{b_1 z^{-1} + b_2 z^{-2} + b_3 z^{-3}}{1 + a_1 z^{-1} + a_2 z^{-2} + a_3 z^{-3}}$$

Homework Problems for Section 22.5

22.5.1   Consider the system

$$\frac{Y(s)}{X(s)} = H(s) = \frac{1}{s(s+1)}$$

Suppose the input signal to this system is generated by a ZOH and the output signal is sampled with a sampling period of T_s. Find the equivalent discrete time system in state space form.

22.5.2   A state space representation of a mass-spring-damper system, as developed in Example 11.4.3, is given by

$$\dot{\vec{q}}(t) = \begin{bmatrix} 0 & 1 \\ -\dfrac{k_{st}}{m_{st}} & -\dfrac{c_{st}}{m_{st}} \end{bmatrix} \begin{bmatrix} q_1(t) \\ q_2(t) \end{bmatrix} + \begin{bmatrix} 0 \\ \dfrac{1}{m_{st}} \end{bmatrix} f_{st}(t),$$

$$y_{st}(t) = \begin{bmatrix} 1 & 0 \end{bmatrix} \vec{q}(t).$$

Let the parameters of this system be $k_{st} = 10$ N/m, $c_{st} = 8$ N/m/sec, and $m_{st} = 1$ kg. Suppose that the input signal is generated by a ZOH and the output signal is sampled with a sampling frequency of ω_s. Determine the poles of the system as the sampling frequency varies between $1 \le \omega_s \le 10$ rad/sec.

22.5.3   Consider the system

$$\dot{\vec{q}}(t) = \begin{bmatrix} 0 & 1 \\ -1 & 0 \end{bmatrix} \vec{q}(t) + \begin{bmatrix} 1 \\ 0 \end{bmatrix} x(t)$$

Suppose the input signal to this system is generated by a sampler followed by a ZOH. The sampling period is T_s.
(a) Find the equivalent discrete state space equations.
(b) Find the poles and plot them in the z-plane relative to the unit circle.
(c) For what values of the sampling period T_s is the input matrix zero?

Chapter 23

Frequency Domain Analysis of Discrete-Time Systems

Chapter Outline

In Chapters 21 and 22 we have developed several system representations and their interrelationship. In Chapter 20 we discussed several representations of a signal. In this chapter we will study the relationship between the input and output signal and the system.

We will adopt the following terminology. Given an input signal the system will produce an output signal. We say the input signal *propagates* through the system. This terminology is consistent with the definition of a system, which defines the system in terms of the input and output signal. This terminology suggests that the relationship between the input and output signals and the system can be studied by comparing the input signal to the output signal. The change between these two signals can be attributed to the system. We shall pursue this idea in this chapter.

We begin by studying the response of the system to a sinusoidal input signal. It turns out that the output signal is also a sinusoid at the same frequency. Furthermore, the amplitude and phase of the output signal can be determined from the transfer function of the system. So the output *signal* of a system is directly related to the *system* representation and the input *signal*. This result is precisely stated in the frequency response theorem which is discussed in this chapter. This theorem is the basis for establishing the relationship between the system and the signal propagating through it. The linkage between the input and output signals and the system in the frequency response theorem is of immense importance in many aspects of signals and systems.

The frequency response theorem identifies the frequency response function as the system property that links the system to the signal propagating through it. This function is identical to the DTFT transfer function, and it can be easily calculated from the z-transform transfer function. It is really the frequency response function that is of central interest in this chapter. The frequency response function is a complex function of a real variable Ω. This function is usually displayed by graphing the magnitude and phase as a function of the real variable Ω. The graph of the frequency response function is the form of this function that is most widely used.

It turns out that the shape of the magnitude of the frequency response function tells us which sinusoidal signals are passed by the system with no attenuation of their amplitude and which sinusoidal signals have their amplitude attenuated significantly (they are blocked). This idea is called *filtering*. Some sinusoidal signals are allowed to pass through the system unattenuated while other signals are blocked, or filtered out. This terminology suggests that we call a system a <u>filter</u>. A system as a filter can be classified depending on the frequencies of the sinusoids that are transmitted by the system and those sinusoids that are blocked. The notion of the bandwidth of the system is introduced to describe the shape of the magnitude of the frequency response function. This concept plays a fundamental role in system theory.

In Chapter 20 we introduced several signal representations including the amplitude and phase spectra, the energy spectral density, and the power spectrum. In this chapter we extend the frequency response theorem to accommodate energy and power signals as input and output signals. By combining these concepts with the frequency response function from the frequency response theorem, we obtain a graphical interpretation of how a signal propagates through a system. These interpretations lend deep insight into how a signal propagates through a system.

These frequency domain concepts lead naturally into the design of a filter. We would like to synthesize a filter such that the frequency response function of the filter satisfies certain given specifications.   Filter synthesis for analog filters is introduced in Section 14.7. That conceptual framework extends directly to digital filters. In fact, one way to design a digital filter is to transform the transfer function of an analog filter into a transfer function of a discrete filter. This approach allows us to incorporate the extensive knowledge of the analog filters into the design of digital filters. A second class of digital filters is based on FIR filters. Since these filters have no equivalent analog filter, we introduce a new class of design methodologies for these filters. For FIR filter design we show how to window the impulse response function of an ideal digital filter to obtain an FIR filter with linear phase. This approach takes place completely in the discrete domain. Hence, we are able to obtain properties of the digital filter that can't be found in analog filters. An introduction to both IIR and FIR digital filter design is given in this chapter.

Summary of Sections

Section 23.1:   We discuss the frequency response theorem.

Section 23.2:   We discuss the relationship between sampling and the frequency response function of a digital filter.

Section 23.3:   We introduce classifications of the frequency response function.

Section 23.4:   We discuss the transformation of analog filters into digital filters.

Section 23.5:   We discuss linear phase filters.

Section 23.6:   We extend the frequency response theorem to energy and power input signals.

Section 23.7:   Chapter summary section.

Coverage of the Text

The results in this chapter require the results in Chapters 21 and 22. Section 23.2 requires a basic knowledge of sampling from Chapter 19. Section 23.4 requires the results from Section 14.7. Section 23.6 requires some results from Sections 20.1 and 20.2.

23.1 FREQUENCY RESPONSE THEOREM FOR DISCRETE SYSTEMS

23.1.1 Introduction

We introduced the concept of the frequency response of a continuous system in Chapter 14, and we showed how this concept is fundamental for the analysis and design of continuous systems in the frequency domain. The most fundamental result that links the interaction between a signal and a system in the frequency domain is the frequency response theorem. The frequency response theorem is stated in two versions for continuous systems in Chapter 14. In one version the Laplace transform is used and in the other version the Fourier transform is used. In this section we will state the frequency response theorem for discrete systems in two versions using the one-sided z-transform and using the DTFT. The discrete-time theorems closely parallel the continuous-time theorems, but the differences between continuous-time and discrete-time results will be noted.

In order to provide a common framework for the two versions of the frequency response theorem we will assume that the system can be represented as

$$(23.1.1)$$

$$y(n) = \sum_{k=-\infty}^{\infty} h(n-k)x(k).$$

We will assume that the input signal is a cosine in this section. We will then derive the relationship between the input signal and the output signal in terms of the parameters of the sinusoid (the amplitude and phase) and the frequency response function of the system. In this section we will introduce the terminology related to this concept and illustrate the graphical interpretation of these results which are similar to continuous-time graphical interpretations of the frequency response function.

23.1.2 The Frequency Response Theorem

In this subsection we will present the two versions of the frequency response theorem. We begin with the Fourier transform version.

Theorem 23.1.1: (Frequency Response Theorem)   Suppose a system can be represented by a DTFT transfer function

$$\frac{Y(\Omega)}{X(\Omega)} = H(\Omega).$$

Let the input signal be

$$x(n) = \cos(\Omega_i n).$$

Then the output signal is given by

$$y_{ss}(n) = |H(\Omega_i)|\cos(\Omega_i n + \angle H(\Omega_i)), \quad -\infty < n < \infty.$$

▲▲

The proof of this theorem parallels the proof of its counterpart in continuous-time, the Fourier version of the frequency response theorem, Theorem 14.2.1.

Proof: Using Euler's identity, the input signal can be written as

$$x(n) = \cos(\Omega_i n) = \frac{1}{2}\left[e^{j\Omega_i n} + e^{-j\Omega_i n}\right] = \frac{1}{2}[x_1(n) + x_2(n)]. \tag{23.1.2}$$

Suppose the input signal to the system is $x_1(n)$. The input signal is bounded and the system is BIBO stable as required for the existence of the DTFT transfer function.[1] So the output signal for the system (23.1.1) exists and it is given by

$$y_1(n) = \sum_{k=-\infty}^{\infty} h(k)x_1(n-k) = \sum_{k=-\infty}^{\infty} h(k)\left(e^{j\Omega_i(n-k)}\right) = (e^{j\Omega_i n}) \sum_{k=-\infty}^{\infty} h(k)\left(e^{-j\Omega_i k}\right). \tag{23.1.3}$$

From Definition 18.4.1 of the DTFT we see that the last term in (23.1.3) is just the DTFT of the impulse response function

$$\sum_{k=-\infty}^{\infty} h(k)\left(e^{-j\Omega_i k}\right) = H(\Omega_i). \tag{23.1.4}$$

By a similar analysis the output signal corresponding to the input signal $y_2(n)$ is

$$y_2(n) = \sum_{k=-\infty}^{\infty} h(k)x_2(n-k) = e^{-j\Omega_i n} \sum_{k=-\infty}^{\infty} h(k)\left(e^{j\Omega_i k}\right) = e^{-j\Omega_i n} H(-\Omega_i) \tag{23.1.5}$$

$$= e^{-j\Omega_i n} \overline{H(\Omega_i)}.$$

Noting that the system is linear and using Euler's identity we can write

$$y(n) = \left(\frac{1}{2}\right)[y_1(n) + y_2(n)] = \left(\frac{1}{2}\right)\left[H(\Omega)e^{j\Omega_i n} + \overline{H(\Omega)}e^{-j\Omega_i n}\right] \tag{23.1.6}$$

$$= |H(\Omega)| \cos(\Omega_i n + \angle H(\Omega)).$$

▲▲

The frequency response theorem also can be stated in terms of the one-sided z-transform. This version of the frequency response theorem for discrete systems parallels the Laplace version of the frequency response theorem for continuous-time systems.

---

[1] Theorem 22.4.2.

Theorem 23.1.2: (Frequency Response Theorem Using One-Sided z-Transforms) Assume that the system is represented by a one-sided z-transform transfer function

$$\frac{Y(z)}{X(z)} = H(z).$$

Also assume that this system is BIBO stable. Let the input signal be

$$x(n) = \left[\cos(\Omega_i n)\right] u_s(n).$$

Then the output of the system $y(n)$ is given by

$$y(n) = y_{tr}(n) + y_{ss}(n)$$

where

$$y_{ss}(n) = |H(\Omega_i)|\cos(\Omega_i n + \angle H(\Omega_i)), \quad n \geq 0,$$

with

$$H(z)\big|_{z=e^{j\Omega_i}} = H(\Omega_i)$$

and

$$y_{tr}(n) = y(n) - y_{ss}(n).$$

Furthermore,

$$\lim_{n\to\infty} y_{tr}(n) = 0.$$

▲▲

The proof of this theorem parallels the proof of the frequency response theorem in Section 14.1. It will not be repeated here.

The statement of Theorems 23.1.1 and 23.1.2 assumes that the input signal to the system is a sinusoid at a single frequency. This observation gives rise to the following definition.

Definition 23.1.3:   The term $y_{ss}(n)$ in Theorems 23.1.1 and 23.1.2 is called the (sinusoidal) steady state response of the system.

▲▲

In Theorem 23.1.2 there is an additional component in the output signal.

Definition 23.1.4: The term $y_{tr}(n) = y(n) - y_{ss}(n)$ is called the transient response of the system.

▲▲

The terminology in Definition 23.1.4 suggests that transient term dies out as time increases.

The primary difference between Theorems 23.1.1 and 23.1.2 is the difference of the input signals. In Theorem 23.1.1, the input signal is present for all time. The system is in steady state. Hence, only the steady state output is present. In Theorem 23.1.2 the input signal is "turned on" at time $n = 0$. This action leads to the transient response. Once the system reaches steady state, the two theorems yield the same result.

In Section 14.3 we discussed the similarities and differences between the Laplace and Fourier versions of the frequency response theorem for continuous-time systems. Those comments apply to the two versions of the frequency response theorem for discrete systems presented in this section.

Implicit in the statements of Theorems 23.1.1 and 23.1.2 are assumptions about the properties of the systems. Both theorems assume that the system can be modeled by a transfer function. This assumption requires the system to be linear and time invariant. This assumption is significant. If the system is neither linear nor time invariant, neither theorem holds. Since this theorem is the basis of much of the theory and design of discrete systems, we are deprived of many design tools for nonlinear, time-varying systems.

Both Theorems 23.1.1 and 23.1.2 require the system to be BIBO stable. In Theorem 23.1.1 it is assumed that the transfer function is derived using the DTFT. This assumption requires the system to be BIBO stable (Theorem 22.4.2). Theorem 23.1.2 uses the one-sided z-transform transfer function. This transfer function exists for unstable systems. Hence, the theorem contains an explicit statement requiring the system to be BIBO stable.

Theorem 23.1.1 allows the impulse response function to be nonzero for negative time. That is, the system may be noncausal. The Fourier version of the frequency response theorem for continuous systems also allows the system to be noncausal, but this type of continuous system never arises in practice. Discrete, noncausal systems are sometimes encountered, however. Hence, this extension is significant.

23.1.3 The Frequency Response Function

There are two basic messages in the frequency response theorem. Both of these messages involve the comparison of the input signal to the steady state response of the system. The first message is that the steady state response is also a sinusoid *at the same frequency* as the sinusoidal input signal. This fact is the most important aspect of the frequency response theorem. The second message is that if we compare the steady state response to the input signal, we see that the output signal's amplitude and phase have been changed relative to the input signal. Furthermore, the change in the amplitude and phase can be predicted from the transfer function of the system evaluated at the frequency of the input sinusoid. By extension, if we can evaluate the transfer function over all frequencies, then we will know how the system responds to all sinusoidal input signals. This information is contained in the following function.

Definition 23.1.5:   Suppose that a system is represented by a convolution representation (23.1.1) or by a transfer function $H(z) = Y(z)/X(z)$. Then the function $H(\Omega)$

$$H(\Omega) = H(z)\big|_{z=e^{j\Omega}} = \mathcal{DTFT}\{h(n)\}$$

is called the <u>frequency response (function)</u> of the system.

▲▲

Terminology: The frequency response function and the DTFT transfer function are one and the same function.

   The interpretation of the frequency response function for discrete systems is exactly the same as the interpretation of the frequency response function for continuous systems. In particular, this function is usually displayed in two graphs.

Definition 23.1.6: Let $H(\Omega)$ be the frequency response function. The graph

$$|H(\Omega)| \text{ vs. frequency}$$

is called the <u>magnitude graph</u> of the frequency response function. The graph

$$\angle H(\Omega) \text{ vs. frequency}$$

is called <u>phase graph</u> of the frequency response function.

▲▲

Like continuous-time systems, these graphs can be plotted with a variety of scales.
   There is one other quantity that is sometimes used to characterize the phase of the frequency response function. To define this quantity let the phase of the frequency response function be given by $\Theta(\Omega) = \angle H(\Omega)$.

Definition 23.1.7:   The <u>group delay</u> of the frequency response function $H(\Omega)$ is defined as

$$\tau(\Omega) = -\frac{d\Theta(\Omega)}{d\Omega}.$$

▲▲

We will return to the group delay when we discuss distortionless filters.
   The frequency response function for a continuous system is computed as

$$H(s)\big|_{s=j\omega} = H(\omega). \tag{23.1.7}$$

The function $H(\omega)$ can be thought of as the transfer function $H(s)$ evaluated along the $j\omega$ axis. The frequency response function for a discrete system is

$$H(z)\big|_{z=e^{j\Omega}} = H(\Omega). \tag{23.1.8}$$

In (23.1.8) as Ω varies from 0 to ∞, the function $e^{j\Omega}$ traces out the unit circle in the z-plane transversing the circle every 2π radians. The frequency response function $H(\Omega)$ can be thought of as the transfer function $H(z)$ evaluated on the unit circle $z = e^{j\Omega}$.

This is the second time we have encountered the unit circle in connection with discrete systems. The first time was in connection with BIBO stability. See Figure 22.3.2. The unit circle is the boundary between the stability regions in the z-plane just as the $j\omega$ axis is the boundary between the stability regions in the s-plane. Now we see that the unit circle is also directly linked to the frequency response function as is the $j\omega$ axis in continuous-time theory. The unit circle in discrete-time system theory plays the role of the $j\omega$ axis in continuous-time system theory.

One of the fundamental differences between the frequency response function of a continuous and a discrete system is that the frequency response function of a discrete system is periodic. The periodic nature of the frequency response function $H(\Omega)$ can be observed from its definition. Definition 23.1.5 says that

$$H(\Omega) = \mathcal{DTFT}\{h(n)\} = \sum_{n=-\infty}^{\infty} h(n)e^{-j\Omega n}. \tag{23.1.9}$$

Here we see that $H(\Omega)$ is a sum of periodic functions whose periods are integer multiples of each other. The periodicity of $H(\Omega)$ is also implied by thinking of this function as $H(z)$ evaluated around the unit circle. As Ω varies from 0 to ∞, the function $e^{j\Omega}$ traces out the unit circle, transversing the circle every 2π radians. Since $H(z)$ is being evaluated along the same path repeatedly, it must be periodic.

The implication of this periodicity is important. We have to work with the frequency response function only over one period, i.e., over a finite interval. The most common intervals are $0 \le \Omega \le 2\pi$, $-\pi \le \Omega \le \pi$, and $0 \le \Omega \le \pi$. This comment is illustrated in the next example.

In the next example we develop the frequency response function of a system, and show its periodic nature. We also illustrate the basic interpretation of the frequency response theorem.

Example 23.1.8: Consider the causal system

$$\frac{Y(z)}{X(z)} = H(z) = \frac{1 - 2r\zeta z^{-1} + z^{-2}}{1 - 2r\zeta z^{-1} + r^2 z^{-2}}. \tag{23.1.10}$$

Suppose that $\zeta = 0.707$ and $r = 0.9$. The poles and zeros of this transfer function are shown in Figure 23.1.1. The frequency response function is shown in Figure 23.1.2. The periodic character of the magnitude of the frequency response function is clearly evident in Figure 23.1.2. Obviously we need only plot the frequency response function on the frequency interval $0 \le \Omega < 2\pi$. (The interval $-\pi < \Omega < \pi$ could also

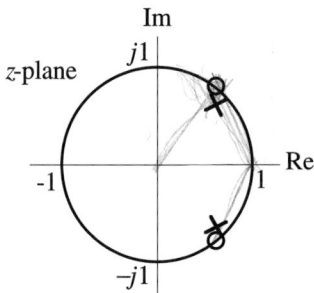

Figure 23.1.1 Poles and Zeros of the Transfer Function in (23.1.10)

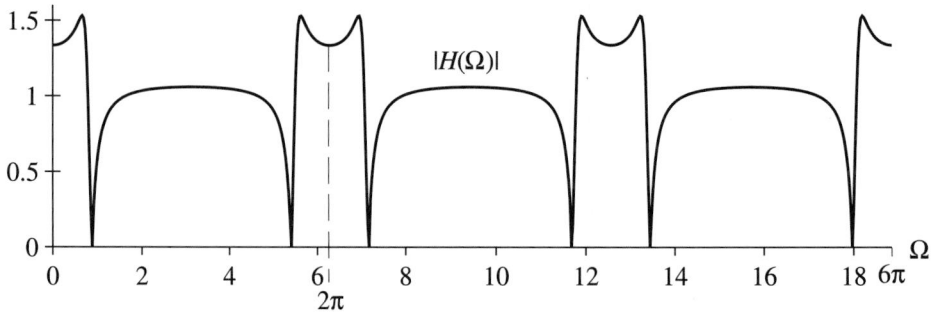

Figure 23.1.2 Frequency Response Function for the System in (23.1.10)

be used.) Notice also that on the $0 \le \Omega < 2\pi$ this function has even symmetry about the frequency $\Omega = \pi$. So we really need only plot this function on half of this interval. The frequency response function on the interval $0 \le \Omega < \pi$ is shown in Figure 23.1.3.

Suppose the input signal is

$$x(n) = \cos(\Omega_i n) = \cos(0.35n), \quad \Omega_i = 0.35. \tag{23.1.11}$$

To determine the steady state response for the input signal (23.1.11) we see from either Theorem 23.1.1 or 23.1.2 that the steady state response is given by

$$y_{ss}(n) = |H(\Omega_i)|\cos(\Omega_i n + \angle H(\Omega_i)). \tag{23.1.12}$$

So we need to evaluate the magnitude and phase of the frequency response function at the frequency of the input signal. These two numbers can be read from the graphs in Figure 23.1.3 as shown. We see that

$$|H(\Omega_i)| = |H(0.35)| = 1.4, \quad \angle H(\Omega_i) = \angle H(0.35) = -0.15. \tag{23.1.13}$$

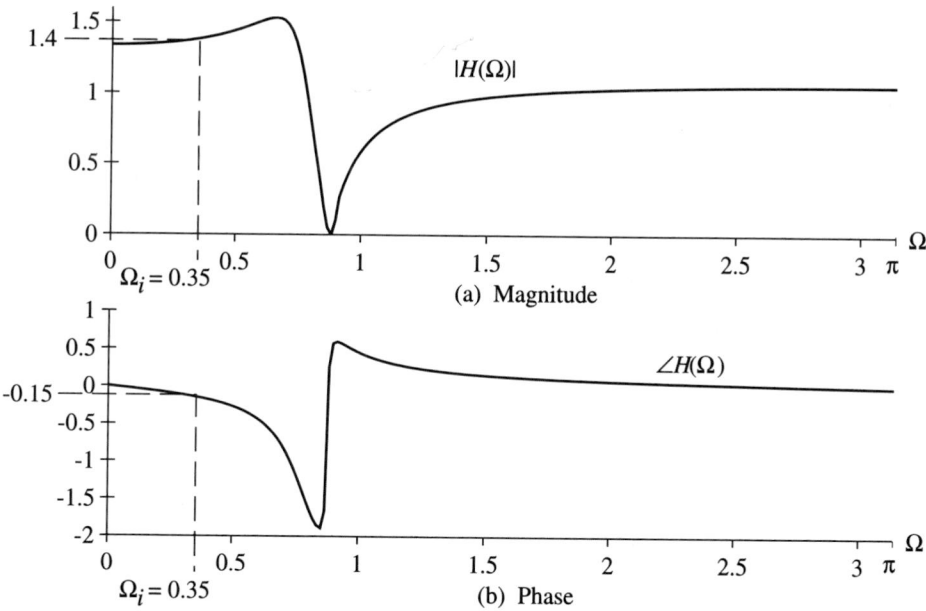

Figure 23.1.3 Frequency Response Function on the Interval $0 \le \Omega < \pi$

Now the steady state response $y_{ss}(n)$ is

$$y_{ss}(n) = |H(\Omega_i)|\cos(\Omega_i n + \angle H(\Omega_i)) = |H(0.35)|\cos(0.35n + \angle H(0.35)) \qquad (23.1.14)$$
$$= 1.4\cos(0.35n - 0.15).$$

The input and output signal are shown in Figure 23.1.4. The output signal in Figure 23.1.4 was computed using Theorem 23.1.2. The transient response is also identified in Figure 23.1.4 along with the steady state response. As predicted by the frequency response theorem, the amplitude of the steady state response in Figure 23.1.4 is

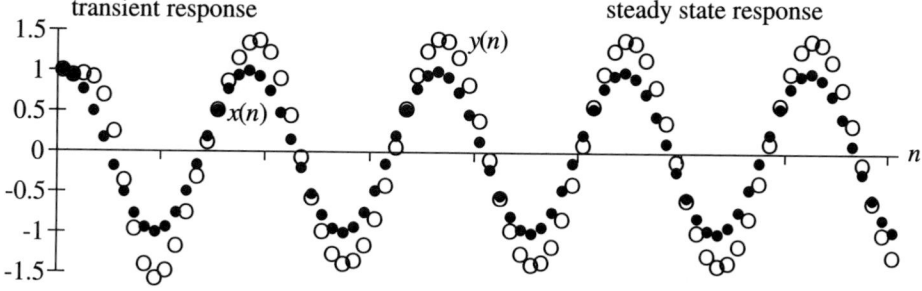

Figure 23.1.4 Input and Output Signal for Example 23.1.8

$|H(\Omega_i)| = 1.4$. It is also seen that the steady state response lags the input signal slightly.

This example shows that the *shape* of the magnitude of the frequency response function in Figure 23.1.3 determines how the system responds to sinusoidal signals. This example shows that an input signal with a frequency of $\Omega_i = 0.9$ is removed from the output signal, because the magnitude of the frequency response function is zero at that frequency. At other frequencies the magnitude of the transfer function is much larger than zero. This filter is a notch filter.

▲▲

Terminology: A BIBO stable, linear, time invariant system allows some sinusoidal signals to pass through it relatively unchanged, while other signals are blocked by the system. Therefore, we call this system a <u>filter</u>. Which sinusoids are passed and which sinusoids are blocked depends on the frequency of the input signal. Therefore, we call this filter a <u>frequency selective</u> filter.

The frequency response theorem for discrete-time systems shows how a discrete signal propagates through a discrete system. This interpretation is identical to the interpretation we gave for continuous-time systems. We can directly apply all of the filtering concepts developed for continuous-time systems in Chapter 14 to discrete-time systems.

Terminology: A discrete system can be implemented in a computer. Because of the finite length of the registers, however, all of the signals in the computer are quantized; they are digital signals. The coefficients of the transfer function are also quantized. Hence, we call a discrete system implemented in a computer a <u>digital</u> filter.

23.1.4 MATLAB Experiments

The following MATLAB M-file was used to generate the graphs in Example 23.1.8. The frequency response function for a discrete system is calculated using the **freqz** command. The input arguments to this command are the system representation and the points around the unit circle where the frequency response function are to be evaluated. The output arguments of this command are the (complex-valued) frequency response function and the frequency vector. Various options exist for determining at which frequency points the frequency response function is calculated.

This command is similar to the **bode** command for continuous systems, except the magnitude and phase of the frequency response function must be calculated separately. The **freqs** command is the continuous-time counterpart of the **freqz** command.

Example 23.1.8: (*continued*)
```
clear
% define system
r = .9;  q = 45;                % define pole locations
zeta = cos((pi/180)*q);
b = [1,-2*r*zeta,1];            % numerator polynomial
a = [1,-2*r*zeta,r^2];          % denominator polynomial
```

```
% Calculate the frequency response
w = linspace(0,pi,400);       % frequency scale
[h,w] = freqz(b,a,w);
figure(1)
plot(w,abs(h))                % plot the magnitude of the frequency response function
xlabel('frequency')
ylabel('magnitude')
title('Frequency Response Function')
figure(2)
plot(w,angle(h))              % plot the phase of the frequency response function
xlabel('frequency')
ylabel('phase')
title('Frequency Response Function')
% Calculate the time responses
n = [0:80];                   % define time axis
wi = 0.35;                    % frequency of the input sinusoid
x = cos(wi*n);               % input signal
y = filter(b,a,x);            % calculate output signal
figure(3)
plot(n,y,'o',n,x,'+')         % plot the input and output signals
xlabel('time')
title('Time Histories')
legend('output signal','input signal')
```

▲▲

Exploratory Exercise 23.1.9: Vary the frequency of the input signal by changing wi, $0 < wi < \pi$. (The time vector n may have to be changed to obtain the desired plots of the input and output signals.) Verify the amplitude of the steady state response from the magnitude of the frequency response function. What is the amplitude of the steady state response when $wi = 0.9$?

▲▲

Exploratory Exercise 23.1.10: Describe the variation of the magnitude and phase of the frequency response function when r is varied between $0 < r < 1$. How do the poles move as r is varied?

▲▲

Exploratory Exercise 23.1.11: For what value of q is the notch located at $\Omega = 2$? Verify this answer with the M-file above.

▲▲

Exploratory Exercise 23.1.12: Plot the magnitude and phase of the frequency response function over an expanded frequency scale. Observe the periodicity of these functions.

▲▲

23.2 RELATIONSHIP TO CONTINUOUS-TIME SIGNALS

23.2.1 Introduction

Quite frequently, digital filters are used to filter continuous signals by sampling the continuous signal and then passing the sampled signal through the digital filter (discrete system). This configuration is shown in Figure 23.2.1. The purpose of filtering, of course, is to alter the spectrum of the continuous signal in a desirable way. To determine the effect of the filter on the signal propagating through it we compare the frequency response of the system to the spectrum of the input signal, and then use the frequency response theorem. This basic analysis is given in the last section. The problem here, however, is that the input signal is in continuous time and the filter operates in discrete time. Also the sampling process plays a role in how the filter responds to the input signal. In this section we will show how to interpret the behavior of the system in Figure 23.2.1 in the frequency domain. This discussion will clarify the role of sampling in the performance of this system.

23.2.2 Analysis

We will analyze the system in Figure 23.2.1 when the input signal is a sinusoid

$$x(t) = \cos(\omega_i t) \tag{23.2.1}$$

where the frequency of the input signal is ω_i rad/sec, a continuous-time frequency. We sample this signal with a sample period of T_s. Then the discrete-time sampled version of $x(t)$ is

$$x(t)\big|_{t=nT_s} = \hat{x}(n) = \cos(\omega_i T_s n) = \cos(\Omega_i n). \tag{23.2.2}$$

From (23.2.2) we see that the fundamental relationship between continuous-time and discrete-time frequency is

$$\omega_i T_s = \Omega_i. \tag{23.2.3}$$

The relationship in (23.2.3) is basic to the interpretation of sampled signal in the discrete-time domain.

In this section we will illustrate the relationship between sampling and discrete system frequency response via an example. Consider the system

$$\frac{Y(z)}{X(z)} = H(z) = \frac{1 - 2r\zeta z^{-1} + z^{-2}}{1 - 2r\zeta z^{-1} + r^2 z^{-2}}. \tag{23.2.4}$$

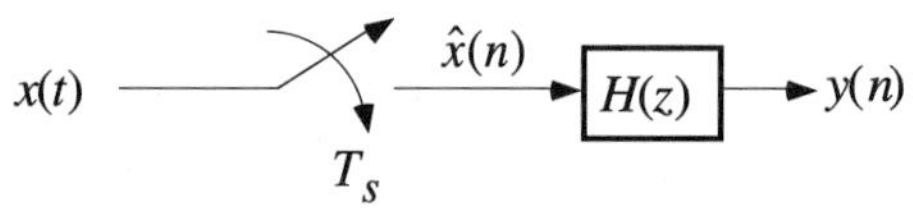

Figure 23.2.1 Sampled Signal as the Input Signal to a Discrete System

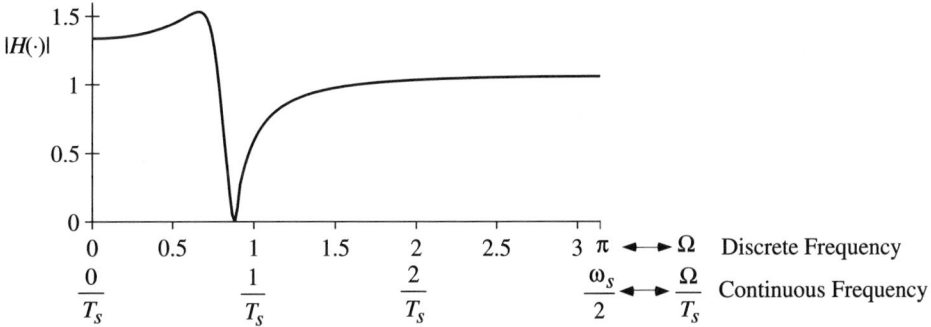

Figure 23.2.2 Magnitude of the Frequency Response Function Labeled in Both Continuous and Discrete Frequency

Suppose that $\zeta = 0.707$ and $r = 0.9$. This system is discussed in Example 23.1.8. The magnitude of the frequency response function of (23.2.4) is shown in Figure 23.2.2. In order to determine the response of the discrete system we must map the continuous frequency of the input signal (23.2.1) into the discrete frequency axis in Figure 23.2.2. This mapping is accomplished using (23.2.3). In fact we can also re-label the frequency axis in continuous frequency using (23.2.3) as shown in Figure 23.2.2. The discrete frequency $\Omega = \pi$ corresponds to the continuous frequency π/T_S which is 1/2 of the sampling frequency $\omega_s = 2\pi/T_S$. That is, the frequency π/T_S is the Nyquist frequency of the sampling frequency $\omega_s = 2\pi/T_S$.

The next example illustrates the calculations necessary to determine the output signal given the input signal, the sampling frequency, and the system.

Example 23.2.1: Suppose the signal

$$x(t) = \cos(2\pi(400)t) \tag{23.2.5}$$

is the input signal to the sampler in Figure 23.2.1 where (23.2.4) is the discrete system. Also suppose that the sample frequency is $f_S = 1$ kHz. Then the discrete-time sampled version of $x(t)$ is

$$\hat{x}(n) = \cos(800\pi(T_S n)) = \cos(800\pi(0.001)n)) = \cos(0.8\pi n). \tag{23.2.6}$$

The magnitude of the frequency response function is shown in Figure 23.2.3 with the frequency scale marked in both continuous and discrete frequency. The steady state response of the filter in (23.2.4) to the input signal (23.2.6) is determined by the frequency response theorem, Theorem 23.1.1, as outlined in the last section. The input frequency is located on the frequency axis in Figure 23.2.3 as shown, and then the magnitude is read off. We have

$$y_{ss}(n) = |H(0.8\pi)|\cos(0.8\pi n + \angle H(0.8\pi)) = (1.05)\cos(0.8\pi n + 0.04). \tag{23.2.7}$$

▲▲

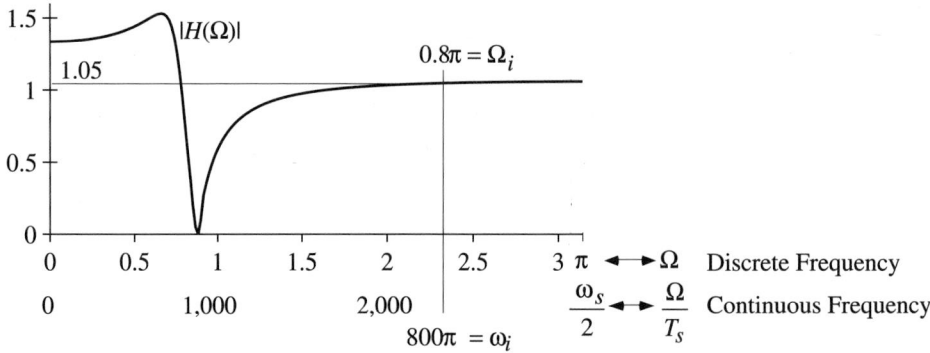

Figure 23.2.3 Magnitude of the Frequency Response Function for the System in (23.2.4) with the Sampling Frequency $f_s = 1$ kHz $(T_s = 0.001)$

By comparing the results of the last example to the results of the next example we see the effects of sampling on the performance of the filter.

Example 23.2.2: Suppose the signal

$$x(t) = \cos(2\pi(400)t) \tag{23.2.8}$$

is the input signal to the sampler in Figure 23.2.1 where (23.2.4) is the discrete system. Also suppose that the sampling frequency is $f_s = 5$ kHz. Then the discrete-time sampled version of $x(t)$ is

$$\hat{x}(n) = \cos(800\pi(T_s n)) = \cos(800\pi(0.0002)n) = \cos(0.16\pi n). \tag{23.2.9}$$

The magnitude of the frequency response function is shown in Figure 23.2.4 with the frequency scale marked in both continuous and discrete frequency. Note that the continuous frequency axis has changed because we have changed the sampling frequency. The discrete frequency axis is unchanged, of course. The steady state response of the system in (23.2.4) to the input signal (23.2.9) is determined as described in the last example. We have

$$y_{ss}(n) = |H(0.16\pi)| \cos(0.16\pi n + \angle H(0.16\pi)) = (1.44)\cos(0.16\pi n - 0.27) \tag{23.2.10}$$

where the phase is obtained from Figure 23.1.3. When this example is compared to Example 23.2.1, the role of the sampling frequency relative to the discrete system is highlighted. By changing the sampling frequency, we have changed the frequency of the discrete sinusoid that we obtained by sampling the continuous-time input signal. Because the frequency of the discrete sinusoidal input signal has changed, the way the filter processes the signal changes.

▲▲

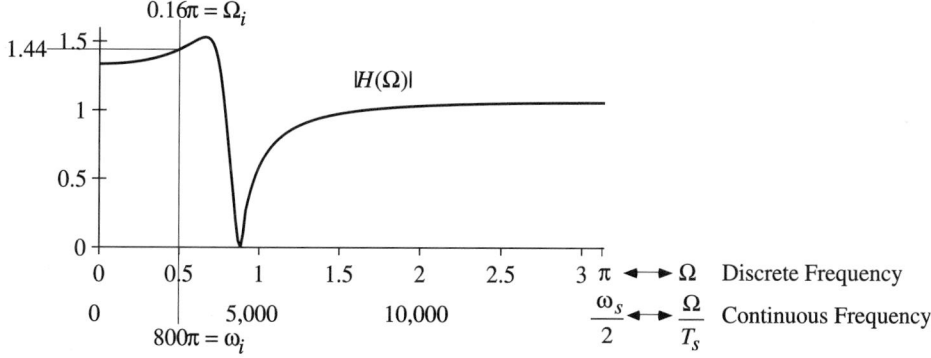

Figure 23.2.4 Magnitude of the Frequency Response Function for the System
in (23.2.4) with the Sampling Frequency $f_S = 5$ kHz ($T_S = 0.002$)

23.2.3 MATLAB Experiments

Exploratory Exercise 23.2.3: The M-file given at the end of the previous section calculates the frequency response of the system used in the examples in this section. Modify that M-file so that the frequency axis of the frequency response functions is plotted in continuous frequency. Use the sampling frequencies in the examples above. Then modify the M-file so that the input signal is a sampled sinusoid as discussed above. That is, the frequency of the input signal should be determined by the sampling frequency and the frequency of the continuous time sinusoid. Plot the output signals (along with the input signals) determined in the examples above. For the input signal in (23.2.5), what sampling frequency will cause the steady state response to be (essentially) zero? Verify your answer through simulation.

▲▲

23.3 CLASSIFICATION OF FREQUENCY RESPONSE FUNCTIONS

23.3.1 Introduction

In Section 23.1 we introduced the frequency response theorem, Theorem 23.1.1, for discrete systems. This theorem showed that the frequency response function characterizes how a system responds to a sinusoidal signal. In particular the *shape* of the frequency response function determines which sinusoids are passed by the system and which sinusoids are blocked by the system. Hence, we can classify a system based on the shape of the frequency response function. In this section we will discuss this classification.

The basic notion in this classification of frequency response functions is the bandwidth of a system. This idea is discussed at length in Sections 14.5 and 14.6 for continuous systems. All of those concepts for continuous systems translate essentially unchanged to discrete systems. Therefore, in this section we will only briefly review these concepts.

23.3.2 Bandwidth of a System

Suppose we are given a (BIBO stable) system

$$\frac{Y(\Omega)}{X(\Omega)} = H(\Omega). \tag{23.3.1}$$

Suppose further the input signal is a sinusoid

$$x(n) = A\cos(\Omega_i n). \tag{23.3.2}$$

Then according to the frequency response theorem, Theorem 23.1.1, the steady state output signal is given by

$$y_{ss}(n) = A|H(\Omega_i)|\cos(\Omega_i n + \angle H(\Omega_i)). \tag{23.3.3}$$

It can be shown that the power[2] of the input signal and output signal are, respectively,

$$P_x = \frac{A^2}{2} \quad \text{and} \quad P_y = \frac{A^2|H(\Omega_i)|^2}{2}. \tag{23.3.4}$$

If we take the ratio of these two powers we see that

$$\frac{P_y}{P_x} = |H(\Omega_i)|^2. \tag{23.3.5}$$

(Recall that the frequency response theorem is fundamentally a comparison of the input signal to the output signal.)   Hence, we see that the frequency response function squared determines how much power of the input signal is transmitted to the output signal.   In particular, when the frequency response function is near unity almost all of the power of the input signal is transmitted to the output signal, while if the frequency response function is near zero almost no power is transmitted to the output signal. This observation leads to the concept of bandwidth.

For the purposes of this discussion, suppose that the magnitude of the frequency response function is as shown in Figure 23.3.1.

Note: The frequency response function in Figure 23.3.1 is plotted on the interval from 0 to π since this frequency response function corresponds to a discrete system. The frequency response function of a discrete system is fundamentally different than a continuous system as explained in Section 23.1. Hence, throughout this chapter we will restrict ourselves to the finite frequency interval $0 < \Omega < \pi$.

In Figure 23.3.1 we have divided the frequency axis into three intervals.

---

[2] See Definition 20.2.7.

Passband: The passband is the frequency interval $0 < \Omega < \Omega_c$ in Figure 23.3.1. In the passband the frequency response function is close to unity. If the frequency of the input sinusoid Ω_i is in frequency interval I_p, the corresponding output signal has approximately the same power as the input signal.

Stopband: The stopband is the frequency interval $\Omega_c < \Omega < \pi$ in Figure 23.3.1. In the stopband the frequency response function is close to zero. If the frequency of the input sinusoid Ω_i is in frequency interval I_s, the corresponding output signal experiences large attenuation and the power of the output signal is significantly reduced over the power of the input signal.

Transition Band: The transition band is sometimes introduced to divide the passband from the stopband, particularly in filter design. This band is not shown in Figure 23.3.1.

Cutoff Frequency: The <u>cutoff frequency</u> Ω_c is the frequency that divides the passband from the stopband. Some filters may have more than one cutoff frequency such as bandpass filters. In realistic filters, such as the one shown in Figure 23.3.1, there may be no clear boundary between the passband and the stopband. In this case, the boundary is established by selecting a number k_1, as shown in Figure 23.3.1, which measures how close the magnitude is to unity. This number can be chosen in several different ways depending on the application at hand. A typical choice for k_1 is $k_1 = 0.707$, as explained in Section 14.5.[3] For this choice of k_1, if the frequency of the input signal is Ω_c, then from (23.3.5) the power of the output signal is one-half of the power of the input signal.

From these specifications we obtain another important parameter of a filter.

Bandwidth (Definition 14.5.3): The <u>bandwidth</u> β of the system is defined as the width of the passband.

▲▲

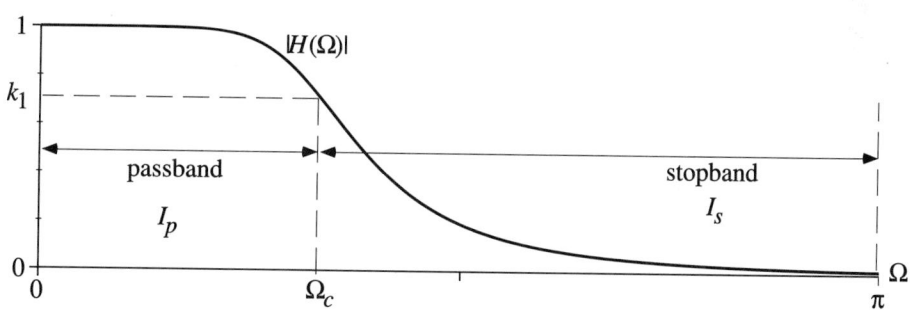

Figure 23.3.1 Magnitude of the Frequency Response Function of (23.3.1)

---

[3] When the maximum magnitude of the frequency response function is not equal to unity, then k_1 is defined by multiplying the maximum value by 0.707.

Terminology: In Figure 23.3.1 the cutoff frequency also defines the bandwidth of this filter. If $k_1 = 0.707$, then the bandwidth is called the <u>half power</u> bandwidth.

23.3.3 Ideal Filters

In the realistic systems we have encountered so far, the frequency response function has been fairly complicated. For the purpose of understanding the basic characteristics of filters and classifying these filters it is often useful to simplify the shape of the frequency response function through idealization. Furthermore, ideal filters often represent the response we would ideally like to obtain from a realistic filter. Ideal filters provide a performance metric against which realistic filters are measured. For the rest of this section, we will study ideal filters.

The most fundamental of the ideal filters is a filter which does nothing to the signal as it propagates through it except introduce a time delay.

Definition 23.3.1: A system is <u>distortionless</u> if the frequency response function is of the form

$$\frac{Y(\Omega)}{X(\Omega)} = H(\Omega) = Ke^{-j\Omega n_d}.$$

▲▲

The frequency response function of a distortionless filter is shown in Figure 23.3.2.

Suppose that the input signal into a distortionless filter is $x(n)$. Then using Property 18.4.9, the time shift property of the DTFT, the output signal is

$$y(n) = Kx(n - n_d). \tag{23.3.6}$$

So we see that if n_d is an integer, then the output signal is a delayed version of the input signal. The definition of a distortionless filter includes a time delay in the passband because all real filters have some phase delay associated with them.

The group delay (Definition 23.1.7) of the distortionless filter is

$$\tau(\Omega) = -\frac{d\Theta(\Omega)}{d\Omega} = -\frac{d(-\Omega n_d)}{d\Omega} = n_d. \tag{23.3.7}$$

The group delay is just the slope of the phase in Figure 23.3.1. Furthermore, it is a constant. Thus, the group delay of a distortionless filter is the time delay between the input and output signal. We obtain this result for this filter because the phase is

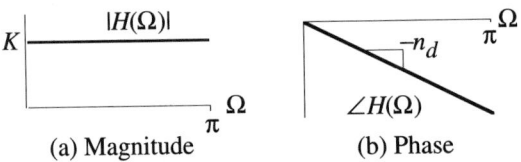

(a) Magnitude   (b) Phase

Figure 23.3.2 Frequency Response Function of a Distortionless Filter

linear. In general, the group delay can be used as a measure of the linearity of the phase of the filter.

Next we use the distortionless filter to define several ideal filters by introducing a passband and a stopband into the distortionless filter. The following definition introduces four classes of such systems based on the shape of their frequency response function. This definition is essentially the same as Definitions 14.6.2–5.

Definition 23.3.2: (a) The frequency response function for an ideal lowpass filter $H_{LP}(\Omega)$ is given by

$$|H_{LP}(\Omega)| = \begin{cases} 1, & 0 \le |\Omega| \le \beta \\ 0, & \beta < |\Omega| < \pi \end{cases}$$

$$\angle H_{LP}(\Omega) = \begin{cases} -\Omega n_d, & 0 \le |\Omega| \le \beta \\ 0, & \beta < |\Omega| < \pi \end{cases}$$

(b) The frequency response function for an ideal bandpass filter $H_{BP}(\Omega)$ is given by

$$|H_{BP}(\Omega)| = \begin{cases} 1, & 0 < \beta_1 \le |\Omega| \le \beta_2 < \pi \\ 0, & \text{otherwise} \end{cases}$$

$$\angle H_{BP}(\Omega) = \begin{cases} -\Omega n_d, & 0 < \beta_1 \le |\Omega| \le \beta_2 < \pi \\ 0, & \text{otherwise} \end{cases}$$

(c) The frequency response function for an ideal highpass filter $H_{HP}(\Omega)$ is given by

$$|H_{HP}(\Omega)| = \begin{cases} 0, & 0 \le |\Omega| \le \beta \\ 1, & \beta < |\Omega| < \pi \end{cases}$$

$$\angle H_{HP}(\Omega) = \begin{cases} 0, & 0 \le |\Omega| \le \beta \\ -\Omega n_d, & \beta < |\Omega| < \pi \end{cases}$$

(d) The frequency response function for an ideal notch filter $H_N(\Omega)$ is given by

$$|H_N(\Omega)| = \begin{cases} 0, & 0 < \beta_1 \le |\Omega| \le \beta_2 < \pi \\ 1, & \text{otherwise} \end{cases}$$

$$\angle H_N(\Omega) = \begin{cases} 0, & 0 < \beta_1 \le |\Omega| \le \beta_2 < \pi \\ -\Omega n_d, & \text{otherwise} \end{cases}$$

▲▲

The magnitude of the frequency response function of each of these filters is shown in Figure 23.3.3. As can be seen from Figure 23.3.3, the frequencies β, β_1, and β_2 are cutoff frequencies. The cutoff frequencies determine the bandwidth of the ideal filters.

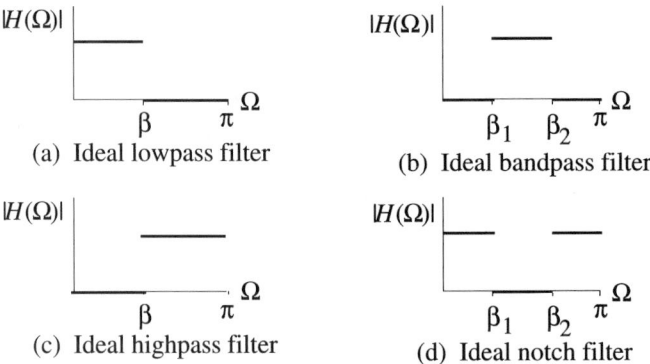

Figure 23.3.3 Ideal Filters

Suppose that the input signal to an ideal lowpass filter is

$$x(n) = \cos(\Omega_1 n) + \cos(\Omega_2 n), \qquad 0 < \Omega_1 < \beta < \Omega_2 < \pi. \tag{23.3.8}$$

Then for Definition 23.3.2(a) the output signal is

$$y(n) = |H_{LP}(\Omega_1)|\cos(\Omega_1 n + \angle H_{LP}(\Omega_1)) + |H_{LP}(\Omega_2)|\cos(\Omega_2 n + \angle H_{LP}(\Omega_2)) \tag{23.3.9}$$
$$= \cos(\Omega_1 n - \Omega_1 n_d)).$$

The frequency of the first cosine term in (23.3.8) falls within the passband of the ideal lowpass filter. This term is passed through the system with only a time delay. The filter is called *ideal* because this term is not attenuated. Also note that this cosine term experiences a pure time delay equal to the group delay of the ideal filter in the passband. The second cosine term in (23.3.8) is eliminated completely from the output signal because the filter is ideal. This frequency falls within the stopband of the filter. This filter is called a *low*pass filter because the passband begins at zero frequency. We can perform a similar analysis for the other filters in Figure 23.3.3.

In Figure 23.3.3 the magnitude of each ideal filter has a discontinuity. In a real filter the magnitude will not change discontinuously. A system is classified by matching its frequency response function to the type of ideal filter it most closely resembles. Also for nonideal filters the cutoff frequencies must be identified according to some criteria as discussed above.

23.4 IIR FILTER DESIGN

23.4.1 Introduction

In Chapter 17 we introduced the idea of a digital filter as a potential replacement for an analog filter. To briefly summarize that discussion, we assume that we are given a BIBO stable continuous-time transfer function

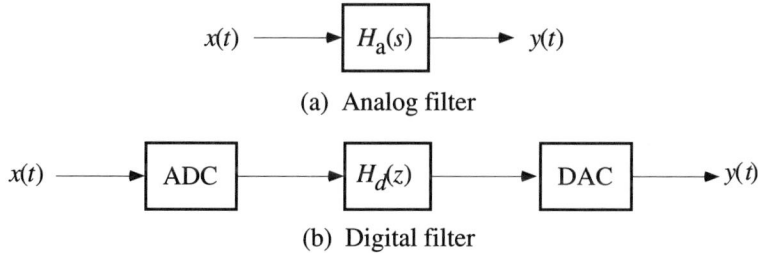

(a) Analog filter

(b) Digital filter

Figure 23.4.1 Equivalent Systems

$$\frac{Y(s)}{X(s)} = H_a(s). \tag{23.4.1}$$

This analog filter is shown in Figure 23.4.1a. We would like to replace the analog filter with the discrete system shown in Figure 23.4.1b. This discrete system contains an ADC to sample the continuous-time input signal and a DAC to convert the discrete-time output signal from the digital filter back into a continuous-time signal. *It is assumed that these devices operate synchronously with a fixed sampling frequency* f_s. We would like to construct a BIBO stable discrete-time transfer function

$$\frac{Y(z)}{X(z)} = H_d(z) \tag{23.4.2}$$

such that the discrete system in Figure 23.4.1b has the same characteristics as the continuous system in (23.4.1). For the purposes of this discussion we will ignore the sampling effects of the ADC and the DAC. Then the problem boils down to finding a digital filter (23.4.2) that has the same characteristics as the analog filter in (23.4.1).

In particular, we would like the frequency response function of the discrete system to match the frequency response function of the continuous system. Because the input signal to the digital system is sampled, the frequency response functions of the two systems can't match above the Nyquist frequency for f_s.[4] Therefore, we must restrict our attention to continuous frequencies below the Nyquist frequency, or, said another way, to the discrete frequency interval $\Omega = [0, \pi]$. The two systems will not be completely equivalent. We will measure how well they match each other by comparing their frequency response functions over the discrete frequency interval $\Omega = [0, \pi]$.

In this section we will assume that the performance specifications on the analog filter are given in terms of the magnitude of the frequency response function. These specifications are discussed in detail in Section 14.7. Also discussed in Section 14.7 are several standard filters including Butterworth, Chebyshev, and elliptical filters. The discussion in this section assumes knowledge of the material in Section 14.7.

---

[4] See Definition 19.3.5. The discussion in Section 23.2 is also generally relevant to this discussion.

23.4.2 Bilinear Transformation

The basic idea is that we want to transform the analog filter (23.4.1), which is a complex function, into the digital filter (23.4.2), which is also a complex function. This transformation is accomplished by mapping the s-plane into the z-plane with a function $s = f(z)$ which is called a <u>conformal mapping</u> or a <u>transformation</u>. Using this transformation the digital filter is given by

$$H_d(z) = H_a(s)\big|_{s=f(z)} = H_a(f(z)). \tag{23.4.3}$$

This conformal mapping should have two properties. First, we want the digital filter to be BIBO stable. Therefore, the mapping should take the poles of the analog transfer function in the LHP of the s-plane into poles of the digital filter that are inside the unit circle in the z-plane. To insure that all of the poles are mapped correctly, the conformal mapping should take the LHP in the s-plane into the unit disk in the z-plane. Second, the conformal mapping should preserve the frequency response function. The frequency response function of the analog filter is obtained by evaluating the analog transfer function (23.4.1) along the $j\omega$-axis in the s-plane. The frequency response function of the discrete system is obtained by evaluating the discrete transfer function around the unit circle in the z-plane. Therefore, the conformal mapping should take the $j\omega$-axis in the s-plane into the unit circle in the z-plane. In particular, the origin of the s-plane should map to the point $(1,0)$ in the z-plane. There are several conformal mappings that possess these properties. Here we will focus on the one that is most widely used.

Definition 23.4.1: The <u>bilinear transformation</u> is a conformal mapping between the s-plane and the z-plane given by

$$s = C\left(\frac{1 - z^{-1}}{1 + z^{-1}}\right).$$

▲▲

The constant C in the bilinear transformation is a design parameter. We will discuss how to choose C below. In the absence of other criteria, the constant C is usually chosen as

$$C = \frac{2}{T_S}. \tag{23.4.4}$$

Terminology: The bilinear transformation with C given in (23.4.4) is also known as <u>Tustin's rule</u>.

This mapping takes the entire LHP in the s-plane into the unit disk in the z-plane. The points at infinity in the s-plane are mapped into the point $(-1,0)$ in the z-plane. The origin in the s-plane is mapped to the point $(1,0)$ in the z-plane. Consequently, all poles of $H_a(s)$ in the left-hand plane are mapped into poles of $H_d(z)$ inside the unit circle. The zeros at infinity of $H_a(s)$ are mapped to zeros at $z = -1$ of $H_d(z)$. A

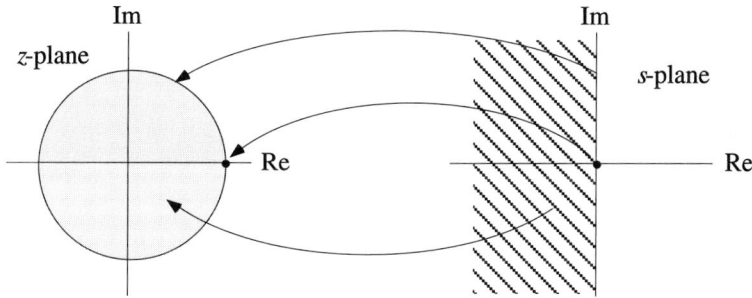

Figure 23.4.2 The Bilinear Mapping

BIBO stable continuous-time system will be mapped to a BIBO stable discrete system. This bilinear mapping is illustrated in Figure 23.4.2. This transformation maps the $j\omega$ axis in the s-plane to the unit circle in the z-plane according to

$$\omega = C \tan\left(\frac{\Omega}{2}\right). \tag{23.4.5}$$

The relationship in (23.4.5) follows by substituting $s = j\omega$ and $z = e^{j\Omega}$ into Definition 23.4.1. Unfortunately, this relationship between the $j\omega$ axis and the unit circle is nonlinear. The entire $j\omega$ axis which is of infinite length is mapped to the unit circle which is of finite length. The zero frequency in continuous frequency is mapped to the zero frequency in discrete frequency. Infinitely large continuous frequencies are mapped to discrete frequencies near $\Omega = \pi$. This compression is called <u>frequency warping</u>. Note that the mapping, and hence the frequency warping, depends on the constant C in (23.4.5).

Example 23.4.2: Suppose that the analog filter in (23.4.1) is a 4th-order Chebyshev Type I filter with a bandwidth of $\beta = 1$ rad/sec. For simplicity assume that the sampling frequency is $f_s = 1$ Hz. A digital filter is constructed from the analog filter using the bilinear command in MATLAB. The magnitude of the frequency response function of the analog filter and the digital filter are shown in Figure 23.4.3. Although the shapes of the two frequency response functions are the same, the effects of the frequency warping are clearly visible in Figure 23.4.3.
▲▲

There is one discrete frequency Ω_d that is mapped to the corresponding continuous frequency ω_d by both mappings $\Omega = T_s\omega$ and (23.4.5). At the frequency Ω_d the frequency response function of the analog filter will match the frequency response function of the digital filter. Where this desired frequency falls depends on the constant C in the bilinear transformation. By choosing C appropriately, we can select the frequency $\omega_d = \Omega_d$ at which the continuous and discrete frequency response functions match. Given the desired continuous-time frequency ω_d the desired coefficient C is calculated from

$$C = \omega_d \left[\tan\left(\frac{\Omega_d}{2}\right) \right]^{-1} = \omega_d \left[\tan\left(\frac{\omega_d T_s}{2}\right) \right]^{-1} . \qquad (23.4.6)$$

Terminology: The processes of selecting C in (23.4.6) is called <u>prewarping</u> of the desired continuous-time frequency ω_d.

Prewarping is used when one frequency can be identified as being of particular importance. One common choice of the desired frequency is the bandwidth of the analog filter.

Example 23.4.2: (*continued*) Consider again the Chebyshev filter in Example 23.4.2. Suppose we select the bandwidth of this filter as the frequency to be prewarped, $\omega_d = \beta = 1$ rad/sec. After prewarping, the magnitude of the frequency response function of both the analog and digital filter are shown in Figure 23.4.4. In general, prewarping will yield a better match between analog and digital frequency response functions, although differences remain as can be seen in Figure 23.4.4. ▲▲

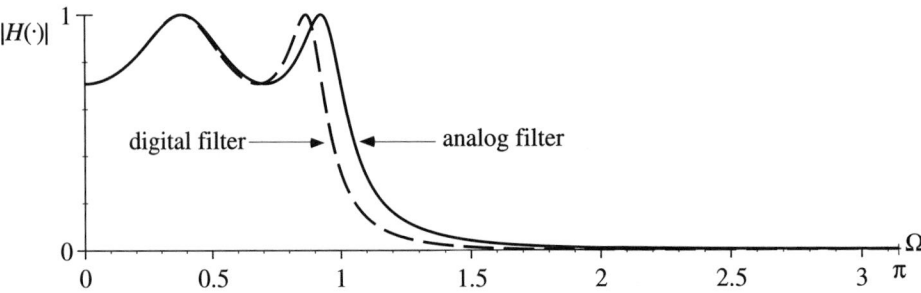

Figure 23.4.3 Magnitude of the Frequency Response Functions of an Analog and Digital Chebyshev Filter

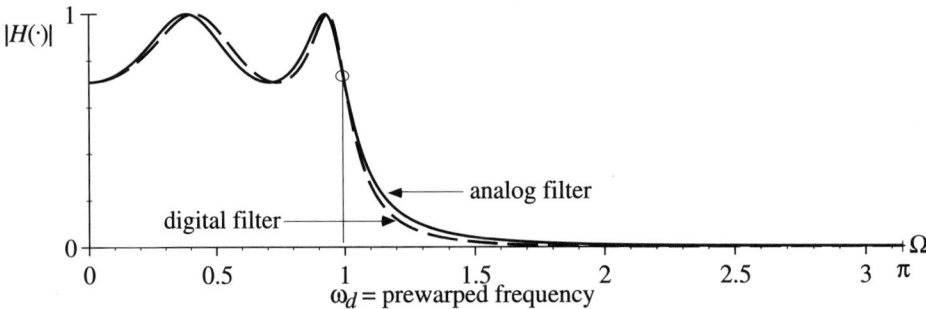

Figure 23.4.4 Magnitude of the Frequency Response Functions of an Analog and Digital Chebyshev Filter After Prewarping

The bilinear transformation may be carried out by using the bilinear transformation to map the poles and zeros of the analog transfer function. Let the analog transfer function be given by

$$H_a(s) = \frac{K(s-\sigma_1)(s-\sigma_2)\cdots(s-\sigma_M)}{(s-s_1)(s-s_2)\cdots(s-s_N)}. \tag{23.4.7}$$

In order to determine the discrete transfer function in (23.4.9) from the analog transfer function in (23.4.7), the poles and zeros of the analog transfer function are mapped into the poles and zeros of the discrete transfer function using the bilinear transformation in Definition 23.4.1. The poles and zero map according to the relationship

$$z_k = \frac{1+\left(\frac{1}{C}\right)\sigma_k}{1-\left(\frac{1}{C}\right)\sigma_k}, \quad p_k = \frac{1+\left(\frac{1}{C}\right)s_k}{1-\left(\frac{1}{C}\right)s_k} \tag{23.4.8}$$

Applying this transformation to (23.4.7) we obtain

$$H_d(z) = b_0(1+z^{-1})^{N-M}\frac{(1-z_1z^{-1})(1-z_2z^{-1})\cdots(1-z_Mz^{-1})}{(1-p_1z^{-1})(1-p_2z^{-1})\cdots(1-p_Nz^{-1})} \tag{23.4.9}$$

The zeros at -1 in $H_d(z)$ account for the zeros at infinity[5] in $H_a(s)$. The gain b_0 in $H_d(z)$ is determined from the relationship $H_a(0) = H_d(1)$. The following example illustrates these calculations.

Example 23.4.3: Suppose we are given an analog transfer function

$$\frac{Y(s)}{X(s)} = H_a(s) = \frac{1}{s+\alpha}, \tag{23.4.10}$$

and a sampling frequency f_s. We wish to construct a digital filter to approximate the analog filter using the bilinear transformation. For prewarping, we select the desired frequency to be the bandwidth of the analog filter $\omega_d = \alpha$. Then using the relationship in (23.4.6) the constant C is given by

$$C = \omega_d\left[\tan\left(\frac{\omega_d T_s}{2}\right)\right]^{-1} = \alpha\left[\tan\left(\frac{\alpha T_s}{2}\right)\right]^{-1}. \tag{23.4.11}$$

Now the poles and zeros of the discrete transfer function are given by

---

[5] See the discussion following Theorem 3.2.7.

$$\text{zero at } \infty \quad \leftrightarrow \quad 1 + z^{-1} \tag{23.4.12}$$

$$s_1 = -\alpha \quad \leftrightarrow \quad p_1 = \frac{1 + \left(\frac{1}{C}\right)s_1}{1 - \left(\frac{1}{C}\right)s_1} = \frac{1 + \left(\frac{1}{C}\right)(-\alpha)}{1 - \left(\frac{1}{C}\right)(-\alpha)}$$

$$b_0 = 1 \quad \leftrightarrow \quad b_0 = H_a(0)\left(\frac{1 - p_1}{2}\right)$$

So the digital filter is given by

$$\frac{Y(z)}{X(z)} = H_d(z) = \frac{b_0(1 + z^{-1})}{1 - p_1 z^{-1}}. \tag{23.4.13}$$

▲▲

23.4.3 Filter Design

In Section 14.7 we discussed the design of analog filters including Butterworth, Chebyshev, and elliptic filters. By combining these analog filters with the bilinear transformation, we obtain a powerful method for designing IIR digital filters. The basic procedure is to first design an analog filter which meets the design specifications, and then transform that filter into a digital filter using the bilinear transformation. This procedure is automated in MATLAB.

This procedure can be used to design all four classes of filters: lowpass, bandpass, notch, and highpass. The actual design procedure consists of the following steps. First, a prototype lowpass analog filter is constructed of the specified type and order. Then an analog transformation is applied to transform the prototype lowpass filter into a lowpass, bandpass, highpass, or notch filter as specified. This second transformation also shifts the passband to the desired frequency range. These transformations are discussed in Section 14.7. Finally, the bilinear transformation is applied to transform the analog filter into a digital filter. The bilinear transformation automatically incorporates prewarping based on the cutoff frequencies.

23.4.4 MATLAB Experiments

The MATLAB command **bilinear** implements the bilinear transformation. The input arguments are the analog filter and the sampling frequency. An optional argument allows the user to specify the frequency for prewarping. The output arguments are the coefficients of the digital filter.

The MATLAB commands **butter**, **cheby1**, **cheby2**, and **ellip** automate completely the design process described above for Butterworth, Chebyshev, and elliptic filters. These commands automatically generate a lowpass analog prototype filter, transform the prototype filter to the desired filter, and use the bilinear transformation to transform the analog filter into a digital filter. The input arguments to these commands are the order of the filter, the cutoff frequencies, and the passband and/or stopband ripple, if appropriate. Because these filters are discrete systems, the cutoff frequencies must be between 0 and π. MATLAB assumes that these cutoff frequencies are normalized to be between 0 and 1. If a cutoff frequency greater than 1 is entered, the filter coefficients are returned, but the filter is meaningless. To create a bandpass filter, the cutoff frequencies are entered as a vector. An additional optional parameter is required when the filter is a highpass or notch filter. The output arguments of these commands are the filter representation in one of the three standard representations. In all cases, the MATLAB help files should be consulted for specifics.

MATLAB uses the bilinear transformation to calculate the digital filter from the prototype lowpass analog filter. Due to the approximations in the filter design process, the frequency response of the digital filter may not meet the frequency response specifications. In this case, the cutoff frequencies in the input argument may have to be adjusted.

The following example M-file illustrates the basics of creating a lowpass Chebyshev Type I filter.

Example 23.4.2: (*continued*)   The following M-file generates a 4th-order Chebyshev Type 1 lowpass filter with 3 dB of ripple in the passband.

```
clear
order = 4;                        % order of the filter
ripple = 3;                       % passband ripple
cutoff = 2;                       % cutoff frequency
normcf = 2/pi;                    % normalized cutoff frequency


% generate filter
[b,a] = cheby1(order,ripple,normcf);
% calculate frequency response
w = linspace(0,pi,600);           % generate frequency vector
h = freqz(b,a,w);                 % calculate frequency response function
mag = abs(h);                     % calculate magnitude
phs = (180/pi)*angle(h);          % calculate phase
% plot frequency response function
subplot(2,1,1), plot(w,mag)
xlabel('discrete frequency')
title('magnitude')
subplot(2,1,2), plot(w,phs)
xlabel('discrete frequency')
ylabel('deg')
title('phase')
```

▲▲

Exploratory Exercise 23.4.4:   Generate Butterworth, Chebyshev Type II, and elliptic filters with the same order and cutoff frequencies. Plot the magnitudes on the same graph. Plot the phases on the same graph. Also plot the step response of each filter. What are the differences between these filters?

▲▲

Exploratory Exercise 23.4.5: Modify the M-file above to generate a bandpass filter with cutoff frequencies $[\beta_1,\beta_2] = [1,2]$. Then modify this bandpass filter into a notch filter.

▲▲

23.5 LINEAR PHASE FIR FILTERS

23.5.1 Introduction

In the previous section we discussed the digital filters as an extension of analog filters. In this section we briefly introduce linear phase FIR filters. These filters have no analog equivalent because they rely on hardware implementations that have no analog equivalent. These FIR filters are attractive because they are inherently BIBO stable, making them easy to implement. We will present one way to design these filters. There are several other design methods as well but they are beyond the scope of this text.

The design specifications for these filters are the same as the specifications we used for the IIR filters in the last section. Hence, the goals of the filter design remain the same. We will just construct a different type of filter to meet these specifications. First, we will discuss a few of the basic properties of linear phase FIR filters and then we will discuss a design methodology for these filters.

23.5.2 Linear Phase Filters

In this section we consider a system

$$\frac{Y(\Omega)}{X(\Omega)} = H(\Omega). \tag{23.5.1}$$

In previous sections we have characterized filters by the shape of the magnitude of their frequency response function. Next we introduce a characterization of filters based on the phase of the frequency response function.

Definition 23.5.1: The system in (23.5.1) is said to have <u>linear phase</u> if the phase of the filter satisfies

$$\angle H(\Omega) = -n_d \Omega,$$

or

$$\angle H(\Omega) = \gamma - n_d \Omega.$$

▲▲

Note that the group delay of both of these filters is a constant $\tau(\Omega) = n_d$.

To understand the linear phase property of these digital filters, consider a discrete-time ideal lowpass filter

$$|H_{LP}(\Omega)| = \begin{cases} 1, & |\Omega| \le \beta \\ 0, & |\Omega| > \beta \end{cases} \tag{23.5.2}$$

$$\angle H_{LP}(\Omega) = \begin{cases} -n_d \Omega, & |\Omega| \le \beta \\ 0, & |\Omega| > \beta \end{cases}$$

By Definition 23.5.1 this filter has linear phase. The impulse response function for this system is computed from the inverse DTFT. We have

$$h_{LP}(n) = \frac{1}{2\pi} \int_{-\pi}^{\pi} H_{LP}(\Omega)e^{j\Omega n} d\Omega = \frac{1}{2\pi} \int_{-\Omega_c}^{\Omega_c} \left(1e^{-jn_d\Omega}\right)e^{j\Omega n} d\Omega. \tag{23.5.3}$$

This integration leads to

$$h_{LP}(n) = \begin{cases} \dfrac{\beta}{\pi} \dfrac{\sin(\beta(n-n_d))}{\beta(n-n_d)}, & n \neq n_d \\[2ex] \dfrac{\beta}{\pi}, & n = n_d \end{cases} \tag{23.5.4}$$

The frequency response function and its impulse response function are shown in Figure 23.5.1.

If the input signal $x(n)$ to the filter in (23.5.2) is bandlimited to frequencies below β, the output signal of such a filter is $y(n) = x(n - n_d)$. The fact that the output signal is just a time delayed version of the input signal is due to the linear

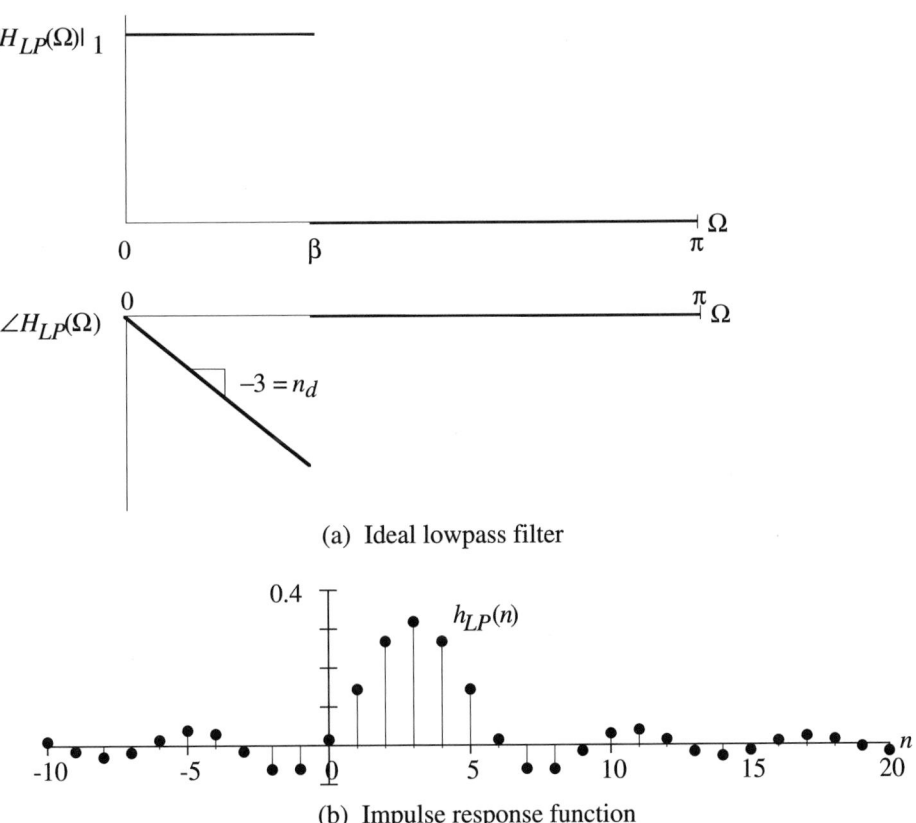

(a) Ideal lowpass filter

(b) Impulse response function

Figure 23.5.1 Ideal Lowpass Filter

phase of the filter. Clearly, such a filter is distortionless in its passband. Linear phase is an important property for a filter where signal distortion is undesirable. This situation often arises in speech processing and high speed digital communication applications.

The impulse response function of the ideal lowpass filter is shown in Figure 23.5.1b. Note that the peak of the sinc function has been shifted away from the origin by an amount equal to the group delay.

23.5.3 FIR Filters

The ideal filter shown in Figure 23.5.1 has the linear phase property, but it is noncausal. Hence, this system can't be used for real-time filtering. It can also be shown that any causal IIR system can't have linear phase. (Since all continuous systems are IIR systems, the linear phase property is not discussed in conjunction with continuous systems.) Causal FIR systems can have linear phase, however. In this section we will introduce several linear phase FIR filters and discuss some of their properties.

We will consider FIR filters whose difference equation is of the form

$$y(n) = b_0 x(n) + b_1 x(n-1) + \cdots + b_{M-1} x(n - M + 1) = \sum_{k=0}^{M-1} b_k x(n-k). \tag{23.5.5}$$

The transfer function of this system is

$$\frac{Y(z)}{X(z)} = H(z) = \sum_{k=0}^{M-1} b_k z^{-k}. \tag{23.5.6}$$

Notation: There are M nonzero coefficients of this FIR filter. This parameter is one of the primary design parameters of FIR filters.

Comparing (23.5.5) to the convolution representation of a system we see that the impulse response function of this system is given by

$$h_{FIR}(n) = \begin{cases} 0, & n < 0 \\ b_n, & 0 \le n \le M - 1 \\ 0, & M \le n \end{cases} \tag{23.5.7}$$

So the system in (23.5.5) is indeed FIR.

For the FIR system (23.5.5) to have linear phase, the coefficients must satisfy special properties. These properties are given in the next theorem.

Theorem 23.5.2: Consider the FIR system in (23.5.5).
(a) If the coefficients of this system satisfy

$$b_k = b_{M-1-k}, \quad \begin{cases} k = 0,1,\ldots,(M-1)/2, & M \text{ odd} \\ k = 0,1,\ldots,(M/2)-1, & M \text{ even} \end{cases}$$

then the phase of the frequency response function will be given by

$$\angle H(\Omega) = -\left(\frac{M-1}{2}\right)\Omega.$$

(b) If the coefficients of the system (23.5.5) satisfy

$$b_k = -b_{M-1-k}, \quad \begin{cases} k = 0,1,\ldots,(M-1)/2, & M \text{ odd} \\ k = 0,1,\ldots,(M/2)-1, & M \text{ even} \end{cases}$$

then the phase of the frequency response function will be given by

$$\angle H(\Omega) = \frac{\pi}{2} - \left(\frac{M-1}{2}\right)\Omega.$$

▲▲

Theorem 23.5.2 states that for an FIR filter to have linear phase, the coefficients must only satisfy a symmetry condition. There are no restrictions on the value of the coefficients. It should be noted that the group delay depends on the order of the filter $M-1$. Condition (a) of Theorem 23.5.2 says that the coefficients must have even symmetry about the midpoint for the system to have linear phase. If the number of coefficients is odd, the point of symmetry falls on a value of impulse response function. If the number of coefficients is even, the point of symmetry falls in between two values of impulse response function. Condition (b) of Theorem 23.5.2 says that the coefficients must have odd symmetry about the midpoint for the system to have linear phase "plus an offset." These coefficients must satisfy the same symmetry conditions.

Hence, Theorem 23.5.2 identifies four different types of FIR filters with linear phase depending on the symmetry of the coefficients and whether the number of coefficients is even or odd. These filter types are shown in Figure 23.5.2. The coefficients not shown in Figure 23.5.2 are zero.

Notation: We label these four types of filters as shown in Table 23.5.1.

23.5.4 Windowing

Compare the Type I filter in Figure 23.5.2a to the impulse response function of the ideal lowpass filter in Figure 23.5.1b. Note that if we truncate the impulse response function in Figure 23.5.1b, setting to zero those values of the impulse response function where $n < 0$, and $n > 6$, we obtain the FIR impulse response function in Figure 23.5.2a. We can formalize this process. Define the function

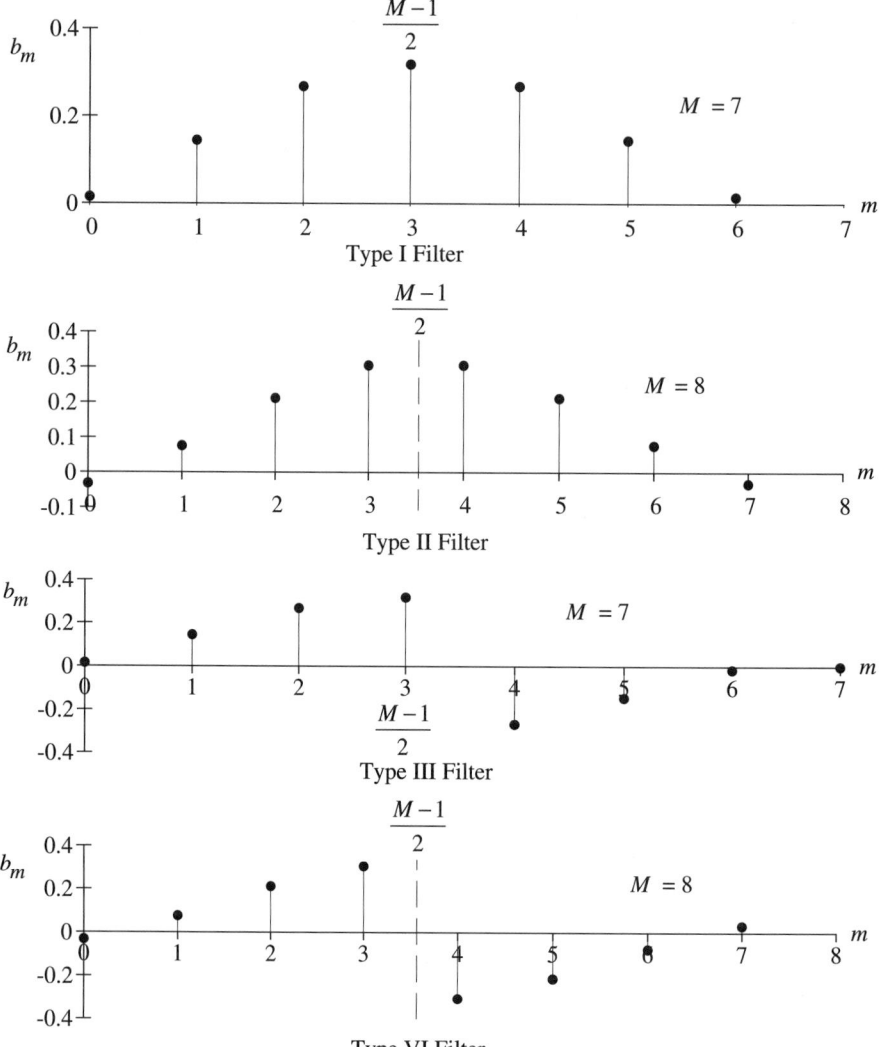

Figure 23.5.2 Four Types of Linear Phase Filters

$$w(n) = u_s(n) - u_s(n - M). \tag{23.5.8}$$

Note that this function is zero outside the time interval $0 \le n < M-1$. Using (23.5.8) with $M = 7$ we can formally truncate the impulse response function in (23.5.4) to obtain the impulse response function in Figure 23.5.2a. We have

$$h_I(n) = h_{LP}(n)w(n). \tag{23.5.9}$$

Table 23.5.1 Linear Phase Filter Types

Type	$H(\Omega)$	Symmetry	M	Phase
I	$H_I(\Omega)$	even	odd	$\angle H_I(\Omega) = -\left(\dfrac{M-1}{2}\right)\Omega$
II	$H_{II}(\Omega)$	even	even	$\angle H_{II}(\Omega) = -\left(\dfrac{M-1}{2}\right)\Omega$
III	$H_{III}(\Omega)$	odd	odd	$\angle H_{III}(\Omega) = \dfrac{\pi}{2} - \left(\dfrac{M-1}{2}\right)\Omega$
IV	$H_{IV}(\Omega)$	odd	even	$\angle H_{IV}(\Omega) = \dfrac{\pi}{2} - \left(\dfrac{M-1}{2}\right)\Omega$

Equation (23.5.9) suggests a method for constructing FIR linear phase filters. An FIR filter can be constructed from an IIR impulse response function by multiplying the IIR impulse response function with any finite length sequence. The resulting FIR impulse response function will have linear phase if the coefficients satisfy the symmetry conditions in Theorem 23.5.2. To insure the symmetry conditions, first, the function $w(n)$ must be symmetric about its midpoint. Second, the midpoint of the impulse response function must be aligned with the midpoint of the function $w(n)$. We will discuss this method of constructing an FIR filter in more detail in the next subsection.

The operation of extracting a finite length sequence from an infinite length sequence using the function $w(n)$ has a name.

Definition 23.5.3: A finite length sequence which is symmetric about its midpoint is called a <u>window</u> function. The process of multiplying an infinite length sequence by a window function is called <u>windowing</u>. The discrete signal created by windowing is said to be a <u>windowed</u> signal.

▲▲

Notation: We will denote the length of a window function by M.

Terminology: The function in (23.5.7) is called a <u>rectangular</u> window. Several other windows are given in Table 23.5.2.

In the following discussion we will use windows to design linear phase filters. Windows have many other uses in signal processing that are not explored here.

23.5.5 FIR Filter Design Using Windows

In this subsection we will discuss a method for FIR linear phase filter design that uses windows. The basic idea is to start with an ideal linear phase filter and then window the impulse response function. In order to insure that the coefficients of the FIR filter satisfy the symmetry conditions of Theorem 23.5.2, the group delay of the ideal filter must be chosen to match the length of the window. The details of this methodology are given next.

Table 23.5.2 Window Functions

Window	Function $n = 0, 1, \ldots, M-1$
Rectangular	$u_S(n) - u_S(n-M)$
Barlett	$1 - \dfrac{2\left\|n - \dfrac{M-1}{2}\right\|}{M-1}$
Hanning	$\dfrac{1}{2}\left[1 - \cos\left(\dfrac{2\pi n}{M-1}\right)\right] = \sin^2\left(\dfrac{\pi n}{M-1}\right)$
Hamming	$0.54 - 0.46\cos\left(\dfrac{2\pi n}{M-1}\right)$
Blackman	$0.42 - 0.5\cos\left(\dfrac{2\pi n}{M-1}\right) + 0.08\cos\left(\dfrac{4\pi n}{M-1}\right)$
Kaiser	$\dfrac{I_0\left(\beta\sqrt{1 - \left[\dfrac{2n}{M-1}\right]^2}\right)}{I_0(\beta)}$

Procedure 23.5.4: **Design of FIR Filters Using Windows** We assume that we are given specifications on the magnitude of the frequency response function.

Step 1 Select an ideal impulse response function from Table 23.5.3 whose frequency response function satisfies the given specifications. That is, choose the cutoff frequencies for the impulse response function.

Step 2 Chose the order of the filter to be $M - 1$. The group delay will be

$$n_d = \frac{M-1}{2}. \tag{23.5.10}$$

Step 3 Select a window of length M, $w(n;M)$. Now the coefficients of the FIR filter are given by

$$b_n = h_{LP}(n;\beta,n_d)w(n;M). \tag{23.5.11}$$

Step 4 Evaluate the frequency response function of the FIR filter. If it meets the specifications, quit. If it doesn't meet the specifications, return to Step 2 and adjust the filter length. Or return to Step 3 and change the window. ▲▲

The FIR filter coefficients calculated according to Procedure 23.5.4 will satisfy the symmetry conditions of Theorem 23.5.2 because: (1) the choice of the group delay in (23.5.10), (2) the shape of the ideal impulse response function, and (3) the length and symmetry of the window function. This filter will have linear phase in the passband. However, the magnitude may not meet the passband and stopband ripple specifications or the transition bandwidth specifications. In this event it is necessary to iterate on the number of coefficients of the filter M and the selection of the window. This iteration is easily accomplished with MATLAB. If a Type III or

Table 23.5.3 Ideal Impulse Response Functions

Filter Type	Impulse Response Function $h_d(n)$	
	$h(n - n_d), n \neq n_d$	$h(n_d)$
Lowpass $h_{LP}(n)$	$\dfrac{\beta}{\pi} \dfrac{\sin(\beta(n - n_d))}{\beta(n - n_d)}$	$\dfrac{\beta}{\pi}$
Bandpass $h_{BP}(n)$	$\dfrac{\beta_2}{\pi} \dfrac{\sin(\beta_2(n - n_d))}{\beta_2(n - n_d)} - \dfrac{\beta_1}{\pi} \dfrac{\sin(\beta_1(n - n_d))}{\beta_1(n - n_d)}$	$\dfrac{\beta_2}{\pi} - \dfrac{\beta_1}{\pi}$
Notch $h_N(n)$	$\dfrac{\beta_1}{\pi} \dfrac{\sin(\beta_1(n - n_d))}{\beta_1(n - n_d)} - \dfrac{\beta_2}{\pi} \dfrac{\sin(\beta_2(n - n_d))}{\beta_2(n - n_d)}$	$1 - \left[\dfrac{\beta_2}{\pi} - \dfrac{\beta_1}{\pi} \right]$
Highpass $h_{HP}(n)$	$- \dfrac{\beta}{\pi} \dfrac{\sin(\beta(n - n_d))}{\beta(n - n_d)}$	$1 - \dfrac{\beta}{\pi}$

Type IV filter is desired, then a Type I or Type II filter is designed and the appropriate coefficients negated.

23.5.6 Selection of Windows

Next we will discuss the effect of the order and window on the frequency response function of the FIR filter in the context of an example. We will assume that we want to design a lowpass filter with the cutoff frequency set to $\beta = 1.5$. The simplest and most obvious window to select in (23.5.11) is the rectangular window. If an FIR filter is designed using a rectangular window, the resulting frequency response function is shown in Figure 23.5.3 for two different values of the filter length.

The frequency response functions in Figure 23.5.3 are typical of the magnitude of a linear phase FIR filter. They exhibit both passband and stopband ripple. The low order filter shows a very slow roll off in the transition band. Generally, as the order increases, the magnitude shows a sharper rolloff in the transition band. As a rule, a very high order FIR filter is required to achieve the same transition bandwidth as a low order IIR filter.

The frequency response functions in Figure 23.5.3 also demonstrate characteristics of a rectangular window. The magnitude exhibits a constant 9%

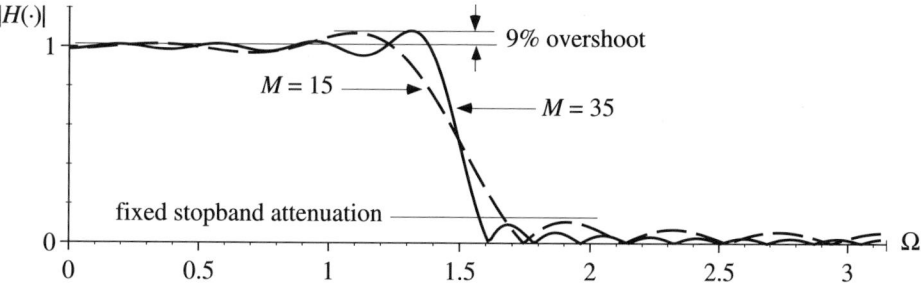

Figure 23.5.3 A Comparison of Rectangular Window Lengths

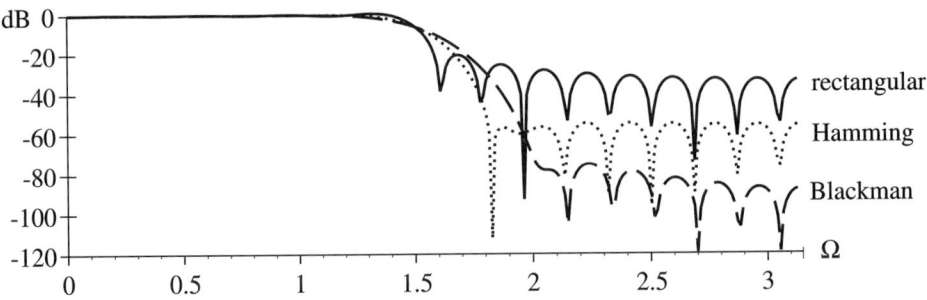

Figure 23.5.4 Comparison of Windows for FIR Filter Design

overshoot in the passband (due to the Gibb's phenomenon) and no more than −21 dB stopband attenuation regardless of the number of coefficients, M. These characteristics are due to the abrupt termination of the window (in the time domain), and they generally preclude the use of this window.

In Procedure 23.5.4, Step 3 the FIR filter is created by windowing a ideal IIR. The shape of the magnitude of the filter depends on the selection of the window. By selecting different windows from Table 23.5.2 the passband ripple can be reduced, and the stopband attenuation increased. To illustrate the influence of the window on the frequency response function, a comparison of two of these windows with the rectangular window is given in Figure 23.5.4 for a filter of fixed length $M = 35$.

The frequency response functions in Figure 23.5.4 show how a window can change the shape of the frequency response function. First, note that both the Hamming and Blackman windows remove the overshoot of the rectangular window. Both of these windows also achieve a better stopband attenuation than the rectangular window. The improved stopband attenuation is accomplished at the expense of an increased transition bandwidth. Indeed, the Blackman window has the best stopband attenuation, but it also has the largest transition bandwidth. This observation can be said to be a general principle in FIR filter design.

23.5.7 Summary

The frequency responses in the three figures above indicate the basic trade-offs involved in this design approach to FIR filters.

1. The properties of the filter improve with increasing order. The complexity of the filter also increases, however.
2. The passband and stopband ripple can be reduced by appropriate selection of the window. As the stopband ripple is reduced, however, the transition bandwidth increases.

The advantages and disadvantages of the window method of design can be summarized as follows.

1. The design procedure is very straightforward and easy to use.
2. The discontinuities of the desired frequency response lead to ripples and overshoot in the frequency response function. Windows can smooth these

ripples and reduce the overshoot, but at the expense of a greater transition bandwidth.

3. The window characteristics completely define the filter characteristics. Hence, there is very little flexibility is shaping the frequency response of the filter.

23.5.8 MATLAB Experiments

MATLAB contains M-file to implement the window functions in Table 23.5.2. These command names are the same as the names given in Table 23.5.2.

The design methodology for linear phase FIR filters is embodied in the MATLAB command **fir1**. The input arguments are the order of the filter and the bandwidth of the filter. The default window in the **fir1** command is the Hamming window. An optional input argument allows the user to specify an alternative window.

The following M-file implements the design methodology discussed in this section for a lowpass filter with resorting to the **fir1** command. Note that the parameters for the impulse response function have been modified so that the **sinc** command can be used.

Example 23.5.5: M-file Implementation of Design Procedure 23.5.4

```
clear
M = 15;                 % define the order of the filter
nd = (M-1)/2;           % group delay
bdwd = 1.5;             % bandwidth of the filter
n = [0:M-1];            % time vector
% calculate the filter coefficients
h = bdwd/pi*sinc((bdwd/pi)*(n-nd));      % calculate the impulse response function
hm = h.*hamming(M)';                     % window impulse response function
% calculate frequency response
[H,w] = freqz(hm,1,256);
mag = 20*log10(abs(H));
% plot magnitude of the frequency response function
plot(w,mag)
xlabel('discrete frequency')
ylabel('dB')
title('Linear Phase Filter')
```

▲▲

Exploratory Exercise 23.5.6: Modify the M-file to recreate Figure 23.5.1. Verify the limitations of the rectangular window by changing the order of the filter.

▲▲

Exploratory Exercise 23.5.7: Modify the M-file to recreate Figure 23.5.2 and Figure 23.5.3. Experiment with the other windows available in MATLAB.

▲▲

Exploratory Exercise 23.5.8: Modify the M-file so that the filter always has unity gain at zero frequency independent of the order of the filter by adjusting the gain of the transfer function. (The MATLAB command **fir1** contains this modification.)
▲▲

Exploratory Exercise 23.5.9: Modify the M-file for the design of bandpass filters.
▲▲

Exploratory Exercise 23.5.10: Rewrite this M-file using the **fir1** command.
▲▲

Exploratory Exercise 23.5.11: Modify the M-file above to generate a Type III filter. Compare the phase of the Type III filter to the phase of the Type I filter.
▲▲

Exploratory Exercise 23.5.12: Plot all of the windows in Table 23.5.2 on the same graph. What are the differences between these functions?
▲▲

23.6 SYSTEM RESPONSE TO ARBITRARY INPUT SIGNALS

23.6.1 Introduction

In this section we will discuss the response of a discrete system to an arbitrary input signal in the frequency domain. The purpose of this discussion is to highlight the relationship between the input and output signal of each filter in terms of the frequency response of the filter, and the amplitude and phase spectra (or energy or power spectrum) of the input and output signals. The context of this discussion is the same as the discussion in Chapter 15. In fact, all of the conclusions in Chapter 15 apply equally well to discrete systems, as we will show here. Herein lies the power of frequency domain analysis.

In the analysis presented in this section we think of the input signal as propagating through the system and the system modifying the signal as it passes through the system. To describe the effects of the system on the signal, we represent both the signals and the system in the frequency domain. The input and output signals are represented by their amplitude and phase spectra or their energy or power spectral density. The relationship between the energy spectral density of the input and output signals can be concisely expressed using the frequency response function of the system. This result is a generalization of the frequency response theorem in Section 23.1.

We then present an extended example of a pulse signal propagating through an IIR filter and a FIR filter. This example will illustrate the interplay between the time and frequency domain descriptions of signals and systems. This example will also illustrate the difference between an IIR filter with nonlinear phase and an FIR linear phase filter.

23.6.2 System Response to Energy and Power Signals

Consider the linear, time invariant BIBO stable system

$$y(n) = \sum_{k=-\infty}^{\infty} h(n-k)x(k). \tag{23.6.1}$$

This system can also be represented by its DTFT transfer function

$$\frac{Y(\Omega)}{X(\Omega)} = H(\Omega). \tag{23.6.2}$$

The frequency response theorem, Theorem 23.1.1, characterizes the output signal of this system when the input signal is a sinusoid. In this subsection we will extend this theorem to include input signals that are energy or power signals.

This extension relies on the definition of the autocorrelation of a signal. Recall that if $x(n)$ is an energy signal, the autocorrelation of this signal is defined as (Definition 20.2.13)

$$r_{xx}(k) = \sum_{n=-\infty}^{\infty} x(n+k)x(n). \tag{23.6.3}$$

If $y(n)$ is power signal, then the autocorrelation of this signal is defined as

$$r_{yy}(k) = \lim_{\kappa \to \infty} \frac{1}{2\kappa+1} \sum_{n=-\kappa}^{\kappa} y(n+k)y(n) \tag{23.6.4}$$

provided the limit exists. From Theorem 20.2.14 we know that the energy (power) spectral density of a signal is the DTFT of the autocorrelation function of that signal. That is,

$$D_x(\Omega) = \mathcal{DTFT}\{r_{xx}(k)\}, \quad \text{and} \quad S_y(\Omega) = \mathcal{DTFT}\{r_{yy}(k)\}. \tag{23.6.5}$$

Autocorrelation functions are convenient for analysis because they exist for both energy and power signals. We will use autocorrelation functions without distinguishing whether we are working with an energy or power signal.

The autocorrelation function can be written in another form using the convolution notation. If we write the convolution summation of the signals $x(n)$ and $x(-n)$ we obtain the autocorrelation of the signals $x(n)$. That is,

$$x(k) * x(-k) = \sum_{n=-\infty}^{\infty} x(n)x(-(k-n)) = \sum_{n=-\infty}^{\infty} x(n)x(n-k) \tag{23.6.6}$$

$$= \sum_{m=-\infty}^{\infty} x(m+k)x(m) = r_{xx}(k)$$

where in the last step in (23.6.6) we introduced a change of variables in the summation. The * notation borrowed from convolution is a very convenient notation to represent correlation functions.

Next we will develop a characterization of the autocorrelation of the output signal of a system in terms of the autocorrelation of the input signal and the impulse response function of the system. We assume that the autocorrelation functions exist. The autocorrelation of the output signal is given by

$$r_{yy}(k) = y(n) * y(-n) = [h(n) * x(n)] * [h(-n) * x(-n)] \tag{23.6.7}$$

$$= [h(n) * h(-n)] * [x(n) * x(-n)] = r_{hh}(k) * r_{xx}(k).$$

If we take the Fourier transform of (23.6.7) we obtain the following theorem.

Theorem 23.6.1:   Let (23.6.1) be a BIBO stable system.   Assume that the autocorrelation function exists for the input signal.
(a) If the input signal has a DTFT, then the DTFT of the output signal exists and it is given by

$$|Y(\Omega)| = |H(\Omega)||X(\Omega)|$$

$$\angle Y(\Omega) = \angle H(\Omega) + \angle X(\Omega)$$

(b) Suppose that the input signal is an energy signal. Then the output signal is an energy signal and the energy spectral density of the output signal is given by

$$D_y(\Omega) = |H(\Omega)|^2 D_x(\Omega).$$

(c) Suppose that the input signal is a power signal. Then the output signal is a power signal and the power spectral density of the output signal is given by

$$S_y(\Omega) = |H(\Omega)|^2 S_x(\Omega).$$

▲▲

This theorem says that the energy spectral density of the output signal can be computed from the energy spectral density of the input signal and the frequency response function. Theorem 23.6.1 parallels exactly Theorem 15.1.1. Essentially all of the frequency domain interpretations in Chapter 15 carry over to discrete systems. The example given next illustrates this theorem.

23.6.3 Filter Comparison

Theorem 23.6.1 is the most important result describing the relationship between the input signal, the output signal, and the system in the frequency domain. This result forms the foundation for digital filter design. This theorem is best understood through graphical interpretation. In the remainder of this section we will give an example that illustrates this interpretation.

The idea behind this example is shown in Figure 23.6.1. We assume that we are given two digital filters: an IIR filter and an FIR linear phase filter. The IIR filter is a 4th-order elliptical filter given by

$$\frac{Y_1(z)}{X_1(z)} = H_1(z) = \frac{0.11 - 0.23z^{-1} + 0.32z^{-2} - 0.23z^{-3} + 0.11z^{-4}}{1 - 2.87z^{-1} + 3.50z^{-2} - 2.05z^{-3} + 0.50z^{-4}}. \tag{23.6.8}$$

The second filter is a Type 1 linear phase FIR filter of the form

$$\frac{Y_2(z)}{X_2(z)} = H_2(z) = \sum_{k=0}^{M-1} b_k z^{-k}. \tag{23.6.9}$$

For this example $M = 41$. The coefficients for this filter can be obtained from the M-file below. In order to compare these two filters, they are chosen to have the same bandwidth, $\beta = 0.2\pi$ rad. The magnitude and phase of the frequency response function is shown in Figure 23.6.2. The FIR filter is linear phase within the bandwidth of the filter while the IIR doesn't have linear phase. Note that the cutoff of the FIR filter are not as sharp as the IIR filter. It is generally true that the order of an FIR filter must be much larger than the order of an IIR filter to achieve roughly the same magnitude of the frequency response function.

Each of these filters receives the same input signal, and we want to compare the output signals. The input signal is chosen as

$$x(n) = u_s(n) - u_s(n - pw) \tag{23.6.10}$$

where pw is the width of the pulse. The input and output signals of the IIR filter (23.6.8) are shown in Figure 23.6.3a. The linear phase property of the FIR filter means that there is a time delay of $(M-1)/2$ in the output signal of the filter. (See the discussion of distortionless filters in Subsection 23.3.3.) Therefore, to fairly compare the input signal pulse with the output signal of the filter, we create a delayed version of the input signal

$$x_d(n) = u_s(n - n_d) - u_s(n - pw - n_d), \qquad n_d = \frac{M-1}{2}. \tag{23.6.11}$$

This delayed input signal and the output signal of the FIR filter are shown in Figure 23.6.3b. (This delayed input signal is used only in this plot. The output signal of the FIR filter is generated with (23.6.10).) The amplitude spectra of the input signal

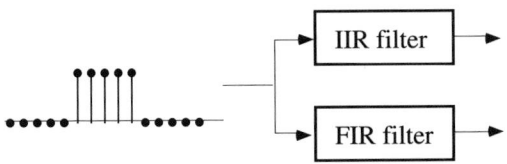

Figure 23.6.1 Filter Comparison with a Pulse Input Signal

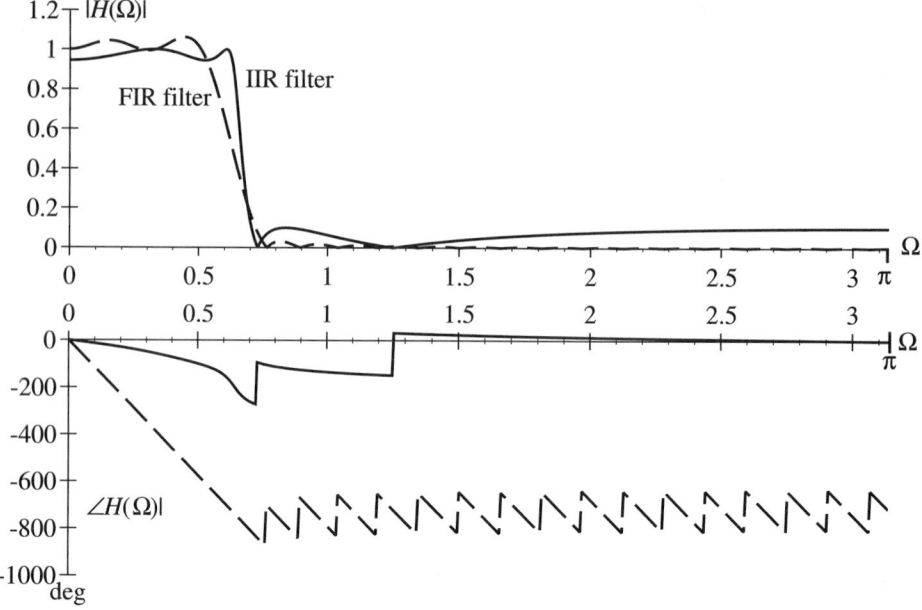

Figure 23.6.2 Magnitude and Phase of the IIR and FIR Filters

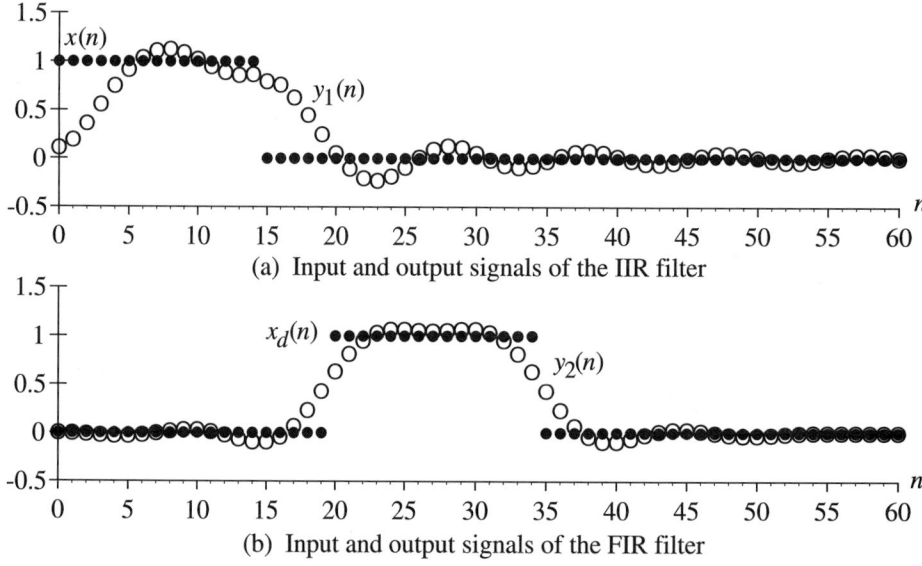

Figure 23.6.3 Response of the IIR and FIR Filters to the Input Signal (23.6.1)

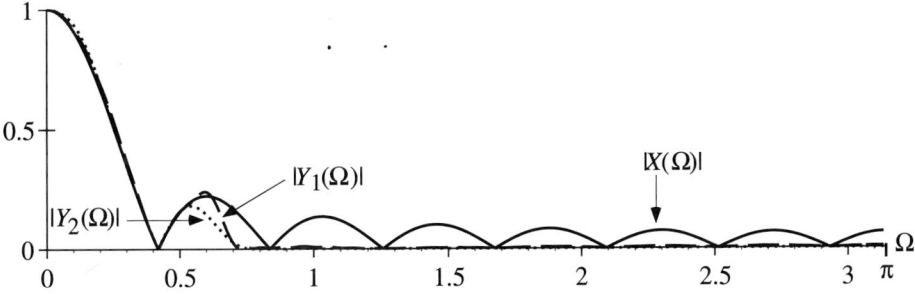

Figure 23.6.4 Amplitude Spectra of the Input and Output Signals

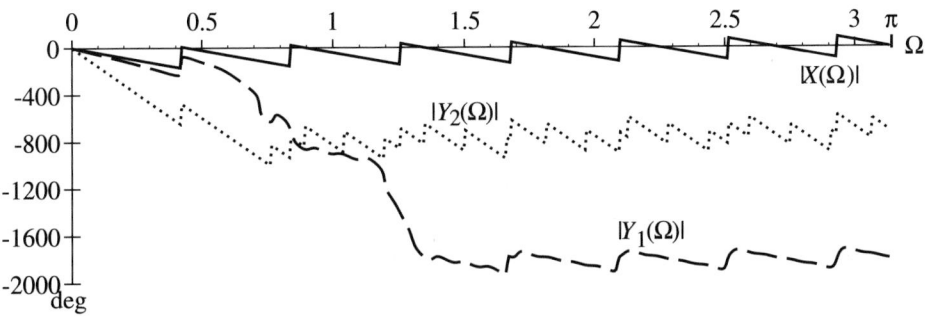

Figure 23.6.5 Phase Spectra of the Input and Output Signals

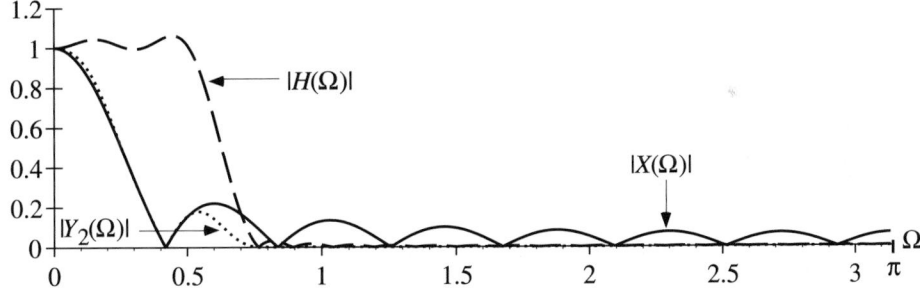

Figure 23.6.6 Magnitude of the FIR Filter along with the Amplitude Spectrum
of the Input and Output Signals

and the two output signals are shown in Figure 23.6.4. The phase spectra of these
signals are shown in Figure 23.6.5.

   Next we undertake a comparison of the output signals of the two filters in both
the time and frequency domain. Consider first the time responses in Figure 23.6.3.
The output signal of the IIR filter in Figure 23.6.3a is recognizable as a pulse;

however, the shape of the pulse is distorted. There is also a significant amount of ringing in the output signal after the termination of the pulse at the input of the filter. The output signal of the FIR filter is shown in Figure 23.6.3b along with the delayed input signal. The shape of the pulse, while slightly distorted, is still quite recognizable as a pulse. Also there is only a small amount of ringing before and after the pulse. The effect of the linear phase of the FIR filter is quite obvious when the two output signals in Figure 23.6.3 are compared.

Next we will apply Theorem 23.6.1 to interpret the propagation of the signal through the system in the frequency domain. To illustrate the analysis methodology the magnitude of the FIR filter along with the amplitude spectrum of its input and output signal are shown in Figure 23.6.6. Theorem 23.6.1 says that the amplitude spectrum of the output signal is obtained by multiplying the amplitude spectrum of the input signal by the magnitude of the frequency response function of the filter. This calculation is clearly shown in Figure 23.6.6. The magnitude has ripples in the passband which cause some ripples in the amplitude spectrum of the output signal. In the stopband of the filter the amplitude spectrum of the output signal is driven to zero. A similar calculation is performed to obtain the amplitude spectrum of the output signal of the IIR filter.

Consider again Figure 23.6.6, comparing the amplitude spectrum of the input signal to the amplitude spectrum of the output signal. We see that there is little difference between these two amplitude spectra since the main lobe of the input signal falls within the passband of the filter. The side lobes of the input signal have been removed from the output signal by the filter. The removal of these side lobes causes the ringing in the output signal. Otherwise we would expect the pulse to roughly retain its shape. This behavior is observed in Figure 23.6.3.

Next consider the amplitude spectra of the two output signals in Figure 23.6.4. Since the magnitude of the two filters is almost the same, the amplitude spectrum of the two output signals is almost the same. Based just on the amplitude spectra of the two output signals, we would expect the two output signal to look the same. Any variation between the two output signals must be due to the way the phase of the output signal of the IIR filter is distorted by the nonlinear phase of the IIR filter.

Next consider the phase of the input and output signals in Figure 23.6.5. According to Theorem 23.6.1, the phase of the output signal in Figure 23.6.5 is the sum of the phase of the input signal in Figure 23.6.5 and the phase of the filter in Figure 23.6.2. Because the phase of the frequency response function of the FIR filter is linear in the passband of the filter, the phase spectrum of the output signal is a delayed version of the phase spectrum of the input signal. The phase spectrum of the output signal of the IIR filter, when compared to the phase spectrum of the input signal, shows the effects of the nonlinear phase of the IIR filter, however. The distortion of the phase spectrum of the output signal of the IIR filter (compared to the input signal) causes the additional distortion observed in the output signal of the IIR filter when compared to the output signal of the FIR filter.

23.6.4 MATLAB Experiments

The following M-file generates the plots in the example above. The IIR filter is an elliptic filter whose parameters are discussed in Section 23.4. The FIR filter is generated using the window method as discussed in Section 23.5. The bandwidth of these filters can be easily changed using the parameter bdwd. The following M-file plots the amplitude spectra of the input and output signals. In order to compare these spectra, they are normalized.

% Comparison of IIR and FIR linear phase filters

```
clear
% generate the filter coefficients
bdwd = .2;                              % bandwidth of filter is 0.2*pi
[b,a] = ellip(4,.5,20,bdwd);            % IIR filter
fltord = 40;                            % order of the FIR filter
h2 = fir1(fltord,bdwd,kaiser(fltord+1,2));  % FIR linear phase filter
% generate the input signal
pw = 15;                                % width of pulse
n = [0:pw*5];                           % time vector
x = zeros(size(n));
x(1:pw) = ones(1,pw);                   % pulse input signal
% calculate the output signals
y1 = filter(b,a,x);                     % IIR filter response
y2 = filter(h2,1,x);                    % FIR filter response
% generate delayed input signal for FIR plot
xshift = zeros(size(n));
delay = ceil((fltord-1)/2);
xshift(delay:delay+pw-1) = ones(1,pw);
% plot the filter responses
figure(1)
subplot(2,1,1),  plot(n,x,'+w',n,y1,'oy')
title('Input and Output Signal of the IIR Filter')
xlabel('time')
legend('+w','input signal','oy','output signal')
subplot(2,1,2),  plot(n,xshift,'+w',n,y2,'ob')
title('Input and Output Signal of the FIR Filter')
xlabel('time')
legend('+w','input signal','ob','output signal')
% calculate the frequency response of the filters
[hmag1,f1] = freqz(b,a);                % IIR frequency response
[hmag2,f2] = freqz(h2,1);               % FIR frequency response
% plot the magnitude and phase of the filters
figure(2)
subplot(2,1,1),  plot(f1,abs(hmag1),'y',f2,abs(hmag2),'--b')
title('Magnitude of Elliptic Filter and FIR Filter')
xlabel('frequency, rad')
legend('-y','IIR filter','--b','FIR filter')
subplot(2,1,2),
plot(f1,(180/pi)*unwrap(angle(hmag1)),'y',f2,(180/pi)*unwrap(angle(hmag2)),'--b')
title('Phase of Elliptic and FIR Filter')
xlabel('frequency, rad')
ylabel('deg')
legend('-y','IIR filter','--b','FIR filter')
% calculate DTFT of the input and output signals
[xspec,fx] = freqz(x,1);                % DTFT of input signal
```

```
[y1spec,fy1] = freqz(y1,1);              % DTFT of output signal
[y2spec,fy2] = freqz(y2,1);              % DTFT of output signal
% normalize amplitude spectrums
xmagn = abs(xspec)./max(abs(xspec));
y1magn = abs(y1spec)./max(abs(y1spec));
y2magn = abs(y2spec)./max(abs(y2spec));
% unwrap phase spectrums
xphs = (180/pi)*unwrap(angle(xspec));
y1phs = (180/pi)*unwrap(angle(y1spec));
y2phs = (180/pi)*unwrap(angle(y2spec));
% plot amplitude and phase spectrums of the input and output signals
figure(3)
subplot(2,1,1),  plot(fx,xmagn,'w',fy1,y1magn,'--y',fy2,y2magn,':b')
xlabel('frequency, rad')
title('Amplitude Spectrum of x, y1 and y2')
legend('-w','input signal','--y','IIR output signal',':b','FIR output signal')
subplot(2,1,2),  plot(fx,xphs,'w',fy1,y1phs,'--y',fy2,y2phs,':b')
xlabel('frequency, rad')
ylabel('deg')
title('Phase of the FFT of x, y1 and y2')
legend('-w','input signal','--y','IIR output signal',':b','FIR output signal')
```

▲▲

Exploratory Exercise 23.6.2:   Vary the filter order of the FIR filter between fltord = 5 and fltord = 80. How does the magnitude and phase of the frequency response function change in comparison to the IIR filter? What are the effects on the output signal of the FIR filter?

▲▲

Exploratory Exercise 23.6.3: Vary the pulse width pw of the input signal. How does the output signal change with this change of the width of the pulse of the input signal? How are these changes reflected in the amplitude spectra of the input and output signals?

▲▲

Exploratory Exercise 23.6.4: Change the type of IIR filter from an elliptic to a Butterworth or a Chebyshev filter. How does changing the type of filter change the behavior of the output signal?

▲▲

Exploratory Exercise 23.6.5: Plot the difference between the phase of the IIR and FIR filter over the bandwidth of the filter.

▲▲

23.7 CHAPTER SUMMARY

In this chapter we have investigated the effect of a system on a signal propagating through it. The most fundamental result is given in the frequency response theorem (Theorem 23.1.1). Suppose a system can be represented by a DTFT transfer function

$$\frac{Y(\Omega)}{X(\Omega)} = H(\Omega). \tag{23.7.1}$$

Let the input signal be

$$x(n) = \cos(\Omega_i n). \tag{23.7.2}$$

Then the output signal is given by

$$y_{ss}(n) = |H(\Omega_i)|\cos(\Omega_i n + \angle H(\Omega_i)), \quad -\infty < n < \infty. \tag{23.7.3}$$

This theorem, which is stated using one-sided z-transforms in Theorem 23.1.2, expresses the output signal as a function of the sinusoidal input signal and the transfer function of the system. The term $y_{ss}(n)$ in Theorem 23.1.1 and 23.1.2 is called the (sinusoidal) steady state response of the system (Definition 23.1.3).

The effect of the system on the sinusoidal signal propagating through it can be determined from the frequency response function $H(\Omega)$ (Definition 23.1.6)

$$H(\Omega) = H(z)\big|_{z=e^{j\Omega}} = \mathcal{DTFT}\{h(n)\}. \tag{23.7.4}$$

Usually, the frequency response function is given in graphical form (Definition 23.1.6) in terms of the magnitude graph

$$|H(\Omega)| \quad \text{vs. frequency} \tag{23.7.5}$$

and the phase graph

$$\angle H(\Omega) \quad \text{vs. frequency} \tag{23.7.6}$$

The frequency response theorem is extended to energy and power input signals by using autocorrelation functions (Theorem 23.6.1). If the input signal has a DTFT, then the DTFT of the output signal exists and it is given by

$$|Y(\Omega)| = |H(\Omega)||X(\Omega)| \quad \text{and} \quad \angle Y(\Omega) = \angle H(\Omega) + \angle X(\Omega). \tag{23.7.7}$$

If the input signal is an energy signal, then the output signal is an energy signal and the energy spectral density of the output signal is given by

$$D_y(\Omega) = |H(\Omega)|^2 D_x(\Omega). \tag{23.7.8}$$

If the input signal is a power signal, then the output signal is a power signal and the power spectral density of the output signal is given by

$$S_y(\Omega) = |H(\Omega)|^2 S_x(\Omega). \tag{23.7.9}$$

Theorem 23.6.1 forms the basis of digital filter design which is discussed in the rest of this chapter. The shape of the frequency response function determines which sinusoidal signals are passed through the system with little attenuation and which sinusoids are essentially blocked by the system. That is, the passband is the frequency interval where the frequency response function is close to unity. The stopband is the frequency interval where the frequency response function is close to zero. The bandwidth of the system is the width of the passband. (See Section 23.3.)

Filter design is motivated by the desire to create an ideal, or distortionless filter (Definition 23.3.1). Such a filter only introduces a time delay into the signal as it propagates through the system. Ideal lowpass, bandpass, highpass, and notch filters are introduced in Definition 23.3.2 as a starting point for the development of filter specifications.

Sections 23.4 and 23.5 contain a brief introduction to digital filter design. Two methods are presented for filter design: one method for IIR filter design and one method for FIR filter design. IIR filter design generally proceeds by transforming a prototype analog filter into a digital filter by using the bilinear transformation (Definition 23.4.1). The bilinear transformation yield digital versions of Butterworth, Chebyshev I and II, and elliptical filters. One method for FIR filter design consists of windowing the impulse response function of an ideal filter. By choosing the appropriate window, the shape of the magnitude of the frequency response function can be tailored to the design specifications. Both of the design methodologies are automated in MATLAB.

23.8 HOMEWORK FOR CHAPTER 23

Homework Problems for Section 23.1

23.1.1 What is the frequency response function of the following systems?

(i) $\dfrac{Y(z)}{X(z)} = 1 + 2z^{-1} + z^{-2}$ (ii) $\dfrac{Y(z)}{X(z)} = 1 + 2z^{-1} - z^{-2}$

23.1.2 Plot the frequency response function of the following system.
$$y(n) + 0.5043y(n-1) + 0.5289y(n-2) - 0.0423y(n-3)$$
$$= 0.2489x(n) + 0.7466x(n-1) + 0.7466x(n-2) + 0.2489x(n-3)$$

23.1.3 A simple FIR notch filter is given by

$$\frac{Y(z)}{X(z)} = H(z) = 1 - 2\cos(\Omega_0)z^{-1} + z^{-2}$$

(a) Find the difference equation that corresponds to this transfer function.

(b) Show that the zeros of this transfer function are given by $z = e^{\pm j\Omega_0}$.

(c) Suppose we want to eliminate the signal $x(n) = \cos(0.3\pi n)$ from the output of this system. Find the appropriate value of Ω_0.

(d) Draw the pole-zero diagram.

(e) Plot the frequency response function for this filter.

(f) Plot the input and output signals to show the performance of the filter.

(g) Draw an all-delay block diagram of this system.

23.1.4   The magnitude and phase of the frequency response function for a BIBO stable system are shown in Figure P23.1.4. For each of the following input signals find the steady state response.

(i) $x(n) = \sin(2n)$ (iii) $x(n) = \cos(n + 0.3)$

(ii) $x(n) = \cos(1.2n)$ (iv) $x(n) = \sin(0.2n + 1.1)$

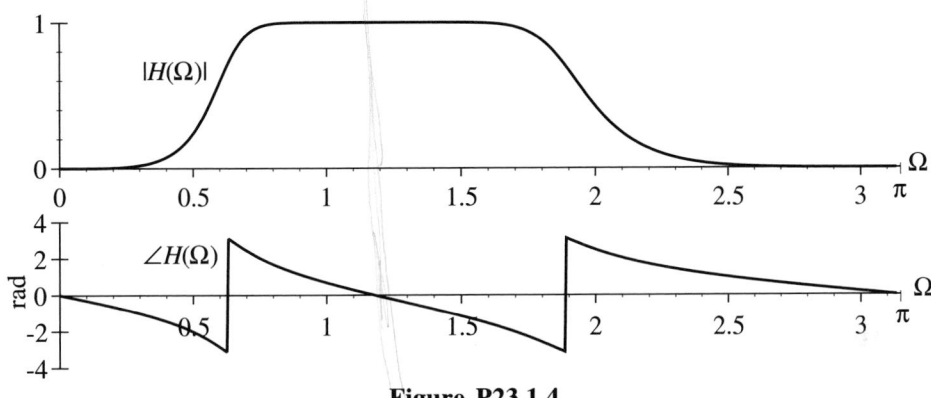

Figure P23.1.4

23.1.5   The pole-zero plots of several digital filters are shown in Figure P23.1.5. Sketch the magnitude of the frequency response function of each filter.

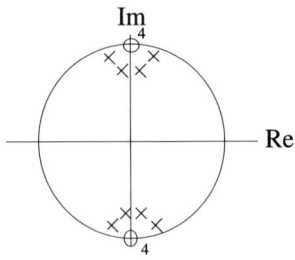

Figure P23.1.5a

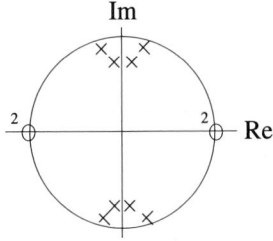

Figure P23.1.5b.

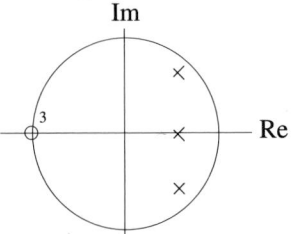

Figure P23.1.5c

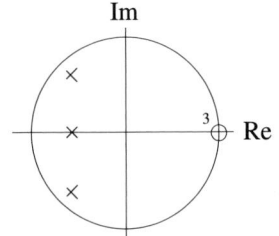

Figure P23.1.5d

23.1.6 Consider the system

$$y(n+1)+0.3y(n)=0.5x(n), \quad y(0)=-1.$$

Suppose the input signal is given by

$$x(n)=\sin(0.1n)u_n(n)$$

(a) Is this system BIBO stable? Why?
(b) Find the output signal using z-transforms. (Use the MATLAB command **residuez**).
(c) Plot the input and output signal on the same graph using the **filter** command. Identify the transient and steady state solutions.

23.1.7 For the following BIBO stable systems:
(a) Find the frequency response function.
(b) Find the magnitude and phase of the frequency response function.
(c) Sketch the magnitude and phase of the frequency response function.
(d) Find the group delay of each filter.

(i) $\dfrac{Y(z)}{X(z)}=\dfrac{z}{z-a}$ $\qquad$ (iii) $\dfrac{Y(z)}{X(z)}=\dfrac{1}{z^3}$

(ii) $\dfrac{Y(z)}{X(z)}=\dfrac{1}{z}$ $\qquad$ (iv) $\dfrac{Y(z)}{X(z)}=\dfrac{1}{z^2+2r\zeta z+r^2}$

23.1.8 Consider the two systems

$$\frac{Y_1(z)}{X(z)}=H(z), \quad \frac{Y_2(z)}{X_2(z)}=H(z).$$

These two systems have the same transfer function but different input and output signals. Let $x(n)$ be a finite length signal. Suppose this signal is the input signal to the first filter. Now define the input signal into the second filter as $x_2(n)=y_1(-n)$. Finally, define the signal $y(n)=y_2(-n)$.

(a) Show that the relationship between the input signal $x(n)$ and the signal $y(n)$ is

$$Y(\Omega)=|H(\Omega)|^2 X(\Omega).$$

This filtering scheme filters the signal $x(n)$ with no phase distortion. It is implemented in the MATLAB command **filtfilt**.

(b) Is this filtering scheme causal?

Homework Problems for Section 23.2

23.2.1   Consider the signal

$$x(t) = \sin(2\pi t)$$

(a) Suppose that this signal is the input signal to a continuous-time ideal distortionless filter with a gain of $K = 1$ and a group delay of $t_d = 1$. Plot the input and output signal of the distortionless filter on the same graph.

(b) Suppose that this signal is sampled with a sampling frequency of $f_s = 10$ Hz and the sampled signal is the input signal to a discrete-time ideal distortionless filter with a gain of $K = 1$ and a group delay of $n_d = t_d$. Plot the input and output signal of the distortionless discrete-time filter on the same graph as the input and output signal of the continuous-time distortionless filter.

(c) Repeat (a) and (b) for a group delay of $n_d = t_d = 1.5$.

This exercise shows how to interpret a fractional group delay for a discrete system.

23.2.2   The discrete-time sampled version of

$$x(t) = \cos(\omega t)$$

is used as the input signal to this system

$$\frac{Y(z)}{X(z)} = H(z) = \frac{0.2z^2}{z^2 - 1.8z + 1}.$$

If $x(t)$ has a bandwidth of 10 kHz, what is the minimum sampling frequency that will result in aliasing in the half-power bandwidth of $H(\Omega)$?

23.2.3   The frequency response of a digital filter is shown in Figure P23.2.2.

(a) Suppose the input signal to this filter is a continuous signal that is sampled with sampling frequency of $f_s = 1$ kHz. Locate the cutoff frequencies of the notch in terms of continuous frequency.

(b) Suppose the input signal to this filter is a continuous signal that is sampled with sampling frequency of $f_s = 4$ kHz. Locate the cutoff frequencies of the notch in terms of continuous frequency.

(c) Suppose the continuous-time input signal to this filter is

$$x(t) = 1.5\cos(400\pi t) + 0.3\cos(800\pi t + 1.2)$$

Find the amplitude of each component of the steady state response of this filter for each of the sampling frequencies in (a) and (b).

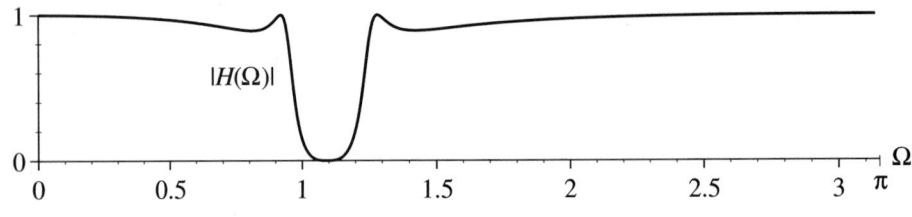

Figure P23.2.3

23.2.4   The discrete-time sampled version of

$$x(t) = \sin(3\pi t)$$

is used as the input signal to a moving average filter with an impulse response function of

$$h(n) = u_s(n) - u_s(n - 5).$$

Find a sampling frequency so that the output signal is zero for $n > 5$.

Homework Problems for Section 23.3

23.3.1   The frequency response functions for a system are shown in Figure P23.1.4.
(a) What type of filter is this system?
(b) What is the bandwidth of this filter?
(c) Repeat (a) and (b) for the frequency response function in Figure P23.2.3.

23.3.2   The magnitude of the frequency response function of a digital filter is shown in Figure P23.3.2.
(a) What type of filter is this system?
(b) What is the bandwidth of this filter in discrete frequency?
(c) Suppose the sampling frequency is $f_s = 10$ Hz. What is the bandwidth of this filter in continuous frequency?

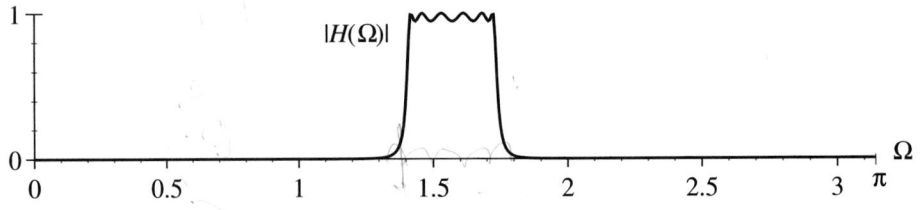

Figure P23.3.2

23.3.3   The magnitude of the frequency response function of a digital filter is shown in Figure P23.3.3.
(a) What type of filter is this system?
(b) What is the bandwidth of this filter in discrete frequency?

(c) Suppose the sampling frequency is $f_s = 30$ kHz. What is the bandwidth of this filter in continuous frequency?

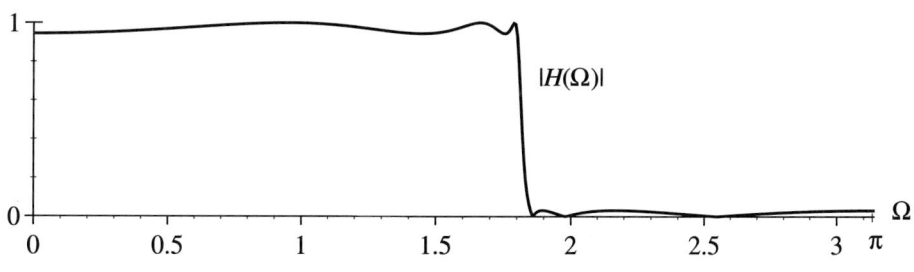

Figure P23.3.3

23.3.4 Consider the system

$$\frac{Y(z)}{X(z)} = \frac{0.057 + 0.1709z^{-1} + 0.1709z^{-2} + 0.057z^{-3}}{1 - 1.1905z^{-1} + 1.0452z^{-2} - 0.3990z^{-3}}$$

(a) Plot the pole-zero diagram.
(b) Plot the frequency response function.
(c) What type of filter is this system?
(d) If the edge of the stopband is defined as -10 dB, what is the stopband?

23.3.5 Suppose that the frequency response function of a filter is given by

$$|H(\Omega)| = K|\Omega|,$$

$$\angle H(\Omega) = \begin{cases} \pi/2, & \Omega < 0 \\ -\pi/2, & \Omega \geq 0 \end{cases}$$

(a) Sketch the magnitude and phase of the frequency response function.
(b) If the input signal is $x(n) = \cos(\Omega_0 n)$, find the output signal.
(c) Describe the function of this filter.

23.3.6 Suppose an ideal continuous-time lowpass filter has bandwidth of 1000 rad/sec and a group delay of 0.2 sec. Suppose this filter is to be implemented as a digital filter with a sampling frequency of $f_s = 5$ kHz.
(a) Sketch the frequency response function of this filter in discrete frequency.
(b) Find an analytical expression for the frequency response function of this system.

23.3.7 Suppose that the cutoff frequency of an ideal highpass filter is 1 rad and a group delay of 2 sec.
(a) Sketch the frequency response function of this filter.
(b) Find an analytical expression for the frequency response function.

23.3.8 Suppose an ideal notch filter has a notch with a width of 2 kHz centered on 30 kHz. Suppose that the group delay of this filter is 3 μsec. Suppose this filter is to be implemented as a digital filter with a sampling frequency of $f_s = 300$ kHz.

(a) Sketch the frequency response function of this filter in discrete frequency.

(b) Find an analytical expression for the frequency response function of this system.

23.3.9 Suppose an ideal bandpass filter has a bandwidth of 300 Hz and a group delay of 4 msec. Suppose this filter is to be implemented as a digital filter with a sampling frequency of $f_s = 10$ kHz.

(a) Sketch the frequency response function of this filter in discrete frequency.

(b) Find an analytical expression for the frequency response function.

Homework Problems for Section 23.4

23.4.1 Consider the analog filter

$$\frac{Y(s)}{X(s)} = \frac{(s^2 + \omega_1^2)}{(s^2 + 2(.2)\omega_1 s + \omega_1^2)}$$

where $\omega_1 = 2\pi(7000)$ rad/sec. This filter is to be implemented digitally using a sampling frequency of:

    (i) $f_s = 60$ kHz

    (ii) $f_s = 300$ kHz

(a) Using the bilinear transformation, find the equivalent digital filter using MATLAB.

(b) Plot the pole-zero diagram of this filter.

(c) Sketch the magnitude of the frequency response function of the filter. Label the axis in both continuous and discrete frequency. Identify the frequency ω_1 in continuous and discrete frequency.

(d) What kind of filter is this system?

23.4.2 Consider the bilinear transformation

$$s = \frac{2}{T_s} \frac{z-1}{z+1}.$$

(a) Using $z = \rho e^{j\Omega}$, show that

$$s = \frac{2}{T_s}\left[\frac{\rho^2 - 1}{1 + \rho^2 + 2\rho\cos\Omega} + j\frac{2\rho\sin\Omega}{1 + \rho^2 + 2\rho\cos\Omega}\right].$$

(b) Show that the bilinear transformation maps the LHP of the s-plane into the unit circle.

(c) Show that the bilinear transformation maps the $j\omega$ axis in the s-plane into the unit circle.

23.4.3 Derive the relationship between the poles and zeros of the continuous and discrete transfer functions given in (23.4.9).

23.4.4 Write a MATLAB M-file to implement directly the bilinear transformation of the single pole filter in Example 23.4.3. Investigate the effects of prewarping.

23.4.5 Suppose that a continuous signal is sampled with a sampling frequency of $f_S = 1$ kHz. This sampled signal is filtered with a digital 4th-order Chebyshev I filter with a bandwidth of $\Omega_c = 0.66\pi$ rad and 1 dB of ripple in the passband.

(a) Generate the digital filter using MATLAB.

(b) Plot the magnitude and phase of the digital filter. Label the frequency axis in both continuous and discrete frequency.

(c) Suppose the input signal to the filter is obtained by sampling the continuous-time signal $x(t) = \sin(1500t)$. What is the amplitude and phase of the discrete-time steady state output signal? Explain your answer using the frequency response function in (b).

23.4.6 The amplitude spectrum of a signal $x(t)$ is shown in Figure P23.4.6. This signal has two main components: a sinusoid at 60 Hz, and another component centered at 200 Hz with a bandwidth of 50 Hz. We wish to pass this signal through a digital filter to remove the 60 Hz component while minimizing the distortion to the other signal component. Specifically, the center frequency of 200 Hz should have less than 0.1 dB attenuation, and the 60 Hz component should be attenuated by more than 40 dB.

(a) The sampling frequency is $f_S = 500$ Hz. Sketch the amplitude spectrum of the sampled signal. Label the sampling frequency.

(b) What type of filter is required?

(c) Design the required filter. What is the transfer function?

(d) Plot the frequency response function.

(e) Plot the poles and zeros of the filter.

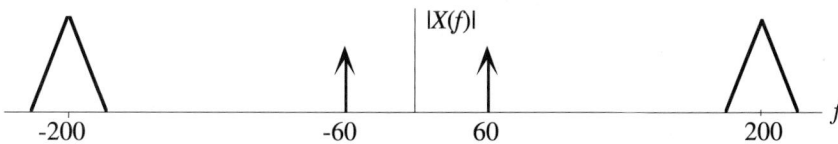

Figure P23.4.6

23.4.7   Suppose we want to design a lowpass filter with the following specifications:

bandwidth	0.3π rad
passband ripple	0.5 dB
stopband cutoff frequency	0.5π rad
stopband ripple	> 30 dB

(a) Design a minimal order Butterworth filter to meet these specifications.
(b) Design a minimal order Chebyshev Type I filter to meet these specifications.
(c) Design a minimal order Chebyshev Type II filter to meet these specifications.
(d) Design a minimal order elliptic filter to meet these specifications.
(e) What is the order of each of these filters?
(f) In each case plot the magnitude and phase of the frequency response function.

23.4.8   We want to design a bandpass filter with the following specifications:

passband cutoff frequencies	4 and 8 kHz
passband ripple	0.1 dB
stopband ripple	30 dB
transition bandwidth	1 kHz
sampling frequency	40 kHz

(a) Design a minimal order Butterworth filter to meet these specifications.
(b) Design a minimal order Chebyshev Type I filter to meet these specifications.
(c) Design a minimal order Chebyshev Type II filter to meet these specifications.
(d) Design a minimal order elliptic filter to meet these specifications.
(e) What is the order of each of these filters?
(f) Plot the magnitude and phase of each frequency response function.

Homework Problems for Section 23.4

23.5.1   Consider the impulse response function of the ideal lowpass filter.
(a) How does this function change if the bandwidth β changes?
(b) How does this function change if the group delay n_d changes?
(c) Verify your results in MATLAB.

23.5.2   (a) Design an FIR filter to reject sinusoidal input signals with frequencies of $\Omega = \frac{\pi}{4}, \frac{3\pi}{4}$ with a DC gain of 1.
(b) What is the minimum order of this filter?
(c) What is the group delay of this filter?
(d) Plot the magnitude of the frequency response function.

23.5.3   Consider the FIR filter $H(z) = b_0 + b_1 z^{-1} + b_0 z^{-2}$.
(a) Show that the frequency response function of this filter is
(b) Let the input signal to this filter be $x(n) = \cos(0.8n) + \cos(1.5n)$. Suppose that we want to block the low frequency cosine but pass the

high frequency cosine. Determine b_0 and b_1.

(c) Write a MATLAB M-file to calculate and plot the output signal. Also calculate and plot the frequency response function.

23.5.4 Find the transfer function of the system in Figure P23.5.4.

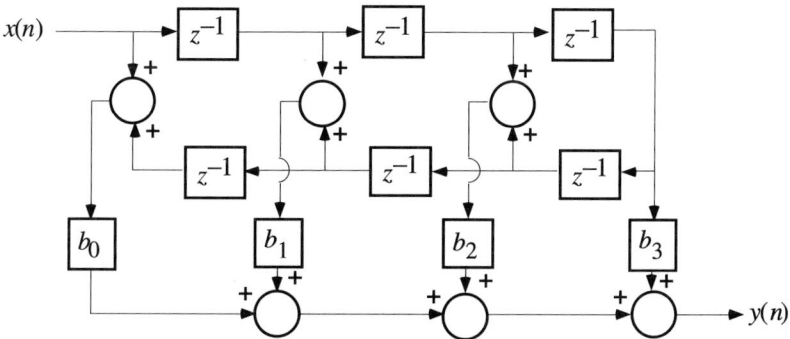

Figure P23.5.4

23.5.5 Design a linear phase, lowpass FIR filter with a bandwidth of $\beta = 2$ rad with $M = 6$ nonzero coefficients in the impulse response function using the window method with a Hamming window.

23.5.6 Find the filter coefficients of a symmetric, lowpass, linear phase FIR filter with $M = 4$ and a bandwidth of $\beta = 0.2$ rad using the window method. Use a Hanning window.

23.5.7 We want to design a lowpass FIR filter with the following specifications:

bandwidth	0.3π rad
passband ripple	0.5 dB
stopband cutoff frequency	0.5π rad
stopband ripple	> 30 dB

(a) Design a minimal order FIR filter to meet these specifications using a Hamming window.

(b) Design a minimal order FIR filter to meet these specifications using a Blackman window.

(c) Design a minimal order FIR filter to meet these specifications using a Kaiser window.

(d) Compare the order and transition bandwidth of each filter.

23.5.8 We want to design a bandpass filter with the following specifications:

passband bandwidth	4–8 kHz
passband ripple	0.1 dB
stopband ripple	30 dB
transition bandwidth	1 kHz

This filter is to be implemented digitally with a sampling frequency of $f_s = 40$ kHz.

(a) Translate the filter specifications into discrete frequency.
(b) Design a minimal order FIR filter to meet these specifications using a Hamming window.
(c) Design a minimal order FIR filter to meet these specifications using a Blackman window.
(d) Design a minimal order FIR filter to meet these specifications using a Kaiser window.
(e) Compare the order of each filter.

23.5.9 The specifications for a filter are shown in Figure P23.5.9. This filter is to be implemented digitally with a sampling frequency of $f_s = 100$ kHz.

(a) Translate the filter specifications into discrete frequency.
(b) Design a Butterworth filter to meet these specifications.
(c) Design a Chebyshev Type I filter to meet these specifications.
(d) Design a Chebyshev Type II filter to meet these specifications.
(e) Design an elliptic filter to meet these specifications.
(f) Design a minimal order FIR filter to meet these specifications using a Hamming window.
(g) Design a minimal order FIR filter to meet these specifications using a Blackman window.
(h) Design a minimal order FIR filter to meet these specifications using a Kaiser window.
(i) In each case plot the magnitude and phase of the frequency response function.
(j) Compare and contrast each of the filter designs

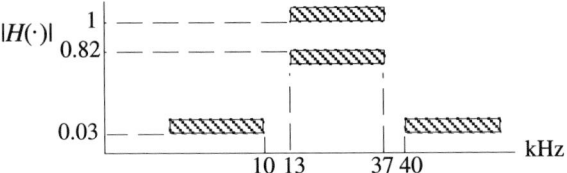

Figure P23.5.9

23.5.10 When a keypad is pushed on a standard telephone keypad two tones (sinusoids at specified frequencies) are generated. The frequencies for each keypad are shown in the Table P23.5.10. Note that each dual tone consists of one low frequency and one high frequency. The problem is to determine which two frequencies are present in the received signal. One approach is to sample the received signal and pass the sampled signal through a bank of narrowband filters. One filter is designed to pass each frequency. The dual tone is detected by determining which two filter output signals are high. Assume that the sampling frequency is 8 kHz.

(a) Develop a set of filter specifications.
(b) Design a Butterworth, Chebyshev II, and elliptic filter to meet the design requirements. In each case minimize the order of the filter. Determine the minimum length of the signal needed to detect the tone.

Compare and contrast your filter designs.

(c) Design an FIR linear phase filter to meet the specifications using the window method with Hamming and Kaiser windows. In each case minimize the order of the filter. Determine the minimum length of the signal needed to detect the tone. Compare and contrast your filter designs.

(d) Compare and contrast the IIR and FIR filter designs.

Table P23.5.10 Dual Tone Frequencies

frequencies, Hz	1209	1336	1477	1633
697	1	2	3	A
770	4	5	6	B
852	7	8	9	C
941	*	0	#	D

Homework Problems for Section 23.6

23.6.1 Suppose that a system is given by

$$y(n) = h(n) * x(n)$$

(a) Show that the crosscorrelation between the input and output signals is given by

$$r_{yx}(k) = h(k) * r_{xx}(k)$$

(b) Show that the cross-energy spectral density of $y(n)$ and $x(n)$ is

$$D_{xy}(\Omega) = H(\Omega)D_{xx}(\Omega)$$

(c) Suppose that the energy spectral density of the input signal is flat; i.e., $D_{xx}(\Omega) = E_x$. Show that

$$h(k) = \frac{1}{E_x} r_{yx}(k)$$

(d) This observation provides a method for finding the impulse response function for an unknown system. What criteria must the system satisfy to be able to apply this identification technique? Write a MATLAB M-file to test out this method.

23.6.2 The magnitude of the frequency response function of a digital filter is shown in Figure P23.6.2. Suppose the sampling frequency is $f_s = 30$ kHz.

(a) Suppose the input signal to this filter is

$$x(t) = [\text{Sa}(5t)]^2 \cos(50t).$$

Plot the energy spectral density of the sampled signal.
(b) Plot the energy spectral density of the output signal.

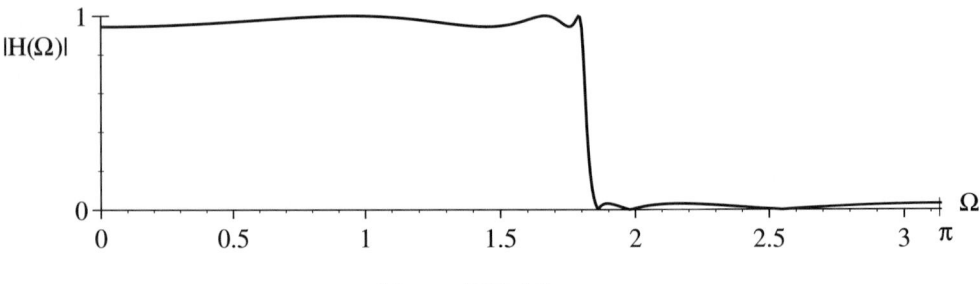

Figure P23.6.2

23.6.3 Consider the signal

$$x(n) = 0.4 \, \text{Sa}(0.4\pi n) \cos(0.5\pi n).$$

(a) Find the energy spectral density of this signal.
(b) Design a 6th-order bandpass Chebyshev I filter with a bandwidth of $\beta = 0.1\pi$ rad whose passband is centered on $\Omega = 0.5\pi$.
(c) If $x(n)$ is the input signal to this filter, plot the input and output signals on the same graph.
(d) Plot the energy spectral density of the input and output signals on the same graph along with the magnitude squared of the frequency response function.
(e) Repeat parts (b) - (d) if the bandwidth of the filter is $\beta = 0.4\pi$ rad.
(f) Repeat parts (b) - (d) if the bandwidth of the filter is $\beta = 0.7\pi$ rad.

23.6.4 The magnitude of the frequency response function of a digital filter is shown in Figure P23.6.4. Assume that the sampling frequency is $f_s = 10$ Hz.
(a) Suppose the input signal to this filter is

$$x(t) = \text{Sa}(5t) \cos(30t).$$

Plot the energy spectral density of the sampled signal.
(b) Sketch the energy spectral density of the output signal.

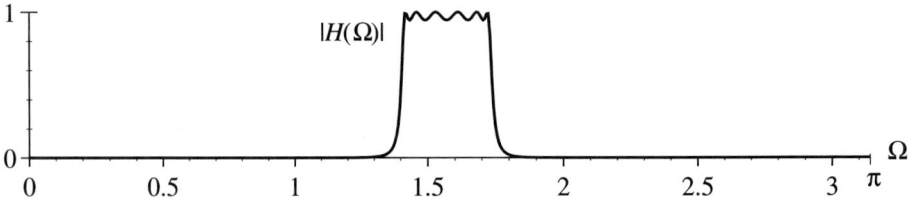

Figure P23.6.4

Nomenclature

Global Notation   There are several variables that occur with great frequency throughout the text. These variables are:

t	continuous time	units are seconds (sec)
ω	(continuous) frequency	units are radians/second (rad/sec)
Ω	(discrete) frequency	units are radians (rad)
f	(continuous) frequency	units are hertz (Hz)

If the units on these variables are not shown, they are the units shown above.

Selected Notation   The following list contains selected notation used throughout the text. This list is organized chapter by chapter with the notation shown when it first appears.

Continuous-Time Signals and Systems

Chapter 2

$\delta(t)$	unit impulse function	Definition 2.2.1
$u_s(t)$	unit step function	Definition 2.2.3
$r_p(t)$	unit ramp function	Definition 2.2.4
$\Pi(t)$	unit pulse function	Definition 2.2.5
$\Lambda(t)$	unit triangle function	Definition 2.2.6
$\ln(t)$	natural logarithm	Definition 2.2.8
$\log(t)$	logarithm base 10	Definition 2.2.9
$\text{Sa}(t)$	Sa function	Definition 2.2.11
$\text{sinc}(t)$	sinc function	Definition 2.2.11
n	discrete-time function	Definition 2.3.1
κ	fixed integer	Definition 2.3.1
$\delta(n)$	discrete unit impulse function	Definition 2.3.3
$u_s(n)$	unit step function	Definition 2.3.5
$\text{Sa}(n)$	discrete Sa function	Definition 2.3.7
$\text{sinc}(n)$	discrete sinc function	Definition 2.3.7
$\Pi_\kappa(n)$	unit pulse function	Definition 2.3.8

Chapter 3

s	complex number	Definition 3.1.1
σ	real part of a complex number	Definition 3.1.2
$j\omega$	imaginary part of a complex number	Definition 3.1.2
ρ	magnitude of a complex number	Definition 3.1.7
θ	phase of a complex number	Definition 3.1.7
$\bar{s}$	complex conjugate	Definition 3.1.12
s-plane	copy of the set of complex numbers	Definition 3.1.13
RHP, LHP	right-and left-half plane	Definition 3.1.14
n	order of a rational function	Definition 3.2.5
ω_n	natural frequency	(3.2.8), Definition 6.4.1
ζ	damping ratio	(3.2.10), Definition 6.4.1

Chapter 5 and 6

$x(t)$	a signal, input signal to a system	Definition 5.1.1
$\mathcal{H}[\cdot]$	abstract notation for a system	(6.1.1)
$y(t)$	a signal, output signal of a system	(6.1.1)
$h(t)$	impulse response function	Definition 6.2.1

Chapter 7

T_0	period of a periodic signal	Definition 7.1.1
$\int_{T_0} dt$	integration over any interval of length T_0	(7.3.2)
$\mathcal{F}\{\cdot\}$	Fourier transform of $x(t)$	Definition 7.4.1

Chapter 8

E_x	energy of a signal $x(t)$	Definition 8.2.1
P_x	power in the signal $x(t)$	Definition 8.2.2
D_x	energy spectral density of $x(t)$	Definition 8.3.2
S_x	power spectral density of $x(t)$	Definition 8.4.2
BL	bandlimit of a signal	Definition 8.3.4
B	bandwidth of a signal	Definition 8.3.5

Chapter 9

$\mathcal{L}\{\cdot\}$	Laplace transform of $x(t)$	Definition 9.1.1
s	Laplace transform variable	Definition 9.1.1

Chapter 10

$H(s)$	Laplace transfer function of a system	Definition 10.1.3
$a(s)$	denominator of $H(s)$	(10.1.18)
$b(s)$	numerator of $H(s)$	(10.1.18)

n	order of the transfer function	Definition 10.1.5
A, B, C, D	state space representation	Definition 10.5.8

Chapter 12

$h_1 * h_2$	convolution of two systems	(12.1.25)
$H(\omega)$	Fourier transfer function	Definition 12.4.1

Chapter 14

dB	decibel	Definition 14.3.5
β	bandwidth of a system	Definition 14.5.3

Discrete-Time Signals and Systems

While most of the notation is standardized throughout the text, there are two changes in notation in Chapters 17–23. The reason for this change is to present the material in the style of the signal processing literature.

First Change It is customary to index the coefficients of a rational function differently in discrete-time than in continuous-time. Recall that a rational transfer function in the Laplace variable s is indexed as

$$\frac{Y(s)}{X(s)} = \frac{b_3 s^3 + b_2 s^2 + b_1 s + b_0}{s^3 + a_2 s^2 + a_1 s + a_0} \tag{1}$$

The index of the coefficient is the same as the power of s.

In Chapters 17–23 we discuss rational transfer functions of the form

$$\frac{Y(z)}{X(z)} = \frac{b_0 + b_1 z^{-1} + b_2 z^{-2} + b_3 z^{-3}}{1 + a_1 z^{-1} + a_2 z^{-2} + a_3 z^{-3}} \cdot \tag{2}$$

Note the change in the indexing of the coefficients. In (2) the index on the coefficient matches the negative of the power of z. Results in Chapters 3–16 can be translated into the results in Chapters 17–23 by an appropriate change in the indexing of the coefficients.

Second Change In Chapter 17–23 discrete-time functions $x(n)$ are discussed. Here the variable n is an integer, and it represents discrete time. In Chapters 3–16 we use n to stand for the order of the denominator polynomial of a rational function. In Chapters 17–23 we will use N to stand for the order of the denominator polynomial of a rational function.

Chapter 17

$x(n)$	discrete-time signal	Section 17.1.2
T_s	sampling period	Definition 17.2.1
f_s	sampling frequency in hertz	Definition 17.2.1
ω_s	sampling frequency in radians	Definition 17.2.1
$\hat{x}(n)$	discrete-time sampled version of $x(t)$	Definition 17.2.1
ZOH	zero-order hold	Definition 17.4.1

Chapter 18

z	z-transform variable	Definition 18.1.1
$\mathcal{Z}\{\cdot\}$	one or two-sided z-transform	Definition 18.1.1, 18.31
$\mathcal{DTFT}\{\cdot\}$	discrete-time Fourier transform	Definition 18.4.1

Chapter 19

$x^*(t)$	continuous-time sampled version of $x(t)$	Definition 19.1.1

Chapter 20

DFT	discrete Fourier transform	Definition 20.3.3
$\tilde{X}(k)$	DFT of $\tilde{x}(n)$	Definition 20.3.3
$\{2 \quad 1\}$ $\uparrow$	see description in text	(20.4.2)

Chapter 21

FIR	finite impulse response	Definition 21.1.4
IIR	infinite impulse response	Definition 21.1.4
N	order of a transfer function	Definition 21.2.1
$H(z)$	(z-transform) transfer function	Definition 21.2.6
$H(\Omega)$	DTFT transfer function	Definition 21.4.1

Index